DESCRIPTIVE SET THEORY

STUDIES IN LOGIC

AND

THE FOUNDATIONS OF MATHEMATICS

VOLUME 100

NORTH-HOLLAND PUBLISHING COMPANY – AMSTERDAM · NEW YORK · OXFORD

DESCRIPTIVE SET THEORY

YIANNIS N. MOSCHOVAKIS
University of California, Los Angeles

1980

NORTH-HOLLAND PUBLISHING COMPANY – AMSTERDAM · NEW YORK · OXFORD

North-Holland ISBN: 0 444 70199 0

Published by:

North-Holland Publishing Company – Amsterdam · New York · Oxford

Sole distributors for the U.S.A. and Canada:

Elsevier North-Holland, Inc.
52 Vanderbilt Avenue
New York, N.Y. 10017

Library of Congress Cataloging in Publication Data

Moschovakis, Yiannis N.
Descriptive set theory.

(Studies in logic and the foundations of mathematics; v. 100)
Bibliography: p.
Includes index.
1. Descriptive set theory. I. Title. II Series,
QA248.M66 511'.3 79-11792
ISBN 0-444-85305-7

PRINTED IN THE NETHERLANDS

This book is dedicated to the memory of my father Nicholas
a good and gentle man.

This page intentionally left blank

PREFACE

This book was conceived in the winter of 1970 when I heard that I was getting a Sloan Fellowship and I thought I would take a year off to write a book. It took a bit longer than that, but I have many good excuses.

I am grateful to the Sloan Foundation, the National Science Foundation and the University of California for their financial support—and to the Mathematics Department at UCLA for the stimulating and pleasant working environment that it provides.

One often sees in prefaces long lists of persons who have contributed to the project in one way or another and I hope I will be forgiven for not complying with tradition; in my case any reasonably complete list would have to start with Lebesgue and increase the size of the book beyond the publisher's indulgence. I will, however, mention my student Chris Freiling who read carefully through the entire final version of the manuscript and corrected all my errors.

My wife Joan is the only person who really knows how much I owe to her and she is too kind to tell. But I know too.

Finally, my deepest feelings of gratitude and appreciation are reserved for the very few friends with whom I have spent so many hours during the last ten years arguing about descriptive set theory; Bob Solovay and Tony Martin in the beginning, Aleko Kechris, Ken Kunen and Leo Harrington a little later. Their influence on my work will be obvious to anyone who glances through this book and I consider them my teachers—although of course, they are all so much younger than me. No doubt I would still work in this field if they were all priests or generals—but I would not enjoy it half as much.

Santa Monica, California December 22, 1978

Added in proof. I am deeply grateful to Dr. Haimanti Sarbadhikari who read the first seven chapters in proof and corrected all the errors missed by Chris Freiling. I am also indebted to Anna and Nicholas Moschovakis for their substantial help in constructing the indexes and to Tony Martin for the sustenance he offered me during the last stages of this work.

TABLE OF CONTENTS

ABOUT THIS BOOK

My aim in this monograph is to give a brief but coherent exposition of the main results and methods of descriptive set theory. I have made no attempt to be complete; in a subject so broad this would degenerate into a long catalog of specialized results which would cover up the main thread. On the contrary, I have tried very hard to be selective, so that the central ideas stand out.

Much of the material is in the exercises. A very few of them are simple, to test the reader's comprehension and a few more give interesting extensions of the theory or sidelines. The vast majority of the exercises are an integral part of the monograph and would be normally billed "theorems." There are extensive "hints" for them, proofs really, with some of the details omitted.

I have tried hard to attribute all the important results and ideas to those who invented them but this was not an easy task and I have undoubtedly made many errors. *There is no suggestion that unattributed results are mine or are published here for the first time.* When I do not give credit for something, the most likely explanation is that I could not determine the correct credit. My own results are immodestly attributed to me, including those which are first published here.

Many of the references are in the historical sections at the end of each chapter. The paragraphs of these sections are numbered and the footnotes in the body of the text refer to these paragraphs—each time meaning the section at the end of the chapter where the reference occurs. In a first reading, it is best to skip looking up these references and read the historical sections as they come after one is familiar with the material in the chapter.

The order of exposition follows roughly the historical development of the subject, simply because this seemed the best way to do it. It goes without saying that the classical results are presented from a modern point of view and using modern notation.

What appeals to me most about descriptive set theory is that to study it you must really understand so many things: you need a little bit of topology, analysis and logic, a good deal of recursive function theory and a great deal of set theory, including constructibility, forcing, large cardinals and determinacy. What makes the writing of a book on the subject so difficult is that you must explain so many things: a little bit of topology, analysis and logic, a good deal of recursive function theory, etc. Of course, one could aim the book at those who already know all the prerequisites, but chances are that these few potential readers already know descriptive set theory. My aim has been to make this material accessible to a mathematician whose particular field of specialization could be anything, but who has an interest in set theory, or at least what used to be called "the theory of pointsets." He certainly knows whatever little topology and analysis are required, because he learned that as an undergraduate, and he has read Halmos' *Naive Set Theory* [1960] or a similar text. Beyond that, what he needs to read this book is patience and a basic interest in the central problem of descriptive set theory and definability theory in general: *to find and study the characteristic properties of definable objects.*

INTRODUCTION

The roots of *Descriptive Set Theory* go back to the work of Borel, Baire and Lebesgue around the turn of the centrury, when the young French analysts were trying to come to grips with the abstract notion of a *function* introduced by Dirichlet and Riemann. A function was to be an arbitrary correspondence between objects, with no regard for any method or procedure by which this correspondence could be established. They had some doubts whether so general a concept should be accepted; in any case, it was obvious that all the specific functions which were studied in practice were determined by simple *analytic expressions*, explicit formulas, infinite series and the like. The problem was to delineate the functions which could be defined by such accepted methods and search for their characteristic properties, presumably nice properties not shared by all functions.

Baire was first to introduce in his Thesis [1899] what we now call *Baire functions* (of several real variables), the smallest set which contains all continuous functions and is closed under the taking of (pointwise) limits. He gave an inductive definition: the continuous functions are *of class* 0 and for each countable ordinal ξ, a function is *of class* ξ if it is the limit of a sequence of functions of smaller classes and is not itself of lower class. Baire, however, concentrated on a detailed study of the functions of class 1 and 2 and he said little about the general notion beyond the definition.

The first systematic study of definable functions was Lebesgue's [1905], *Sur les fonctions représentables analytiquement.* This beautiful and seminal paper truly started the subject of descriptive set theory.

Lebesgue defined the collection of *analytically representable functions as the smallest set which contains all constants and projections* $(x_1, x_2, \ldots, x_n) \mapsto x_i$ and which is closed under sums, products and the taking of limits. It is easy to verify that these are precisely the Baire functions. Lebesgue then showed that there exist Baire functions of every countable class and that there exist definable functions which are not analytically representable. He also defined the *Borel measurable* functions and showed that they too coincide with the Baire functions. In fact he proved a much stronger theorem along these lines which relates the *hierarchy* of Baire functions with a natural hierarchy of the Borel measurable sets at each level.

Today we recognize Lebesgue [1905] as a classic work in the *theory of definability*. It introduced and studied systematically several natural notions of definable functions and sets and it established the first important hierarchy theorems and structure results for collections of definable objects. In it we can find the origins of many standard tools and techniques that we use today, for example *universal sets* and applications of the Cantor *diagonal method* to questions of definability.

One of Lebesgue's results in [1905] identified the *implicity analytically definable* functions with the Baire functions. To take a simple case, suppose that $f:\mathcal{R}^2 \to \mathcal{R}$ is analytically representable and for each x,

$$f(x, y) = 0$$

has exactly one solution in y. This equation then defines y implicitly as a function of x; Lebesgue showed that it is an analytically representable function of x, by an argument which was "simple, short but false." The wrong step in the proof was hidden in a lemma taken as trivial, that a set in the line which is the projection of a Borel measurable set in the plane is itself Borel measurable.

Ten years later the error was spotted by Suslin, then a young student of Lusin at the University of Moscow, who rushed to tell his professor in a scene charmingly described in Sierpinski [1950].

Suslin called the projections of Borel sets *analytic* and showed that indeed there are analytic sets which are not Borel measurable. Together with Lusin they quickly established most of the basic properties of analytic sets and they announced their results in two short notes in the Comptes Rendus, Suslin [1917] and Lusin [1917].

The class of analytic sets is rich and complicated but the sets in it are nice. They are measurable in the sense of Lebesgue, they have the property of Baire and they satisfy the continuum hypothesis, i.e. every uncountable analytic set is equinumerous with the set of all real numbers. The best result in Suslin [1917] is a characterization of the Borel measurable sets as precisely those analytic sets which have analytic complements. Lusin [1917] announced another basic theorem which implied that Lebesgue's contention about implicitly analytically definable functions is true, despite the error in the original proof.

Suslin died in 1919 and the study of analytic sets was continued mostly by Lusin and his students in Moscow and by Sierpinski in Warsaw. Because of what Lusin delicately called "difficulties of international communication" those years, they were isolated from each other and from the wider mathematical community, and there were very few publications in western journals in the early twenties.

The next significant step was the introduction of *projective sets* by Lusin and Sierpinski in 1925: a set is projective if it can be constructed starting with Borel measurable sets and iterating the operations of projection and complementation. Using later terminology, let us call analytic sets A sets, analytic complements CA sets, projections of CA sets PCA sets, complements of these $CPCA$ sets, etc. Lusin in his [1925a], [1925b], [1925c] and Sierpinski [1925] showed that these classes of sets are all distinct and they established their elementary properties. But it was clear from the very beginning that the theory of projective sets was not easy. There was no obvious way to extend to these more complicated sets the regularity properties of Borel and analytic sets; for example, it was an open problem whether analytic complements satisfy the continuum hypothesis or whether PCA sets are Lebesgue measurable.

Another fundamental and difficult problem was posed in Lusin [1930a]. Suppose P is a subset of the plane; a subset P^* of P *uniformizes* P if P^* is the graph of a function and it has the same projection on the line as P.

The natural question is whether definable sets admit definable uniformizations and it comes up often, for example when we seek "canonical" solutions for y in terms of x in an equation

$$f(x, y) = 0.$$

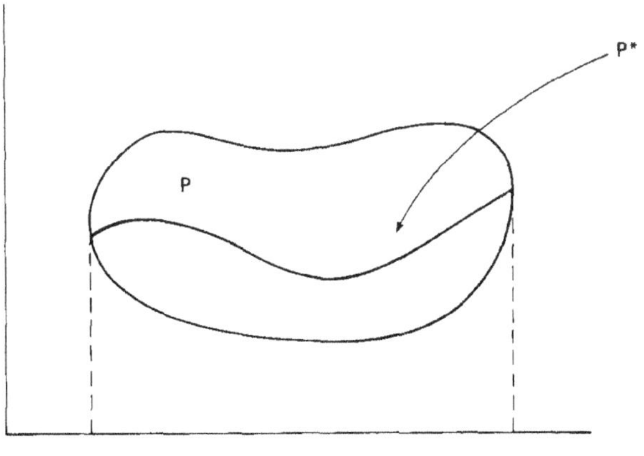

Lusin and Sierpinski showed that Borel sets can be uniformized by analytic complements and Lusin also verified that analytic sets can be projectively uniformized. In a fundamental advance in the subject, Kondo [1938] completed earlier work of Novikov and proved that analytic complements and *PCA* sets can be uniformized by sets in the same classes. Again, there was no clear method for extending the known techniques to solve the uniformization problem for the higher projective classes.

As it turned out, the "difficulties of the theory of projective sets" which bothered Lusin from his very first publications in the subject could not be overcome by ingenuity alone. There was an insurmountable technical obstruction to answering the central open questions in the field, since *all of them were independent of the axioms of classical set theory*. It goes without saying that the researchers in descriptive set theory were formulating and trying to prove their assertions within axiomatic Zermelo–Fraenkel set theory, as all mathematicians still do, consciously or not.

The first independence results were proved by Gödel, in fact they were by-products of his famous consistency proof of the continuum hypothesis. He announced in his [1938] that in the model L of constructible sets there is a *PCA* set which is not Lebesgue measurable: it follows that one cannot establish in Zermelo–Frankel set theory (with the axiom of choice and even the continuum hypothesis) that all *PCA* sets are Lebesgue measurable. His results were followed up by some people, notably Mostowski and Kuratowski, but that was another period of "difficulties of international communication" and nothing was published until the late forties. Addison [1959b] gave the first exposition in print of the consistency and independence results that are obtained by analysing Gödel's L.

The independence of the continuum hypothesis was proved by Cohen [1963], whose powerful method of *forcing* was soon after applied to independence questions in descriptive set theory. One of the most significant papers in forcing was Solovay [1970], where it is shown (among other things) that one can consistently assume the axioms of Zermelo–Frankel set theory (with choice and even the continuum hypothesis) together with the proposition that all projective sets are Lebesgue measurable; from this and Gödel's work it follows that *in classical set theory we can neither prove nor disprove the Lebesgue measurability of PCA sets.*

Similar consistency and independence results were obtained about all the central problems left open in the classical period of descriptive set theory, say up to 1940. It says something about the power of the mathematicians working in the field those years, that in almost every instance they obtained the best theorems that could be proved from the axioms they were assuming.

So the logicians entered the picture in their usual style, as spoilers. There was, however, another parallel development which brought them in more substantially and in a friendlier role. Before going into that, let us make a few remarks about the appropriate context for studying problems of definability of functions and sets.

We have been recounting the development of descriptive set theory on the real numbers, but it is obvious that the basic notions are topological in nature and can be formulated in the context of more general topological spaces. All the important results can be extended easily to *complete, separable, metric spaces.* In fact, it was noticed early on that the theory assumes a particularly simple form on *Baire space*

$$\mathcal{N} = {}^{\omega}\omega,$$

the set of all infinite sequences of natural numbers, topologized with the product topology (taking ω discrete). The key fact about $\mathcal{N}$ is that it is homeomorphis with its own square $\mathcal{N} \times \mathcal{N}$, so that irrelevant problems of dimension do not come up. Results in the theory are often proved just for $\mathcal{N}$, with the (suitable) generalizations to other spaces and the reals in particular left for the reader or simply stated without proof.

Let us now go back to a discussion of the impact of logic and logicians on descriptive set theory.

The fundamental work of Gödel [1931] on incompleteness phenomena in formal systems suggested that it would be profitable to delineate and study those functions (of several variables) on the set ω of natural numbers which are *effectively computable.* A great deal of work was done on this problem in the thirties by Church, Kleene, Turing, Post and Gödel among others, from which emerged a coherent and beautiful theory of *computability* or *recursion.* The class of *recursive functions* (of several variables) *on* ω was characterized as the smallest set which contains all the constants, the successor and the projections $(x_1, x_2, \ldots, x_n) \mapsto x_i$ and which is closed under composition, a form of simple definition by induction (primitive recursion) and minimalization, where g is defined from f by the equation

$$g(x_1, x_2, \ldots, x_n) = \text{least } w \text{ such that } f(x_1, x_2, \ldots, x_n, w) = 0,$$

assuming that for each $x_1, x_2, \ldots, x_n$ there is a root to the equation $f(x_1, x_2, \ldots, x_n, w) = 0$. Church [1936] proposed the *thesis* (hypothesis) that all number theoretic functions which can be computed effectively by some algorithm are in fact recursive, and to this date no serious evidence has been presented to dispute this.

Kleene [1952a], [1952b] extended the theory of recursion to functions

$$f : \omega^n \times \mathcal{N}^k \rightarrow \omega$$

with domain some finite cartesian product of copies of the natural numbers and Baire space. For example, a function $f : \omega \times \mathcal{N} \rightarrow \omega$ is recursive (by the natural extension of Church's Thesis) if there is an algorithm which will compute $f(n, \alpha)$ given n and a sufficiently long initial segment of the infinite sequence α.

A set $A \subseteq \omega^n \times \mathcal{N}^k$ is recursive if its characteristic function is recursive. By Church's Thesis again, these are the *decidable* sets for which we have (at least in principle) an algorithm for testing membership.

Using recursion theory as his main tool, Kleene developed a rich and intricate theory of definability on the natural numbers in the sequence of papers [1943], [1955a], [1955b], [1955c].

The class of *arithmetical* sets is the smallest family which contains all recursive sets and is closed under complementation and projection on ω. The *analytical* sets are defined similarly, starting with the arithmetical sets and iterating any finite number of times the operations of complementation, projection on ω and projection on $\mathcal{N}$. Both of these classes are naturally ramified into subclasses, much like the subclasses $A, CA, PCA, \ldots$ of projective sets of reals. Notice that the definitions make sense for subsets of an arbitrary product space of the form $\omega^n \times \mathcal{N}^k$. Kleene, however, was primarily interested in classifying definable sets of natural numbers and he stated his ultimate results just for them. The more complicated product spaces were brought in only so projection on $\mathcal{N}$ could be utilized to define complicated subsets of ω.

Kleene studied a third notion (discovered independently by Davis [1950] and Mostowski [1951]) which is substantially more complicated. The class of *hyperarithmetical* sets of natural numbers is the smallest family of subsets of ω which contains the recursive sets and is closed under complementation and "recursive" countable union, suitably defined. The precise definition is quite intricate and the proofs of the main results are subtle, often depending on delicate estimates of the complexity of explicit and inductive definitions.

Using later terminology, let us call Σ_1^1 the simplest analytical sets of numbers, those which are projections to ω of arithmetical subsets of $\omega \times \mathcal{N}$. The most significant result of Kleene [1955c] (and the whole theory for that matter) was a characterization of the hyperarithmetical sets as precisely those Σ_1^1 sets which have Σ_1^1 complements.

Now this is clearly reminiscent of Suslin's characterization of the Borel sets. A closer look at specific results reveals a deep resemblance between these two fundamental theorems and suggests the following *analogy* between the classical theory and Kleene's definability theory for subsets of ω:

$\mathcal{R}$ or $\mathcal{N}$	ω
continuous functions	recursive functions
Borel sets	hyperarithmetical sets
analytic sets	Σ_1^1 sets
projective sets	analytical sets

In fact the theories of the corresponding classes of objects in this table are so similar, that one naturally conjectures that Kleene was consciously trying to create an "effective analog" on the space ω of classical descriptive set theory.

As it happened, Kleene did not know the classical theory, since he was a logician by trade and at the time that was considered part of topology. Mostowski knew it, being Polish, and he first used classical methods in his [1946], where he obtained independently many of the results of Kleene [1943]. More significantly, Mostowski introduced the hyperarithmetical sets following closely the classical approach to Borel sets, as opposed to Kleene's initial rather different definition in [1955b].

First to establish firmly the analogies in the table above was Addison, in his Thesis [1954] and later in his [1959a]. Over the years and with the work of many people, what was first conceived as "analogies" developed into a general theory which yields in a unified manner both the classical results and the theorems of the recursion theorists; more precisely, this effective theory yields *refinements* of the classical results and *extensions* of the theorems of the recursion theorists.

It is this extended, *effective descriptive set theory* which concerns us here.

Powerful as they are, the methods from logic and recursion theory cannot solve the "difficulties of the theory of projective sets," since they too are restricted by the limitations of Zermelo–Fraenkel set theory. The natural next step was taken in the fundamental paper Solovay [1969], where for the first time strong set theoretic hypotheses were shown to imply significant results about projective sets.

Solovay proved that if there exist measurable cardinals, then *PCA* sets are Lebesgue measurable, they have the property of Baire and they satisfy the continuum hypothesis. Later, he and Martin proved a difficult uniformization theorem about *CPCA* sets in their joint [1969] and Martin [19?a] established several deep properties of *CPCA* sets, all under the same hypothesis, that there exist measurable cardinals.

For our purposes here, it is not important to know exactly what measurable cardinals are. Suffice it to say that their existence cannot be shown in Zermelo–Fraenkel set theory and that if they exist, they are terribly large sets: bigger than the continuum, bigger than the first strongly inaccessible cardinal, bigger than the first Mahlo cardinal, etc. It is also fair to add that few people are willing to buy their existence after a casual look at their definition. Nevertheless, no one has shown that they do not exist, and it was known from previous work of Scott, Gaifman Rowbottom and Silver that the existence of measurable cardinals implies new and interesting propositions about sets, even about real numbers. These, however, were metamathematical results, the kind that only logicians can love. Solovay's chief contribution was that he used this new and strange hypothesis to solve problems posed by Lusin more than forty years earlier.

Unfortunately, measurable cardinals were not a panacea. Soon after Solovay's original work it was shown by himself, Martin and Silver among others that they do not resolve the open questions about projective sets beyond the *CPCA* class, except for some isolated results about *PCPCA* sets.

The next step was quite unexpected, even by those actively searching for strong hypotheses to settle the old open problems. Blackwell [1967] published a new, short and elegant proof of an old result of Lusin's about analytic sets, using the determinancy of open games.

Briefly, an *infinite game* (of perfect information) is described by an arbitrary subset $A \subseteq \mathcal{N}$ of Baire space. We imagine two players I and II successively choosing natural numbers, with I choosing k_0, then II choosing k_1, then I choosing k_2, etc; after an infinite sequence

$$\alpha = (k_0, k_1, \ldots)$$

has been specified in this manner, we say that I wins if $\alpha \in A$, II wins if

$\alpha \notin A$. The game (or the set A which describes it) is *determined*, if one of the two players has a *winning strategy*, a method of playing against arbitrary moves of his opponent which will always produce a sequence winnning for him.

It was known that open games are determined and Blackwell's proof hinged on that fact. It was also known that one could prove the existence of non-determined games using the axiom of choice, but no definable non-determined game had ever been produced.

Working independently, Addison and Martin realized that Blackwell's proof could be lifted to yield new results about the third class of projective sets, if only one *assumed* the hypothesis that enough projective sets are determined. Soon after, Martin and Moschovakis again independently used the *hypothesis of projective determinacy* to settle a whole slew of old questions about all levels of the projective hierarchy, see Addison-Moschovakis [1968] and Martin [1968]. Three years later the uniformization problem was solved on the same hypothesis in Moschovakis [1971a] and the methods introduced there led quickly to an almost complete structure theory for the classes of projective sets, see especially Kechris [1973], [1974], [1975], Martin [19?a] and Moschovakis [1973], [1974c].

This is where matters stand today.

This page intentionally left blank

CHAPTER 1

THE BASIC CLASSICAL NOTIONS

Let $\omega = \{0, 1, 2,...\}$ be the set of (nonnegative) *integers* and let $\mathcal{R}$ be the set of *real numbers*. The main business of *Descriptive Set Theory* is the study of ω, $\mathcal{R}$ and their subsets, with particular emphasis on the *definable* sets of integers and reals. Another fair name for it is *Definability Theory for the Continuum*.

In this first chapter we will introduce some of the basic notions of the subject and we will establish the elementary facts about them.

1A. Perfect Polish spaces

Instead of working specifically with the reals, we will frame our results in the wider context of *complete, separable metric spaces* (*Polish spaces*) *with no isolated points* (*perfect*). One of the reasons for doing this is the wider applicability of the theory thus developed. More than that, we often need to look at more complicated spaces in order to prove results about $\mathcal{R}$.[(1–5)]

Of course $\mathcal{R}$ is a perfect Polish space and so is the real n-space $\mathcal{R}^n$ for each $n \geq 2$. There are two other important examples of such spaces which will play a key role in the sequel.

Baire space is the set of all infinite sequences of integers,

$$\mathcal{N} = {}^{\omega}\omega$$

with the natural product topology, taking ω discrete. The basic neighborhoods are of the form

$$N(k_0,..., k_n) = \{\alpha \in \mathcal{N}: \alpha(0) = k_0,..., \alpha(n) = k_n\},$$

one for each tuple $k_0,..., k_n$. We picture $\mathcal{N}$ as (the set of infinite branches of) a tree, where each node splits into countably many one-point extensions, Figure 1A.1.

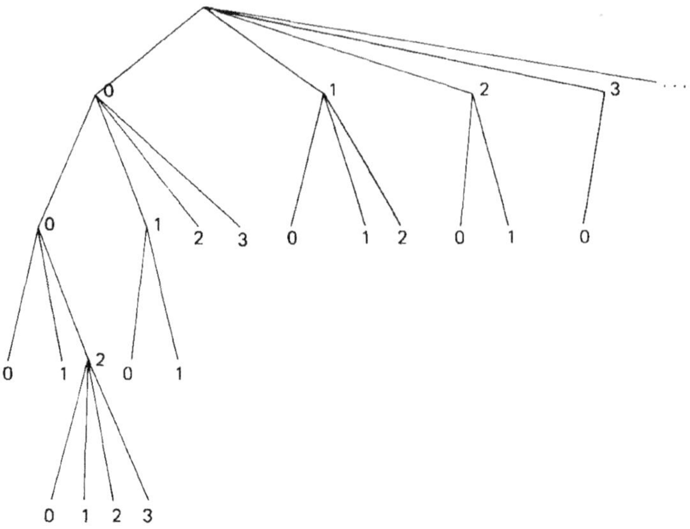

Figure 1A.1. Picturing $\mathcal{N}$ as a tree.

It is easy to verify that the topology on $\mathcal{N}$ is generated by the metric

$$d(\alpha,\beta)=\begin{cases}0, & \text{if } \alpha=\beta,\\ \dfrac{1}{\textit{least}\ n[\alpha(n)\neq\beta(n)]+1}, & \text{if } \alpha\neq\beta.\end{cases}$$

Also, $\mathcal{N}$ is complete with this metric and the set of *ultimately constant sequences* is countable and dense in $\mathcal{N}$, so $\mathcal{N}$ is a perfect Polish space.

One can show that $\mathcal{N}$ is homeomorphic with the set of irrational numbers, topologized as a subspace of $\mathcal{R}$. The proof appeals to some basic properties of continued fractions and does not concern us here—it can be found in any good book on number theory, for example Hardy and Wright [1960]. Although we will never use this result, we will find it convenient to call the members of $\mathcal{N}$ *irrationals*.

Notice that Baire space is *totally disconnected*, i.e. the neighborhood base given above consists of clopen sets.

1A.1. THEOREM. *For every Polish space* $\mathfrak{M}$, *there is a continuous surjection*

$$\pi:\mathcal{N}\twoheadrightarrow\mathfrak{M}$$

of Baire space onto $\mathfrak{M}$.

PROOF. Fix a countable dense subset

$$D = \{r_0, r_1, r_2, \ldots\}$$

of $\mathfrak{M}$ and to each $\alpha \in \mathcal{N}$ *assign the sequence* $\{x_n^\alpha\} = \{x_n\}$ by the induction

$$x_0 = r_{\alpha(0)},$$

$$x_{n+1} = \begin{cases} r_{\alpha(n+1)} & \text{if } d(x_n, r_{\alpha(n+1)}) < 2^{-n}, \\ x_n & \text{if } d(x_n, r_{\alpha(n+1)}) \geq 2^{-n}. \end{cases}$$

Now for each n,

$$d(x_n, x_{n+1}) < 2^{-n},$$

so $\{x_n^\alpha\}$ is Cauchy and we can set

$$\pi(\alpha) = \mathit{limit}_{n \to \infty}\, x_n^\alpha.$$

It is obvious that π is continuous since

$$\alpha(0) = \beta(0), \ldots, \alpha(n) = \beta(n) \Longrightarrow x_0^\alpha = x_0^\beta, \ldots, x_n^\alpha = x_n^\beta$$

from which it follows immediately that

$$\begin{aligned} d(\pi(\alpha), \pi(\beta)) &\leq d(\pi(\alpha), x_n^\alpha) + d(x_n^\beta, \pi(\beta)) \\ &\leq 2^{-n+1} + 2^{-n+1} = 2^{-n+2}. \end{aligned}$$

On the other hand, for each $x \in \mathfrak{M}$ let

$$\alpha(n) = \textit{least } k \textit{ such that } d(x, r_k) < 2^{-n-1}$$

and check that $\pi(\alpha) = \mathit{limit}_n\, r_{\alpha(n)} = x$. ⊣

Another very useful perfect Polish space is the set of all infinite binary sequences

$$\mathcal{C} = {}^\omega 2,$$

again with the product topology. This is a *compact* subspace of $\mathcal{N}$ naturally represented by the complete binary tree. It is obviously homeomorphic with the classical *Cantor set* obtained from the closed interval [0, 1] on the line by successively removing the open middle third, as in Figure 1A.2. Again we will abuse terminology a bit by calling $\mathcal{C}$ *the Cantor set.*

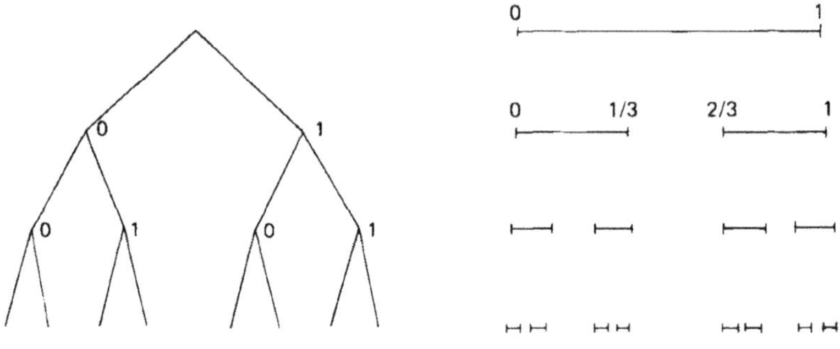

Figure 1A.2. The Cantor set.

With each perfect Polish space $\mathfrak{M}$ we can associate a fixed enumeration

$$N(\mathfrak{M}, 0),\ N(\mathfrak{M}, 1),\ N(\mathfrak{M}, 2),\ldots$$

of a countable set of open nbhds which generates the topology. When $\mathfrak{M}$ is clearly understood by the context we will use the simpler notation

$$N_0,\ N_1,\ N_2,\ldots\ .$$

Of course we may assume that the N_i's are open balls. There are situations, however, when this is not convenient. For example, if $\mathfrak{M} = X_1 \times X_2$ is the product of two spaces, it is often preferable to work with the nbhds of the form $B_1 \times B_2$, where B_1 and B_2 are chosen from bases in X_1 and X_2.

We will leave open the possibility that the N_i's are not open balls. However, we will assume that with each N_i we have associated a *center* x_i and a *radius* p_i such that the following hold:

(1) $x_i \in N_i$, if $N_i \neq \emptyset$.

(2) If $x \in N_i$, then $d(x, x_i) < p_i$.

(3) If x is any point, then for every n we can find some N_i such that $x \in N_i$ and $radius(N_i) < 2^{-n}$.

For any set $P \subseteq \mathfrak{M}$, let

$$\bar{P} = \text{closure of } P,$$

so that $\bar{N}_s = \bar{N}(\mathfrak{M}, s)$ is the closure of the s'th nbhd in the fixed base for the topology of $\mathfrak{M}$.

The simple construction in the next result will be useful in many situations beyond the corollary following it.

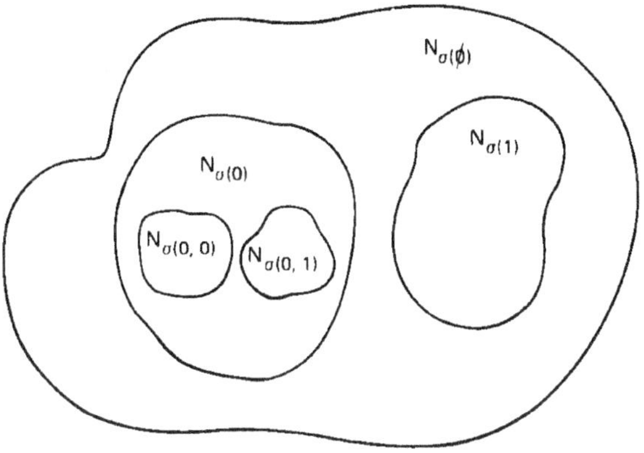

Figure 1A.3.

1A.2. THEOREM. *Let* $\mathfrak{M}$ *be a perfect, Polish space. We can assign to each finite binary sequence* $u = (t_0, \ldots, t_{n-1})$ $(t_i \leq 1)$ *an open* nbhd $N_{\sigma(u)} \neq \emptyset$ *in* $\mathfrak{M}$ *so that*

(i) *if* u *is an initial segment of* v, *then* $\bar{N}_{\sigma(v)} \subseteq N_{\sigma(u)}$,

(ii) *if* u *and* v *are incompatible, then*

$$\bar{N}_{\sigma(u)} \cap \bar{N}_{\sigma(v)} = \emptyset,$$

(iii) *if* $u = (t_0, \ldots, t_{n-1})$ *has length* n, *then radius*$(N_{\sigma(u)}) \leq 2^{-n}$. (See Figure 1A.3.)

PROOF. Two sequences $u = (t_0, \ldots, t_{n-1})$, $v = (s_0, \ldots, s_{k-1})$ are incompatible if for some $i < n$, $i < k$ we have $t_i \neq s_i$.

We define $N_{\sigma(u)}$ by induction on the length n of the binary sequence $u = (t_0, \ldots, t_{n-1})$, starting with some $N_{\sigma(\emptyset)}$ of radius $\leq 1 = 2^{-0}$ that we assign to the empty sequence.

Given $u = (t_0, \ldots, t_{n-1})$ and assuming that $N_{\sigma(u)}$ has already been defined, we know that there must be infinitely many points in $N_{\sigma(u)}$, or else the center of this nbhd would be isolated. Choose then $x \neq y$ in $N_{\sigma(u)}$ and find open balls B_x, B_y with centers x and y respectively and such that

$$\bar{B}_x \subseteq N_{\sigma(u)}, \qquad \bar{B}_y \subseteq N_{\sigma(u)},$$
$$\bar{B}_x \cap \bar{B}_y = \emptyset,$$

as in Figure 1A.4.

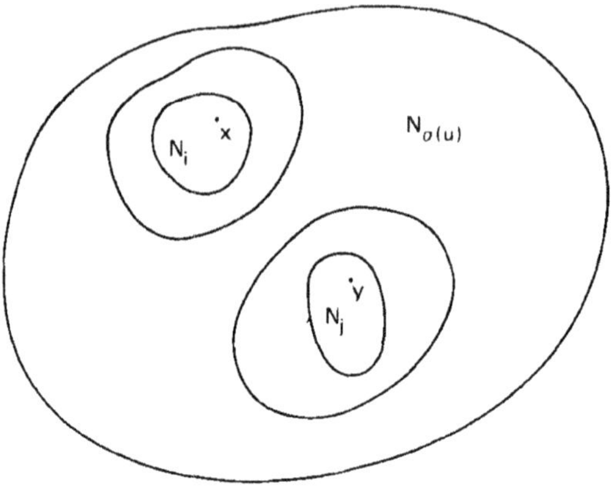

Figure 1A.4.

It is now enough to choose i, j such that $N_i \subseteq B_x$, $N_j \subseteq B_y$ and N_i, N_j have radii $\leq 2^{-n-1}$ and set

$$\sigma(t_0, \ldots, t_{n-1}, 0) = i, \qquad \sigma(t_0, \ldots, t_{n-1}, 1) = j.$$

Verification of (i), (ii), (iii) with this definition of σ is trivial. ⊣

1A.3. COROLLARY. *For every perfect Polish space $\mathcal{M}$, there is a continuous injection*

$$\pi : \mathcal{C} \rightarrowtail \mathcal{M}$$

of the Cantor set into $\mathcal{M}$.

PROOF. Given an infinite binary sequence α, put

$$x_n^\alpha = \textit{the center of } N(\mathcal{M}, \sigma(\alpha(0), \ldots, \alpha(n-1)))$$

and let

$$\pi(\alpha) = \textit{limit}_{n\to\infty}\, x_n^\alpha.$$

It is immediate that π is an injection (one-to-one). That π is continuous can be proved by verifying

$$\pi^{-1}[N_s] = \bigcup_n \{\alpha : N(\mathcal{M}, \sigma(\alpha(0), \ldots, \alpha(n-1))) \subseteq N_s\}.$$ ⊣

EXERCISES

1A.4. For each compact Polish space X, let $C[X]$ be the set of all continuous functions on X to $\mathcal{R}$ with the usual supnorm distance,

$$d(f, g) = \textit{supremum}\{|f(x) - g(x)| : x \in X\}.$$

Prove that $C[X]$ is a perfect Polish space. ⊣

1A.5. For each perfect Polish space X, let $H[X]$ be the set of all *compact* non-empty subsets of X. If $x \in X$ and $A \in H[X]$, put

$$d(x, A) = \textit{infimum}\{d(x, y) : y \in A\}$$

where on the right d is the distance function on X. The *Hausdorff distance* between two compact sets is defined by

$$d(A, B) = \textit{maximum}\{\textit{supremum}\{d(x, B) : x \in A\}, \\ \textit{supremum}\{d(y, A) : y \in B\}\}.$$

Prove that this is metric on $H[X]$ and that $H[X]$ is a perfect Polish space.

Hint. The set of all finite subsets of any dense subset of X is dense in $H[X]$. ⊣

1B. The Borel pointclasses of finite order

In order to study the subsets of a perfect Polish space $\mathcal{M}$, it will be necessary to consider other space related to $\mathcal{M}$, e.g. the products $\mathcal{M} \times \mathcal{M}$, $\mathcal{N} \times \mathcal{M}$, $\omega \times \mathcal{M}$. Let us first establish notation and terminology which make these detours easy.

We fix once and for all a collection $\mathfrak{F}$ of metric spaces with the following properties:

(1) The discrete space ω, the reals $\mathcal{R}$, Baire space $\mathcal{N}$ and the Cantor set $\mathcal{C}$ are in $\mathfrak{F}$.

(2) Every space in $\mathfrak{F}$ other than ω is a perfect Polish space.

Except for these restrictions we can leave membership in $\mathfrak{F}$ open—e.g. one might take ω, $\mathcal{R}$, $\mathcal{N}$ and $\mathcal{C}$ to be the only spaces in $\mathfrak{F}$. The idea is that we put in $\mathfrak{F}$ all the perfect Polish spaces in which we are interested.

The members of $\mathfrak{F}$ are the *basic* spaces. A *product space* (by definition) is any cartesian product

$$\mathfrak{X} = X_1 \times \cdots \times X_k,$$

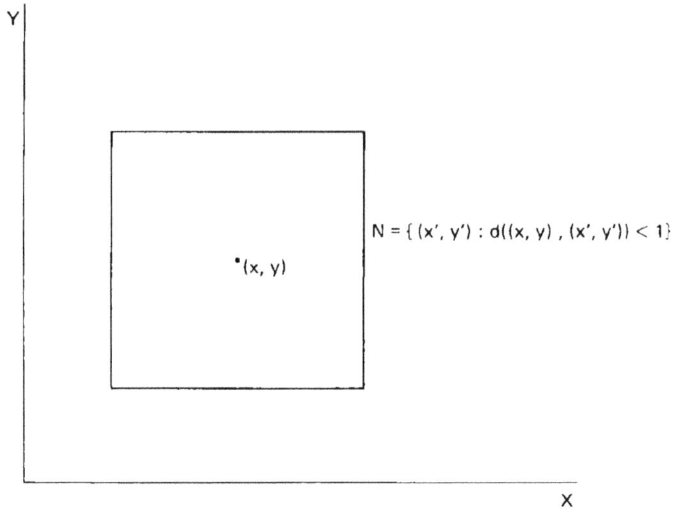

Figure 1B.1. The unit ball in a product space.

where each X_i is basic. Basic spaces count as product spaces by allowing $k = 1$ here. We naturally topologize $X_1 \times \cdots \times X_k$ as a product, i.e. with basic nbhds of the form

$$N = B_1 \times \cdots \times B_k,$$

where each B_i is a nbhd in X_i. It is easy to verify that this topology on $\mathfrak{X}$ is induced by the metric

$$d((x_1, \ldots, x_k), (y_1, \ldots, y_k)) = maximum\{d_1(x_1, y_1), \ldots, d_k(x_k, y_k)\},$$

where each d_i is the given metric on X_i. (See Figure 1B.1.)

Two product spaces $\mathfrak{X} = X_1 \times \cdots \times X_k$ and $\mathcal{Y} = Y_1 \times \cdots \times Y_l$ are *equal* if $k = l$ and $X_1 = Y_1, \ldots, X_k = Y_k$. We then define *products of product spaces* by going back to the basic factors, i.e. if

$$\mathfrak{X} = X_1 \times \cdots \times X_k$$

and

$$\mathcal{Y} = Y_1 \times \cdots \times Y_l,$$

then (by definition)

$$\mathfrak{X} \times \mathcal{Y} = X_1 \times \cdots \times X_k \times Y_1 \times \cdots \times Y_l.$$

Thus

$$\mathfrak{X} \times (\mathfrak{Y} \times \mathfrak{Z}) = (\mathfrak{X} \times \mathfrak{Y}) \times \mathfrak{Z} = \mathfrak{X} \times \mathfrak{Y} \times \mathfrak{Z}.$$

We call the tuples in these product spaces *points* and the subsets of these spaces *pointsets.*

If $x = (x_1, \ldots, x_k)$ and $y = (y_1, \ldots, y_l)$, then (by definition)

$$(x, y) = (x_1, \ldots, x_k, y_1, \ldots, y_l).$$

As with products of product spaces, this pairing operation is associative,

$$(x, (y, z)) = ((x, y), z) = (x, y, z).$$

We can think of pointsets as *sets* or as *relations* with arguments in the basic spaces. Both points of view are useful and we will use interchangeably the customary notations for these, i.e. for $P \subseteq \mathfrak{X}$,

$$x \in P \Leftrightarrow P(x).$$

Of course, we will not be studying individual pointsets so much as *collections of pointsets,* call them *pointclasses.* Thus a pointclass Λ is a collection of sets such that each P in Λ *is a subset of some product space* $\mathfrak{X}$. For example, we may have

$$\begin{aligned}\Lambda &= \text{all open pointsets}\\ &= \xi\{P\colon P \subseteq \mathfrak{X} \text{ for some product space } \mathfrak{X} \text{ and } P \text{ is open}\}.\end{aligned}$$

In definability theory we typically start with a small pointclass Λ and certain operations on pointsets and then we study the sets which can be constructed by applying (once or repeatedly) the given operations to the members of Λ. For *the Borel sets of finite order* we start with the open sets and we apply repeatedly the operations of *complementation* or *negation* ($\neg$) and *projection along* ω or *existential number quantification* ($\exists^\omega$).

More precisely, if $P \subseteq \mathfrak{X}$ is any pointset, put

$$\neg P = \mathfrak{X} - P.$$

For a pointclass Λ, let

$$\neg \Lambda = \{\neg P\colon P \in \Lambda\}$$

be *the dual pointclass.*

Similarly, if $P \in \mathfrak{X} \times \omega$ for some $\mathfrak{X}$, let

$$\begin{aligned}\exists^\omega P &= \{x \in \mathfrak{X}\colon \text{for some } n, P(x, n)\}\\ &= \{x \in \mathfrak{X}\colon (\exists n) P(x, n)\}\end{aligned}$$

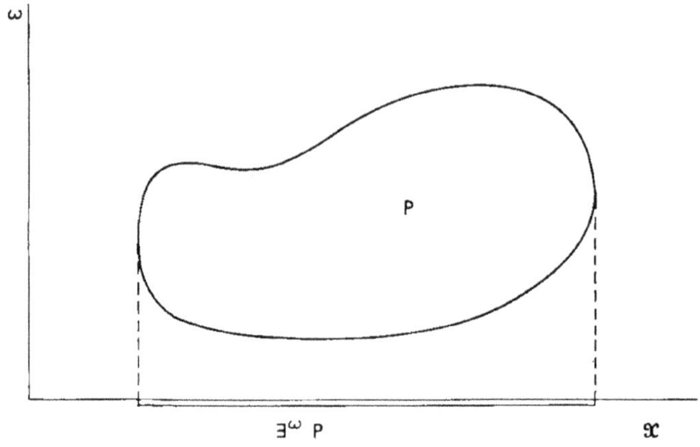

Figure 1B.2. Projection along ω.

and for a pointclass Λ, put

$$\exists^{\omega}\Lambda = \{\exists^{\omega}P \colon P \in \Lambda, P \subseteq \mathfrak{X} \times \omega \text{ for some } \mathfrak{X}\};$$

see Figure 1B.2.

The *Borel pointclasses of finite order* $\boldsymbol{\Sigma}^0_n (n \geq 1)$ are defined by the induction

$$\boldsymbol{\Sigma}^0_1 = \textit{all open pointsets},$$

$$\boldsymbol{\Sigma}^0_{n+1} = \exists^{\omega} \neg \boldsymbol{\Sigma}^0_n;$$

the *dual Borel pointclasses* $\boldsymbol{\Pi}^0_n$ are defined by

$$\boldsymbol{\Pi}^0_n = \neg \boldsymbol{\Sigma}^0_n;$$

finally, the *ambiguous Borel pointclasses* $\boldsymbol{\Delta}^0_n$ are given by

$$\boldsymbol{\Delta}^0_n = \boldsymbol{\Sigma}^0_n \cap \boldsymbol{\Pi}^0_n.^{(13)}$$

Thus, $\boldsymbol{\Pi}^0_1$ consists of all closed pointsets, $\boldsymbol{\Delta}^0_1$ is the class of all clopen sets, $\boldsymbol{\Sigma}^0_2$ is the class of all projections along ω of closed sets, etc. Put another way, a set $P \subseteq \mathfrak{X}$ is $\boldsymbol{\Sigma}^0_2$ if there is a closed $F \subseteq \mathfrak{X} \times \omega$ such that for all x,

$$P(x) \Leftrightarrow (\exists t) F(x, t).$$

Similarly, P is $\boldsymbol{\Sigma}^0_3$ if there is a closed F such that

$$\begin{aligned} P(x) &\Leftrightarrow (\exists t) \neg (\exists s) F(x, t, s) \\ &\Leftrightarrow (\exists t)(\forall s) \neg F(x, t, s), \end{aligned}$$

i.e. P is $\mathbf{\Sigma}^0_3$ if there is an open pointset G such that

$$P(x) \Leftrightarrow (\exists t_1)(\forall t_2)G(x, t_1, t_2).$$

Similar *normal forms* can be computed for the pointclasses $\mathbf{\Pi}^0_n$, e.g. P is $\mathbf{\Pi}^0_4$ if there is some open G such that for all x,

$$P(x) \Leftrightarrow (\forall t_1)(\exists t_2)(\forall t_3)G(x, t_1, t_2, t_3).$$

In the classical terminology, $\mathbf{\Sigma}^0_2$ sets are called F_σ *sets*, $\mathbf{\Pi}^0_2$ sets are G_δ, $\mathbf{\Sigma}^0_3$ sets are $G_{\delta\sigma}$, $\mathbf{\Pi}^0_3$ sets are $F_{\sigma\delta}$, etc. It is a cumbersome notation and we will not use it, except for an occasional reference to F_σ's and G_δ's.

1B.1. THEOREM. *The following diagram of inclusions holds among the Borel pointclasses of finite order*:

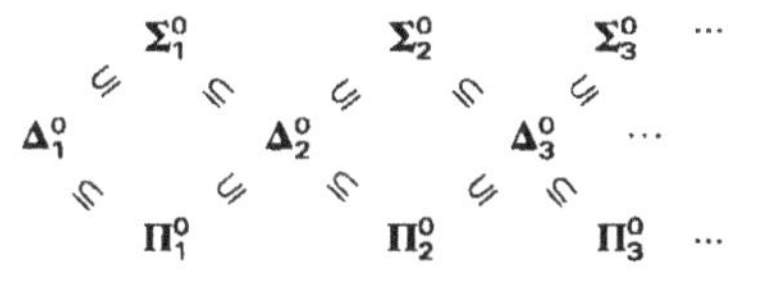

Diagram 1B.3. The Borel pointclasses of finite order.

PROOF. The inclusions

$$\mathbf{\Sigma}^0_n \subseteq \mathbf{\Pi}^0_{n+1}$$

are almost immediate from the definitions. Taking $n = 3$ to simplify notation, if P is $\mathbf{\Sigma}^0_3$, then

$$P(x) \Leftrightarrow (\exists t_1)(\forall t_2)G(x, t_1, t_2)$$

with some open $G \subseteq \mathcal{X} \times \omega \times \omega$. We can rewrite this as

$$(*) \qquad P(x) \Leftrightarrow (\forall s)(\exists t_1)(\forall t_2)G(x, t_1, t_2)$$

since the addition of the *vacuous quantifier* $(\forall s)$ does not affect the meaning of the equivalence. Now define

$$G'(x, s, t_1, t_2) \Leftrightarrow G(x, t_1, t_2)$$

and notice that G' is (trivially) open, so equivalence $(*)$ above establishes that P is $\mathbf{\Pi}^0_4$.

To prove the inclusions

$$\mathbf{\Sigma}^0_n \subseteq \mathbf{\Sigma}^0_{n+1},$$

recall that in a separable metric space every open set is a countable union of closed sets. If $G \subseteq \mathfrak{X}$ and

$$G = \bigcup_t F_t$$

with each F_t closed, define $F \subseteq \mathfrak{X} \times \omega$ by

$$F(x, t) \Leftrightarrow x \in F_t$$

and notice that F is closed and

$$G(x) \Leftrightarrow (\exists t) F(x, t).$$

Thus G is $\boldsymbol{\Sigma}^0_2$, and since G was arbitrary open,

$$\boldsymbol{\Sigma}^0_1 \subseteq \boldsymbol{\Sigma}^0_2.$$

Hence $\boldsymbol{\Sigma}^0_2 = \exists^\omega \neg \boldsymbol{\Sigma}^0_1 \subseteq \exists^\omega \neg \boldsymbol{\Sigma}^0_2 = \boldsymbol{\Sigma}^0_3$ and inductively, $\boldsymbol{\Sigma}^0_n \subseteq \boldsymbol{\Sigma}^0_{n+1}$.

This establishes

$$\boldsymbol{\Sigma}^0_n \subseteq \boldsymbol{\Delta}^0_{n+1}$$

for every n, so taking negations,

$$\boldsymbol{\Pi}^0_n \subseteq \boldsymbol{\Delta}^0_{n+1}$$

and the remaining inclusions in the diagram are trivial. ⊣

EXERCISES

1B.2. Prove that if $\mathfrak{X} = X_1 \times \cdots \times X_k$ is a product space with at least one factor $X_i = \mathcal{N}$ and every X_j either ω or $\mathcal{N}$, then $\mathfrak{X}$ is homeomorphic with $\mathcal{N}$.

Hint. Construct homeomorphisms of $\omega \times \mathcal{N}$ and $\mathcal{N} \times \mathcal{N}$ with $\mathcal{N}$ and then use induction on k. ⊣

1B.3. Prove that if $\mathfrak{X} = X_1 \times \cdots \times X_k$ is a product space with at least one factor X_i not ω, then $\mathfrak{X}$ is a perfect Polish space. ⊣

1B.4. Prove that a pointset P is $\boldsymbol{\Sigma}^0_2$ if and only if

$$P = \bigcup_{i=0}^{\infty} F_i,$$

with each F_i closed.

Similarly, P is $\boldsymbol{\Pi}^0_2$, if and only if

$$P = \bigcap_{i=0}^{\infty} G_i$$

with each G_i open. ⊣

This is the classical definition of F_σ and G_δ sets. These occur quite often in analysis, for example consider the following problem.

1B.5. Let $f:\mathcal{R}\to\mathcal{R}$ be an arbitrary function on the line. Prove that the set

$$A=\{x\in\mathcal{R}: f \text{ is continuous at } x\}$$

is a G_δ.

Hint. Define the *variation* of f on an interval (a, b) by

$$V(a,b)=\mathit{supremum}\{f(x): a<x<b\} \\ -\mathit{infimum}\{f(x): a<x<b\},$$

where the value may be ∞ or $-\infty$. The *local variation* of f is given by

$$v(x)=\mathit{limit}_{n\to\infty}\, V\left(x-\frac{1}{n}, x+\frac{1}{n}\right)$$

and it is clear that f is continuous at x just in case $v(x)=0$. Show that for each n, the set

$$A_n=\left\{x: v(x)<\frac{1}{n}\right\}$$

is open and $A=\bigcap_n A_n$. ⊣

1B.6. Prove that if $n\geq 3$ is odd, then P is $\boldsymbol{\Sigma}^0_n$ if and only if there is an open set G such that

$$P(x)\Leftrightarrow(\exists t_1)(\forall t_2)(\exists t_3)(\forall t_4)\cdots(\forall t_{n-1})G(x,t_1,\ldots,t_{n-1}).$$

Similarly, if n is even, then P is $\boldsymbol{\Sigma}^0_n$ if and only if there is a closed set F such that

$$P(x)\Leftrightarrow(\exists t_1)(\forall t_2)(\exists t_3)\cdots(\exists t_{n-1})F(x,t_1,\ldots,t_{n-1}).$$

Find similar normal forms for the $\boldsymbol{\Pi}^0_n$ pointsets. ⊣

1B.7. Prove that if $\mathfrak{X}$ is a product of copies of ω and $\mathcal{N}$ and P is $\boldsymbol{\Sigma}^0_n$ with n odd, then there exists a clopen set R such that

$$P(x)\Leftrightarrow(\exists t_1)(\forall t_2)\cdots(\forall t_{n-1})(\exists t_n)R(x,t_1,\ldots,t_n);$$

similarly for even n, with the last quantifier $\forall$.

Hint. If $A\subseteq\mathfrak{X}$ is open, then $A=\bigcup_n R_n$ with clopen R_n in these spaces and we can take

$$R(x,n)\Leftrightarrow x\in R_n.$$

⊣

1C. Computing with relations; closure properties

The relational notation for pointsets is particularly convenient in putting down compact expressions for complicated definitions. For example, suppose that $Q \subseteq \mathfrak{X} \times \mathcal{N}$, $R \subseteq \mathfrak{X} \times \mathcal{N} \times \omega$ and let

$$P(x) \Leftrightarrow (\forall \alpha)[Q(x, \alpha) \Rightarrow (\exists i) R(x, \alpha, i)].$$

Here the logical symbols are taken with their customary meaning, as we have been using them all along: $\forall$ (for all), $\exists$ (there exists), $\Rightarrow$ (implies), & (and), $\vee$ (or), $\neg$ (not).

We will also use customarily Greek variables α, β, γ,... from the beginning of the alphabet over $\mathcal{N}$ and i, j, k, l, m, n, s, t, over ω. This will save us having to specify explicitly the range of the quantifiers in each definition.

One can view the logical symbols as denoting operations on pointsets. In general, a k-*ary pointset operation* is a function Φ with domain some set of k-tuples of pointsets and pointsets as values.

With this terminology, *conjunction* & is the binary pointset operation which assigns to every pair P, Q *of subsets of the same space* $\mathfrak{X}$ the set P & Q,

$$x \in (P \,\&\, Q) \Leftrightarrow P(x) \,\&\, Q(x).$$

Of course

$$P \,\&\, Q = P \cap Q.$$

We will, however keep the symbol $\cap$ for denoting the general set theoretic operation of intersection, with $A \cap B$ defined for arbitrary sets A, B.

Similarly, the *disjunction* $P \vee Q$ of two pointsets is defined when P and Q are subsets of the same $\mathfrak{X}$ and

$$P \vee Q = P \cup Q = \{x : P(x) \vee Q(x)\}.$$

Negation is most conveniently regarded as a collection of operations $\neg_{\mathfrak{X}}$, one for each product space $\mathfrak{X}$, with $\neg_{\mathfrak{X}} P$ defined when $P \subseteq \mathfrak{X}$:

$$\neg_{\mathfrak{X}} P = \mathfrak{X} - P = \{x \in \mathfrak{X} : \neg P(x)\}.$$

In practice we will always write $\neg P$ for $\neg_{\mathfrak{X}} P$, as $\mathfrak{X}$ is clear from the context.

From these we can construct more pointset operations by *composition*, e.g. the *implication* $P \Rightarrow Q$ of P and Q is defined by

$$(P \Rightarrow Q) = \neg P \vee Q.$$

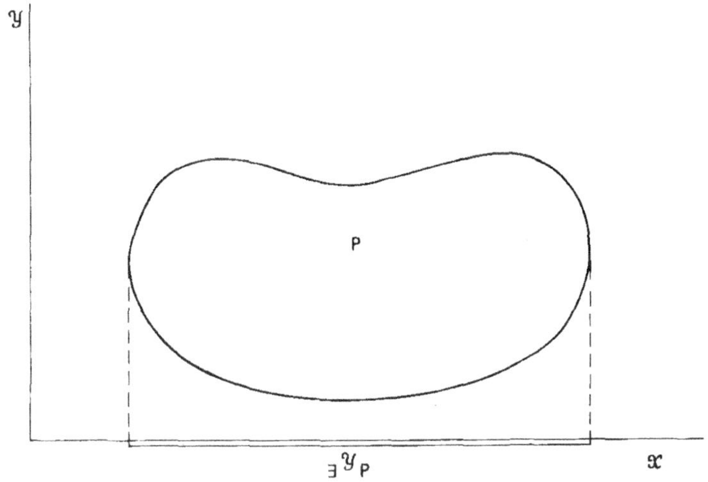

Figure 1C.1. Projection along $\mathcal{Y}$.

More interesting than these *propositional* pointset operations are the *projections* and *dual projections* or *quantifiers.* If $P \subseteq \mathcal{X} \times \mathcal{Y}$, put

$$\exists^{\mathcal{Y}} P = \{x \in \mathcal{X} \colon (\exists y) P(x, y)\}$$

as in Figure 1C.1.

For each fixed product space $\mathcal{Y}$, we call the operation $\exists^{\mathcal{Y}}$ *projection along* $\mathcal{Y}$ or *existential quantification on* $\mathcal{Y}$. Clearly $\exists^{\mathcal{Y}} P$ is defined when $P \subseteq \mathcal{X} \times \mathcal{Y}$ for some $\mathcal{X}$, in which case $\exists^{\mathcal{Y}} P \subseteq \mathcal{X}$.

We have already used projection along ω, $\exists^{\omega}$.

Only the projections along basic spaces are fundamental, since all the others can be obtained from these by composition; for example, if $\mathcal{Y} = \omega \times \mathcal{N}$, then for each $P \subseteq \mathcal{X} \times \omega \times \mathcal{N}$,

$$\exists^{\mathcal{Y}} P = \exists^{\omega} \exists^{\mathcal{N}} P,$$

i.e. in relational notation,

$$(\exists y \in \mathcal{Y}) P(x, y) \Leftrightarrow (\exists n)(\exists \alpha) P(x, n, \alpha).$$

If $P \subseteq \mathcal{X} \times \mathcal{Y}$, put

$$\forall^{\mathcal{Y}} P = \neg \exists^{\mathcal{Y}} \neg P,$$

i.e.

$$\begin{aligned} x \in \forall^{\mathcal{Y}} P &\Leftrightarrow \neg (\exists y \in \mathcal{Y}) \neg P(x, y) \\ &\Leftrightarrow (\forall y \in \mathcal{Y}) P(x, y). \end{aligned}$$

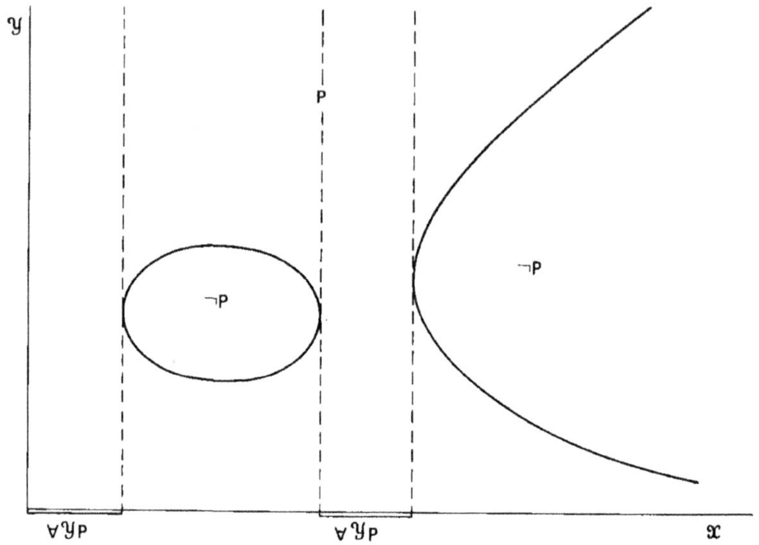

Figure 1C.2. Universal quantification on 𝔜.

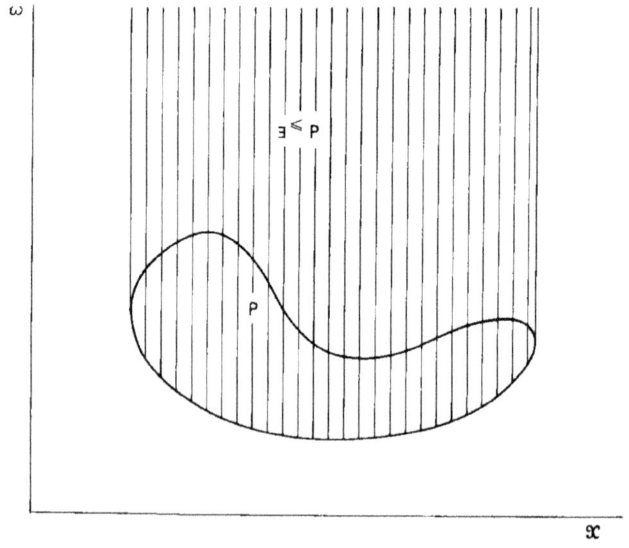

Figure 1C.3. Bounded number quantification.

Fixing $\mathcal{Y}$, we call the operation $\forall^{\mathcal{Y}}$ *dual projection along* $\mathcal{Y}$ or *universal quantification on* $\mathcal{Y}$. Again, $\forall^{\mathcal{Y}}P$ is defined when $P \subseteq \mathcal{X} \times \mathcal{Y}$ for some $\mathcal{X}$ and then $\forall^{\mathcal{Y}}P \subseteq \mathcal{X}$, see Figure 1C.2.

In addition to the operations $\exists^{\omega}$, $\forall^{\omega}$, the *bounded number quantifiers* will prove useful,

$$(x, n) \in \exists^{\leq} P \Leftrightarrow (\exists m \leq n) P(x, m),$$

$$(x, n) \in \forall^{\leq} P \Leftrightarrow (\forall m \leq n) P(x, m),$$

see Figure 1C.3

Clearly $\exists^{\leq}P$, $\forall^{\leq}P$ are defined when $P \subseteq \mathcal{X} \times \omega$ for some $\mathcal{X}$, in which case both $\exists^{\leq}P$ and $\forall^{\leq}P$ are subsets of $\mathcal{X} \times \omega$.

A pointclass Λ *is closed under* a k-ary pointset operation Φ if whenever $P_1, \ldots, P_k$ are in Λ and $\Phi(P_1, \ldots, P_k)$ is defined, then $\Phi(P_1, \ldots, P_k)$ is also in Λ. For example, Λ is closed under conjunction if whenever P and Q are subsets of the same space $\mathcal{X}$ and both are in Λ, then $P \,\&\, Q = P \cap Q$ is in Λ.

Similarly, Λ is closed under negation $\neg$, if $\neg\Lambda \subseteq \Lambda$, i.e. if for every $P \in \Lambda$, $P \subseteq \mathcal{X}$ we have $\mathcal{X} - P \in \Lambda$.

We say that Λ is closed under *continuous substitution* if for every continuous function $f : \mathcal{X} \to \mathcal{Y}$ and every $P \in \Lambda$, $P \subseteq \mathcal{Y}$, $f^{-1}[P] \in \Lambda$. Here of course

$$x \in f^{-1}[P] \Leftrightarrow f(x) \in P \Leftrightarrow P(f(x)).$$

It is worth putting down a very useful alternate version of this closure property.

1C.1. LEMMA. *Suppose Λ is a pointclass closed under continuous substitution, let $f_1 : \mathcal{X} \to \mathcal{Y}_1, \ldots, f_m : \mathcal{X} \to \mathcal{Y}_m$ be continuous functions and assume that $Q \subseteq \mathcal{Y}_1 \times \cdots \times \mathcal{Y}_m$ is a pointset in Λ. If*

$$P(x) \Leftrightarrow Q(f_1(x), \ldots, f_m(x)),$$

then P is also in Λ.

PROOF. The function $g : \mathcal{X} \to \mathcal{Y}_1 \times \cdots \times \mathcal{Y}_m$ defined by

$$g(x) = (f_1(x), \ldots, f_m(x))$$

is continuous and

$$P(x) \Leftrightarrow Q(g(x)).$$

For example, suppose

$$P(x, y) \Leftrightarrow Q(y, x) \& R(x, y, y),$$

where $Q \subseteq \mathcal{Y} \times \mathcal{X}$, $R \subseteq \mathcal{X} \times \mathcal{Y} \times \mathcal{Y}$ and both Q and R are in some pointclass Λ closed under continuous substitution and &. Then P too is in Λ, since

$$P(x, y) \Leftrightarrow Q'(x, y) \& R'(x, y),$$

where

$$Q'(x, y) \Leftrightarrow Q(f_1(x, y), f_2(x, y)),$$
$$R'(x, y) \Leftrightarrow R(f_2(x, y), f_1(x, y), f_1(x, y))$$

with

$$f_1(x, y) = y, \qquad f_2(x, y) = x.$$

In effect, closure under continuous substitution allows us to *permute* or *identify* variables in a relation and stay in the pointclass we are working with.

After these preliminary remarks we can state concisely the elementary closure properties of the Borel classes. To prove them, we will need functions that code finite sequences of integers by single integers.

Let

$$p(i) = p_i = \textit{the } i\textit{'th prime},$$

with $p_0 = 2$, and for each n, put

$$\langle t_0, \ldots, t_{n-1} \rangle = p_0^{t_0+1} \cdot \cdots \cdot p_{n-1}^{t_{n-1}+1}.$$

By convention the empty product is 1, so that

$$\langle \emptyset \rangle = 1,$$

and 1 is the code of the empty sequence. With this particular coding of tuples we associate the natural decoding functions and relations

$$\begin{aligned} Seq(u) &\Leftrightarrow u = 1 \text{ or } u = \langle t_0, \ldots, t_{n-1} \rangle \quad \text{for some } t_0, \ldots, t_{n-1}, \\ lh(u) &= n \quad \text{if } u = \langle t_0, \ldots, t_{n-1} \rangle \quad \text{for some } t_0, \ldots, t_{n-1}, \\ &= 0 \quad \text{otherwise}, \\ (u)_i &= t_i \quad \text{if } u = \langle t_0, \ldots, t_{n-1} \rangle \quad \text{for some} \\ &\qquad t_0, \ldots, t_{n-1} \text{ and } i < n, \\ &= 0 \quad \text{otherwise.} \end{aligned}$$

It is often convenient to index a finite sequence starting with 1 rather than 0. Notice that if

$$u = \langle t_1, \ldots, t_n \rangle,$$

then for $i < n$, $(u)_i = t_{i+1}$.

1C.2. THEOREM. *Each Borel pointclass* $\mathbf{\Sigma}^0_n$ $(n \geq 1)$ *is closed under continuous substitution,* $\vee$, $\&$, $\exists^{\leq}$, $\forall^{\leq}$ *and* $\exists^{\omega}$.

Each dual pointclass $\mathbf{\Pi}^0_n$ *is closed under continuous substitution,* $\vee$, $\&$, $\exists^{\leq}$, $\forall^{\leq}$ *and* $\forall^{\omega}$.

Each ambiguous Borel pointclass $\mathbf{\Delta}^0_n$ *is closed under continuous substitution,* $\neg$, $\vee$, $\&$, $\exists^{\leq}$ *and* $\forall^{\leq}$.

PROOF. The results about $\mathbf{\Pi}^0_n$ and $\mathbf{\Delta}^0_n$ follow immediately from those about $\mathbf{\Sigma}^0_n$. Also the closure properties of $\mathbf{\Sigma}^0_1$ are trivial, except perhaps for closure under $\exists^{\leq}$ and $\forall^{\leq}$ which follow easily from the equations

$$\exists^{\leq} P = \bigcup_n \{(x, n) : (\exists m \leq n) P(x, m)\},$$
$$\forall^{\leq} P = \bigcup_n \{(x, n) : (\forall m \leq n) P(x, m)\}.$$

Assume now that $\mathbf{\Sigma}^0_n$ has all the right closure properties—we will show the same for $\mathbf{\Sigma}^0_{n+1}$.

Suppose first that Q is a typical $\mathbf{\Sigma}^0_{n+1}$ subset of $\mathcal{Y}$, i.e.

$$Q(y) \Leftrightarrow (\exists m) \neg P(y, m),$$

with P some $\mathbf{\Sigma}^0_n$ subset of $\mathcal{Y} \times \omega$. Assume also that $f : \mathcal{X} \to \mathcal{Y}$ is continuous. Now

$$Q(f(x)) \Leftrightarrow (\exists m) \neg P(f(x), m)$$
$$\Leftrightarrow (\exists m) \neg P'(x, m)$$

with

$$P'(x, m) \Leftrightarrow P(f(x), m).$$

Since P' is $\mathbf{\Sigma}^0_n$ by 1C.1 and the induction hypothesis, $f^{-1}[Q]$ is $\mathbf{\Sigma}^0_{n+1}$. Hence $\mathbf{\Sigma}^0_{n+1}$ is closed under continuous substitution.

To prove closure of $\mathbf{\Sigma}^0_{n+1}$ under $\&$, compute

$$R(x) \Leftrightarrow (\exists s) \neg P(x, s) \,\&\, (\exists t) \neg Q(x, t)$$
$$\Leftrightarrow (\exists u)[\neg P(x, (u)_0) \,\&\, \neg Q(x, (u)_1)]$$
$$\Leftrightarrow (\exists u) \neg [P(x, (u)_0) \vee Q(x, (u)_1)].$$

If P and Q are Σ^0_n, then

$$P'(x, u) \Leftrightarrow P(x, (u)_0) \vee Q(x, (u)_1)$$

is also Σ^0_n by closure under continuous substitution and $\vee$, so R is Σ^0_{n+1}.

This method of proof goes by the fancy name of *like quantifiei contraction* and yields equally trivial proofs of closure of Σ^0_{n+1} under $\vee$ and $\exists^\omega$. For closure under $\forall^\leq$ we need a slightly more elaborate contraction of finitely many quantifiers.

Suppose

$$R(x, n) \Leftrightarrow (\forall m \leq n)(\exists s) \neg P(x, m, s)$$

with P in Σ^0_n and compute,

$$\begin{aligned} R(x, n) &\Leftrightarrow (\exists u)(\forall m \leq n) \neg P(x, m, (u)_m) \\ &\Leftrightarrow (\exists u) \neg (\exists m \leq n) P(x, m, (u)_m). \end{aligned}$$

Again

$$P'(x, n, u) \Leftrightarrow (\exists m \leq n) P(x, m, (u)_m)$$

is Σ^0_n by closure of this class under continuous substitution and $\exists^\leq$, so R is Σ^0_{n+1}.

Proof of closure of Σ^0_{n+1} under $\exists^\leq$ is trivial. ⊣

This simple argument is a good illustration of the advantage of relational (or logical) notation, i.e. writing

$$R(x, n) \Leftrightarrow (\forall m \leq n)(\exists s) \neg P(x, m, s)$$

rather than

$$R = \forall^\leq \exists^\omega \neg P.$$

In fact the whole proof rested on some *quantifier manipulation rules* whose truth is transparent in logical notation. We list them here for reference, but we will apply them in the future without much ado.

$$(\exists s)(\exists t) P(s, t) \Leftrightarrow (\exists u) P((u)_0, (u)_1),$$

$$(\forall s)(\forall t) P(s, t) \Leftrightarrow (\forall u) P((u)_0, (u)_1),$$

$$(\forall m \leq n)(\exists s) P(m, s) \Leftrightarrow (\exists u)(\forall m \leq n) P(m, (u)_m),$$

$$(\exists m \leq n)(\forall s) P(m, s) \Leftrightarrow (\forall u)(\exists m \leq n) P(m, (u)_m).$$

These rules are useful because they allow us to simplify the *quantifier prefix* of a complicated logical expression by introducing *continuous substitutions* in the *matrix*.

To see how we can use the closure properties of a pointclass, suppose that

$$P(x) \Leftrightarrow (\exists t)(\exists s)\{Q(x, s) \Rightarrow (\exists u)[R(u, f(x, u), t) \vee S(u, x, s)]\},$$

where Q, R, S are Σ^0_n, f is continuous and t, s, u range over ω. We will argue that P is Σ^0_{n+1}.

First put

$$Q'(x, t, s) \Leftrightarrow Q(x, s),$$
$$R'(x, t, s, u) \Leftrightarrow R(u, f(x, u), t),$$
$$S'(x, t, s, u) \Leftrightarrow S(u, x, s),$$

and notice that Q', R', S' are Σ^0_n by closure of this pointclass under continuous substitution. Now

$$P(x) \Leftrightarrow (\exists t)(\exists s)\{\neg Q'(x, t, s) \vee (\exists u)[R'(x, t, s, u) \vee S'(x, t, s, u)]\}$$
$$\Leftrightarrow (\exists t)(\exists s)\{\neg Q'(x, t, s) \vee (\exists u)T(x, t, s, u)\}$$
$$\Leftrightarrow (\exists t)(\exists s)\{\neg Q'(x, t, s) \vee T'(x, t, s)\}$$
$$\Leftrightarrow (\exists t)(\exists s)T''(x, t, s)$$

where T, T', T'' are defined by

$$T(x, t, s, u) \Leftrightarrow R'(x, t, s, u) \vee S'(x, t, s, u),$$
$$T'(x, t, s) \Leftrightarrow (\exists u)T(x, t, s, u),$$
$$T''(x, t, s) \Leftrightarrow \neg Q(x, t, s) \vee T'(x, t, s).$$

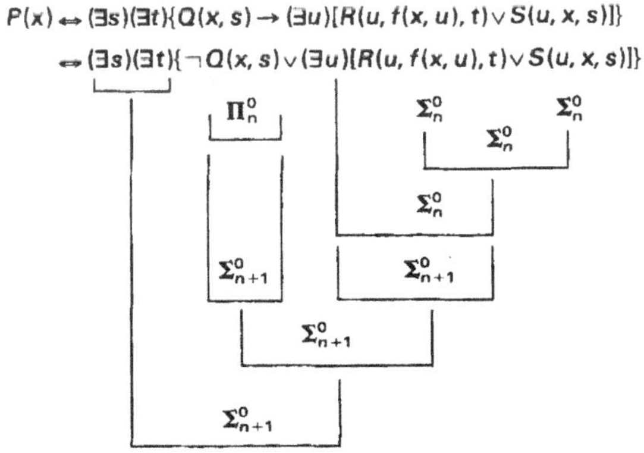

Diagram 1C.4.

Clearly T and T' are $\mathbf{\Sigma}^0_n$ by the closure properties of this pointclass. Hence T' is $\mathbf{\Sigma}^0_{n+1}$ by 1B.1 and since $\neg Q$ is also $\mathbf{\Sigma}^0_{n+1}$, T'' is $\mathbf{\Sigma}^0_{n+1}$. Finally P is $\mathbf{\Sigma}^0_{n+1}$ by two applications of closure of this pointclass under $\exists^\omega$.

This kind of computation is so simple that we will not usually bother to put it down. One way to make computations of this type with a minimum of writing is to use a diagram like 1C.4 above which shows step-by-step the properties of the relevant pointclasses that we use.

EXERCISES

1C.3. Let $f:\mathcal{R}\to\mathcal{R}$ be a continuous function on the line. Prove that the relations

$$P(x, y) \Leftrightarrow f'(x) = y,$$

$$Q(x) \Leftrightarrow f'(x) \text{ exists}$$

are both $\mathbf{\Pi}^0_3$.

Hint. Let $r_0, r_1, \ldots$ be an enumeration of all rational numbers and put

$$R(x, y, s, k, m) \Leftrightarrow r_m \neq 0 \;\&\; \left|\frac{1}{r_m}\{f(x+r_m)-f(x)\}-y\right| \leq \frac{1}{s+1}.$$

Clearly R is a closed relation. It is easy to verify that

$$P(x, y) \Leftrightarrow (\forall s)(\exists k)(\forall m)\left\{0<|r_m|<\frac{1}{k+1} \Rightarrow R(x, y, s, k, m)\right\}.$$

The second assertion is proved similarly, starting with the relation

$$S(x, s, k, m, n) \Leftrightarrow r_m \neq 0 \;\&\; r_n \neq 0$$

$$\&\; \left|\frac{1}{r_m}\{f(x+r_m)-f(x)\}-\frac{1}{r_n}\{f(x+r_n)-f(x)\}\right| \leq \frac{1}{s+1}. \qquad \dashv$$

1C.4. Let $C[0, 1]$ be the space of continuous real functions on the unit interval and define $Q \subseteq C[0, 1]\times\mathcal{R}$ by

$$Q(f, x) \Leftrightarrow 0<x<1 \;\&\; f'(x) \text{ exists}.$$

Prove that Q is $\mathbf{\Pi}^0_3$. $\dashv$

1C.5. Prove that if $P\subseteq\mathcal{X}$ *and* $Q\subseteq\mathcal{Y}$ are $\mathbf{\Sigma}^0_n$, then the product $P\times Q\subseteq \mathcal{X}\times\mathcal{Y}$ is $\mathbf{\Sigma}^0_n$.

Hint. Use closure under continuous substitution. $\dashv$

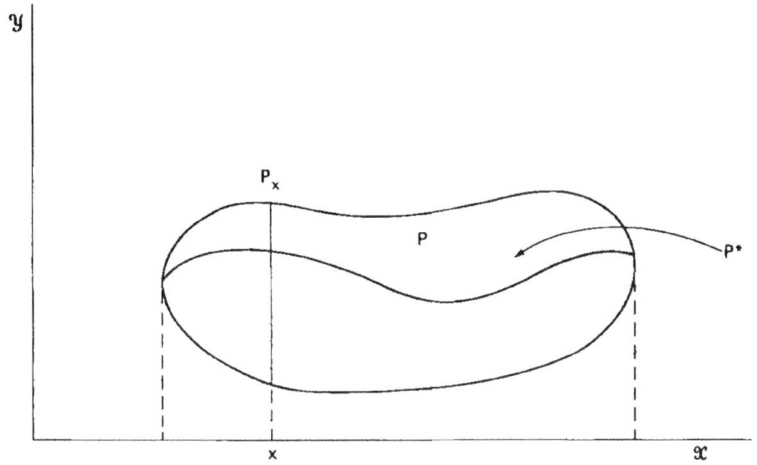

Figure 1C.5. Uniformization.

For the next exercise we introduce the basic problem of *uniformization.*[(17)]

Suppose $P \subseteq \mathfrak{X} \times \mathfrak{Y}$ is a subset of the product $\mathfrak{X} \times \mathfrak{Y}$. We say that P^* *uniformizes* P, if $P^* \subseteq P$ and P^* is the graph of a function with domain the projection $\exists^{\mathfrak{Y}} P$. Intuitively, P^* assigns to each point in $\exists^{\mathfrak{Y}} P$ just one member of the *section* or *fiber*

$$P_x = \{y : P(x, y)\}$$

as in Figure 1C.5.

It follows from the axiom of choice that each P can be uniformized by some P^*; on the other hand, it is often very difficult to find a definable uniformizing set, even if the given set is very simple.

The next exercise solves the uniformization problem in a very simple situation, but we will see later that even this easy result is useful.

1C.6. Prove that for each $n > 1$, if $P \subseteq \mathfrak{X} \times \omega$ is in $\mathbf{\Sigma}^0_n$, then there is some P^* also in $\mathbf{\Sigma}^0_n$ which uniformizes P.

Hint. Suppose

$$P(x, m) \Leftrightarrow (\exists i) Q(x, m, i)$$

with Q in $\mathbf{\Pi}^0_{n-1}$. Put

$$R(x, s) \Leftrightarrow Q(x, (s)_0, (s)_1) \ \& \ (\forall t < s) \neg Q(x, (t)_0, (t)_1),$$

$$P^*(x, m) \Leftrightarrow (\exists i) R(x, \langle m, i \rangle). \qquad \dashv$$

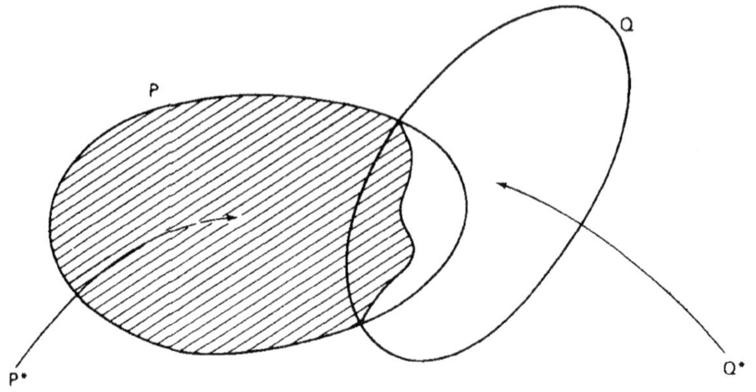

Figure 1C.6. Reduction.

Suppose P and Q are subsets of the same space $\mathfrak{X}$. We say that the pair P^*, Q^* *reduces* the pair P, Q if the following hold:

$$P^* \subseteq P, \qquad Q^* \subseteq Q,$$
$$P \cup Q = P^* \cup Q^*,$$
$$P^* \cap Q^* = \emptyset.$$ [16]

(See Figure 1C.6.)

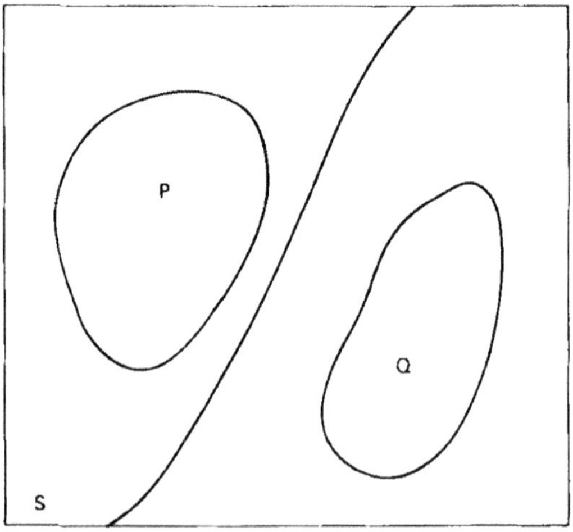

Figure 1C.7. Separation.

1C.7. Prove that for each $n > 1$, every pair of sets P, Q in $\boldsymbol{\Sigma}^0_n$ is reducible by a pair P^*, Q^* in $\boldsymbol{\Sigma}^0_n$.

Hint. Uniformize the set R defined by

$$R(x, m) \Leftrightarrow \{P(x) \ \& \ m = 0\} \vee \{Q(x) \ \& \ m = 1\}. \qquad \dashv$$

Suppose that P and Q are disjoint subsets of the same space $\mathcal{X}$. We say that the set S *separates* P from Q if

$$P \subseteq S, \qquad Q \cap S = \emptyset.^{(16)}$$

(See Figure 1C.7.)

1C.8. Prove that for each $n > 1$, every disjoint pair of sets P, Q in $\boldsymbol{\Pi}^0_n$ can be separated by a set in $\boldsymbol{\Delta}^0_n$.

Hint. To separate P from Q, reduce the pair $\mathcal{X} - P$, $\mathcal{X} - Q$. $\dashv$

1D. Parametrization and hierarchy theorems

In the most general situation, a *parametrization* of a set $\mathcal{S}$ *on* I (*with code set* I) is any surjection

$$\pi : I \twoheadrightarrow \mathcal{S}$$

of I onto $\mathcal{S}$. Often we need parametrizations which are "nice"—e.g. we may want π to be definable or to reflect some given structure on $\mathcal{S}$.

Here we are interested in the case when $\mathcal{S}$ is *the restriction of a given pointclass* Γ *to some product space* $\mathcal{X}$,

$$\Gamma{\restriction}\mathcal{X} = \{P \subseteq \mathcal{X} : P \in \Gamma\}.$$

In fact we seek parametrizations of $\Gamma{\restriction}\mathcal{X}$ on product spaces.

If $P \subseteq \mathcal{Y} \times \mathcal{X}$ and $y \in \mathcal{Y}$, let P_y be the y-*section of* P,

$$P_y = \{x \in \mathcal{X} : P(y, x)\},$$

as in Figure 1D.1.

A pointset $G \subseteq \mathcal{Y} \times \mathcal{X}$ is *universal for* $\Gamma{\restriction}\mathcal{X}$, if G is in Γ and the map

$$y \mapsto G_y$$

is a parametrization of $\Gamma{\restriction}\mathcal{X}$ on $\mathcal{Y}$, i.e. for $P \subseteq \mathcal{X}$,

$$P \in \Gamma \Leftrightarrow \textit{for some } y \in \mathcal{Y},\ P = G_y.^{(15)}$$

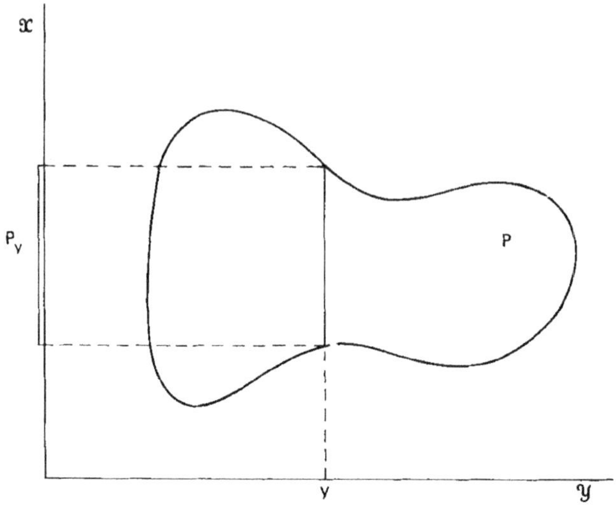

Figure 1D.1. The section above y.

A pointclass Γ is $\mathcal{Y}$*-parametrized* if for every product space $\mathfrak{X}$ there is some $G \subseteq \mathcal{Y} \times \mathfrak{X}$ which is universal for $\Gamma \restriction \mathfrak{X}$.

Let

$$N_0, N_1, N_2, \ldots$$

be an enumeration of a basis for the topology of some product space $\mathfrak{X}$ and define $O \subseteq \mathcal{N} \times \mathfrak{X}$ by

$$O(\varepsilon, x) \Leftrightarrow (\exists n)[x \in N_{\varepsilon(n)}].$$

Clearly O is open and each open set $P \subseteq \mathfrak{X}$ is of the form

$$P = O_\varepsilon = \bigcup_n N_{\varepsilon(n)}$$

for some $\varepsilon \in \mathcal{N}$, so that O is universal for $\mathbf{\Sigma}^0_1 \restriction \mathfrak{X}$. Thus $\mathbf{\Sigma}^0_1$ is $\mathcal{N}$-parametrized and it is trivial to prove from this that all the Borel pointclasses $\mathbf{\Sigma}^0_n$ and their duals $\mathbf{\Pi}^0_n$ are $\mathcal{N}$-parametrized. The next theorem establishes a little more.

1D.1. THE PARAMETRIZATION THEOREM FOR $\mathbf{\Sigma}^0_1$. *For every perfect product space* $\mathcal{Y}$, $\mathbf{\Sigma}^0_1$ *is* $\mathcal{Y}$*-parametrized.*[15]

PROOF. Let $N(\mathcal{Y}, 0), N(\mathcal{Y}, 1),\ldots$ and $N(\mathcal{X}, 0), N(\mathcal{X}, 1),\ldots$ enumerate bases for the topology of $\mathcal{Y}$ and a fixed product space $\mathcal{X}$ respectively. Recall from Theorem 1A.2 that there is a function σ which assigns to each finite binary sequence u a nbhd $N(\mathcal{Y}, \sigma(u))$ in $\mathcal{Y}$ such that (i), (ii) and (iii) of 1A.2 hold. Using this σ, define $G \subseteq \mathcal{Y} \times \mathcal{X}$ by

$$G(y, x) \Leftrightarrow \textit{there exists a finite binary sequence}$$
$$u = (t_0,\ldots, t_n) \textit{ such that } t_n = 0,\ y \in N(\mathcal{Y}, \sigma(u))$$
$$\textit{and } x \in N(\mathcal{X}, n).$$

It is immediate that G is open and hence every section $G_y \subseteq \mathcal{X}$ is open. The proof will be complete if we show that every open subset of $\mathcal{X}$ is a section of G, since then G will be universal for $\mathbf{\Sigma}^0_1 \restriction \mathcal{X}$ and $\mathcal{X}$ was arbitrary.

If $P \subseteq \mathcal{X}$ is open, then there is a set of integers A such that

$$x \in P \Leftrightarrow (\exists n)[n \in A \ \&\ x \in N(\mathcal{X}, n)].$$

Put

$$t_n = \begin{cases} 0 & \text{if } n \in A, \\ 1 & \text{if } n \notin A \end{cases}$$

and as in the proof of 1A.3 define the sequence $\{y_n\}$ in $\mathcal{Y}$ by

$$y_n = \textit{the center of } N(\mathcal{Y}, \sigma(t_0,\ldots, t_n)).$$

The properties of σ imply that $\{y_n\}$ is Cauchy, so let

$$y = \textit{limit}_{n\to\infty}\, y_n.$$

We claim that for this y,

$$G(y, x) \Leftrightarrow x \in P.$$

If $x \in P$, then for some n we have $t_n = 0$ and $x \in N(\mathcal{X}, n)$, and by the properties of σ, $y \in N\sigma(t_0,\ldots, t_n))$, so by the definition of G we have $G(y, x)$.

Conversely, if $G(y, x)$, then there is some $u = (t'_0, t'_1,\ldots, t'_n)$ such that $y \in N(\mathcal{Y}, \sigma(u))$ and $t'_n = 0$ and $x \in N(\mathcal{X}, n)$. Since $y \in N(\mathcal{Y}, \sigma(t_0, t_1,\ldots, t_n))$, the sequences $(t_0,\ldots, t_n)$ and $(t'_0,\ldots, t'_n)$ are compatible by the properties of σ. But binary sequences of the same length are compatible only when they are identical, so $t_0 = t'_0,\ldots, t_n = t'_n = 0$, hence $t_n = 0$ and $x \in N(\mathcal{X}, n)$, so $x \in P$. ⊣

1D.2. THEOREM. *If a pointclass Γ is $\mathcal{Y}$-parametrized, then so are the pointclasses $\neg\Gamma$ and $\exists^{\mathcal{Z}}\Gamma$, where $\mathcal{Z}$ is any product space. In particular all* *the Borel pointclasses* $\mathbf{\Sigma}^0_n$ *and their duals* $\mathbf{\Pi}^0_n$ *are* $\mathscr{Y}$*-parametrized, where* $\mathscr{Y}$ *is any perfect product space.*

PROOF. If $G \subseteq \mathscr{Y} \times \mathscr{X}$ is universal for $\Gamma \restriction \mathscr{X}$, then $\neg G = \mathscr{Y} \times \mathscr{X} - G$ is obviously universal for $\neg \Gamma \restriction \mathscr{X}$. Similarly, if $G \subseteq \mathscr{Y} \times \mathscr{X} \times \mathscr{Z}$ is universal for $\Gamma \restriction (\mathscr{X} \times \mathscr{Z})$, define $H \subseteq \mathscr{Y} \times \mathscr{X}$ by

$$H(y, x) \Leftrightarrow (\exists z) G(y, x, z)$$

and verify immediately that H is universal for $\exists^{\mathscr{Z}} \Gamma \restriction \mathscr{X}$. ⊣

The significance of parametrizations is evident in the next result which we formulate in a very general setting.

1D.3. THE HIERARCHY LEMMA. *Let* Γ *be a pointclass such that for every product space* $\mathscr{X}$ *and every pointset* $P \subseteq \mathscr{X} \times \mathscr{X}$ *in* Γ, *the diagonal* $P' = \{x : P(x, x)\}$ *is also in* Γ. *If* Γ *is* $\mathscr{Y}$*-parametrized, then some* $P \subseteq \mathscr{Y}$ *is in* Γ *but not in* $\neg \Gamma$.(15)

PROOF. Let $G \subseteq \mathscr{Y} \times \mathscr{Y}$ be universal for $\Gamma \restriction \mathscr{Y}$ and take

$$P = \{y : G(y, y)\}.$$

By hypothesis $P \in \Gamma$. If $\neg P \in \Gamma$, then for some fixed $y^* \in \mathscr{Y}$ we would have

$$G(y^*, y) \Leftrightarrow \neg P(y) \Leftrightarrow \neg G(y, y)$$

which is absurd for $y = y^*$. ⊣

1D.4. THE HIERARCHY THEOREM FOR THE BOREL POINTCLASSES OF FINITE ORDER. *If* $\mathscr{X}$ *is any perfect product space, then the following diagram of proper inclusions holds:*

$$\begin{array}{ccccccc}
 & & \mathbf{\Sigma}^0_1\restriction\mathscr{X} & & & & \mathbf{\Sigma}^0_2\restriction\mathscr{X} \quad \cdots \\
 & \subsetneq & & \subsetneq & & \subsetneq & \\
\mathbf{\Delta}^0_1\restriction\mathscr{X} & & & & \mathbf{\Delta}^0_2\restriction\mathscr{X} & & \quad \cdots \\
 & \subsetneq & & \subsetneq & & \subsetneq & \\
 & & \mathbf{\Pi}^0_1\restriction\mathscr{X} & & & & \mathbf{\Pi}^0_2\restriction\mathscr{X} \quad \cdots
\end{array}$$

Diagram 1D.2. The Borel pointclasses of finite order.

PROOF. We have the inclusions from 1B.1, so it is enough to prove that they are proper.

From 1D.2 we know that $\mathbf{\Sigma}^0_n$ and $\mathbf{\Pi}^0_n$ are $\mathfrak{X}$-parametrized, hence by 1D.3 there is some $P \subseteq \mathfrak{X}$, $P \in \mathbf{\Sigma}^0_n$, $P \notin \mathbf{\Pi}^0_n$. Thus $\mathbf{\Delta}^0_n \restriction \mathfrak{X} \subsetneq \mathbf{\Sigma}^0_n \restriction \mathfrak{X}$ and similarly $\mathbf{\Delta}^0_n \restriction \mathfrak{X} \subsetneq \mathbf{\Pi}^0_n \restriction \mathfrak{X}$. On the other hand, if $\mathbf{\Sigma}^0_n \restriction \mathfrak{X} = \mathbf{\Delta}^0_{n+1} \restriction \mathfrak{X}$, then $\mathbf{\Sigma}^0_n \restriction \mathfrak{X}$ would be closed under $\neg$, so $\mathbf{\Pi}^0_n \restriction \mathfrak{X} \subseteq \mathbf{\Sigma}^0_n \restriction \mathfrak{X}$ contradicting $P \in \mathbf{\Sigma}^0_n - \mathbf{\Pi}^0_n$. ⊣

1E. The projective sets

We now introduce a second hierarchy of pointclasses by applying repeatedly the operations of negation and projection along $\mathcal{N}$.

For each pointclass Λ let

$$\begin{aligned} \exists^{\mathcal{N}} \Lambda &= \{\exists^{\mathcal{N}} \mathrm{P} : \mathrm{P} \in \Lambda\} \\ &= \{\exists^{\mathcal{N}} \mathrm{P} : \mathrm{P} \in \Lambda \restriction (\mathfrak{X} \times \mathcal{N}) \text{ for some } \mathfrak{X}\}. \end{aligned}$$

The *Lusin pointclasses* $\mathbf{\Sigma}^1_n$ $(n \geq 1)$ are defined by the induction

$$\begin{aligned} \mathbf{\Sigma}^1_1 &= \exists^{\mathcal{N}} \mathbf{\Pi}^0_1, \\ \mathbf{\Sigma}^1_{n+1} &= \exists^{\mathcal{N}} \neg \mathbf{\Sigma}^1_n, \end{aligned}$$

and as with the Borel pointclasses we define *the dual* and *ambiguous* Lusin pointclasses by

$$\begin{aligned} \mathbf{\Pi}^1_n &= \neg \mathbf{\Sigma}^1_n, \\ \mathbf{\Delta}^1_n &= \mathbf{\Sigma}^1_n \cap \mathbf{\Pi}^1_n. \end{aligned}$$ (11,12)

Thus a pointset $P \subseteq \mathfrak{X}$ is $\mathbf{\Sigma}^1_1$ if there is a closed $F \subseteq \mathfrak{X} \times \mathcal{N}$ such that for all x

$$P(x) \Leftrightarrow (\exists \alpha) F(x, \alpha),$$

P is $\mathbf{\Sigma}^1_2$ if there is an open $G \subseteq \mathfrak{X} \times \mathcal{N} \times \mathcal{N}$ such that

$$P(x) \Leftrightarrow (\exists \alpha_1)(\forall \alpha_2) G(x, \alpha_1, \alpha_2),$$

etc. Similarly, P is $\mathbf{\Pi}^1_1$ if there is an open G such that

$$P(x) \Leftrightarrow (\forall \alpha) G(x, \alpha),$$

P is $\mathbf{\Pi}^1_2$ if there is a closed F such that

$$P(x) \Leftrightarrow (\forall \alpha_1)(\exists \alpha_2) F(x, \alpha_1, \alpha_2),$$

etc.

The pointsets that occur in these Lusin pointclasses are *the projective sets*, the chief objects of our study.

1E.1. THEOREM. *The following diagram of inclusions holds among the Lusin pointclasses:*

$$
\begin{array}{ccccccc}
 & \boldsymbol{\Sigma}^1_1 & & \boldsymbol{\Sigma}^1_2 & & \cdots \\
\rotatebox{45}{$\subseteq$} & & \rotatebox{-45}{$\subseteq$} \quad \rotatebox{45}{$\subseteq$} & & & \\
\boldsymbol{\Delta}^1_1 & & \boldsymbol{\Delta}^1_2 & & \cdots & \\
\rotatebox{-45}{$\subseteq$} & & \rotatebox{45}{$\subseteq$} \quad \rotatebox{-45}{$\subseteq$} & & & \\
 & \boldsymbol{\Pi}^1_1 & & \boldsymbol{\Pi}^1_2 & & \cdots
\end{array}
$$

Diagram 1E.1. The Lusin pointclasses.

PROOF. The inclusions $\boldsymbol{\Sigma}^1_n \subseteq \boldsymbol{\Pi}^1_{n+1}$ are proved by vacuous quantification, the same way we showed $\boldsymbol{\Sigma}^0_n \subseteq \boldsymbol{\Pi}^0_{n+1}$ in 1B.1.

If F is a closed set, then F is $\boldsymbol{\Pi}^0_2$ by 1B.1, so for some open G,

$$
\begin{aligned}
F(x) &\Leftrightarrow (\forall t)G(x, t) \\
&\Leftrightarrow (\forall \alpha)G(x, \alpha(0)).
\end{aligned}
$$

Now the set $G' \subseteq \mathscr{X} \times \mathscr{N}$ defined by

$$G'(x, \alpha) \Leftrightarrow G(x, \alpha(0))$$

is also open since G is and the map

$$(x, \alpha) \mapsto (x, \alpha(0))$$

is continuous, hence F is $\boldsymbol{\Pi}^1_1$. Thus every closed set is $\boldsymbol{\Pi}^1_1$ and then, by definition, every $\boldsymbol{\Sigma}^1_1$ set is $\boldsymbol{\Sigma}^1_2$, from which

$$\boldsymbol{\Sigma}^1_n \subseteq \boldsymbol{\Sigma}^1_{n+1}$$

follows immediately by induction.

The remaining inclusions in the diagram are trivial. $\dashv$

To prove the closure properties of the Lusin pointclasses we need maps that allow us to code infinite sequences of irrationals by single irrationals. Put

$$(\alpha)_i = (t \mapsto \alpha(\langle i, t\rangle)),$$

i.e.

$$(\alpha)_i = \beta, \quad \text{where } \beta(t) = \alpha(\langle i, t\rangle).$$

There is a k-ary inverse of this function for each $k \geq 1$,

$$
\begin{aligned}
\langle \alpha_0, \ldots, \alpha_{k-1}\rangle(\langle i, t\rangle) &= \alpha_i(t) && \text{if } i < k, \\
\langle \alpha_0, \ldots, \alpha_{k-1}\rangle(n) &= 0 && \text{if } n \neq \langle i, t\rangle \text{ for all } t \text{ and } i < k.
\end{aligned}
$$

The maps

$$(\alpha, i) \mapsto (\alpha)_i,$$
$$(\alpha_0, \ldots, \alpha_{k-1}) \mapsto \langle \alpha_0, \ldots, \alpha_{k-1} \rangle$$

are obviously continuous and for each k and $i < k$,

$$(\langle \alpha_0, \ldots, \alpha_{k-1} \rangle)_i = \alpha_i.$$

It will also be useful to have a notation for the *shift* map,

$$\alpha^\star = (t \mapsto \alpha(t+1)).$$

Again, $\alpha \mapsto \alpha^\star$ is continuous.

1E.2. THEOREM. *Each Lusin pointclass $\mathbf{\Sigma}_n^1$ is closed under continuous substitution, $\vee$, &, $\exists^\leq$, $\forall^\leq$, $\forall^\omega$ and $\exists^{\mathcal{Y}}$ for every product space $\mathcal{Y}$.*

Each dual Lusin pointclass $\mathbf{\Pi}_n^1$ is closed under continuous substitution, $\vee$, &, $\exists^\leq$, $\forall^\leq$, $\exists^\omega$ and $\forall^{\mathcal{Y}}$ for every product space $\mathcal{Y}$.

Each ambiguous Lusin pointclass $\mathbf{\Delta}_n^1$ is closed under $\neg$, $\vee$, &, $\exists^\leq$, $\forall^\leq$, $\exists^\omega$ and $\forall^\omega$.

In particular, every pointset of finite Borel order is $\mathbf{\Delta}_1^1$.

PROOF. The results about $\mathbf{\Pi}_n^1$ and $\mathbf{\Delta}_n^1$ follow immediately from those about $\mathbf{\Sigma}_n^1$ and the last assertion is a trivial consequence of the closure properties of $\mathbf{\Delta}_1^1$.

Closure of $\mathbf{\Sigma}_1^1$ under continuous substitution follows from the closure of $\mathbf{\Pi}_1^0$ under continuous substitution.

To prove closure of $\mathbf{\Sigma}_1^1$ under $\vee$, &, $\exists^\leq$, $\forall^\leq$ and $\exists^{\mathcal{Y}}$ we use quantifier contractions. For example, to prove closure under $\exists^{\mathcal{N}}$, assume that

$$P(x, \alpha) \Leftrightarrow (\exists \beta) F(x, \alpha, \beta)$$

with F closed. Then

$$\begin{aligned}(\exists \alpha) P(x, \alpha) &\Leftrightarrow (\exists \alpha)(\exists \beta) F(x, \alpha, \beta)\\ &\Leftrightarrow (\exists \gamma) F(x, (\gamma)_0, (\gamma)_1)\end{aligned}$$

and $\exists^{\mathcal{Y}} P$ is $\mathbf{\Sigma}_1^1$ by closure of $\mathbf{\Pi}_1^0$ under continuous substitution.

To take one more example, suppose

$$P(x, m) \Leftrightarrow (\exists \beta) F(x, m, \beta).$$

Then

$$\begin{aligned}(\forall m \leq n) P(x, m) &\Leftrightarrow (\forall m \leq n)(\exists \beta) F(x, m, \beta)\\ &\Leftrightarrow (\exists \gamma)(\forall m \leq n) F(x, m, (\gamma)_m)\end{aligned}$$

and again $\forall^{\leq}P$ is Σ_1^1 by closure of Π_1^0 under continuous substitution and $\forall^{\leq}$.

Closure of Σ_1^1 under $\exists^\omega$ follows immediately from the equivalence

$$(\exists t)(\exists\alpha)Q(x, t, \alpha) \Leftrightarrow (\exists\gamma)Q(x, \gamma(0), \gamma^\star).$$

For every product space $\mathcal{Y}$, there is a continuous surjection

$$\pi : \mathcal{N} \twoheadrightarrow \mathcal{Y}$$

of $\mathcal{N}$ onto $\mathcal{Y}$ by 1A.1. Thus if $P \subseteq \mathcal{X} \times \mathcal{Y}$, *then*

$$(\exists y)P(x, y) \Leftrightarrow (\exists\alpha)P(x, \pi(\alpha))$$

and closure of Σ_1^1 under $\exists^{\mathcal{Y}}$ follows from closure under continuous substitution and $\exists^{\mathcal{N}}$.

Finally, to prove closure of Σ_1^1 under $\forall^\omega$, suppose

$$P(x, t) \Leftrightarrow (\exists\alpha)F(x, t, \alpha)$$

with F in Π_1^0. Then

$$\begin{aligned}(\forall t)P(x, t) &\Leftrightarrow (\forall t)(\exists\alpha)F(x, t, \alpha)\\ &\Leftrightarrow (\exists\gamma)(\forall t)F(x, t, (\gamma)_t),\end{aligned}$$

so $\forall^\omega P$ is Σ_1^1 by closure of Π_1^0 under continuous substitution and $\forall^\omega$.

The closure properties of Σ_n^1 for $n > 1$ follow by induction, using the same quantifier manipulations that we used for the case of Σ_1^1. ⊣

In addition to the obvious quantifier contractions

$$(\exists\alpha)(\exists\beta)P(\alpha, \beta) \Leftrightarrow (\exists\gamma)P((\gamma)_0, (\gamma)_1),$$

$$(\forall\alpha)(\forall\beta)P(\alpha, \beta) \Leftrightarrow (\forall\gamma)P((\gamma)_0, (\gamma)_1),$$

we also used in this proof the equivalence

$$(\forall t)(\exists\alpha)P(t, \alpha) \Leftrightarrow (\exists\gamma)(\forall t)P(t, (\gamma)_t).$$

This expresses the *countable axiom of choice* for pointsets. The dual equivalence

$$(\exists t)(\forall\alpha)P(t, \alpha) \Leftrightarrow (\forall\gamma)(\exists t)P(t, (\gamma)_t)$$

looks a bit mysterious at first sight. We prove it by taking the negation of each side in the countable axiom of choice.

Theorems 1D.1–1D.4 and 1E.1 yield immediately the following result.

1E.3. THE PARAMETRIZATION AND HIERARCHY PROPERTIES OF THE LUSIN POINTCLASSES. For each $n \geq 1$ and for each perfect product space $\mathcal{Y}$, the pointclasses $\boldsymbol{\Sigma}^1_n$, $\boldsymbol{\Pi}^1_n$ are $\mathcal{Y}$-parametrized. Hence they satisfy the following diagram of proper inclusions, where $\mathcal{X}$ is any perfect product space:[(12)]

$$\begin{array}{ccccccc}
 & & \boldsymbol{\Sigma}^1_1{\restriction}\mathcal{X} & & & & \boldsymbol{\Sigma}^1_2{\restriction}\mathcal{X} \quad \cdots \\
 & \subsetneq & & \supsetneq & & \subsetneq & \\
\boldsymbol{\Delta}^1_1{\restriction}\mathcal{X} & & & & \boldsymbol{\Delta}^1_2{\restriction}\mathcal{X} & & \cdots \\
 & \supsetneq & & \subsetneq & & \supsetneq & \\
 & & \boldsymbol{\Pi}^1_1{\restriction}\mathcal{X} & & & & \boldsymbol{\Pi}^1_2{\restriction}\mathcal{X} \quad \cdots
\end{array}$$

Diagram 1E.2. The Lusin pointclasses. ⊣

In the classical terminology the $\boldsymbol{\Sigma}^1_1$ pointsets are called *analytic* or *A*-sets. They include most of the sets one encounters in hard analysis. The $\boldsymbol{\Pi}^1_1$ sets are *co-analytic* or *CA*-sets, the $\boldsymbol{\Sigma}^1_2$ sets are *PCA*-sets, the $\boldsymbol{\Pi}^1_2$ sets are *CPCA*-sets, etc.

EXERCISES

1E.4. If $f : \mathcal{X} \to \mathcal{Y}$, let

$$Graph(f) = \{(x, y) : f(x) = y\}.$$

Prove that if f is continuous, then $Graph(f)$ is closed. ⊣

1E.5. Prove that if $f : \mathcal{X} \to \mathcal{Y}$ is continuous and P is a $\boldsymbol{\Sigma}^1_n$ subset of $\mathcal{X}$, then $f[P] = \{f(x) : P(x)\}$ is $\boldsymbol{\Sigma}^1_n$. ⊣

1E.6. Prove that for every pointset $P \subseteq \mathcal{X}$,

$$P \text{ is } \boldsymbol{\Sigma}^1_1 \Leftrightarrow P = f[\mathcal{N}] \text{ for some continuous } f,$$

$$P \text{ is } \boldsymbol{\Sigma}^1_{n+1} \Leftrightarrow P = f[Q] \text{ for some } \boldsymbol{\Pi}^1_n \text{ set } Q \subseteq \mathcal{N} \text{ and some continuous } f.$$

Hint. For the first assertion, suppose P is the projection of some closed subset C of $\mathcal{X} \times \mathcal{N}$. Consider C as a metric space with the metric it inherits from $\mathcal{X} \times \mathcal{N}$; it is easily separable and complete, so by 1A.1, there is a continuous surjection $f : \mathcal{N} \twoheadrightarrow C$. Now P is the image of $\mathcal{N}$ under f followed by the continuous projection function. ⊣

We cannot replace $\mathcal{N}$ by an arbitrary perfect product space in this result, because of the next exercise. However, see 1G.12 for a related characterization of $\boldsymbol{\Sigma}^1_1$.

1E.7. Prove that if $f : \mathcal{R} \to \mathfrak{X}$ is continuous and F is a closed set of reals, then $f[F]$ is $\mathbf{\Sigma}^0_2$.

Hint. $\mathcal{R}$ is a countable union of compact sets. ⊣

Practically every specific pointset which comes up in the usual constructions of analysis and topology is easily shown to be projective—in fact, almost always, it is $\mathbf{\Sigma}^1_1$ or $\mathbf{\Pi}^1_1$. We only mention a couple of simple examples here, since we will meet several interesting projective pointsets later on.

1E.8. On the space $C[0, 1]$ of continuous real functions on the unit interval, put

$$Q(f) \Leftrightarrow f \text{ is differentiable on } [0, 1],$$

$$R(f) \Leftrightarrow f \text{ is continuously differentiable on } [0, 1],$$

where at the endpoints we naturally take the one-sided derivatives. Prove that Q is $\mathbf{\Pi}^1_1$ and R is $\mathbf{\Sigma}^1_1$. ⊣

1F. Countable operations and the transfinite Borel pointclasses

A *countable pointset operation* is any function Φ with domain some set of infinite sequences of pointsets and pointsets as values. We will often use the notation

$$\Phi_i P_i = \Phi(P_0, P_1, P_2, \ldots).$$

The most obvious countable operations are *countable conjunction*, $\bigwedge^\omega$, and countable disjunction, $\bigvee^\omega$. Here $\bigwedge^\omega_i P_i$ and $\bigvee^\omega_i P_i$ are defined when all the P_i are subsets of the same space $\mathfrak{X}$ and

$$x \in \textstyle\bigwedge^\omega_i P_i \Leftrightarrow \bigwedge_i P_i(x) \Leftrightarrow \text{for all } i \in \omega,\ P_i(x),$$

$$x \in \textstyle\bigvee^\omega_i P_i \Leftrightarrow \bigvee_i P_i(x) \Leftrightarrow \text{for some } i \in \omega,\ P_i(x).$$

In set theoretic notation

$$\textstyle\bigwedge^\omega_i P_i = \bigcap_i P_i, \qquad \bigvee^\omega_i P_i = \bigcup_i P_i$$

whenever all the P_i are subsets of the same space.

A pointclass Λ is a *closed under a countable operation* Φ, if whenever $P_0, P_1, \ldots$ are all in Λ and $\Phi_i P_i$ is defined, then $\Phi_i P_i$ is also in Λ.

1F.1. THEOREM. *Let Γ be an $\mathcal{N}$-parametrized pointclass which is closed under continuous substitution. If Γ is closed under $\exists^\omega$, then it is closed under $\bigvee^\omega$ and if Γ is closed under $\forall^\omega$, then it is also closed under $\bigwedge^\omega$.*

PROOF. Suppose $P_i \subseteq \mathfrak{X}$, $P_i \in \Gamma$, let $G \subseteq \mathcal{N} \times \mathfrak{X}$ be universal and choose irrationals ε_i such that

$$P_i = G_{\varepsilon_i} = \{x \in \mathfrak{X}: G(\varepsilon_i, x)\}.$$

Now pick ε so that for every i,

$$(\varepsilon)_i = \varepsilon_i$$

and set

$$x \in P \Leftrightarrow (\exists i) G((\varepsilon)_i, x).$$

Clearly $P \in \Gamma$ by closure under continuous substitution and $\exists^\omega$ and $P = \bigcup_i P_i$.

The argument about $\forall^\omega$ is similar. ⊣

1F.2. COROLLARY. *Each $\mathbf{\Sigma}^0_n$ is closed under $\bigvee^\omega$, each $\mathbf{\Pi}^0_n$ is closed under $\bigwedge^\omega$ and all $\mathbf{\Sigma}^1_n$, $\mathbf{\Pi}^1_n$, $\mathbf{\Delta}^1_n$ are closed under both $\bigvee^\omega$ and $\bigwedge^\omega$.*(12) ⊣

If Φ is a k-ary or countable set operation and Λ is a pointclass, put

$$\Phi\Lambda = \{\Phi(P_0, P_1, \ldots): P_0, P_1, \ldots \in \Lambda \text{ and } \Phi(P_0, P_1, \ldots) \text{ is defined}\}.$$

We have already used this notation in connection with $\exists^\omega$ and $\exists^{\mathcal{N}}$.

It is trivial to verify that if Λ is closed under continuous substitution, then

$$\exists^\omega \Lambda \subseteq \bigvee^\omega \Lambda,$$

i.e. every projection along ω of a set in Λ can be written as a countable union of sets in Λ. This together with 1F.2 give us a new inductive characterization of the finite Borel pointclasses,

$$\mathbf{\Sigma}^0_1 = \textit{all open sets},$$
$$\mathbf{\Sigma}^0_{n+1} = \bigvee^\omega \neg \mathbf{\Sigma}^0_n.$$

Now the class of all pointsets of finite Borel order is closed under $\exists^\omega$ but it is not closed under $\bigvee^\omega$; for example, choose $G_n \subseteq \mathcal{N}$ to be in $\mathbf{\Sigma}^0_n - \mathbf{\Pi}^0_n$ and verify that

$$G = \bigcup_n \{(n, \alpha): \alpha \in G_n\}$$

is not in any $\mathbf{\Sigma}^0_n$. This suggests an extension of the finite Borel hierarchy into the transfinite as follows.

Take

$$\boldsymbol{\Sigma}^0_1 = \textit{all open pointsets}$$

and for each ordinal number $\xi > 1$, let

$$\boldsymbol{\Sigma}^0_\xi = \bigvee^\omega \neg(\bigcup_{\eta<\xi} \boldsymbol{\Sigma}^0_\eta).$$

Unscrambling this, P is in $\boldsymbol{\Sigma}^0_\xi$ if there are pointsets $P_0, P_1, \ldots$, with each P_i in some $\boldsymbol{\Sigma}^0_\eta$, $\eta < \xi$, such that

$$P = \bigcup_i (\mathcal{X} - P_i).$$

We call $\boldsymbol{\Sigma}^0_\xi$ the *Borel pointclass of order* ξ. The *dual* and *ambiguous* Borel pointclasses are defined in the obvious way,

$$\boldsymbol{\Pi}^0_\xi = \neg\boldsymbol{\Sigma}^0_\xi,$$
$$\boldsymbol{\Delta}^0_\xi = \boldsymbol{\Sigma}^0_\xi \cap \boldsymbol{\Pi}^0_\xi.^{(8,9)}$$

For finite ξ this definition yields the pointclasses $\boldsymbol{\Sigma}^0_\xi$ as we know them, so there is no conflict in notation.

It is very easy to extend the basic properties of the finite Borel pointclasses to all Borel pointclasses and we will leave this for the exercises. We only state here the basic characterization of the class $\mathbf{B}$ of *Borel sets*,

$$\mathbf{B} = \bigcup_\xi \boldsymbol{\Sigma}^0_\xi.^{(8)}$$

1F.3. THEOREM. *For each product space $\mathcal{X}$, the class $\mathbf{B}\restriction\mathcal{X}$ of Borel subsets of $\mathcal{X}$ is the smallest collection of subsets of $\mathcal{X}$ which contains the open sets and is closed under complementation and countable union; similarly, $\mathbf{B}\restriction\mathcal{X}$ is the smallest collection of subsets of $\mathcal{X}$ which contains the open (or the closed) sets and is closed under countable union and countable intersection.*

PROOF. If P is Borel, then P is in $\boldsymbol{\Sigma}^0_\xi$ for some ξ, so $\neg P = \mathcal{X} - P \in \boldsymbol{\Sigma}^0_{\xi+1}$, in particular, $\neg P$ is Borel. Also, if P_i is Borel for every i, $P_i \subseteq \mathcal{X}$, then $P_i \in \boldsymbol{\Sigma}^0_{\xi_i}$ for some ξ_i so $\neg P_i \in \boldsymbol{\Sigma}^0_{\xi_i+1}$ and taking

$$\xi = \textit{supremum}\{\xi_i + 2 \colon i = 0, 1, 2, \ldots\},$$

we have $P \in \boldsymbol{\Sigma}^0_\xi$, since

$$P = \bigcup_i P_i = \bigcup_i (\mathcal{X} - (\mathcal{X} - P_i)).$$

Thus the class of Borel subsets of $\mathcal{X}$ is closed under $\neg$ and $\bigvee^\omega$.

Conversely, if $\mathcal{S}$ is any collection of subsets of $\mathcal{X}$ which is closed under $\neg$ and $\bigvee^\omega$, then $\mathcal{S}$ clearly contains all open subsets of $\mathcal{X}$ and an easy induction on ξ shows that

$$P \subseteq \mathcal{X}, P \in \mathbf{\Sigma}^0_\xi \Longrightarrow P \in \mathcal{S}.$$

For the second assertion, notice first that $\mathbf{B}\restriction\mathcal{X}$ is easily closed under countable intersection, since

$$\bigcap_i P_i = \mathcal{X} - (\bigcup_i (\mathcal{X} - P_i)).$$

Conversely, if $\mathcal{S}$ contains all the open subsets of $\mathcal{X}$ and is closed under both countable unions and countable intersection, then each $P \subseteq \mathcal{X}$ which is either $\mathbf{\Sigma}^0_\xi$ or $\mathbf{\Pi}^0_\xi$ is in $\mathcal{S}$ by a trivial induction on ξ; because closed sets are countable intersections of open sets and in general

$$P \in \mathbf{\Sigma}^0_\xi \Longrightarrow P = \bigcup_i P_i \quad \text{with each} \quad P_i \in \mathbf{\Pi}^0_{\eta_i}, \qquad \eta_i < \xi,$$
$$P \in \mathbf{\Pi}^0_\xi \Longrightarrow P = \bigcap_i P_i \quad \text{with each} \quad P_i \in \mathbf{\Sigma}^0_{\eta_i}, \qquad \eta_i < \xi.$$

It is immediate from the definition of the Borel sets and the closure properties of $\mathbf{\Delta}^1_1$ that $\mathbf{B} \subseteq \mathbf{\Delta}^1_1$. Actually,

$$\mathbf{B} = \mathbf{\Delta}^1_1.$$

This is one of the central results of the theory—we will prove it in Chapter 2.

EXERCISES

1F.4. Prove that each Borel pointclass $\mathbf{\Sigma}^0_\xi$ is closed under continuous substitution, $\vee$, &, $\exists^\leq$, $\forall^\leq$, $\exists^\omega$ and $\bigvee^\omega$. State and prove the natural closure properties of the pointclasses $\mathbf{\Pi}^0_\xi$ and $\mathbf{\Delta}^0_\xi$.

Hint. Use induction on ξ. One way to arrange the computations is to show the following lemmas, for Λ's which contain all clopen sets:

(1) If Λ is closed under continuous substitution, $\vee$ and &, then $\bigvee^\omega \neg \Lambda$ is also closed under these operations and $\bigvee^\omega$.

(2) If Λ is closed under continuous substitution, $\vee$, & and $\bigvee^\omega$, then Λ is closed under $\exists^\leq$, $\forall^\leq$ and $\exists^\omega$.

(3) If $\Lambda_0 \subseteq \Lambda_1 \subseteq \cdots$ is an increasing sequence of pointclasses, each closed under continuous substitution, $\vee$ and &, then $\bigcup_i \Lambda_i$ has the same properties. ⊣

1F.5. Prove that

$$\mathbf{B} = \bigcup_{\xi<\aleph_1} \mathbf{\Sigma}^0_\xi,$$

i.e. every Borel set occurs in some Borel class of countable order. ⊣

1F.6. Prove that for each countable ξ, $\mathbf{\Sigma}^0_\xi$ is $\mathcal{N}$-parametrized. Infer that

$$\mathbf{\Delta}^0_\xi{\restriction}\mathcal{N} \subsetneq \mathbf{\Sigma}^0_\xi{\restriction}\mathcal{N} \subsetneq \mathbf{\Delta}^0_{\xi+1}{\restriction}\mathcal{N},$$

so in particular no Borel class $\mathbf{\Sigma}^0_\xi$ of countable order exhausts the Borel sets.

Hint. The result is known for finite ξ, so we proceed by induction. Choose $\xi_0, \xi_1,\ldots$ so that *supremum*$\{\xi_i+1 : i = 0, 1,\ldots\} = \xi$ and let $G_i \subseteq \mathcal{N} \times \mathfrak{X}$ be universal for $\mathbf{\Sigma}^0_{\xi_i}{\restriction}\mathfrak{X}$. Put

$$G(\alpha, x) \Leftrightarrow \bigvee_i \neg G_i((\alpha)_i, x)$$

and show that G is universal for $\mathbf{\Sigma}^0_\xi{\restriction}\mathfrak{X}$ by verifying that each P in $\mathbf{\Sigma}^0_\xi{\restriction}\mathfrak{X}$ satisfies

$$P(x) \Leftrightarrow \bigvee_i \neg P_i(x)$$

with each P_i in $\mathbf{\Sigma}^0_{\xi_i}$. ⊣

In the exercises of the next section we will extend this result to show that each $\mathbf{\Sigma}^0_\xi$ is $\mathcal{Y}$-parametrized for each perfect product space $\mathcal{Y}$.

1F.7. Suppose $R_0, R_1,\ldots$ are all subsets of the same space $\mathfrak{X}$ and we take

$$R(x, s) \Leftrightarrow R_s(x).$$

Prove that if each R_s is in $\mathbf{\Sigma}^0_\xi$, then so is R.

Hint. For each i, put

$$P_i(x, s) \Leftrightarrow R_i(x) \,\&\, s = i$$

and notice that P_i is in $\mathbf{\Sigma}^0_\xi$ by the closure properties of this pointclass. Now

$$R(x, s) \Leftrightarrow \bigvee_i P_i(x, s). \qquad ⊣$$

Recall the definitions of *parametrization*, *reduction* and *separation* given in the exercises of 1C.

1F.8. Prove that for each $\xi > 1$, if $P \subseteq \mathfrak{X} \times \omega$ is in $\mathbf{\Sigma}^0_\xi$, then there is a P^* also in $\mathbf{\Sigma}^0_\xi$ which uniformizes P.

Hint. We follow the same argument as in 1C.6, but now we deal with the infinitary operation $\bigvee^\omega$ instead of projection on ω, $\exists^\omega$. If

$$P(x, m) \Leftrightarrow \bigvee_i Q_i(x, m),$$

where each Q_i is in some $\mathbf{\Pi}^0_{\xi_i}$ with $\xi_i < \xi$, put

$$R_s(x) \Leftrightarrow Q_{(s)_1}(x, (s)_0) \,\&\, \bigwedge_{t<s} \neg Q_{(t)_1}(x, (t)_0)$$

and check that each R_s is in $\mathbf{\Pi}^0_{\xi_i}$, so that by the preceding exercise,

$$R(x, s) \Leftrightarrow R_s(x)$$

is in $\mathbf{\Sigma}^0_\xi$. Take

$$P^*(x, m) \Leftrightarrow (\exists i) R(x, \langle m, i\rangle). \qquad \dashv$$

1F.9. Prove that for each $\xi > 1$, every pair of sets P, Q in $\mathbf{\Sigma}^0_\xi$ is reducible by a pair P^*, Q^* in $\mathbf{\Sigma}^0_\xi$.[(16)] ⊣

1F.10. Prove that for each $\xi > 1$, every disjoint pair of sets in $\mathbf{\Pi}^0_\xi$ can be separated by a set in $\mathbf{\Delta}^0_\xi$.[(16)] ⊣

1G. Borel functions and isomorphisms

Let Λ be a fixed pointclass and let

$$f : \mathcal{X} \to \mathcal{Y}$$

a function. We say that F is *Λ-measurable*, if for each basic nbhd $N_s \subseteq \mathcal{Y}$, the inverse image $f^{-1}[N_s]$ is in Λ. This notion is due to Lebesgue.[(10)]

Here we are mostly interested in *Borel measurable* or simply *Borel functions*. A *Borel isomorphism* between two spaces is a bijection

$$f : \mathcal{X} \rightarrowtail\!\!\!\!\twoheadrightarrow \mathcal{Y}$$

such that both f and its inverse are Borel measurable.

The main result of this section is that every perfect product space is both Borel isomorphic with $\mathcal{N}$ and the continuous one-to-one image of some closed subset of $\mathcal{N}$. We will also show that the Lusin pointclasses are closed under Borel substitution. Thus in studying projective sets we can often simplify proofs by assuming that the space under consideration is $\mathcal{N}$.

We will leave for the exercises some very interesting results about $\mathbf{\Sigma}^0_\xi$-measurable functions.

Let us first dispose of the easy result.

1G.1. THEOREM. *If $f:\mathcal{X}\to\mathcal{Y}$ is a Borel function and $P\subseteq\mathcal{Y}$ is in any of the pointclasses* $\mathbf{B}$, $\mathbf{\Delta}^1_n$, $\mathbf{\Sigma}^1_n$, $\mathbf{\Pi}^1_n$, *then $f^{-1}[P]$ is in the same pointclass.*

In particular, the collection of Borel functions is closed under composition.

PROOF. A simple induction of ξ shows that if f is Borel and P is $\mathbf{\Sigma}^0_\xi$, then $f^{-1}[P]$ is Borel. Thus $\mathbf{B}$ is closed under Borel substitution. Also, if $f:\mathcal{X}\to\mathcal{Y}$, $g:\mathcal{Y}\to\mathcal{Z}$ are both Borel and $h:\mathcal{X}\to\mathcal{Z}$ is the composition,

$$h(x)=g(f(x)),$$

then for each open set $P\subseteq\mathcal{Z}$,

$$h^{-1}[P]=f^{-1}[g^{-1}[P]],$$

so $h^{-1}[P]$ is Borel and h is Borel measurable.

For the rest, notice that

$$f(x)=y\Leftrightarrow\bigwedge\nolimits_s[y\in N_s\Rightarrow f(x)\in N_s],$$

so that *the graph* of f

$$Graph(f)=\{(x,y):f(x)=y\}$$

is Borel. Now for any $P\subseteq\mathcal{Y}$,

$$\begin{aligned}P(f(x))&\Leftrightarrow(\exists y)[P(y)\ \&\ f(x)=y]\\&\Leftrightarrow(\forall y)[P(y)\vee f(x)\neq y].\end{aligned}$$

These equivalences, the fact that $\mathbf{B}\subseteq\mathbf{\Delta}^1_1$ and the closure properties of the pointclasses $\mathbf{\Delta}^1_n$, $\mathbf{\Sigma}^1_n$, $\mathbf{\Pi}^1_n$ imply immediately that if P is in one of them, then so if $f^{-1}[P]$. ⊣

We now go to the *transfer theorems* which often allow us to study just subsets of $\mathcal{N}$ instead of arbitrary pointsets. The first of these is a more refined statement of Theorem 1A.1.[(18)]

1G.2. THEOREM. *For every product space $\mathcal{X}$ there is a continuous surjection*

$$\pi:\mathcal{N}\twoheadrightarrow\mathcal{X}$$

and a closed set $A\subseteq\mathcal{N}$ such that π is one-to-one on A and $\pi[A]=\mathcal{X}$. Moreover, there is a Borel injection

$$f:\mathcal{X}\rightarrowtail\mathcal{N}$$

which is precisely the inverse of π restricted to A, i.e. for all $\alpha \in A$, $f(\pi(\alpha)) = \alpha$ and for all $x \in \mathfrak{X}$, $f(x) \in A$ and $\pi(f(x)) = x$.

PROOF. To begin with, let

$$\rho : \mathcal{N} \twoheadrightarrow \mathfrak{X}$$

be the surjection defined in the proof of 1A.1 and for $x \in \mathfrak{X}$, put

$$g(x) = \alpha,$$

where

$$\alpha(n) = \text{least } k \text{ such that } d(x, r_k) \leq 2^{-n-2}$$

It is very simple to check that for all $x \in \mathfrak{X}$, $\rho(g(x)) = x$, so g is an injection. Moreover, if we put

$$B = g[\mathfrak{X}],$$

then g is precisely the inverse of ρ restricted to B, since

$$\begin{aligned} \alpha \in B &\Rightarrow \alpha = g(x) \quad \text{for some } x, \\ &\Rightarrow g(\rho(\alpha)) = g(\rho(g(x))) = g(x) = \alpha. \end{aligned}$$

If $g(x) = \alpha$, then

$$\alpha(n) = k \Leftrightarrow d(x, r_k) \leq 2^{-n-2} \;\&\; (\forall s < k)[d(x, r_s) > 2^{-n-2}].$$

Thus if

$$B_{n,k} = \{\alpha : \alpha(n) = k\},$$

each $g^{-1}[B_{n,k}]$ is a Borel subset of $\mathfrak{X}$. Hence for each basic nbhd $N = \{\alpha : \alpha(0) = k_0, \ldots, \alpha(n-1) = k_{n-1}\}$ in $\mathcal{N}$, the set

$$g^{-1}[N] = g^{-1}[B_{0,k_0}] \cap \cdots \cap g^{-1}[B_{n-1,k_{n-1}}]$$

is Borel and g is a Borel function.

Now, easily

$$\alpha \in B \Leftrightarrow (\forall n)[d(\rho(\alpha), r_{\alpha(n)}) \leq 2^{-n-2} \;\&\; (\forall k < \alpha(n))[d(\rho(\alpha), r_k) > 2^{-n-2}]],$$

so B is a $\mathbf{\Pi}^0_2$ subset of $\mathcal{N}$. We must refine the construction a bit to get π and A with the same properties, with A a closed set.

Put B in normal form

$$\alpha \in B \Leftrightarrow (\forall n)(\exists s) R(\alpha, n, s),$$

where R is a clopen pointset by 1B.7 and define $A \subseteq \mathcal{N} \times \mathcal{N}$ by

$$(\alpha, \beta) \in A \Leftrightarrow (\forall n)[R(\alpha, n, \beta(n)) \;\&\; (\forall k < \beta(n)) \neg R(\alpha, n, k)].$$

Clearly A is closed. Moreover, the projection $\sigma : \mathcal{N} \times \mathcal{N} \to \mathcal{N}$, $\sigma(\alpha, \beta) = \alpha$ takes A onto B and is one-to-one on A, since

$$(\alpha, \beta) \in A \Rightarrow \beta(n) = \text{least } k \text{ such that } R(\alpha, n, k).$$

Hence the composition $\pi = \rho \circ \sigma$ takes A onto $\mathfrak{X}$ and is continuous, one-to-one.

It is trivial to check that the inverse of π

$$f(x) = (g(x), n \mapsto \text{least } k \text{ such that } R(g(x), n, k))$$

is Borel. The proof is completed by carrying A to $\mathcal{N}$ via some trivial homeomorphism of $\mathcal{N}$ with $\mathcal{N} \times \mathcal{N}$, e.g. the map

$$(n_0, n_1, n_2, \ldots) \mapsto ((n_0, n_2, n_4, \ldots), (n_1, n_3, n_5, \ldots)). \qquad \dashv$$

The function f of this proof is an example of an interesting class of functions. Let us temporarily call a function

$$f : \mathfrak{X} \rightarrowtail \mathcal{Y}$$

a good Borel injection if

(1) f is a Borel injection,

(2) there is a Borel surjection

$$g : \mathcal{Y} \twoheadrightarrow \mathfrak{X}$$

such that $g \circ f$ is the identity on $\mathfrak{X}$, i.e.

$$g(f(x)) = x \quad (x \in \mathfrak{X}).$$

We refer to any such g as *a Borel inverse* of f.

It will turn out that every Borel injection is a good Borel injection. This is a special case of a fairly difficult theorem which we will prove in 2E and again in Chapter 4. Here we only need show that enough good Borel injections exist.

Notice that if $f : \mathfrak{X} \rightarrowtail \mathcal{Y}$ is a good Borel injection, then

$$y \in f[\mathfrak{X}] \Leftrightarrow f(g(y)) = y$$

with g any Borel inverse of f, so $f[\mathfrak{X}]$ is a Borel set. Moreover, if P is any Borel subset of $\mathfrak{X}$, then

$$y \in f[P] \Leftrightarrow y \in f[\mathfrak{X}] \;\&\; g(y) \in P,$$

so that $f[P]$ is Borel. Thus *the image of a Borel set by a good Borel injection is Borel.*

It is also immediate that the class of good Borel injections is closed under composition.

1G.3. LEMMA. *For every perfect product space* $\mathfrak{X}$, *there are good Borel injections*

$$f : \mathfrak{X} \rightarrowtail \mathcal{N},$$
$$h : \mathcal{N} \rightarrowtail \mathfrak{X}.$$

PROOF. We have already constructed f in 1G.2.

To construct h, define first $h_1 : \mathcal{N} \rightarrowtail \mathcal{C}$ by

$$h_1(\alpha) = \beta,$$

where

$$\beta(n) = \begin{cases} 0 & \text{if } \alpha((n)_0) = (n)_1, \\ 1 & \text{if } \alpha((n)_0) \neq (n)_1. \end{cases}$$

It is trivial to verify that h_1 is a Borel function, and

$$\begin{aligned} \beta \in h_1[\mathcal{N}] \Leftrightarrow & (\forall n)[\beta(n) = \beta(\langle (n)_0, (n)_1 \rangle) \\ & \& (\forall n)(\forall k)[[\beta(n) = 0 \; \& \; \beta(k) = 0 \; \& \; (n)_0 = (k)_0] \Rightarrow (n)_1 = (k)_1] \\ & \& (\forall n)(\exists k)[\beta(\langle n, k \rangle) = 0]], \end{aligned}$$

so that $h_1[\mathcal{N}]$ is Borel. Define now $g_1 : \mathcal{C} \twoheadrightarrow \mathcal{N}$ by

$$g_1(\beta) = \text{the constant function } 0 \quad \text{if } \beta \notin h_1[\mathcal{N}],$$
$$g_1(\beta) = \alpha \quad \text{if } \beta \in h_1[\mathcal{N}],$$

where

$$\alpha(n) = \textit{the unique } m \textit{ such that } \beta(\langle n, m \rangle) = 0$$

and verify easily that g_1 is a Borel inverse of h_1, so that h_1 is a good Borel injection.

Now let

$$\pi : \mathcal{C} \rightarrowtail \mathfrak{X}$$

be the continuous injection constructed in 1A.3 with $\mathfrak{M} = \mathfrak{X}$. Since $\mathcal{C}$ is compact and π is a continuous injection, we know that $\pi[\mathcal{C}]$ is compact; in any case, we can compute $\pi[\mathcal{C}]$ using the function σ of 1A.2,

$$\begin{aligned} x \in \pi[\mathcal{C}] \Leftrightarrow \textstyle\bigwedge_n \bigvee_u [u = (t_0, \dots, t_{n-1}) & \text{ for some } t_0, \dots, t_{n-1} \\ & \& \; x \in N_{\sigma(u)}]. \end{aligned}$$

For an inverse to π, take

$$\rho(x) = \begin{cases} \text{the constant 0 function} & \text{if } x \notin \pi[\mathcal{C}], \\ \text{the unique } \alpha \in \mathcal{C} \text{ such that } \pi(\alpha) = x & \text{if } x \in \pi[\mathcal{C}]. \end{cases}$$

If

$$B = \{\alpha : \alpha(0) = k_0, \ldots, \alpha(n) = k_n\}$$

is a typical nbhd in $\mathcal{C}$, then

$$\rho(x) \in B \Leftrightarrow \rho(x)(0) = k_0 \;\&\; \cdots \;\&\; \rho(x)(n) = k_n,$$

so to prove that ρ is Borel it is enough to show that for each n, the relation

$$P_n(x) \Leftrightarrow \rho(x)(n) = 0$$

is Borel. This is true, since

$$P_n(x) \Leftrightarrow x \notin \pi[\mathcal{C}] \vee \bigvee_u [u = (t_0, \ldots, t_{n-1}) \text{ for some } t_0, \ldots, t_{n-1}$$
$$\& \; t_{n-1} = 0 \;\&\; x \in N_{\sigma(u)}].$$

Now $h = \pi \circ h_1$ is a good Borel injection of $\mathcal{N}$ into $\mathfrak{X}$. ⊣

1G.4. THEOREM. Every perfect product space is Borel isomorphic with $\mathcal{N}$.[(18)]

PROOF. Recall the classical Schroeder–Bernstein Theorem, that from given injections $h : \mathcal{N} \rightarrowtail \mathfrak{X}$ and $f : \mathfrak{X} \rightarrowtail \mathcal{N}$ we can construct a bijection $g : \mathcal{N} \twoheadrightarrow \mathfrak{X}$. We will verify that if h, f are good Borel injections, then the resulting bijection is a Borel isomorphism.

Define the sequences of sets $\mathcal{N}_0, \mathcal{N}_1, \ldots, \mathfrak{X}_0, \mathfrak{X}_1, \ldots$ inductively as follows (see Diagram 1G.1 on top of page 55).

An easy induction shows that

$$\mathcal{N}_n \supseteq f[\mathfrak{X}_n] \supseteq \mathcal{N}_{n+1},$$
$$\mathfrak{X}_n \supseteq h[\mathcal{N}_n] \supseteq \mathfrak{X}_{n+1},$$

so that

$$\mathcal{N} = \mathcal{N}_0 \supseteq f[\mathfrak{X}_0] \supseteq \mathcal{N}_1 \supseteq f[\mathfrak{X}_1] \supseteq \mathcal{N}_2 \supseteq f[\mathfrak{X}_2] \supseteq \cdots,$$
$$\mathfrak{X} = \mathfrak{X}_0 \supseteq h[\mathcal{N}_0] \supseteq \mathfrak{X}_1 \supseteq h[\mathcal{N}_1] \supseteq \mathfrak{X}_2 \supseteq h[\mathcal{Y}_2] \supseteq \cdots.$$

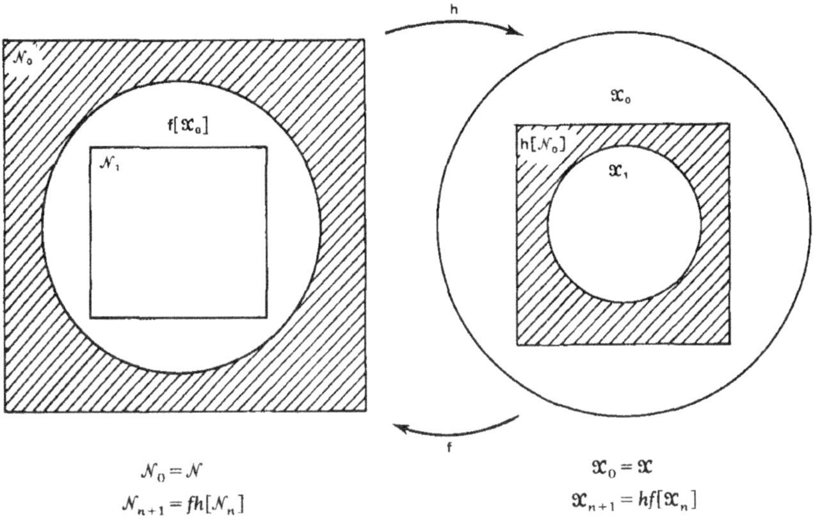

Diagram 1G.1.

Put also

$$\mathcal{N}^* = \bigcap_n \mathcal{N}_n, \qquad \mathfrak{X}^* = \bigcap_n \mathfrak{X}_n$$

and notice that

$$\mathfrak{X}^* = \bigcap_n \mathfrak{X}_n \supseteq \bigcap_n h[\mathcal{N}_n] \supseteq \bigcap_n \mathfrak{X}_{n+1} = \mathfrak{X}^*,$$

and since h is an injection,

$$h[\mathcal{N}^*] = h(\cap_n \mathcal{N}_n] = \bigcap_n h[\mathcal{N}_n] = \mathfrak{X}^*.$$

Thus h gives a bijection of $\mathcal{N}^*$ with $\mathfrak{X}^*$. On the other hand,

$$\mathcal{N} = (\mathcal{N}_0 - f[\mathfrak{X}_0]) \cup (f[\mathfrak{X}_0] - \mathcal{N}_1) \cup (\mathcal{N}_1 - f[\mathfrak{X}_1]) \cup (f[\mathfrak{X}_1] - \mathcal{N}_2) \cup \cdots \cup \mathcal{N}^*$$

$$\mathfrak{X} = (\mathfrak{X}_0 - h[\mathcal{N}_0]) \cup (h[\mathcal{N}_0] - \mathfrak{X}_1) \cup (\mathfrak{X}_1 - h[\mathcal{N}_1]) \cup (h[\mathcal{N}_1] - \mathfrak{X}_2) \cup \cdots \cup \mathfrak{X}^*,$$

where the sets in these unions are disjoint. Moreover, h is a bijection of $\mathcal{N}_n - f[\mathfrak{X}_n]$ with $h[\mathcal{N}_n] - \mathfrak{X}_{n+1}$, since h is an injection and $f[\mathfrak{X}_n] \subseteq \mathcal{N}_n$, so that

$$h[\mathcal{N}_n - f[\mathfrak{X}_n]] = h[\mathcal{N}_n] - hf[\mathfrak{X}_n] = h[\mathcal{N}_n] - \mathfrak{X}_{n+1},$$

and similarly, f is a bijection of $\mathfrak{X}_n - h[\mathfrak{X}_n]$ with $f[\mathfrak{X}_n] - \mathcal{N}_{n+1}$. So we have a bijection of $\mathcal{Y}$ with $\mathfrak{X}$,

$$g(\alpha)=\begin{cases} h(\alpha) & \text{if } \alpha\in\mathcal{N}^* \text{ or } \alpha\in\mathcal{N}_n - f[\mathfrak{X}_n] \text{ for some } n,\\ f^{-1}(\alpha) & \text{if } \alpha\notin\mathcal{N}^* \text{ and } \alpha\in f[\mathfrak{X}_n]-\mathcal{N}_{n+1} \text{ for some } n.\end{cases}$$

It remains to verify that g is Borel.

Recall that good Borel injections map Borel sets onto Borel sets. This implies that all the sets $\mathcal{N}_n$, $\mathfrak{X}_n$ are Borel, hence $\mathcal{N}^*$, $\mathfrak{X}^*$ and all the differences $\mathcal{N}_n - f[\mathfrak{X}_n]$, $f[\mathfrak{X}_n]-\mathcal{N}_{n+1}$ are Borel. From this it follows immediately that g is Borel. ⊣

EXERCISES

Let us start with a very simple representation of Borel sets which comes out of 1G.2.

1G.5. Prove that every Borel set is the continuous, injective image of a closed set of irrationals; i.e. if $P\subseteq\mathfrak{X}$ is Borel, then there exists a continuous $\pi:\mathcal{N}\to\mathfrak{X}$ and a closed $B\subseteq\mathcal{N}$ such that π is one-to-one on B and $\pi[B]=P$.

Hint. Suppose first P is closed. Let $\pi:\mathcal{N}\twoheadrightarrow\mathfrak{X}$ and A be as in 1G.2 and take $B=\pi^{-1}[P]\cap A$.

For each finite sequence $k_0,\ldots, k_{n-1}$, let

$$N(k_0,\ldots, k_{n-1})=\{\alpha: \alpha(0)=k_0,\ldots, \alpha(n-1)=k_{n-1}\}.$$

Each $N(k_0,\ldots, k_{n-1})$ is trivially homeomorphic with $\mathcal{N}$.

Suppose $P=\bigcup_n P_n$ and each P_n satisfies the result. We may assume then that there are closed sets $B_n\subseteq N(n)$ and continuous maps $\pi_n: N(n)\to\mathfrak{X}$ such that $\pi_n[B_n]=P_n$ and π_n is injective on B_n. Take

$$B=\{\alpha: \alpha^{\star}\in B_{\alpha(0)}\},$$

with $\alpha^{\star}=(t\mapsto\alpha(t+1))$ and $\pi(\alpha)=\pi_{\alpha(0)}(\alpha^{\star})$.

Suppose $P=\bigcap_n P_n$ with B_n, π_n again as above. Let $(\alpha)_i$ be defined as in 1E and put

$$\begin{aligned}\alpha\in B \Leftrightarrow\ & (\forall n)[(\alpha)_n\in B_n]\\ & \&\ (\forall n)(\forall m)[\pi_n((\alpha)_n)=\pi_m((\alpha)_m)]\\ & \&\ (\forall n)(\forall s)(\forall t)[t\neq\langle n, s\rangle \Rightarrow \alpha(t)=0].\end{aligned}$$

Clearly B is closed. Let

$$\pi(\alpha) = \pi_0((\alpha)_0)$$

and verify that π is one-to-one on B and $\pi[B] = \bigcap_n P_n$. ⊣

We will prove in 2E.7 and 2E.8 that this is actually a *characterization* of Borel sets, i.e. every continuous injective image of a closed set is Borel. In 4A.7 we will also give a very different proof of this result.

By our basic definition, a function $f : \mathcal{X} \to \mathcal{Y}$ is $\boldsymbol{\Sigma}^0_\xi$-measurable if $f^{-1}[P]$ is in $\boldsymbol{\Sigma}^0_\xi$ for each open $P \subseteq \mathcal{Y}$. Clearly, the $\boldsymbol{\Sigma}^0_1$-measurable functions are precisely the continuous functions.

1G.6. Suppose that $f : \mathcal{X} \to \mathcal{Y}$, $g : \mathcal{Y} \to \mathcal{Z}$ and $h : \mathcal{X} \to \mathcal{Z}$ is the composition of f and g,

$$h(x) = (g \circ f)(x) = g(f(x)).$$

Prove that if one of the two given functions is continuous and the other is $\boldsymbol{\Sigma}^0_\xi$-measurable, then h is $\boldsymbol{\Sigma}^0_\xi$-measurable.

Hint. For the case when g is continuous, use the closure of $\boldsymbol{\Sigma}^0_\xi$ under continuous substitution. ⊣

1G.7. Prove that if $f : \mathcal{X} \to \mathcal{Y}$ is $\boldsymbol{\Sigma}^0_\xi$-measurable and P is a $\boldsymbol{\Sigma}^0_\eta$ subset of $\mathcal{Y}$, then $f^{-1}[P]$ is $\boldsymbol{\Sigma}^0_{\xi+\eta}$.

Hint. Use induction on η; notice that $\xi + \eta$ denotes the ordinal sum of ξ and η so that

$$\textit{supremum}\{\xi + \eta_i : i = 0, 1, \ldots\} = \xi + \textit{supremum}\{\eta_i : i = 0, 1, \ldots\}.$$ ⊣

1G.8. Prove that if $f : \mathcal{X} \to \mathcal{Y}$ is $\boldsymbol{\Sigma}^0_\xi$-measurable and $g : \mathcal{Y} \to \mathcal{Z}$ is $\boldsymbol{\Sigma}^0_\eta$-measurable, then the composition $g \circ f : \mathcal{X} \to \mathcal{Z}$,

$$(g \circ f)(x) = g(f(x))$$

is $\boldsymbol{\Sigma}^0_{\xi+\eta}$-measurable. ⊣

With each function

$$f : \mathcal{X} \to \mathcal{N}$$

we associate the function

$$f^* : \mathcal{X} \times \omega \to \omega$$

defined by

$$f^*(x, n) = f(x)(n).$$

1G.9. Prove that for each $\mathfrak{X}$ and each countable ξ, $f:\mathfrak{X}\to\mathcal{N}$ is $\mathbf{\Sigma}^0_\xi$-measurable if and only if the associated function $f^*:\mathfrak{X}\times\omega\to\omega$ is $\mathbf{\Sigma}^0_\xi$-measurable. ⊣

1G.10. Prove that for each perfect product space $\mathfrak{X}$ there is a $\mathbf{\Sigma}^0_2$-measurable surjection

$$f:\mathfrak{X}\twoheadrightarrow\mathcal{N}$$

of $\mathfrak{X}$ onto Baire space.

Hint. Let σ be the function on binary sequences associated with $\mathfrak{X}$ in 1A.2 and put

$$P(x,n,w)\Leftrightarrow \bigvee_u[u=\langle t_0,t_1,\ldots,t_{\langle n,w\rangle}\rangle \;\&\; t_{\langle n,w\rangle}=0 \;\&\; x\in N(\mathfrak{X},\sigma(u))]$$
$$\vee \bigwedge_u \bigwedge_k[(u=\langle t_0,\ldots,t_{\langle n,k\rangle}\rangle \;\&\; t_{\langle n,k\rangle}=0)\Rightarrow x\notin N(\mathfrak{X},\sigma(u))].$$

Clearly P is $\mathbf{\Sigma}^0_2$ and for each x, n there is some w such that $P(x,n,w)$. Consider P as a subset of $(\mathfrak{X}\times\omega)\times\omega$ and choose a $\mathbf{\Sigma}^0_2$ set P^* which uniformizes P by 1C.6. Now P^* is the graph of a $\mathbf{\Sigma}^0_2$-measurable function $f^*:\mathfrak{X}\times\omega\to\omega$, so the associated $f:\mathfrak{X}\to\mathcal{N}$,

$$f(x)=n\mapsto f^*(x,n)$$

is also $\mathbf{\Sigma}^0_2$-measurable. To show that f is onto $\mathcal{N}$, given α let

$$\beta(s)=\begin{cases}0 & \text{if } s=\langle n,\alpha(n)\rangle \text{ for some } n,\\ 1 & \text{otherwise}\end{cases}$$

and take $x=\pi(\beta)$, where π is the canonical injection of $\mathcal{C}$ into $\mathfrak{X}$ defined in 1A.3. It is easy to check that $f(x)=\alpha$. ⊣

1G.11. Prove that for each countable ξ and each perfect product space $\mathcal{Y}$, $\mathbf{\Sigma}^0_\xi$ is $\mathcal{Y}$-parametrized. Infer that for each countable ξ and each perfect $\mathfrak{X}$,

$$\mathbf{\Delta}^0_\xi{\restriction}\mathfrak{X}\subsetneq\mathbf{\Sigma}^0_\xi{\restriction}\mathfrak{X}\subsetneq\mathbf{\Delta}^0_{\xi+1}{\restriction}\mathfrak{X}.^{(15)}$$

Hint. Use 1F.6, 1G.7 and 1G.10. ⊣

1G.11 also yields an alternative characterization of $\mathbf{\Sigma}^1_1$ sets which is worth pointing out.

1G.12. Suppose $\mathfrak{X}$ is perfect and $P\subseteq\mathcal{Y}$. Prove that P is $\mathbf{\Sigma}^1_1$ if and only if there is a $\mathbf{\Sigma}^0_2$-measurable $f:\mathfrak{X}\to\mathcal{Y}$ such that $P=f[\mathfrak{X}]$. Similarly, P is $\mathbf{\Sigma}^1_1$ if and only if P is the projection of some $\mathbf{\Pi}^0_2$ subset Q of $\mathcal{Y}\times\mathfrak{X}$.

In particular, every Σ_1^1 set of reals is the projection of a G_δ set in the plane.

Hint. For the first assertion use 1G.10 and 1E.6. For the second assertion, let $C \subseteq \mathcal{Y} \times \mathcal{N}$ be closed with projection P, let $f : \mathcal{X} \twoheadrightarrow \mathcal{N}$ be Σ_2^0-measurable and take $Q = \{(y, x) : C(y, f(x))\}$. ⊣

1G.13. Prove that the function f which we defined in the proof of 1G.2 is actually Σ_n^0-measurable for a small, finite n. ⊣

1G.14. Prove that for each perfect product space $\mathcal{X}$ there is a bijection

$$f : \mathcal{N} \twoheadrightarrow \mathcal{X}$$

such that both f and its inverse are $\Sigma_{\omega+1}^0$-measurable. ⊣

1G.15. Prove that a function $f : \mathcal{X} \to \mathcal{Y}$ is Borel measurable if and only if it is Σ_ξ^0-measurable for some countable ordinal ξ. ⊣

1G.16. Prove that for each perfect product space $\mathcal{X}$, for each product space $\mathcal{Y}$ and for each countable ordinal ξ, there exists a function $f : \mathcal{X} \to \mathcal{Y}$ which is Σ_ξ^0-measurable but is not Σ_η^0-measurable for any $\eta < \xi$.[(10)]

Hint. Choose a subset P of $\mathcal{X}$ which is Δ_ξ^0 but not Σ_η^0 for any $\eta < \xi$, let y_0 and y_1 be distinct points in $\mathcal{Y}$ and take

$$f(x) = \begin{cases} y_0 & \text{if } x \in P, \\ y_1 & \text{if } x \notin P. \end{cases}$$ ⊣

A function $f : \mathcal{X} \to \mathcal{Y}$ is of *Baire class* 0 if it is continuous; it is of *Baire class* 1 if it is not continuous but it is Σ_2^0-measurable. Proceeding inductively, for each countable ordinal ξ, a function $f : \mathcal{X} \to \mathcal{Y}$ is of *Baire class* ξ if it is not of Baire class $\eta < \xi$ and there exists a sequence $f_0, f_1, \ldots$ such that each f_n is of Baire class $<\xi$ and

$$f = \mathit{limit}_{n \to \infty} f_n \quad \text{(pointwise)},$$

i.e. for each $x \in \mathcal{X}$,

$$f(x) = \mathit{limit}_{n \to \infty} f_n(x).$$

1G.17. Prove that if $f : \mathcal{X} \to \mathcal{Y}$ is of Baire class ξ, then f is $\Sigma_{\xi+1}^0$-measurable.[(10)]

Hint. Use induction on ξ, taking cases on whether ξ is a successor or a limit ordinal. The key equivalence is the following, where $f = \mathit{limit}_{n \to \infty} f_n$

and the typical open ball P is written as a countable union of closed balls with the same center,

$$P = \bigcup_i F_i;$$

$$f(x) \in P \leftrightarrow (\bigvee_n)(\bigvee_i)(\bigwedge_{m \geq n})[f_m(x) \in F_i]. \quad \dashv$$

We first establish the converse of this elegant characterization in the simple cases when $\mathcal{Y}$ is ω or $\mathcal{N}$.

1G.18. Prove that if $f: \mathfrak{X} \to \omega$ is $\mathbf{\Sigma}^0_{\xi+1}$-measurable, then f is of Baire class $\leq \xi$.

Hint. Use induction on ξ. Given some $\mathbf{\Sigma}^0_{\xi+1}$-measurable function f with $\xi + 1 \geq 3$, we clearly have

$$f(x) = w \leftrightarrow \bigvee_n \bigwedge_m [x \in P_{n,m,w}]$$

where each $P_{n,m,w}$ is $\mathbf{\Sigma}^0_{\eta_i}$ for some $\eta_i < \xi$. Put

$$g_s(x) = \text{the least } t \text{ such that } (\forall m \leq s)(x \in P_{(t)_0, m, (t)_1}],$$

$$f_s(x) = (g_s(x))_1$$

and verify easily that $f = \mathit{limit}_{s\to\infty} f_s$. Now each f_s is easily $\mathbf{\Sigma}^0_{\eta+1}$-measurable for some $\eta < \xi$, so by induction hypothesis, each f_s is of Baire class $<\xi$; hence f is of Baire class $\leq \xi$. $\dashv$

1G.19. Prove that if $f: \mathfrak{X} \to \mathcal{N}$ is $\mathbf{\Sigma}^0_{\xi+1}$-measurable, then f is of Baire class $\leq \xi$.

Hint. Use 1G.9 above and the corresponding result for functions of Baire class ξ. $\dashv$

1G.20. Prove that if $f: \mathfrak{X} \to \mathcal{Y}$ is $\mathbf{\Sigma}^0_{\xi+1}$-measurable, then f is of Baire class $\leq \xi$.[(10)]

Hint. Put

$$P(x, n, i) \leftrightarrow d(f(x), r_i) < 2^{-n-1},$$

where $r_0, r_1, \ldots$ is the fixed dense set in $\mathcal{Y}$ and notice that P is a $\mathbf{\Sigma}^0_{\xi+1}$ subset of $(\mathfrak{X} \times \omega) \times \omega$. By 1F.8, choose P^* to uniformize P, P^* also in $\mathbf{\Sigma}^0_{\xi+1}$ and notice that P^* is the graph of a function $g^*: \mathfrak{X} \times \omega \to \omega$. Put

$$g(x) = n \mapsto g^*(x, n),$$

so that both g and g^* are $\mathbf{\Sigma}^0_{\xi+1}$-measurable, and check that

$$f(x) = \pi(g(x)),$$

where π is the canonical continuous surjection of $\mathcal{N}$ onto $\mathcal{Y}$ defined in 1A.1. Finally, verify that the collection of functions of Baire class $\leq\xi$ is closed under composition with continuous functions and apply 1G.19. ⊣

If $\mathcal{X}=\mathcal{N}$ or $\mathcal{Y}=\mathcal{R}$, we can extend this characterization of $\boldsymbol{\Sigma}^0_{\xi+1}$-measurability to $\boldsymbol{\Sigma}^0_2$. We only state here the result for the case $\mathcal{X}=\mathcal{N}$, since it can be established easily by the methods we have been using.

1G.21. Prove that a function $f:\mathcal{N}\to\mathcal{Y}$ is $\boldsymbol{\Sigma}^0_2$-measurable if and only if there is a sequence $f_0, f_1, \ldots$ of continuous functions on $\mathcal{N}$ to $\mathcal{Y}$, such that $f=limit_{n\to\infty} f_n$.[(10)]

Hint. Use the method of 1G.18, together with the fact that every closed subset of $\mathcal{N}\times\omega$ is a countable intersection of clopen sets. ⊣

1H. Historical and other remarks

[1] The early papers in descriptive set theory were all concerned with sets and functions in real n-space. It was quickly recognized, however, that most results generalized easily at least to Polish spaces, and soon two tendencies developed: one was to stick with the reals or the irrationals and prove the strongest possible results, the other to aim for the widest context in which the basic facts can be established.

[2] Lusin works in the irrationals in his classic [1930b] and Sierpinski [1950] gives a brief exposition of the theory for the reals. Among the general books in set theory and topology which cover descriptive set theory, the three best references are Hausdorff [1957], Sierpinski [1956] and Kuratowski [1966]. Kuratowski's book is by far the most comprehensive of the three and serves as the standard reference for the classical theory.

[3] In this book we are mostly interested in the *theory of definable sets of real numbers.* To study this, however, we must consider the irrationals and finite products of copies of $\mathcal{R}$, $\mathcal{N}$ and ω; as it happens, it is no harder and a bit neater to develop the theory for *finite products of perfect Polish spaces and copies of* ω.

[4] There is no real restriction in taking the basic spaces *perfect,* since every Polish space X is a closed subset of the perfect Polish space $X\times\mathcal{N}$ and results about X can be easily read off the results about $X\times\mathcal{N}$. On the other hand, there are some definite technical advantages to our convention, particularly in the *effective theory* which we will study starting with Chapter 3.

[5] We should point out that a large part of the theory can be developed in a very general context, in fact many of the basic results have been extended recently to nonseparable and even nonmetrizable spaces. These extensions are important and significant for the applications of descriptive set theory to set-theoretic topology, functional analysis and potential theory. The interested reader should consult Christensen [1974], Hoffman-Jørgensen [1970] and the further references given there.

[6] In citing classical references, we will not always specify the context in which the results were first proved or give credit for the subsequent generalizations, unless these involved genuinely new ideas.

[7] As we mentioned in the introduction, the earliest notions of descriptive set theory were the *Baire classes of functions* on $\mathcal{R}^n$ to $\mathcal{R}$, defined in Baire [1899] and studied extensively in Lebesgue [1905].

[8] Lebesgue [1905] also introduced the class of *Borel measurable sets* and defined the first hierarchy on $\mathbf{B}$. According to Lebesgue, a set $P \subseteq \mathcal{R}^n$ is *open of class* ξ if there is a function $f: \mathcal{R}^n \to \mathcal{R}$ of Baire class ξ and an interval (a, b) in the line, such that

$$P = f^{-1}(a, b) = \{x : a < f(x) < b\};$$

a set is *closed of class* ξ if its complement is open of class ξ and a set is *of row* ξ is it can be written as a countable intersection of sets which are closed of class $<\xi$ and is not itself closed of class $<\xi$. Lebesgue then proved (in our notation) that

$$P \text{ is open of class} \leq \xi \Leftrightarrow P \text{ is } \mathbf{\Sigma}^0_{\xi+1},$$
$$\text{for limit } \xi, P \text{ is of row} \leq \xi \Leftrightarrow P \text{ is } \mathbf{\Pi}^0_\xi.$$

[9] Our own approach of taking the classes $\mathbf{\Sigma}^0_\xi$ and $\mathbf{\Pi}^0_\xi$ as the basic definitions traces back to Hausdorff [1919]. We mention the Lebesgue definitions here because they were often used in the early papers, through the 1920's. Another notion often taken as basic is that of set of *Baire-de la Vallee-Poussin class* $\leq \xi$, $\mathbf{\Delta}^0_{\xi+1}$ set in our terminology.

[10] Lebesgue [1905] defined the general notion of Λ*-measurability* and established that the Baire functions coincide with the Borel measurable functions, as well as the step-by-step characterization of 1G.17, 1G.20 and 1G.21 with $\mathcal{X} = \mathcal{R}^n$, $\mathcal{Y} = \mathcal{R}$. The more general result about arbitrary Polish spaces is due to Banach [1931]. It appears that the hints to these exercises outline a new proof of Banach's result—whether new or not, it is a simple proof which illustrates the value of having the trivial space ω available as a factor in our product spaces.

[11] *Analytic* (Σ^1_1) *sets* were introduced in Suslin [1917]. Suslin's definition was in terms of the operation $\mathscr{A}$ which we will study in the next chapter, but he characterized the analytic sets in $\mathcal{R}$ as precisely the projections of Borel (or G_δ) sets in the plane. He also proved the key result that there are analytic sets which are not Borel measurable, as well as all the simple closure properties of analytic sets, including closure under projection. In a companion note, Lusin [1917] (essentially) characterized Σ^1_1 sets of reals as the images of $\mathcal{R}$ by Σ^0_2-measurable functions (1G.12) and proved that every Borel set is the continuous, injective image of a closed set of irrationals (1G.5).

[12] *Projective sets* were introduced by Lusin [1925a, b, c] and (apparently independently) by Sierpinski [1925]. The main result in both these papers is the *hierarchy property for the Lusin pointclasses* on the reals. Somewhat later, Sierpinski [1928] showed the closure of these classes under countable unions and intersections.

[13] The finite Borel pointclasses were not studied separately from the transfinite ones in the classical theory, so it was not noticed that they can be defined using projection on ω in a manner analogous to the definition of the projective classes. Our approach here derives from the work of the recursion theorists, Kleene [1943] and Mostowski [1946].

[14] Another major difference between the approach to the subject in the early papers and the present theory is our heavy use of the *operations of logic* both in stating and in establishing the closure properties of the various pointclasses. A good many of the quantifier rules and logical transformations which we used in 1C and 1E were first applied in the fundamental papers Kuratowski–Tarski [1931] and Kuratowski [1931], where the connection between descriptive set theory and logic was first noticed. The use of *codings of finite sequences* and continuous substitutions to prove closure properties is essentially due to Kleene.

[15] *Universal sets* were introduced by Lusin [1925d], who credits them to Lebesgue [1905]. The reference is not entirely accurate, as Lebesgue had what we called in 1*D parametrizations* (on collections of functions) rather than universal sets, although he certainly initiated the use of the diagonal method to prove hierarchy results. Many papers were written on universal sets, proving their existence for various pointclasses in diverse spaces or constructing specific universal sets which somehow appeared to be "natural." On the other hand, the simple construction in 1D.1 and its Corollaries 1D.2, 1E.3 and 1G.11 seem to have been missed—the strongest result mentioned in Kuratowski [1966] is that the Borel and

Lusin pointclasses are $\mathcal{N}$-parametrized. Sierpinski [1950] has a general hierarchy lemma, very similar to our 1D.3.

[16] It is not entirely clear who introduced first the notion of *separation* for sets, probably Lusin. The separability of disjoint $\mathbf{\Pi}^0_\xi$ sets by a $\mathbf{\Delta}^0_\xi$ set was apparently first proved by Sierpinski [1924] and independently by Lavrentieff [1925]. Kuratowski [1936] defined the *reduction* property and showed that $\mathbf{\Sigma}^0_\xi$ sets can be reduced by $\mathbf{\Sigma}^0_\xi$ sets.

[17] The more fundamental *uniformization* property was introduced in Lusin [1930a]. Lusin established there some difficult uniformization theorems and introduced the difficult problem of uniformization for $\mathbf{\Pi}^1_1$ sets. Of course, the question of uniformizing subsets of $\mathfrak{X} \times \omega$ was not considered in the classical theory, since they never studied the trivial space ω. It comes up naturally in the effective theory and we will come back to it in the sequel. Notice how useful the trivial 1F.8 is in proving 1F.9, 1G.10 and 1G.20.

[18] There is a large number of *transfer theorems* like our 1G.2 and 1G.4, some of them stronger than the simple ones we have established. Kuratowski [1966] is an excellent source for results of this kind and for references to the original sources.

[19] Finally, a word about the "logical" notation of the $\mathbf{\Sigma}$'s $\mathbf{\Pi}$'s and $\mathbf{\Delta}$'s which we have adopted for the Borel and the projective pointclasses. This was introduced by Addison [1959a] and Shoenfield [1961] and was quickly accepted by the logicians, though not by all the topologists and set theorists who were working in descriptive set theory. No other comprehensive system of notation has gained wide acceptance and it seems that all the reasons given by Addison [1959a] for adopting this one are still valid today.

CHAPTER 2

$\varkappa$-SUSLIN AND λ-BOREL

One of the chief motivations for studying projective sets is that we can settle for them many questions which seem intractable for arbitrary sets. The hierarchy of the Lusin pointclasses is important because it is often the case that a simple observation about open sets turn into a deep theorem about $\boldsymbol{\Sigma}^1_1$ sets, an elegant generalization about $\boldsymbol{\Sigma}^1_2$ sets and a very difficult problem about $\boldsymbol{\Sigma}^1_3$ or $\boldsymbol{\Sigma}^1_4$ sets.

For example, consider the central question of set theory, *the continuum problem:* must each uncountable pointset be equinumerous with $\mathcal{R}$? Gödel and Cohen have shown that both answers are consistent with the currently accepted axioms of set theory, but it is still open whether the question may be settled on the basis of generally acceptable properties of sets. In any case, we can try to settle it for specific pointclasses, preferably large pointclasses that contain most sets encountered in traditional mathematics.

One of the first important results of descriptive set theory was that every uncountable $\boldsymbol{\Sigma}^1_1$ set is equinumerous with $\mathcal{R}$. More recently, this has been extended by Solovay and Mansfield to $\boldsymbol{\Sigma}^1_2$ and $\boldsymbol{\Sigma}^1_3$ sets respectively, granting some strong set theoretic axioms that are unprovable in Zermelo–Fraenkel set theory. The situation is a bit murkier for the higher Lusin pointclasses, but there are (very strong) plausible hypotheses which imply that every uncountable projective set is equinumerous with $\mathcal{R}$.

The same situation occurs with several other *regularity properties* of sets. For example, every $\boldsymbol{\Sigma}^1_1$ set is absolutely measurable and has the property of Baire. There are again suitable generalizations of these results to the higher Lusin pointclasses, if we assume strong set theoretic hyotheses.

The central classical result of the theory is Suslin's Theorem: *every $\boldsymbol{\Delta}^1_1$ set is Borel.* More than a regularity property, this is a *construction principle,* since it yields a reduction of the complicated projection operation (in this simple instance) to an iteration of the more elementary operations of countable union and complementation. A somewhat weaker construction principle is Sierpinski's theorem that every $\mathbf{\Sigma}^1_2$ set is the union of $\aleph_1$ Borel sets.

In this chapter we will establish some of the basic classical structure results about $\mathbf{\Sigma}^1_1$ and $\mathbf{\Sigma}^1_2$ pointsets.

Actually we will work with the wider classes of $\varkappa$-*Suslin* sets, where $\varkappa$ is any infinite cardinal number—this will ease extension of this theorem to the higher Lusin pointclasses. The $\mathbf{\Sigma}^1_1$ sets are precisely the $\aleph_0$-Suslin sets and every $\mathbf{\Sigma}^1_2$ set is $\aleph_1$-Suslin.

2A. The Cantor–Bendixson theorem

For any set A, let

$$card(A) = \textit{the cardinal number of } A.$$

By 1G.4, every perfect Polish space $\mathfrak{M}$ is equinumerous with $\mathcal{N}$, hence

$$card(\mathfrak{M}) = card(\mathcal{N}) = 2^{\aleph_0}.$$

If P is a *perfect subset* of $\mathfrak{M}$ (i.e. P is closed and has no isolated points) and $P \neq \emptyset$, then P is a perfect Polish subspace of $\mathfrak{M}$, so again $card(P) = 2^{\aleph_0}$. This settles the continuum problem for perfect sets.

2A.1. THE CANTOR–BENDIXSON THEOREM. *If A is a closed pointset, then*

$$A = P \cup S$$

where P is perfect, S is countable and $P \cap S = \emptyset$. Moreover, there is only one such decomposition of A into two disjoint sets, one perfect the other countable.

PROOF. A point x is a *condensation point* of A if every nbhd of x intersects A in an uncountable set. Put

$$P = \{x : x \text{ is a condensation point of } A\},$$
$$S = A - P.$$

Since condensation points are clearly limit points and A is closed, we have $P \subseteq A$, and by definition $P \cap S = \emptyset$, $A = P \cup S$.

We will show that S is countable, P is perfect and if $A = P' \cup S'$ with P' perfect, S' countable and $P' \cap S' = \emptyset$, then $P' = P$, $S' = S$.

To each $y \in S$ we can assign some basic nbhd N^y such that $N^y \cap A$ is countable. Since there are only countably many basic nbhds altogether, there is a countable sequence $N^0, N^1, \ldots$ such that

$$S \subseteq \bigcup_{i \in \omega} (N^i \cap A)$$

with each $N^i \cap A$ countable, so S is countable.

To prove that P is closed, let x be a limit point of P, N any nbhd of x. Then some $x' \in N \cap P$, hence N contains some nbhd N' of x' which in turn contains uncountably many points of A; hence $x \in P$. To prove P perfect, if $x \in P$, then every nbhd of x contains uncountably many points of A of which only countably many can be in S—hence at least two are in P.

Finally, assume that $A = P' \cup S'$ with P', S' as above. If $x \in P'$ and N is any nbhd of x, choose some nbhd N_1 of x with $N_1 \subseteq N$ and check that $\overline{N_1 \cap P'}$ is a perfect subset of $\overline{N_1 \cap P'} \subseteq N \cap P'$, so $N \cap P'$ is uncountable and hence $x \in P$; this proves $P' \subseteq P$. On the other hand, if $y \in S'$, then there is some nbhd N of y such that $N \cap P' = \emptyset$, since P' is closed; hence $N \cap A = N \cap S'$, i.e. $N \cap A$ is countable and $y \in S$. ⊣

In this canonical decomposition

$$A = P \cup S$$

of a closed pointset, we call P the *(perfect) kernel* and S the *scattered part* of A.

It is worth putting down explicitly the corollary about the size of closed sets.

2A.2. COROLLARY. *Every uncountable closed pointset contains a perfect subset and hence has cardinality* $2^{\aleph_0}$. ⊣

We can think of the Cantor–Bendixson Theorem as a construction principle, since it gives us a method of building up the closed sets from the apparently simpler perfect sets and countable sets.

EXERCISES

2A.3. A point x is an isolated point of the pointset A if $x \in A$ and x is not a limit point of A. Prove that a pointset has at most countably many isolated points. ⊣

2A.4. Define the *derivative* A' of a pointset A by

$$A' = \{x \in A : x \text{ is a limit point of } A\}.$$

For fixed closed A, define by transfinite recursion the sets

$$A_0 = A,$$
$$A_{\xi+1} = (A_\xi)'$$
$$A_\lambda = \bigcap_{\xi<\lambda} A_\xi \quad \text{if } \lambda \text{ is a limit ordinal.}$$

Prove that $\bigcap_\xi A_\xi$ is perfect (perhaps empty), that $\bigcap_\xi A_\xi = A_\lambda$ for a countable ordinal λ and that $A - A_\lambda$ is countable. (This is an alternative proof of the Cantor–Bendixson Theorem.) ⊣

2B. $\varkappa$-Suslin sets

Let $\varkappa$ be an infinite cardinal number. A pointset $P \subseteq \mathfrak{X}$ is *$\varkappa$-Suslin* if there is a closed set $C \subseteq \mathfrak{X} \times {}^\omega\varkappa$ such that

$$P = \mathfrak{p}C = \text{the projection of } C \text{ along } {}^\omega\varkappa,$$

i.e.

$$x \in P \Leftrightarrow (\exists f \in {}^\omega\varkappa)(x, f) \in C.$$

Here we naturally topologize ${}^\omega\varkappa$ with the product topology, taking $\varkappa$ discrete, the typical nbhds being determined by finite sequences from $\varkappa$,

$$N(\xi_0, \ldots, \xi_n) = \{f \in {}^\omega\varkappa : f(0) = \xi_0, \ldots, f(n) = \xi_n\}.$$

This makes ${}^\omega\varkappa$ into a metric space which is perfect but of course not separable if $\varkappa > \aleph_0$. The set of ultimately constant $f \in {}^\omega\varkappa$ is dense and has cardinality $\varkappa$.

It is immediate from the definitions that the $\mathbf{\Sigma}^1_1$ pointsets are precisely the $\aleph_0$-Suslin or simply *Suslin sets*. Many of their properties can be proved just as easily for $\varkappa$-Suslin sets with $\varkappa > \aleph_0$ and there are applications of these more general results.

In this section we will establish the elementary properties of $\varkappa$-Suslin sets, starting with the equivalence of the definition above with two very useful and seemingly unrelated conditions.[1-5]

Let us take up first the representation of $\varkappa$-Suslin sets in terms of a pointset operation. A *$\varkappa$-Suslin system* is a mapping

$$u \mapsto P_u$$

which assigns to each finite sequence $u = (\xi_0, \ldots, \xi_{n-1})$ from $\varkappa$ a subset P_u of some fixed product space $\mathfrak{X}$. The operation $\mathcal{A}^\varkappa$ is defined on such

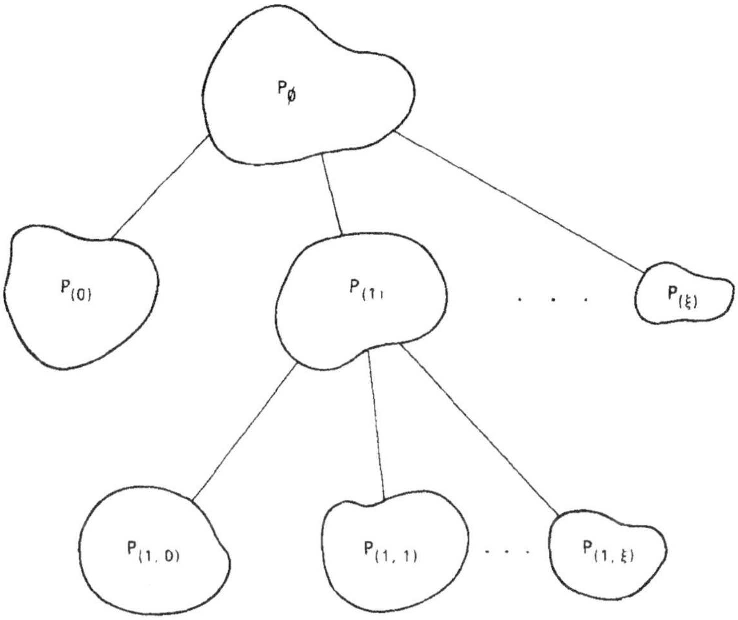

Figure 2B.1. The operation $\mathscr{A}$.

systems by

$$\mathscr{A}^{\varkappa}_{u}P_u = \bigcup_f \bigcap_n P_{f\upharpoonright n},$$

where f varies over ${}^{\omega}\varkappa$ (cf. Figure 2B.1). Thus

$$x \in \mathscr{A}^{\varkappa}_{u}P_u \Leftrightarrow \bigvee_f \bigwedge_n [x \in P_{f\upharpoonright n}].$$

It turns out that the $\varkappa$-Suslin pointsets are precisely the sets of the form $\mathscr{A}^{\varkappa}_{u}P_u$, where each P_u is closed. This is easy enough to prove directly, but we might as well take up the third condition we will need and prove the equivalence of all three round-robin style.

A *norm* on a pointset P is any function

$$\varphi : P \to \text{Ordinals}$$

which assigns an ordinal number $\varphi(x)$ to every $x \in P$. If for every $x \in P$ we have $\varphi(x) < \lambda$, we call φ a λ-*norm.*

A *semiscale* on P is a sequence

$$\bar{\varphi} = \{\varphi_n\}_{n \in \omega}$$

of norms on P such that the following limit condition holds: *if $x_0, x_1, x_2,\ldots$ are in P and $\text{limit}_{i\to\infty} x_i = x$ and if for each n the sequence of ordinals*

$$\varphi_n(x_0), \varphi_n(x_1), \varphi_n(x_2),\ldots$$

is ultimately constant, then $x \in P$.

We call $\bar{\varphi} = \{\varphi_n\}_{n\in\omega}$ a *λ-semiscale* if every norm φ_n is a λ-norm.

A ϰ-Suslin system $u \mapsto P_u$ is *regular* if the following conditions hold:

(i) Each P_u is the closure $\bar{N}_s$ of some basic nbhd, perhaps $\bar{N}_s = \emptyset$.

(ii) If the sequence u is an initial segment of the sequence v, then $P_u \supseteq P_v$.

(iii) If $u = (\xi_0,\ldots,\xi_{n-1})$ is a sequence of length n and $P_u = \bar{N}_s$, then $\textit{radius}(N_s) \leq 2^{-n+1}$.

2B.1. THEOREM. *For every infinite cardinal ϰ and every pointset $P \subseteq \mathfrak{X}$, the following conditions are equivalent.*

(i) *P is ϰ-Suslin, i.e.*

$$P = \mathfrak{p}C = \{x : (\exists f \in {}^{\omega}\varkappa)C(x, f)\}$$

with a closed $C \subseteq \mathfrak{X} \times {}^{\omega}\varkappa$.

(ii) *P admits a ϰ-semiscale.*

(iii) *$P = \mathscr{A}^{\varkappa}_{u}P_u$, where the ϰ-Suslin system $u \mapsto P_u$ is regular.*

(iv) *$P = \mathscr{A}^{\varkappa}_{u}P_u$ with a ϰ-Suslin system $u \mapsto P_u$ where each P_u is closed.*(1–5)

PROOF. (i) ⟹ (ii). For each $x \in P$, choose some $f_x \in {}^{\omega}\varkappa$ such that $(x, f_x) \in C$ and put

$$\varphi_n(x) = f_x(n).$$

To prove that the sequence $\bar{\varphi} = \{\varphi_n\}_{n\in\omega}$ of ϰ-norms on P is a semiscale, assume that $x_0, x_1,\ldots$ are in P, that $\textit{limit}_{i\to\infty} x_i = x$ and that for each n and all large i,

$$\varphi_n(x_i) = f_{x_i}(n) = \xi_n.$$

Let

$$f(n) = \xi_n.$$

Clearly

$$\text{limit}_{i\to\infty}(x_i, f_{x_i}) = (x, f)$$

and since for each i, $(x_i, f_{x_i}) \in C$ and C is closed, we have $(x, f) \in C$, i.e. $x \in P$.

(ii) $\Rightarrow$ (iii). Let $\bar{\varphi} = \{\varphi_n\}_{n\in\omega}$ be a $\varkappa$-semiscale on P. Choose a bijection

$$\pi : \varkappa \rightarrowtail\!\!\!\!\twoheadrightarrow \omega \times \varkappa$$

of $\varkappa$ with all pairs of integers and ordinals below $\varkappa$,

$$\pi(\xi) = (\pi_1(\xi), \pi_2(\xi)).$$

For convenience in notation let

$$N(s) = N_s$$

be the s'th basic nbhd of the space $\mathfrak{X}$ in which P lies.

Define now

$$\begin{aligned} P_{(\xi_0,\ldots,\xi_{n-1})} = \{x:\ & \bar{N}(\pi_1(\xi_0)) \supseteq \bar{N}(\pi_1(\xi_1)) \supseteq \cdots \supseteq \bar{N}(\pi_1(\xi_{n-1})) \\ & \&\ \text{for } i = 0,\ldots, n-1,\ radius(N(\pi_1(\xi_i))) \leq 2^{-i} \\ & \&\ \text{for some } y \in N(\pi_1(\xi_{n-1})),\ y \in P \text{ and} \\ & \quad \varphi_0(y) = \pi_2(\xi_0),\ \varphi_1(y) = \pi_2(\xi_1),\ldots,\ \varphi_{n-1}(y) = \pi_2(\xi_{n-1}), \\ & \&\ x \in \bar{N}(\pi_1(\xi_{n-1}))\}, \end{aligned}$$

so that $P_{(\xi_0,\ldots,\xi_{n-1})}$ is either $\emptyset$ or $\bar{N}(\pi_1(\xi_{n-1}))$ and the system $u \mapsto P_u$ is clearly regular. We will show that

$$P = \mathscr{A}^{\varkappa}_{u} P_u.$$

Suppose first that $x \in P$. Choose nbhds $\bar{N}(s_0) \supseteq \bar{N}(s_1) \supseteq \cdots$ of x such that $radius(N(s_i)) \leq 2^{-i}$ and for each i let ξ_i be the ordinal below $\varkappa$ such that

$$\pi_1(\xi_i) = s_i, \qquad \pi_2(\xi_i) = \varphi_i(x).$$

We obviously have $x \in P_{(\xi_0,\ldots,\xi_{n-1})}$ for every n, so $x \in \mathscr{A}^{\varkappa}_{u} P_u$.

Conversely, assume that there is a sequence $\xi_0, \xi_1, \ldots$ of ordinals such that $x \in P_{(\xi_0,\ldots,\xi_{n-1})}$ for every n. By the definition then,

$$\bar{N}(\pi_1(\xi_0)) \supseteq \bar{N}(\pi_1(\xi_1)) \supseteq \ldots,$$

$x \in \bar{N}(\pi_1(\xi_{n-1}))$ for each n, the radius of $N(\pi_1(\xi_{n-1}))$ shrinks to 0 as $n \to \infty$ and for every n we have some $y_n \in N(\pi_1(\xi_{n-1}))$ such that $\varphi_0(y_n) = \xi_0, \ldots, \varphi_{n-1}(y_n) = \xi_{n-1}$. In particular, $\text{limit}_{n\to\infty}\, y_n = x$ and for $n > i$, $\varphi_i(y_n) = \xi_i$, so by the basic property of a semiscale we have $x \in P$.

(iii) $\Rightarrow$ (iv) is trivial.

(iv) $\Rightarrow$ (i). Assume $P = \mathscr{A}^{\varkappa}_u P_u$ with each P_u closed and put

$$C(x, f) \Leftrightarrow \bigwedge_n [x \in P_{f \restriction n}].$$

Clearly C is closed and

$$\begin{aligned} x \in \mathrm{p}C &\Leftrightarrow (\exists f) C(x, f) \\ &\Leftrightarrow (\exists f)(\forall n)[x \in P_{f \restriction n}] \\ &\Leftrightarrow x \in P. \end{aligned}$$ ⊣

Suslin's original definition of analytic sets was via *the operation* $\mathscr{A}$,

$$\mathscr{A} = \mathscr{A}^{\aleph_0}$$

and the essential content of the equivalences (i) $\Leftrightarrow$ (iii) $\Leftrightarrow$ (iv) was already announced in the basic papers Suslin [1917], Lusin [1917].

Let $\mathbf{S}(\varkappa) = \mathbf{S}_\varkappa$ be the pointclass of all $\varkappa$-Suslin sets, so in particular

$$\mathbf{S}(\aleph_0) = \mathbf{\Sigma}^1_1.$$

2B.2. THEOREM. *For each cardinal* $\varkappa \geq \aleph_0$, *the pointclass* $\mathbf{S}_\varkappa$ *is closed under Borel substitution,* $\exists^{\mathcal{Y}}$ *for every product space* $\mathcal{Y}$, *countable conjunction,* $\bigwedge^\omega$, *disjunction of length* $\varkappa$, $\bigvee^\varkappa$ *and the operation* $\mathscr{A}^\varkappa$.

Moreover, if $\lambda \leq \varkappa$, *then* $\mathbf{S}_\lambda \subseteq \mathbf{S}_\varkappa$; *in particular every* $\mathbf{\Sigma}^1_1$ *pointset is* $\varkappa$*-Suslin.*

PROOF. Closure of $\mathbf{S}_\varkappa$ under continuous substitution is immediate, so we can use it in the arguments below.

To prove closure under $\exists^{\mathcal{Y}}$, suppose $C \subseteq \mathcal{X} \times \mathcal{N} \times {}^\omega\varkappa$ is closed and

$$P(x, \alpha) \Leftrightarrow (\exists f) C(x, \alpha, f),$$

so that

$$(\exists \alpha) P(x, \alpha) \Leftrightarrow (\exists \alpha)(\exists f) C(x, \alpha, f).$$

Let

$$\pi(\xi) = (\pi_1(\xi), \pi_2(\xi))$$

be a bijection of $\varkappa$ with $\omega \times \varkappa$ as in the proof of 2B.1 and notice that the mapping

$$\rho(g) = (g_1, g_2)$$

where

$$\begin{aligned} g_1(n) &= \pi_1(g(n)), \\ g_2(n) &= \pi_2(g(n)), \end{aligned}$$

is a homemorphism of ${}^{\omega}\varkappa$ with ${}^{\omega}\omega \times {}^{\omega}\varkappa = \mathcal{N} \times {}^{\omega}\varkappa$. Thus if we define

$$C^*(x, g) \Leftrightarrow C(x, g_1, g_2),$$

the set C^* is closed in ${}^{\omega}\varkappa$ and

$$(\exists\alpha)P(x, \alpha) \Leftrightarrow (\exists g)C^*(x, g),$$

so $\exists^{\mathcal{N}}P$ is $\varkappa$-Suslin.

We can now prove closure of $\mathbf{S}_\varkappa$ under $\exists^{\mathcal{Y}}$ using closure under $\exists^{\mathcal{N}}$ and the fact that every $\mathcal{Y}$ is a continuous image of $\mathcal{N}$.

If $P_\xi = \mathfrak{p}C_\xi$ for each $\xi < \varkappa$, put

$$C(x, f) \Leftrightarrow C_{f(0)}(x, f^\star),$$

where by definition

$$f^\star(n) = f(n+1).$$

Clearly C is closed and

$$\begin{aligned} \bigvee_{\xi<\varkappa} P_\xi(x) &\Leftrightarrow \bigvee_{\xi<\varkappa} (\exists f)C_\xi(x, f) \\ &\Leftrightarrow (\bigvee_f)C_{f(0)}(x, f^\star) \\ &\Leftrightarrow (\exists f)C(x, f), \end{aligned}$$

hence $\bigvee_\xi^\varkappa P_\xi$ is $\varkappa$-Suslin.

Similarly, if $P_m = \mathfrak{p}C_m$ for each $m \in \omega$, put

$$C(x, f) \Leftrightarrow \bigwedge_m C_m(x, f_m),$$

where

$$f_m(n) = f(\langle m, n\rangle)$$

and notice that C is closed and

$$\begin{aligned} \bigwedge_m P_m(x) &\Leftrightarrow \bigwedge_m (\exists f)C_m(x, f) \\ &\Leftrightarrow (\exists f) \bigwedge_m C_m(x, f_m) \\ &\Leftrightarrow (\exists f)C(x, f), \end{aligned}$$

so that $\bigwedge_m^\omega P_m$ is $\varkappa$-Suslin.

To prove closure of $\mathbf{S}_\varkappa$ under the operation $\mathcal{A}^\varkappa$, suppose $u \mapsto P_u$ is a $\varkappa$-Suslin system where for each u,

$$P_u(x) \Leftrightarrow (\exists g)C_u(x, g),$$

with C_u closed. Then

$$\begin{aligned} x \in \mathscr{A}^{\varkappa}_u P_u &\Leftrightarrow (\bigvee_f)(\bigwedge_n) P_{f \upharpoonright n}(x) \\ &\Leftrightarrow (\bigvee_f)(\bigwedge_n)(\exists g) C_{f\upharpoonright n}(x, g) \\ &\Leftrightarrow (\bigvee_f)(\exists g)(\bigwedge_n) C_{f\upharpoonright n}(x, g_n) \end{aligned}$$

where (as above) $g_n(m) = g(\langle n, m\rangle)$. Now the set

$$C(x, f, g) \Leftrightarrow \bigwedge_n C_{f\upharpoonright n}(x, g_n)$$

is obviously closed in $\mathfrak{X} \times {}^{\omega}\varkappa \times {}^{\omega}\varkappa$, and

$$x \in \mathscr{A}^{\varkappa}_u P_u \Leftrightarrow (\exists f)(\exists g) C(x, f, g).$$

We can easily find a closed C^* in $\mathfrak{X} \times {}^{\omega}\varkappa$, such that $\mathscr{A}^{\varkappa}_u P_u = \mathfrak{p}C^*$ using the obvious fact that ${}^{\omega}\varkappa \times {}^{\omega}\varkappa$ is homeomorphic with ${}^{\omega}\varkappa$.

If $\lambda \leq \varkappa$, then every λ-semiscale on a pointset P is also a $\varkappa$-semiscale. Hence every λ-Suslin set is $\varkappa$-Suslin.

Closure under Borel substitution follows immediately, since if $f: \mathfrak{X} \to \mathfrak{Y}$ is Borel and $P = \mathscr{A}^{\varkappa}_u P_u$ with each P_u closed, then

$$f^{-1}[P] = \mathscr{A}^{\varkappa}_u f^{-1}[P_u]$$

with each $f^{-1}[P_u]$ Borel, hence $\varkappa$-Suslin. Thus $f^{-1}[P]$ is $\varkappa$-Suslin by closure under $\mathscr{A}^{\varkappa}$. ⊣

Since every perfect product space is Borel isomorphic with $\mathcal{N}$, the closure of $\mathbf{S}_{\varkappa}$ under Borel substitution reduces the study of this pointclass to the study of $\varkappa$-Suslin sets of irrationals. This is often a useful reduction, especially because of a simple characterization of such sets in terms of *trees* which we will establish in the next section.

EXERCISES

2B.3. Suppose $u \mapsto P_u$ is a $\varkappa$-Suslin system such that

(i) if u is an initial segment of v, then $P_u \supseteq P_v$,

(ii) if u, v are distinct sequences of the same length n, then

$$P_u \cap P_v = \emptyset.$$

Prove that

$$\mathscr{A}^{\varkappa}_u P_u = \bigcap_n \bigcup_{\xi_0, \ldots, \xi_{n-1}} P_{(\xi_0, \ldots, \xi_{n-1})}.$$

Hint. If $x \in \bigcap_n \bigcup_{\xi_0,\ldots,\xi_{n-1}} P_{(\xi_0,\ldots,\xi_{n-1})}$, then by (ii), for each n there is exactly one sequence u of length n such that $x \in P_u$ and by (i) there is some $f \in {}^\omega\varkappa$ so that this $u = f \upharpoonright n$. ⊣

2B.4. Suppose $\varkappa$ is a cardinal of cofinality $>\omega$, i.e. if $\xi_0, \xi_1, \ldots$ are all $<\varkappa$, then $supremum\{\xi_n : n = 0, 1, 2, \ldots\} < \varkappa$. Prove that a pointset $P \subseteq \mathfrak{X}$ is $\varkappa$-Suslin if and only if

$$P = \bigcup_{\xi<\varkappa} P_\xi,$$

where each P_ξ is λ-Suslin for some cardinal $\lambda < \varkappa$ (Martin [19?a]).

Hint. Suppose

$$P(x) \Leftrightarrow (\exists f \in {}^\omega\varkappa) C(x, f)$$

with C closed and for each $\xi < \varkappa$, put

$$P_\xi(x) \Leftrightarrow (\exists f \in {}^\omega\xi) C(x, f).$$

Clearly $P = \bigcup_{\xi<\varkappa} P_\xi$, using $\mathrm{cf}(\varkappa) > \omega$. On the other hand, letting $\lambda = card(\xi)$ and $\pi : \lambda \twoheadrightarrow \xi$ any bijection, the set

$$C_\xi(x, f) \Leftrightarrow C(x, n \mapsto \pi(f(n)))$$

is obviously closed in $\mathfrak{X} \times {}^\omega\lambda$ and $P_\xi = \mathfrak{p}C_\xi$, so that P_ξ is λ-Suslin. ⊣

Despite this result, it is often useful to consider $\varkappa$-Suslin sets with $\varkappa$ of cofinality $>\omega$.

2B.5. Prove that if $n \geq 2$, then the pointclasses $\mathbf{\Sigma}^1_n$, $\mathbf{\Pi}^1_n$, $\mathbf{\Delta}^1_n$ are all closed under the operation $\mathcal{A} = \mathcal{A}^{\aleph_0}$ (Kantorovitch–Livenson [1932]).[6]

Hint. It is easy to check that if each P_u is $\mathbf{\Sigma}^1_n$, then $\mathcal{A}_u P_u$ is $\mathbf{\Sigma}^1_n$ $(n \geq 2)$. Assume that each P_u is $\mathbf{\Pi}^1_n$,

$$P_u(x) \Leftrightarrow (\forall \alpha) Q_u(x, \alpha) \quad (Q_u \text{ in } \mathbf{\Sigma}^1_{n-1}, n \geq 2)$$

and let

$$P(x) \Leftrightarrow \mathcal{A}_u P_u(x) \Leftrightarrow (\bigvee_\beta)(\bigwedge_t)(\forall \alpha) Q_{\beta \upharpoonright t}(x, \alpha)$$

so that

$$\neg P(x) \Leftrightarrow (\bigwedge_\beta)(\bigvee_t)(\forall \alpha) \neg Q_{\beta \upharpoonright t}(x, \alpha).$$

Now, only countably many α's are needed to verify the right hand side, hence

$$\neg P(x) \Leftrightarrow (\exists \alpha)(\bigwedge_\beta)(\bigvee_t)(\exists m) \neg Q_{\beta \upharpoonright t}(x, (\alpha)_m).$$

From this the result follows easily, by verifying that the relation

$$Q'(x, \alpha, \beta, t, m) \Leftrightarrow \neg Q_{\beta \upharpoonright t}(x, (\alpha)_m)$$

is $\boldsymbol{\Pi}^1_{n-1}$. ⊣

2C. Trees and the perfect set theorem

The main result of this section is that the continuum hypothesis holds for $\boldsymbol{\Sigma}^1_1$ sets—in fact every uncountable $\boldsymbol{\Sigma}^1_1$ set has a non-empty perfect subset.

For our purposes, a *tree* on a (non-empty) set X is a set T of finite sequences of members of X such that if $u \in T$ and v is an initial segment of u, then $v \in T$.

We often call the members of T *nodes* or *finite paths*. By definition, the empty sequence $\emptyset$ is a node of every non-empty tree—we call it the *root*. The terminology is motivated by the standard picture of a tree, see Figure 2C.1.

A function $f \in {}^{\omega}\mathfrak{X}$ is an *infinite branch* (or *path*) of a tree T, if for every n,

$$f \upharpoonright n = (f(0), \dots, f(n-1)) \in T.$$

We let

$$[T] = \{f \in {}^{\omega}X : f \quad \text{is an infinite branch of} \quad T\}$$

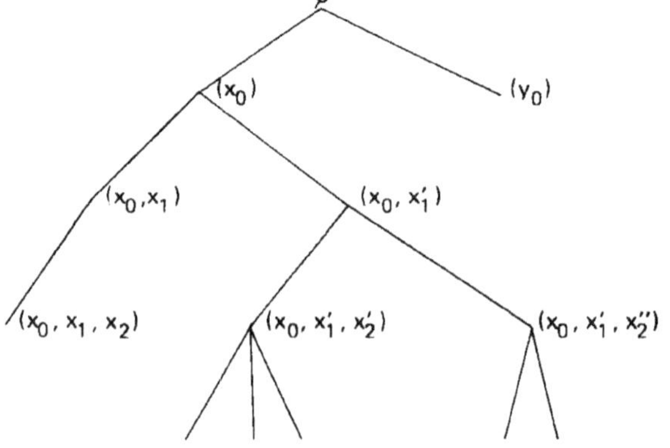

Figure 2C.1. A tree.

be *the body of* T, the subset of ${}^{\omega}X$ naturally associated with T.

We are particularly interested in *trees of pairs*, where we take

$$X = \omega \times \varkappa$$

with some infinite cardinal $\varkappa$. There is an obvious bijection of ${}^{\omega}(\omega \times \varkappa)$ with ${}^{\omega}\omega \times {}^{\omega}\varkappa = \mathcal{N} \times {}^{\omega}\varkappa$ which sends $g \in {}^{\omega}(\omega \times \varkappa)$ to (α, f), where

$$(*) \qquad g(n) = (\alpha(n), f(n)).$$

Let us agree that when T is a tree on $\omega \times \varkappa$ for some $\varkappa$, then we will take the body of T to be the obvious subset of $\mathcal{N} \times {}^{\omega}\varkappa$,

$$[T] = \{(\alpha, f)\text{: for all } n, ((\alpha(0), f(0)), \ldots, (\alpha(n-1), f(n-1))) \in T\}.$$

One could raise a pedantic objection to this ambiguous use of the symbol $[T]$, but it will cause no problems. It will always be clear from the context when we consider T to be a tree of pairs.

We will simplify notation further by denoting an arbitrary sequence

$$((t_0, \xi_0), \ldots, (t_{n-1}, \xi_{n-1}))$$

in $\omega \times \varkappa$ by

$$(t_0, \xi_0, \ldots, t_{n-1}, \xi_{n-1}).$$

There is no point in putting down all these parentheses. Thus if T is a tree on $\omega \times \varkappa$,

$$[T] = \{(\alpha, f)\text{: for all } n, (\alpha(0), f(0), \ldots, \alpha(n-1), f(n-1)) \in T\}.$$

2C.1. THEOREM. *For each non-empty set* X, *put the product topology on* ${}^{\omega}X$, *taking* X *discrete; then a set* $C \subseteq {}^{\omega}X$ *is closed if and only if there is a tree* T *on* X *such that* C *is the body of* T,

$$C = [T].$$

Similarly, for each cardinal $\varkappa \geq \aleph_0$, *a set* $C \subseteq \mathcal{N} \times {}^{\omega}\varkappa$ *is closed if and only if there is a tree* T *of pairs on* $\omega \times \varkappa$ *such that*

$$C = [T] = \{(\alpha, f)\text{: for all } n, (\alpha(0), f(0), \ldots, \alpha(n-1), f(n-1)) \in T\};$$

hence a set of irrationals

$$P \subseteq \mathcal{N}$$

is $\varkappa$*-Suslin if and only if there is a tree* T *on* $\omega \times \varkappa$ *such that*

$$P = \mathrm{p}[T] = \{\alpha\text{: } (\exists f)(\forall n)[(\alpha(0), f(0), \ldots, \alpha(n-1), f(n-1)) \in T]\}.$$

PROOF. It is enough to prove the first assertion, from which the second follows immediately, by the definition of $\varkappa$-Suslin sets and the obvious fact that the map $g \mapsto (\alpha, f)$ defined by $(*)$ above is a homeomorphism of ${}^{\omega}(\omega \times \varkappa)$ with $\mathcal{N} \times {}^{\omega}\varkappa$.

Suppose T is a tree on X and $f \notin [T]$; then for some n, $(f(0),\dots, f(n-1)) \notin T$, so that the basic nbhd $\{g : g(0) = f(0),\dots, g(n-1) = f(n-1)\}$ of ${}^{\omega}X$ is disjoint from $[T]$ and hence the complement of $[T]$ is open.

Conversely, if $C \subseteq {}^{\omega}X$ is closed, put

$$T = \{(f(0),\dots, f(n-1)) : f \in C\};$$

clearly $C \subseteq [T]$ and C is dense in $[T]$, so $C = [T]$. ⊣

Two finite sequences u and v from X are *compatible* if they have *a common extension*—i.e. if there is some w such that both u and v are initial segments of w. This simply means that either $u = v$ or one of u and v is an initial segment of the other.

For each tree T on X and each finite sequence u from X, let (cf. Figure 2C.2).

$$T_u = \{v \in T : v \textit{ is compatible with } u\}.$$

Evidently T_u is always a tree, the result of pruning all the side branches of

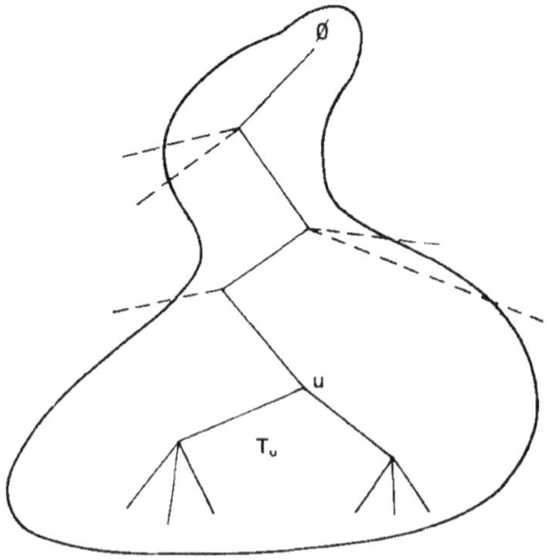

Figure 2C.2. The truncation of a tree.

T below u. In particular,

$$T_{\varnothing} = T.$$

Notice that if $u = (x_0, \ldots, x_{n-1})$ is a sequence of length n, then

$$[T_u] = [T] \cap \{f \in {}^{\omega}X : f \upharpoonright n = u\}$$
$$= \bigcup_{x \in X} [T_{u^\frown(x)}],$$

where of course for each $x \in X$,

$$u^\frown(x) = (x_0, \ldots, x_{n-1})^\frown(x) = (x_0, \ldots, x_{n-1}, x).$$

In connection with *projections of trees of pairs*, notice that if $u = (t_0, \xi_0, \ldots, t_{n-1}, \xi_{n-1})$ is a finite sequence from $\omega \times \kappa$, then

$$[T_u] = \bigcup_{t<\omega, \xi<\kappa} [T_{u^\frown(t,\xi)}],$$

so that

$$\mathrm{p}[T_u] = \bigcup_{t<\omega, \xi<\kappa} \mathrm{p}[T_{u^\frown(t,\xi)}].$$

We could prove the next result by an adaptation of the topological argument we used to establish the Cantor–Bendixson Theorem 2A.1. It will be more informative, however, to extend the argument of 2A.4. The use of trees is not essential in this instance, but they do make the proof neater.

2C.2. THE PERFECT SET THEOREM. *Let κ be an infinite cardinal and assume that P is a κ-Suslin pointset with more than κ elements. Then P has a non-empty perfect subset (Suslin, Mansfield).*[7]

PROOF. Suppose we know the result for every subset of $\mathcal{N}$ and P is a κ-Suslin subset of some perfect product space $\mathfrak{X}$. Let

$$\pi : \mathcal{N} \twoheadrightarrow \mathfrak{X}$$

be the continuous surjection of $\mathcal{N}$ onto $\mathfrak{X}$ guaranteed by 1G.2 and such that for some $\boldsymbol{\Delta}^0_1$ set $A \subseteq \mathcal{N}$, π is one-to-one on A and $\pi[A] = \mathfrak{X}$. Take $P' = \pi^{-1}[P] \cap A$. Now P' is a κ-Suslin subset of $\mathcal{N}$ with more than κ-elements, so it has a perfect subset Q. This Q must contain a non-empty perfect compact set Q_0—to see this apply 1A.3 with $\mathfrak{M} = Q$, considered as a subspace of $\mathcal{N}$. Hence $\pi[Q_0]$ is a perfect subset of P, since the continuous one-to-one image of a perfect compact set is easily perfect.

To establish the result for subsets of $\mathcal{N}$, let $P \subseteq \mathcal{N}$ be $\varkappa$-Suslin and choose a tree T on $\omega \times \varkappa$ such that

$$P = \mathrm{p}[T] = \{\alpha : (\exists f \in {}^{\omega}\varkappa)(\alpha, f) \in [T]\}.$$

Define by transfinite recursion the sets $T^{\xi} \subseteq T$,

$$T^0 = T,$$

$$T^{\xi+1} = \{u \in T^{\xi} : \mathrm{p}[T^{\xi}_u] \text{ has more than one (irrational) element}\},$$

$$T^{\lambda} = \bigcap_{\xi<\lambda} T^{\xi}, \text{ if } \lambda \text{ is a limit ordinal.}$$

It is immediate that each T^{ξ} is a tree and

$$\eta < \xi \Rightarrow T^{\eta} \supseteq T^{\xi}.$$

There are at most $\varkappa$ nodes in T, so there must be some ordinal λ of cardinality $\varkappa$ $(\lambda < \varkappa^+)$ such that

$$T^{\lambda+1} = T^{\lambda}.$$

Choose the least such λ and put

$$S = T^{\lambda}.$$

The heart of the proof is the following simple lemma about S.

LEMMA. $S \neq \emptyset$.

PROOF. Assume $S = \emptyset$, towards a contradiction.

For each $\alpha \in P = \mathrm{p}[T]$ choose $f \in {}^{\omega}\varkappa$ so that $(\alpha, f) \in [T]$ and notice that there must exist some $\xi < \lambda$ such that

$$(\alpha, f) \in [T^{\xi}] - [T^{\xi+1}];$$

this is because $(\alpha, f) \notin [T^{\lambda}]$ and for limit ζ,

$$(\alpha, f) \in [T^{\eta}], \quad \text{all } \eta < \zeta \Rightarrow (\alpha, f) \in [T^{\zeta}].$$

It follows that for some n,

$$u = (\alpha(0), f(0), \alpha(1), f(1), \ldots, \alpha(n-1), f(n-1)) \notin T^{\xi+1},$$

i.e. by definition

$$\mathrm{p}[T^{\xi}_u] \quad \text{has at most one element.}$$

Thus we have shown that

$$P \subseteq \bigcup \{\mathrm{p}[T^{\xi}_u] : \xi \leq \lambda, u \in T^{\xi} - T^{\xi+1}\}$$

which is absurd since the set on the right is the union of at most κ singletons and P has cardinality greater than κ. This proves the lemma.

If $u=(t_0, \xi_0, \ldots, t_{n-1}, \xi_{n-1})$ and $v=(s_0, \zeta_0, \ldots, s_{m-1}, \zeta_{m-1})$, call u and v *incompatible in the first coordinate* just in case $t_i \neq s_i$ for some $i<n$, $i<m$. It is immediate that every u in S has extensions u', u'' which are incompatible in the first coordinate—otherwise $\mathfrak{p}[S_u]$ would have at most one irrational in it and $u \notin T^{\lambda+1}=T^{\lambda}=S$.

We now imitate the proof of 1A.2. For each $u \in S$, let $l(u)$, $r(u)$ be extensions of u in S which are incompatible in the first coordinate and for each $f \in {}^{\omega}2$ define the sequence $u_0^f, u_1^f, \ldots$ of nodes in S by the induction

$$u_0^f=\emptyset,$$

$$u_{n+1}^f=\begin{cases} l(u_n^f) & \text{if } f(n)=0, \\ r(u_n^f) & \text{if } f(n)=1. \end{cases}$$

Let J be the set of all initial segments of all sequences u_n^f, $f \in {}^{\omega}2$. Clearly J is a tree, $J \subseteq S$ *and every two incompatible sequences in* J *are incompatible in the first coordinate*. The set $[J]$ is perfect (and compact) in $\mathcal{N} \times {}^{\omega}\kappa$ and since the projection mapping $\mathfrak{p}$ is continuous and one-to-one on $[J]$, $\mathfrak{p}[J]$ is perfect—this is the desired perfect subset of $P=\mathfrak{p}[T]$. ⊣

2C.3. COROLLARY. *Every uncountable* $\boldsymbol{\Sigma}_1^1$ *set has a perfect subset* (*Suslin*).[7] ⊣

We will see in Chapter 5 that this result cannot be extended to $\boldsymbol{\Sigma}_2^1$ sets (or even $\boldsymbol{\Delta}_1^1$ sets) in the context of Zermelo–Fraenkel set theory. On the other hand, there are better results that follow from strong set theoretic assumptions, as we mentioned in the introduction to this chapter.

EXERCISES

2C.4. Prove that we can decompose the reals into two disjoint sets

$$\mathcal{R}=A \cup B,$$

such that both A and B are uncountable and every non-empty perfect set intersects both A and B. In particular, A is an uncountable set which has no perfect subset other than $\emptyset$.

Hint. You need the axiom of choice for this. First argue that there are exactly $2^{\aleph_0}$ non-empty perfect sets. Wellorder $\mathcal{R}=\{x_\xi : \xi<2^{\aleph_0}\}$ and the collection of non-empty perfect sets $\mathcal{P}=\{P_\xi : \xi<2^{\aleph_0}\}$ and define by transfinite recursion set A_ξ, B_ξ such that $A_\xi \cap B_\xi=\emptyset$, $A_\xi \cap P_\xi \neq \emptyset$, $B_\xi \cap P_\xi \neq \emptyset$, $\eta \leq \xi \Rightarrow A_\eta \subseteq A_\xi$, $B_\eta \subseteq B_\xi$ and $card(A_\xi)<2^{\aleph_0}$, $card(B_\xi)<2^{\aleph_0}$. ⊣

2C.5. Prove that if $P \subseteq \mathfrak{X}$ and $Q \subseteq \mathcal{Y}$ are Borel sets such that

$$card(P) = card(Q), \qquad card(\mathfrak{X} - P) = card(\mathcal{Y} - Q),$$

then there exists a Borel isomorphism $f : \mathfrak{X} \rightarrowtail\!\!\!\!\twoheadrightarrow \mathcal{Y}$ such that $f[P] = Q$.

Hint. Use 2C.3 and the method of 1G.3. ⊣

2D. Wellfounded trees

A tree T on X is *wellfounded* if $[T] = \emptyset$, i.e. if T has no infinite branches. The name comes from considering the relation of *proper extension* of finite sequences from X,

$$u \succ v \Leftrightarrow u \textit{ is a proper initial segment of } v.$$

Clearly T is wellfounded if and only if the restriction of $\succ$ to T has no infinite descending chains.

Here we discuss briefly proof and definition by *backwards induction* on a wellfounded tree, which we will need in the next section. We will also introduce *rank functions* on wellfounded trees and use them to prove that $\mathbf{\Sigma}^1_2$ pointsets are $\aleph_1$-Suslin.

Let T be a wellfounded tree on X and suppose P is a relation on the finite sequences from X such that:

(∗) *if $P(u^\frown(x))$ holds for every $u^\frown(x) \in T$, then $P(u)$ holds.*

It follows that $P(u)$ must hold for every sequence u in X; otherwise there is some u_0 such that $\neg P(u_0)$, hence there is some x_0 with $u_0^\frown(x_0) \in T$ and $\neg P(u_0^\frown(x_0))$, hence there is some x_1 with $u_0^\frown(x_0, x_1) \in T$ and $\neg P(u_0^\frown(x_0, x_1))$, etc., so we get an infinite branch $u_0^\frown(x_0, x_1, x_2, \ldots)$ in T contradicting $[T] = \emptyset$. This method of proof is called *backwards* or *bar induction* on T.

In the same way we can justify *definition by backwards* or *bar recursion* on a wellfounded tree T: in order to define $F(u)$ for every finite sequence u from X, it is enough to show how to compute $F(u)$ if we know $F(u^\frown(x))$ for every $u^\frown(x) \in T$.

Formally, a function $F(u)$ is defined by bar recursion if we are given an equation of the form

$$F(u) = G(u, \{(x, F(u^\frown(x))) : u^\frown(x) \in T\})$$

with G a given function. We then put

$$R(u, z) \Leftrightarrow \text{there is some function } f \text{ such that } u \in Domain(f) \,\&\, f(u) = z$$
$$\& (\forall u', x)[[u' \in Domain(f) \,\&\, u'^\frown(x) \in T] \Rightarrow u'^\frown(x) \in Domain(f)]$$
$$\& (\forall u' \in Domain(f))[f(u') = G(u', \{(x, f(u'^\frown(x))) : u'^\frown(x) \in T\})]$$

and show by bar induction on T that for every u there is exactly one z such that $R(u, z)$, so we can set

$$F(u) = \textit{the unique } z \textit{ such that } R(u, z).$$

This clearly satisfies the given equation. Another simple bar induction shows that no other F' can satisfy the determining equation.

A *rank function* for a tree T on X is any mapping ρ defined on all the finite sequences from X, with ordinal values, such that

$$\textit{if } u^\frown(x) \in T, \textit{ then } \rho(u) > \rho(u^\frown(x)).$$

The next result is trivial but useful enough to deserve billing as a theorem.

2D.1. THEOREM. *A tree T on X is wellfounded if and only if it admits a rank function. Moreover, if $card(X) = \varkappa$ and T is wellfounded, then T admits a rank function ρ such that for every u, $\rho(u) < \varkappa^+$.*

PROOF. If T admits a rank function ρ, then T is obviously wellfounded since any infinite branch

$$f = (x_0, x_1, x_2, \ldots)$$

in T would define the infinite decreasing sequence of ordinals

$$\rho(x_0) > \rho(x_0, x_1) > \rho(x_0, x_1, x_2) > \ldots .$$

Conversely, if T is wellfounded, we can define ρ on T by bar recursion,

$$\rho(u) = 0 \quad \textit{if } u \textit{ is terminal in } T \textit{ or } u \notin T,$$
$$\rho(u) = supremum\{\rho(u^\frown(x)) + 1 : u^\frown(x) \in T\} \quad \textit{if } u \textit{ is not terminal.}$$

Actually the second equation suffices if we adopt the useful convention

$$supremum(\emptyset) = 0.$$

It is immediate that ρ is a rank function on T.

Fix now this canonical ρ associated with a wellfounded tree $T \neq \emptyset$. A trivial induction on ξ shows that

$$\text{if } \rho(u) = \xi, \quad \text{then for every } \zeta < \xi \text{ there is some } v \text{ which extends } u \text{ such that } \rho(v) = \zeta.$$

Thus the range of ρ is an initial segment of ordinals, i.e. ρ is *onto* $\lambda = \rho(\emptyset) + 1$. For each $\xi < \lambda$, choose some $u_\xi \in T$ such that $\rho(u_\xi) = \xi$. Now the map

$$\xi \mapsto u_\xi$$

establishes a one-to-one correspondence of λ with a subset of T, which has cardinality $\varkappa$, so that $\lambda < \varkappa^+$. ⊣

We will sometimes distinguish the rank function ρ associated with a wellfounded tree T in this proof and call it *the rank function of* T,

$$\rho = \rho^T.$$

The *length* of T is defined by

$$|T| = \textit{supremum}\{\rho^T(u) \colon u \in T\}.$$

If $T \neq \emptyset$, then clearly $|T| = \rho^T(\emptyset)$.

The notion of a wellfounded tree gives an alternative way of putting down the characterization of $\varkappa$-Suslin sets of irrationals of 2C.1. If T is a tree on $\omega \times \varkappa$, and $\alpha \in \mathcal{N}$, put

$$T(\alpha) = \{(\xi_0, \ldots, \xi_{n-1}) \colon (\alpha(0), \xi_0, \alpha(1), \xi_1, \ldots, \alpha(n-1), \xi_{n-1}) \in T\}.$$

Evidently $T(\alpha)$ is a tree on $\varkappa$. It is important to notice that with this notation, whether $(\xi_0, \ldots, \xi_{n-1})$ is in $T(\alpha)$ or not depends only on the first n values of α, i.e.

$$\alpha \upharpoonright n = \beta \upharpoonright n \Rightarrow [(\xi_0, \ldots, \xi_{n-1}) \in T(\alpha) \Leftrightarrow (\xi_0, \ldots, \xi_{n-1}) \in T(\beta)].$$

2D.2. THEOREM. *Let $\varkappa$ be an infinite cardinal and $P \subseteq \mathcal{N}$ a set of irrationals. Then P is $\varkappa$-Suslin if and only if there is a tree T on $\omega \times \varkappa$ such that*

$$P(\alpha) \Leftrightarrow T(\alpha) \text{ is not wellfounded.} \quad ⊣$$

We now put these two results to good use.

2D.3. THEOREM (Shoenfield).[5] *Every $\mathbf{\Sigma}^1_2$ pointset is $\aleph_1$-Suslin.*

PROOF. By 2B.2, it is enough to show that every $\mathbf{\Pi}^1_1$ set of irrationals is $\aleph_1$-Suslin.

Assume then that T is a tree on $\omega\times\omega$ and

$$P(\alpha) \Leftrightarrow T(\alpha) \textit{ is wellfounded}$$
$$\Leftrightarrow T(\alpha) \textit{ admits a rank function into } \aleph_1.$$

By 2D.2 and 2D.1 every Π^1_1 set $P \subseteq \mathcal{N}$ can be represented in this way. The idea of the proof is to define a tree S on $\omega\times\aleph_1$ such that every infinite branch of $S(\alpha)$ codes a rank function of $T(\alpha)$.

Let $u_0, u_1, u_2, \ldots$ be an enumeration of all finite sequences from ω such that

$$length(u_n) \leq n.$$

This is easy to arrange. For each n then, $u_n = (s_0, \ldots, s_{k-1})$ with some $k \leq n$.

Call $u = (s_0, \ldots, s_{k-1})$ T-*compatible with* $(t_0, \ldots, t_{m-1})$, if $k \leq m$ and $(t_0, s_0, \ldots, t_{k-1}, s_{k-1}) \in T$.

Put

$$(t_0, \xi_0, \ldots, t_{n-1}, \xi_{n-1}) \in S \Leftrightarrow \text{for every } i, j < n, \text{ if } u_i, u_j \text{ are}$$
$$T\text{-compatible with } (t_0, \ldots, t_{n-1}) \text{ and}$$
$$u_i \text{ is an initial segment of } u_j, \text{ then } \xi_i > \xi_j.$$

Easily S is a tree on $\omega\times\aleph_1$. The claim is that

$$P(\alpha) \Leftrightarrow S(\alpha) \textit{ is not wellfounded.}$$

Notice that for any fixed α and $u = (s_0, \ldots, s_{k-1})$,

$$u \text{ is } T\text{-compatible with } (\alpha(0), \ldots, \alpha(n-1))$$
$$\Leftrightarrow k \leq n \,\&\, (\alpha(0), s_0, \ldots, \alpha(k-1), s_{k-1}) \in T$$
$$\Leftrightarrow k \leq n \,\&\, u \in T(\alpha).$$

Using the condition length $(u_n) \leq n$, we then have

$$(\xi_0, \ldots, \xi_{n-1}) \in S(\alpha) \Leftrightarrow (\alpha(0), \xi_0, \ldots, \alpha(n-1), \xi_{n-1}) \in S$$
$$\Leftrightarrow \text{for every } i, j < n, \text{ if } u_i, u_j \text{ are}$$
$$\text{in } T(\alpha) \text{ and } u_i \text{ is an initial segment}$$
$$\text{of } u_j, \text{ then } \xi_i > \xi_j.$$

This observation implies immediately that if $(\xi_0, \xi_1, \ldots)$ is an infinite branch of $S(\alpha)$, then the mapping

$$u_i \mapsto \xi_i$$

is a rank function on $T(\alpha)$, so that $T(\alpha)$ is wellfounded. Conversely, if $T(\alpha)$ is wellfounded, let ρ be a rank function on $T(\alpha)$, put $\xi_i = \rho(u_i)$ and check immediately that $(\xi_0, \xi_1, \ldots)$ is an infinite branch of $S(\alpha)$, so that $S(\alpha)$ is not wellfounded. ⊣

We will prove later much better representation theorems for $\mathbf{\Pi}^1_1$ and $\mathbf{\Sigma}^1_2$ along these lines. However this result implies already that a $\mathbf{\Sigma}^1_2$ set with more than $\aleph_1$ elements has a perfect subset.

2E. The Suslin theorem

Fix a space $\mathcal{X}$ and an ordinal $\lambda > \omega$. A collection $\mathcal{C}$ of subsets of $\mathcal{X}$ is a *λ-algebra* if $\emptyset \in \mathcal{C}$ and $\mathcal{C}$ is closed under complementation and unions of length less than λ, i.e.

$$\xi < \lambda \text{ and for all } \eta < \xi,\ A_\eta \in \mathcal{C}$$
$$\Longrightarrow \bigcup\nolimits_{\eta<\xi} A_\eta \in \mathcal{C}.$$

The collection $\mathbf{B}_\lambda \restriction \mathcal{X}$ of *λ-Borel* subsets of $\mathcal{X}$ is the least λ-algebra on $\mathcal{X}$ which contains all open sets and $\mathbf{B}_\lambda$ is the pointclass of λ-Borel pointsets in all product spaces.

Clearly, $\mathbf{B}_{\omega+1}$ is the pointclass of Borel measurable sets as we defined them in 1F. Also, $\mathbf{B}_{\omega+1} = \mathbf{B}_{\aleph_1}$, and in general, if λ is not a cardinal, then $\mathbf{B}_\lambda = \mathbf{B}_{\lambda^+}$. It will be convenient to have $\mathbf{B}_\lambda$ defined for all $\lambda > \omega$.

Let $\mathbf{B}'_\lambda$ be the collection of all λ-Borel pointsets $P \subseteq \mathcal{Y}$, such that for every Borel function $f : \mathcal{X} \to \mathcal{Y}$, $f^{-1}[P]$ is λ-Borel. Clearly $\mathbf{B}'_\lambda$ contains all open sets and is closed under $\neg$ and unions of length less than λ, hence $\mathbf{B}'_\lambda = \mathbf{B}_\lambda$ and $\mathbf{B}_\lambda$ is closed under Borel substitution. We will leave for the exercises the remaining easy closure properties of $\mathbf{B}_\lambda$. Here we want to concentrate on the Strong Separation Theorem and its corollary, the Suslin Theorem which is the chief construction principle of classical descriptive set theory.

Recall from 1C that a set C *separates* A from B if $A \subseteq C$, $B \cap C = \emptyset$ (see Figure 2E.1).

2E.1. THE STRONG SEPARATION THEOREM. *Let ϰ be an infinite cardinal, assume that A and B are disjoint ϰ-Suslin subsets of some perfect product space 𝒳. There exists a (ϰ+1)-Borel set C which separates A from B (Lusin).*[(8)]

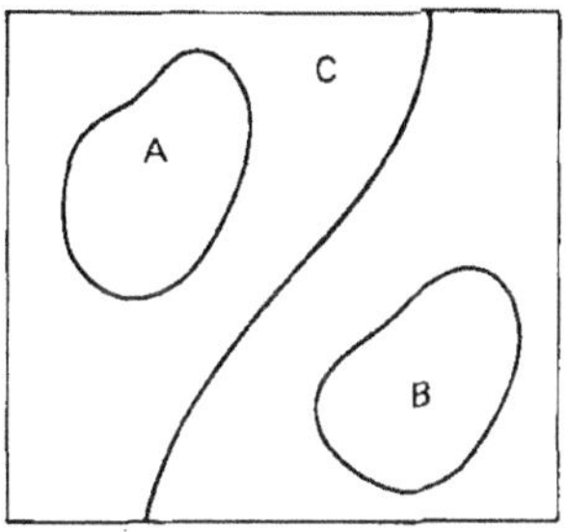

Figure 2E.1. Separation.

PROOF. We may assume that A, B are subsets of $\mathcal{N}$, since $\mathfrak{X}$ is Borel isomorphic with $\mathcal{N}$ and both $\mathbf{S}_\varkappa$ and $\mathbf{B}_{\varkappa+1}$ are closed under Borel substitution.

The key to the proof is the following simple combinatorial fact about separating sets. Suppose

$$A = \bigcup_{i \in I} A_i, \qquad B = \bigcup_{j \in J} B_j$$

are unions of sets, where the index sets I, J are quite arbitrary, suppose that for each $i \in I$, $j \in J$ there is a set $C_{i,j}$ which separates A_i from B_j. Then the set

$$C = \bigcup_{i \in I} \bigcap_{j \in J} C_{i,j}$$

separates A from B. To prove this, notice that for each i, j, $A_i \subseteq C_{i,j}$, hence $A_i \subseteq \bigcap_{j \in J} C_{i,j}$, hence $A = \bigcup_{i \in I} A_i \subseteq \bigcup_{i \in I} \bigcap_{j \in J} C_{i,j} = C$. On the other hand, for each i, j, $B_j \subseteq \mathcal{N} - C_{i,j}$, hence $B = \bigcup_{j \in J} B_j \subseteq \bigcup_{j \in J} (\mathcal{N} - C_{i,j})$ and since this holds for arbitrary i,

$$\begin{aligned} B \subseteq \bigcap_{i \in I} \bigcup_{j \in J} (\mathcal{N} - C_{i,j}) &= \bigcap_{i \in I} (\mathcal{N} - \bigcap_{j \in J} C_{i,j}) \\ &= \mathcal{N} - \bigcup_{i \in I}, \bigcap_{j \in J} C_{i,j} = \mathcal{N} - C. \end{aligned}$$

Suppose now that A and B are disjoint $\varkappa$-Suslin sets of irrationals, so there are trees T and S on $\omega \times \varkappa$ and

$$A = \mathfrak{p}[T], \qquad B = \mathfrak{p}[S].$$

We give two proofs of the result—first a simple argument by contradiction and then a constructive proof which actually exhibits a $(\varkappa + 1)$-Borel set C that separates A from B.

Proof by contradiction. Assume that A cannot be separated from B by a $(\varkappa+1)$-Borel set. Since

$$A = \mathfrak{p}[T] = \bigcup_{t\in\omega,\xi<\varkappa} \mathfrak{p}[T_{(t,\xi)}],$$
$$B = \mathfrak{p}[S] = \bigcup_{s\in\omega,\eta<\varkappa} \mathfrak{p}[S_{(s,\eta)}],$$

by the remarks above there must be some t_0, ξ_0, s_0, η_0 such that $\mathfrak{p}[T_{(t_0,\xi_0)}]$ and $\mathfrak{p}[S_{(s_0,\eta_0)}]$ cannot be separated. This implies that $t_0 = s_0$, or else we can take

$$C = \{\alpha : \alpha(0) = t_0\},$$

which surely separates these two sets. Hence $(t_0, \xi_0) \in T$, $(t_0, \eta_0) \in S$ and $\mathfrak{p}[T_{(t_0,\xi_0)}]$, $\mathfrak{p}[S_{(t_0,\eta_0)}]$ cannot be separated by a $(\varkappa+1)$-Borel set.

Proceeding inductively, we find $t_0, t_1, t_2, \ldots, \xi_0, \xi_1, \xi_2, \ldots, \eta_0, \eta_1, \eta_2, \ldots$ such that for each n, $u = (t_0, \xi_0, \ldots, t_{n-1}, \xi_{n-1}) \in T$, $v = (t_0, \eta_0, \ldots, t_{n-1}, \eta_{n-1}) \in S$ and $\mathfrak{p}[T_u]$, $\mathfrak{p}[S_v]$ cannot be separated by a $(\varkappa+1)$-Borel set. However this is absurd, since then $\alpha = (t_0, t_1, \ldots)$ is in both A and B and these sets were assumed disjoint.

Constructive proof. Define the tree J on $\omega \times \varkappa \times \varkappa$ by

$$((t_0, \xi_0, \eta_0), \ldots, (t_{n-1}, \xi_{n-1}, \eta_{n-1})) \in J$$
$$\Leftrightarrow (t_0, \xi_0, \ldots, t_{n-1}, \xi_{n-1}) \in T \,\&\, (t_0, \eta_0, \ldots, t_{n-1}, \eta_{n-1}) \in S.$$

We will omit the parentheses in writing the nodes of J as we have been doing for trees of pairs,

$$(t_0, \xi_0, \eta_0, t_1, \xi_1, \eta_1, \ldots, t_{n-1}, \xi_{n-1}, \eta_{n-1}) = ((t_0, \xi_0, \eta_0), \ldots, (t_{n-1}, \xi_{n-1}, \eta_{n-1})).$$

Any infinite branch $(t_0, \xi_0, \eta_0, t_1, \xi_1, \eta_1, \ldots)$ in J would determine infinite branches $(t_0, \xi_0, t_1, \xi_1, \ldots)$ in T and $(t_0, \eta_0, t_1, \eta_1, \ldots)$ in S with the same irrational part $\alpha = (t_0, t_1, \ldots)$, so that $\alpha \in A \cap B$ contrary to hypothesis. Hence J is a wellfounded tree.

To simplify notation, assign to each sequence

$$u = (t_0, \xi_0, \eta_0, \ldots, t_{n-1}, \xi_{n-1}, \eta_{n-1})$$

from $\omega \times \varkappa \times \varkappa$ the two sequences that it determines in $\omega \times \varkappa$

$$\tau(u) = (t_0, \xi_0, \ldots, t_{n-1}, \eta_{n-1}),$$
$$\sigma(u) = (t_0, \eta_0, \ldots, t_{n-1}, \eta_{n-1}).$$

By the usual convention,

$$\tau(\emptyset) = \sigma(\emptyset) = \emptyset.$$

Now

$$J = \{u : \tau(u) \in T \quad \text{and} \quad \sigma(u) \in S\}.$$

If v is a sequence from $\omega \times \varkappa$ put

$$A_v = \mathrm{p}[T_v], \qquad B_v = \mathrm{p}[S_v].$$

We will define by bar recursion on the wellfounded tree J a function

$$u \mapsto C_u$$

such that for each sequence u in $\omega \times \varkappa \times \varkappa$,

(a) C_u is $\varkappa + 1$-Borel,

(b) C_u separates $A_{\tau(u)}$ from $B_{\sigma(u)}$.

This will complete the proof since $A_{\tau(\emptyset)} = A_\emptyset = A$ and $B_{\sigma(\emptyset)} = B_\emptyset = B$, so $C = C_\emptyset$ will be the required set.

We have for each u

$$A_{\tau(u)} = [T_{\tau(u)}] = \bigcup_{t<\omega, \xi<\varkappa} A_{\tau(u)^\frown(t,\xi)},$$

$$B_{\sigma(u)} = [S_{\sigma(u)}] = \bigcup_{s<\omega, \eta<\varkappa} B_{\sigma(u)^\frown(s,\eta)};$$

hence by the remarks at the beginning of this proof, it is enough to define sets $D_{t,\xi,s,\eta}$ such that

(c) $D_{t,\xi,s,\eta}$ is $(\varkappa+1)$-Borel,

(d) $D_{t,\xi,s,\eta}$ separates $A_{\tau(u)^\frown(t,\xi)}$ from $B_{\sigma(u)^\frown(s,\eta)}$, since then the set

$$C_u = \bigcup_{t<\omega,\xi<\varkappa} \bigcap_{s<\omega,\eta<\varkappa} D_{t,\xi,s,\eta}$$

will surely be $(\varkappa+1)$-Borel and separate $A_{\tau(u)}$ from $B_{\sigma(u)}$.

If $t = s$ and $u^\frown(t, \xi, \eta) \in J$, we can take

$$D_{t,\xi,s,\eta} = C_{u^\frown(t,\xi,\eta)},$$

since by the induction hypothesis of the bar recursion we can assume that $C_{u^\frown(t,\xi,\eta)}$ has been defined, it is $(\varkappa+1)$-Borel and it separates $A_{\tau(u)^\frown(t,\xi)}$ from $B_{\sigma(u)^\frown(s,\eta)}$. Hence it is enough to define $D_{t,\xi,s,\eta}$ when $t \neq s$ or $t = s$ but $u^\frown(t, \xi, \eta) \notin J$.

If $t \neq s$, take

$$D_{t,\xi,s,\eta} = \{\alpha : \alpha(n) = t\},$$

where n is the length of the sequence u, so that

$$\alpha \in A_{\tau(u)^\frown(t,\xi)} \Longrightarrow \alpha(n) = t;$$

clearly $A_{\tau(u)^\frown(t,\xi)} \subseteq \{\alpha : \alpha(n) = t\}$, while

$$B_{\sigma(u)^\frown(s,\eta)} \cap \{\alpha : \alpha(n) = t\} = \emptyset.$$

If $t = s$ but $u^\frown(t, \xi, \eta) \notin J$, there are two cases.

Case 1. $\tau(u)^\frown(t, \xi) \notin T$. In this case $A_{\tau(u)^\frown(t,\xi)} = \emptyset$ and we can take $D_{t,\xi,s,\eta} = \emptyset$.

Case 2. $\sigma(u)^\frown(s, \eta) \notin S$. In this case, $B_{\sigma(u)^\frown(s,\eta)} = \emptyset$ and we can take $D_{t,\xi,s,\eta} = \mathcal{N}$. ⊣

The constructive argument in this proof is somehow more satisfying, since it actually shows us how to build a separating set C from the trees T and S that determine the given sets. More than that, there is additional information about C that is implicit in the proof and which we will extract and utilize later on.

2E.2. THE SUSLIN THEOREM. *Let $\varkappa$ be an infinite cardinal. If a pointset $A \subseteq \mathfrak{X}$ and its complement $\mathfrak{X} - A$ are both $\varkappa$-Suslin, then A is $(\varkappa + 1)$-Borel.*

In particular, a set is Borel if and only if it is $\mathbf{\Delta}^1_1$.(8)

PROOF is immediate from 2E.1, taking $B = \mathfrak{X} - A$. ⊣

The theorem of Suslin is the standard *construction principle*, the result we always try to imitate or extend to more general situations. It reduces the fairly complex operation of projection along $\mathcal{N}$ to an iteration of complementation and countable union; this, of course, only in the special circumstance when we know that both the given set A and its complement can be defined by projecting closed sets. It will become clear as we go on that projection along $\mathcal{N}$ applied to more complicated sets is a very complex operation. In the general situation, it produces sets much more difficult to understand than those we apply it to.

EXERCISES

2E.3. Prove that for each ordinal $\lambda > \omega$, the pointclass $\mathbf{B}_\lambda$ is closed under continuous substitution, $\neg$, &, $\vee$, $\exists^\omega$, $\forall^\omega$ and $\bigvee^\xi, \bigwedge^\xi$ for every $\xi < \lambda$. ⊣

2E.4. Prove that if $f : \mathfrak{X} \to \mathfrak{Y}$ and $Graph(f) = \{(x, y) : f(x) = y\}$ is $\mathbf{\Sigma}^1_1$, then f is Borel.

Hint. $$f(x) \in N_s \Leftrightarrow (\exists y)[f(x) = y \;\&\; y \in N_s] \Leftrightarrow (\forall y)[f(x) = y \Rightarrow y \in N_s].$$ ⊣

2E.5. Suppose $f:\mathfrak{X}\twoheadrightarrow \mathcal{R}$ is a Borel function and for each $x\in\mathfrak{X}$ there is exactly one solution y of the equation

$$f(x, y)=0,$$

so that this equation determines y as a function of x,

$$y=g(x).$$

Prove that g is a Borel function.

In particular, if $f:\mathfrak{X}\twoheadrightarrow\mathcal{Y}$ is a Borel bijection of $\mathfrak{X}$ with $\mathcal{Y}$, then f^{-1} is also Borel, so f is a Borel isomorphism.[9]

Hint. $g(x)=y \Leftrightarrow f(x, y)=0.$ ⊣

In general, it is not true that every $(\varkappa+1)$-Borel set is $\varkappa$-Suslin. This extra fact allows much stronger results to be proved in the case $\varkappa=\aleph_0$.

2E.6. Let $A_0, A_1,\ldots$ be a sequence of pairwise disjoint $\mathbf{\Sigma}^1_1$ subsets of some perfect product space $\mathfrak{X}$. Prove that there exists a sequence $C_0, C_1,\ldots$ of pairwise disjoint Borel subsets of $\mathfrak{X}$ such that $A_0\subseteq C_0$, $A_1\subseteq C_1,\ldots$.

Hint. Choose C_0 to separate A_0 from $\bigcup_{n\geq 1} A_n$, then choose C_1 to separate A_1 from $C_0\cup\bigcup_{n\geq 2} A_n$, then choose C_2 to separate A_2 from $C_0\cup C_1\cup\bigcup_{n\geq 3} A_n$, etc. ⊣

We can use this extension of 2E.1 to establish a very important theorem apparently due to Lusin and Suslin.

2E.7. Suppose $f:\mathfrak{X}\to\mathcal{Y}$ is continuous, $A\subseteq\mathfrak{X}$ is Borel and f is one-to-one on A; then the image $f[A]$ is Borel.[9]

Hint. By 1G.5, it is enough to consider the case $f:\mathcal{N}\to\mathcal{Y}$, with A a *closed* subset of $\mathcal{N}$. As before, let

$$N(k_0,\ldots, k_{n-1})=\{\alpha: \alpha(0)=k_0,\ldots, \alpha(n-1)=k_{n-1}\}$$

and put

$$A(k_0,\ldots, k_{n-1})=f[A\cap N(k_0,\ldots, k_{n-1})].$$

Each $A(k_0,\ldots, k_{n-1})$ is $\mathbf{\Sigma}^1_1$ and these sets are pairwise disjoint for fixed n since f is injective on A; hence by 2E.6 there exist Borel sets $B(k_0,\ldots, k_{n-1})$, pairwise disjoint for each fixed n, so that

$$A(k_0,\ldots, k_{n-1})\subseteq B(k_0,\ldots, k_{n-1}).$$

Let us get a better separating sequence by putting

$$B^*(k) = B(k) \cap \overline{A(k)},$$

$$B^*(k_0, k_1) = B(k_0, k_1) \cap \overline{A(k_0, k_1)} \cap B^*(k_0),$$

and in general

$$B^*(k_0,\dots, k_n) = B(k_0,\dots, k_n) \cap \overline{A(k_0,\dots, k_n)} \cap B^*(k_0,\dots, k_{n-1}).$$

Easily,

$$A(k_0,\dots, k_{n-1}) \subseteq B^*(k_0,\dots, k_{n-1}) \subseteq \overline{A(k_0,\dots, k_{n-1})}$$

and it is not hard to check that

$$f[A] = \mathcal{A}_u B^*(u);$$

this is because if $x \in \bigcap_n B^*(\alpha(0),\dots, \alpha(n-1))$, then $\alpha \in A$ and $x = f(\alpha)$. Moreover, the system $u \mapsto B^*(u)$ satisfies the conditions of 2B.3, hence $\mathcal{A}_u B^*(u)$ is Borel. ⊣

Together with 1G.5, we now have Lusin's favorite characterization of Borel sets:

2E.8. Prove that a set $P \subseteq \mathfrak{X}$ is Borel if and only if P is the continuous, injective image of a closed set in $\mathcal{N}$.(9) ⊣

The result in 2E.7 extends easily to Borel functions.

2E.9. Prove that if $f: \mathfrak{X} \to \mathfrak{Y}$ is a Borel function, $A \subseteq \mathfrak{X}$ is a Borel set and f is injective on A, then $f[A]$ is Borel.(9)

Hint. The set $B = \{(x, y): x \in A \;\&\; y = f(x)\}$ is Borel and $f[A]$ is a continuous, injective image of B, via the projection $(x, y) \mapsto x$. ⊣

In Chapter 4 we will prove by an entirely different method some important generalizations of 2E.7–2E.9.

2E.10. Prove that every Borel injection is a good Borel injection (in the sense of 1G).(9) ⊣

2E.11. Suppose $f: \mathfrak{X} \times \mathfrak{Y} \to \mathcal{R}$ is a Borel function such that for each x, there is at most one y such that $f(x, y)=0$. Prove that the set

$$D=\{x: (\exists y)[f(x, y)=0]\}$$

is a Borel set and there exists a Borel function $g: \mathfrak{X} \to \mathcal{Y}$ such that for $x \in D$,

$$f(x, g(x))=0.^{(9)}$$

Hint. The set $A=\{(x, y): f(x)=y\}$ is Borel and the map $(x, y) \mapsto x$ is an injection of A onto D, so D is Borel by 2E.7. Now define g so it has Borel graph and use 2E.4. ⊣

2F. Inductive analysis of projections of trees

The chief result of this section is that $\boldsymbol{\Sigma}^1_1$ sets can be expressed as both the union and the intersection of $\aleph_1$ Borel sets. This will be an easy corollary of a general structure result about projections of trees.

In 2D we associated a canonical rank function $\rho=\rho^T$ with every wellfounded tree T on some X. It is convenient to have a rank function for T even when T is not wellfounded—we will simply put $\rho(u)=\infty$ if u is not in the wellfounded part of T.

To be precise, if T is a tree on X, the *wellfounded part* of T is defined by

$$WF(T)=\{u: u\notin T \text{ or } u\in T \text{ but there is no infinite sequence } x_0, x_1,\ldots \text{ such that for every } n,\ u^\frown(x_0,\ldots, x_{n-1})\in T\}.$$

Putting into $WF(T)$ the sequences outside T is of course only a matter of convenience. Now $WF(T)$ is not a tree, but it is clear that we can define functions by bar recursion on $WF(T)$ exactly as we do on all the sequences from X when T is wellfounded. In that case, of course, $WF(T)=\{u: u \textit{ is a sequence from } X\}$. Put then

$$\rho(u)=\textit{supremum}\{\rho(u^\frown(x)+1): u^\frown(x)\in T\} \quad \text{if } u\in WF(T),$$
$$\rho(u)=\infty \quad \text{if } u\notin WF(T),$$

where ∞ is assumed greater than all ordinals in the equations below. If there is need to identify the tree with which we are working we write

$$\rho(u)=\rho^T(u)=\rho(T, u).$$

It follows exactly as in 2D.1 that if $card(X) = \varkappa$ and T is a tree on X, then

$$u \in WF(T) \Rightarrow \rho(T, u) < \varkappa^+.$$

2F.1. THEOREM (Sierpinski).[10] *Let $\varkappa$ be an infinite cardinal, let T be a tree on $\omega \times \varkappa$ and put*

$$A = \mathfrak{p}[T], \qquad B = \mathcal{N} - A.$$

For each sequence $u = (\xi_0, \ldots, \xi_{n-1})$ from $\varkappa$ and each $\lambda < \varkappa^+$ put

$$B_u^\lambda = \{\alpha : \rho(T(\alpha), u) \leq \lambda\}.$$

Then

$$B_u^0 = \bigcap_{\xi<\varkappa} \{\alpha : (\alpha(0), \xi_0, \ldots, \alpha(n-1), \xi_{n-1}, \alpha(n), \xi) \notin T\},$$
$$B_u^\lambda = \bigcap_{\xi<\varkappa} \bigcup_{\zeta<\lambda} B_{u^\frown(\xi)}^\zeta \quad \text{if} \quad \lambda > 0,$$

and

$$B = \bigcup_{\lambda<\varkappa^+} B_\emptyset^\lambda.$$

PROOF. We compute:

$$\begin{aligned}\alpha \in B_u^0 &\Leftrightarrow \rho(T(\alpha), u) = 0\\ &\Leftrightarrow (\forall \xi < \varkappa)[u^\frown(\xi) \notin T(\alpha)]\\ &\Leftrightarrow (\forall \xi < \varkappa)[(\alpha(0), \xi_0, \ldots, \alpha(n-1), \xi_{n-1}, \alpha(n), \xi) \notin T].\end{aligned}$$

For $\lambda > 0$,

$$\begin{aligned}\alpha \in B_u^\lambda &\Leftrightarrow \mathit{supremum}\{\rho(T(\alpha), u^\frown(\xi)) + 1 : u^\frown(\xi) \in T(\alpha)\} \leq \lambda.\\ &\Leftrightarrow (\forall \xi < \varkappa)(\exists \zeta < \lambda)\{u^\frown(\xi) \in T(\alpha) \Rightarrow \rho(T(\alpha), u^\frown(\xi)) \leq \zeta\}\\ &\Leftrightarrow (\forall \xi < \varkappa)(\exists \zeta < \lambda)[\rho(T(\alpha), u^\frown(\xi)) \leq \zeta]\\ &\Leftrightarrow (\forall \xi < \varkappa)(\exists \zeta < \lambda)[\alpha \in B_{u^\frown(\xi)}^\zeta].\end{aligned}$$

The last assertion follows from 2D.2 since

$$\begin{aligned}\alpha \in B &\Leftrightarrow T(\alpha) \quad \text{is wellfounded}\\ &\Leftrightarrow \rho(T(\alpha), \emptyset) \quad \text{is defined}\\ &\Leftrightarrow \rho(T(\alpha), \emptyset) < \varkappa^+.\end{aligned}$$

⊣

2F.2. THEOREM. *If $\varkappa$ is an infinite cardinal and A is a $\varkappa$-Suslin pointset,* *then*

$$A = \bigcup_{\lambda<\kappa^+} C_\lambda,$$
$$A = \bigcap_{\lambda<\kappa^+} D_\lambda,$$

where the sets C_λ, D_λ are $(\kappa+1)$-Borel. In particular, every $\boldsymbol{\Sigma}^1_1$ set is both a union and an intersection of $\aleph_1$ Borel sets.[(10)]

PROOF. It is enough to prove the result for $A \subseteq \mathcal{N}$. If $A = \mathrm{p}[T]$ with T a tree on $\omega \times \kappa$ and $B = \mathcal{N} - A$, then by 2F.1

$$B = \bigcup_{\lambda<\kappa^+} B^\lambda_\emptyset,$$

hence

$$A = \bigcap_{\lambda<\kappa^+} (\mathcal{N} - B^\lambda_\emptyset).$$

This shows that A is the intersection of κ^+ sets which are $(\kappa+1)$-Borel, since it is evident from 2F.1 that every B^λ_u is $(\kappa+1)$-Borel.

With the same notation and for $\lambda < \kappa^+$, put

$$E_\lambda = \{\alpha : \rho(T(\alpha), \emptyset) \le \lambda\} \cup \{\alpha : (\exists u)[\rho(T(\alpha), u) = \lambda]\},$$

where u varies over all sequences from κ. Again by 2F.1, each E_λ is $(\kappa+1)$-Borel, since

$$E_\lambda = B^\lambda_\emptyset \cup \bigcup_u [B^\lambda_u - \bigcup_{\xi<\lambda} B^\xi_u].$$

We claim that

$$B = \bigcap_{\lambda<\kappa^+} E_\lambda;$$

proof of this claim will be sufficient, since then

$$A = \bigcup_{\lambda<\kappa^+} (\mathcal{N} - E_\lambda).$$

Assume first that $\alpha \in B$ so that $T(\alpha)$ is wellfounded. The mapping

$$u \mapsto \rho(T(\alpha), u)$$

take the sequences from κ onto the initial segment of κ^+ bounded by $\rho(T(\alpha), \emptyset)$. Thus for each $\lambda < \kappa^+$, either $\rho(T(\alpha), \emptyset) \le \lambda$ or $\lambda < \rho(T(\alpha), \emptyset)$, in which case $\lambda = \rho(T(\alpha), u)$ for some u. In either case, $\alpha \in E_\lambda$.

In the other direction, assume towards a contradiction that $\alpha \in \bigcap_{\lambda<\kappa^+} E_\lambda$ but $\alpha \notin B$. Now $T(\alpha)$ is not wellfounded so $\rho(T(\alpha), \emptyset) \le \lambda$ is false for every λ, hence for every λ there must be some u_λ with $\rho(T(\alpha), u_\lambda) = \lambda$. This establishes a mapping $u_\lambda \mapsto \lambda$ from the sequences of κ onto κ^+ which is absurd. ⊣

The result implies that Σ^1_2 sets can also be written as unions of $\aleph_1$ Borel sets.

2F.3. THEOREM (Sierpinski).[10] *For each Σ^1_2 pointset P there are Borel sets B_ξ, $\xi<\aleph_1$, such that*

$$P=\bigcup_{\xi<\aleph_1} B_\xi.$$

PROOF. There is a Π^1_1 set Q such that

$$P(x)\Leftrightarrow(\exists\alpha)Q(x,\alpha),$$

and by 2F.2 there are Borel sets C_ξ such that

$$Q(x,\alpha)\Leftrightarrow(\exists\xi<\aleph_1)C_\xi(x,\alpha).$$

Thus

$$\begin{aligned}P(x)&\Leftrightarrow(\exists\alpha)(\exists\xi<\aleph_1)C_\xi(x,\alpha)\\&\Leftrightarrow(\exists\xi<\aleph_1)(\exists\alpha)C_\xi(x,\alpha).\end{aligned}$$

Now put

$$D_\xi=\{x:(\exists\alpha)C_\xi(x,\alpha)\}$$

and notice that for each ξ, D_ξ is Σ^1_1. Hence by 2F.2 again

$$D_\xi=\bigcup_\eta E_{\xi,\eta}$$

with each $E_{\xi,\eta}$ Borel and

$$P=\bigcup_{\xi<\aleph_1}\bigcup_{\eta<\aleph_1}E_{\xi,\eta}.$$

This expresses P as a union of $\aleph_1$ Borel sets. (Notice the use of the axiom of choice in this argument.) ⊣

This is as far as results of this type can go, even if we go to set theories stronger than Zermelo–Fraenkel. One of the exciting modern results is the theorem of Martin that with strong hypotheses, Π^1_2 (and hence Σ^1_3) sets are the unions of $\aleph_2$ Borel sets!

Theorems 2F.2 and 2F.3 are trivial if one assumes the continuum hypothesis, that $2^{\aleph_0}=\aleph_1$, since then every subset of $\mathcal{N}$ is the union of $\aleph_1$ singletons (which are Borel sets) and the intersection of $\aleph_1$ complements of singletons (which are Borel sets). If, on the other hand we think of $2^{\aleph_0}$ as very large compared to $\aleph_1$, then 2F.3 can be considered a construction principle for Σ^1_2. Because surely Borel sets are very simple compared to Σ^1_2 sets and we need just a few ($\aleph_1$) of them to build up any given Σ^1_2 set.

EXERCISES

Let us first use Theorem 2F.2 to get a simple characterization of $\aleph_n$-Suslin sets for $n = 1, 2, \ldots$.

2F.4. Prove that a pointset $P \subseteq \mathfrak{X}$ is $\aleph_n$-Suslin ($n \geq 1$) if and only if

$$P = \bigcup_{\xi < \aleph_n} P_\xi$$

where each P_ξ is Borel (Martin [19?a]).

Hint. One way comes directly from the closure properties of $\varkappa$-Suslin sets. For the converse suppose first P is $\aleph_1$-Suslin; then by 2B.4,

$$P = \bigcup_{\xi < \aleph_1} P_\xi,$$

where each P_ξ is $\aleph_0$-Suslin, i.e. Σ^1_1; each P_ξ in turn is the union of $\aleph_1$ Borel sets by 2F.2, so P is the union of $\aleph_1 \cdot \aleph_1 = \aleph_1$ Borel sets. The result follows by induction on n.

(Notice the use of the axiom of choice in this proof.) ⊣

The next two exercises outline a new and interesting proof of the Strong Separation Theorem.[11]

Suppose $B \subseteq \mathcal{N}$ is the complement of some $\varkappa$-Suslin set $A = \mathrm{p}[T]$, where T is a tree on $\omega \times \varkappa$. For each set $C \subseteq B$ and each $u = (k_0, \xi_0, \ldots, k_{n-1}, \xi_{n-1})$, put

$$Index(C, T, u) = supremum\{\rho(T(\alpha), (\xi_0, \ldots, \xi_{n-1})) : \alpha \in C \;\&\; \alpha(0) = k_0, \ldots, \alpha(n-1) = k_{n-1}\},$$

and let

$$Index(C, T) = Index(C, T, \emptyset).$$

2F.5. In the notation just introduced, prove that if C is a $\varkappa$-Suslin subset of B, then

$$Index(C, T) < \varkappa^+.$$

Hint. Assume towards a contradiction that

$$Index(C, T, \emptyset) = \varkappa^+$$

and

$$C = \mathrm{p}[S],$$

where S is a tree on $\omega \times \varkappa$. Now

$$Index(C, T, \emptyset) = supremum_\alpha\{\rho(T(\alpha), \emptyset): \alpha \in C\}$$

and for each $\alpha \in C$,

$$\begin{aligned}\rho(T(\alpha), \emptyset) &= supremum_\xi\{\rho(T(\alpha), (\xi)) + 1: (\xi) \in T(\alpha)\}\\ &= supremum_\xi\{\rho(T(\alpha), (\xi)) + 1: (\alpha(0), \xi) \in T\},\end{aligned}$$

from which we easily get

$$Index(C, T, \emptyset) = supremum_{\xi,n}\{supremum_\alpha\{\rho(T(\alpha), (\xi)) + 1: \alpha \in C \;\&\; \alpha(0) = n\}\}.$$

Now using the fact that $\varkappa^+$ is regular, we infer that for some n_0, ξ_0

$$supremum\{\rho(T(\alpha), (\xi_0)) + 1: \alpha \in C \;\&\; \alpha(0) = n_0\} = \varkappa^+$$

which implies

$$Index(C, T, (n_0, \xi_0)) = \varkappa^+.$$

Also

$$C = \bigcup_{m,\eta} \mathfrak{p}[S_{(m,\eta)}],$$

so for some fixed (m_0, η_0),

$$Index(\mathfrak{p}[S_{(m_0,\eta_0)}], T, (n_0, \xi_0)) = \varkappa^+.$$

Argue that we must have $m_0 = n_0$, so that

$$Index(\mathfrak{p}[S_{(n_0,\eta_0)}], T, (n_0, \xi_0)) = \kappa^+,$$

and then repeat the construction to obtain $n_0, n_1, \ldots, \xi_0, \xi_1, \ldots, \eta_0, \eta_1, \ldots,$ so that for each k,

$$Index(\mathfrak{p}[S_{(n_0,\, \eta_0, \ldots,\, n_{k-1},\, \eta_{k-1})}], T, (n_0, \xi_0, \ldots, n_{k-1}, \xi_{k-1})) = \varkappa^+.$$

Now let

$$\alpha = (n_0, n_1, \ldots)$$

and notice that $\alpha \in \mathfrak{p}[S] \cap \mathfrak{p}[T] = A \cap C$, contrary to the hypothesis $C \subseteq B$. ⊣

2F.6. Prove that if $B \subseteq \mathfrak{X}$ is the complement of a $\varkappa$-Suslin set A, then

$$B = \bigcup_{\lambda < \varkappa^+} B^\lambda$$

where each B^λ is $(\varkappa+1)$-Borel and if $C\subseteq B$ in $\varkappa$-Suslin, then $C\subseteq B^\lambda$ for some λ. Use this to get a new proof of the Strong Separation Theorem.[11] ⊣

2G. The Kunen–Martin theorem

One can easily prove by classical methods that every $\mathbf{\Sigma}_1^1$ wellfounded relation has countable length; in particular, there cannot be a $\mathbf{\Sigma}_1^1$ wellordering of the continuum. We will prove here a much more general recent result due independently to Kunen and Martin.

One of the consequences of the Kunen–Martin Theorem is that $\mathbf{\Sigma}_2^1$ wellfounded relations have length less than $\aleph_2$; in particular, if there is a $\mathbf{\Sigma}_2^1$ wellordering of $\mathcal{R}$, then the continuum hypothesis holds. This was proved by Martin (before the general result) in one of the first spectacular demonstrations of what modern set theoretic techniques can do for the classical theory.

We will give Kunen's proof of the Kunen–Martin Theorem since it is very simple and in the spirit of the methods we have been using in this chapter.

With each binary relation $R(x, y)$ on a set S we associate its *strict part* $<_R$,

$$x <_R y \Leftrightarrow R(x, y)\ \&\ \neg R(y, x).$$

Of course it may be that $<_R = R$ if R is already *strict*, i.e. if

$$R(x, y) \Rightarrow \neg R(y, x).$$

We call R *wellfounded* if every nonempty subset of S has a $<_R$-minimal element, i.e.

$$\emptyset \subsetneq A \subseteq S \Rightarrow \text{for some } x\in A \text{ and all } y\in A,\ \neg y <_R x.$$

It is easy to verify that this is equivalent to the condition that there are no $<_R$-*infinite descending chains*, i.e. there is no sequence

$$x_0 >_R x_1 >_R x_1 >_R \cdots.$$

It is common to study wellfounded relations with various additional properties, like *transitivity* or *reflexiveness*—see the exercises for a statement of these conditions. Many results, however, go through without such restrictions and it is convenient to prove them in this generality. Since only the strict part of a relation comes into the definition of wellfoundedness, we often restrict attention to *strict, wellfounded* relations which we denote by symbole like $<$, $\prec$, etc.

We may justify *proof by induction* and *definition by recursion* on a wellfounded relation exactly as we did for wellfounded trees in 2D. In particular, each wellfounded relation R on S admits a *rank function*

$$\rho : S \twoheadrightarrow \lambda = |R|,$$

where ρ is determined by the recursion

$$\rho(x) = \mathit{supremum}\{\rho(y)+1 : y <_R x\}$$

and $|R| = |<_R|$ is the *length* of R,

$$|R| = \mathit{supremum}\{\rho(x)+1 : x \in S\}.$$

Notice that

$$\rho(x) = 0 \Leftrightarrow \text{for every } y, \ \neg y <_R x.$$

It will be convenient for the proof of the Kunen–Martin Theorem to introduce the notion of a *good semiscale.*

A sequence $\bar{\varphi} = \{\varphi_n\}_{n\in\omega}$ of $\varkappa$-norms on a pointset P is a *good* $\varkappa$-*semiscale* if whenever $x_0, x_1, \ldots$ are in P and for each fixed n the sequence of ordinals

$$\varphi_n(x_0), \varphi_n(x_1), \varphi_n(x_2), \ldots$$

is ultimately constant, then there is some $x \in P$ such that $\mathit{limit}_{i\to\infty} x_i = x$.

2G.1. LEMMA. *If $\varkappa$ is an infinite cardinal and a pointset P admits a $\varkappa$-semiscale, then P admits a good $\varkappa$-semiscale.*

PROOF. Let

$$\pi : \varkappa \times \omega \rightarrowtail\!\!\twoheadrightarrow \varkappa$$

be a bijection of $\varkappa \times \omega$ with $\varkappa$ and for each x, choose $q(x, i)$ such that

$$x \in N_{q(x,i)}, \qquad \mathit{radius}(N_{q(x,i)}) \leq 2^{-i}.$$

Here of course

$$N_0, N_1, \ldots$$

is a basis for the open sets in the product space $\mathfrak{X}$ which contains P.

Given a $\varkappa$-semiscale $\bar{\varphi} = \{\varphi_n\}_{n\in\omega}$ on P, put

$$\psi_n(x) = \pi(\varphi_n(x), q(x, n)).$$

If $x_0, x_1, \ldots$ are in P and for each n, $\psi_n(x_i)$ is ultimately constant, then clearly for each n the sequences

$$q(x_0, n), q(x_1, n), q(x_2, n), \ldots$$
$$\varphi_n(x_0), \varphi_n(x_1), \varphi_n(x_2), \ldots$$

are ultimately constant. In particular, for each n the sequence $x_0, x_1, \ldots$ is ultimately trapped in the nbhd $N_{q(x_m,n)}$ of radius $\leq 2^{-n}$, so there is a point x to which $x_0, x_1, x_2, \ldots$ converges. Now the fact that $\bar{\varphi}$ is a $\varkappa$-semiscale on P implies that $x \in P$, so $\bar{\psi} = \{\psi_n\}_{n\in\omega}$ is a good $\varkappa$-semiscale on P. ⊣

If $u = (x_0, \ldots, x_{n-1})$ and $v = (y_0, \ldots, y_{m-1})$ are sequences in some set X, put

$$\begin{aligned} u > v &\Leftrightarrow u \text{ is a proper initial segment of } v \\ &\Leftrightarrow n < m \quad \text{and} \quad x_0 = y_0, \ldots, x_{n-1} = y_{n-1}. \end{aligned}$$

2G.2. THE KUNEN–MARTIN THEOREM. *Let $\varkappa$ be an infinite cardinal, suppose $\prec$ is a strict wellfounded relation on a subset of some perfect product space $\mathfrak{X}$, suppose further that (as a subject of $\mathfrak{X} \times \mathfrak{X}$), $\prec$ is $\varkappa$-Suslin. Then the length of $\prec$ is less than $\varkappa^+$.*[12]

PROOF. Consider the tree T on $\mathfrak{X}$ defined by

$$T = \{(x_0, x_1, \ldots, x_{n-1}) : x_0 \succ x_1 \succ \cdots \succ x_{n-1}\}.$$

Clearly T is wellfounded. If ρ is the rank function of $\prec$, it is easy to check by induction on x that

$$(x_0, \ldots, x_{n-1}, x) \in T \Leftrightarrow \rho(T, (x_0, \ldots, x_{n-1}, x)) = \rho(x),$$

where $\rho(T, u)$ is the rank of u in T. Hence

$$\begin{aligned} |\prec| &= \textit{supremum}\{\rho(x) + 1 : x \in \textit{field}(\prec)\} \\ &= \textit{supremum}\{\rho(T, x) + 1 : (x) \in T\} \\ &= \rho(T, \emptyset), \end{aligned}$$

i.e.

$$|\prec| = |T| = \textit{the length of } T$$

and it is enough to prove that $|T| < \varkappa^+$.

The method of proof is to define a wellfounded tree S on $\varkappa$ and a mapping

$$\sigma : T \to S$$

which preserves the relation of *proper extension* on finite sequences,

$$u > v \Rightarrow \sigma(u) > \sigma(v).$$

In these circumstances it is immediate by bar induction on T that

$$\rho(T, u) \leq \rho(S, \sigma(u));$$

hence $|T| = \rho(T, \emptyset) \leq \rho(S, \sigma(\emptyset)) \leq \rho(S, \emptyset) = |S|$ and $|S| < \varkappa^+$ by 2D.1 since S is a tree on $\varkappa$.

To define S and σ let $\bar{\psi} = \{\bar{\psi}_n\}_{n \in \omega}$ be a good $\varkappa$-semiscale on $\{(x, y) : x \rightarrowtail y\}$. Here this means that if $x_0 \rightarrowtail y_0$, $x_1 \rightarrowtail y_1, \ldots$ and if for each n the sequence of ordinals

$$\psi_n(x_0, y_0), \psi_n(x_1, y_1), \ldots$$

is ultimately constant, then $limit_{i \to \infty} x_i = x$, $limit_{i \to \infty} y_i = y$ for some x, y and $x \rightarrowtail y$.

We now define σ directly—S will be the set of all initial segments of sequences $\sigma(u)$ with u in T. Put

$$\sigma(\emptyset) = \emptyset,$$

$$\sigma((x_0)) = (0),$$

$$\sigma((x_0, x_1)) = (0, \psi_0(x_0, x_1)),$$

$$\sigma((x_0, x_1, x_2)) = (0, \psi_0(x_0, x_1), \psi_0(x_1, x_2), \psi_1(x_1, x_2), \psi_1(x_0, x_1))$$

and in general for $n \geq 2$,

$$\sigma((x_0, \ldots, x_n)) = \sigma((x_0, \ldots, x_{n-1}))^\frown(\psi_0(x_{n-1}, x_n), \psi_1(x_{n-1}, x_n), \ldots, \psi_{n-1}(x_{n-1}, x_n),$$
$$\psi_{n-1}(x_{n-2}, x_{n-1}), \psi_{n-1}(x_{n-3}, x_{n-2}), \ldots, \psi_{n-1}(x_0, x_1)).$$

The idea is to include in $\sigma((x_0, \ldots, x_n))$ all ordinals $\psi_j(x_i, x_{i+1})$ for $i < n$, $j < n$. The sequence in which we do this is clear from Diagram 2G.1.

	ψ_0	ψ_1	ψ_2	$\psi_3 \cdots$
(x_0, x_1)	1	4	9	
(x_1, x_2)	2	3	8	
(x_2, x_3)	5	6	7	

Diagram 2G.1.

It is immediate that σ preserves the relation of proper extension on finite sequences, so it will be enough to verify that the tree

$$S = \{v\colon \text{for some } u \text{ in } T,\ v \succ \sigma(u)\}$$

is wellfounded.

Towards a contradiction assume that in some sequence

$$\sigma((x_0^0, x_1^0)) \succ \sigma((x_0^1, x_1^1, x_2^1)) \succ \sigma((x_0^2, x_1^2, x_2^2, x_3^2)) \succ \cdots$$

each term is a proper initial segment of the next, where of course we have $x_0^0 \succ\!\!- x_1^0$, $x_0^1 \succ\!\!- x_1^1 \succ\!\!- x_2^1$, $x_0^2 \succ\!\!- x_1^2 \succ\!\!- x_2^2 \succ\!\!- x_3^2$, etc. Then in the diagram

$$\begin{array}{lllll}
\psi_0(x_0^0, x_1^0), & & & & \\
\psi_0(x_0^1, x_1^1), & \psi_0(x_1^1, x_2^1), & \psi_1(x_1^1, x_2^1), & \psi_1(x_0^1, x_1^1), & \\
\psi_0(x_0^2, x_1^2), & \psi_0(x_1^2, x_2^2), & \psi_1(x_1^2, x_2^2), & \psi_1(x_0^2, x_1^2), & \psi_0(x_2^2, x_3^2), \ldots \\
\vdots & \vdots & \vdots & \vdots & \vdots
\end{array}$$

each column consists of identical ordinals. Hence for each fixed j and each fixed n the sequence

$$\psi_n(x_j^0, x_{j+1}^0),\ \psi_n(x_j^1, x_{j+1}^1),\ \psi_n(x_j^2, x_{j+1}^2), \ldots$$

is ultimately constant, so by the limit property of $\{\psi_n\}_{n\in\omega}$, each $limit_{i\to\infty}\, x_j^i$ exists, call it x_j and we have

$$x_0 \succ\!\!- x_1 \succ\!\!- x_2 \succ\!\!- \cdots$$

which is absurd. ⊣

This is the key tool for computing the length of wellfounded projective relations and it will be used again and again in the sequel. Here we only draw the conclusions mentioned in the beginning of this section.

2G.3. COROLLARY. *Every strict Σ_1^1 wellfounded relation has countable length.*[(12)] ⊣

2G.4. COROLLARY. *Every strict Σ_2^1 wellfounded relation has length less than $\aleph_2$. In particular, if $2^{\aleph_0} > \aleph_1$, then there is no Σ_2^1 wellordering of the continuum (Martin [19?a]).*[(12)]

PROOF. Use 2D.3 and the Kunen–Martin Theorem. ⊣

EXERCISES

2G.5. Prove that a binary relation $R(x, y)$ on a set S is wellfounded if and only if there are no infinite $<_R$-descending chains.

Hint. If $A \neq \emptyset$ and A has no $<_R$-minimal element, then we can successively choose $x_0 \in A_1$, $x_1 <_R x_0$, $x_2 <_R x_1, \ldots$, and get an infinite $<_R$-descending chain. ⊣

Consider the following conditions on a binary relation $\leq$ on a set S.

(a) $\leq$ is *transitive*, i.e.

$$x \leq y \;\&\; y \leq z \Rightarrow x \leq z.$$

(b) $\leq$ is *reflexive*, i.e. for all $x \in S$, $x \leq x$.

(c) $\leq$ is *antisymmetric*, i.e.

$$x \leq y \;\&\; y \leq x \Rightarrow x = y.$$

(d) $\leq$ is *connected*, i.e. for every x, y in S, $x \leq y$ or $y \leq x$.

(e) $\leq$ is *wellfounded*.

There are various names attached to relations that satisfy some of these conditions and we put them down here for the record.

(1) $\leq$ is a *partial ordering* if it is transitive, reflexive and antisymmetric.

(2) $\leq$ is an *ordering* if it is a connected partial ordering.

(3) $\leq$ is a *wellordering* if it is a wellfounded ordering.

(4) $\leq$ is a *prewellordering* if it is transitive, reflexive, connected and wellfounded—i.e. if $\leq$ has all the properties of a wellordering except for antisymmetry.

The strict part of a $\mathbf{\Sigma}_1^1$ relation need not be $\mathbf{\Sigma}_1^1$, so corollaries 2G.3, 2G.4 do not apply to arbitrary wellfounded relations. The best we can do here is state the trivial consequence of these results for $\mathbf{\Delta}_1^1$ and $\mathbf{\Delta}_2^1$ relations.

2G.6. Prove that every Borel wellfounded relation has countable length and every $\mathbf{\Delta}_2^1$ wellfounded relation has length less than $\aleph_2$. ⊣

There is a simple but useful characterization of the rank function implicit in its definition.

2G.7. Let R be a wellfounded relation on S with rank function ρ, let

$f: S \to Ordinals$ be any order preserving function, i.e.

$$x <_R y \Longrightarrow f(x) < f(y).$$

Prove that for every x in S, $\rho(x) \leq f(x)$. ⊣

A norm φ on a set S is *regular* if $\varphi: S \twoheadrightarrow \lambda$ is *onto* some ordinal λ, i.e.

$$\varphi(x) = \xi \ \&\ \eta < \xi \Longrightarrow \text{for some } y, \varphi(y) = \eta.$$

With each norm φ on S we associate the binary relation $\leq^{\varphi}$,

$$x \leq^{\varphi} y \Longleftrightarrow \varphi(x) \leq \varphi(y).$$

2G.8. Prove that a binary relation $\lesssim$ on a set S is a prewellordering if and only if there is a norm φ on S such that $\lesssim = \leq^{\varphi}$. Moreover, if $\lesssim$ is a prewellordering, then there is a unique regular φ on S such that $\lesssim = \leq^{\varphi}$.

Hint. Given $\lesssim$, take $\varphi = \rho$ to be the rank function of $\lesssim$. ⊣

2H. Category and measure

We have proved that not every $\boldsymbol{\Sigma}^1_1$ set is Borel. There are times, however, when it is useful to know that a pointset P is *approximately* equal to some Borel set P^*, in the sense that the *symmetric difference*

$$P \triangle P^* = (P - P^*) \cup (P^* - P)$$

is *small*. We establish here a general, set theoretic result about approximations of $\varkappa$-Suslin sets by $(\varkappa+1)$-Borel sets modulo a given $\varkappa$-ideal. This will imply, in particular, that $\boldsymbol{\Sigma}^1_1$ sets are Lebesgue measurable and have the property of Baire.

Fix a perfect product space $\mathfrak{X}$. A collection J of subsets of $\mathfrak{X}$ is a *$\varkappa$-ideal* ($\varkappa$ an infinite cardinal) if J is closed under subsets and unions of $\varkappa$ elements, i.e.

$$A \subseteq B \ \&\ B \in J \Longrightarrow A \in J,$$

$$\text{for each } \xi < \varkappa, A_\xi \in J \Longrightarrow \bigcup_{\xi<\varkappa} A_\xi \in J.$$

If $\varkappa = \aleph_0$, instead of $\aleph_0$-ideals we talk of *σ-ideals*.

Suppose $\mathcal{C}$ is a fixed λ-algebra of subsets of $\mathfrak{X}$. We say that $P \subseteq \mathfrak{X}$ *is in $\mathcal{C}$ modulo J* if there is some $P^* \in \mathcal{C}$ such that $P \triangle P^* \in J$. In particular, P is $(\varkappa+1)$-Borel modulo J if $P \triangle P^* \in J$ for some $(\varkappa+1)$-Borel P^*.

Recall that a pointset A is *meager* if $A = \bigcup_n A_n$ with each A_n *nowhere dense*, i.e. such that the closure $\bar{A}_n$ contains no open set. The collection M of all meager subsets of $\mathfrak{X}$ is obviously a σ-ideal.

Suppose μ is a *σ-finite Borel measure on* $\mathfrak{X}$, i.e. a countably additive function on the Borel subsets of $\mathfrak{X}$ with values real numbers ≥ 0 or ∞ and such that we can write

$$\mathfrak{X} = \bigcup_n^\omega A_n$$

with $\mu(A_n) < \infty$ for each n. Let Z_μ be the collection of *sets of measure* 0 (in the completed measure), i.e.

$$A \in Z_\mu \Leftrightarrow \text{there exists a Borel set } B$$
$$\text{such that } A \subseteq B \text{ and } \mu(B) = 0.$$

Again it is clear that Z_μ is a σ-ideal.

These are the two standard examples which we want covered by the approximation theorem. They satisfy an additional hypothesis which will be crucial to the proof.

Suppose again J is a $\varkappa$-ideal on $\mathfrak{X}$ and $\mathcal{C}$ is a $(\varkappa+1)$-algebra of subsets of $\mathfrak{X}$. We say that J is *regular from above relative to* $\mathcal{C}$ if for every $P \subseteq \mathfrak{X}$ there is some $\tilde{P} \in \mathcal{C}$ such that (see Figure 2H.1)

(1) $P \subseteq \tilde{P}$,

(2) if $A \subseteq \tilde{P} - P$ and $A \in \mathcal{C}$, then $A \in J$.

We will outline proofs in the exercises that the σ-ideals of meager and null sets are regular from above relative to the Borel sets.

2H.1. THE APPROXIMATION THEOREM. *Let $\varkappa$ be an infinite cardinal, suppose J is a $\varkappa$-ideal on some perfect product space $\mathfrak{X}$, assume that J is*

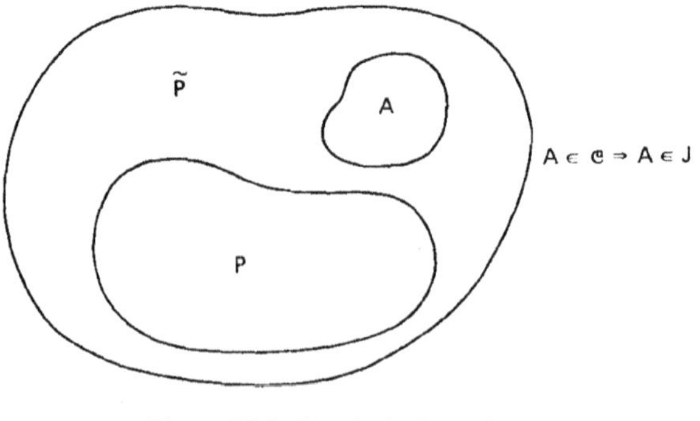

Figure 2H.1. Regularity from above.

regular from above relative to some $(\varkappa+1)$-algebra of sets $\mathcal{C}$. Then the collection of sets which are in $\mathcal{C}$ modulo J is closed under complementation, unions of length $\varkappa$ and the operation $\mathcal{A}^{\varkappa}$.

In particular, every $\varkappa$-Suslin subset of $\mathfrak{X}$ is $(\varkappa+1)$-Borel modulo J, taking $\mathcal{C}=\mathbf{B}_{\varkappa+1}\restriction\mathfrak{X}$.[13]

PROOF. If $P_\xi \triangle P^*_\xi \in J$ for all $\xi<\varkappa$, then

$$(\bigcup_{\xi<\varkappa} P_\xi) \triangle (\bigcup_{\xi<\varkappa} P^*_\xi) \subseteq \bigcup_{\xi<\varkappa} (P_\xi \triangle P^*_\xi) \in J.$$

Similarly, if $P \triangle P^* \in J$, then $(\mathfrak{X}-P) \triangle (\mathfrak{X}-P^*) = P \triangle P^* \in J$. Thus the collection of subsets of $\mathfrak{X}$ which are in $\mathcal{C}$ modulo J is closed under complementation and unions of length $\varkappa$.

Assume now that

$$P = \mathcal{A}^{\varkappa}_u P_u$$

where each P_u is in $\mathcal{C}$. For each sequence $u=(\xi_0,\ldots,\xi_{n-1})$ in $\varkappa$, put

$$Q_u = \mathcal{A}^{\varkappa}_v P_{u^\frown v} = \bigcup_f \bigcap_n P_{(\xi_0,\ldots,\xi_{n-1})^\frown f\restriction n},$$

so that

$$Q_\emptyset = P,$$

and for each u,

$$Q_u \subseteq P_u.$$

Notice also that by the definition,

$$Q_u = \bigcup_{\xi<\varkappa} Q_{u^\frown(\xi)}.$$

Since J is regular from above, we can find in $\mathcal{C}$ sets Q^*_u such that (see Figure 2H.2)

(i) $$Q_u \subseteq Q^*_u,$$

(ii) if $A \subseteq Q^*_u - Q_u$ is in $\mathcal{C}$, then $A \in J$.

We may also assume that

(iii) $$Q^*_u \subseteq P_u,$$

since in any case the sets $Q^*_u \cap P_u$ are in $\mathcal{C}$ and satisfy the crucial properties (i), (ii).

We now claim that

$$Q^*_\emptyset - P \subseteq \bigcup_u (Q^*_u - \bigcup_{\xi<\varkappa} Q^*_{u^\frown(\xi)}).$$

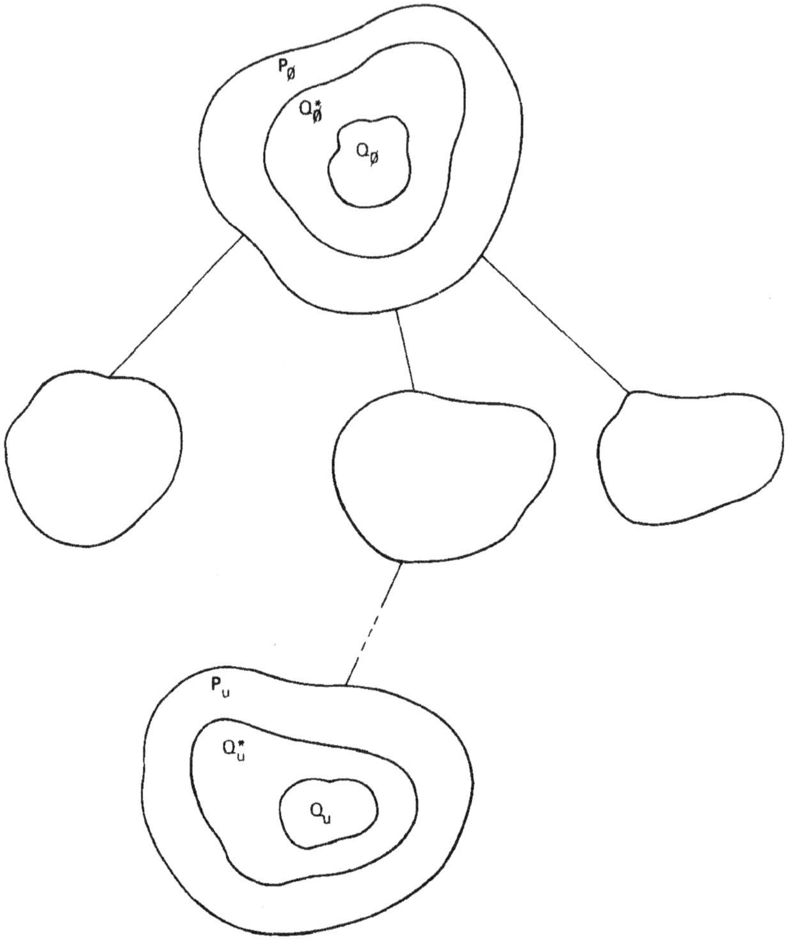

Figure 2H.2. The approximation theorem.

To prove this by contradiction, assume that

$$x \in Q^*_\emptyset - P$$

but for every sequence u in $\varkappa$,

$$x \notin Q^*_u \quad \text{or for some} \quad \xi < \varkappa,\, x \in Q^*_{u^\frown(\xi)}.$$

Taking $u = \emptyset$, this means that there is some ξ_0 so that

$$x \in Q^*_{(\xi_0)}.$$

Taking $u=(\xi_0)$ now, there must be some ξ_1 so that

$$x \in Q^*_{(\xi_0,\xi_1)},$$

and proceeding inductively we define some $f \in {}^\omega\varkappa$ such that

$$x \in \bigcap_n Q^*_{f\upharpoonright n}.$$

Since $Q^*_{f\upharpoonright n} \subseteq P_{f\upharpoonright n}$, we thus have

$$x \in \bigcap_n P_{f\upharpoonright n},$$

so that $x \in \mathcal{A}_u P_u = P$ contradicting $x \notin P$.

For each u, clearly $Q^*_u - \bigcup_{\xi<\varkappa} Q^*_{u^\frown(\xi)}$ is in $\mathfrak{C}$ and

$$Q^*_u - \bigcup_{\xi<\varkappa} Q^*_{u^\frown(\xi)} \subseteq Q^*_u - \bigcup_{\xi<\varkappa} Q_{u^\frown(\xi)} = Q^*_u - Q_u,$$

so that $Q^*_u - \bigcup_{\xi<\varkappa} Q^*_{u^\frown(\xi)} \in J$. Since there are only $\varkappa$ finite sequences of elements of $\varkappa$, $\bigcup_u (Q^*_u - \bigcup_{\xi<\varkappa} Q^*_{u^\frown(\xi)}) \in J$ and hence $P \triangle Q^*_\emptyset = Q^*_\emptyset - P \in J$.

This argument proves that if $P = \mathcal{A}^\varkappa_u P_u$ with each P_u in $\mathfrak{C}$, then P is in $\mathfrak{C}$ modulo J.

For the more general assertion, assume that

$$P = \mathcal{A}^\varkappa_u P_u$$

with each P_u in $\mathfrak{C}$ modulo J and choose sets P^*_u in $\mathfrak{C}$ such that

$$P_u \triangle P^*_u \in J.$$

To prove that P is in $\mathfrak{C}$ modulo J, it is enough to show that

(iv) $$\mathcal{A}^\varkappa_u P_u \triangle \mathcal{A}^\varkappa_u P^*_u \subseteq \bigcup_u (P_u \triangle P^*_u)$$

since the set on the right is the union of $\varkappa$ sets in J.

Assume then that

$$x \in \mathcal{A}^\varkappa_u P_u - \mathcal{A}^\varkappa_u P^*_u,$$

i.e.

$$(\exists f)(\forall n)P_{f\upharpoonright n}(x) \;\&\; (\forall f)(\exists n)\neg P^*_{f\upharpoonright n}(x),$$

and choose f so that $(\forall n)P_{f\upharpoonright n}(x)$. Then there is some n such that $\neg P^*_{f\upharpoonright n}(x)$, so with $u = f\upharpoonright n$ we have $x \in P_u - P^*_u$. A symmetric argument shows that if $x \in \mathcal{A}^\varkappa_u P^*_u - \mathcal{A}^\varkappa_u P_u$, then for some u, $x \in P^*_u - P_u$. Thus (iv) is established and the proof is complete. ⊣

EXERCISES

2H.2. Prove that in a complete metric space no open ball is meager (The Baire Category Theorem).

Hint. Assume $B \subseteq \bigcup_n \bar{A}_n$, where each $\bar{A}_n$ is closed and nowhere dense. Choose an open ball B_1 so that $\bar{B}_1 \subseteq B - \bar{A}_1$ and $radius(B_1) < 1$, choose an open ball B_2 so that $\bar{B}_2 \subseteq B_1 - \bar{A}_2$ and $radius(B_2) < \frac{1}{2}$, etc. Show that if $x \in \bigcap_n B_n$, then $x \notin \bigcup_n \bar{A}_n$, which is absurd. ⊣

A pointset P has *the property of Baire* if there is some open set P^* such that $P \triangle P^*$ is meager.

2H.3. Prove that every Borel pointset has the property of Baire.

Hint. Open sets clearly have the property of Baire. If P is closed, let $P^* = \text{Interior}(P) = \{x \in P\colon \text{for some nbhd } N \text{ of } x, N \subseteq P\}$. Show that $P - P^*$ is nowhere dense, so $P \triangle P^*$ is meager. Notice that $(P' \triangle Q) = (P \triangle Q')$, where $'$ denotes the complement, and use this to show that if P has the property of Baire, so does P'. Show finally that if each P_n has the property of Baire, so does $\bigcup_n P_n$. ⊣

2H.4. Prove that for every pointset $P \subseteq \mathfrak{X}$, there is an F_σ set $\tilde{P} \supseteq P$ such that if $A \subseteq \tilde{P} - P$ is any Borel set, then A is meager.

Hint. Let

$$D(P) = \{x\colon \text{for every nbhd } N \text{ of } x, N \cap P \text{ is not meager}\}.$$

Show that $D(P)$ is closed and that $P - D(P)$ is meager, so $P - D(P) \subseteq W$ for some meager F_σ set W. Take $\tilde{P} = D(P) \cup W$. If $A \subseteq \tilde{P} - P = (D(P) \cup W) - P$ and A is Borel but not meager, choose an open N such that $N - A = V$ is meager, so $N \subseteq A \cup V \subseteq (D(P) \cup W) - P \cup V \subseteq (D(P) - P) \cup Y$, where Y is meager. Now $N \cap P \subseteq Y$, so $N \cap P$ is meager, hence $N \cap D(P) = \emptyset$, hence $N \subseteq Y$ which contradicts the Baire Category Theorem. ⊣

2H.5. Prove that the collection of pointsets with the property of Baire is closed under the operation $\mathscr{A}$; in particular $\boldsymbol{\Sigma}^1_1$ sets have the property of Baire.(13) ⊣

2H.6. Prove that there are sets of real numbers which do not have the property of Baire.

Hint. This needs the axiom of choice. One way to do it is by a construction similar to that of 2C.4.

First argue that there are exactly $2^{\aleph_0}$ pairs (G, F), where G is open and F is a meager F_σ. Wellorder $\mathbb{R} = \{x_\xi\colon \xi < 2^{\aleph_0}\}$ and these pairs,

$\{(G_\xi, F_\xi): \xi < 2^{\aleph_0}\}$. You want to construct a set A such that the inclusion

$$A \triangle G_\xi = (A - G_\xi) \cup (G_\xi - A) \subseteq F_\xi$$

fails for every ξ. Define by recursion disjoint sets A_ξ, B_ξ, such that $card(A_\xi) < 2^{\aleph_0}$, $card(B_\xi) < 2^{\aleph_0}$ and $(A_\xi - G_\xi) \cup (G_\xi \cap B_\xi) \subseteq F_\xi$ fails. At the ξ'th step, either $\mathcal{R} - (G_\xi \cup F_\xi)$ is uncountable, hence of cardinality $2^{\aleph_0}$ and we can throw in A_ξ some element of this set; or $\mathcal{R} = G_\xi \cup F'_\xi$ where F'_ξ is meager, F_σ and $F'_\xi \supseteq F_\xi$, hence $G_\xi - F'_\xi$ is uncountable and has cardinality $2^{\aleph_0}$ and we can throw in B_ξ an element of this set. ⊣

2H.7. Let μ be a σ-finite Borel measure on some product space $\mathfrak{X}$. Prove that the σ-ideal Z_μ of sets of measure 0 is regular from above relative to the Borel sets.

Hint. Suppose first that P is contained in some Q with $\mu(Q) < \infty$ and put

$$x = infimum\{\mu(Q): Q \quad \text{Borel}, \quad P \subseteq Q\}.$$

Choose a decreasing sequence $Q_1 \supseteq Q_2 \supseteq \cdots$ of Borel sets, $Q_n \supseteq P$, such that $limit_{n \to \infty} \mu(Q_n) = x$ and take $\tilde{P} = \bigcap_n Q_n$. If P is large, let $\mathfrak{X} = \bigcup_n A_n$ with each A_n Borel, $\mu(A_n) < \infty$ and $A_n \cap A_m = \emptyset$ for $n \neq m$, and use the result on each $P \cap A_n$. ⊣

A set $P \subseteq \mathfrak{X}$ is *measurable* relative to a σ-finite Borel measure μ on $\mathfrak{X}$ if there are Borel sets $\tilde{P}$ and Q such that $P \triangle \tilde{P} \subseteq Q$ and $\mu(Q) = 0$. We let $\mu(P) = \mu(\tilde{P})$ for any $\tilde{P}$ with this property—$\mu(\tilde{P})$ is obviously independent of the choice of $\tilde{P}$.

2H.8. Prove that the collection of sets measurable relative to a σ-finite Borel measure μ on $\mathfrak{X}$ contains all $\mathbf{\Sigma}^1_1$ and $\mathbf{\Pi}^1_1$ sets and is closed under complementation, countable unions and the operation $\mathcal{A}$.(13) ⊣

Recall that if A is a set of reals, then the *Lebesgue outer measure* of A is defined by

$$\mu^*(A) = infimum\{\textstyle\sum_{i=0}^{\infty} (b_i - a_i): A \subseteq \bigcup_i^{\infty} (a_i, b_i)\},$$

where of course (a_i, b_i) is the open interval from a_i to b_i. We call A Lebesgue *measurable* if for every closed interval $[a, b]$

$$\mu^*(A \cap [a, b]) + \mu^*([a, b] - A) = b - a.$$

It is proved in every textbook in real variables that the collection of Lebesgue measurable sets contains all open sets and is closed under both complementation and countable union; hence every Borel set is Lebesgue measurable. Moreover, μ^* is a measure on the measurable sets, in particular μ^* restricted to the Borel sets is a Borel measure. The definition of measurability we gave above for arbitrary σ-finite Borel measures is consistent with this definition of Lebesgue measurability.

2H.9. Prove that there is a set of reals A contained in the unit interval $[0, 1]$ such that $\mu^*(A) = \mu^*([0, 1] - A) = 1$. In particular, A is not Lebesgue measurable.

Hint. This is one more construction by transfinite recursion and choice. First argue that there are $2^{\aleph_0}$ open coverings $\bigcup_i^\infty (a_i, b_i)$ with $\sum_{i=0}^\infty (b_i - a_i) < 1$, wellorder them $\{G_\xi : \xi < 2^{\aleph_0}\}$. Now build A_ξ, B_ξ so that $A_\xi \cap B_\xi = \emptyset$, $card(A_\xi) < 2^{\aleph_0}$, $card(B_\xi) < 2^{\aleph_0}$ and each $[0, 1]$-G_ξ intersects both A_ξ and B_ξ. The key observation is that each $[0, 1]$-G_ξ has cardinality $2^{\aleph_0}$. ⊣

We will see later that these results about category and measure are best possible in the context of Zermelo–Fraenkel set theory. One cannot prove in this theory that $\mathbf{\Delta}_2^1$ sets of reals have the property of Baire or are Lebesgue measurable. There are, however, natural strong axioms of set theory which imply that all $\mathbf{\Sigma}_2^1$ sets have these regularity properties and still stronger axioms which allow us to establish that all projective sets are Lebesgue measurable and have the property of Baire.

By the basic definition of Λ-measurability in Section 1G, we call $f : \mathcal{X} \to \mathcal{Y}$ *Baire-measurable* if for each basic nbhd $N_s \subseteq \mathcal{Y}$, $f^{-1}[N_s]$ has the property of Baire. Similarly, if μ is a σ-finite Borel measure on $\mathcal{X}$, then $f : \mathcal{X} \to \mathcal{Y}$ is μ-measurable if each $f^{-1}[N_s]$ is measurable relative to μ. We say that $f : \mathcal{X} \to \mathcal{Y}$ is *absolutely measurable* if f is μ-measurable relative to every σ-finite Borel measure μ on $\mathcal{X}$.

These functions come up often in the applications of descriptive set theory to analysis. Here we will confine ourselves to a simple but useful remark about them.

2H.10. Prove that if $f : \mathcal{X} \to \mathcal{Y}$ is Baire-measurable, then there exists a G_δ set $P \subseteq \mathcal{X}$ which is *comeager* (i.e. $\mathcal{X} - P$ is meager) and such that the restriction $f \upharpoonright P$ of f to P is continuous.

Similarly, if $f : \mathcal{X} \to \mathcal{Y}$ is μ-measurable relative to a σ-finite Borel measure μ, then there exists an F_σ set $P \subseteq \mathcal{X}$ and a Borel function

$f^*: \mathfrak{X} \to \mathcal{Y}$, such that $\mathfrak{X} - P$ has measure 0 and

$$x \in P \Longrightarrow f(x) = f^*(x).$$

Hint. For each basic nbhd $N_s \subseteq \mathcal{Y}$ let G_s be open in $\mathfrak{X}$ such that $f^{-1}[N_s] \triangle G_s$ is meager, choose a meager F_σ-set Q_s so that $f^{-1}[N_s] \triangle G_s \subseteq Q_s$ and take $Q = \bigcup_s Q_s$, $P = \mathfrak{X} - Q$. The argument for measure is similar. ⊣

2I. Historical remarks

[1] As we have already noted, the operation $\mathcal{A} = \mathcal{A}^{\aleph_0}$ was introduced in the basic paper Suslin [1917], although some similar ideas can be found in Alexandroff [1916] and Hausdorff [1916]. Suslin [1917] and Lusin [1917] also contain, at least implicitly, the characterization of $\mathbf{\Sigma}^1_1$ sets of reals as projections of closed sets in $\mathbb{R} \times \mathcal{N}$.

[2] The more general operation $\mathcal{A}^\kappa$ for any cardinal κ was introduced by Maximoff [1940], who also defined what we have called here λ-Borel sets. Maximoff worked in large, non-separable spaces and defined "κ-Suslin" to be the sets obtained via the operation $\mathcal{A}^\kappa$ applied to Suslin systems $u \mapsto P_u$, where each P_u is $(\kappa+1)$-Borel; this is a much larger class of sets than our κ-Suslin sets. Stone [1962] studies the present notion of κ-Suslin, but he relates these to the usual Borel ($\aleph_1$-Borel) sets, again in non-separable spaces.

[3] Our own approach here has been to use these general notions of κ-Suslin and λ-Borel sets as tools for obtaining specific information about projective pointsets. There is some anticipation of this in Sierpinski [1927], where he shows that his "hyperborelian" sets of reals must have cardinality $\leq \aleph_1$ or $2^{\aleph_0}$; these turn out to be precisely the $\aleph_1$-Suslin sets, although Sierpinski defined them differently. The modern approach is due to Mansfield [1970] who used trees and particularly Martin [19?a] who saw most clearly its potentialities.

[4] *Semiscales* are quite modern and come from the *scales* introduced in Moschovakis [1971] to study uniformization problems. We will look at these closely in Chapter 4.

[5] In Chapters 6, 7 and 8 it will become obvious why we developed here the theory of κ-Suslin sets rather than concentrate on the classical $\mathbf{\Sigma}^1_1$ sets. The proofs for the special case are no simpler than the ones we gave. A hint for the kind of applications that are forthcoming shows in 2D.3, the

fact that $\mathbf{\Sigma}^1_2$ pointsets are $\aleph_1$-Suslin. This important result is implicit in Shoenfield [1961].

[6] The closure of the projective classes $\mathbf{\Sigma}^1_n$, $\mathbf{\Pi}^1_n$, $\mathbf{\Delta}^1_n$ under the operation $\mathscr{A}$ (2B.4) was established by Kantorovitch and Livenson [1932] and later, (by the simple proof we gave) by Addison and Kleene [1957].

[7] The *Perfect Set Theorem* 2C.2 is due to Mansfield [1970] in its full generality, but of course there were several similar earlier results. The specific application to $\mathbf{\Sigma}^1_1$ sets is due to Suslin—it was announced in Lusin [1917]. The proof we gave is due to Solovay.

[8] Suslin [1917] announced the *Suslin Theorem* (2E.2, for $\varkappa = \aleph_0$ of course) but gave no hint of its proof. The first published proof is in Lusin–Sierpinski [1918]—this is the argument outlined in 2F.4 and 2F.5. Another proof was given in Lusin–Sierpinski [1923]. Lusin [1927] established the more general *Separation Theorem* 2E.1, but the proofs in the preceding two papers could certainly have been used for this too. Lusin [1930b] gives both the argument by contradiction and a constructive proof, as we did.

[9] Among the immediate corollaries of the Suslin theorem, perhaps the most significant are 2E.5, 2E.8, 2E.9, 2E.10 and 2E.11. The characterizations of Borel sets as the continuous, injective images of closed subsets of $\mathscr{N}$ is already stated (in somewhat different form) in Lusin [1917]; Lusin says that it can be proved "using a method of Suslin," so it is quite likely a joint result. The same is probably true of 2E.5, 2E.9, 2E.10 and 2E.11 which were considered particularly important, since they showed that the claims of Lebesgue [1905] about implicitly defined functions were correct, even though Lebesgue's proof was wrong. These results were all treated in detail in Lusin [1927].

[10] Lusin–Sierpinski [1918] established that $\mathbf{\Pi}^1_1$ sets are unions of $\aleph_1$ Borel sets and Lusin–Sierpinski [1923] proved the same representation for $\mathbf{\Sigma}^1_1$ sets. This representation for $\mathbf{\Sigma}^1_2$ sets (2F.3) is due to Sierpinski [1925] who also established the elegant equations of 2F.1 in his [1926].

[11] The exercises of 2F are directly from Lusin–Sierpinski [1918]. There are many applications of the so-called *Lusin–Sierpinski index* which we will study in Chapter 4, in a general setting.

[12] Logicians interested in descriptive set theory often refer to "the classical result" that $\mathbf{\Sigma}^1_1$ wellfounded relations have countable length. This was apparently never put down on paper, but it is certainly easy to show by classical methods. Martin showed in 1968 that $\mathbf{\Sigma}^1_2$ wellfounded relations have length below $\aleph_2$ (2G.4) by a sophisticated argument, using

forcing. The more general (and simpler) *Kunen–Martin Theorem* 2G.2 was proved independently in 1971 by its two authors and was not published until Martin [19?a].

[13] According to Kuratowski [1966], the *Approximation Theorem* 2H.1 is due to Szpilrajn–Marcewski who published it in Polish in 1929. The specific corollaries were established earlier as follows: $\boldsymbol{\Sigma}^1_1$ sets have the property of Baire (Lusin–Sierpinski [1923]); the collection of sets with the property of Baire is closed under the operation $\mathscr{A}$ (Nikodym [1925]); $\boldsymbol{\Sigma}^1_1$ sets are Lebesgue measurable (Lusin [1917]); the collection of Lebesgue measurable sets is closed under the operation $\mathscr{A}$ (Lusin–Sierpinski [1918]).

[14] The fact that the collection of μ-measurable sets is closed under the operation $\mathscr{A}$ (2H.8) has been extended by Choquet [1955] from measures to *capacities*, roughly "subadditive measures." A very simple and elegant exposition of this important theorem can be found in Carleson [1967].

CHAPTER 3

BASIC NOTIONS OF THE EFFECTIVE THEORY

Our choice of basic notions in Chapter 1 was based on the implicit assumption that open sets are somehow "simple." They are just given at the very start, and then we build more complicated sets from them. Let us try here to analyze this view.

Suppose G is an open set of reals, say

$$G = \bigcup_n (a_n, b_n),$$

where each (a_n, b_n) is an open interval with rational endpoints. Given a real number x, we may attempt to find out if $x \in G$ by searching for some n such that $a_n < x < b_n$. One natural way to be "given" x is via a sequence of rationals converging to it with a known modulus of convergence, say

$$x = limit_{i \to \infty} q_i,$$

where for each i,

$$|x - q_i| \leq 2^{-i}.$$

We now search for some n and i such that

$$(*) \qquad a_n + 2^{-i} < q_i < b_n - 2^{-i};$$

if and when we find them, we will know that $x \in G$.

We have described a *semieffective membership test* for G which will verify that $x \in G$ if this is true. If $x \notin G$, this procedure will not terminate—we will simply not be able to find n and i such that $(*)$ holds.

It seems improbable that we can discover a genuine *effective membership test* which will decide by a finite computation whether an arbitrary given x is or is not a member of G. In fact, even if $G = (0, 1)$ and it just so happens that $x = 1$, we will never be able to assert with certainty that $x \notin G$ by looking at the approximations q_i.

This argument suggests that open sets are "simple" because they are "semieffective."

One factor we did not consider is the complexity of the function

$$n \mapsto (a_n, b_n).$$

Suppose, for example, that $(p_0, q_0), (p_1, q_1), \ldots$ is an effective enumeration of all open intervals with rational endpoints and put

$$(a_n, b_n) = \begin{cases} (p_n, q_n) & \text{if} \quad f(n, x_1, \ldots, x_k) \neq 0 \quad \text{for all integers } x_1, \ldots, x_k, \\ (0, 1) & \text{otherwise.} \end{cases}$$

Suppose further that $f(n, x_1, \ldots, x_k) = 0$ is a hopelessly complicated Diophantine equation which cannot (apparently) be solved by any of the standard methods. To verify that $a_n < x < b_n$, we must first find out if $f(n, x_1, \ldots, x_k) = 0$ has solutions, or else we do not even know whether $(a_n, b_n) = (p_n, q_n)$ or $(a_n, b_n) = (0, 1)$. Here the "semieffective" membership test for $G = \bigcup_n (a_n, b_n)$ breaks down at the very beginning—we do not know for what intervals (a, b) we should attempt to verify that $a < x < b$.

Of course this is a perverse example. The open sets that occur naturally in mathematical practice are almost always of the form $\bigcup_n (a_n, b_n)$ where the function $n \mapsto (a_n, b_n)$ can be computed by some explicit or inductive procedure.

An open set G is *semirecursive* if $G = \bigcup_n (a_n, b_n)$, where the function $n \mapsto (a_n, b_n)$ is computable. To make this precise, we will appeal to *Church's Thesis*, one of the central discoveries of modern mathematical logic. This identifies the intuitive notion of a *computable function on the integers* with the precise, mathematical concept of a *recursive function.*

The semirecursive pointsets are just the effectively described open sets, those open sets for which the procedure described above can in fact be carried out. They include almost all open sets one is likely to encounter in analysis or topology. Starting with them, we will define effective Borel and Lusin pointclasses and develop an interesting and non-trivial *refinement* of the theory in Chapter 1.

Using semirecursive pointsets one can also introduce in a natural way, *recursive functions*

$$f : \mathcal{X} \to \mathcal{Y}$$

on product spaces. Intuitively, f is recursive if we have an algorithm which given (sufficiently close approximations to) x produces (arbitrarily accurate approximations to) $f(x)$. Every recursive function is continuous, but not vice versa. Again, every special continuous function that one is likely to meet in ordinary mathematical practice is in fact recursive.

It is obvious from these remarks that we will study *recursion theory* as an effective version, a *refinement* of *pointset topology*.

One of the most fascinating aspects of this approach is that it leads naturally to an *effective descriptive set theory* on the space ω. Contrary to our promise in the introduction to Chapter 1, we have said nothing about definable sets of integers. Every subset of ω is open, so the Borel and Lusin pointclasses trivialize on this space. On the other hand, there are only countably many semirecursive pointsets and recursive functions. It turns out that the effective Borel and Lusin pointclasses yield interesting and non-trivial hierarchies of subsets of ω.

As a matter of fact, the theory for ω was developed by Kleene in the period 1940–1955 (roughly) entirely independently of classical descriptive set theory. Similarities and "analogies" between the two theories were then noticed, particularly by Addison who initiated the development of the unified treatment we are presenting here.

It should be emphasized that the effective theory is not only interesting in its own right—it is also a powerful tool for studying the classical Borel and Lusin pointclasses. Some of the most important recent results about projective sets depend essentially on the use of recursion theoretic concepts and techniques.

The development in this chapter is brief but totally self-contained, i.e. it presupposes no knowledge of logic or recursion theory. Consequently, the reader who is well versed in these subjects should skip much of it, particularly Sections 3A and 3F which establish some of the standard results about recursion on ω. On the other hand, the reader with no experience in recursion theory should go carefully over 3A and do all the exercises. These give a stock of recursive functions which we then use constantly and without apologies or special reference.

3A. Recursive functions on the integers[(1)]

Consider the following "constructive" schemes for defining a function f with integer arguments from given functions.

Composition. Given $g_1(x), g_2(x), \ldots, g_m(x)$ and $h(n_1, \ldots, n_m)$, define f by

$$f(x) = h(g_1(x), g_2(x), \ldots, g_m(x)).$$

Here and in the equations below x varies over ω^k,

$$x = (x_1, \ldots, x_k) \in \omega^k$$

and all functions take integers as values.

Primitive recursion. Given $g(x)$ and $h(u, n, x)$, define $f(n, x)$ by the recursion

$$\begin{cases} f(0, x) = g(x), \\ f(n+1, x) = h(f(n, x), n, x). \end{cases}$$

It is clear that f is determined by these two equations if g and h are given. One example of primitive recursion is the usual definition of the addition function,

$$\begin{cases} f(0, m) = m, \\ f(n+1, m) = f(n, m) + 1. \end{cases}$$

One proves easily by induction on n that for all m,

$$f(n, m) = n + m.$$

The definition can be brought to the standard form of primitive recursion that we listed above if we take

$$g(m) = m$$

$$h(u, n, m) = u + 1.$$

Another example is the usual definition of multiplication from addition,

$$\begin{cases} f(0, m) = 0 \\ f(n+1, m) = f(n, m) + m. \end{cases}$$

Again it is obvious that

$$f(n, m) = n \cdot m$$

and we can put this recursion into the form above by choosing

$$g(m) = 0$$

$$h(u, n, m) = u + m.$$

There is a simpler kind of *primitive recursion* appropriate for defining functions of one variable,

$$\begin{cases} f(0) = w_0 \\ f(n+1) = h(f(n), n). \end{cases}$$

For example the *predecessor* function

$$pd(n) = \begin{cases} n - 1 & \text{if} \quad n > 0, \\ 0 & \text{if} \quad n = 0 \end{cases}$$

can be defined this way,

$$\begin{cases} pd(0) = 0 \\ pd(n+1) = n. \end{cases}$$

We will include this simple scheme when we talk of definition by primitive recursion.

Minimalization. Suppose $g(n, x)$ is such that

for every x there is some n such that $g(n, x) = 0$.

Put

$$\begin{aligned} f(x) &= \mu n[g(n, x) = 0] \\ &= \textit{the least number n such that } g(n, x) = 0. \end{aligned}$$

We called these schemes *constructive* because they give us a direct way of computing the values of the new function f in terms of the values of the given functions. For example, if f is defined from g and h by primitive recursion, to compute $f(2, x)$ we successively compute

$$\begin{aligned} f(0, x) &= g(x) = w_0, \\ f(1, x) &= h(w_0, 0, x) = w_1, \\ f(2, x) &= h(w_1, 1, x) = w_2. \end{aligned}$$

Similarly, if f is defined from g by minimalization, to compute $f(x)$ we successively compute

$$g(0, x), g(1, x), g(2, x), \ldots$$

until we find some w such that $g(w, x) = 0$; we set

$$f(x) = w$$

for the first such w.

The intention is to call a number theoretic function *recursive* (or *computable*) if we can define it by successive applications of these three simple schemes. Of course we must have some simple functions to start with, and for these we choose the following completely trivial functions.

$S(n) = n + 1$	*successor*
$C^k_w(x_1, \ldots, x_k) = w$	*constant* w, as a function of k arguments
$P^k_i(x_1, \ldots, x_k) = x_i$	*projection* in the i'th component, $1 \leq i \leq k$.

Of these, the projection P^1_1 would be better named the *identity* function, $P^1_1(n) = n$.

Now, a function is *recursive* if it can be defined by successive applications of composition, primitive recursion and minimalization starting with the functions S, C^k_w, P^k_i. More precisely, *the class of recursive functions is the smallest collection of number theoretic functions which contains the successor S, all constants C^k_w and projections P^k_k and which is closed under composition, primitive recursion and minimalization.*

For example, to prove that addition

$$f(n, m) = n + m$$

is recursive, it is enough to show that g and h are recursive, where

$$g(m) = m$$
$$h(u, n, m) = u + 1,$$

by the argument above. But $g = P^1_1$ and

$$h(u, n, m) = S(P^3_1(u, n, m)),$$

so h is recursive as the composition of recursive functions.

Similarly, to show that multiplication

$$f(n, m) = n \cdot m$$

is recursive, it is enough to show that g, h are recursive, where

$$g(m) = 0,$$
$$h(u, n, m) = u + m.$$

Again, $g = C^1_0$ and

$$h(u, n, m) = P^3_1(u, n, m) + P^3_3(u, n, m),$$

so h is recursive as the composition of $+$, P^3_1 and P^3_3.

Church's Thesis is the metamathematical claim that every number theoretic function which is intuitively computable is in fact recursive. By "intuitively computable" we mean that there is an effective, uniform method for computing $f(x)$ once we are given x.

To justify Church's Thesis, one must make a deep and detailed study of the class of recursive functions as well as a careful analysis of the notion of "effective method" or "algorithm." Books on recursion theory take great pains to do this carefully. We will not do it here, as it would take us far afield from our central interest in the study of pointsets.

From the strictly technical point of view, Church's Thesis is irrelevant—one always works with the precise concept of recursiveness rather than the vague notion of intuitive computability. After all, no one takes great pains in the classsical theory to justify starting with the open sets—it is taken for granted that these are the simplest sets we can think of. Church's Thesis becomes important when we attempt to draw foundational or philosophical inferences from technical results—and in those instances one should explicitly bring it in as a consideration.

We will need to know that a great many functions are recursive. This is the point of the lengthy list of exercises in this section. One should look at these problems much as one looks at the basic limit theorems in Calculus—the limit of a sum is the sum of the limits, etc. They are mostly used to prove that various functions are continuous. After a while, one gets a certain intuitive understanding of continuity and seldom bothers to give a detailed $\varepsilon-\delta$ argument. Here too, after these exercises, we will often assert that "obviously f is recursive" without a proof. The implication is that the recursiveness of f can be established routinely by the methods of this section.

A k-ary relation ω, $P \subseteq \omega^k$ is *recursive* if its characteristic function χ_P is recursive, where

$$\chi_P(x) = \begin{cases} 1 & \text{if} \quad P(x), \\ 0 & \text{if} \quad \neg P(x). \end{cases}$$

Intuitively, P is recursive if we have an effective way of deciding for each x whether $P(x)$ or $\neg P(x)$ holds—we simply compute $\chi_P(x)$. Recursive relations are also a good tool to use in proving that various functions are recursive.

EXERCISES

3A.1. Prove that if $g(x)$ is recursive, where x varies over ω^k and $f(x, y)$ is defined by

$$f(x, y) = g(x),$$

with y varying over ω^l, then f is recursive. (*Addition of inert variables.*)

Prove that if π is a permutation of $\{1,\dots, k\}$ and $g(x_1,\dots, x_k)$ is recursive, then so is f defined by

$$f(x_1,\dots, x_k) = g(x_{\pi(1)},\dots, x_{\pi(k)}).$$

(*Permutation of variables.*)

Hint. Use composition and the projection functions. ⊣

3A.2. Prove that the following functions are recursive.

*1. $$f(k, n) = k^n \quad (=1 \quad \text{if} \quad k = n = 0).$$

Hint. By 3A.1 it is enough to show that

$$g(n, k) = k^n$$

is recursive and for this we have the primitive recursion

$$g(0, k) = 1,$$
$$g(n+1, k) = g(n, k) \cdot k = h(g(n, k), n, k),$$

where

$$h(u, n, k) = u \cdot k$$

is recursive by 3A.1, since multiplication is recursive.

We will not bother to indicate the necessary applications of 3A.1 in the hints below.

*2. $$k \dot{-} n = \begin{cases} k - n & \text{if} \quad k \geq n \\ 0 & \text{if} \quad k < n. \end{cases} \quad \textit{(arithmetic subtraction)}$$

Hint.

$$\begin{cases} k \dot{-} 0 = k \\ k \dot{-} (n+1) = pd(k \dot{-} n). \end{cases}$$

*3. $$max(x_1, \ldots, x_k) = \textit{the largest of } x_1, \ldots, x_k.$$

Hint. Usc induction on k to prove that each of these functions is recursive.

$$max(x_1, x_2) = (x_1 \dot{-} x_2) + x_2,$$
$$max(x_1, \ldots, x_k, x_{k+1}) = max(max(x_1, \ldots, x_k), x_{k+1}).$$

*4. $$min(x_1, \ldots, x_k) = \textit{the smallest of } x_1, \ldots, x_k.$$

Hint. As in *4, starting with

$$min(x_1, x_2) = x_1 + x_2 \dot{-} max(x_1, x_2).$$

*5. $$sg(n) = \begin{cases} 0 & \text{if} \quad n = 0 \\ 1 & \text{if} \quad n > 0. \end{cases}$$

Hint. $sg(n) = 1 \dot{-} (1 \dot{-} n)$.

*6.
$$\overline{sg}(n) = \begin{cases} 1 & \text{if } n = 0, \\ 0 & \text{if } n > 0. \end{cases}$$

Hint. $\overline{sg}(n) = 1 \dot{-} n$.

*7.
$$\begin{aligned} |n - k| &= \textit{absolute value of the difference of } n, k \\ &= (n \dot{-} k) + (k \dot{-} n). \end{aligned}$$

*8.
$$\begin{aligned} [n/k] &= \textit{the unique } q \textit{ such that for some } r < k, \\ & \quad n = qk + r \quad \textit{if} \quad n \geq k > 0, \\ &= 0 \quad \textit{if} \quad n < k \quad \textit{or} \quad k = 0. \end{aligned}$$

Hint. $[n/k] = sg(k) \cdot \mu q[(n \cdot qk) \dot{-} (k \dot{-} 1) = 0] \cdot \overline{sg}(k \dot{-} n)$.

*9.
$$\begin{aligned} rm(n, k) &= \textit{the unique } r < k \textit{ such that for some } q, \\ & \quad n = qk + r \quad \textit{if} \quad n, k > 0 \\ &= 0 \quad \textit{if} \quad n = 0 \quad \textit{or} \quad k = 0. \end{aligned}$$

Hint. $rm(n, k) = sg(k) \cdot \overline{sg}(k \dot{-} n) \cdot [n \dot{-} [n/k] \cdot k] + sg(k \dot{-} n) \cdot n$. ⊣

3A.3. Prove that the relations

$$n = m, \quad n \leq m, \quad n < m$$

are recursive and that the class of recursive relations is closed under the operations $\neg$, &, $\vee$, $\rightarrow$, $\exists^{\leq}$, $\forall^{\leq}$ and substitution of recursive functions.

Hint. The first assertion is trivial, e.g. the characteristic function of $=$ is

$$\chi_{=}(n, m) = \overline{sg}\,|n - m|.$$

Closure under the propositional operations is also easy, e.g. if $P(x)$, $Q(x)$ are given and

$$R(x) \Leftrightarrow P(x) \,\&\, Q(x), \; S(x) \Leftrightarrow P(x) \vee Q(x),$$

then

$$\begin{aligned} \chi_R(x) &= \chi_P(x) \cdot \chi_Q(x) \\ \chi_S(x) &= sg(\chi_P(x) + \chi_Q(x)). \end{aligned}$$

If $P = \exists^{\leq} Q$, so that

$$P(x, n) \Leftrightarrow (\exists m \leq n) Q(x, m),$$

define $\chi_P(x, n)$ by the recursion

$$\chi_P(x, 0) = \chi_Q(x, 0)$$
$$\chi_P(x, n+1) = sg(\chi_P(x, n) + \chi_Q(x, n+1)).$$

Finally, if

$$P(x) \Leftrightarrow Q(f_1(x), \ldots, f_m(x))$$

with the f_i recursive, then

$$\chi_P(x) = \chi_Q(f_1(x), \ldots, f_m(x)). \qquad \dashv$$

3A.4. Prove that if $P_1(x), \ldots, P_m(x)$ are recursive relations and $f_1(x), \ldots, f_m(x)$, $f_{m+1}(x)$ are recursive functions, then f defined below *by cases* is recursive.

$$f(x) = \begin{cases} f_1(x) & \text{if } P_1(x), \\ f_2(x) & \text{if } \neg P_1(x) \,\&\, P_2(x), \\ \cdots & \cdots \\ f_m(x) & \text{if } \neg P_1(x) \,\&\, \neg P_2(x) \,\&\, \cdots \,\&\, \neg P_{m-1}(x) \,\&\, P_m(x), \\ f_{m+1}(x) & \text{otherwise.} \end{cases}$$

Hint. Taking $m = 1$ with $P = P_1$,

$$f(x) = f_1(x) sg\chi_P(x) + f_2(x)\overline{sg}\chi_P(x). \qquad \dashv$$

3A.5. Prove that $f : \mathfrak{X} \to \omega$ is recursive if and only if the graph of f, $Graph(f) = \{(x, n) : f(x) = n\}$ is recursive. $\dashv$

This trivial observation is a very useful tool for proving the recursiveness of functions using 3A.3 above.

3A.6. Prove that the following functions and relations are recursive.

*10. $\quad$ *Divides*$(m, n) \Leftrightarrow$ *n divides m.*

Hint. *Divides*$(m, n) \Leftrightarrow rm(m, n) = 0$.

*11. $\quad$ *Prime*$(m) \Leftrightarrow$ *m is a prime number.*

Hint. *Prime*$(m) \Leftrightarrow m > 1 \,\&\, (\forall n \le m)[\neg Divides(m, n) \vee n = 0 \vee n = 1 \vee n = m]$. The relation

$$P(m, n) \Leftrightarrow \neg Divides(m, n) \vee n = 0 \vee n = 1 \vee n = m$$

is obviously recursive, hence so is $Q(m, k)$ defined by

$$Q(m, k) \Leftrightarrow (\forall n \le k) P(m, n).$$

Now

$$Prime(m) \Leftrightarrow m > 1 \;\&\; Q(m, m).$$

This is the standard way of treating restricted quantifiers which are applied simultaneously with various substitutions.

*12. $$p(i) = p_i = \textit{the } i\textit{'th prime}.$$

Hint.

$$\begin{cases} p_0 = 2 \\ p_{i+1} = \mu n[Prime(n) \;\&\; n \ge p_i + 1]. \end{cases}$$

*13. $$\langle t_0, \dots, t_{n-1} \rangle = p_0^{t_0+1} \cdot \cdots \cdot p_{n-1}^{t_{n-1}+1}.$$

Recall that this is defined even when $n = 0$,

$$\langle\ \rangle = 1.$$

Hint. There are infinitely many functions here, one for each n. Show by induction on n that each is recursive,

$$\langle t_0, \dots, t_n \rangle = \langle t_0, \dots, t_{n-1} \rangle \cdot p_n^{t_n+1}.$$

*14. $$Seq(u) \Leftrightarrow \text{for some } t_0, \dots, t_{n-1},\ u = \langle t_0, \dots, t_{n-1} \rangle.$$

Hint.

$$\begin{aligned} Seq(u) \Leftrightarrow u > 0 & \\ & \& \; (\forall m \le u)(\forall s \le m)\{[Prime(m) \;\&\; Prime(s) \\ & \& \; Divides(u, m)] \Rightarrow Divides(u, s)\}. \end{aligned}$$

*15. $$\begin{aligned} lh(u) &= n \quad \text{if} \quad u = \langle t_0, \dots, t_{n-1} \rangle \quad \text{for some} \quad n \ge 1, \\ &= 0 \quad \text{otherwise.} \end{aligned}$$

Hint.

$$\begin{aligned} lh(u) = n \Leftrightarrow & [(u = 1 \vee \neg Seq(u)) \;\&\; n = 0] \\ & \vee [Seq(u) \;\&\; u > 1 \;\&\; (\forall m < n) Divides(u, p_m) \\ & \quad \& \; \neg Divides(u, p_n)]. \end{aligned}$$

*16. $$\begin{aligned} (u)_i = t_i \quad & \text{if} \quad u = \langle t_0, \dots, t_{n-1} \rangle \quad \text{for some} \quad t_0, \dots, t_{n-1} \\ & \text{and} \quad 0 \le i \le n-1, \\ = 0 \quad & \text{otherwise.} \end{aligned}$$

Hint.

$$(u)_i = t \Leftrightarrow [Seq(u) \& i < lh(u) \& Divides(u, p_i^{t+1}) \\ \& \neg Divides(u, p_i^{t+2})] \\ \vee [(\neg Seq(u) \vee i \geq lh(u)) \& t = 0].$$

*17. $$u \upharpoonright t = \langle (u)_0, \ldots, (u)_{t-1} \rangle.$$

Hint.

$$u \upharpoonright t = z \Leftrightarrow Seq(z) \& lh(z) = t \\ \& (\forall i < t)[(z)_i = (u)_i].$$

*18. $$u * v = \langle (u)_0, \ldots, (u)_{lh(u) \dot{-} 1}, (v)_0, \ldots, (v)_{lh(v) \dot{-} 1} \rangle.$$

Hint. Show that the graph is recursive. ⊣

The functions in *13–*18 allow us to deal effectively with finite sequences of integers.

3A.7. Suppose $g(u, x)$ is recursive and $f(n, x)$ satisfies the equation

$$f(n, x) = g(\langle f(0, x), \ldots, f(n-1, x) \rangle, x),$$

where for $n = 0$, $\langle\ \rangle = 1$ by convention, so

$$f(0, x) = g(1, x).$$

Prove that f is recursive. (*Definition by complete recursion.*)

Hint. Define $h(n, x)$ by the primitive recursion

$$h(0, x) = \langle g(1, x) \rangle$$
$$h(n+1, x) = h(n, x) * \langle g(h(n, x), x \rangle$$

and verify that

$$f(n, x) = (h(n, x))_n.$$ ⊣

3A.8. Suppose g_1, h_1, g_2, h_2 are all recursive and f_1, f_2 are defined by the *simultaneous recursion*

$$f_1(0, x) = g_1(x), \qquad f_2(0, x) = g_2(x),$$
$$f_1(n+1, x) = h_1(f_1(n, x), f_2(n, x), n, x),$$
$$f_2(n+1, x) = h_2(f_1(n+1, x), f_2(n, x), n, x).$$

Prove that both f_1 and f_2 are recursive.

Hint. Show that the function

$$f(n, x) = \langle f_1(n, x), f_2(n, x)\rangle$$

is recursive. ⊣

3A.9. Enumerate the rational numbers by the function

$$r_i = (-1)^{(i)_0} \cdot \frac{(i)_1}{(i)_2 + 1}.$$

Prove that *addition is recursive* in this coding of the rationals, i.e. there is a recursive $f(i, j)$ such that

$$r_i + r_j = r_{f(i,j)}.$$

Do the same for subtraction, multiplication and division, where for simplicity

$$\frac{r_i}{0} = 0.$$ ⊣

3B. Recursive presentations

Suppose $\mathfrak{M}$ is a separable metric space with distance function d. A *recursive presentation* of $\mathfrak{M}$ is any sequence

$$\{r_0, r_1, \ldots\}$$

of points in $\mathfrak{M}$ satisfying the following two conditions.

(1) The set $\{r_0, r_1, \ldots\}$ is dense in $\mathcal{M}$.

(2) The relations

$$P(i, j, m, k) \Leftrightarrow d(r_i, r_j) \leq \frac{m}{k+1},$$

$$Q(i, j, m, k) \Leftrightarrow d(r_i, r_j) < \frac{m}{k+1}$$

are recursive.

Not every separable metric space admits a recursive presentation—but every interesting space certainly does. Consider first the basic examples ω, $\mathcal{R}$, $\mathcal{N}$, $\mathcal{C}$.

In the case of ω we have the trivial distance function

$$d(i, j) = \begin{cases} 0 & \text{if } i = j, \\ 1 & \text{if } i \neq j. \end{cases}$$

Take $r_i = i$ so that

$$d(i, j) \leq \frac{m}{k+1} \Leftrightarrow [i = j \vee k+1 \leq m],$$

$$d(i, j) < \frac{m}{k+1} \Leftrightarrow [i = j \vee k+1 < m].$$

For the real numbers, choose any effective enumeration of the rationals, where repetitions are allowed, e.g.

$$r_i = (-1)^{(i)_0} \cdot \frac{(i)_1}{(i)_2 + 1}.$$

Proof that this is a recursive presentation is routine by the methods of 3A.

For Baire space, recall that

$$d(\alpha, \beta) = \begin{cases} 0 & \text{if } \alpha = \beta, \\ \dfrac{1}{\mu n[\alpha(n) \neq \beta(n)] + 1} & \text{if } \alpha \neq \beta. \end{cases}$$

Here we need an effective enumertion of all ultimately zero sequences of integers, e.g.

$$r_i(n) = (i)_n.$$

Again, the fact that this is a recursive presentation is easy.

Similarly, for the Cantor set $\mathscr{C}$ we take all ultimately 0 binary sequences,

$$r_i(n) = sg((i)_n).$$

(For the definition of sg see *5 of 3A.2.)

Recall that in Section 1B we fixed once and for all a collection $\mathfrak{F}$ of basic spaces including ω, $\mathscr{R}$, $\mathscr{N}$ and $\mathscr{C}$. *We now assume further that we are given a fixed recursive presentation*

$$\{r_0^{\mathfrak{M}}, r_1^{\mathfrak{M}}, \ldots\}$$

for each basic space $\mathfrak{M}$. For ω, $\mathscr{R}$, $\mathscr{N}$ and $\mathscr{C}$ we take the presentations given above.

Suppose $\mathfrak{X} = X_1 \times \cdots \times X_k$ and we are given metrics $d_1, \ldots, d_k$ and recursive presentations $\{r_0^1, r_1^1, \ldots\}$, $\{r_0^2, r_1^2, \ldots\}, \ldots$ of the spaces $X_1, \ldots, X_k$. It is well-known that the function

$$d((x_1,\ldots,x_k),(x_1',\ldots x_k')) = maximum\{d_1(x_1,x_1'),\ldots,d_k(x_k,x_u')\}$$

is a metric on $\mathfrak{X}$ which generates the natural product topology. For each $i \in \omega$, put

$$r_i = (r^1_{(i)_1},\ldots,r^k_{(i)_k});$$

we leave it for 3B.3 that $\{r_0, r_1,\ldots\}$ is a recursive presentation of $\mathfrak{X}$ with this metric.

We now have a fixed recursive presentation for each product space. If X is a basic space, $x_0 \in X$ and p is any rational number ≥ 0, let $B(x_0, p)$ be the open ball with center x_0 and radius p,

$$B(x_0, p) = \{x \in X: d(x, x_0) < p\}.$$

Taking $\{r_0, r_1, \ldots\}$ to be the fixed recursive presentation of X, put for each $s \in \omega$

$$B_s = B(X, s) = B\left(r_{(s)_0}, \frac{(s)_1}{(s)_2+1}\right).$$

Clearly $B_0, B_1,\ldots$ is an effective enumeration of a basis for the topology of X. Notice that the empty set occurs in this enumeration. In fact, $B_s = \emptyset$ whenever $(s)_1 = 0$, in particular

$$B_0 = \emptyset.$$

For product spaces it is easier to work directly with the natural nbhd basis for the product topology. For each $\mathfrak{X} = X_1 \times \cdots \times X_k$ and each $s \in \omega$, put

$$N_s = N(\mathfrak{X}, s) = B(X_1,(s)_1) \times B(X_2,(s)_2) \times \cdots \times B(X_k,(s)_k)$$
$$= \{(x_1,\ldots,x_k): x_1 \in B_{(s)_1} \,\&\, \cdots \,\&\, x_k \in B_{(s)_k}\}.$$

Now $N_0, N_1,\ldots$ is an effective enumeration of a nbhd basis for the topology of $\mathfrak{X}$. Notice again,

$$N_0 = \emptyset.$$

In several constructions in the first two chapters we used some enumeration of a nbhd basis for the topology of a product space. *We now fix once and for all the canonical basis of nbhds*

$$N(\mathfrak{X}, 0), N(\mathfrak{X}, 1),\ldots$$

associated with the fixed recursive presentation of $\mathfrak{X}$.

Sometimes we needed a *center* and a *radius* for the basic nbhds, as we described and used these in Section 1A. We naturally put

$$center(N_s) = r_i = (r^1_{(i)_1}, \ldots, r^k_{(i)_k}),$$

where

$$(i)_1 = ((s)_1)_0, \ldots, (i)_k = ((s)_k)_0,$$

and

$$radius(N_s) = maximum\{p_1, \ldots, p_k\},$$

where

$$p_1 = \frac{((s)_1)_1}{((s)_1)_2 + 1}, \ldots, p_k = \frac{((s)_k)_1}{((s)_k)_2 + 1}.$$

There is one slightly annoying technical detail we should clear up here. We have identified each basic space X with the "product space" $\mathfrak{X} = X$ whose only factor is X. Now we have described two bases for the topology of X, the sequence

$$B(X, 0), B(X, 1), \ldots$$

of open balls, thinking of X as basic, and the sequence

$$N(\mathfrak{X}, 0), N(\mathfrak{X}, 1), \ldots$$

of "products of one factor,"

$$N(X, s) = B(X, (s)_1).$$

Of course the bases are identical, but the *enumeration* is different as the last displayed equation plainly shows. Notice also that

$$B(X, s) = N(X, \langle 0, s \rangle).$$

We need two simple lemmas to deal effectively with these codings.

3B.1. LEMMA. *For each two product spaces $\mathfrak{X}$, $\mathfrak{Y}$, there are recusive functions f, g, h such that*

$$N(\mathfrak{X}, s) \times N(\mathfrak{Y}, t) = N(\mathfrak{X} \times \mathfrak{Y}, f(s, t)),$$
$$N(\mathfrak{X} \times \mathfrak{Y}, s) = N(\mathfrak{X}, g(s)) \times N(\mathfrak{Y}, h(s)).$$

PROOF. If $\mathfrak{X} = X_1 \times \cdots \times X_k$ and $\mathfrak{Y} = Y_1 \times \cdots \times Y_l$, then

$$N(\mathfrak{X} \times \mathfrak{Y}, s) = B(X_1, (s)_1) \times \cdots \times B(X_k, (s)_k) \times B(Y_1, (s)_{k+1}) \times \cdots \times B(Y_l, (s)_{k+l});$$

from this follows immediately that we can take

$$f(s,t)=\langle 0,(s)_1,\dots,(s)_k,(t)_1,\dots,(t)_l\rangle,$$
$$g(s)=\langle 0,(s)_1,\dots,(s)_k\rangle,$$
$$h(s)=\langle 0,(s)_{k+1},\dots,(s)_{k+l}\rangle. \quad \dashv$$

The second one is a bit messier.

3B.2. LEMMA. *For each product space $\mathfrak{X}$, there is a recursive function f such that*

$$N(\mathfrak{X},s)\cap N(\mathfrak{X},t)=\bigcup_n N(\mathfrak{X},f(s,t,n)).$$

Similarly, there is a recursive function g, such that

$$\bigcap_{i\le m} N(\mathfrak{X},(u)_i)=\bigcup_n N(\mathfrak{X},g(u,m,n)).$$

PROOF. We show the second assertion first.

Let X be a basic space with the recursive presentation $\{r_0, r_1,\dots\}$ and let $B(x_0,p_0),\dots,B(x_m,p_m)$ be $m+1$ open balls in X. Then

$$x\in B(x_0,p_0)\cap\cdots\cap B(x_m,p_m)\leftrightarrow(\exists i)(\exists k)\Big\{d(r_i,x)<\frac{(k)_1}{(k)_2+1}$$
$$\&\ d(x_0,r_i)<p_0-\frac{(k)_1}{(k)_2+1}$$
$$\&\ \cdots$$
$$\&\ d(x_m,r_i)<p_m-\frac{(k)_1}{(k)_2+1}\Big\};$$

the implication from right to left is trivial, while if the left-handside holds, then

$$A=\Big\{z:(\exists k)\Big[d(z,x)<\frac{(k)_1}{(k)_2+1}\ \&\ d(x_0,z)<p_0-\frac{(k)_1}{(k)_2+1}\ \&\cdots$$
$$\&\ d(x_m,z)<p_m-\frac{(k)_1}{(k)_2+1}\Big]\Big\}$$

is open and nonempty (since $x\in A$), so A must contain some r_i. Using this equivalence and the definition of a recursive presentation, it is easy to see that there is a recursive relation $P(s, m, n)$ such that

$$x \in B(X, s_0) \cap \cdots \cap B(x, s_m) \leftrightarrow (\exists n)[x \in B(X, n) \;\&\; P(\langle s_0, \ldots, s_m\rangle, m, n)],$$

i.e.

$$(*) \qquad \bigcap\nolimits_{i \le m} B(X, s_i) = \bigcup\nolimits_n \{B(X, n) : P(\langle s_0, \ldots, s_m\rangle, m, n)\}.$$

Suppose now that

$$\mathfrak{X} = X_1 \times \cdots \times X_k$$

and let $P_1, \ldots, P_k$ be recursive relations so that $(*)$ holds with $X_1, \ldots, X_k$ respectively. Using the definition of the coding, we compute:

$$\begin{aligned}\bigcap\nolimits_{i \le m} N(\mathfrak{X}, (u)_i) &= \bigcap\nolimits_{i \le m} [B(X_1, ((u)_i)_1) \times \cdots \times B(X_k, ((u)_i)_k)] \\ &= [\bigcap\nolimits_{i \le m} B(X_1, ((u)_i)_1] \times \cdots \times [\bigcap\nolimits_{i \le m} B(X_k, ((u)_i)_k)].\end{aligned}$$

Now for each $j = 1, \ldots, k$,

$$\begin{aligned}\bigcap\nolimits_{i \le m} B(X_j, ((u)_i)_j) &= \bigcup\nolimits_n \{B(X_j, n) : P_j(\langle((u)_0)_j, \ldots, ((u)_m)_j\rangle, m, n)\} \\ &= \bigcup\nolimits_n \{B(X_j, n) : P_j^*(u, m, n)\}\end{aligned}$$

with an obvious recursive P_j^*, hence

$$\begin{aligned}\bigcap\nolimits_{i \le m} N(\mathfrak{X}, (u)_i) &= [\bigcup\nolimits_n \{B(X_1, n) : P_1^*(u, m, n)\}] \\ &\qquad \times \cdots \times [\bigcup\nolimits_n \{B(X_k, n) : P_k^*(u, m, n)\}] \\ &= \bigcup\nolimits_n \{B(X_1, (n)_1) \times \cdots \times B(X_k, (n)_k) : \\ &\qquad P_1^*(u, m)(n)_1) \;\&\; \cdots \;\&\; P_k^*(u, m, (n)_k)\} \\ &= \bigcup\nolimits_n \{N(\mathfrak{X}, n) : P^*(u, m, n)\},\end{aligned}$$

with some recursive P^*. The result follows by setting

$$g(u, m, n) = \begin{cases} n & \textit{if} \quad P^*(u, m, n), \\ 0 & \textit{otherwise.}\end{cases}$$

The first assertion follows immediately, taking

$$f(s, t, n) = g(\langle s, t\rangle, 1, n). \qquad \dashv$$

EXERCISES

3B.3. Prove that if $\{r_0^j, r_1^j, \ldots\}$ is a recursive presentation of X_j for $j = 1, \ldots, k$, then the sequence

$$r_i = (r^1_{(i)_1}, r^2_{(i)_2}, \ldots, r^k_{(i)_k})$$

is a recursive presentation of $\mathfrak{X} = X_1 \times \cdots \times X_k$.

Hint. You must show that the relation

$$P(i, j, m, l) \Leftrightarrow d(r_i, r_j) \leq \frac{m}{l+1}$$
$$\Leftrightarrow maximum\{d_1(r^1_{(i)_1}, r^1_{(j)_1}, \ldots, d_k(r^k_{(i)_k}, r^k_{(j)_k})\} \leq \frac{m}{l+1}$$

is recursive, and similarly with $<$. See 3A.9. ⊣

3B.4. Let $\mathfrak{X} = X_1 \times \cdots \times X_k$ be a product space and let $N_s = N(\mathfrak{X}, s)$ be the basic nbhd with code $s \in \omega$. Prove that the relations

$$P(s, m, l) \Leftrightarrow radius(N_s) \leq \frac{m}{l+1},$$

$$Q(s, m, l) \Leftrightarrow radius(N_s) < \frac{m}{l+1}$$

are both recursive.

Prove also that

$$center(N_s) = r_{f(s)}$$

where $\{r_0, r_1, \ldots\}$ is the recursive presentation of $\mathfrak{X}$ and f is some recursive function. ⊣

3B.5. Prove that there are recursive functions $g : \omega \to \omega$ and $h : \omega^2 \to \omega$ such that

$$\alpha \in N(\mathcal{N}, s) \Leftrightarrow ((s)_1)_1 \neq 0 \;\&\; (\forall i < g(s))[\alpha(i) = h(s, i)].$$

Hint. The idea is that by the definitions,

$$B(\mathcal{N}, t) = \emptyset \quad \text{if} \quad (t)_1 = 0$$

and for $(t)_1 \neq 0$,

$$B(\mathcal{N}, t) = \{\alpha : \alpha(0) = k_0, \alpha(1) = k_1, \ldots, \alpha(l-1) = k_{l-1}\}$$

with $l, k_0, \ldots, k_{l-1}$ effectively computable from t—where in the case $l = 0$, we have $B(\mathcal{N}, t) = \mathcal{N}$. Write

$$l = l(t),$$
$$k_i = k(t, i)$$

with suitable recursive functions and take

$$g(s) = l((s)_1),$$
$$h(s, i) = k((s)_1, i).$$

3B.6. Find a recursive presentation for your favorite perfect Polish space, e.g. C[0, 1] or $H[0, 1]$, as we defined these in the exercises of 1A. ⊣

3C. Semirecursive pointsets[2,3]

A pointset $G \subseteq \mathcal{X}$ is *semirecursive* if

$$G = \bigcup_n N(\mathcal{X}, \varepsilon(n))$$

with some *recursive irrational* ε, i.e. with an irrational ε such that the function

$$n \mapsto \varepsilon(n)$$

is recursive. Intuitively, G is semirecursive if it can be written as *a recursive union of basic nbhds.*

The definition suggests that the pointclass of semirecursive sets depends on the particular *coding* of basic nbhds by integers which we fixed. We will see in the exercises that this is not so. It is obvious, however, that the notion of semirecursion depends on the particular *recursive presentations* of the basic spaces which we adopted.

It is natural to consider the family of all semircursive subsets of $\mathcal{X}$ as a *recursive topology* on $\mathcal{X}$. It is not closed under arbitrary unions, but it has strong closure properties, as we now proceed to show.

If $\mathcal{X} = X_1 \times \cdots \times X_k$, let

$$\pi_i : \mathcal{X} \to X_i$$

be *the projection function,*

$$\pi_i(x_1, \ldots, x_k) = x_i.$$

A function

$$f : \mathcal{X} \to \mathcal{Y} = Y_1 \times \cdots \times Y_l$$

is *trivial* if there are projection functions

$$f_1 : \mathcal{X} \to Y_1, \ldots, f_l : \mathcal{X} \to Y_l$$

such that

$$f(x) = (f_1(x), \ldots, f_l(x)).$$

For example, the map

$$(x_1, x_2, x_3, x_4) \mapsto (x_2, x_1, x_1)$$

of $X_1 \times X_2 \times X_3 \times X_4$ into $X_2 \times X_1 \times X_1$ is trivial.

3C.1. THEOREM. *The pointclass of semirecursive sets contains the empty set, every product space* $\mathfrak{X}$, *every recursive relation on* ω^k, *every basic nbhd* $N(\mathfrak{X}, s)$ *and the basic nbhd relation*

$$\{(x, s): x \in N(\mathfrak{X}, s)\}$$

for each $\mathfrak{X}$; *moreover, it is closed under substitution of trivial functions,* &, $\vee$, $\exists^{\leq}$, $\forall^{\leq}$ *and* $\exists^{\omega}$.

PROOF. Notice first that if

$$P = \bigcup \{N(\mathfrak{X}, f(n)): P^*(n)\}$$

where f is a recursive function and P^* a recursive relation, then P is semirecursive, since

$$P = \bigcup_n N(\mathfrak{X}, \varepsilon(n))$$

with

$$\varepsilon(n) = \begin{cases} f(n) & \text{if} \quad P^*(n), \\ 0 & \text{otherwise.} \end{cases}$$

This will help a little in the computations below.

Clearly

$$\emptyset = \bigcup_n N(\mathfrak{X}, 0),$$
$$\mathfrak{X} = \bigcup_n N(\mathfrak{X}, n),$$

so both $\emptyset$ and $\mathfrak{X}$ are semirecursive.

By the definition of the coding of nbhds in ω,

$$B(\omega, \langle i, 1, 1\rangle) = \left\{m: d(m, i) < \frac{1}{1+1}\right\} = \{i\},$$

hence each singleton is a basic nbhd in ω^k:

$$\{(n_1, \ldots, n_k)\} = N(\omega^k, \langle 0, \langle n_1, 1, 1\rangle, \ldots, \langle n_k, 1, 1\rangle\rangle).$$

If $R \subseteq \omega^k$ is recursive, then

$$R = \bigcup \{N(\omega^k, \langle 0, \langle (n)_1, 1, 1\rangle, \ldots, \langle (n)_k, 1, 1\rangle\rangle): R((n)_1, \ldots, (n)_k)\},$$

so R is semirecursive.

Again,

$$N(\mathfrak{X}, s) = \bigcup_n N(\mathfrak{X}, s),$$

so $N(\mathfrak{X}, s)$ is semirecursive since the constant function $n \mapsto s$ is recursive.

To check that $\{(x, s): x \in N(\mathfrak{X}, s)\}$ is semirecursive, notice that

$$\begin{aligned} N(\mathfrak{X}, s)\times\{s\} &= N(\mathfrak{X}, s)\times N(\omega, \langle 0, \langle s, 1, 1\rangle\rangle) \\ &= N(\mathfrak{X}\times\omega, f(s, \langle 0, \langle s, 1, 1\rangle\rangle)) \end{aligned}$$

using the recursive function f of 3B.1, so that

$$\{(x, s): x \in N(\mathfrak{X}, s)\} = \bigcup_s N(\mathfrak{X}\times\omega, f(s, \langle 0, \langle s, 1, 1\rangle\rangle)).$$

Going to the closure properties, suppose first that

$$P = \bigcup_n N(\mathfrak{X}, \alpha(n)),\ Q = \bigcup_m N(\mathfrak{X}, \beta(m))$$

with both α and β recursive. Then

$$[\bigcup_n N(\mathfrak{X}, \alpha(n))]\cup[\bigcup_m N(\mathfrak{X}, \beta(m))] = \bigcup_t N(\mathfrak{X}, \gamma(t)),$$

where γ enumerates the union of the sets enumerated by α and β,

$$\begin{aligned} \gamma(2k) &= \alpha(k), \\ \gamma(2k+1) &= \beta(k). \end{aligned}$$

Similarly,

$$\begin{aligned} P\cap Q &= [\bigcup_n N(\mathfrak{X}, \alpha(n))]\cap[\bigcup_m N(\mathfrak{X}, \beta(m))] \\ &= \bigcup_{n,m} [N(\mathfrak{X}, \alpha(n))\cap N(\mathfrak{X}, \beta(m))] \\ &= \bigcup_{n,m,s} N(\mathfrak{X}, f(\alpha(n), \beta(m), s)) \end{aligned}$$

where f is the recursive function of 3B.2; thus

$$P\cap Q = \bigcup_t N(\mathfrak{X}, f(\alpha((t)_0), \beta((t)_1), (t)_2)),$$

and $P\cap Q$ is semirecursive. This establishes closure under $\vee$ and &.

To prove closure under $\exists^\omega$, suppose

$$P(x) \Leftrightarrow (\exists m)Q(x, m)$$

and

$$Q = \bigcup_n N(\mathfrak{X}\times\omega, \varepsilon(n));$$

then

$$\begin{aligned} P(x) &\Leftrightarrow (\exists m)(\exists n)[(x, m)\in N(\mathfrak{X}\times\omega, \varepsilon(n))] \\ &\Leftrightarrow (\exists m)(\exists n)[x\in N(\mathfrak{X}, g(\varepsilon(n)))\ \&\ m\in N(\omega, h(\varepsilon(n)))] \end{aligned}$$

where g, h are recursive by 3B.1. The relation

$$R(m, n) \Leftrightarrow m\in N(\omega, h(\varepsilon(n)))$$

is easily proved recursive, so we have

$$P(x) \Leftrightarrow (\exists t)[x \in N(\mathfrak{X}, g(\varepsilon((t)_0))) \mathbin{\&} R((t)_1, (t)_0)]$$

and P is semirecursive.

Suppose now that

$$f: X_1 \times \cdots \times X_k \to \mathcal{Y}$$

is trivial, that is

$$f(x_1,\ldots, x_k) = (x_{i_1},\ldots, x_{i_l})$$

where the numbers $i_1,\ldots, i_l$ are between 1 and k. If

$$Q = \bigcup_n N(\mathcal{Y}, \varepsilon(n))$$

and

$$P(x) \Leftrightarrow Q(f(x)),$$

then:

$$\begin{aligned} P(x_1,\ldots, x_k) &\Leftrightarrow (\exists n)[(x_{i_1},\ldots, x_{i_l}) \in N(\mathcal{Y}, \varepsilon(n))] \\ &\Leftrightarrow (\exists n)[x_{i_1} \in B(X_{i_1}, (\varepsilon(n))_1) \mathbin{\&} \cdots \mathbin{\&} X_{i_l} \in B(X_{i_l}, (\varepsilon(n))_l)]. \end{aligned}$$

For fixed j, easily

$$\begin{aligned} x_j \in B(X_j, m) &\Leftrightarrow (\exists t)[x_1 \in B(X_1, (t)_1) \mathbin{\&} \cdots \mathbin{\&} x_j \in B(X_j, m) \\ &\qquad\qquad \mathbin{\&} \cdots \mathbin{\&} x_k \in B(X_k, (t)_k)] \\ &\Leftrightarrow (\exists t)[(x_1,\ldots, x_k) \in N(\mathfrak{X}, f_j(m, t))] \end{aligned}$$

with a recursive function f_j; using the argument which established that $\{(x, s) : x \in N(\mathfrak{X}, s)\}$ is semirecursive, it is easy to verify that each relation

$$R_j(x_1,\ldots, x_k, m, t) \Leftrightarrow (x_1,\ldots, x_k) \in N(\mathfrak{X}, f_j(m, t))$$

is semirecursive, so by closure under $\exists^\omega$ we have

$$P(x_1,\ldots, x_k) \Leftrightarrow (\exists n)[R^*_{i_1}(x_1,\ldots, x_k, n) \mathbin{\&} \cdots \mathbin{\&} R^*_{i_l}(x_1,\ldots, x_k, n)]$$

with suitable semirecursive $R^*_{i_1},\ldots, R^*_{i_l}$, and P is semirecursive by closure under & and $\exists^\omega$.

At this point it becomes easy to combine the results we already have to prove closure under $\exists^\leq$, $\forall^\leq$.

If

$$P(x, n) \Leftrightarrow (\exists i \leq n) Q(x, i)$$

with Q semirecursive, then

$$\begin{aligned} P(x, n) &\Leftrightarrow (\exists i)[i \leq n \mathbin{\&} Q(x, i)] \\ &\Leftrightarrow (\exists i)[R(x, n, i) \mathbin{\&} S(x, n, i)] \end{aligned}$$

where

$$R(x, n, i) \Leftrightarrow i \leq n,$$

$$S(x, n, i) \Leftrightarrow Q(x, i),$$

are both semirecursive by closure under the trivial substitutions $(x, n, i) \mapsto (i, n)$, $(x, n, i) \mapsto (x, i)$ and the semirecursiveness of $\leq$ and Q. Now use closure under & and $\exists^\omega$.

Similarly, if

$$P(x, n) \Leftrightarrow (\forall i \leq n) Q(x, i)$$

with

$$Q = \bigcup_m N(\mathcal{X} \times \omega, \varepsilon(m)),$$

then

$$\begin{aligned} P(x, n) &\Leftrightarrow (\forall i \leq n)(\exists m)[(x, i) \in N(\mathcal{X} \times \omega, \varepsilon(m))] \\ &\Leftrightarrow (\exists s)(\forall i \leq n)[(x, i) \in N(\mathcal{X} \times \omega, \varepsilon((s)_i))] \\ &\Leftrightarrow (\exists s)(\forall i \leq n)[x \in N(\mathcal{X}, f_1(s, i)) \;\&\; i \in N(\omega, f_2(s, i))] \end{aligned}$$

with f_1, f_2 recursive by 3B.1; thus

$$\begin{aligned} P(x, n) \Leftrightarrow (\exists s)(\exists u)\{&(\forall i \leq n)[f_1(s, i) = (u)_i] \\ &\& \; (\forall i \leq n)[x \in N(\mathcal{X}, (u)_i)] \\ &\& \; (\forall i \leq n)[i \in N(\omega, f_2(s, i)]\}. \end{aligned}$$

Now using 3B.2 and rearranging,

$$P(x, n) \Leftrightarrow (\exists u)(\exists s)[(\exists t)[x \in N(\mathcal{X}, g(u, n, t))] \;\&\; R(n, s, u)]$$

with a recursive function g and a recursive R, i.e.

$$\begin{aligned} P(x, n) \Leftrightarrow (\exists u)(\exists s)\{(\exists t)(\exists m)[m = g(u, n, t) \;\&\; x \in N(\mathcal{X}, m)] \\ \& \; R(n, s, u)\} \end{aligned}$$

so P is semirecursive by the closure properties we have established already. ⊣

The proof of 3C.1 was messy, because we were forced to deal directly with the coding of nbhds. In the sequel, we will be able to give fairly simple proofs of semirecursiveness by applying 3C.1 and the remaining results of this section.

It is worth emphasizing the usefulness of closure under trivial substitutions, which allows us to identify, permute or introduce new arguments in relations. For example, suppose

$$P(x, y, x) \Leftrightarrow (\exists n)\{Q(x, n) \,\&\, R(n, y, z)\}$$

with Q, R semirecursive; then

$$P(x, y, z) \Leftrightarrow (\exists n)\{Q^*(x, y, z, n) \,\&\, R^*(x, y, z, n)\}$$

with

$$Q^*(x, y, z, n) \Leftrightarrow Q(x, n),$$
$$R^*(x, y, z, n) \Leftrightarrow R(n, y, z)$$

and since both Q^* and R^* are semirecursive by closure under trivial substitutions, so is P by closure under & and $\exists^\omega$. In Section 1C we were appealing to closure of the finite Borel classes under continuous substitution to do this kind of computation.

We will see in 3D.2 that the pointclass of semirecursive sets is closed under substitution of *recursive functions* on product spaces, which include the trivial functions.

Let us call a product space $\mathfrak{X}$ *of type* 0 if $\mathfrak{X} = \omega^k$ for some k. These are *the discrete* product spaces. A pointset P is *of type* 0 if $P \subseteq \omega^k$ for some k.

3C.2. THEOREM. *A pointset $P \subseteq \omega^k$ of type 0 is semirecursive if and only if there is a recursive relation $R \subseteq \omega^{k+1}$ such that*

$$P(x) \Leftrightarrow (\exists n)R(x, n).$$

Moreover, P is recursive if and only if both P and $\neg P = \omega^k - P$ are semirecursive.

PROOF. Assume first that $P \subseteq \omega^k$ is semirecursive,

$$P = \bigcup_n N(\omega^k, \varepsilon(n)),$$

with ε recursive. Then

$$P(x_1, \ldots, x_k) \Leftrightarrow (\exists n)(\exists m)[\varepsilon(n) = m \,\&\, (x_1, \ldots, x_k) \in N(\omega^k, m)]$$

so P is of the required form, since the relation

$$S(x_1, \ldots, x_k, m) \Leftrightarrow (x_1, \ldots, x_k) \in N(\omega^k, m)$$

is easily proved recursive by direct computation.

Conversely, if

$$P(x_1, \ldots, x_k) \Leftrightarrow (\exists n)R(x_1, \ldots, x_k, n),$$

then

$$P = \bigcup_n \{N(\omega^k, \langle 0, \langle (n)_1, 1, 1\rangle, \ldots, \langle (n)_k, 1, 1\rangle\rangle) : R((n)_1, \ldots, (n)_k, (n)_0)\},$$

so P is semirecursive.

If both P and $\neg P$ are semirecursive, then

$$P(x_1, \ldots, x_k) \Leftrightarrow (\exists n) R(x_1, \ldots, x_k, n),$$
$$\neg P(x_1, \ldots, x_k) \Leftrightarrow (\exists m) S(x_1, \ldots, x_k, m),$$

with both R and S recursive. It follows that for each $x_1, \ldots, x_k$ there is some n such that

$$R(x_1, \ldots, x_k, n) \vee S(x_1, \ldots, x_k, n),$$

so that the function

$$f(x_1, \ldots, x_k) = \mu n[R(x_1, \ldots, x_k, n) \vee S(x_1, \ldots, x_k, n)]$$

is recursive. Now easily

$$P(x_1, \ldots, x_k) \Leftrightarrow R(x_1, \ldots, x_k, f(x_1, \ldots, x_k)),$$

so P is recursive by 3A.3. ⊣

A pointset $P \subseteq \mathcal{X}$ is *recursive* if both P and $\neg P = \mathcal{X} - P$ are semirecursive—by 3C.2 this definition agrees with the one we already have for pointsets of type 0.

When we define recursive functions on arbitrary product spaces in 3D, we will verify that P is recursive precisely when its characteristic function χ_P is recursive. These pointsets are *clopen*, so they are trivial in connected spaces like the reals. They are very important in studying products of ω and $\mathcal{N}$.

A space $\mathcal{X} = X_1 \times \cdots \times X_k$ is *of type* 1 if each X_i is either ω or $\mathcal{N}$ and at least one X_i is $\mathcal{N}$. Again, P is of type 1 if P is a subset of some space $\mathcal{X}$ of type 1.

3C.3. THEOREM. *The pointclass of recursive sets contains the empty set, every product space $\mathcal{X}$, every recursive relation on ω, the pointset*

$$\{(\alpha, n, w) : \alpha(n) = w\}$$

and for each space $\mathcal{X}$ of type 0 or 1 every basic nbhd $N(\mathcal{X}, s)$ as well as the basic nbhd relation $\{(x, s) : x \in N(\mathcal{X}, s)\}$; moreover, it is closed under substitution of trivial functions, $\neg$, &, $\vee$, $\exists^{\leq}$ and $\forall^{\leq}$.

PROOF. The closure properties are immediate from 3C.1 and so are the facts that $\emptyset$, each $\mathfrak{X}$ and each recursive relation on ω are recursive.

Recall from 3B.5 that there are recursive functions g and h such that

$$\alpha \in N(\mathcal{N}, s) \leftrightarrow ((s)_1)_1 \neq 0 \;\&\; (\forall i < g(s))[\alpha(i) = h(s, i)].$$

Thus

$$\alpha(n) = w \Leftrightarrow (\exists s)[\alpha \in N(\mathcal{N}, s) \;\&\; n < g(s) \;\&\; h(s, n) = w],$$

so $\{(\alpha, n, w): \alpha(n) = w\}$ is semirecursive by 3C.1. It is also recursive, since

$$\alpha(n) \neq w \leftrightarrow (\exists m)[m \neq w \;\&\; \alpha(n) = m].$$

Using again 3B.5,

$$\alpha \notin N(\mathcal{N}, s) \Leftrightarrow ((s)_1)_1 = 0 \vee (\exists i < g(s))(\exists w)[\alpha(i) = w \;\&\; w \neq h(s, i)],$$

so $\{(\alpha, s): \alpha \notin N(\mathcal{N}, s)\}$ is semirecursive and hence recursive by 3C.1. The corresponding set for ω is trivially recursive, and then by 3B.1 and closure under &, $\{(x, s): x \in N(\mathfrak{X}, s)\}$ is recursive for every space $\mathfrak{X}$ of type 0 or 1. ⊣

The characterization of 3C.2 extends to pointsets of type 1.

3C.4. THEOREM. *A pointset $P \subseteq \mathfrak{X}$ of type 0 or 1 is semirecursive if and only if there is a recursive $R \subseteq \mathfrak{X} \times \omega$ such that*

$$P(x) \leftrightarrow (\exists n) R(x, n).$$

PROOF. One way is immediate by 3C.1. For the converse, assume that $P \subseteq \mathfrak{X}$ is semirecursive with $\mathfrak{X}$ of type 0 or 1, so that

$$\begin{aligned} P(x) &\leftrightarrow (\exists n)[x \in N(\mathfrak{X}, \varepsilon(n))] \\ &\leftrightarrow (\exists n)(\exists m)[\varepsilon(n) = m \;\&\; x \in N(\mathfrak{X}, m)] \\ &\leftrightarrow (\exists t)[\varepsilon((t)_0) = (t)_1 \;\&\; x \in N(\mathfrak{X}, (t)_1)] \end{aligned}$$

with a recursive ε. Thus it is enough to show that the relation

$$S(x, t) \leftrightarrow x \in N(\mathfrak{X}, (t)_1)$$

is recursive when $\mathfrak{X}$ is of type 0 or 1; it is by 3C.3, since

$$S(x, t) \leftrightarrow (\exists m)[(t)_1 = m \;\&\; x \in N(\mathfrak{X}, m)],$$

$$\neg S(x, t) \leftrightarrow (\exists m)[(t)_1 = m \;\&\; x \notin N(\mathfrak{X}, m)].$$ ⊣

This simple characterization cannot be extended to arbitrary spaces, since it implies that a semirecursive set is a countable union of clopen sets,

$$\{x: (\exists n)R(x, n)\} = \textstyle\bigcup_n \{x: R(x, n)\}.$$

We list for the record some similar, simple normal forms in arbitrary product spaces which are sometimes useful.

3C.5. THEOREM. *A pointset $P \subseteq \mathfrak{X}$ is semirecursive if and only if there is a semirecursive $P^* \subseteq \omega$ such that*

$$P(x) \Leftrightarrow (\exists s)\{x \in N(\mathfrak{X}, s) \;\&\; P^*(s)\}.$$

More generally, $P \subseteq \mathfrak{X} \times \mathfrak{Y}$ is semirecursive if and only if there is some semirecursive $P^* \subseteq \omega^2$ such that

$$P(x, y) \Leftrightarrow (\exists s)(\exists t)\{x \in N(\mathfrak{X}, s) \;\&\; y \in N(\mathfrak{Y}, t) \;\&\; P^*(s, t)\}.$$

More specifically, $P \subseteq \omega \times \mathfrak{X}$ is semirecursive if and only if there is a semirecursive $P^* \subseteq \omega^2$ such that

$$P(n, x) \Leftrightarrow (\exists s)\{x \in N(\mathfrak{X}, s) \;\&\; P^*(n, s)\}.$$

PROOF is implicit in many of the constructions we have been making, particularly in the proof of 3C.1. To take just the last assertion, it is obvious by 3C.1 that any P satisfying such an equivalence is semirecursive. Conversely, if

$$P(n, x) \Leftrightarrow (\exists m)[(n, x) \in N(\omega \times \mathfrak{X}, \varepsilon(m))]$$

with a recursive ε, then by 3B.1

$$P(n, x) \Leftrightarrow (\exists m)\{n \in N(\omega, g(\varepsilon(m))) \;\&\; x \in N(\mathfrak{X}, h(\varepsilon(m)))\}$$

with recursive g, h, so

$$P(n, x) \Leftrightarrow (\exists s)\{x \in N(\mathfrak{X}, s) \;\&\; (\exists m)[s = h(\varepsilon(m)) \;\&\; n \in N(\omega, g(\varepsilon(m)))]\}$$

which is the required representation, since the second conjunct within the braces is obviously semirecursive. ⊣

EXERCISES

3C.6. Prove that for each $\mathfrak{X}$, the relation $Q \subseteq \mathfrak{X} \times \mathfrak{X}$,

$$Q(x, y) \Leftrightarrow x \neq y$$

is semirecursive.

Hint.

$$x \neq y \Leftrightarrow (\exists s)(\exists t)\{x \in N_s \ \&\ y \in N_t \ \&\ radius(N_s) + radius(N_t) < d(center(N_s), center(N_t))\}. \quad \dashv$$

3C.7. Prove that if $\mathfrak{X}$ is of type 0 and $f : \mathfrak{X} \to \omega$ is a function, then f is recursive if and only if the graph of f, $Graph(f) = \{(x, n) : f(x) = m\}$ is semirecursive.

Hint. If f is recursive, use 3C.1 and 3A.5. If $Graph(f)$ is semirecursive, then by 3C.2

$$f(x) = m \Leftrightarrow (\exists n)R(x, m, n)$$

with R recursive and

$$f(x) = (\mu n R(x, (n)_0, (n)_1))_0. \quad \dashv$$

This exercise is often a useful tool for proving that specific functions are recursive.

3C.8. Prove that a set $A \subseteq \omega$ is semirecursive if and only if $A = \emptyset$ or there exists a recursive function $f : \omega \to \omega$ which *enumerates* A, i.e.

$$A = \{f(0), f(1), f(2), \ldots\}. \quad \dashv$$

Because of this result, semirecursive subsets of ω are usually called *recursively enumerable.*

3C.9. Let $\mathfrak{X}$ be a product space with fixed recursive presentation $\{r_0, r_1, \ldots\}$. Prove that the relations

$$P(x, i, m, k) \Leftrightarrow d(x, r_i) < \frac{m}{k+1},$$

$$Q(x, i, m, k) \Leftrightarrow \frac{m}{k+1} < d(x, r_i),$$

are both semirecursive.

Hint.

$$d(x, r_i) < \frac{m}{k+1} \Leftrightarrow (\exists s)\left[x \in N_s \ \&\ center(N_s) = r_i \ \&\ radius(N_s) < \frac{m}{k+1}\right],$$

$$d(x, r_i) > \frac{m}{k+1} \Leftrightarrow (\exists s)\left[x \in N_s \ \&\ \frac{m}{k+1} + radius(N_s) < d(r_i, center(N_s))\right].$$

3C.10. Let $\mathfrak{X}$ be a product space. Prove that the relations

$$P(x, y, m, k) \Leftrightarrow d(x, y) < \frac{m}{k+1},$$

$$Q(x, y, m, k) \Leftrightarrow d(x, y) > \frac{m}{k+1},$$

are both semirecursive. ⊣

3C.11. Prove that the relation $x < y$ on the reals is semirecursive. ⊣

3C.12. Prove that the collection of semirecursive sets is the smallest pointclass which contains all recursive pointsets of type 0 and the relation $P^X \subseteq X \times \omega^3$ for each basic space X,

$$P^X(x, i, m, k) \Leftrightarrow d(r_i, x) < \frac{m}{k+1}$$

and which is closed under trivial substitutions, &, $\vee$, $\exists^{\leq}$, $\forall^{\leq}$ and $\exists^{\omega}$.

Hint. It is enough to show that for each $\mathfrak{X}$, $\{(x, s): x \in N(\mathfrak{X}, s)\}$ must belong to every pointclass with these properties:

$$\begin{aligned} x \in N(\mathfrak{X}, s) &\Leftrightarrow x_1 \in B(X_1, (s)_1) \,\&\, \cdots \,\&\, x_k \in B(X_k, (s)_k) \\ &\Leftrightarrow P^{X_1}(x_1, ((s)_1)_0, ((s)_1)_1, ((s)_1)_2) \\ &\qquad \& \cdots \\ &\qquad \& \, P^{X_k}(x_k, ((s)_k)_0, ((s)_k)_1, ((s)_k)_2). \end{aligned}$$ ⊣

This problem shows that the definition of semirecursive sets does not depend in any essential way on the coding of nbhds we chose.

3C.13. Prove that a pointset $P \subseteq \mathfrak{X}$ is open if and only if there is a semirecursive $Q \subseteq \mathcal{N} \times \mathfrak{X}$ and an irrational ε such that

$$P(x) \Leftrightarrow Q(\varepsilon, x).$$ ⊣

3C.14. Prove that the pointclass of semirecursive sets is closed under $\exists^{\mathcal{Y}}$ for every product space $\mathcal{Y}$.

Hint. Suppose $P \subseteq \mathfrak{X} \times Y$ with Y basic and P is semirecursive,

$$P(x, y) \Leftrightarrow (\exists n)[(x, y) \in N(\mathfrak{X} \times Y, \varepsilon(n))]$$

with a recursive ε. Using 3B.1, there are then recursive functions f_1, f_2 such that

$$P(x, y) \Leftrightarrow (\exists n)[x \in N(\mathcal{X}, f_1(n)) \; \& \; y \in N(Y, f_2(n))],$$

so that

$$(\exists y)P(x, y) \Leftrightarrow (\exists n)[x \in N(\mathcal{X}, f_1(n)) \; \& \; (\exists y)[y \in N(Y, f_2(n))]].$$

But

$$(\exists y)[y \in N(Y, f_2(n))] \Leftrightarrow (\exists y)[y \in B(Y, (f_2(n))_1]$$
$$\Leftrightarrow radius(B(Y, (f_2(n))_1) > 0,$$

so this relation is recursive.

Show closure under $\exists^{\mathcal{Y}}$ for arbitrary $\mathcal{Y}$ by iterating closure under $\exists^{Y}$ for basic Y. ⊣

3D. Recursive and Γ-recursive functions[(2,3)]

With each function

$$f : \mathcal{X} \to \mathcal{Y}$$

we associate the *nbhd diagram* $G^f \subseteq \mathcal{X} \times \omega$ of f,

$$G^f(x, s) \Leftrightarrow f(x) \in N(\mathcal{Y}, s).$$

Clearly G^f determines f uniquely.

We say that f is *recursive* if G^f is semirecursive; more generally, for each pointclass Γ, we say that f is *Γ-recursive* if its nbhd diagram is in Γ.

If f is recursive, then we can effectively compute arbitrarily good approximations to $f(x)$: given n, simply search for some s such that $radius(N_s) < 2^{-n}$ and $f(x) \in N_s$.

It is clear that this notion of recursiveness is an *effective refinement* of continuity. We should point out at the outset, however, that not all "simple" continuous functions are recursive and that some of the most elementary properties of continuous functions do not carry over to recursive functions.

Not all constant functions are recursive—only those whose constant value can be effectively approximated to any desired degree of accuracy. It makes no sense to ask if "f is recursive at x." Similarly, it makes no sense to ask whether a function $f : \mathbb{R}^2 \to \mathbb{R}$ of two variables is "separately recursive."

The more general notion of Γ-recursiveness is an effective refinement of Lebesgue's *Γ-measurability* which we introduced in 1G. We will see in the exercises that for suitable Γ, a function is Γ-recursive precisely when it is Γ-measurable.

To study profitably Γ-recursive functions we must restrict ourselves to pointclasses which satisfy the closure properties of the semirecursive sets. Call Γ a *Σ-pointclass* if it contains all semirecursive pointsets and it is closed under trivial substitutions, &, $\vee$, $\exists^{\leq}$, $\forall^{\leq}$ and $\exists^{\omega}$. By 3C.1 then, the collection of semirecursive sets is the smallest Σ-pointclass and we introduce the notation

$$\Sigma^0_1 = \text{all semirecursive sets.}$$

Notice the *lightface* font which distinguishes Σ^0_1 from the pointclass $\mathbf{\Sigma}^0_1$ of open sets.

All $\mathbf{\Sigma}^0_\xi$ are Σ-pointclasses, as are all $\mathbf{\Sigma}^1_n$, $\mathbf{\Pi}^1_n$, $\mathbf{\Delta}^1_n$. The multiplicative Borel classes $\mathbf{\Pi}^0_\xi$ are not Σ-pointclasses.

The first simple but very useful lemma is due to Dellacherie.

3D.1. LEMMA. (Dellacherie). *Let Γ be a Σ-pointclass. A function*

$$f : \mathfrak{X} \to \mathcal{Y}$$

is Γ-recursive if and only if for every semirecursive $P \subseteq \omega \times \mathcal{Y}$, the set $P^f \subseteq \omega \times \mathfrak{X}$ defined by

$$P^f(n, x) \Leftrightarrow P(n, f(x))$$

is in Γ.

PROOF. Suppose first that $f : \mathfrak{X} \to \mathcal{Y}$ is Γ-recursive and $P \subseteq \omega \times \mathcal{Y}$ is Σ^0_1. By 3C.5 then, there is a semirecursive $P^* \subseteq \omega^2$ such that

$$P(n, y) \Leftrightarrow (\exists s)\{y \in N(\mathcal{Y}, s) \ \&\ P^*(n, s)\},$$

so that

$$\begin{aligned} P^f(n, x) &\Leftrightarrow (\exists s)\{f(x) \in N(\mathcal{Y}, s) \ \&\ P^*(n, s)\} \\ &\Leftrightarrow (\exists s)\{G^f(x, s) \ \&\ P^*(n, s)\}; \end{aligned}$$

thus P^f is in Γ by the closure properties of a Σ-pointclass.

The converse is trivial, taking

$$P(n, y) \Leftrightarrow y \in N(\mathcal{Y}, n). \qquad \dashv$$

Call a sequence $P_0, P_1, P_2, \ldots$ of subsets of some $\mathfrak{X}$ *Γ-enumerable* if the relation

$$P(n, x) \Leftrightarrow P_n(x)$$

is in Γ. Then 3D.1 says that $f : \mathfrak{X} \to \mathfrak{Y}$ is Γ-recursive exactly when the inverse image

$$f^{-1}[P_0], f^{-1}[P_1], \ldots$$

of every Σ^0_1-enumerable sequence

$$P_0, P_1, \ldots$$

of subsets of $\mathfrak{Y}$ is a Γ-enumerable sequence of subsets of $\mathfrak{X}$.

3D.2. THEOREM. *Let Γ be a Σ-pointclass.*

(i) *A function $f : \mathfrak{X} \to \omega$ is Γ-recursive if and only if $Graph(f) = \{(x, n) : f(x) = n\}$ is in Γ.*

(ii) *If $\mathfrak{X}$ is of type 0 and $f : \mathfrak{X} \to \omega$, then f is recursive in the present sense (i.e. Σ^0_1-recursive) exactly when it is recursive in the sense of* 3A.

(iii) *Suppose $Q \subseteq Y_1 \times \cdots \times Y_l$ and*

$$P(x) \Leftrightarrow Q(f_1(x), \ldots, f_l(x)),$$

where each f_i is trivial or Γ-recursive into ω. If Q is in Γ, then so is P.

PROOF. (i). Assume first that f is Γ-recursive and take

$$R(m, n) \Leftrightarrow m = n;$$

then R is in Σ^0_1 and

$$f(x) = m \Leftrightarrow R^f(m, x),$$

so $Graph(f)$ is in Γ.

Conversely, for any $P \subseteq \omega \times \omega$ in Σ^0_1,

$$\begin{aligned} P^f(n, x) &\Leftrightarrow P(n, f(x)) \\ &\Leftrightarrow (\exists m)\{f(x) = m \;\&\; P(n, m)\}, \end{aligned}$$

so P^f is in Γ by the closure properties of a Σ-pointclass.

(ii) is immediate from (i) and 3C.7.

(iii) To simplify notation, suppose

$$P(x_1, x_2, x_3) \Leftrightarrow Q(x_2, g(x_1, x_2, x_3)),$$

where Q is in Γ and $g:\mathfrak{X}_1 \times \mathfrak{X}_2 \times \mathfrak{X}_3 \to \omega$ is Γ-recursive. Then

$$P(x_1, x_2, x_3) \Leftrightarrow (\exists m)\{g(x_1, x_2, x_3) = m \ \& \ Q(x_2, m)\}$$

and P is in Γ by (i) and closure of Γ under trivial substitutions, & and $\exists^\omega$. ⊣

This simple result is very useful and we will use it constantly without explicit reference. For example, if

$$P(x, n, m) \Leftrightarrow (\exists t)\{Q(x, \langle n, m\rangle) \ \& \ f(x, (n)_0) = (t)_1\}$$

with Q in Γ and f a Γ-recursive function into ω, then P is also in Γ, since

$$P(x, n, m) \Leftrightarrow (\exists t)\{Q'(x, n, m, t) \ \& \ R'(x, n, m, t)\},$$

with Q', R' obtained from Q and

$$R(x, n, t) \Leftrightarrow f(x, n) = t$$

by suitable substitutions of trivial functions and recursive (hence Γ-recursive) functions into ω.

Recall from 1G that with each

$$f:\mathfrak{X} \to \mathcal{N}$$

we have associated the function

$$f^*:\mathfrak{X} \times \omega \to \omega$$

defined by

$$f^*(x, n) = f(x)(n).$$

3D.3. THEOREM. *Let Γ be a Σ-pointclass.*

(i) *A function $f:\mathfrak{X} \to \mathcal{N}$ is Γ-recursive if and only if the associated $f^*:\mathfrak{X} \times \omega \to \omega$ is Γ-recursive.*

(ii) *A function $f:\mathfrak{X} \to \mathcal{Y} = Y_1 \times \cdots \times Y_l$ is Γ-recursive if and only if*

$$f(x) = (f_1(x), \ldots, f_l(x))$$

with suitable Γ-recursive functions $f_1, \ldots, f_l$.

PROOF. Assume first that $f:\mathfrak{X} \to \mathcal{N}$ is Γ-recursive and take

$$R(u, \alpha) \Leftrightarrow \alpha((u)_0) = (u)_1,$$

so that

$$R^f(u, x) \Leftrightarrow f(x)((u)_0) = (u)_1.$$

Clearly R is Σ_1^0 and

$$f^*(x, n) = m \Leftrightarrow R^f(\langle n, m\rangle, x),$$

so f^* is Γ-recursive by 3D.2.

To prove the converse we appeal to 3B.5 according to which there are recursive functions g and h such that

$$\alpha \in N(\mathcal{N}, s) \Leftrightarrow ((s)_1)_1 \neq 0 \;\&\; (\forall i < g(s))[\alpha(i) = h(s, i)].$$

Hence for $f : \mathfrak{X} \to \mathcal{N}$,

$$\begin{aligned} G^f(x, s) &\Leftrightarrow f(x) \in N(\mathcal{N}, s) \\ &\Leftrightarrow ((s)_1)_1 \neq 0 \;\&\; (\forall i < g(s))[f(x)(i) = h(s, i)] \\ &\Leftrightarrow ((s)_1)_1 \neq 0 \;\&\; (\forall i < g(s))[f^*(x, i) = h(s, i)] \end{aligned}$$

so that if f^* is Γ-recursive, then G^f is in Γ by 3D.2 again and the closure properties of Γ.

(ii) is trivial. ⊣

This result gives the easiest method for proving that a function into a space of type 0 or 1 is Γ-recursive.

3D.4. THEOREM. *Let Γ be a Σ-pointclass.*

(i) *Every trivial function $f : \mathfrak{X} \to \mathfrak{Y}$ is recursive.*

(ii) *If $f : \mathfrak{X} \to \mathfrak{Y}$ is Γ-recursive and $g : \mathfrak{Y} \to \mathfrak{Z}$ is recursive, then the composition $h : \mathfrak{X} \to \mathfrak{Z}$,*

$$h(x) = g(f(x))$$

is Γ-recursive.

In particular, the class of recursive functions is closed under composition.

PROOF. (i) is completely trivial. To prove (ii), notice that

$$\begin{aligned} g(f(x)) \in N(\mathfrak{Z}, s) &\Leftrightarrow f(x) \in \{y : g(y) \in N(\mathfrak{Z}, s)\} \\ &\Leftrightarrow G^g(f(x), s). \end{aligned}$$

Now G^g is Σ_1^0 since g is recursive and hence so is

$$P(s, y) \Leftrightarrow G^g(y, s),$$

so that G^h is in Γ by 3D.1. ⊣

It is not always true that the Γ-recursive functions are closed under composition. For example, all $\mathbf{\Sigma}_\xi^0$ are Σ-pointclasses and the $\mathbf{\Sigma}_\xi^0$-recursive functions coincide with the $\mathbf{\Sigma}^0_\xi$-measurable functions, see 3D.21; these are not closed under composition for $\xi > 1$.

It is also not true in general that a Σ-pointclass is closed under substitution of recursive functions. This is a useful special fact about Σ^0_1.

3D.5. THEOREM. *The pointclass Σ^0_1 of semirecursive sets is closed under recursive substitution.*

PROOF. Suppose $P \subseteq \mathcal{Y}$ is Σ^0_1, so

$$P(y) \Leftrightarrow (\exists s)[y \in N(\mathcal{Y}, s) \ \&\ P^*(s)]$$

with a semirecursive P^* by 3C.5, let $f : \mathcal{X} \to \mathcal{Y}$ be recursive. Then

$$Q(x) \Leftrightarrow P(f(x)) \Leftrightarrow (\exists s)[f(x) \in N(\mathcal{Y}, s) \ \&\ P^*(s)],$$

and this is obviously semirecursive. ⊣

With each pointclass Γ we associate the *ambiguous* part of Γ,

$$\Delta = \Gamma \cap \neg\Gamma.$$

3D.6. THEOREM. *Let Γ be a Σ-pointclass. A set $P \subseteq \mathcal{X}$ is in Δ if and only if its characteristic function χ_P is Γ-recursive.*

In particular, P is recursive if and only if χ_P is recursive.

PROOF. On the one hand

$$\chi_P(x) = m \Leftrightarrow [P(x) \ \&\ m = 1] \vee [\neg P(x) \ \&\ m = 0],$$

so if P is in Δ, then χ_P is Γ-recursive by 3D.2.

On the other hand,

$$P(x) \Leftrightarrow \chi_P(x) = 1,$$
$$\neg P(x) \Leftrightarrow \chi_P(x) = 0,$$

so if χ_P is Γ-recursive both P and $\neg P$ are in Γ easily by 3D.3. ⊣

With each pointclass Γ and each point $z \in \mathcal{Z}$ we associate the *relativization* $\Gamma(z)$ *of* Γ *to* z: $P \subseteq \mathcal{X}$ is in $\Gamma(z)$ if there exists some $Q \subseteq \mathcal{Z} \times \mathcal{X}$ in Γ such that

$$P(x) \Leftrightarrow Q(z, x).$$

In particular, the sets in $\Sigma^0_1(z)$ are called *semirecursive in* z and the functions which are $\Sigma^0_1(z)$-recursive are called *recursive in* z.

A point $x \in \mathfrak{X}$ is *Γ-recursive* if the set of codes of nbhds of x is in Γ, i.e. if

$$\mathfrak{U}(x) = \{s : x \in N(\mathfrak{X}, s)\}$$

is in Γ. We often call these simply *the points in Γ*, in fact we will on occasion consider them as members of Γ,

$$x \in \Gamma \Leftrightarrow x \text{ is } \Gamma\text{-recursive}.$$

The points in Σ^0_1 are called *recursive*, the points in $\Sigma^0_1(z)$ are called *recursive in z*.

3D.7. THEOREM. *Let Γ be a Σ-pointclass.*

(i) *For each point $z, \Gamma(z)$ is a Σ-pointclass.*

(ii) *A point x is Γ-recursive if and only if for each $\mathcal{Y}$, the constant function $y \mapsto x$ is Γ-recursive.*

(iii) *If x is recursive in y and y is Γ-recursive, then x is Γ-recursive.*

(iv) *If $f : \mathfrak{X} \to \mathcal{Y}$ is Γ-recursive, then for each $x \in \mathfrak{X}$, $f(x)$ is $\Gamma(x)$-recursive.*

In particular, if $f : \mathfrak{X} \to \mathcal{Y}$ is recursive and x is recursive, then $f(x)$ is also recursive.

PROOF. (i) is very easy and (ii) is trivial, since if $f : \mathcal{Y} \to \mathfrak{X}$ is the constant function $y \mapsto x$, then

$$G^f(y, s) \Leftrightarrow x \in N(\mathfrak{X}, s) \Leftrightarrow s \in \mathfrak{U}(x).$$

(iii) If x is recursive in y, then $\mathfrak{U}(x)$ is in $\Sigma^0_1(y)$, i.e. there is a Σ^0_1 set $P \subseteq \mathcal{Y} \times \omega$ such that

$$s \in \mathfrak{U}(x) \Leftrightarrow P(y, s);$$

since y is Γ-recursive, the constant map $n \mapsto y$ is Γ-recursive, so that for each semirecursive Q, the relation

$$Q^y(s, n) \Leftrightarrow Q(s, y)$$

is in Γ. The result follows by taking $Q = \{(s, y) : P(y, s)\}$ so that

$$s \in \mathfrak{U}(x) \Leftrightarrow Q^y(s, s).$$

(iv) is immediate since

$$s \in \mathfrak{U}(f(x)) \Leftrightarrow G^f(x, s)$$

and the last assertion comes from (iv) and (iii). ⊣

We leave for the exercises some of the other interesting properties of these simple notions.

EXERCISES

3D.8. Prove that the following functions are recursive. (We are continuing the numbering from 3A.6.)

*19. $$f(\alpha, n) = \alpha(n).$$

*20. $$f(\alpha, n) = \bar{\alpha}(n) = \langle \alpha(0), \ldots, \alpha(n-1) \rangle$$

Hint. $\bar{\alpha}(n) = u \Leftrightarrow Seq(u)\ \&\ lh(u) = n\ \&\ (\forall i < lh(u))[\alpha(i) = (u)_i]$.

*21. $$f(\alpha, i) = (\alpha)_i,$$

where for each t, $(\alpha)_i(t) = \alpha(\langle i, t \rangle)$.

*22. $$f(\alpha) = \alpha^\star = (t \quad \alpha(t+1))$$

*23. $$f(\alpha_0, \ldots, \alpha_{k-1}) = \langle \alpha_0, \ldots, \alpha_{k-1} \rangle,$$

where

$$\langle \alpha_0, \ldots, \alpha_{k-1} \rangle(\langle i, t \rangle) = \alpha_i(t) \quad \text{for} \quad i = 0, \ldots, k-1,$$
$$\langle \alpha_0, \ldots, \alpha_{k-1} \rangle(n) = 0 \quad \text{if} \quad n \neq \langle i, t \rangle \quad \text{for all} \quad t, i < k.$$

*24. $$f(i) = r_i,$$

where $f : \omega \to \mathfrak{X}$ and $\{r_0, r_1, \ldots\}$ is the fixed recursive presentation of $\mathfrak{X}$.

*25. $$f(x, y) = x + y \quad (x, y \in \mathcal{R}).$$

*26. $$f(x, y) = x \cdot y \quad (x, y \in \mathcal{R}).$$ ⊣

3D.9. Prove that a function $f : \mathfrak{X} \to \mathcal{R}$ into the reals is recursive if and only if the relation

$$P(x, u, v) \Leftrightarrow (-1)^{(u)_0} \cdot \frac{(u)_1}{(u)_2 + 1} < f(x) < (-1)^{(v)_0} \cdot \frac{(v)_1}{(v)_2 + 1}$$

is semirecursive. ⊣

3D.10. Prove that for each $\mathfrak{X}$, the distance function $f : \mathfrak{X} \times \mathfrak{X} \to \mathcal{R}$,

$$f(x, y) = d(x, y)$$

is recursive. ⊣

3D.11. Assume that Γ is a Σ-pointclass, $g:\omega\times\mathcal{X}\to\omega$ is Γ-recursive and for each x, there is some n such that $g(n,x)=0$. Prove that the function f defined by minimalization

$$f(x)=\mu n[g(n,x)=0]$$

is Γ-recursive. ⊣

Recall from 1C the definition of a *uniformizing* subset $P^*\subseteq P$, for $P\subseteq\mathcal{X}\times\mathcal{Y}$.

3D.12. Prove that if $\mathcal{X}$ is of type 0 or 1 and $\mathcal{Y}$ is of type 0, then every semirecursive subset $P\subseteq\mathcal{X}\times\mathcal{Y}$ can be uniformized by some semirecursive $P^*\subseteq P$.

Infer that with the same hypotheses on $\mathcal{X}$, $\mathcal{Y}$, P, if in addition $(\forall x)(\exists y)P(x,y)$ holds, then there is a recursive function $f:\mathcal{X}\to\mathcal{Y}$ such that $(\forall x)P(x,f(x))$. (*The Selection Principle.*)

Hint. Use 3C.4. ⊣

A homeomorphism $\pi:\mathcal{X}\rightarrowtail\!\!\!\!\twoheadrightarrow\mathcal{Y}$ is *recursive* if both π and its inverse π^{-1} are recursive functions.

3D.13. Prove that if $\mathcal{X}$ and $\mathcal{Y}$ are of the same type 0 or 1, then they are recursively homeomorphic.

Hint. For $\mathcal{X}$ of type 0 the result is trivial. For type 1, take $\mathcal{X}=\mathcal{N}$ and use induction on the number of factors in $\mathcal{Y}$ after producing trivial homeomorphisms of $\omega\times\mathcal{N}$ and $\mathcal{N}\times\mathcal{N}$ with $\mathcal{N}$. ⊣

3D.14. Prove that for every product space $\mathcal{X}$ there is a recursive surjection

$$\pi:\mathcal{N}\twoheadrightarrow\mathcal{X}$$

of Baire space onto $\mathcal{X}$.

Hint. Use the map of 1.A.1. ⊣

3D.15. Prove that for each perfect product space $\mathcal{X}$ there is a recursive function σ which assigns to each code u of a binary sequence $((u)_0,\dots,(u)_{n-1})$ a basic nbhd $N_{\sigma(u)}$ of $\mathcal{X}$ such that conditions (i), (ii), (iii) of Theorem 1A.2 hold. Infer that there is a recursive injection

$$\pi:\mathcal{C}\rightarrowtail\mathcal{X}.$$

Hint. Put

$$
\begin{aligned}
P(n, i, j) \Leftrightarrow\ & center(N_i) \in N_n \;\&\; center(N_j) \in N_n \\
& \&\; radius(N_i) + d(center(N_i), center(N_n)) < \tfrac{1}{2}\, radius(N_n) \\
& \&\; radius(N_j) + d(center(N_j), center(N_n)) < \tfrac{1}{2}\, radius(N_n) \\
& \&\; radius(N_i) + radius(N_j) < d(center(N_i), center(N_j)),
\end{aligned}
$$

where $N_s = N(\mathfrak{X}, s)$. Now $(\forall n)(\exists i)(\exists j)P(n, i, j)$ and P is semirecursive (in fact recursive), so by 3D.12. there is a recursive $f(n) = (g(n), h(n))$ so that $(\forall n)P(n, g(n), h(n))$. Fix some z_0 so that radius $(N(\mathfrak{X}, z_0)) \leq 1$ and put

$$
\begin{aligned}
Q(u, m) \Leftrightarrow\ & Seq(u) \;\&\; (\forall i < lh(u))[(u)_i \leq 1] \\
& \&\; (\exists z)[(z)_0 = z_0 \;\&\; (\forall i < lh(u))\{[(u)_i = 0 \Rightarrow (z)_{i+1} = g((z)_i)] \\
& \&\; [(u)_i = 1 \Rightarrow (z)_{i+1} = h((z)_i)]\} \;\&\; (z)_{lh(u)} = m].
\end{aligned}
$$

Clearly Q is semirecursive and with $A = \{u : Seq(u) \;\&\; (\forall i < lh(u))[(u)_i \leq 1]\}$ we have $(\forall u \in A)(\exists m)Q(u, m)$, so by 3D.12 again there is a recursive σ such that $(\forall u \in A)Q(u, \sigma(u))$. ⊣

3D.16. Prove that every integer is a recursive point, an irrational ε is recursive if and only if the function $n \mapsto \varepsilon(n)$ is recursive and $(x_1,\ldots, x_k)$ is recursive if and only if $x_1,\ldots, x_k$ are all recursive. ⊣

3D.17. Prove that a point $x \in \mathfrak{X}$ is recursive if and only if there is a recursive irrational ε such that

$$\text{limit}_{i \to \infty}\, r_{\varepsilon(i)} = x$$

and for each i,

$$d(r_{\varepsilon(i)}, r_{\varepsilon(i+1)}) < 2^{-i}.$$

(Here $\{r_0, r_1,\ldots\}$ is the recursive persentation of $\mathfrak{X}$.)

Hint. The "if" part is easy. To prove the "only if" part, put

$$P(i, j) \Leftrightarrow d(x, r_j) < 2^{-i-1},$$

show that P is semirecursive and use the Selection Principle, 3D.12. ⊣

3D.18. Prove that a real number x is recursive if and only if

$$\{m : (-1)^{(m)_0} \cdot \frac{(m)_1}{(m)_2 + 1} < x\}$$

is recursive.

Hint. Take cases on whether x is rational or not. ⊣

3D.19. Prove that the set of recursive real numbers is a field.

Hint. Use the characterization of 3D.18. ⊣

Put

$$x \leq_T y \Leftrightarrow x \text{ is recursive in } y$$
$$\Leftrightarrow x \text{ is } \Sigma_1^0(y)\text{-recursive}$$

and

$$x \equiv_T y \Leftrightarrow x \leq_T y \;\&\; y \leq_T x.$$

The subscript T stands for *Turing* and the relations $\leq_T$, $\equiv_T$ are often called *Turing redubility* and *Turing equivalence.*

3D.20. Prove that $\equiv_T$ is an equivalence relation on the set of all points and $\leq_T$ induces a partial ordering on the set of equivalence classes of $\equiv_T$. ⊣

The equivalence classes of irrationals in $\equiv_T$ are often called *Turing degrees.*

3D.21. Prove that if a function $f: \mathfrak{X} \to \mathcal{Y}$ is recursive in some z, then f is continuous; conversely, every continuous function is recursive in some $\varepsilon \in \mathcal{N}$. ⊣

3D.22. Prove that if Γ is a Σ-pointclass closed under countable conjunction $\bigvee^\omega$, then the Γ-recursive functions are precisely the Γ-measurable functions.

Hint. Every Γ-recursive function is trivially Γ-measurable. For the converse, assume that $f: \mathfrak{X} \to \mathcal{Y}$ is Γ-measurable and $P \subseteq \omega \times \mathcal{Y}$, notice that

$$P^f = \bigcup_n (\{n\} \times f^{-1}[P_n])$$

where $P_n = \{y : P(n, y)\}$ and show that $P^f \in \Gamma$. ⊣

3E. The Kleene pointclasses

We now introduce the *Kleene pointclasses* by effectivizing the definitions in 1B and 1E. These are also called the *lightface* classes—notice the lightface font in the symbols we use to denote them. By analogy, the Borel and Lusin pointclasses are often called *boldface*.

Put

$$\Sigma^0_1 = \text{all semirecursive pointsets,}$$
$$\Sigma^0_{n+1} = \exists^\omega \neg\Sigma^0_n,$$
$$\Pi^0_n = \neg\Sigma^0_n,$$
$$\Delta^0_n = \Sigma^0_n \cap \Pi^0_n.^{(4)}$$

Similarly,

$$\Sigma^1_1 = \exists^{\mathcal{N}}\Pi^0_1,$$
$$\Sigma^1_{n+1} = \exists^{\mathcal{N}} \neg\Sigma^1_n,$$
$$\Pi^1_n = \neg\Sigma^1_n,$$
$$\Delta^1_n = \Sigma^1_n \cap \Pi^1_n.^{(5,6)}$$

For reasons that will become clear later, we call the pointsets in $\bigcup_n \Sigma^0_n$ *arithmetical* and the pointsets in $\bigcup_n \Sigma^1_n$ *analytical*. They are the effective versions of the finite Borel and the projective sets respectively.

Notice the possible source of conflict between the term "*analytical*" and the classical name "*analytic*" for $\mathbf{\Sigma}^1_1$ sets. One way to avoid confusion is to observe scrupulously the difference in suffix between these two words; people have been careless with this pedantic distinction, especially in some early papers in recursion theory. It is prudent to say "$\mathbf{\Sigma}^1_1$" rather than "analytic" in contexts where analytical sets are also discussed.

We can also define the *relativized Kleene pointclasses* $\Sigma^0_n(z)$, $\Pi^0_n(z)$, $\Sigma^1_n(z)$, $\Pi^1_n(z)$ by the general process we described in 3D. Again,

$$\Delta^0_n(z) = \Sigma^0_n(z) \cap \Pi^0_n(z),$$
$$\Delta^1_n(z) = \Sigma^1_n(z) \cap \Pi^1_n(z).$$

One should be careful with this notation, since *it is not the case that* $\Delta^0_n(z)$ *is the relativization of* Δ^0_n *to* z, see 3F.9.

The sets in $\bigcup_n \Sigma^0_n(z)$ are called *arithmetical in* z and the sets in $\bigcup_n \Sigma^1_n(z)$ are called *analytical in* z. We willl not always bother to state explicitly results about these relativized notions since they are similar to those about the absolute pointclasses and they are obtained (usually) by the same arguments.

The basic properties of the Kleene pointclasses can be established very easily, simply by copying the proofs in Chapter 1 and substituting "semirecursive" for "open" and "recursive function" for "continuous function." We will do this somewhat more generally, so we will not have to repeat it when we introduce new pointclasses later on.

The normal forms for the finite Borel and Lusin pointclasses carry over to the Kleene pointclasses immediately. For example, P is Σ^0_2 if there is a Π^0_1 set F so that

$$P(x) \Leftrightarrow (\exists t)F(x, t),$$

P is Σ^0_3 if there is a Σ^0_1 (semirecursive) set G so that

$$P(x) \Leftrightarrow (\exists t_1)(\forall t_2)G(x, t_1, t_2),$$

etc. Similarly, P is Σ^1_1 if there is a Π^0_1 set F such that

$$P(x) \Leftrightarrow (\exists \alpha)F(x, \alpha),$$

P is Σ^1_2 if there is a semirecursive G such that

$$P(x) \Leftrightarrow (\exists \alpha_1)(\forall \alpha_2)G(x, \alpha_1, \alpha_2),$$

etc.

These forms become a bit simpler for spaces of type 0 or 1 because of the characterization in 3C.4. Thus, if P is a pointset of type 0 or 1, then P is Π^0_1 if there is a recursive R such that

$$P(x) \Leftrightarrow (\forall t)R(x, t),$$

P is Σ^0_2 if there is a recursive R such that

$$P(x) \Leftrightarrow (\exists t_1)(\forall t_2)R(x, t_1, t_2),$$

P is Σ^1_1 if there is a recursive R such that

$$P(x) \Leftrightarrow (\exists \alpha)(\forall t)R(x, \alpha, t),$$

etc.

The key to the closure properties of the Kleene pointclasses are the closure properties of Σ^0_1 given in 3C.1 and 3D.5.

To simplify statements of results, let us call a pointclass Γ *adequate* if it contains all recursive pointsets and is closed under recursive substitution, &, $\vee$, $\exists^{\leq}$ and $\forall^{\leq}$. Clearly Σ^0_1 is adequate and closed under $\exists^{\omega}$; Δ^0_1 is adequate and closed under $\neg$.

Recall the notation we introduced in 1F, where for a pointclass Λ and a pointset operation Φ,

$$\Phi\Lambda = \{\Phi(P_0, P_1, \ldots): P_0, P_1, \ldots, \in \Lambda \quad \text{and} \quad \Phi(P_0, P_1, \ldots) \text{ is defined}\}.$$

3E.1. THEOREM. *If Λ is an adequate pointclass, then so are $\neg\Lambda$, $\exists^{\omega}\Lambda$, $\forall^{\omega}\Lambda$, $\exists^{\mathcal{N}}\Lambda$, $\forall^{\mathcal{Y}}\Lambda$.*

Moreover, $\exists^{\omega}\Lambda$ is closed under $\exists^{\omega}$, $\forall^{\omega}\Lambda$ is closed under $\forall^{\omega}$, $\exists^{\mathcal{N}}\Lambda$ is closed under $\exists^{\mathcal{Y}}$ for all product spaces $\mathcal{Y}$ and $\forall^{\mathcal{N}}\Lambda$ is closed under $\forall^{\mathcal{Y}}$ for all product spaces $\mathcal{Y}$.

PROOF. The arguments in 1C.2 and 1E.2 suffice here too if we notice that the continuous substitutions we used there are in fact recursive. We omit the details. ⊣

3E.2. COROLLARY. *All Kleene pointclasses are adequate. Moreover, Σ^0_n is closed under $\exists^{\omega}$, Π^0_n is closed under $\forall^{\omega}$, Σ^1_n is closed under $\forall^{\omega}$ and $\exists^{\mathcal{Y}}$ for all $\mathcal{Y}$ and Π^1_n is closed under $\exists^{\omega}$ and $\forall^{\mathcal{Y}}$ for every $\mathcal{Y}$.*

PROOF. Use 3E.1 and induction on n. To prove closure of Σ^1_n under $\forall^{\omega}$ and the dual closure of Π^1_n under $\exists^{\omega}$ look up the proof of 1E.2. ⊣

3E.3. THEOREM. *The following diagram of inclusions holds for the Kleene pointclasses.*

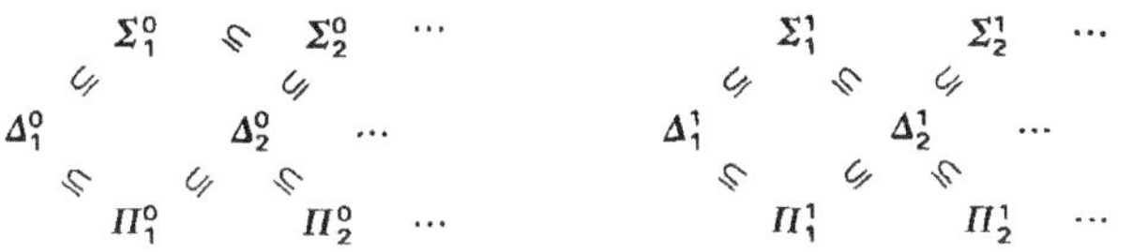

Diagram 3E.1. The Kleene pointclasses.

Moreover, every arithmetical pointset is Δ^1_1.

PROOF. The second assertion follows immediately by the closure properties of Δ^1_1.

To prove the inclusion diagrams we imitate the proofs of 1B.1 and 1E.1. The only new ingredient we need is a proof of

$$\Sigma^0_1 \subseteq \Sigma^0_2,$$

since the proof of $\mathbf{\Sigma}^0_1 \subseteq \mathbf{\Sigma}^0_2$ given in 1B.1 does not immediately yield the lightface version.

All recursive pointsets of type 0 are clearly in Σ^0_2 (by vacuous quantification). Also, Σ^0_2 is closed under trivial substitutions, &, $\vee$, $\exists^{\leq}$, $\forall^{\leq}$ and $\exists^{\omega}$ by 3E.2; hence to show that $\Sigma^0_1 \subseteq \Sigma^0_2$ by applying 3C.12 it is enough to verify that for each basic space X the relation

$$P^X(x, i, m, k) \Leftrightarrow d(r_i, x) < \frac{m}{k+1}$$

is in Σ^0_2. Clearly,

$$P^X(x, i, m, k) \Leftrightarrow (\exists m')(\exists k')\left\{\frac{m'}{k'+1} < \frac{m}{k+1} \;\&\; \neg\left(\frac{m'}{k'+1} < d(r_i, x)\right)\right\},$$

and then P^X is in Σ^0_2 by 3C.9 and the closure properties. ⊣

Let us state for the record the rather obvious relationship between the lightface and the boldface pointclasses.

3E.4. THEOREM. *Let Γ be Σ^0_n, Π^0_n, Σ^1_n or Π^1_n,* let $\mathbf{\Gamma}$ *be the corresponding boldface pointclass,* $\mathbf{\Sigma}^0_n$, $\mathbf{\Pi}^0_n$, $\mathbf{\Sigma}^1_n$ or $\mathbf{\Pi}^1_n$. *For each product space $\mathfrak{X}$, there is a pointset $G \subseteq \mathcal{N} \times \mathfrak{X}$ in Γ which is universal for $\mathbf{\Gamma} \restriction \mathfrak{X}$, the class of subsets of $\mathfrak{X}$ in $\mathbf{\Gamma}$.*

In particular, $P \subseteq \mathfrak{X}$ is in $\mathbf{\Gamma}$ if and only if P is in $\Gamma(\varepsilon)$ for some $\varepsilon \in \mathcal{N}$ i.e. if and only if

$$P(x) \Leftrightarrow P^*(\varepsilon, x)$$

for some fixed $\varepsilon \in \mathcal{N}$ and some P^* in Γ.

Also for the ambiguous pointclasses, P is $\mathbf{\Delta}^0_n$ (or $\mathbf{\Delta}^1_n$) if and only if P is $\Delta^0_n(\varepsilon)$ (or $\Delta^1_n(\varepsilon)$) *for some ε.*

PROOF. For Σ^0_1, take

$$G(\varepsilon, x) \Leftrightarrow (\exists n)[x \in N(\mathfrak{X}, \varepsilon(n))].$$

This is obviously Σ^0_1 and universal for $\mathbf{\Sigma}^0_1 \restriction \mathfrak{X}$. The result follows by a trivial induction.

The second statement is immediate from the first.

For the ambiguous pointclasses, if $P \subseteq \mathfrak{X}$ is in $\mathbf{\Delta}^0_n$ (say), choose ε_1, ε_2 and $P_1 \subseteq \mathcal{N} \times \mathfrak{X}$, $P_2 \subseteq \mathcal{N} \times \mathfrak{X}$ in Σ^0_n and Π^0_n respectively so that

$$P(x) \Leftrightarrow P_1(\varepsilon_1, x),$$
$$P(x) \Leftrightarrow P_2(\varepsilon_2, x),$$

choose ε such that $(\varepsilon)_1 = \varepsilon_1$, $(\varepsilon)_2 = \varepsilon_2$ and notice that P is $\Delta^0_n(\varepsilon)$, since

$$P(x) \Leftrightarrow P_1((\varepsilon)_1, x),$$
$$P(x) \Leftrightarrow P_2((\varepsilon)_2, x). \qquad ⊣$$

We leave for the exercises similar strong versions of the parametrization and hierarchy theorems 1D.1–1D.4.

Clearly all Σ^0_n are Σ-pointclasses, as are all Σ^1_n, Π^1_n and Δ^1_n, so we can study recursion theory for them. The case of Σ^0_n-recursion is somewhat interesting and some of the exercises will deal with it.

Here we are interested in *Δ^1_1-recursion*, the effective refinement of *Borel measurability*.

3E.5. THEOREM. (i) *The following four conditions on a function $f:\mathfrak{X}\to\mathfrak{Y}$ are equivalent*:

(1) *f is Δ^1_1-recursive.*

(2) *f is Σ^1_1-recursive.*

(3) *$Graph(f)=\{(x,y): f(x)=y\}$ is Σ^1_1.*

(4) *$Graph(f)$ is Δ^1_1.*

(ii) *All the analytical pointclasses Σ^1_n, Π^1_n, Δ^1_n are closed under substitution of Δ^1_1-recursive functions.*

PROOF. (i) (1) $\Rightarrow$ (2) is immediate and (2) $\Rightarrow$ (3) follows from the equivalence

$$f(x)=y \Leftrightarrow (\forall s)\{y\in N(\mathfrak{Y},s)\Rightarrow f(x)\in N(\mathfrak{Y},s)\}.$$

To prove (3) $\Rightarrow$ (4), notice that

$$f(x)\neq y \Leftrightarrow (\exists z)\{f(x)=z\ \&\ z\neq y\}$$

and for (4) $\Rightarrow$ (1) use

$$\begin{aligned} f(x)\in N(\mathfrak{Y},s) &\Leftrightarrow (\exists y)\{f(x)=y\ \&\ y\in N(\mathfrak{Y},s)\}\\ &\Leftrightarrow (\forall y)\{f(x)\neq y \vee y\in N(\mathfrak{Y},s)\}.\end{aligned}$$

(ii) is immediate from the equivalences

$$\begin{aligned} P(f(x)) &\Leftrightarrow (\exists y)\{f(x)=y\ \&\ P(y)\}\\ &\Leftrightarrow (\forall y)\{f(x)\neq y\vee P(y)\}. \end{aligned}$$ ⊣

We often say "*Δ^1_1 function*" instead of "Δ^1_1-recursive function." For the moment, we think of these as the effective Borel functions simply because Borel measurability is the same as $\mathbf{\Delta}^1_1$-measurability by Suslin's Theorem. In Chapter 7 we will look at some deeper reasons for the analogy.

We now state effective versions of the "Transfer Theorems" 1G.2 and 1G.3. These are very important as they allow us to reducce the study of the analytical Kleene pointclasses to the study of analytical sets of irrationals.

3E.6. THEOREM. *For every product space $\mathfrak{X}$ there is a recursive surjection*

$$\pi : \mathcal{N} \twoheadrightarrow \mathfrak{X}$$

and a Π^0_1 set $A \subseteq \mathcal{N}$ such that π is one-to-one on A and $\pi[A] = \mathfrak{X}$. Moreover, there is a Δ^1_1 injection

$$f : \mathfrak{X} \to \mathcal{N}$$

which is precisely the inverse of π restricted to A, i.e. for all $\alpha \in A$, $f(\pi(\alpha)) = \alpha$ and for all x, $f(x) \in A$ and $\pi(f(x)) = x$.

PROOF is exactly that of 1G.2. ⊣

3E.7. THEOREM. *For every perfect product space $\mathfrak{X}$ there is a Δ^1_1 bijection*

$$g : \mathcal{N} \rightarrowtail\!\!\!\!\twoheadrightarrow \mathfrak{X}$$

whose inverse g^{-1} is also Δ^1_1.

PROOF. As in 1G, we call an injection

$$f : \mathfrak{X} \rightarrowtail \mathcal{Y}$$

a *good Δ^1_1 injection* if there is a Δ^1_1 surjection

$$f^* : \mathcal{Y} \twoheadrightarrow \mathfrak{X}$$

such that

$$f^* f(x) = x.$$

Using the proof of 1G.4 and 3D.15 we can easily show the existence of good Δ^1_1 injections

$$h : \mathcal{N} \rightarrowtail \mathfrak{X},$$

$$f : \mathfrak{X} \rightarrowtail \mathcal{N}.$$

Define now the sets $\mathcal{N}_n$, $\mathfrak{X}_n$ exactly as in the proof of 1G.4. It is enough to prove that the four relations

$$\alpha \in \mathcal{N}_n, \quad x \in \mathfrak{X}_n, \quad \alpha \in f[\mathfrak{X}_n], \quad x \in h[\mathcal{N}_n]$$

are Δ^1_1. From this it is immediate that the bijection g defined in the proof of 1G.4 has Δ^1_1 graph and is therefore Δ^1_1 by 3E.5.

Let us concentrate on the relation $\alpha \in \mathcal{N}_n$. To begin with, it is almost trivial that

(1) $$\alpha \in \mathcal{N}_n \Leftrightarrow (\exists\beta)\{(\forall i < n)[(\beta)_{i+1} = fh((\beta)_i)] \;\&\; \alpha = (\beta)_n\}.$$

We prove direction ($\Rightarrow$) of (1) by choosing $\beta_0, \beta_1, \ldots, \beta_n$ so that $\beta_1 = fh(\beta_0)$, $\beta_2 = fh(\beta_1), \ldots, \beta_n = fh(\beta_{n-1}) = \alpha$ and then picking β so that for $i < n$, $(\beta)_i = \beta_i$. For the direction ($\Leftarrow$), choose any β which satisfies the matrix on the right of (1) and verify by induction on $i < n$ that $(\beta)_{i+1} \in \mathcal{N}_{i+1}$, so that $\alpha = (\beta)_n \in \mathcal{N}_n$.

Equivalence (1) establishes that the relation $\alpha \in \mathcal{N}_n$ is Σ^1_1. To show that this relation is also Π^1_1, we need the slightly less perspicuous equivalence

(2) $$\alpha \in \mathcal{N}_n \Leftrightarrow (\forall\beta)\{(\forall i < n)[(\beta)_i = h^*f^*((\beta)_{i+1}) \;\&\; (\beta)_n = \alpha] \Rightarrow (\forall i < n)[(\beta)_{i+1} = fh((\beta)_i)]\},$$

where f^* and h^* are Δ^1_1 "inverses" of the good Δ^1_1 injections f and h.

Proof of direction ($\Rightarrow$) *of* (2) is by induction on n. For $n = 0$, $\alpha \in \mathcal{N}_0 = \mathcal{N}$ and the right hand side is vacuously true. Assume $\alpha \in \mathcal{N}_{n+1}$, so that for some $\gamma \in \mathcal{N}_n$ we have

$$\alpha = fh(\gamma)$$

and therefore

$$h^*f^*(\alpha) = \gamma.$$

Any β which satisfies

$$(\forall i < n+1)[(\beta)_i = h^*f^*((\beta)_{i+1}) \;\&\; (\beta)_{n+1} = \alpha]$$

obviously satisfies

$$(\forall i < n)[(\beta)_i = h^*f^*((\beta)_{i+1}) \;\&\; (\beta)_n = \gamma],$$

so by the induction hypothesis applied to $\gamma \in \mathcal{N}_n$,

$$(\forall i < n)[(\beta)_{i+1} = fh((\beta)_i)].$$

Since also

$$(\beta)_{n+1} = \alpha = fh(\gamma) = fh((\beta)_n),$$

we have

$$(\forall i < n+1)[(\beta)_{i+1} = fh((\beta)_i)]$$

and we have shown the right hand side of (2) for $n+1$.

Proof of direction ($\Leftarrow$) *of* (2). Given α so that the right hand side of (2) holds, choose β so that $(\beta)_n = \alpha$, $(\beta)_{n-1} = h^*f^*((\beta)_n)$, $(\beta)_{n-2} = h^*f^*((\beta)_{n-1}), \ldots, (\beta)_0 = h^*f^*((\beta)_1)$. We then have that $(\forall i<n)$ $[(\beta)_{i+1} = fh((\beta)_i)]$ from which it follows immediately that for each $i<n$, $(\beta)_{i+1} \in \mathcal{N}_{i+1}$, so that $\alpha = (\beta)_n \in \mathcal{N}_n$.

A symmetric argument establishes that the relation $x \in \mathfrak{X}_n$ is Δ^1_1. Finally,

$$\begin{aligned} \alpha \in f[\mathfrak{X}_n] &\Leftrightarrow \alpha \in f[\mathfrak{X}] \;\&\; f^*(\alpha) \in \mathfrak{X}_n \\ &\Leftrightarrow ff^*(\alpha) = \alpha \;\&\; f^*(\alpha) \in \mathfrak{X}_n \end{aligned}$$

so that $\alpha \in f[\mathfrak{X}_n]$ is Δ^1_1 and similarly for the relation $x \in h[\mathcal{N}_n]$. ⊣

The analytical pointclasses are not closed under the infinitary operations of countable union and intersection. Because of this, it is not obvious what is the correct effective version of the Suslin Theorem. Actually, there is a beautiful result of Kleene which characterizes Δ^1_1 as the smallest pointclass containing all semirecursive sets and closed under "effective" countable union and complementation. This is quite difficult and we will leave it for Chapter 7.

EXERCISES

3E.8. Prove that if $\mathfrak{X}$ is of type 1, then for each n, there is some pointset $P \subseteq \mathfrak{X}$ in $\Sigma^0_n - \boldsymbol{\Pi}^0_n$. Similarly, the differences $\Sigma^1_n{\restriction}\mathfrak{X} - \boldsymbol{\Pi}^1_n{\restriction}\mathfrak{X}$, $\Pi^0_n{\restriction}\mathfrak{X} - \boldsymbol{\Sigma}^0_n{\restriction}\mathfrak{X}$, $\Pi^1_n{\restriction}\mathfrak{X} - \boldsymbol{\Sigma}^1_n{\restriction}\mathfrak{X}$ are all nonempty.

Hint. Use the fact that $\mathcal{N} \times \mathfrak{X}$ is recursively homeomorphic with $\mathfrak{X}$ and that some $G \subseteq \mathcal{N} \times \mathfrak{X}$ in Σ^0_n parametrizes $\boldsymbol{\Sigma}^0_n{\restriction}\mathfrak{X}$. ⊣

3E.9. Prove that if $\mathfrak{Y}$ is a prerfect product space, then for each $\mathfrak{X}$ there is a Σ^0_1 set $G \subseteq \mathfrak{Y} \times \mathfrak{X}$ which is universal for $\boldsymbol{\Sigma}^0_1{\restriction}\mathfrak{X}$.

Similarly with Σ^0_n, Σ^1_n in place of Σ^0_1.

Infer that for each perfect product space $\mathfrak{X}$ the differences $\Sigma^0_n{\restriction}\mathfrak{X} - \boldsymbol{\Pi}^0_n{\restriction}\mathfrak{X}$, $\Sigma^1_n{\restriction}\mathfrak{X} - \boldsymbol{\Pi}^1_n{\restriction}\mathfrak{X}$, $\Pi^0_n{\restriction}\mathfrak{X} - \boldsymbol{\Sigma}^0_n{\restriction}\mathfrak{X}$, $\Pi^1_n{\restriction}\mathfrak{X} - \boldsymbol{\Sigma}^1_n{\restriction}\mathfrak{X}$ are nonempty.

Hint. Follow the proofs of 1D.1–1D.4 and 3D.15. ⊣

3E.10. Prove that for each $n>1$, every Σ^0_n set $P \subseteq \mathfrak{X} \times \omega$ can be uniformized by some Σ^0_n subset $P^* \subseteq P$.

Hint. See 1C.6. ⊣

3E.11. Prove that for each perfect product space $\mathfrak{X}$, there is a Σ^0_2-recursive surjection

$$f:\mathfrak{X} \twoheadrightarrow \mathcal{N}.$$

Hint. See 1G.8 and 1G.10. ⊣

We will see in 4D.10 that 1E.6 does not have an effective version—it is not true that the Σ^1_1 subsets of a space $\mathfrak{X}$ are precisely the recursive images of $\mathcal{N}$. Similarly, only part of 1G.12 holds for Σ^1_1.

3E.12. Suppose $\mathfrak{X}$ is perfect and $P \subseteq \mathcal{Y}$. Prove that P is Σ^1_1 if and only if P is the projection of some Π^0_2 subset of $\mathcal{Y} \times \mathfrak{X}$.

In particular, every Σ^1_1 set of reals is the projection of a Π^0_2 subset of the plane.

Hint. See the hint to 1G.12. ⊣

Call a function $f:\mathfrak{X} \to \mathcal{Y}$ of *effective Baire class* 0 if it is recursive, of *effective Baire class* 1 if it is Σ^0_2-recursive but not recursive and, inductively, of *effective Baire class* $n+1 \geq 2$ if it is not of effective Baire class $\leq n$ and there exists a function

$$g:\omega \times \mathfrak{X} \to \mathcal{Y},$$

of effective Baire class n such that

$$f(x) = limit_{m\to\infty}\, g(m, x).$$

3E.13. Prove that a function $f:\mathfrak{X} \to \mathcal{Y}$ is of effective Baire class $\leq n$ if and only if f is Σ^0_{n+1}-recursive.

Hint. Study the proofs of 1G.16–1G.19. ⊣

3E.14. Prove that a function $f:\mathcal{N} \to \mathcal{Y}$ is Σ^0_2-recursive if and only if there is a recursive $g:\omega \times \mathcal{N} \to \mathcal{Y}$ such that

$$f(\alpha) = limit_{m\to\infty}\, g(m, \alpha). \quad ⊣$$

This last result holds also for functions $f:\mathfrak{X} \to \mathcal{R}$.

It is worth putting down a few exercises on the relativized Kleene pointclasses, $\Sigma^0_n(z)$, $\Pi^0_n(z)$, etc. Their theory is very similar to that of the *absolute* Kleene pointclasses Σ^0_1, Π^0_1, etc.

3E.15. Prove that if z is recursive in w, then for every Kleene pointclass Γ, $\Gamma(z) \subseteq \Gamma(w)$.

Hint. If z is recursive in w, then the constant function $x \mapsto z$ is $\Sigma^0_1(w)$-recursive. ⊣

The next result is completely trivial but very useful and we often tend to use it without citing.

3E.16. Prove that if the singleton $\{x_0\}$ is $\Sigma^1_n(z)$, then x_0 is in $\Delta^1_n(z)$ (i.e. x_0 is $\Delta^1_n(z)$-recursive).

Hint. Let $\{x_0\} = P \subseteq \mathfrak{X}$. Then

$$\begin{aligned} x_0 \in N(\mathfrak{X}, s) &\Leftrightarrow (\exists x)[P(x) \ \&\ x \in N(\mathfrak{X}, s)] \\ &\Leftrightarrow (\forall x)[P(x) \Rightarrow x \in N(\mathfrak{X}, s)]. \end{aligned}$$ ⊣

3E.17. Prove that if x is $\Delta^1_n(z, y)$-recursive and y is $\Delta^1_n(z)$-recursive, then x is $\Delta^1_n(z)$-recursive.

Hint. We have Σ^1_n relations P and Q and Π^1_n relations P', Q', so that

$$\begin{aligned} x \in N_s &\Leftrightarrow P(z, y, s) \Leftrightarrow P'(z, y, s), \\ y \in N_t &\Leftrightarrow Q(z, t) \Leftrightarrow Q'(z, t); \end{aligned}$$

now

$$\begin{aligned} x \in N_s &\Leftrightarrow (\exists y')\{(\forall t)[y' \in N_t \Rightarrow Q(z, t)] \ \&\ P(z, y', s)\} \\ &\Leftrightarrow (\forall y')\{(\forall t)[y' \in N_t \Rightarrow Q(z, t)] \Rightarrow P'(z, y', s)\}. \end{aligned}$$ ⊣

3F. Universal sets for the Kleene pointclasses[1]

It is almost obvious from the definitions that there are only countably many recursive functions. Here we will prove the much stronger result that Σ^0_1 is ω-parameterized.

The reader with some knowledge of basic recursion theory will want to skip this section, after he peruses the final results 3F.6 and 3F.7.

Let us go back to the definition of the class of number theoretic recursive functions in 3A and analyse it.

A *recursive derivation* is a sequence

$$f_0, f_1, \ldots, f_n$$

of (number theoretic) functions such that each f_i is the successor function S, one of the constants C^k_w or the projections P^k_i or else can be defined by composition, primitive recursion or minimalization from functions preceding it in the sequence $f_0, \ldots, f_n$. We can think of a recursive derivation as a *proof* that f_n is recursive.

3F.1. THEOREM. *A function* $f : \omega^k \to \omega$ *is recursive if and only if there is a recursive derivation* $f_0, \ldots, f_n$ with $f_n = f$.

PROOF. If $f_0, \ldots, f_n$ is a recursive derivation, then each f_i is recursive by induction on i. On the other hand, the collection of all functions which occur in recursive derivations obviously contains S, all C^k_w, P^k_i and is closed under composition, primitive recursion and minimalization, so it contains every recursive function. ⊣

To verify that a given sequence of functions $f_0, \ldots, f_n$ is a recursive derivation, we must give a *justification* for including each f_j in the list—because f_j is S or f_j is defined from functions listed before it by composition, etc. We now give a formal coding of such justifications by finite sequences of *numbers*.

Let a sequence $f_0, \ldots, f_n$ of *functions* and a sequence $\hat{f}_0, \ldots, \hat{f}_n$ of *numbers* be given. We say that $\hat{f}_0, \ldots, \hat{f}_n$ is a *justification* for $f_0, \ldots, f_n$, if one of the following conditions holds for each $j \leq n$.

Case 1. $f_j = S$ and $\hat{f}_j = \langle 1, 1 \rangle$.

Case 2. $f_j = C^k_w$ and $\hat{f}_j = \langle 2, k, w \rangle$.

Case 3. $f_j = P^k_i \; (1 \leq i \leq k)$ and $\hat{f}_j = \langle 3, k, i \rangle$.

Case 4. $f_j(x_1, \ldots, x_k) = h(g_1(x), \ldots, g_m(x))$ where the functions $h, g_1, \ldots, g_m$ precede f_j in the list $f_0, \ldots, f_n$ and

$$\hat{f}_j = \langle 4, k, \langle \hat{h}, \hat{g}_1, \ldots, \hat{g}_m \rangle \rangle.$$

By this of course we mean that $h = f_{j_0}$, $g_1 = f_{j_1}, \ldots, g_m = f_{j_m}$ with $j_0, j_1, \ldots, j_m < j$ and

$$\hat{f}_j = \langle 4, k, \langle \hat{f}_{j_0}, \hat{f}_{j_1}, \ldots, \hat{f}_{j_m} \rangle \rangle,$$

and similarly for the next two cases.

Case 5(i).

$$f_j(0, x) = g(x), \qquad f_j(n+1, x) = h(f_j(n, x), n, x),$$

where g and h precede f_j in the list and

$$\hat{f}_j = \langle 5, k+1, \langle \hat{g}, \hat{h} \rangle \rangle.$$

Case 5(ii).

$$f_j(0)=w_0, \qquad f_j(n+1)=h(f_j(n),n),$$

where h precedes f_j in the list and

$$\hat{f}_j=\langle 5,1,\langle w_0,\hat{h}\rangle\rangle.$$

Case 6. $f_j(x)=\mu m[g(m,x)=0]$, where g precedes f_j in the list and

$$\hat{f}_j=\langle 6,k,\hat{g}\rangle.$$

It is now immediate that every recursive derivation has a justification and that if $f_0,\ldots,f_n$ has a justification, then $f_0,\ldots,f_n$ is a recursive derivation. All we have done is code with finite sequences of numbers all canonical proofs that particular sequences of functions are recursive derivations.

More than that, our coding is one-to-one in a very strong sense.

3F.2. THEOREM. *Suppose $\hat{f}_0,\ldots,\hat{f}_n$ is a justification for $f_0,\ldots,f_n$, $\hat{g}_0,\ldots,\hat{g}_m$ is a justification for $g_0,\ldots,g_m$ and for some $j\leq n$, $i\leq m$ we have $\hat{f}_j=\hat{g}_i$. Then $f_j=g_i$.*

PROOF is by induction on the number $\hat{f}_j$. There are six cases to the proof corresponding to the definition of a justification, but looking at just two of them will be sufficient to give the idea.

Case 2. $\hat{f}_j=\langle 2,k,w\rangle$ for some k, w. Then $f_j=C^k_w$ and since also $\hat{g}_i=\langle 2,k,w\rangle$, we have $g_i=C^k_w=f_j$.

Case 6. $\hat{f}_j=\langle 6,k,z\rangle$ for some numbers $k\geq 1$, z. Now by the definition of a justification, it follows that f_j is a k-ary function and that for some f_l with $l<j$, f_l is $(k+1)$-ary, $z=\hat{f}_l$ and

$$f_j(x)=\mu t[f_l(t,x)=0].$$

By the same argument, there is some g_s with $s<i$ such that $z=\hat{g}_s$ and

$$g_i(x)=\mu t[g_s(t,x)=0].$$

Now $z=(\hat{f}_j)_2<\hat{f}_j$, so by induction hypothesis $f_l=g_s$ and hence $f_j=g_i$. ⊣

Put

$$C=\{z\colon \text{there exists some sequence of integers } z_0,\ldots,z_n=z \text{ which is a justification for a recursive derivation}\}.$$

It follows from 3F.2, that if $z \in C$, then in every justification in which z occurs it "codes" the same function, call it f_z. We call C the set of *codes* of recursive functions. Since the map $z \mapsto f_z$ takes C onto the recursive functions, we have proved the first main result of this section.

3F.3. THEOREM. *There are only countably many recursive number theoretic functions.* ⊣

Our coding is such that if $z \in C$, then f_z is k-ary with $k = (z)_1$. For fixed $x_1, \ldots, x_k$, m, let us think of the number

$$\langle z, \langle x_1, \ldots, x_k \rangle, m \rangle$$

as a code of the *assertion* that

$$f_z(x_1, \ldots, x_k) = m.$$

Of course this assertion may be true or false. We now construct a semirecursive set A such that for every $z \in C$, A contains all codes of true assertions $\langle z, \langle x_1, \ldots, x_k \rangle, m \rangle$ and no codes of false assertions about f_z. Notice that A will contain many members of the form $\langle z, \langle x_1, \ldots, x_k \rangle, m \rangle$ where $z \notin C$—these will give us no information about recursive functions.

The set A is the smallest set of integers satisfying (1)–(6), where the symbols $\hat{f}$, $\hat{g}$, $\hat{h}$ are used just as *variables* over ω.

(1) If $\hat{f} = \langle 1, 1 \rangle$, then

$$\text{for every } n, \quad \langle \hat{f}, n, n+1 \rangle \in A.$$

(2) If $\hat{f} = \langle 2, k, w \rangle$ for some $k \geq 1$ and some w, then

$$\text{for every } x_1, \ldots, x_k, \quad \langle \hat{f}, \langle x_1, \ldots, x_k \rangle, w \rangle \in A.$$

(3) If $\hat{f} = \langle 3, k, i \rangle$ for some $k \geq 1$ and $1 \leq i \leq k$, then

$$\text{for every } x_1, \ldots, x_k, \quad \langle \hat{f}, \langle x_1, \ldots, x_k \rangle, x_i \rangle \in A.$$

(4) If $\hat{f} = \langle 4, k, \langle \hat{h}, \hat{g}_1, \ldots, \hat{g}_m \rangle \rangle$ for some $\hat{h}, \hat{g}_1, \ldots, \hat{g}_m$ and if for some $x_1, \ldots, x_k, w_1, \ldots, w_m, w$ we have

$$\langle \hat{g}_1, \langle x_1, \ldots, x_k \rangle, w_1 \rangle \in A, \ldots, \langle \hat{g}_m, \langle x_1, \ldots, x_k \rangle, w_m \rangle \in A,$$
$$\langle \hat{h}, \langle w_1, \ldots, w_m \rangle, w \rangle \in A,$$

then

$$\langle \hat{f}, \langle x_1, \ldots, x_k \rangle, w \rangle \in A.$$

(5) If $\hat{f} = \langle 5, k+1, \langle \hat{g}, \hat{h} \rangle \rangle$ for some $\hat{g}$, $\hat{h}$ and for some $x_1, \ldots, x_k$, w_0 we have $\langle \hat{g}, \langle x_1, \ldots, x_k \rangle, w_0 \rangle \in A$, then

$$\langle \hat{f}, \langle 0, x_1, \ldots, x_k \rangle, w_0 \rangle \in A.$$

(5′) If $\hat{f} = \langle 5, k+1, \langle \hat{g}, \hat{h} \rangle \rangle$ for some $\hat{g}$, $\hat{h}$ and for some $x_1, \ldots, x_k$, w_n, w, n, $\langle \hat{f}, \langle n, x \rangle, w_n \rangle \in A$ and $\langle \hat{h}, \langle w_n, n, x \rangle, w \rangle \in A$, then

$$\langle \hat{f}, \langle n+1, x_1, \ldots, x_k \rangle, w \rangle \in A.$$

(There are two similar clauses which come from Case 5(ii) in the definition of justifications and which we will omit here.)

(6) If $\hat{f} = \langle 6, k, \hat{g} \rangle$ for some $\hat{g}$ and if for some $w_0 \neq 0$, $w_1 \neq 0, \ldots, w_{m-1} \neq 0$ and some $x_1, \ldots, x_k$ we have

$$\langle \hat{g}, \langle 0, x_1, \ldots, x_k \rangle, w_0 \rangle \in A, \ldots, \langle \hat{g}, \langle m-1, x_1, \ldots, x_k \rangle, w_{m-1} \rangle \in A,$$
$$\langle \hat{g}, \langle m, x_1, \ldots, x_k \rangle, 0 \rangle \in A,$$

then

$$\langle \hat{f}, \langle x_1, \ldots, x_k \rangle, m \rangle \in A.$$

(In this clause we allow $m = 0$ in which case the sequence $w_0, w_1, \ldots, w_{m-1}$ is empty.)

3F.4. THEOREM. *The set A is semirecursive. Moreover, for each code z of a recursive function f_z,*

$$f_z(x_1, \ldots, x_k) = w \Leftrightarrow \langle z, \langle x_1, \ldots, x_k \rangle, w \rangle \in A.$$

PROOF. We show first the second assertion, *which holds only for $z \in C$*, i.e. there is no suggestion that whenever $\langle z, \langle x_1, \ldots, x_k \rangle, w \rangle \in A$, then we must have $z \in C$.

Proof of the implication

$$f_z(x_1, \ldots, x_k) = w \Rightarrow \langle z, \langle x_1, \ldots, x_k \rangle, w \rangle \in A$$

is by induction on the code $z \in C$ and it is easy. For example, if $z = \langle 6, k, e \rangle$, then we know that e codes a $(k+1)$-ary function f_e and

$$f_z(x_1, \ldots, x_k) = \mu m[f_e(m, x_1, \ldots, x_k) = 0].$$

Thus if $f_z(x_1, \ldots, x_k) = m$, there are numbers $w_0, \ldots, w_{m-1}$ all $\neq 0$ such that $f_e(0, x_1, \ldots, x_k) = w_0, \ldots, f_e(m-1, x_1, \ldots, x_k) = w_{m-1}$ and $f_e(m, x_1, \ldots, x_k) = 0$. Now $e = (z)_2 < z$, so we can apply the induction hypothesis on these assertions, whence by clause (6) in the defining conditions for A, $\langle z, \langle x_1, \ldots, x_k \rangle, m \rangle \in A$.

We will omit the other cases of this induction, but it is worth pointing out that in case (5), when $z = \langle 5, k+1, \langle e, u\rangle\rangle$ for some k, e, u, the implication

$$f_z(n, x_1,\ldots, x_k) = w \Longrightarrow \langle z, \langle n, x_1,\ldots, x_k\rangle, w\rangle \in A$$

is proved by induction on n.

To prove the converse implication, notice that if $v \in A$, then v must be of the form in one of the conclusions of the clauses (1)–(6) and it must satisfy the corresponding hypothesis, e.g. $v = \langle \hat{f}, \langle 0, x_1,\ldots, x_k\rangle, w_0\rangle$ for some $x_1,\ldots, x_k, w_0$ and $\hat{f} = \langle 5, k+1, \langle \hat{g}, \hat{h}\rangle\rangle$ and $\langle \hat{g}, \langle x_1,\ldots, x_k\rangle, w_0\rangle \in A$; because if v is not of the proper form or does not satisfy the corresponding hypothesis, then $A - \{v\}$ satisfies all (1)–(6) and hence $A - \{v\} \supseteq A$ by the definition of A, i.e. $v \notin A$. Now the implication

$$\langle z, \langle x_1,\ldots, x_k\rangle, w\rangle \in A \Longrightarrow f_z(x_1,\ldots, x_k) = w$$

can be proved for every *code* z by induction on z easily, just as the converse implication was proved.

We now outline a proof of the first assertion of the theorem, that A is semirecursive. The idea is to analyze the inductive definition of A in the same way that we analyzed the inductive definition of the class of recursive functions. Thus an *A-derivation* is a finite sequence of numbers

$$u_0, u_1,\ldots, u_n$$

which *proves* that $u_n \in A$, i.e. each u_i is in A either by virtue of clauses (1), (2), (3) or by virtue of one of the remaining clauses and the fact that certain u_j's with $j < i$ are of a certain form. Once this is written down explicitly, it is trivial to check that the relation

$$P(u) \Longleftrightarrow Seq(u) \mathbin{\&} (\exists n \leq u)[lh(u) = n \mathbin{\&} (u)_0,\ldots, (u)_{n-1} \textit{ is an } A\textit{-derivation}]$$

is recursive. But

$$v \in A \Longleftrightarrow (\exists u)(\exists n)[P(u) \mathbin{\&} lh(u) = n \mathbin{\&} v = (u)_{n-1}],$$

so that A is semirecursive. ⊣

3F.5. THEOREM (Kleene [1943]). *For each $k \geq 1$, there is a semirecursive $G \subseteq \omega \times \omega^k$ which parameterizes the semirecursive subsets of ω^k.*

PROOF. Put

$$G(z, x_1,\ldots, x_k) \Longleftrightarrow (\exists t)[\langle z, \langle x_1,\ldots, x_k, t\rangle, 1\rangle \in A].$$

Clearly G is semirecursive.

If $P \subseteq \omega^k$ is semirecursive, then for some recursive R,

$$\begin{aligned} P(x_1,\dots,x_k) &\Leftrightarrow (\exists t)R(x_1,\dots,x_k,t) \\ &\Leftrightarrow (\exists t)[\chi_R(x_1,\dots,x_k,t)=1] \\ &\Leftrightarrow (\exists t)[f_z(x_1,\dots,x_k,t)=1], \end{aligned}$$

with z any code of χ_R. By 3F.4 then,

$$\begin{aligned} P(x_1,\dots,x_k) &\Leftrightarrow (\exists t)[\langle z, \langle x_1,\dots,x_k,t\rangle, 1\rangle \in A] \\ &\Leftrightarrow G(z,x_1,\dots,x_k). \end{aligned}$$ ⊣

This theorem is usually called the *Enumeration Theorem* for semirecursive (or *recursively enumerable*) relations on ω. It is one of the basic results of recursion theory. From it we can prove easily the key result of this section.

3F.6. *The Parametrization Theorem for the Kleene Pointclasses.* For each product space $\mathcal{Y}$, the Kleene pointclasses Σ^0_n, Π^0_n, Σ^1_n, Π^1_n are $\mathcal{Y}$-parametrized.

PROOF. Take first the simple case $\mathcal{Y}=\omega$. Given $\mathcal{X}$, choose $G \subseteq \omega\times\omega$ to be semirecursive and parametrize the semirecursive subsets of ω and put

$$H(z,x) \Leftrightarrow (\exists s)[x \in N(\mathcal{X},s) \;\&\; G(z,s)].$$

By 3C.5, H is universal for $\Sigma^0_1{\restriction}\mathcal{X}$.

If $\mathcal{Y}=\omega^n$ for some n, take

$$H'(z_1,\dots,z_n,x) \Leftrightarrow H(z_1,x)$$

with this H.

If $\mathcal{Y}$ is not of type 0, then $\mathcal{Y}$ is perfect. Choose a recursive σ which satisfies Lemma 1A.2 by 3D.15 for the space $\mathcal{Y}$. Given a space $\mathcal{X}$, let G be as above and put

$$\begin{aligned} H(y,x) \Leftrightarrow (\exists u)(\exists z)[&Seq(u) \;\&\; lh(u)=z+1 \\ &\&\; (\forall i<z)[(u)_i=0] \;\&\; (u)_z=1 \;\&\; y \in N(\mathcal{Y},\sigma(u)) \\ &\&\; (\exists s)[x\in N(\mathcal{X},s) \;\&\; G(z,s)]]. \end{aligned}$$

Clearly H is semirecursive.

Fix $y \in \mathcal{Y}$ and suppose that for some x, $H(y,x)$ holds. It follows easily from the properties of σ that there is a unique sequence code u of the form $\langle 0, 0, \ldots, 0, 1\rangle$ and length $z+1$ such that for all x,

$$H(y, x) \Leftrightarrow (\exists s)[x \in N(\mathcal{X}, s) \ \&\ G(z, s)].$$

Thus

$$H_y = \{x : H(y, x)\}$$

is a semirecursive subset of $\mathcal{X}$.

Conversely, if

$$P(x) \Leftrightarrow (\exists s)[x \in N(\mathcal{X}, s) \ \&\ G(z, s)]$$

is any semirecursive subset of $\mathcal{X}$, take u with $lh(u) = z+1$, $u = \langle 0, 0, \ldots, 0, 1\rangle$, choose any $y \in N(\mathcal{Y}, \sigma(u))$ and verify easily that

$$\begin{aligned} H(y, x) &\Leftrightarrow (\exists s)[x \in N(\mathcal{X}, s) \ \&\ G(z, s)] \\ &\Leftrightarrow P(x), \end{aligned}$$

so that $P = H_y$.

We have now shown that Σ^0_1 is $\mathcal{Y}$-parametrized for every $\mathcal{Y}$ and the theorem follows by 1D.2. ⊣

3F.7. *The Hierarchy Theorem for the Kleene Pointclasses.* For every product space $\mathcal{X}$ the following diagrams of proper inclusions hold:

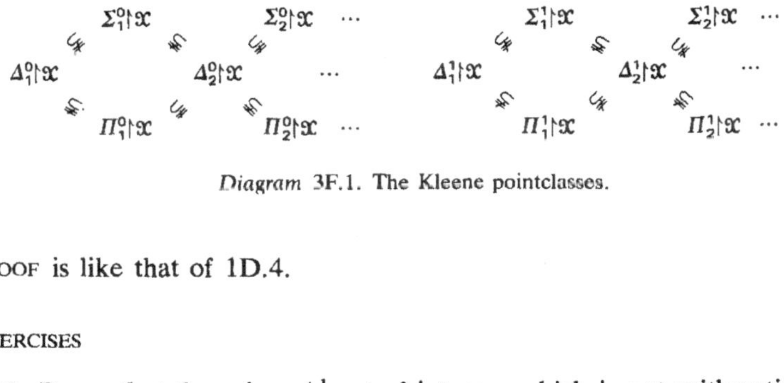

Diagram 3F.1. The Kleene pointclasses.

PROOF is like that of 1D.4. ⊣

EXERCISES

3F.8. Prove that there is a Δ^1_1 set of integers which is not arithmetical.

Hint. Define $G_n \subseteq \omega \times \omega$ in a canonical way to be universal for $\Sigma^0_n{\restriction}\omega$ and take

$$H = \{\langle n, e, t\rangle : G_n(e, t)\}.$$ ⊣

3F.9. Prove that for each n, there is a set $A \subseteq \mathcal{N}$ in $\mathbf{\Delta}^0_n$ such that for every product space $\mathcal{Z}$ and for every Δ^0_n set $Q \subseteq \mathcal{Z} \times \mathcal{N}$ and for every $z \in \mathcal{Z}$,

$$A \neq Q_z = \{\alpha : Q(z, \alpha)\}.$$

Infer that the following plausible sounding conjecture is false: P is $\Delta^0_n(z)$ if and only if there is some Δ^0_n set Q such that $P(x) \Leftrightarrow Q(z, x)$.

Similarly with $\mathbf{\Delta}^1_n$, Δ^1_n in place of $\mathbf{\Delta}^0_n$, Δ^0_n throughout.

Hint. Let $Q^0, Q^1, \ldots$ be an enumeration of all the Δ^0_n subsets of $\mathcal{N} \times \mathcal{N}$ and take

$$A = \{\alpha : \neg Q^{\alpha(0)}(\alpha^\star, \alpha)\},$$

where $\alpha^\star = t \mapsto \alpha(t+1)$. It is immediate that $A \neq Q_\varepsilon$ for every Δ^0_n subset Q of $\mathcal{N} \times \mathcal{N}$ and every ε. Using a recursive surjection $\pi : \mathcal{N} \twoheadrightarrow \mathcal{Z}$, show that $A \neq Q_z$ for every Δ^0_n set $Q \subseteq \mathcal{Z} \times \mathcal{N}$ and every $z \in \mathcal{Z}$. ⊣

In the case of $\mathbf{\Delta}^0_1$, this construction gives a clopen $A \subseteq \mathcal{N}$ which is not Q_z for any recursive $Q \subseteq \mathcal{Z} \times \mathcal{N}$, any z. Its characteristic function χ_A is continuous and cannot be obtained from any recursive function by fixing one of the arguments.

3G. Partial functions and the substitution property

The notion of Γ-recursion is an effective refinement of the classical notion of Γ-measurability and its basic properties can be established easily when Γ is a Σ-pointclass, as we saw in 3D. To obtain a smoother theory of Γ-recursion which refines the classical theory of *continuous functions* we must impose an additional condition on Γ which will insure (in particular) that Γ is closed under substitution of Γ-recursive functions. As it turns out, the correct formulation of this *substitution property* involves *partial functions* in a natural way.

A *partial function on* $\mathfrak{X}$ *to* $\mathcal{Y}$ is simply a (total) function

$$f : D \to \mathcal{Y}$$

with domain some subset D of $\mathfrak{X}$. We will use the notation

$$f : \mathfrak{X} \rightharpoonup \mathcal{Y}$$

for these objects, even when the domain D of f is a proper subset of $\mathfrak{X}$—even in the extreme case when $D = \emptyset$! We also write

$$f(x)\downarrow \Leftrightarrow f \text{ is defined at } x,$$

so that

$$D = Domain(f) = \{x \in \mathcal{X} : f(x)\downarrow\}.$$

The *domain of the composition* of two partial functions is defined in the natural way:

$$f(g(x)\downarrow \Rightarrow g(x))\downarrow \ \&\ [\text{if } g(x) = y, \text{ then } f(y)\downarrow].$$

In particular, if

$$f(x) = (f_1(x), \dots, f_l(x))$$

with $f_1, \dots, f_l$ partial, then

$$f(x)\downarrow \Leftrightarrow f_1(x)\downarrow \ \&\ \cdots\ \&\ f_l(x)\downarrow.$$

We should emphasize that when we call $f : \mathcal{X} \to \mathcal{Y}$ a *partial function*, f *could* be total; but when we call f a *function*, then in fact f *must* be total, i.e. $Domain(f) = \mathcal{X}$.

If $f : \mathcal{X} \to \mathcal{Y}$ is a partial function, $D \subseteq Domain(f)$ and $P \subseteq \mathcal{X} \times \omega$ is in some pointclass Γ, we say that *P computes f on D* if

$$x \in D \Rightarrow (\forall s)[f(x) \in N_s \Leftrightarrow P(x, s)].$$

In the notation we introduced in 3D, we can rewrite this as

$$x \in D \Rightarrow \mathcal{U}(f(x)) = \{s : P(x, s)\} = P_x.$$

A partial function $f : \mathcal{X} \to \mathcal{Y}$ is *Γ-recursive on D* if some P in Γ computes f on D. Most often we will be looking at partial functions which are *Γ-recursive on their domain*; if f is Γ-recursive on $D = Domain(f)$ and in addition $Domain(f)$ is in Γ, we say that f is *Γ-recursive.* If $\Gamma = \Sigma^0_1$, we say *recursive* for Σ^0_1-recursive.

The class of recursive partial functions has been studied extensively on ω, but in the wider context of product spaces it is difficult to control the domains of partial functions and the weaker notion of *recursion on a set D* is more useful.

It is clear that if $f : \mathcal{X} \to \mathcal{Y}$ is total, then f is *Γ-recursive* (*on* $\mathcal{X}$) in the present sense exactly when it is Γ-recursive in the sense of 3D.

The condition we want to impose on a pointclass Γ is (roughly) that it be closed under substitution of partial functions which are Γ-recursive on some set D, at least when we restrict these substitutions to the points in D. Precisely: a pointclass Γ has *the substitution property* if for each $Q \subseteq \mathcal{Y}$ in Γ and for each partial function $f : \mathcal{X} \to \mathcal{Y}$ which is Γ-recursive on its domain, there is some $Q^* \subseteq \mathfrak{X}$ in Γ such that for all $x \in \mathfrak{X}$,

$$f(x)\downarrow \Rightarrow [Q^*(x) \Leftrightarrow Q(f(x))].$$

Σ-pointclasses with the substitution property carry a very reasonable recursion theory, particularly if they are ω-parametrized (as is Σ_1^0). We will come back to this in Chapter 7, but we will first put down here the few facts that we need in the interim.

3G.1. THEOREM. *If Γ is a Σ-pointclass with the substitution property, then the collection of partial functions which are Γ-recursive on their domain is closed under composition; moreover, Γ is closed under the substitution of Γ-recursive (total) functions, so in particular, Γ is adequate.*

PROOF. The second assertion is immediate.

To prove the first assertion, suppose $g : \mathfrak{X} \to \mathfrak{Y}$, $f : \mathfrak{Y} \to \mathfrak{Z}$ are both Γ-recursive on their domain computed by $P \subseteq \mathfrak{X} \times \omega$, $Q \subseteq \mathfrak{Y} \times \omega$ respectively and let $h(x) = f(g(x))$. Now if $h(x)\downarrow$, then $g(x)\downarrow$, say $g(x) = y$ and $f(y)\downarrow$. Since Q computes f, we have

$$(\forall s)[f(y) \in N_s \Leftrightarrow Q(y, s)]$$

i.e.

$$(\forall s)[f(g(x)) \in N_s \Leftrightarrow Q(g(x), s)].$$

Since g is Γ-recursive on its domain, so is (easily) the map

$$(x, s) \mapsto (g(x), s);$$

thus by the substitution property, there is some Q^* in Γ so that

$$g(x)\downarrow \Rightarrow [Q(g(x), s) \Leftrightarrow Q^*(x, s)].$$

It follows immediately that

$$h(x)\downarrow \Rightarrow (\forall s)[h(x) \in N_s \Leftrightarrow Q^*(x, s)]$$

so that h is Γ-recursive on its domain. ⊣

Let us now verify that the substitution property is easy to establish.

3G.2. THEOREM. (i) *Σ_1^0 has the substitution property.*

(ii) *If Γ is a Σ-pointclass with the substitution property, then so is each relativization $\Gamma(w)$.*

(iii) *If Γ is a Σ-pointclass closed under $\forall^\omega$ and either $\exists^{\mathcal{Y}}$ or $\forall^{\mathcal{Y}}$, then Γ has the substitution property; in particular, Σ^1_n, Π^1_n all do.*

PROOF. (i) Suppose $Q \subseteq \mathcal{Y}$ is semirecursive, so that by 3C.5

$$Q(y) \Leftrightarrow (\exists s)[y \in N_s \ \&\ Q'(s)],$$

with a semirecursive Q'. If $f : \mathcal{X} \to \mathcal{Y}$ is partial and computed on its domain by some semirecursive $P \subseteq \mathcal{X} \times \omega$, put

$$Q^*(x) \Leftrightarrow (\exists s)[P(x, s) \ \&\ Q'(s)];$$

now if $f(x)\downarrow$, then

$$f(x) \in N_s \Leftrightarrow P(x, s),$$

so that

$$\begin{aligned} Q^*(x) &\Leftrightarrow (\exists s)[f(x) \in N_s \ \&\ Q'(s)] \\ &\Leftrightarrow Q(f(x)). \end{aligned}$$

(ii) Suppose $Q \subseteq \mathcal{Y}$ is in $\Gamma(w)$, so that

$$Q(y) \Leftrightarrow Q'(w, y)$$

for some Q' in Γ and suppose that $f : \mathcal{X} \to \mathcal{Y}$ is computed on its domain by some $P \subseteq \mathcal{X} \times \omega$ in $\Gamma(w)$; again,

$$P(x, s) \Leftrightarrow P'(w, x, s)$$

for some P' in Γ. Now P' computes on its domain the partial function $f' : \mathcal{W} \times \mathcal{X} \to \mathcal{Y}$ defined by

$$\begin{aligned} &f'(w', x)\downarrow \Leftrightarrow \text{for some } y, \mathcal{U}(y) = \{s : P'(w', x, s)\}, \\ &f'(w', x)\downarrow \Rightarrow (\forall s)[f'(w', x) \in N_s \Leftrightarrow P'(w', x, s)]. \end{aligned}$$

Notice that for the specific fixed w we have

$$\begin{aligned} &f(x)\downarrow \Rightarrow f'(w, x)\downarrow, \\ &f(x) = f'(w, x). \end{aligned}$$

The partial function

$$g(w', x) = (w', f'(w', x))$$

is Γ-recursive on its domain, so by the substitution property for Γ, there is some $Q'' \subseteq \mathcal{W} \times \mathcal{X}$ in Γ so that

$$g(w', x)\downarrow \Rightarrow [Q''(w', x) \Leftrightarrow Q'(w', f'(w', x))];$$

setting $w' = w$ then, we have

$$f(x)\downarrow \Rightarrow [Q''(w, x) \Leftrightarrow Q'(w, f'(w, x))]$$
$$\Rightarrow [Q''(w, x) \Leftrightarrow Q(f(x))]$$

and we can take

$$Q^*(x) \Leftrightarrow Q''(w, x)$$

to satisfy the substitution property for $\Gamma(w)$.

(iii) Suppose the partial function $f: \mathfrak{X} \to \mathcal{Y}$ is computed on its domain by $P \subseteq \mathfrak{X} \times \omega$ in Γ, $Q \subseteq \mathcal{Y}$ is in Γ and Γ is closed under $\forall^\omega$ and $\exists^{\mathcal{Y}}$. Take

$$Q^*(x) \Leftrightarrow (\exists y)[Q(y) \,\&\, (\forall s)[y \in N_s \Rightarrow P(x, s)]].$$

This is easily in Γ and if $f(x)\downarrow$, then for any y,

$$(\forall s)[y \in N_s \Rightarrow P(x, s)] \Rightarrow (\forall s)[y \in N_s \Rightarrow f(x) \in N_s]$$
$$\Rightarrow y = f(x),$$

so that

$$Q^*(x) \Leftrightarrow Q(f(x)).$$

Similarly, if Γ is closed under $\forall^{\mathcal{Y}}$, take

$$Q^*(x) \Leftrightarrow (\forall y)[Q(y) \vee (\exists s)[P(x, s) \,\&\, y \notin N_s]]. \qquad \dashv$$

To appreciate the need for using partial functions in formulating the substitution property one should spend a few minutes trying to prove that if a Σ-pointclass Γ is closed under (total) Γ-recursive substitutions, then each relativization $\Gamma(w)$ is closed under $\Gamma(w)$-recursive substitutions.

EXERCISES

3G.3. Show that not every Σ-pointclass is adequate.

Hint. Choose a measure μ_X on each basic space X, with $\mu_{\mathcal{R}} =$ Lebesgue measure and $\mu_\omega =$ the trivial (counting) measure, let $\mu_{\mathfrak{X}}$ be the (completed) product measure on each product space and take

$$\Gamma = \text{all measurable pointsets.}$$

It is easy to check that Γ is a Σ-pointclass. If $A \subseteq \mathcal{R}$ is not Lebesgue measurable with characteristic function χ_A, then

$$B = \{(x, \chi_A(x)) : x \in \mathcal{R}\}$$

is measurable (with measure 0) in the plane, since it is a subset of two lines. But

$$x \in A \Leftrightarrow (x, 1) \in B,$$

so if Γ were closed under recursive substitution then A would be in Γ, since $x \mapsto (x, 1)$ is recursive. ⊣

3G.4. Assume that Γ is a Σ-pointclass with the substitution property and prove the following.

(i) A partial function $f : \mathfrak{X} \rightharpoonup \omega$ is Γ-recursive on its domain if and only if there is some $Q \subseteq \mathfrak{X} \times \omega$ in Γ so that

$$f(x)\downarrow \Rightarrow [f(x) = w \Leftrightarrow Q(x, w)].$$

(ii) A partial function $f : \mathfrak{X} \rightharpoonup \mathcal{N}$ is Γ-recursive on its domain if and only if the function $f^* : \mathfrak{X} \times \omega \rightharpoonup \omega$ is Γ-recursive on its domain, where

$$f^*(x, i)\downarrow \Leftrightarrow f(x)\downarrow,$$
$$f(x)\downarrow \Rightarrow [f^*(x, i) = f(x)(i)].$$

(iii) A partial function $f : \mathfrak{X} \rightharpoonup \mathcal{Y}_1 \times \cdots \times \mathcal{Y}_l$ is Γ-recursive on its domain if and only if

$$f(x) = (f_1(x), \ldots, f_l(x))$$

with $f_1, \ldots, f_l$ Γ-recursive on their (common) domain. ⊣

The most useful property of Γ-recursion is embodied in the following completely trivial result, which we put down for the record.

3G.5. If Γ is a Σ-pointclass with the substitution property and $f : \mathfrak{X} \rightharpoonup \mathcal{Y}$ is Γ-recursive on its domain, then for each x,

$$f(x)\downarrow \Rightarrow f(x) \text{ is } \Gamma(x)\text{-recursive};$$

if in addition $\Pi^0_1 \subseteq \Gamma$, then

$$f(x)\downarrow \Rightarrow f(x) \text{ is } \Delta(x)\text{-recursive}.$$

Hint. By the definitions,

$$f(x)\downarrow \Rightarrow \mathfrak{U}(f(x)) = \{s : P(x, s)\}$$

where $P \in \Gamma$ computes f on its domain.

For the second assertion, the relation

$$Q(y, s) \Leftrightarrow y \notin N_s$$

is Π^0_1, hence in Γ. Since the partial function $(x, s) \rightharpoonup (f(x), s)$ is obviously Γ-recursive on *Domain*(f), by the substitution property there is some Q^* in Γ so that

$$f(x)\downarrow \Longrightarrow [f(x) \notin N_s \Leftrightarrow Q^*(x, s)]$$
$$\Longrightarrow \mathfrak{U}(f(x)) = \{s: \neg Q^*(x, s)\},$$

so that $f(x)$ is also $\neg\Gamma(x)$-recursive, hence $\Delta(x)$-recursive. ⊣

3H. Codings, uniformity and good parametrizations

A coding for a set A is any surjection

$$\pi : D \twoheadrightarrow A$$

of a set $D = Domain(\pi)$ onto A. If $a \in A$ and $\pi(c) = a$, we call c *a code for* a in the coding π.

In our case we will always have $D \subseteq \mathfrak{X}$ for some product space $\mathfrak{X}$ (usually $\mathcal{N}$), so we can think of a *coding in* $\mathfrak{X}$ as a partial function

$$\pi : \mathfrak{X} \twoheadrightarrow A,$$

with $D = Domain(\pi)$.

Let us look at some examples.

(1) The set ω codes the collection of basic nbhds of a product space $\mathfrak{X}$ by the map

$$\pi(s) = N_s.$$

(2) The set of integers C defined in 3F codes the collection of recursive functions on ω by the map

$$\pi(z) = f_z \quad (z \in C).$$

(3) Suppose Γ is $\mathcal{N}$-parametrized and $G \subseteq \mathcal{N} \times \mathfrak{X}$ is universal for the Γ-subsets of $\mathfrak{X}$. The map

$$\alpha \rightharpoonup G_\alpha = \{x: G(\alpha, x)\}$$

codes the Γ-subsets of $\mathfrak{X}$ in $\mathcal{N}$.

(4) With Γ and G as above, let

$$C = \{\alpha: G_{(\alpha)_0} \text{ is the complement of } G_{(\alpha)_1}\}.$$

Now each $\alpha \in C$ determines a set $G_{(\alpha)_0}$ in Δ, so that the partial map

$$\pi(\alpha) = G_{(\alpha)_0} \quad (\alpha \in C)$$

is a coding of $\Delta \upharpoonright \mathfrak{X}$.

(5) Fix a space $\mathfrak{X}$ and fix an open set $G \subseteq \mathcal{N} \times \mathfrak{X}$ which is universal for the open subsets of $\mathfrak{X}$. For each countable ordinal ξ define the set B_ξ of codes for $\mathbf{\Sigma}^0_\xi$ by the recursion:

$$B_0 = \{\alpha : \alpha(0) = 0\}$$
$$B_\xi = \{\alpha : \alpha(0) = 1 \;\&\; (\forall n)[(\alpha^\star)_n \in \bigcup_{\eta<\xi} B_\eta]\},$$

where

$$\alpha^\star(t) = \alpha(t+1).$$

Define maps

$$\pi_\xi : B_\xi \twoheadrightarrow \mathbf{\Sigma}^0_\xi \upharpoonright \mathfrak{X}$$

by the recursion

$$\pi_0(\alpha) = G_{\alpha^\star},$$
$$\pi_\xi(\alpha) = \bigcup_n (\mathfrak{X} - \pi_{\eta(n)}((\alpha^\star)_n))$$

where

$$\eta(n) = \text{least } \eta \text{ so that } (\alpha^\star)_n \in B_\eta.$$

It is obvious that each π_ξ is a coding for the $\mathbf{\Sigma}^0_\xi$ subsets of $\mathfrak{X}$. One can also check by an easy induction on ξ that

$$0 < \eta \leq \xi \Rightarrow [B_\eta \subseteq B_\xi \;\&\; \pi_\eta = \pi_\xi \upharpoonright B_\eta]$$

so that the limit function

$$\pi = \bigcup_\xi \pi_\xi$$

is a coding for the Borel subsets of $\mathfrak{X}$ with domain

$$B = \bigcup_\xi B_\xi.$$

If we take $\Gamma = \mathbf{\Sigma}^1_1$ in example (4), we get a natural coding of the $\mathbf{\Delta}^1_1$ subsets of $\mathfrak{X}$. In (5) we defined a coding for the Borel subsets of $\mathfrak{X}$, and by the Suslin theorem *every* $\mathbf{\Delta}^1_1$ *subset of* $\mathfrak{X}$ *is Borel.* We will show in Chapter 7 that the Suslin theorem holds *uniformly in the codings* in the following sense: *there is a partial function*

$$f : \mathcal{N} \to \mathcal{N}$$

which is (Σ^0_1-)*recursive on the set* C (*in particular* $C \subseteq Domain(f)$) *and such that if* α *is a* $\mathbf{\Delta}^1_1$-*code of* $A \subseteq \mathfrak{X}$, *then* $f(\alpha)$ *is a Borel code of* A. This

Suslin-Kleene Theorem is one of the central results of the effective theory.

To define this important notion of *uniformity* in general, suppose that

$$\pi : \mathfrak{X} \twoheadrightarrow A, \qquad \rho : \mathfrak{Y} \twoheadrightarrow B$$

are codings for the sets A, B (π, ρ are partial functions), suppose $R \subseteq A \times B$ is a relation on $A \times B$ and suppose that Γ is a fixed Σ-pointclass with the substitution property. We say that the assertion

$$(*) \qquad (\forall a \in A)(\exists b \in B) R(a, b)$$

holds Γ-uniformly (in the codings π, ρ) if there exists a partial function $f : \mathfrak{X} \to \mathfrak{Y}$ which is Γ-recursive on $Domain(\pi)$ and such that

$$(**) \qquad \pi(x)\downarrow \Rightarrow R(\pi(x), \rho(f(x)));$$

i.e. whenever x codes some $a \in A$, then $f(x)$ gives us a code of some $b \in B$ so that $R(a, b)$.

In the important case $\Gamma = \Sigma^0_1$, we talk of *recursive uniformity* or simply *uniformity*.

To take a trivial example, let $\Delta = \Gamma \cap \neg\Gamma$ be coded in $\mathcal{N}$ as in (4) and consider the assertion

$$\Delta \text{ is closed under complementation,}$$

i.e.

$$(\forall P \in \Delta)(\exists Q \in \Delta)[Q \text{ is the complement of } P];$$

this holds uniformly because the function

$$\alpha \mapsto \langle (\alpha)_1, (\alpha)_0 \rangle$$

is recursive (and hence recursive on the set C of codes).

For a slightly more interesting example, take the assertion: "if $g, h_1, \ldots, h_m$ are recursive functions on ω with the proper number of arguments, then the composition

$$f(x) = g(h_1(x), \ldots, h_m(x))$$

is also recursive." Using the coding (2) above for recursive functions on ω we can easily show that this statement holds uniformly: i.e. there is a recursive function $u(\hat{g}, \hat{h}_1, \ldots, \hat{h}_m)$ so that whenever $\hat{g}, \hat{h}_1, \ldots, \hat{h}_m$ code (in C) recursive functions with the appropriate numbers of variables, then $u(\hat{g}, \hat{h}_1, \ldots, \hat{h}_m)$ codes their composition.

We will often call any partial function f which satisfies $(**)$ above a *uniformity* (which establishes that $(*)$ holds Γ-uniformly).

Starting with the next chapter, we will meet several situations in which codings and uniformities come up naturally and non-trivially. Here we will confine ourselves to one simple but very useful preparatory result.

With each pointclass Γ we associate *the boldface pointclass* $\mathbf{\Gamma}$, where for $P \subseteq \mathfrak{X}$,

$$P \in \mathbf{\Gamma} \Leftrightarrow \text{there is some } P^* \subseteq \mathcal{N} \times \mathfrak{X},\ P^* \in \Gamma \text{ and some } \varepsilon \in \mathcal{N} \text{ so that } P = P^*_\varepsilon = \{x : P^*(\varepsilon, x)\}.$$

As usually

$$\mathbf{\Delta} = \mathbf{\Gamma} \cap \neg\mathbf{\Gamma}.$$

3H.1. THE GOOD PARAMETRIZATION LEMMA. *Suppose Γ is ω-parametrized and closed under recursive substitutions. Then we can associate with each space $\mathfrak{X}$ a set $G^{\mathfrak{X}} \subseteq \mathcal{N} \times \mathfrak{X}$ in Γ which is universal for $\mathbf{\Gamma} \restriction \mathfrak{X}$ and so that the following properties hold:*

(i) *For $P \subseteq \mathfrak{X}$,*

$$P \in \Gamma \Leftrightarrow P = G^{\mathfrak{X}}_\varepsilon \quad \textit{with a recursive } \varepsilon \in \mathcal{N}.$$

(ii) *For each space $\mathfrak{X}$ of type 0 or 1 and each $\mathfrak{Y}$, there is a recursive function*

$$S^{\mathfrak{X},\mathfrak{Y}} = S : \mathcal{N} \times \mathfrak{X} \to \mathcal{N}$$

so that

$$G^{\mathfrak{X}\times\mathfrak{Y}}(\varepsilon, x, y) \Leftrightarrow G^{\mathfrak{Y}}(S(\varepsilon, x), y).$$

PROOF. If $G \subseteq \omega \times \mathcal{N} \times \mathfrak{X}$ parametrizes $\Gamma \restriction (\mathcal{N} \times \mathfrak{X})$, take

$$G^*(\varepsilon, x) \Leftrightarrow G(\varepsilon(0), \varepsilon^\star, x)$$

with $\varepsilon^\star(t) = \varepsilon(t+1)$ and check easily that G^* parametrizes $\mathbf{\Gamma} \restriction \mathfrak{X}$ so that (i) holds. Thus we may assume that we are given parametrizations of $\mathbf{\Gamma}$ in Γ which satisfy (i)—we must obtain a new system of parametrizations which also satisfies (ii).

Call a space $\mathfrak{Y}$ (for this proof) *simple* if $\mathfrak{Y}$ has no initial factor of type 0 or 1, i.e. if it is impossible to write

$$\mathfrak{Y} = \mathfrak{Y}_1 \times \mathfrak{Y}_2$$

with $\mathfrak{Y}_1$ of type 0 or 1. We will first construct suitable parametrizations for all spaces of the form $\mathfrak{X} \times \mathfrak{Y}$ with $\mathfrak{X}$ of type 0 or 1 together with a fixed simple space $\mathfrak{Y}$.

For each space $\mathcal{X}$ of type 0 or 1 then fix a recursive homeomorphism

$$\pi_{\mathcal{X}} : \mathcal{N} \times \mathcal{X} \rightarrowtail\!\!\!\twoheadrightarrow \mathcal{N}$$

and let

$$V \subseteq \mathcal{N} \times (\mathcal{N} \times \mathcal{N} \times \mathcal{Y})$$

be in Γ and universal for the $\mathbf{\Gamma}$-subsets of $\mathcal{N} \times \mathcal{N} \times \mathcal{Y}$ so that (i) holds. Define $G \subseteq \mathcal{N} \times \mathcal{X} \times \mathcal{Y}$ by

$$G(\varepsilon, x, y) \Leftrightarrow V((\varepsilon)_0, (\varepsilon)_1, \pi_{\mathcal{X}}((\varepsilon)_2, x), y).$$

Clearly G is in Γ.

To prove that G is universal for $\mathbf{\Gamma} \upharpoonright \mathcal{X} \times \mathcal{Y}$, suppose that $Q \subseteq \mathcal{X} \times \mathcal{Y}$ is in $\mathbf{\Gamma}$ and set

$$Q'(\alpha, \beta, y) \Leftrightarrow Q(\mathrm{p}_{\mathcal{X}}(\pi_{\mathcal{X}}^{-1}(\beta)), y)$$

where

$$\mathrm{p}_{\mathcal{X}} : \mathcal{N} \times \mathcal{X} \rightarrow \mathcal{X}$$

is the projection map on $\mathcal{X}$. Now $Q' \in \mathbf{\Gamma}$, so that for some $\varepsilon \in \mathcal{N}$,

$$Q'(\alpha, \beta, y) \Leftrightarrow V(\varepsilon, \alpha, \beta, y)$$

and for any α, taking $\beta = \pi_{\mathcal{X}}(\alpha, x)$ we have

$$\begin{aligned} Q(x, y) &\Leftrightarrow Q'(\alpha, \pi_{\mathcal{X}}(\alpha, x), y) \\ &\Leftrightarrow V(\varepsilon, \alpha, \pi_{\mathcal{X}}(\alpha, x), y) \\ &\Leftrightarrow G(\langle \varepsilon, \alpha, \alpha \rangle, x, y). \end{aligned}$$

Choosing a recursive α, say $t \mapsto 0$ and $\varepsilon^* = \langle \varepsilon, \alpha, \alpha \rangle$, we have

$$Q(x, y) \Leftrightarrow G(\varepsilon^*, x, y).$$

It is also immediate that the universal set G satisfies (i).

Fix spaces $\mathcal{X}$, $\mathcal{W}$ of type 0 or 1; we must construct a recursive function

$$S : \mathcal{N} \times \mathcal{X} \rightarrow \mathcal{N}$$

so that

$$G(\varepsilon, x, w, y) \Leftrightarrow G(S(\varepsilon, x), w, y) \tag{1}$$

(where of course the G on the left stands for a different relation than that on the right—a pedantic notation here would introduce a lot of superscripts).

Put

$$P(\alpha, \beta, y) \Leftrightarrow G((\alpha)_0, \mathrm{p}_{\mathcal{X}}(\pi_{\mathcal{X}}^{-1}((\alpha)_1)), \mathrm{p}_{\mathcal{W}}(\pi_{\mathcal{W}}^{-1}(\beta)), y) \tag{2}$$

where G is the universal subset of $\mathcal{N}\times\mathcal{X}\times\mathcal{W}\times\mathcal{Y}$ we just defined and the recursive homeomorphisms and projections $\pi_{\mathcal{X}}$, $\pi_{\mathcal{W}}$, $\mathrm{p}_{\mathcal{X}}$, $\mathrm{p}_{\mathcal{W}}$ are as above. Now P is in Γ, so for a fixed recursive ε^*,

$$(3) \qquad P(\alpha, \beta, y) \Leftrightarrow V(\varepsilon^*, \alpha, \beta, y).$$

For arbitrary ε, x, w, let

$$\alpha = \langle \varepsilon, \pi_{\mathcal{X}}(\varepsilon, x)\rangle, \qquad \beta = \pi_{\mathcal{W}}(\varepsilon, w);$$

substituting in (2) and (3) we get

$$\begin{aligned} G(\varepsilon, x, w, y) &\Leftrightarrow P(\langle \varepsilon, \pi_{\mathcal{X}}(\varepsilon, x)\rangle, \pi_{\mathcal{W}}(\varepsilon, w), y) \\ &\Leftrightarrow V(\varepsilon^*, \langle \varepsilon, \pi_{\mathcal{X}}(\varepsilon, x)\rangle, \pi_{\mathcal{W}}(\varepsilon, w), y) \end{aligned}$$

and then by the definition of the universal sets,

$$G(\varepsilon, x, w, y) \Leftrightarrow G(\langle \varepsilon^*, \langle \varepsilon, \pi_{\mathcal{X}}(\varepsilon, x)\rangle, \varepsilon\rangle, w, y).$$

so that (1) holds with

$$S(\varepsilon, x) = \langle \varepsilon^*, \langle \varepsilon, \pi_{\mathcal{X}}(\varepsilon, x)\rangle, \varepsilon\rangle.$$

The construction is similar for spaces of type 0 or 1 (simply skip the $\mathcal{Y}$ above) and the universal sets we constructed work for the simple spaces (skip the $\mathcal{W}$ in the proof). ⊣

A system of sets $G^{\mathcal{X}} \subseteq \mathcal{N}\times\mathcal{X}$ in Γ which are $\mathbf{\Gamma}$-universal and satisfy (i) and (ii) of the theorem will be called a *good parametrization* (in Γ for $\mathbf{\Gamma}$). We will often simply say "*let* $G \subseteq \mathcal{N}\times\mathcal{X}$ *be a good universal set*" meaning that G belongs to a good parametrization when Γ is clear from the context. We will also tend to be a bit sloppy with notation and avoid all superscripts, so that the basic property of good parametrizations reads

$$G(\varepsilon, x, y) \Leftrightarrow G(S(\varepsilon, x), y).$$

Fix a good parametrization for each ω-parametrized, adequate pointclass Γ and consider the natural coding for $\mathbf{\Gamma}$ determined by this parametrization as in (3) above, by the map

$$\alpha \mapsto G_\alpha = \{x \in \mathcal{X}: G(\alpha, x)\}.$$

The restriction of these maps to recursive α's gives a coding for Γ.

Similarly, $\mathbf{\Delta}$ is coded by

$$\alpha \mapsto G_{(\alpha)_0},$$

on the set of codes $\{\alpha: G_{(\alpha)_0}$ is the complement of $G_{(\alpha)_1}\}$.

When we mention **Γ**-*codes* or **Δ**-*codes* of sets, we will refer to these fixed, *canonical* codings—we will do this quite frequently for $\mathbf{\Sigma}^1_n$-codes or $\mathbf{\Delta}^1_n$-codes, for example.

Lemma 3H.1 says that *the operation of passing to a section at a point of type* 0 *or* 1 is uniform, for an ω-parametrized, adequate Γ: i.e. if ε codes $P \subseteq \mathcal{X} \times \mathcal{Y}$ in $\mathbf{\Gamma}$ with $\mathcal{X}$ of type 0 or 1, then $S(\varepsilon, x)$ codes $P_x = \{y: P(x, y)\}$ in $\mathbf{\Gamma}$. But the lemma actually implies much more.

3H.2. THE UNIFORM CLOSURE THEOREM. *Suppose Γ is an ω-parametrized, adequate pointclass; if Γ is closed under any of the operations* &, $\vee$, $\exists^{\leq}$, $\forall^{\leq}$, $\exists^{\mathcal{Y}}$, $\forall^{\mathcal{Y}}$, then **Γ** *is uniformly closed under the same operation (in the codings induced by a good parametrization).*

PROOF. Suppose for example that Γ is closed under &. We must show that there is a recursive function $\mathsf{u}(\alpha, \beta)$ such that if

$$P(x) \Leftrightarrow G(\alpha, x)$$

$$Q(x) \Leftrightarrow G(\beta, x),$$

then

$$P(x) \,\&\, Q(x) \Leftrightarrow G(\mathsf{u}(\alpha, \beta), x).$$

To check this, put

$$R(\alpha, \beta, x) \Leftrightarrow G(\alpha, x) \,\&\, G(\beta, x).$$

By closure under recursive substitution and &, R is in Γ, so for a fixed recursive $\varepsilon^* \in \mathcal{N}$,

$$\begin{aligned} R(\alpha, \beta, x) &\Leftrightarrow G(\varepsilon^*, \alpha, \beta, x) \\ &\Leftrightarrow G(S(\varepsilon^*, \alpha, \beta), x) \end{aligned}$$

by the good parametrization lemma. Thus we can take

$$\mathsf{u}(\alpha, \beta) = S(\varepsilon^*, \alpha, \beta).$$

The argument for the other cases is similar. ⊣

There is another somewhat tricky corollary of the good parametrization lemma which can be viewed as *fixed-point theorem for parametrized pointclasses*. We put it down here because we need it for an important application in the next chapter, but its full significance will not be appreciated until Chapter 7.

3H.3. KLEENE'S RECURSION THEOREM FOR RELATIONS. *Suppose Γ is ω-parametrized and closed under recursive substitutions and suppose $R \subseteq \mathcal{N} \times \mathfrak{X}$ is in Γ; then we can find a recursive $\varepsilon^* \in \mathcal{N}$ so that the section*

$$R_{\varepsilon^*} = \{x : R(\varepsilon^*, x)\}$$

has Γ-code ε^*, i.e.

$$R(\varepsilon^*, x) \Leftrightarrow G(\varepsilon^*, x)$$

where G is the fixed good universal set for $\Gamma \upharpoonright \mathfrak{X}$.

PROOF. Let

$$P(\alpha, x) \Leftrightarrow R(S(\alpha, \alpha), x)$$

where S is recursive by 3H.1 and such that for all ε, α, x,

$$G(\varepsilon, \alpha, x) \Leftrightarrow G(S(\varepsilon, \alpha), x).$$

Since P is in Γ, there is a fixed recursive ε_0 so that

$$\begin{aligned} P(\alpha, x) &\Leftrightarrow G(\varepsilon_0, \alpha, x) \\ &\Leftrightarrow G(S(\varepsilon_0, \alpha), x) \end{aligned}$$

and hence for all α, x,

$$G(S(\varepsilon_0, \alpha), x) \Leftrightarrow R(S(\alpha, \alpha), x).$$

Now set $\alpha = \varepsilon_0$ in this equivalence and take $\varepsilon^* = S(\varepsilon_0, \varepsilon_0)$. ⊣

EXERCISES

We formulated 3H.1 for parametrizations in $\mathcal{N}$, since this version is most directly applicable. However, there is a similar result for parametrizations in ω which is occcassionally useful.

3H.4. Suppose Γ is ω-parametrized and closed under recursive substitutions. Then we can associate with each space $\mathfrak{X}$ a set $G^{\mathfrak{X}} \subseteq \omega \times \mathfrak{X}$ in Γ which is universal for $\Gamma \upharpoonright \mathfrak{X}$ and so that the following property holds: for each space $\mathfrak{X}$ of type 0 and each $\mathcal{Y}$ there is a recursive function

$$S^{\mathfrak{X} \times \mathcal{Y}} = S : \omega \times \mathfrak{X} \to \omega$$

so that

$$G^{\mathfrak{X} \times \mathcal{Y}}(e, x, y) \Leftrightarrow G^{\mathcal{Y}}(S(e, x), y).$$

Moreover, for any $\mathfrak{X}$, if $R \subseteq \omega \times \mathfrak{X}$ is in Γ, then we can find some $e^* \in \omega$ so that

$$R(e^*, x) \Leftrightarrow G^{\mathfrak{X}}(e^*, x). \qquad \dashv$$

3I. Historical remarks

[1] The class of *recursive functions* on the integers was introduced and studied in the mid-thirties in various ways and by several mathematicians, particularly Church, Kleene, Turing, Post and (later) Markov. We will not attempt to trace its history here since this is done in some detail in the classical monograph on the subject Kleene [1952a]. Our development in 3A follows very closely the approach of Kleene.

[2] The generalization of recursion theory to spaces of type 0 or 1 is (at least) implicit in Kleene [1952a] and more explicit in Kleene [1952b]. Alternative approaches to this theory were given later by Kleene [1959a] and Kreisel [1959].

[3] There were also several attempts to develop the theory of recursive functions on the reals, of which the most direct and successful was Lacombe [1955]. The later paper Lacombe [1959] is more in the spirit of what we are doing here, in fact it develops recursion theory in a more general context. The specific definitions we gave in this chapter are new and perhaps simpler than previous developments, but we have no significant new results here.

[4] The *arithmetical pointclasses* on ω were introduced in Kleene [1943] and later (independently) in Mostowski [1946]. They were studied extensively in Kleene [1952a].

[5] Taking again ω as his basic space, Kleene [1955b] introduced and studied the *analytical pontclasses*. The main aim of that paper was the study of the *hyperarithmetical* relations on ω which had been introduced independently (earlier) by Davis [1950] and Mostowski [1951]. These coincide with the Δ^1_1 relations, but the proof of this is quite difficult—it was first given in Kleene [1955c]. We will postpone studying the hyperarithmetical relations until Chapter 7.

[6] In his original development of the theory Kleene introduced specific parametrizations for the arithmetical and analytical pointclasses which he proved to be "good" in the sense of 3H.1. The fact that good parametrizations can be constructed given arbitrary ω-parametrizations was discovered by several people independently perhaps first by Friedman [1971].

[7] As we mentioned in the introduction to this book, the similarities and "analogies" between the effective theory on ω developed (mostly) by Kleene and classical descriptive set theory on $\mathcal{R}$ were noticed first by Mostowski and Addison. The present, unified *effective descriptive set theory* which we are studying here is the end result of a long process of generalization and abstraction which started with Addison [1954] and [1959a] and involved the work of many people.

CHAPTER 4

STRUCTURE THEORY FOR POINTCLASSES

We are now ready to plunge into a systematic study of the structure of Π^1_1 and Σ^1_2.

In many ways, this chapter is a continuation of Chapter 2; here too we will establish various interesting properties of $\mathbf{\Sigma}^1_1$ and $\mathbf{\Sigma}^1_2$ sets, in fact we will answer several natural questions left open there. What is new and different is that we will use systematically the methods of the effective theory which we developed in the preceding chapter.

It turns out that this infusion of ideas from recursion theory creates a more radical change in the flavor of the subject than one might think. It is not just a case of obtaining "finer" results about the lightface pointclasses with a little more computation, as we did in Chapter 3. Even when we prove theorems which are significant only for the boldface pointclasses, we will use recursion theory to great advantage.

The most important results of the chapter are *uniformization theorems*, particularly the Novikov–Kondo–Addison Theorem 4E.4 and the Δ-Uniformization Criterion 4D.4. The latter implies that in many special circumstances we can uniformize a Borel set by a Borel set.

As in Chapter 2, we will formulate many of the results of this chapter in a general setting, to ease extension to the higher projective pointclasses. This will lead us naturally to the axiomatic definition of a *Spector pointclass*, one of the key notions of the subject. Specifically for uniformization results, the notion of a *scale* will also prove very important.

Perhaps this is the most important chapter of this book, because it is the most characteristic of our subject. One could say that Chapter 1 was mostly topology, Chapter 2 was set theory and Chapter 3 was recursion theory; but this chapter would be out of place in anything but a book on descriptive set theory.

4A. The basic representation theorem for Π^1_1 sets

Most of the results of Chapter 2 depended directly on the fact that $\mathbf{\Sigma}^1_1$ sets are $\aleph_0$-Suslin. Here we will first formulate an effective version of this fact and then refine it to a representation theorem for Π^1_1 sets which is the key to the structure properties of this pointclass.

Recall from 3D.8 (*20) that

$$\bar{\alpha}(n) = \langle \alpha(0), \ldots, \alpha(n-1) \rangle.$$

This is a recursive function of α and n.

4A.1. THEOREM. (i) *A pointset* $P \subseteq \mathfrak{X} \times \mathcal{N}^l$ $(l \geq 1)$ *is* Σ^0_1 *if and only if there is a* Σ^0_1 *set* $Q \subseteq \mathfrak{X} \times \omega^l$ *such that*

$$P(x, \alpha_1, \ldots, \alpha_l) \Leftrightarrow (\exists t) Q(x, \bar{\alpha}_1(t), \ldots, \bar{\alpha}_l(t))$$

and

$$[Q(x, \bar{\alpha}_1(t), \ldots, \bar{\alpha}_l(t)) \;\&\; t < s] \Rightarrow Q(x, \bar{\alpha}_1(s), \ldots, \bar{\alpha}_l(s)).$$

Moreover, if $\mathfrak{X}$ *is of type* 0 *or* 1, *then* Q *may be chosen to be recursive.*

(*ii*) *A pointset* $P \subseteq \mathfrak{X}$ *is* Π^1_1, *if and only if there is a* Σ^0_1 *set* $Q \subseteq \mathfrak{X} \times \omega$ *such that*

$$P(x) \Leftrightarrow (\forall \alpha)(\exists t) Q(x, \bar{\alpha}(t))$$

and

$$[Q(x, \bar{\alpha}(t)) \;\&\; t < s] \Rightarrow Q(x, \bar{\alpha}(s)).$$

Moreover, if $\mathfrak{X}$ *is of type* 0 *or* 1, *then* Q *may be chosen to be recursive.*

PROOF. (ii) follows immediately from (i).

To prove (i), take $l = 1$ for simplicity of notation and suppose by 3C.5 that

$$P(x, \alpha) \Leftrightarrow (\exists u)(\exists v)\{x \in N(\mathfrak{X}, u) \;\&\; \alpha \in N(\mathcal{N}, v) \;\&\; P^*(u, v)\}$$

with P^* semirecursive, so there is a recursive R such that

$$P(x, \alpha) \Leftrightarrow (\exists u)(\exists v)(\exists n)\{x \in N(\mathfrak{X}, u) \;\&\; \alpha \in N(\mathcal{N}, v) \;\&\; R(u, v, n)\}.$$

By 3B.5, there are recursive functions g, h such that

$$\alpha \in N(\mathcal{N}, v) \Leftrightarrow ((v)_1)_1 \neq 0 \;\&\; (\forall i < g(v))[\alpha(i) = h(v, i)],$$

so that whenever $t \geq g(v)$, we easily have

$$\alpha \in N(\mathcal{N}, v) \Leftrightarrow ((v)_1)_1 \neq 0 \;\&\; (\forall i < g(v))[(\bar{\alpha}(t))_i = h(v, i)].$$

Now put

$$
\begin{aligned}
Q(x, w) \Leftrightarrow\ & Seq(w) \\
& \& (\exists u \le lh(w))(\exists v \le lh(w))(\exists n \le lh(w))\{x \in N(\mathfrak{X}, u) \\
& \& g(v) \le lh(w) \& (\forall i < g(v))[(w)_i = h(v, i)] \\
& \& R(u, v, n)\}
\end{aligned}
$$

and verify easily that

$$P(x, \alpha) \Leftrightarrow (\exists t)Q(x, \bar{\alpha}(t)).$$

If $\mathfrak{X}$ is of type 0 or 1, then Q is recursive since $\{(x, u): x \in N(\mathfrak{X}, u)\}$ is recursive by 3C.3. ⊣

With each irrational α we associate the binary relation on ω

$$\le_\alpha = \{(n, m): \alpha(\langle n, m\rangle) = 1\}$$

and we put

$$
\begin{aligned}
\alpha \in LO \Leftrightarrow\ & \le_\alpha \text{ is a linear ordering} \\
\Leftrightarrow\ & (\forall n)(\forall m)[n \le_\alpha m \Rightarrow (n \le_\alpha n \ \&\ m \le_\alpha m)] \\
& \& (\forall n)(\forall m)[(n \le_\alpha m \ \&\ m \le_\alpha n) \Rightarrow n = m] \\
& \& (\forall n)(\forall m)(\forall k)[(n \le_\alpha m \ \&\ m \le_\alpha k) \Rightarrow n \le_\alpha k] \\
& \& (\forall n)(\forall m)[(n \le_\alpha n \ \&\ m \le_\alpha m) \Rightarrow (n \le_\alpha m \vee m \le_\alpha n)],
\end{aligned}
$$

$$
\begin{aligned}
\alpha \in WO \Leftrightarrow\ & \le_\alpha \text{ is a wellordering} \\
\Leftrightarrow\ & \alpha \in LO \ \&\ <_\alpha \text{ has no infinite descending chains} \\
\Leftrightarrow\ & \alpha \in LO \ \&\ (\forall \beta)[(\forall n)[\beta(n+1) \le_\alpha \beta(n)] \Rightarrow (\exists n)[\beta(n+1) = \beta(n)]].
\end{aligned}
$$

If $\alpha \in LO$, let

$$|\alpha| = \text{order type of } \le_\alpha.$$

In particular, the mapping

$$\alpha \mapsto |\alpha|$$

takes WO onto the set of countable ordinals and provides a coding for this set in the sense of 3H.

4A.2. THEOREM. *The set WO of ordinal codes is Π^1_1. Moreover, there are* *relations $\leq_\Pi$, $\leq_\Sigma$ in Π_1^1 and Σ_1^1 respectively, such that*

$$\beta \in WO \Rightarrow \{\alpha \leq_\Pi \beta \Leftrightarrow \alpha \leq_\Sigma \beta \Leftrightarrow [\alpha \in WO \;\&\; |\alpha| \leq |\beta|]\}.^{(5)}$$

PROOF. That WO is Π_1^1 is obvious from the formulas above. To prove the second assertion, take first

$$\begin{aligned} \alpha \leq_\Sigma \beta &\Leftrightarrow \alpha \in LO \;\&\; (\exists \gamma)[\gamma \text{ maps } \leq_\alpha \text{ into } \leq_\beta \text{ in a one-to-one} \\ &\qquad\qquad\qquad \text{order-preserving manner}] \\ &\Leftrightarrow \alpha \in LO \;\&\; (\exists\gamma)(\forall n)(\forall m)[n <_\alpha m \Rightarrow \gamma(n) <_\beta \gamma(m)] \end{aligned}$$

It is immediate that $\leq_\Sigma$ is Σ_1^1 and for $\beta \in WO$,

$$\alpha \leq_\Sigma \beta \Leftrightarrow [\alpha \in WO \;\&\; |\alpha| \leq |\beta|].$$

For the relation $\leq_\Pi$, take

$$\begin{aligned} \alpha \leq_\Pi \beta \Leftrightarrow\; & \alpha \in WO \;\&\; \text{there is no order-preserving map of} \\ & \leq_\beta \text{ onto an initial segment of } \leq_\alpha \\ \Leftrightarrow\; & \alpha \in WO \;\&\; (\forall\gamma)\neg(\exists k)(\forall n)(\forall m)[n \leq_\beta m \Leftrightarrow [\gamma(n) \leq_\alpha \gamma(m) <_\alpha k]], \end{aligned}$$

where of course we abbreviate

$$s <_\alpha t \Leftrightarrow s \leq_\alpha t \;\&\; s \neq t.$$

4A.3. THE BASIC REPRESENTATION THEOREM FOR Π_1^1 (Lusin–Sierpinski, Kleene$^{(1-11)}$). *A pointset $P \subseteq \mathfrak{X}$ is Π_1^1 if and only if there is a Δ_1^1 function $f : \mathfrak{X} \to \mathcal{N}$ such that for all x, $f(x) \in LO$ and*

$$(*) \qquad P(x) \Leftrightarrow f(x) \in WO.$$

In fact, if P is Π_1^1, then we can choose $f : \mathfrak{X} \to \mathcal{N}$ so that $(*)$ holds and the relation

$$R(x, n, m) \Leftrightarrow f(x)(n) = m$$

is arithmetical; if in addition $\mathfrak{X}$ is of type 0 or 1, then $(*)$ holds with a recursive f.

Similarly, P is Π_1^1 if and only if $(*)$ holds with a Borel f, or with a continuous f if $\mathfrak{X}$ is of type 0 or 1.

PROOF. This is an effective and improved version of 2D.2, the representation of complements of $\varkappa$-Suslin sets of irrationals in the form

$$P(\alpha) \Leftrightarrow T(\alpha) \text{ is wellfounded},$$

where T is a tree on $\omega \times \varkappa$. We might as well give here a direct proof for subsets of an arbitrary space $\mathcal{X}$.

The last assertion clearly follows from the claims preceding it.

Assume then by 4A.1 that

$$P(x) \Leftrightarrow (\forall \alpha)(\exists t) R(x, \bar{\alpha}(t))$$

with R semirecursive, or R recursive if $\mathcal{X}$ is of type 0 or 1, where

$$R(x, \bar{\alpha}(t)) \;\&\; t < s \Rightarrow R(x, \bar{\alpha}(s)).$$

For each x, put

$$T(x) = \{(u_0, \dots, u_{t-1}) : \neg R(x, \langle u_0, \dots, u_{t-1} \rangle)\}$$

so that $T(x)$ is a tree on ω and clearly

$$P(x) \Leftrightarrow T(x) \text{ is wellfounded.}$$

What we must do is replace $T(x)$ by a linear ordering on ω which will be wellfounded precisely when $T(x)$ is.

Put

$$\begin{aligned}(v_0, \dots, v_{s-1}) >^x (u_0, \dots, u_{t-1}) \Leftrightarrow\; & (v_0, \dots, v_{s-1}), (u_0, \dots, u_{t-1}) \in T(x) \\ & \& \{v_0 > u_0 \vee [v_0 = u_0 \;\&\; v_1 > u_1] \vee [v_0 = u_0 \;\&\; v_1 = u_1 \;\&\; v_2 > u_2] \\ & \vee [v_0 = u_0 \;\&\; v_1 = u_1 \;\&\; \cdots \;\&\; v_{s-1} = u_{s-1} \;\&\; s < t]\}\end{aligned}$$

where $>$ on the right is the usual "greater than" in ω.

It is immediate that if $(v_0, \dots, v_{s-1})$, $(u_0, \dots, u_{t-1})$ are both in $T(x)$ and $(v_0, \dots, v_{s-1})$ is an initial segment of $(u_0, \dots, u_{t-1})$, then $(v_0, \dots, v_{s-1}) >^x (u_0, \dots, u_{t-1})$; thus if $T(x)$ has an infinite branch, then $>^x$ has an infinite descending chain.

Assume now that $>^x$ has an infinite descending chain, say

$$v^0 >^x v^1 >^x v^2 >^x \dots,$$

where

$$v^i = (v_0^i, v_1^i, \dots, v_{s_i-1}^i),$$

and consider the following array:

$$\begin{aligned} v^0 &= (v_0^0, v_1^0, \dots, v_{s_0-1}^0) \\ v^1 &= (v_0^1, v_1^1, \dots, v_{s_1-1}^1) \\ &\dots \qquad \dots \\ v^i &= (v_0^i, v_1^i, \dots, \dots, v_{s_i-1}^i) \\ &\dots \qquad \dots \end{aligned}$$

The definition of $>^x$ implies immediately that

$$v_0^0 \geq v_0^1 \geq v_0^2 \geq \ldots,$$

i.e. the first column is a nonincreasing sequence of integers. Hence after a while they all are the same, say

$$v_0^i = k_0 \quad \text{for } i \geq i_0.$$

Now the second column is nonincreasing below level i_0, so that for some i_1, k_1

$$v_1^i = k_1 \quad \text{for } i \geq i_1.$$

Proceeding in the same way we find an infinite sequence

$$k_0, k_1, \ldots$$

such that for each s, $(k_0, \ldots, k_{s-1}) \in T(x)$, so $T(x)$ is not wellfounded. Thus we have shown,

$$\begin{aligned} P(x) &\Leftrightarrow T(x) \text{ is wellfounded} \\ &\Leftrightarrow\ >^x \text{ has no infinite descending chains.} \end{aligned}$$

Finally put

$$\begin{aligned} u \leq^x v \Leftrightarrow (\exists t \leq u)(\exists s \leq v)[&Seq(u) \ \&\ lh(u) = t \ \&\ Seq(v) \ \&\ lh(v) = s \\ &\& \ [u = v \vee ((v)_0, \ldots, (v)_{s-1}) >^x ((u)_0, \ldots, (u)_{t-1})]] \end{aligned}$$

and notice that $\leq^x$ is always a linear ordering and

$$P(x) \Leftrightarrow\ \leq^x \text{ is a wellordering.}$$

Moreover, the relation

$$P(x, u, v) \Leftrightarrow u \leq^x v$$

is easily arithmetical for arbitrary $\mathfrak{X}$ and recursive if $\mathfrak{X}$ is of type 0 or 1.

The proof is completed by taking

$$f(x)(n) = \begin{cases} 1 & \text{if } (n)_0 \leq^x (n)_1, \\ 0 & \text{otherwise.} \end{cases}$$

⊣

The linear ordering $\leq^x$ which we used in this proof is variously known in the literature as the *Lusin–Sierpinski* or the *Kleene–Brouwer* ordering.[9,10]

Let us prove here just one very useful corollary of this basic result. Put

$$\omega_1^{CK} = \textit{supremum}\{|\alpha| : \alpha \in WO \text{ and } \alpha \text{ is recursive}\}.$$

One may think of ω_1^{CK} as an "effective analog" of the least uncountable ordinal $\aleph_1$; ω_1^{CK} is the least ordinal which cannot be realized by a recursive wellordering with field in ω.[6]

4A.4. THE BOUNDENESS THEOREM FOR Π_1^1 (Lusin–Sierpinski, Spector[1–11]). *Suppose $P \subseteq \mathfrak{X}$ and P satisfies the equivalence*

$$P(x) \Leftrightarrow f(x) \in WO$$

with some Δ_1^1 function f. Then P is Δ_1^1 if and only if

$$\textit{supremum}\{|f(x)|: P(x)\} < \omega_1^{CK}.$$

Similarly, suppose

$$P(x) \Leftrightarrow f(x) \in WO$$

with some Borel function f. Then P is Borel if and only if

$$\textit{supremum}\{|f(x)|: P(x)\} < \aleph_1.$$

PROOF. Assume first that for all x, if $P(x)$, then $|f(x)| \leq |\alpha|$, where $\alpha \in WO$ and α is recursive. By 4A.2 then,

$$P(x) \Leftrightarrow f(x) \leq_{\Sigma} \alpha,$$

so P is Σ_1^1 and since it is evidently Π_1^1, it is Δ_1^1.

Conversely, suppose *supremum*$\{|f(x)|: P(x)\} \geq \omega_1^{CK}$. Let $Q \subseteq \omega$ be any Π_1^1 relation on ω, so by the basic representation theorem 4A.3 there is a recursive $g: \omega \to \mathcal{N}$ and

$$Q(n) \Leftrightarrow g(n) \in WO.$$

Notice that for every n, $g(n)$ is a recursive irrational by (iv) of 3D.7. Hence

$$\begin{aligned} Q(n) &\Leftrightarrow g(n) \in WO \;\&\; |g(n)| < \omega_1^{CK} \\ &\Leftrightarrow (\exists x)\{P(x) \;\&\; g(n) \leq_{\Sigma} f(x)\} \end{aligned}$$

which implies that if P is Σ_1^1, then so is Q. But Q was arbitrary Π_1^1 on ω and need not be Σ_1^1, so P is not Σ_1^1.

Proof of the boldface result is a bit simpler. ⊣

EXERCISES

Put

$$\delta_1^1 = \textit{supremum}\{|\alpha|: \alpha \in WO \text{ and } \alpha \text{ is } \Delta_1^1\},$$

where α is Δ^1_1 if $\{(n, m): \alpha(n) = m\}$ is Δ^1_1.

4A.5. Prove that $\delta^1_1 = \omega_1^{CK}$. (Spector [1955].[7])

Hint. It is enough to establish that

$$\delta^1_1 \leq \textit{supremum}\{|\alpha|: \alpha \in WO, \alpha \text{ recursive}\},$$

so assume towards a contradiction that there is some $\beta \in WO$, β is Δ^1_1 and for every recursive α, if $\alpha \in WO$, then $|\alpha| < |\beta|$. Choose $P \subseteq \omega$ which is Π^1_1 but not Δ^1_1 and by 4A.3 choose a recursive f such that

$$P(n) \Leftrightarrow f(n) \in WO.$$

Now each $f(n)$ is a recursive irrational, so the assumption above implies

$$P(n) \Leftrightarrow f(n) \in WO \;\&\; |f(n)| \leq |\beta|$$

which via 4A.2 shows P to be Δ^1_1, contrary to hypothesis. ⊣

This result is rather surprising, as one might expect to get longer wellorderings in the complicated pointclass Δ^1_1 than one gets in Δ^0_1.

4A.6. Prove that if A is a $\mathbf{\Sigma}^1_1$ subset of WO, then there is a countable ξ such that

$$\alpha \in A \Rightarrow |\alpha| < \xi.$$

Hint. If not, then every $\mathbf{\Pi}^1_1$ relation P would satisfy

$$P(x) \Leftrightarrow (\exists \alpha)[\alpha \in A \;\&\; f(x) \leq_\Sigma \alpha]$$

with a Borel f and would be $\mathbf{\Delta}^1_1$. ⊣

The next exercise is an effective version of 1G.5.

4A.7. Prove that for each Δ^1_1 pointset $P \subseteq \mathfrak{X}$ there is a recursive function $\pi: \mathcal{N} \to \mathfrak{X}$ and a Π^0_1 set $A \subseteq \mathcal{N}$, such that π is one-to-one on A and $\pi[A] = P$.

Similarly, if P is $\mathbf{\Delta}^1_1$, then there is a continuous $\pi: \mathcal{Y} \to \mathcal{N}$ and a closed $A \subseteq \mathcal{N}$ such that π is one-to-one on A and $\pi[A] = P$ (this is a restatement of 1G.5).

Hint. By 3E.6, we may assume $\mathfrak{X} = \mathcal{N}$. By 4A.3 then, there is a recursive $f: \mathcal{N} \to \mathcal{N}$ such that

$$P(\alpha) \Leftrightarrow f(\alpha) \in WO$$

and by 4A.4, there is a recursive $\beta \in WO$ so that

$$P(\alpha) \Leftrightarrow f(\alpha) \leq_\Sigma \beta.$$

Put

$$\begin{aligned}Q(\gamma,\alpha)&\Leftrightarrow \gamma \text{ maps } \leq_\alpha \text{ onto an initial segment of } \leq_\beta \text{ in an}\\ &\qquad \text{order preserving fashion and } \gamma=0 \text{ outside the field of } \leq_\alpha\\ &\Leftrightarrow (\forall n)\{[\alpha(\langle n,n\rangle)\neq 1 \Rightarrow \gamma(n)=0]\ \&\ [\alpha(\langle n,n\rangle)=1\\ &\Rightarrow \beta(\langle\gamma(n),\gamma(n)\rangle)=1\}\ \&\ (\forall n)(\forall m)\{[\alpha(\langle n,n\rangle)=1\ \&\ \alpha(\langle m,m\rangle)=1]\\ &\Rightarrow [\alpha(\langle n,m\rangle)=1 \Leftrightarrow \beta(\langle\gamma(n),\gamma(m)\rangle)=1]\}\\ &\quad \&\ (\forall n)(\forall m)\{[\alpha(\langle n,n\rangle)=1\ \&\ \beta(\langle m,\gamma(n)\rangle)=1]\\ &\Rightarrow (\exists s)[\alpha(\langle s,n\rangle)=1\ \&\ \gamma(s)=m]\}.\end{aligned}$$

Clearly Q is Π^0_2 and hence so is Q^*,

$$Q^*(\gamma,\alpha)\Leftrightarrow Q(\gamma,f(\alpha)).$$

Moreover, easily

$$\begin{aligned}P(\alpha)&\Leftrightarrow(\exists\gamma)Q^*(\gamma,\alpha)\\ &\Leftrightarrow \text{there exists exactly one } \gamma \text{ such that } Q^*(\gamma,\alpha).\end{aligned}$$

Bring Q^* to normal form

$$Q^*(\gamma,\alpha)\Leftrightarrow(\forall n)(\exists m)R(\gamma,\alpha,n,m)$$

and let

$$S(\delta,\gamma,\alpha)\Leftrightarrow(\forall n)[R(\gamma,\alpha,n,\delta(n))\ \&\ (\forall m<\delta(n))\neg R(\gamma,\alpha,n,m)].$$

Now S is a Π^0_1 subset of $\mathcal{N}\times\mathcal{N}\times\mathcal{N}$ and the recursive map $(\delta,\gamma,\alpha)\mapsto\alpha$ takes S onto P and is one-to-one on S. The result follows because $\mathcal{N}\times\mathcal{N}\times\mathcal{N}$ is recursively homeomorphic with $\mathcal{N}$.

The assertion about $\mathbf{\Delta}^1_1$ sets follows by the same proof, starting with a continuous f such that

$$P(\alpha)\Leftrightarrow f(\alpha)\in WO. \qquad \dashv$$

This result is important, particularly because we will prove later that every injective, recursive image of a Δ^1_1 set is Δ^1_1—see 4D.7.

We can also obtain from this result an interesting partial converse to 3E.16.

4A.8. Prove that a point x_0 is Δ^1_1 if and only if there is a Π^0_1 singleton $\{\alpha_0\}\subseteq\mathcal{N}$ such that x_0 is recursive in α_0.

Hint. If x_0 is recursive in some α_0 with $\{\alpha_0\}$ in Π_1^0, then x_0 is easily Δ_1^1. If x_0 is Δ_1^1, then the pointset

$$P(s, i) \Leftrightarrow [x_0 \in N_s \ \&\ i = 1] \vee [x_0 \notin N_s \ \&\ i = 0]$$

is also Δ_1^1, so by 4A.3 choose a Π_1^0 set $A \subseteq \mathcal{N}$ and a recursive $\pi : \mathcal{N} \to \omega \times \omega$, which is injective on A and such that $\pi[A] = P$. Put

$$\begin{aligned}(\beta, \gamma) \in B \Leftrightarrow (\forall s)[(\beta)_s \in A \ \&\ \pi((\beta)_s) = (s, \gamma(s))] \\ \&\ (\forall u)[\text{if } u \text{ is not of the form } \langle s, i\rangle, \\ \text{then } \beta(u) = 0]\end{aligned}$$

and check easily that $B \subseteq \mathcal{N} \times \mathcal{N}$ is a Π_1^0 singleton and x_0 is recursive in the unique $(\beta_0, \gamma_0) \in B$. Now use the recursive homeomorphism of $\mathcal{N} \times \mathcal{N}$ with $\mathcal{N}$. ⊣

It is not true that every Δ_1^1 point is a Π_1^0 (or even an arithmetical) singleton—this has been shown by Feferman [1965].

4A.9. Prove that for each countable ordinal ξ, the set

$$I_\xi = \{\alpha : \alpha \in WO \ \&\ |\alpha| \leq \xi\}$$

is $\mathbf{\Delta}_1^1$, uniformly in the coding for ordinals determined by WO and the canonical coding for $\mathbf{\Delta}_1^1$.

Hint. We must show that there is a partial function $\mathfrak{u} : \mathcal{N} \to \mathcal{N}$ which is recursive on WO and such that

$$\beta \in WO \Rightarrow \mathfrak{u}(\beta) \text{ is a } \mathbf{\Delta}_1^1\text{-code of } \{\alpha : \alpha \in WO \ \&\ |\alpha| \leq |\beta|\}.$$

Choose recursive irrationals ε_1, ε_0 so that

$$\alpha \leq_\Pi \beta \Leftrightarrow G_1(\varepsilon_1, \beta, \alpha)$$
$$\alpha \leq_\Sigma \beta \Leftrightarrow G_2(\varepsilon_2, \beta, \alpha)$$

where G_1, G_2 are good universal sets in Π_1^1 and Σ_1^1 respectively by 3H.1 and let

$$\mathfrak{u}(\beta) = \langle S(\varepsilon_1, \beta), S(\varepsilon_2, \beta)\rangle. \quad ⊣$$

Of course this exercise is nothing but a restatement of 4A.2 using codings.

4B. The prewellordering property[12,13]

The basic representation theorem implies easily the so-called *prewellordering property for* Π_1^1, which in turn implies directly many of the nice structural properties of this pointclass. This property can be established for Σ_2^1 and many other pointclasses more complicated than Π_1^1, so it is worth studying its consequences in a general setting.

Recall from 2B that a norm on a set P is any function

$$\varphi : P \to \mathit{Ordinals}$$

taking P into the ordinals. There is a simple correspondence between norms and prewellorderings on P established in 2G.8, where with each φ we associate the prewellordering $\leq^{\varphi}$ on P,

$$x \leq^{\varphi} y \Leftrightarrow \varphi(x) \leq \varphi(y).$$

Conversely, if $\prec$ is a prewellordering on P, then $\prec \; = \; \leq^{\varphi}$ for some norm φ; moreover φ is uniquely determined if we insist that it be *regular*, i.e. that φ maps P *onto* some ordinal λ.

Let us call two norms φ and ψ on P *equivalent* if $\leq^{\varphi} \; = \; \leq^{\psi}$, i.e.

$$\varphi(x) \leq \varphi(y) \Leftrightarrow \psi(x) \leq \psi(y).$$

Clearly every norm is equivalent to a unique regular norm.

There are many trivial norms on a set, e.g. the constant 0 function, but the concept becomes nontrivial if we impose definability conditions on a norm in the following way.

Let Γ be a pointclass, $\varphi : P \to \lambda$ a norm on some pointset P. We call φ a *Γ-norm* if there exist relations $\geq_{\Gamma}^{\varphi}$, $\leq_{\neg\Gamma}^{\varphi}$ in Γ and $\neg\Gamma$ respectively such that for every y,

$$(*) \qquad P(y) \Rightarrow (\forall x)\{[P(x) \;\&\; \varphi(x) \leq \varphi(y)] \Leftrightarrow x \leq_{\Gamma}^{\varphi} y \Leftrightarrow x \leq_{\neg\Gamma}^{\varphi} y\}.$$

It is important for the applications that the definition of a Γ-norm be precisely that given by $(*)$. Notice that if Γ is adequate and $P \in \Gamma$, then $(*)$ is stronger than simply requiring that the associated prewellordering $\leq^{\varphi}$ be in Γ but weaker than insisting that $\leq^{\varphi}$ be in $\Gamma \cap \neg\Gamma$.

In addition to the prewellordering $\leq^{\varphi}$, there are two other relations that are naturally associated with a norm φ. Put

$$x \leq_{\varphi}^{*} y \Leftrightarrow P(x) \;\&\; [\neg P(y) \vee \varphi(x) \leq \varphi(y)],$$
$$x <_{\varphi}^{*} y \Leftrightarrow P(x) \;\&\; [\neg P(y) \vee \varphi(x) < \varphi(y)].$$

The meaning of these relations becomes clear if we extend the norm φ on $P \subseteq \mathfrak{X}$ to all of $\mathfrak{X}$ by

$$\varphi(x) = \infty \quad \text{if } \neg P(x),$$

where ∞ is assumed larger than all the ordinals. Then obviously, with this extended φ,

$$x \leq^*_\varphi y \Leftrightarrow P(x) \,\&\, \varphi(x) \leq \varphi(y)$$
$$x <^*_\varphi y \Leftrightarrow P(x) \,\&\, \varphi(x) < \varphi(y).$$

4B.1. THEOREM. *Let Γ be an adequate pointclass and let φ a norm on some P in Γ; then φ is a Γ-norm if and only if both $\leq^*_\varphi$, $<^*_\varphi$ are in Γ.*

PROOF. If $\leq^*_\varphi$, $<^*_\varphi$ are in Γ, we can take

$$x \leq^\varphi_\Gamma y \Leftrightarrow x \leq^*_\varphi y,$$
$$x \leq^\varphi_{\check{\Gamma}} y \Leftrightarrow \neg(y <^*_\varphi x),$$

and verify easily that they prove φ to be a Γ-norm. On the other hand, given such relations $\leq^\varphi_\Gamma$, $\leq^\varphi_{\check{\Gamma}}$, notice that

$$x \leq^*_\varphi y \Leftrightarrow P(x) \,\&\, [x \leq^\varphi_\Gamma y \vee \neg y \leq^\varphi_{\check{\Gamma}} x],$$
$$x <^*_\varphi y \Leftrightarrow P(x) \,\&\, \neg y \leq^\varphi_{\check{\Gamma}} x,$$

so that both $\leq^*_\varphi$ and $<^*_\varphi$ are in Γ. ⊣

A pointclass Γ is *normed* or *has the prewellordering property* if every pointset P in Γ admits a Γ-norm.

4B.2. THEOREM. *Both Π^1_1 and $\boldsymbol{\Pi}^1_1$ are normed.*(12,13)

PROOF. Given P in Π^1_1, choose a Δ^1_1 function f by 4A.3 such that

$$P(x) \Leftrightarrow f(x) \in WO$$

and for $x \in P$, put

$$\varphi(x) = |f(x)|.$$

Using the notation of 4A.2, we can take

$$x \leq^\varphi_\Pi y \Leftrightarrow f(x) \leq_\Pi f(y),$$
$$x \leq^\varphi_\Sigma y \Leftrightarrow f(x) \leq_\Sigma f(y)$$

and verify easily that φ is a Π^1_1-norm.

The same proof works for $\boldsymbol{\Pi}^1_1$, taking a Borel f. ⊣

4B.3. THEOREM (Novikov, Moschovakis[12,13]). *If Γ is adequate, $P \in \Gamma$, $P \subseteq \mathfrak{X} \times \mathcal{N}$ and P admits a Γ-norm, then $\exists^{\mathcal{N}} P$ admits an $\exists^{\mathcal{N}}\forall^{\mathcal{N}}$ Γ-norm.*

Hence, if Γ is adequate closed under $\forall^{\mathcal{N}}$ and normed, then $\exists^{\mathcal{N}}\Gamma$ is normed. In particular, Σ^1_2, $\boldsymbol{\Sigma}^1_2$ are normed.

PROOF. It is enough to establish the first assertion. Assume that

$$Q(x) \Leftrightarrow (\exists\alpha)P(x, \alpha)$$

with P in Γ, let φ be a Γ-norm on P and define ψ on Q by

$$\psi(x) = \mathit{infimum}\{\varphi(x, \alpha) : P(x, \alpha)\}.$$

Proof that ψ is an $\exists^{\mathcal{N}}\forall^{\mathcal{N}}$ Γ-norm is immediate from the equivalences

$$x \leq^*_\psi y \Leftrightarrow (\exists\alpha)(\forall\beta)[(x, \alpha) \leq^*_\varphi (y, \beta)],$$

$$x <^*_\psi y \Leftrightarrow (\exists\alpha)(\forall\beta)[(x, \alpha) <^*_\varphi (y, \beta)]. \qquad \dashv$$

This result is typical of the kind of abstract setting in which the notion of a Γ-norm proves useful. There will be several opportunities for applying 4B.3 in its full generality.

We will study many consequences of the prewellordering property in the next two sections. Here we concentrate on just a few facts which are simple, useful and indicative of the power of this hyothesis about a pointclass.

Recall the definition of a uniformizing set $P^* \subseteq P \subseteq \mathfrak{X} \times \mathcal{Y}$.

4B.4. THE EASY UNIFORMIZATION THEOREM (Kreisel [1962]). *Suppose Γ is an adequate pointclass, $\mathcal{Y}$ is a space of type 0, $P \subseteq \mathfrak{X} \times \mathcal{Y}$ is in Γ and P admits a Γ-norm. Then P can be uniformized by some P^* in $\forall^\omega\Gamma$.*

In particular, if Γ is adequate, normed and closed under $\forall^\omega$, then every $P \subseteq \mathfrak{X} \times \mathcal{Y}$ in Γ with $\mathcal{Y}$ of type 0 can be uniformized by some P^* in Γ.

PROOF. It is obviously enough to prove the result with $\mathcal{Y} = \omega$. Assume then that $P \subseteq \mathfrak{X} \times \omega$ is in Γ, let φ be a Γ-norm on P and put

$$P^*(x, n) \Leftrightarrow P(x, n) \mathbin{\&} (\forall m)[(x, n) \leq^*_\varphi (x, m)]$$
$$\mathbin{\&} (\forall m)[(x, n) <^*_\varphi (x, m) \vee n \leq m],$$

or in other words

$$P^*(x, n) \Leftrightarrow P(x, n) \mathbin{\&} \varphi(x, n) = \mathit{infimum}\{\varphi(x, m) : P(x, m)\}$$
$$\mathbin{\&} n = \mathit{infimum}\{m : P(x, m) \mathbin{\&} \varphi(x, m) = \varphi(x, n)\}.$$

Clearly P^* is in $\forall^\omega\Gamma$ and

$$\begin{aligned} P^*(x, n) \,\&\, P^*(x, n') &\Rightarrow P(x, n) \,\&\, P(x, n') \\ &\quad \& \,\varphi(x, n) = \varphi(x, n') \\ &\quad \& \, n \leq n' \,\&\, n' \leq n \\ &\Rightarrow n = n', \end{aligned}$$

so P^* is the graph of a function. If $(\exists n)P(x, n)$, let

$$\begin{aligned} \xi &= \mathit{infimum}\{\varphi(x, n) : P(x, n)\}, \\ n &= \mathit{infimum}\{m : P(x, m) \,\&\, \varphi(x, m) = \xi\}, \end{aligned}$$

and verify easily that $P^*(x, n)$. Thus P^* uniformizes P. ⊣

The problem of uniformizing subsets of $\mathfrak{X} \times \mathcal{Y}$ for arbitrary product spaces $\mathcal{Y}$ is much harder and cannot be settled using only the prewellordering property. We will deal with it in 4E.

Theorem 4B.4 is most often used in the form of the following easy corollary.

4B.5. The Δ-selection Principle (Kreisel [1962]). *Let Γ be adequate, normed and closed under $\exists^\omega$, $\forall^\omega$, let $P \subseteq \mathfrak{X} \times \mathcal{Y}$ be in Γ with $\mathcal{Y}$ of type* 0, *assume that $A \subseteq \mathfrak{X}$ is in $\Delta = \Gamma \cap \neg\Gamma$ and*

$$(\forall x \in A)(\exists y)P(x, y).$$

Then there exists a Δ-recursive function $f : \mathfrak{X} \to \mathcal{Y}$ such that

$$(\forall x \in A)P(x, f(x)).$$

Proof. Put

$$Q(x, y) \Leftrightarrow x \notin A \vee [x \in A \,\&\, P(x, y)]$$

and choose $Q^* \subseteq Q$ by 4B.4 which is in Γ and uniformizes Q. Clearly Q^* is the graph of a function $f : \mathfrak{X} \to \mathcal{Y}$ and $(\forall x \in A)P(x, f(x))$. Since $\mathcal{Y}$ is of type 0, f is Γ-recursive by 3D.2; now f is Δ-recursive since

$$f(x) \neq y \Leftrightarrow (\exists y')[f(x) = y' \,\&\, y \neq y'].$$ ⊣

Exercises

4B.6. Let Γ be an adequate pointclass. Prove that a norm φ on some P in Γ is a Γ-norm if and only if the unique regular norm ψ which is equivalent to φ is a Γ-norm. Prove also that if φ is a Γ-norm, then there are relations $<^{\varphi}_{\Gamma}$, $<^{\varphi}_{\check{\Gamma}}$ in Γ and $\neg\Gamma$ respectively such that for every y,

$$P(y) \Rightarrow (\forall x)\{[P(x) \,\&\, \varphi(x) < \varphi(y)] \Leftrightarrow x <^{\varphi}_{\Gamma} y \Leftrightarrow x <^{\varphi}_{\check{\Gamma}} y\}. \qquad \dashv$$

4B.7. Prove that if Γ is adequate and normed, then the associated boldface class $\mathbf{\Gamma}$ is also normed. $\dashv$

4B.8. Prove that for $n \geq 2$, the pointclasses Σ^0_n, $\mathbf{\Sigma}^0_n$ are normed. Prove also that every Σ^0_1 (or $\mathbf{\Sigma}^0_1$) pointset of type 0 or 1 admits a Σ^0_1 (or $\mathbf{\Sigma}^0_1$) norm. Show that the latter result fails for sets of reals.

Hint. Given P in Σ^0_n so that

$$P(x) \Leftrightarrow (\exists m) Q(x, m)$$

with Q in Π^0_{n-1}, put

$$\varphi(x) = \text{least } m \text{ such that } Q(x, m). \qquad \dashv$$

4B.9. Suppose Γ is a Σ-pointclass closed under Δ^1_1 substitution. Prove that if every pointset of type 1 in Γ admits a Γ-norm, *then Γ is normed.*

Hint. Use 3E.6. $\dashv$

Recall the definition of *reduction* from 1C. A pointclass Γ has the *reduction property* if every pair P, Q of sets in Γ can be reduced by a pair P^*, Q^* in Γ.

4B.10. Prove that if Γ is adequate and normed, then Γ has the reduction property; in particular, Π^1_1, $\mathbf{\Pi}^1_1$, Σ^1_2 and $\mathbf{\Sigma}^1_2$ have the reduction property. (Kuratowski, Addison.[(13)])

Hint. Given P, Q in Γ, put

$$R(x, n) \Leftrightarrow [P(x) \,\&\, n = 0] \vee [Q(x) \,\&\, n = 1],$$

let φ be a Γ-norm on R and take

$$P^*(x) \Leftrightarrow (x, 0) \leq^*_{\varphi} (x, 1),$$
$$Q^*(x) \Leftrightarrow (x, 1) <^*_{\varphi} (x, 0). \qquad \dashv$$

A pointclass Γ has the *separation property* if when P, Q are in Γ, $P \cap Q = \emptyset$, then there is some R in $\Delta = \Gamma \cap \neg\Gamma$ which separates P from Q. We have already proved in 2E.1 that $\mathbf{\Sigma}^1_1$ has the separation property.

4B.11. Prove that if Γ is adequate and has the reduction property, then

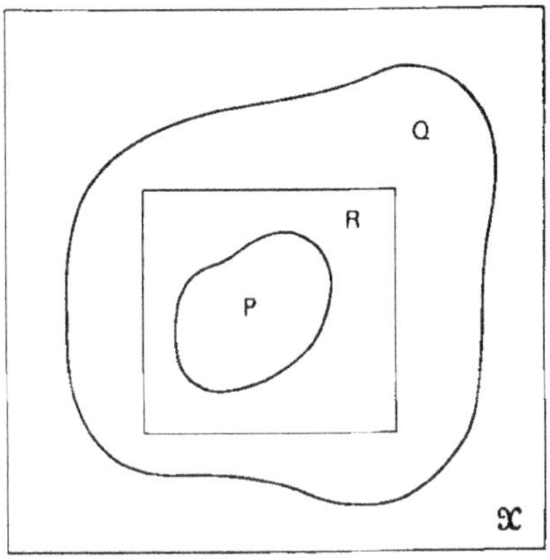

Figure 4B.1. Separation of $P \subseteq Q$ from $\mathfrak{X} - Q$.

the dual class $\neg\Gamma$ has the separation property; in particular, Σ^1_1, $\mathbf{\Sigma}^1_1$, Π^1_2, $\mathbf{\Pi}^1_1$ have the separation property. (Lusin, Novikov, Addison.[(13)])

Hint. Given P, Q in $\neg\Gamma$, both subsets of $\mathfrak{X}$, let $P_1 = \mathfrak{X} - P$, $Q_1 = \mathfrak{X} - Q$, choose P_1^*, Q_1^* to reduce P_1, Q_1 and prove that $P_1^* \cup Q_1^* = \mathfrak{X}$. Take $R = Q_1^*$. ⊣

Many times we use the separation property in the following form: if P is in Γ, Q is in $\neg\Gamma$ and $P \subseteq Q$, then there exists some $R \in \Delta$ so that (see Figure 4B.1)

$$P \subseteq R \subseteq Q.$$

To see this separate P from $\mathfrak{X} - Q$.

4B.12. Prove that if Γ is adequate, ω-parametrized and has the reduction property, then Γ does not have the separation property.

Similarly, if Γ is adequate, $\mathcal{N}$-parametrized and has the reduction property, then Γ does not have the separation property.

In particular, Σ^0_n, Π^1_1, Σ^1_2, $\mathbf{\Sigma}^0_\xi$, $\mathbf{\Pi}^1_1$, $\mathbf{\Sigma}^1_2$ do not have the separation property. (Novikov, Kleene, Addison.[(13)])

Hint. Let $G \subseteq \omega \times \omega$ be universal for $\Gamma{\restriction}\omega$ and put

$$P(n) \Leftrightarrow ((n)_0, n) \in G, \qquad Q(n) \Leftrightarrow ((n)_1, n) \in G.$$

Choose P^*, Q^* in Γ which reduces P, Q and assume towards a

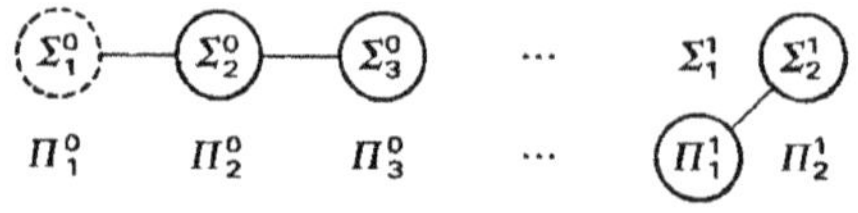

Diagram 4B.2. Normed Kleene pointclasses.

contradiction that some R in Δ separates P^* from Q^*, i.e.

$$P^* \subseteq R, \qquad R \cap Q^* = \emptyset.$$

Choose integers e, m such that

$$R(n) \Leftrightarrow (e, n) \in G, \qquad \neg R(n) \Leftrightarrow (m, n) \in G$$

and let $t = \langle m, e \rangle$. Now show that both assumptions $t \in R$, $t \notin R$ lead to contradictions.

The second assertion is proved similarly. ⊣

4B.13. Prove that if Γ is adequate and ω-parametrized, then at most one of the pointclasses Γ, $\neg\Gamma$ is normed. ⊣

It follows from the results of this section that the Kleene pointclasses which are normed are exactly those circled in Diagram 4B.2.

The circle around Σ^0_1 is dotted, since only Σ^0_1 pointsets of type 0 or 1 admit Σ^0_1-norms.

The diagram for the boldface classes is identical.

We have not included here Σ^1_3, Π^1_3 and the higher analytical pointclasses, as it is not clear at this point which of Σ^1_3 or Π^1_3 is normed, if any.

Many of the results in this section have uniform versions which are easy to establish using the methods of 3H. We put down one theorem of this type as an example.

4B.14. If Γ is ω-parametrized, adequate and has the reduction property, then Γ has the uniform reduction property: i.e. for each $\mathfrak{X}$, there are recursive functions $\mathbf{u}_1(\alpha, \beta)$, $\mathbf{u}_2(\alpha, \beta)$ such that whenever α, β code subsets P, Q of $\mathfrak{X}$ respectively in $\mathbf{\Gamma}$, then $\mathbf{u}_1(\alpha, \beta)$, $\mathbf{u}_2(\alpha, \beta)$ code sets P^*, Q^* respectively which reduce the pair P, Q.

Hint. All codings are relative to a good parametrization of course, so the hypothesis (for example) means that

$$P(x) \Leftrightarrow G(\alpha, x), \; Q(x) \Leftrightarrow G(\beta, x)$$

with G a good universal set.

Define

$$U_1(\alpha, \beta, x) \Leftrightarrow G(\alpha, x),$$
$$U_2(\alpha, \beta, x) \Leftrightarrow G(\beta, x)$$

so that both U_1 and U_2 are in Γ and let U_1^*, U_2^* reduce the pair U_1, U_2 in Γ. There are then recursive irrationals ε_1, ε_2 so that

$$U_1^*(\alpha, \beta, x) \Leftrightarrow G(\varepsilon_1, \alpha, \beta, x) \Leftrightarrow G(S(\varepsilon_1, \alpha, \beta), x)$$
$$U_2^*(\alpha, \beta, x) \Leftrightarrow G(\varepsilon_2, \alpha, \beta, x) \Leftrightarrow G(S(\varepsilon_2, \alpha, \beta), x),$$

where we have used the Good Parametrization Theorem 3H.1. It is easy to check that the recursive functions

$$u_1(\alpha, \beta) = S(\varepsilon_1, \alpha, \beta)$$
$$u_2(\alpha, \beta) = S(\varepsilon_2, \alpha, \beta)$$

have the required properties. ⊣

4C. Spector pointclasses[14-16]

The consequences of the prewellordering property which we proved in 4B depended on several side conditions on a pointclass Γ, e.g. closure under various operations or parametrization. Here we will isolate the most commonly used hypotheses into the basic notion of a *Spector pointclass*. The simplest Spector pointclasses are Π_1^1 and Σ_2^1—in fact Π_1^1 is *the least* Spector pointclass. There are, however, many other examples which make it worth axiomatizing the theory.

A *Spector pointclass* is a collection of pointsets Γ which satisfies the following conditions:

(1) Γ is a Σ-pointclass with the substitution property and closed under $\forall^\omega$.

(2) Γ is ω-parametrized.

(3) Γ is normed.

Recall that (1) implies $\Sigma_1^0 \subseteq \Gamma$ and Γ is closed under &, $\vee$, $\exists^{\leq}$ and $\exists^\omega$, and by 3G.1, Γ is also adequate.

All the Kleene pointclasses Σ_n^1, Π_n^1 satisfy (1) and (2), so to prove that one of these is a Spector pointclass we need only verify the prewellordering property. It is also trivial to check that each relativization $\Gamma(z)$ of a Spector pointclass Γ is a Spector pointclass, see 4C.4. Thus Π_1^1, Σ_2^1, $\Pi_1^1(z)$,

$\Sigma^1_2(z)$ are Spector pointclasses—they are the only ones we know at this time.

In Chapters 5 and 6 we will prove using strong set theoretic hypotheses that some of the higher Kleene pointclasses are also normed and in Chapters 6 and 7 we will introduce many more examples of Spector pointclasses. Here we concentrate on consequences of (1)–(3) above which give us new results about Π^1_1 and Σ^1_2.

First, let us prove a strong closure property of Spector pointclasses which implies that every one of them contains every Π^1_1 relation.

Suppose $Q(x, w)$ is given and we define $P(x, w)$ by

$$(*) \qquad P(x, w) \Leftrightarrow (\forall \alpha)(\exists t)Q(x, w * \bar{\alpha}(t)),$$

where $w * v$ codes the concatenation of the sequences coded by w and v, if $Seq(w)$, $Seq(v)$, see *18 of 3A.6. For each countable ordinal ξ, define the set $P^\xi \subseteq \mathfrak{X} \times \omega$ by the recursion

$$(**) \qquad P^\xi(x, w) \Leftrightarrow Q(x, w) \vee (\forall s)(\exists \eta < \xi)P^\eta(x, w * \langle s \rangle).$$

It is easy to verify by induction on ξ, that

$$P^\xi(x, w) \Rightarrow P(x, w);$$

conversely, if $(\forall \xi)\neg P^\xi(x, w)$ then $\neg Q(x, w)$ and for some $s = s_0$, all ξ, $\neg P^\xi(x, w * \langle s_0 \rangle)$, so again $\neg Q(x, w * \langle s_0 \rangle)$ and for some $s = s_1$, all ξ, $\neg P^\xi(x, w * \langle s_0, s_1 \rangle)$, etc., so finally with $\alpha = (s_0, s_1, \ldots)$ we have $(\forall t)\neg Q(x, w * \bar{\alpha}(t))$, i.e. $\neg P(x, w)$. Thus

$$P = \bigcup_\xi P_\xi.$$

Now define a norm

$$\varphi : P \to ordinals$$

by

$$\varphi(x, w) = \text{least } \xi \text{ such that } P^\xi(x, w);$$

it is immediate from $(**)$ that P satisfies the equivalence

$$P(x, w) \Leftrightarrow Q(x, w) \vee (\forall s)[(x, w * \langle s \rangle) <^*_\varphi (x, w)].$$

It is perhaps a bit surprizing that this equivalence completely determines P.

4C.1. LEMMA. *Suppose $Q(x, w)$, $P(x, w)$ are given and P admits a norm φ*

such that

$$P(x, w) \Leftrightarrow Q(x, w) \vee (\forall s)[(x, w * \langle s \rangle) <^*_\varphi (x, w)];$$

then

$$P(x, w) \Leftrightarrow (\forall \alpha)(\exists t) Q(x, w * \bar{\alpha}(t)).$$

PROOF. First we prove by induction on $\varphi(x, w)$ that

$$P(x, w) \Rightarrow (\forall \alpha)(\exists t) Q(x, w * \bar{\alpha}(t)).$$

Assuming this for all $(x, u) \in P$ with $\varphi(x, u) < \varphi(x, w)$ and supposing that $P(x, w)$ holds, we have by the hypothesis

$$Q(x, w) \vee (\forall s)[(x, w * \langle s \rangle) <^*_\varphi (x, w)].$$

If $Q(x, w)$ holds, then easily $(\forall \alpha) Q(x, w * \bar{\alpha}(0))$ since $w * \bar{\alpha}(0) = w$. Otherwise, we have

$$(\forall s)[(x, w * \langle s \rangle) <^*_\varphi (x, w)],$$

so that for each s, $P(x, w * \langle s \rangle)$ and $\varphi(x, w * \langle s \rangle) < \varphi(x, w)$. By the induction hypothesis then,

$$(\forall s)(\forall \alpha)(\exists t) Q(x, w * \langle s \rangle * \bar{\alpha}(t))$$

from which $(\forall \alpha)(\exists t) Q(x, w * \bar{\alpha}(t))$ follows immediately.

Conversely, assuming $\neg P(x, w)$, we have $\neg Q(x, w)$ and for some $s = s_0$, $\neg(x, w * \langle s_0 \rangle) <^*_\varphi (x, w)$; this means $\neg P(x, w * \langle s_0 \rangle)$, since $P(x, w * \langle s_0 \rangle)$ and $\neg P(x, w)$ implies $(x, w * \langle s_0 \rangle) <^*_\varphi (x, w)$. Again, $\neg Q(x, w * \langle s_0 \rangle)$ and for some $s = s_1$, $\neg(x, w * \langle s_0, s_1 \rangle) <^*_\varphi (x, w * \langle s_0 \rangle)$ etc., so we get some $\alpha = (s_0, s_1, \ldots)$ such that $(\forall t) \neg Q(x, w * \bar{\alpha}(t))$. ⊣

There is a bit of trickery in this proof which will not become completely clear until we look carefully at *inductive definability* in Chapter 7. For now, we can simply view this lemma as a tool for establishing the next very useful result.

4C.2. THEOREM (Moschovakis[15]). *Let Γ be a Spector pointclass, suppose $Q \subseteq \mathfrak{X} \times \omega$ is in Γ and P is defined by*

$$P(x) \Leftrightarrow (\forall \alpha)(\exists t) Q(x, \bar{\alpha}(t));$$

then P is in Γ.

In particular, Π^1_1 is the smallest Spector pointclass and Σ^1_2 is the smallest Spector pointclass closed under $\exists^{\mathcal{N}}$.

PROOF. The second assertion follows immediately from the first by (ii) of 4A.1.

To prove the first assertion using the lemma, it is enough to find some $R^* \subseteq \mathfrak{X} \times \omega$ in Γ which admits some norm φ so that

$$R^*(x, w) \Leftrightarrow Q(x, w) \vee (\forall s)[(x, w * \langle s \rangle) <_\varphi^* (x, w)],$$

since we then have

$$P(x) \Leftrightarrow R^*(x, 1).$$

Here is where we will use Kleene's recursion theorem for relations, 3H.3.

Let $G \subseteq \mathcal{N} \times \mathfrak{X} \times \omega$ be a good universal set in Γ for $\boldsymbol{\Gamma}$, let $\psi : G \to$ *ordinals* be a Γ-norm on G and define

$$R(\alpha, x, w) \Leftrightarrow Q(x, w) \vee (\forall s)[(\alpha, x, w * \langle s \rangle) <_\psi^* (\alpha, x, w)].$$

Now R is in Γ so by 3H.3 there is a fixed recursive ε^* so that

$$R(\varepsilon^*, x, w) \Leftrightarrow G(\varepsilon^*, x, w).$$

Put

$$R^*(x, w) \Leftrightarrow R(\varepsilon^*, x, w)$$

and on R^* put the norm

$$\varphi(x, w) = \psi(\varepsilon^*, x, w).$$

Computing,

$$\begin{aligned} R^*(x, w) &\Leftrightarrow R(\varepsilon^*, x, w) \\ &\Leftrightarrow Q(x, w) \vee (\forall s)[(\varepsilon^*, x, w * \langle s \rangle) <_\psi^* (\varepsilon^*, x, w)] \\ &\Leftrightarrow Q(x, w) \vee (\forall s)[(x, w * \langle s \rangle) <_\varphi^* (x, w)] \end{aligned}$$

so that R^* has the required property. ⊣

This theorem is interesting partly because it gives an intrinsic *structural characterization* of Π_1^1. Of course Π_1^1 can be easily characterized by its closure properties, e.g. it is the smallest Σ-pointclass closed under $\forall^\omega$ and $\forall^{\mathcal{N}}$. But nothing very deep can be proved in general about Σ-pointclasses closed under $\forall^\omega$ and $\forall^{\mathcal{N}}$. We will see that Spector pointclasses have a rich structure theory, much of it giving new results even when we specialize it to Π_1^1.

There is another practical corollary of 4C.2 which we list together with some simple properties of total functions recursive in a Spector pointclass.

4C.3. THEOREM. *Let Γ be a Spector pointclass, suppose $f:\mathfrak{X}\to\mathfrak{Y}$ is total and Γ-recursive; then f is Δ-recursive, Graph$(f)=\{(x,y): f(x)=y\}$ is in Δ and for every x,*

$$f(x)\in\Delta(x)=\Gamma(x)\cap\neg\Gamma(x),$$

i.e. $f(x)$ is a $\Delta(x)$-recursive point.

Moreover, every Δ^1_1 function is Γ-recursive, so in particular, Γ is closed under substitution of Δ^1_1 functions.

PROOF. The first assertion is easy and uses only the fact that Γ is a Σ-pointclass closed under $\forall^\omega$. Thus, $\{(y,s): y\notin N_s\}$ is in Γ since it is Π^0_1 and Γ contains all Π^0_2 sets, hence $\{(x,s): f(x)\notin N_s\}$ is in Γ by closure under substitution of Γ-recursive functions; thus f is Δ-recursive. From this follows trivially that $f(x)\in\Delta(x)$. As for the graph,

$$\begin{aligned} f(x)=y &\Leftrightarrow (\forall s)[y\in N_s \Rightarrow f(x)\in N_s] \\ &\Leftrightarrow (\forall s)[f(x)\in N_s \Rightarrow y\in N_s]. \end{aligned}$$

Now if $f:\mathfrak{X}\to\mathfrak{Y}$ is Δ^1_1, then $\{(x,s): f(x)\in N_s\}$ is Δ^1_1, hence in Γ by 4C.2, hence f is Γ-recursive. ⊣

EXERCISES

4C.4. Prove that if Γ is a Spector pointclass, then so is each relativization $\Gamma(z)$. ⊣

With each pointclass Γ we have associated the *boldface* pointclass $\mathbf{\Gamma}$, where for $P\subseteq\mathfrak{X}$,

$$\begin{aligned} P\in\mathbf{\Gamma} \Leftrightarrow &\text{ for some } P^*\subseteq\mathcal{N}\times\mathfrak{X} \text{ in } \Gamma \\ &\text{ and some } \varepsilon\in\mathcal{N},\ P=P^*_\varepsilon, \end{aligned}$$

i.e.

$$\mathbf{\Gamma}=\textstyle\bigcup_\varepsilon \Gamma(\varepsilon).$$

As usual,

$$\mathbf{\Delta}=\mathbf{\Gamma}\cap\neg\mathbf{\Gamma}.$$

4C.5. Prove that if Γ is a Spector pointclass, then $\mathbf{\Gamma}$ contains $\mathbf{\Pi}^1_1$ and is closed under Borel substitutions, $\exists^\omega$, $\forall^\omega$, $\bigvee^\omega$, $\bigwedge^\omega$, it is $\mathcal{N}$-parametrized and it is normed.

Moreover, every $\mathbf{\Gamma}$-measurable function is $\mathbf{\Delta}$-measurable (in fact $\mathbf{\Delta}$-recursive by 3D.22) and has a graph in $\mathbf{\Delta}$. The pointclass $\mathbf{\Gamma}$ is closed under substitution of $\mathbf{\Gamma}$-measurable functions. ⊣

4C.6. Prove that if Γ is a Spector pointclass, then $\neg\mathbf{\Gamma}$ is closed under the operation $\mathscr{A}$. ⊣

If $\varphi : P \twoheadrightarrow \lambda$ is a regular norm, we call λ *the length of* φ,

$$|\varphi| = \lambda.$$

The length $|\varphi|$ of an arbitrary norm is (by definition) the length of the unique regular norm equivalent to φ.

If $\varphi : P \twoheadrightarrow |\varphi|$ is a regular norm, then for each $\xi < |\varphi|$, *the ξ'th resolvent* of P is defined by

$$P^\xi = \{x : \varphi(x) \leq \xi\}.$$

Clearly

$$P = \bigcup_{\xi < |\varphi|} P^\xi.$$

4C.7. Let Γ be a Spector pointclass and let $\varphi : P \twoheadrightarrow |\varphi|$ be a regular Γ-norm on a pointset P in Γ, where P is of type 0. Prove that for every $\xi < |\varphi|$, the resolvent P^ξ is in Δ.

Similarly, if $\varphi : P \twoheadrightarrow |\varphi|$ is a regular $\mathbf{\Gamma}$-norm on some P in $\mathbf{\Gamma}$, then for every $\xi < |\varphi|$, the resolvent P^ξ is in $\mathbf{\Delta} = \mathbf{\Gamma} \cap \neg\mathbf{\Gamma}$. In particular,

$$P = \bigcup_{\xi < |\varphi|} P^\xi,$$

with each P^ξ in $\mathbf{\Delta}$.

Hint. Choose some $y \in P$ such that $\varphi(y) = \xi$ and notice that

$$\begin{aligned} x \in P^\xi &\Leftrightarrow x \leq^*_\varphi y \\ &\Leftrightarrow \neg(y <^*_\varphi x). \end{aligned}$$ ⊣

This result is more useful if we get an estimate on the length $|\varphi|$ of a Γ-norm. Given a pointclass Γ (which need not be a Spector class), put

$$\delta = \textit{supremum}\{|<| : < \text{ is a prewellordering of } \omega, < \text{ in } \Delta\},$$

$$\boldsymbol{\delta} = \textit{supremum}\{|<| : < \text{ is a prewellordering of } \mathcal{N}, < \text{ in } \mathbf{\Delta}\}.$$

Clearly δ is a countable ordinal, but $\boldsymbol{\delta}$ may well be uncountable—the only obvious bound is

$$\boldsymbol{\delta} < (2^{\aleph_0})^+ = \text{least cardinal} > 2^{\aleph_0}.$$

4C.8. Let Γ be an adequate pointclass closed under $\exists^{\omega}$, $\forall^{\omega}$. Prove that

$$\delta = supremum\{|\leq|: \leq \text{ is a wellordering on } \omega, < \text{ in } \Delta\},$$

δ is a limit ordinal and for every Γ-norm φ on a pointset P of type 0 in Γ, $|\varphi| \leq \delta$. ⊣

4C.9. Prove that if Γ is an adequate pointclass, then for every $\mathbf{\Gamma}$-norm φ on a pointset P in $\mathbf{\Gamma}$, $|\varphi| \leq \boldsymbol{\delta}$.

If Γ is a Spector pointclass, then $\boldsymbol{\delta}$ is an ordinal of cofinality $>\omega$ and every pointset in $\mathbf{\Gamma}$ is the union of $\boldsymbol{\delta}$ sets in $\boldsymbol{\Delta}$.

Hint. An ordinal λ has cofinality $>\omega$ if for every increasing sequence $\xi_0 < \xi_1 < \cdots < \lambda$, $limit_n\, \xi_n < \lambda$. This follows here from closure of $\boldsymbol{\Delta}$ under $\bigvee^{\omega}$. ⊣

This is obviously a "soft" version of part of 2F.2, with a very different proof. To get "hard" corollaries of this exercise we must establish a construction principle for the specific $\boldsymbol{\Delta}$ and also get an estimate of the size of $\boldsymbol{\delta}$. Both of these often turn out to be very hard.

The traditional notation for δ and $\boldsymbol{\delta}$ when Γ is Σ^1_n (or Π^1_n) is δ^1_n and $\boldsymbol{\delta}^1_n$. Similarly, for the relativized class $\Sigma^1_n(z)$ (or $\Pi^1_n(z)$), its ordinal is $\delta^1_n(z)$. (It is trivial to establish that the boldface class corresponding to $\Sigma^1_n(z)$ is $\boldsymbol{\Sigma}^1_n$, so the boldface ordinal of $\Sigma^1_n(z)$ is again $\boldsymbol{\delta}^1_n$.)

From the Kunen–Martin Theorem 2G.2 we know that

$$\boldsymbol{\delta}^1_1 = \aleph_1, \qquad \boldsymbol{\delta}^1_2 \leq \aleph_2.$$

This is about all that can be proved about these ordinals in classical set theory, except for 4A.5, that

$$\delta^1_1 = \omega^{CK}_1 = \text{least nonrecursive ordinal}.$$

The next exercise gives an interesting generalization of the Boundedness Theorem 4A.4 to arbitrary $\boldsymbol{\Pi}^1_1$-norms.

4C.10. Suppose $P \subseteq \mathfrak{X}$ is $\boldsymbol{\Pi}^1_1$ and $\varphi : P \to Ordinals$ is a regular, $\boldsymbol{\Pi}^1_1$-norm on P. Prove that P is Borel if and only if $|\varphi| < \aleph_1$. (*The Boundedness Theorem for* $\boldsymbol{\Pi}^1_1$*-norms.*)

Hint. Let $\lambda = |\varphi|$ and assume first that $\lambda < \aleph$. Now

$$P = \bigcup_{\xi<\lambda} P^{\xi}$$

and each P^{ξ} is Borel by 4C.7, so P is a countable union of Borel sets and hence Borel.

Conversely, if P is Borel, then the prewellordering

$$x \leq y \Leftrightarrow P(x) \;\&\; P(y) \;\&\; \varphi(x) \leq \varphi(y)$$

of length λ is easily Borel and hence $\lambda < \aleph_1$. ⊣

This result is often useful in conjunction with the covering lemma which admits a general version.

4C.11. Let Γ be a Spector pointclass, let φ be a regular Γ-norm on some $P \subseteq \mathfrak{X}$ in $\Gamma - \Delta$, let Q be in $\neg\Gamma$ and assume that either $\mathfrak{X}$ is of type 0 or Γ is closed under $\forall^{\mathcal{N}}$. Prove that

$$Q \subseteq P \Rightarrow \text{for some } \xi < |\varphi|, \quad Q \subseteq P^{\xi} = \{x \in P : \varphi(x) \leq \xi\}.$$

Similarly, let $\boldsymbol{\Gamma}$ be a Spector pointclass closed under $\forall^{\mathcal{N}}$, let φ be a regular $\boldsymbol{\Gamma}$-norm on some $P \subseteq \mathfrak{X}$ in $\boldsymbol{\Gamma} - \boldsymbol{\Delta}$ and let Q be in $\neg\boldsymbol{\Gamma}$. Prove again that

$$Q \subseteq P \Rightarrow \text{for some } \xi < |\varphi|, \quad Q \subseteq P^{\xi}.$$

In particular, if Γ is a Spector pointclass closed under $\forall^{\mathcal{N}}$, $G \subseteq \mathcal{N} \times \mathfrak{X}$ is universal in Γ and $\varphi : G \to \mathit{Ordinals}$ is a Γ-norm on G, then a pointset $P \subseteq \mathfrak{X}$ is in $\boldsymbol{\Delta}$ if and only if there are irrationals ε, ε_0 and some $x_0 \in \mathfrak{X}$ such that $G(\varepsilon_0, x_0)$ and

$$P = \{x \in \mathfrak{X} : G(\varepsilon, x) \;\&\; \varphi(\varepsilon, x) \leq \varphi(\varepsilon_0, x_0)\}$$

(*The Covering Lemma*; see Figure 4C.1).

Hint. By contradiction, see the proof of 4A.4. ⊣

The next result is a simple but interesting extension of the Δ-Selection principle.

4C.12. Let Γ be a Spector pointclass, suppose $P \subseteq \mathfrak{X} \times \mathfrak{Y} \times \mathfrak{Y}$ is in Γ with $\mathfrak{X}$, $\mathfrak{Y}$ of type 0, assume that

$$(\forall x)(\forall y)(\exists y')P(x, y, y').$$

Prove that for each fixed y_0, there is a function $f : \mathfrak{X} \times \omega \to \mathfrak{Y}$ which is Δ-recursive and such that

$$f(x, 0) = y_0,$$
$$(\forall n)P(x, f(x, n), f(x, n+1)).$$

(*The Principle of Γ-dependent choices.*)

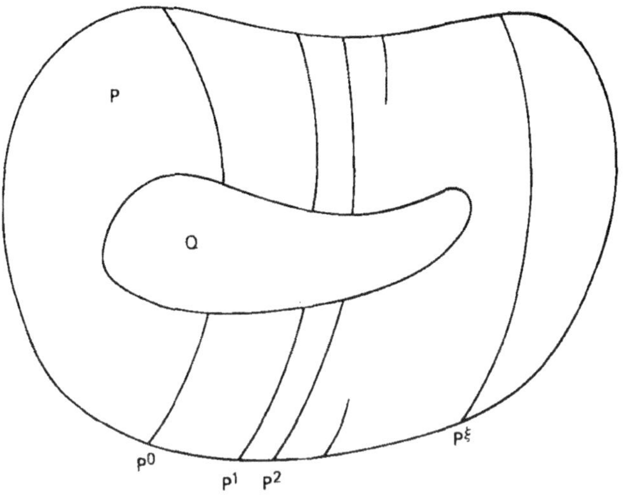

Figure 4C.1. The covering lemma.

Hint. By hypothesis and the Δ-Selection principle 4B.5, there is a Δ-recursive $g : \mathfrak{X} \times \mathcal{Y} \to \mathcal{Y}$ such that $(\forall x)(\forall y)P(x, y, g(x, y))$. Define f by the recursion

$$f(x, 0) = y_0,$$
$$f(x, n+1) = g(x, f(x, n)). \qquad \dashv$$

Another simple but interesting application of the Δ-selection principle comes up in the next result. This is essentially a representation theorem for Δ sets which happen to be open—we will need it in the exercises of 4F.

4C.13. Let Γ be a Spector pointclass closed under $\forall^{\mathcal{N}}$, let $G \subseteq \mathfrak{X}$ be a pointset in Δ which is open. Prove that there is some irrational ε in Δ such that

$$P = \bigcup_n N(\mathfrak{X}, \varepsilon(n))$$

and for each n,

$$\bar{N}(\mathfrak{X}, \varepsilon(n)) \subseteq P.$$

($\bar{N}(\mathfrak{X}, s)$ is the closure of $N(\mathfrak{X}, s)$.)

In particular, under these hypotheses, P is semirecursive in some $\varepsilon \in \Delta$.[25]

Hint. Put

$$Q(x, s) \Leftrightarrow P(x) \;\&\; x \in N_s \;\&\; (\forall y)[y \in \bar{N}_s \Rightarrow P(y)].$$

Clearly Q is in Γ and $(\forall x \in P)(\exists s)Q(x, s)$, so by 4B.5, there is a Δ-recursive $f: \mathfrak{X} \to \omega$ such that $(\forall x)Q(x, f(x))$. The set

$$A = \{s: (\exists x \in P)[f(x) = s]\}$$

is in $\neg\Gamma$ and it is disjoint from

$$B = \{s: (\exists y \notin P)[y \in \bar{N}_s]\},$$

since for each $x \in P$, $\bar{N}_{f(x)} \subseteq P$. By the separation property for $\neg\Gamma$, there is some C in Δ,

$$A \subseteq C, \quad B \cap C = \emptyset.$$

Now it is immediate that

$$P = \cup\{N_s: s \in C\},$$

and for each $s \in C$, $\bar{N}_s \subseteq P$. Take

$$\varepsilon(s) = \begin{cases} s & \text{if } s \in C, \\ 0 & \text{if } s \notin C. \end{cases} \qquad \dashv$$

The last exercise is an interesting generalization of the fact that $\mathbf{\Sigma}^1_1$ relations have countable rank whose proof uses Kleene's recursion theorem for relations, 3H.3 as did the proof of 4C.2.

4C.14. Let Γ be a Spector pointclass closed under $\forall^{\mathcal{N}}$, suppose $<$ is a (strict) wellfounded relation on the perfect product space $\mathfrak{X}$ which is in $\neg\Gamma$, let $G \subseteq \mathcal{N} \times \mathfrak{X}$ be a good universal set in Γ and let $\varphi: G \to$ *Ordinals* be a Γ-norm on G. Then there exists a recursive function

$$f: \mathfrak{X} \to \mathcal{N} \times \mathfrak{X}$$

which is order-preserving from $<$ into φ, i.e.

$$x < y \Rightarrow f(x), f(y) \in G \;\&\; \varphi(f(x)) < \varphi(f(y)).$$

It follows that if φ is any regular Γ-norm on the good universal set G, then

$$|\varphi| = \boldsymbol{\delta}.$$

(Moschovakis [1970].)

Hint. Put

$$Q(\alpha, x) \Leftrightarrow (\forall y)[y < x \Rightarrow (\alpha, y) <^*_\varphi (\alpha, x)]$$

so that Q is in Γ and by 3H.3 there is a recursive $\varepsilon^* \in \mathcal{N}$ satisfying

$$Q(\varepsilon^*, x) \Leftrightarrow G(\varepsilon^*, x).$$

Put

$$f(x) = (\varepsilon^*, x)$$

and check by a trivial induction that if x is in the field of $<$, then

$$f(x) \in G \ \&\ (\forall y)[y < x \Rightarrow \varphi(f(y)) < \varphi(f(x))].$$

Applying this to each relativized pointclass $\Gamma(w)$, we show that $|\varphi|$ exceeds the rank of every strict, wellfounded relation in $\neg\boldsymbol{\Gamma}$ on $\mathfrak{X}$, whence $|\varphi| = \boldsymbol{\delta}$ follows immediately by the fact that every two perfect product spaces are Δ^1_1-isomorphic and 4C.9. ⊣

4D. The parametrization theorem for $\Delta \cap \mathfrak{X}$

Most of the results in 4C follow quite directly from the definitions and depend on only few of the axioms for a Spector pointclass. Here we will consider somewhat deeper propositions whose proofs make essential use of the full set of axioms, including ω-parametrization.

Recall that a partial function

$$f : \mathfrak{X} \to \mathfrak{Y}$$

is *Γ-recursive* if *Domain*(f) is in Γ and f is Γ-recursive on its domain; if Γ is closed under &, this amounts to saying that the relation

$$G^f(x, s) \Leftrightarrow f(x)\downarrow \ \&\ f(x) \in N_s$$

is in Γ. These partial functions are very useful when Γ is a Spector pointclass. We summarize some of their properties in the next result.

4D.1. THEOREM. *Let Γ be a Spector pointclass, let $f : \mathfrak{X} \to \mathfrak{Y}$ be a Γ-recursive partial function.*

(*i*) *The relations*

$$P(x) \Leftrightarrow f(x)\downarrow,$$
$$Q(x,s) \Leftrightarrow f(x)\downarrow \,\&\, f(x) \notin N_s,$$
$$R(x,y) \Leftrightarrow f(x) = y,$$
$$\Leftrightarrow f(x)\downarrow \,\&\, f(x) = y,$$
$$S(x,y) \Leftrightarrow f(x)\downarrow \,\&\, f(x) \neq y,$$

are all in Γ.

(*ii*) *If* $Q \subseteq \mathcal{Y}$ *is in* Γ *and*

$$R(x) \Leftrightarrow f(x)\downarrow \,\&\, Q(f(x))$$

then R *is in* Γ.

(*iii*) *For each* $x \in \mathcal{X}$, *if* $f(x)\downarrow$, *then* $f(x) \in \Delta(x)$, *i.e.* $f(x)$ *is* $\Delta(x)$*-recursive.*

PROOF. The set $\{(y,s) : y \notin N_s\}$ is Π^0_1 and hence in Γ and the partial function $(x,s) \rightharpoonup (f(x), s)$ is easily Γ-recursive, so by the substitution property there is some $Q^*(x,)$ in Γ such that

$$f(x)\downarrow \Rightarrow [Q^*(x,s) \Leftrightarrow f(x) \notin N_s];$$

thus

$$f(x)\downarrow \,\&\, f(x) \notin N_s \Leftrightarrow f(x)\downarrow \,\&\, Q^*(x,s)$$

and this is in Γ.

The other claims are easier:

$$f(x) = y \Leftrightarrow (\forall s)\{y \in N_s \Rightarrow [f(x)\downarrow \,\&\, f(x) \in N_s]\},$$
$$f(x)\downarrow \,\&\, f(x) \neq y \Leftrightarrow (\exists s)\{[f(x)\downarrow \,\&\, f(x) \in N_s] \,\&\, y \notin N_s\}.$$

(ii) Given $Q \subseteq \mathcal{Y}$ in Γ, choose $Q^* \subseteq \mathcal{X}$ in Γ by the substitution property so that

$$f(x)\downarrow \Rightarrow [Q^*(x) \Leftrightarrow Q(f(x))]$$

and notice that

$$R(x) \Leftrightarrow f(x)\downarrow \,\&\, Q^*(x).$$

(iii) Use 3G.5. ⊣

We have been using and will continue to use the handy abbreviation

$$y \in \Delta \Leftrightarrow y \text{ is } \Delta\text{-recursive}$$
$$\Leftrightarrow \mathcal{U}(y) \text{ is in } \Delta,$$

and similarly for $\Delta(x)$. It is also convenient for any pointclass Λ to put

$$\Lambda \cap \mathfrak{X} = \{x \in \mathfrak{X}: x \text{ is } \Lambda\text{-recursive}\}.$$

For example $\Sigma^0_1 \cap \mathcal{R} = \Delta^0_1 \cap \mathcal{R} =$ the set of recursive real numbers.

Using partial functions we can formulate simply an easy to prove but very powerful parametrization theorem for the points in a Spector pointclass.

4D.2. THE PARAMETRIZATION THEOREM FOR THE POINTS IN Δ, $\Delta(x)$. *Let Γ be a Spector pointclass. For each product space $\mathcal{Y}$, there is a Γ-recursive partial function*

$$\mathbf{d}: \omega \rightarrow \mathcal{Y}$$

such that for every $y \in \mathcal{Y}$,

$$y \in \Delta \Leftrightarrow \text{for some } i, \mathbf{d}(i)\downarrow \;\&\; \mathbf{d}(i) = y.$$

Similarly, for each $\mathfrak{X}$, $\mathcal{Y}$ there is a Γ-recursive partial function

$$\mathbf{d}: \omega \times \mathfrak{X} \rightarrow \mathcal{Y}$$

such that for all x, y,

$$y \in \Delta(x) \Leftrightarrow \text{for some } i, \mathbf{d}(i, x)\downarrow \;\&\; \mathbf{d}(i, x) = y.^{(17)}$$

PROOF. Take first the case $\mathcal{Y} = \mathcal{N}$. We prove the second assertion, the first being simpler.

Choose a set $G \subseteq \omega \times \mathfrak{X} \times \omega \times \omega$ which is universal for $\Gamma \restriction (\mathfrak{X} \times \omega \times \omega)$ and let $G^* \subseteq G$ be in Γ and uniformize G by the Easy Uniformization Theorem 4B.4. Here we are thinking of G as a subset of $(\omega \times \mathfrak{X} \times \omega) \times \omega$, i.e. we uniformize only on the last variable. Now put

$$\mathbf{d}(i, x)\downarrow \Leftrightarrow (\forall n)(\exists m) G^*(i, x, n, m)$$

and if $\mathbf{d}(i, x)\downarrow$, let

$$\mathbf{d}(i, x) = \alpha$$

where for all n, m

$$\alpha(n) = m \Leftrightarrow G^*(i, x, n, m).$$

We omit the trivial computation which establishes that $\mathbf{d}$ is in Γ. From this it follows that each $\mathbf{d}(i, x)$ is in $\Delta(x)$ by 4D.1. Conversely, if $\alpha \in \Delta(x)$, choose i so that

$$\alpha(n) = m \Leftrightarrow G(i, x, n, m)$$

so that

$$\alpha(n) = m \Leftrightarrow G^*(i, x, n, m)$$

and hence $\mathbf{d}(i, x)\downarrow \& \ \mathbf{d}(i, x) = \alpha$.

If $\mathcal{Y}$ is of type 0, the result is trivial. Otherwise, there is a Δ_1^1 bijection

$$\pi : \mathcal{N} \rightarrowtail\!\!\!\!\twoheadrightarrow \mathcal{Y}$$

with Δ_1^1 inverse π^{-1} by 3E.7, so let $\mathbf{d}$ as above parametrize the $\Delta(x)$ points in $\mathcal{N}$ and define $\mathbf{d}^* : \omega \times \mathfrak{X} \to \mathcal{Y}$ by

$$\mathbf{d}^*(i, x) = \pi(\mathbf{d}(i, x));$$

clearly

$$\mathbf{d}^*(i, x)\downarrow \& \ \mathbf{d}^*(i, x) \in N_s \Leftrightarrow \mathbf{d}(i, x)\downarrow \& \ \pi(\mathbf{d}(i, x)) \in N_s$$

so $\mathbf{d}^*$ is Γ-recursive. In particular, each $\mathbf{d}^*(i, x)$ is in $\Delta(i, x) = \Delta(x)$. Conversely, if y is in $\Delta(x)$, then $\alpha = \pi^{-1}(y)$ is in $\Delta(x)$ since π^{-1} is Δ_1^1 and hence Γ-recursive, hence $\alpha = \mathbf{d}(i, x)$ for some i and $y = \pi(\alpha) = \mathbf{d}^*(i, x)$. ⊣

There are many interesting corollaries of this theorem and we will leave most of them for the exercises. Two deserve special billing.

4D.3. THE THEOREM ON RESTRICTED QUANTIFICATION (Kleene [1959b][(18)]). *Let Γ be a Spector pointclass, assume that $Q \subseteq \mathfrak{X} \times \mathcal{Y}$ is in Γ and put*

$$P(x) \Leftrightarrow (\exists y \in \Delta) Q(x, y).$$

Then P is in Γ.

Similarly, if $Q \subseteq \mathfrak{X} \times \mathbf{Z} \times \mathcal{Y}$ is in Γ and

$$P(x, z) \Leftrightarrow (\exists y \in \Delta(z)) Q(x, z, y),$$

then P is in Γ.

PROOF. Taking the second case,

$$P(x, z) \Leftrightarrow (\exists i)\{\mathbf{d}(i, z)\downarrow \& \ Q(x, z, \mathbf{d}(i, z))\},$$

so P is in Γ by (ii) of 4D.1. ⊣

The next result gives a very powerful method for uniformizing Borel sets by Borel sets in the special circumstances when this is possible.

4D.4. THE Δ-UNIFORMIZATION CRITERION. Let Γ be a Spector pointclass closed under $\forall^{\mathcal{N}}$, let $P \subseteq \mathfrak{X} \times \mathcal{Y}$ be in Δ and assume that each section $P_x = \{y: P(x, y)\}$ is either $\emptyset$ or contains some points in $\Delta(x) \cap \mathcal{Y}$, i.e.

$$(*) \qquad (\exists y)P(x, y) \Leftrightarrow (\exists y \in \Delta(x))P(x, y).$$

Then the projection $\exists^{\mathcal{Y}} P$ is in Δ and P can be uniformized by some P^* in Δ.

Conversely, if $P \subseteq \mathfrak{X} \times \mathcal{Y}$ is in Δ and can be uniformized by some P^* in Δ, then each non-empty section P_x has some point in $\Delta(x)$.[(24)]

PROOF. Assume $(*)$ and let $Q = \exists^{\mathcal{Y}} P$, i.e.

$$\begin{aligned} Q(x) &\Leftrightarrow (\exists y)P(x, y) \\ &\Leftrightarrow (\exists y \in \Delta(x))P(x, y). \end{aligned}$$

Clearly Q is in Δ by closure of $\neg\Gamma$ under $\exists^{\mathcal{N}}$ and 4D.3.

Now put

$$R(x, i) \Leftrightarrow P(x, \mathbf{d}(i, x))$$

where $\mathbf{d}$ parametrizes $\Delta(x) \cap \mathcal{Y}$ by 4D.2. By the Δ selection principle 4B.5, since $(\forall x \in Q)(\exists i)R(x, i)$, there must be some $g: \mathfrak{X} \to \omega$ in Δ such that $(\forall x \in Q)R(x, g(x))$. Put

$$P^*(x, y) \Leftrightarrow Q(x) \;\&\; \mathbf{d}(g(x), x) = y.$$

It is immediate that P^* uniformizes P and that it is in Δ follows by 4D.1, since

$$P^*(x, y) \Leftrightarrow Q(x) \;\&\; (\exists i)[\mathbf{d}(i, x)\downarrow \;\&\; \mathbf{d}(i, x) = y \;\&\; g(x) = i],$$

$$\neg P^*(x, y) \Leftrightarrow \neg Q(x) \vee (\exists i)[\mathbf{d}(i, x)\downarrow \;\&\; \mathbf{d}(i, x) \neq y \;\&\; g(x) = i].$$

For the converse, suppose $P^* \subseteq P$ is in Δ and uniformizes P and assume that $(\exists y)P(x, y)$. Then there is a unique y^* such that $P^*(x, y^*)$ and

$$\begin{aligned} y^* \in N_s &\Leftrightarrow (\exists y)[P^*(x, y) \;\&\; y \in N_s] \\ &\Leftrightarrow (\forall y)[P^*(x, y) \Rightarrow y \in N_s], \end{aligned}$$

so $y^* \in \Delta(x)$. ⊣

We leave the application of this result for the exercises of this and the next two sections.

EXERCISES

4D.5. Let Γ be a Spector pointclass. Prove that a partial function $f:\mathfrak{X}\to\omega$ is Γ-recursive exactly when its graph, $\{(x,i):f(x)\downarrow \;\&\; f(x)=i\}$ is in Γ. Similarly, a partial function $f:\mathfrak{X}\to\mathcal{N}$ is Γ-recursive exactly when the associated $f^*:\mathfrak{X}\times\omega\to\omega$ is Γ-recursive, where

$$f^*(x,n)=w \Leftrightarrow f(x)\downarrow \;\&\; f(x)(n)=w.$$

Prove also that the collection of Γ-recursive partial functions is closed under composition.

Hint. For the last assertion, compute:

$$f(g(x))\downarrow \;\&\; f(g(x))\in N_s \Leftrightarrow (\exists y\in\Delta(x))[g(x)\downarrow \;\&\; g(x)=y \;\&\; f(y)\downarrow \;\&\; f(y)\in N_s].$$

Use 4D.1 and 4D.3. ⊣

4D.6. Let $P\subseteq\mathfrak{X}\times\mathcal{Y}$ be a pointset in some Spector pointclass Γ. Prove that there exists a Γ-recursive partial function

$$f:\mathfrak{X}\to\mathcal{Y}$$

such that

(i) $f(x)\downarrow \Leftrightarrow (\exists y\in\Delta(x))P(x,y)$,
(ii) $(\exists y\in\Delta(x))P(x,y)\Longrightarrow P(x,f(x))$.

(The Strong Δ-Selection Principle.)

Hint. Put

$$Q(x,i)\Leftrightarrow P(x,\mathbf{d}(i,x))$$

where $\mathbf{d}$ parametrizes $\Delta(x)\cap\mathcal{Y}$ by 4D.2 and let $Q^*\subseteq Q$ uniformize Q in Γ by 4B.4. Now Q^* is the graph of a Γ-recursive partial function $g:\mathfrak{X}\to\omega$ by 4D.5 and the partial function we need is given by

$$f(x)=\mathbf{d}(g(x),x). \qquad ⊣$$

4D.7. Let Γ be a Spector pointclass closed under $\forall^{\mathcal{N}}$, let $P\subseteq\mathfrak{X}$ be in Δ and assume that $f:\mathfrak{X}\to\mathcal{Y}$ is Δ-recursive and one-to-one on P. Prove that $f[P]$ is in Δ and that there is a Δ-recursive function $g:\mathcal{Y}\to\mathfrak{X}$ which agrees with the inverse function f^{-1} on $f[P]$.

Hint. If $P(x)\;\&\; f(x)=y$, then x is the unique point in P whose image is y; hence

$$
\begin{aligned}
s \in \mathfrak{U}(x) &\Leftrightarrow x \in N_s \\
&\Leftrightarrow (\exists x')[f(x') = y \ \& \ P(x') \ \& \ x' \in N_s] \\
&\Leftrightarrow (\forall x')[f(x') \neq y \vee \neg P(x') \vee x' \in N_s]
\end{aligned}
$$

and $\mathfrak{U}(x)$ is in $\Delta(y)$, i.e. $x \in \Delta(y)$. Hence

$$
\begin{aligned}
y \in f[P] &\Leftrightarrow (\exists x)[P(x) \ \& \ y = f(x)] \\
&\Leftrightarrow (\exists x \in \Delta(y))[P(x) \ \& \ y = f(x)]
\end{aligned}
$$

and $f[P]$ is in Δ by closure of $\neg\Gamma$ under $\exists^{\mathcal{N}}$ and 4D.3.

To get the inverse function, notice that $(\forall y \in f[P])(\exists x \in \Delta(y))[f(x) = y]$ and apply the strong Δ-selection principle, 4D.6. ⊣

Taking $\Gamma = \Pi_1^1$, this is a lightface version of 2E.7 with a very different proof. The classical result follows easily from this, by "relativization."

4D.8. Let Γ be a Spector pointclass closed under $\forall^{\mathcal{N}}$, let $P \subseteq \mathfrak{X}$ be in $\boldsymbol{\Delta}$ and assume that $f : \mathfrak{X} \to \mathfrak{Y}$ is $\boldsymbol{\Delta}$-measurable and one-to-one on P. Prove that $f[P]$ is in $\boldsymbol{\Delta}$ and there is a $\boldsymbol{\Delta}$-measurable function $g : \mathfrak{Y} \to \mathfrak{X}$ which agrees with the inverse f^{-1} on $f[P]$.

Hint. If P is in $\boldsymbol{\Delta}$, then P is in $\Gamma(\varepsilon_0)$ and in $\neg\Gamma(\varepsilon_1)$ for some ε_0, ε_1 in $\mathcal{N}$, so easily P is in $\Delta(\varepsilon)$ for some ε, say with $(\varepsilon)_0 = \varepsilon_0$, $(\varepsilon)_1 = \varepsilon_1$. Similarly, if f is $\boldsymbol{\Delta}$-measurable, then f is $\Delta(\varepsilon')$-recursive for any ε' such that $\{(x, s) : f(x) \in N_s\}$ is in $\Delta(\varepsilon')$. Thus we can find some ε^* such that P is in $\Delta(\varepsilon^*)$ and f is $\Delta(\varepsilon^*)$-recursive and apply 4D.7 to $\Gamma^* = \Gamma(\varepsilon^*)$; it follows that $f[P]$ is in $\Delta(\varepsilon^*) \subseteq \boldsymbol{\Delta}$ and similarly for the inverse. ⊣

This technique of obtaining boldface results from lightface, finer theorems is very easy. We will not always bother to put down the boldface consequences, unless they give well-known classical theorems and we want them to stand out.

It is worth putting down for the record the characterization of Δ_1^1 which follows from 4D.7 and 4A.7.

4D.9. Prove that a set $Q \subseteq \mathfrak{X}$ is Δ_1^1 if and only if Q is the recursive, injective image of some Π_1^0 set $P \subseteq \mathcal{N}$. ⊣

Before using 4D.4 to establish some interesting uniformization results, we point out that not every Borel set can be uniformized by a Borel set.

First a lemma which is interesting in its own right.

4D.10. Prove that there is a Π^0_1 set $A \subseteq \mathcal{N}$, such that $A \neq \emptyset$ but A has no Δ^1_1-recursive member; similarly, for each x, there is a $\Pi^0_1(x)$ set $A \subseteq \mathcal{N}$, $A \neq \emptyset$, such that A has no Δ^1_1-recursive member. (Kleene [1955c].[19,20])

Infer that not every Π^0_1 set $A \subseteq \mathcal{N}$ is a recursive image of $\mathcal{N}$.

Hint. Towards a contradiction, assume that every non-empty Π^0_1 set has a member in Δ^1_1 and let $P(n)$ be a Σ^1_1 relation on ω which is not Π^1_1. There is a Π^0_1 set $Q(n, \alpha)$ such that

$$P(n) \Leftrightarrow (\exists\alpha)Q(n, \alpha)$$

and by our assumption, we then have

$$P(n) \Leftrightarrow (\exists\alpha \in \Delta^1_1)Q(n, \alpha)$$

which implies that P is in Π^1_1 by 4D.3.

If $A = f[\mathcal{N}]$ with a recursive f, then A would have recursive members, namely any $f(\alpha)$ with recursive α. ⊣

4D.11. Prove that there is a Π^0_1 set $P \subseteq \mathcal{N} \times \mathcal{N}$ which cannot be uniformized by any $\boldsymbol{\Sigma}^1_1$ set.[19,20]

Hint. Assume the contrary and let $G(n, \varepsilon, \alpha)$ be a universal Π^0_1 subset of $\omega \times \mathcal{N} \times \mathcal{N}$. Since $\omega \times \mathcal{N}$ is recursively homeomorphic with $\mathcal{N}$, the assumption implies that G can be uniformized by some $\boldsymbol{\Sigma}^1_1$ set $G^* \subseteq G$, say G^* is $\Sigma^1_1(\varepsilon^*)$ for a fixed ε^*. Now every $\Pi^0_1(\varepsilon^*)$ set $A \subseteq \mathcal{N}$ is of the form

$$A = \{\alpha : G(n, \varepsilon^*, \alpha)\}$$

with a fixed n; if $A \neq \emptyset$, then $(\exists\alpha)G(n, \varepsilon^*, \alpha)$, so A contains the unique α^* such that $G^*(n, \varepsilon^*, \alpha^*)$. But this α^* is in $\Delta^1_1(\varepsilon^*)$, since

$$\begin{aligned}\alpha^*(t) = w &\Leftrightarrow (\exists\alpha)[G^*(n, \varepsilon^*, \alpha) \;\&\; (t) = w]\\ &\Leftrightarrow (\forall\alpha)[G^*(n, \varepsilon^*, \alpha) \Rightarrow \alpha(t) = w],\end{aligned}$$

contradicting 4D.10. ⊣

Nevertheless, there are many special circumstances in which Borel sets can be uniformized by Borel sets. The next exercise gives a simple topological condition which is often easy to verify and implies the more subtle definability condition of 4D.4.

4D.12. Let Γ be a Spector pointclass closed under $\forall^{\mathcal{N}}$, let $P \subseteq \mathfrak{X} \times \mathfrak{Y}$ be in Δ and assume that for each x, the section P_x has at least one isolated point—e.g. it may be that each P_x is finite, or countable and closed. Prove that P can be uniformized by some P^* in Δ. Infer the same result for P in $\mathbf{\Delta}$, with P^* in $\mathbf{\Delta}$.

Hint. If y is isolated in P_x, then for some s, $P_x \cap N_s = \{y\}$, so that the singleton $\{y\}$ is in $\Delta(x)$ and y is easily $\Delta(x)$-recursive. For the second assertion recall that each P in $\mathbf{\Delta}$ is in some $\Delta(\varepsilon^*)$ and use the result on the Spector pointclass $\Gamma(\varepsilon^*)$. ⊣

In 4F we will improve this result substantially by showing that it is enough to assume each P_x to be a countable union of compact sets.

The next exercise is simple but amusing.

4D.13. Prove that if $P \subseteq \mathbb{R}^n \times \mathbb{R}^m$ is a *convex* Borel set, then the projection

$$Q = \{x \in \mathbb{R}^n : (\exists y \in \mathbb{R}^m) P(x, y)\}$$

is Borel and P can be uniformized by a Borel set.

Hint. For $m = 1$, each section P_x is either a singleton or contains a whole line segment. Use induction on m. ⊣

We now establish some interesting definability results about $\Delta \cap \mathfrak{X}$.

4D.14. Prove that if Γ is a Spector pointclass, then for each $\mathfrak{X}$ the set $\Delta \cap \mathfrak{X}$ is in Γ. Similarly, the relation $\{(x, y) : x \in \Delta(y)\}$ is in Γ. (*Upper classification of* Δ.)(18)

Hint. $x \in \Delta \Leftrightarrow (\exists i)\{\mathbf{d}(i)\downarrow \;\&\; \mathbf{d}(i) = x\}$. ⊣

4D.15. Let Γ be a Spector pointclass and let $\mathbf{d} : \omega \rightarrow \mathfrak{X}$ be a Γ-recursive partial function which parametrizes $\Delta \cap \mathfrak{X} = \{x \in \mathfrak{X} : x \text{ is } \Delta\text{-recursive}\}$. Prove that there is a Γ-recursive partial function

$$\mathbf{c} : \mathfrak{X} \rightarrow \omega$$

such that

$$\mathbf{c}(x)\downarrow \Leftrightarrow x \in \Delta$$

and for $x \in \Delta$, $\mathbf{d}(\mathbf{c}(x)) = x$.

Hint. Use the Easy Uniformization Theorem 4B.4 or 4D.6. ⊣

4D.16. Prove that if Γ is a Spector pointclass closed under either $\forall^{\mathcal{N}}$ or $\exists^{\mathcal{N}}$, then for every perfect product space $\mathfrak{X}$ the set $\Delta \cap \mathfrak{X}$ is not in $\neg\Gamma$. (*Lower classification of* Δ.)(18)

In particular, $\Delta_1^1 \cap \mathfrak{X}$ is not Σ_1^1 and $\Delta_2^1 \cap \mathfrak{X}$ is not Π_2^1.

Hint. If $\pi : \mathcal{N} \rightarrowtail\!\!\!\!\twoheadrightarrow \mathfrak{X}$ is a Δ_1^1 isomorphism, then clearly

$$x \in \Delta \Leftrightarrow \pi^{-1}(x) \in \Delta,$$

so it is enough to prove the result for $\mathcal{N}$. For simplicity in notation put

$$\mathcal{D} = \Delta \cap \mathcal{N}.$$

Case 1. *Γ is closed under* $\forall^{\mathcal{N}}$. Let

$$\begin{aligned} j \in J &\Leftrightarrow (\exists \alpha)[\alpha \in \mathfrak{D} \;\&\; \mathbf{c}(\alpha) = j] \\ &\Leftrightarrow (\exists i)[\mathbf{d}(i)\!\downarrow \;\&\; \mathbf{c}(\mathbf{d}(i)) = j], \end{aligned}$$

so J is in Γ. Also

$$j \notin J \Leftrightarrow (\forall \alpha)[\alpha \notin \mathcal{D} \vee [\mathbf{c}(\alpha)\!\downarrow \;\&\; \mathbf{c}(\alpha) \neq j]],$$

so that if $\mathfrak{D}$ were in Δ, then J would be in Δ. Then the irrational

$$\alpha(j) = \begin{cases} \mathbf{d}(j)(j) + 1 & \text{if } j \in J, \\ 0 & \text{if } j \notin J, \end{cases}$$

would be in Δ and different from all $\mathbf{d}(j)$.

Case 2. *Γ is closed under* $\exists^{\mathcal{N}}$. Let

$$i \in I \Leftrightarrow \mathbf{d}(i)\!\downarrow$$

and let φ be a Γ-norm on I. Put

$$\begin{aligned} P(\alpha) \Leftrightarrow\; &(\forall i)[\alpha(i) = 0 \Rightarrow i \in I] \\ &\& \;(\forall i)(\forall j)[(\alpha(j) = 0 \;\&\; i \leq_{\Gamma}^{\varphi} j) \Rightarrow \alpha(i) = 0]. \end{aligned}$$

Clearly P is in Γ and

$$P(\alpha) \Leftrightarrow \{i : \alpha(i) = 0\} = I \vee (\exists j)[j \in I \;\&\; \{i : \alpha(i) = 0\} = \{i : \varphi(i) < \varphi(j)\}].$$

Since $I \notin \Delta$, or else we get a contradiction as before, we have

$$i \notin I \Leftrightarrow (\exists \alpha)[\alpha \notin \mathcal{D} \;\&\; P(\alpha) \;\&\; \alpha(i) \neq 0]$$

which proves $I \in \Delta$ and yields a contradiction. ⊣

The definition of a Spector pointclass was a bit complicated, because it involved the subtle substitution property. We give here an elegant characterization of Spector pointclasses in terms of a closure property much simpler than substitution.

4D.17. Let Γ be a Σ-pointclass closed under $\forall^\omega$, ω-parametrized and normed. Prove that Γ is a Spector pointclass if and only if it satisfies the following property of *closure under restricted quantification*: if $Q \subseteq \mathcal{X} \times \mathcal{Z} \times \mathcal{Y}$ is in Γ and

$$P(x, z) \Leftrightarrow (\exists y \in \Delta(z))Q(x, z, y),$$

then P is also in Γ. (Kechris.)

Hint. Spector pointclasses are closed under restricted quantification by 4D.3.

Conversely, to establish the substitution property for some Γ satisfying the hypotheses above, suppose $Q \subseteq \mathcal{Y}$ is in Γ and $f : \mathcal{X} \to \mathcal{Y}$ is computed on its domain by some $P \subseteq \mathcal{X} \times \omega$ in Γ. Put

$$Q^*(x) \Leftrightarrow (\exists y \in \Delta(x))[Q(y) \;\&\; (\forall s)[y \in N_s \Rightarrow P(x, s)]];$$

clearly Q^* is in Γ and

$$f(x)\downarrow \;\&\; f(x) \in \Delta(x) \Rightarrow [Q^*(x) \Leftrightarrow Q(f(x))].$$

Thus to complete the proof it will be sufficient to check that under the conditions on Γ,

$$f(x)\downarrow \Rightarrow f(x) \in \Delta(x).$$

Suppose $f(x) = y$ and put

$$S(n, s) \Leftrightarrow P(x, s) \;\&\; radius(N_s) \leq 2^{-n}.$$

Clearly S is in $\Gamma(x)$, which is adequate, closed under $\exists^\omega$, $\forall^\omega$ and normed. Also $(\forall n)(\exists s)S(n, s)$, so by the Δ-selection principle 4B.5 there is a $\Delta(x)$-recursive function $g : \omega \to \omega$ such that $(\forall n)S(n, g(n))$. It is now immediate that

$$y \in N_s \Leftrightarrow (\exists n)[y \in N(\mathcal{Y}, g(n)) \;\&\; N(\mathcal{Y}, g(n)) \subseteq N_s]$$

so that $y \in \Delta(x)$. ⊣

Unfortunately this elegant characterization is not useful in practice since it is usually much easier to establish that a given Γ satisfies the substitution property rather than prove directly closure under restricted quantification.

4E. The uniformization theorem for Π^1_1, Σ^1_2 (19–22)

We now proceed to establish one of the central results in the subject, that Π^1_1 sets can be uniformized by Π^1_1 sets. The key tool for the proof is the notion of a *scale*.

A *scale* on a pointset P is a sequence $\bar{\varphi}=\{\varphi_n\}_{n\in\omega}$ of norms on P such that the following limit condition holds: if $x_0, x_1, x_2, \ldots$ are in P and $limit_{i\to\infty} x_i = x$ and if for each n, the sequence of ordinals

$$\varphi_n(x_0), \varphi_n(x_1), \varphi_n(x_2), \ldots$$

is ultimately constant, say

$$\varphi_n(x_i)=\lambda_n$$

for all large i, then $P(x)$ and for every n,

$$\varphi_n(x)\leq\lambda_n.$$

Thus a scale is just a semiscale (in the sense of 2B) which satisfies an additional *lower semicontinuity property*.

As with norms, there are many trivial scales on a pointset, at least if we use the axiom of choice: choose a one-to-one norm $\varphi: P \twoheadrightarrow \kappa$ and set for each n, $\varphi_n(x)=\varphi(x)$. Again as with norms, we get a nontrivial concept by imposing definability conditions.

Let Γ be a pointclass and $\bar{\varphi}=\{\varphi_n\}_{n\in\omega}$ a scale on some set P. We call $\bar{\varphi}$ a *Γ-scale* if there are relations $S_\Gamma(n, x, y)$, $S_{\check{\Gamma}}(n, x, y)$ in Γ and $\neg\Gamma$ respectively such that for every y,

$$(*)\qquad P(y) \Rightarrow (\forall n)(\forall x)\{[P(x)\ \&\ \varphi_n(x)\leq\varphi_n(y)] \Leftrightarrow S_\Gamma(n, x, y) \Leftrightarrow S_{\check{\Gamma}}(n, x, y)\},$$

In other words, $\bar{\varphi}$ is a Γ-scale if all the norms φ_n are Γ-norms, *uniformly in n.*

It is trivial to verify as in 4B.1 that if Γ is adequate and $\bar{\varphi}$ is a scale on some P in Γ, then $\bar{\varphi}$ is a Γ-scale exactly when the relations

$$R(n, x, y) \Leftrightarrow x \leq^*_{\varphi_n} y,$$
$$S(n, x, y) \Leftrightarrow x <^*_{\varphi_n} y,$$

are in Γ.

A pointclass Γ is *scaled* or *has the scale property* if every pointset in Γ admits a Γ-scale. It is often sufficient for our purposes to prove that pointsets of type 1 in Γ admit Γ-scales, whether or not the stronger scale property holds in Γ (see 4E.6).

4E.1. THEOREM. *Every Π^1_1 pointset of type 1 admits a Π^1_1-scale; similarly, every $\mathbf{\Pi}^1_1$ pointset of type 1 admits a $\mathbf{\Pi}^1_1$-scale.*

PROOF. Let us first develop a bit of notation. If α codes a linear ordering $\leq_\alpha$, i.e. $\alpha \in LO$ as we defined this in 4A, then for every integer n put

$$\begin{aligned}\leq_\alpha \restriction n &= \{(s, t) : s \leq_\alpha t \;\&\; t <_\alpha n\}\\ &= \{(s, t) : \alpha(\langle s, t\rangle) = 1 \;\&\; \alpha(\langle t, n\rangle) = 1\\ &\qquad\qquad \& \; \alpha(\langle n, t\rangle) \neq 1\}.\end{aligned}$$

Clearly $\leq_\alpha \restriction n$ is also a linear ordering—it is the initial segment of $\leq_\alpha$ with top n, if n is in the field of $\leq_\alpha$ and it is the empty relation otherwise. If $\leq_\alpha$ is a wellordering with rank function ρ, then for each n, $\leq_\alpha \restriction n$ is a wellordering and

$$\rho(n) = |\leq_\alpha \restriction n|.$$

In particular, for n, m in the field of $\leq_\alpha$,

$$n \leq_\alpha m \Leftrightarrow |\leq_\alpha \restriction n| \leq |\leq_\alpha \restriction m|.$$

Given a pointset $P \subseteq \mathfrak{X}$ of type 1 in Π^1_1, choose a recursive $f : \mathfrak{X} \to \mathcal{N}$ by 4A.3 such that for every x, $f(x) \in LO$ and

$$P(x) \Leftrightarrow f(x) \in WO.$$

Let

$$(\xi, \eta) \mapsto \langle \xi, \eta\rangle$$

be an order-preserving map of $\aleph_1 \times \aleph_1$ (ordered lexicographically) into the ordinals, i.e.

$$\langle \xi, \eta\rangle \leq \langle \xi', \eta'\rangle \Leftrightarrow [\xi < \xi'] \vee [\xi = \xi' \;\&\; \eta \leq \eta'].$$

Finally, for $x \in P$ put

$$\varphi_n(x) = \langle |\leq_{f(x)}|, |\leq_{f(x)} \restriction n|\rangle.$$

We claim that $\bar{\varphi} = \{\varphi_n\}_{n \in \omega}$ is a Π^1_1-scale on P.

To prove this, assume that $limit_i\, x_i = x$ with $x_0, x_1, \ldots$ in P and that for each n and all large i,

$$\varphi_n(x_i) = \langle \lambda, \lambda_n\rangle.$$

This implies immediately that for each n and all large i,

$$|\leq_{f(x_i)} \restriction n| = \lambda_n.$$

The key to the proof is the fact that f is continuous, being recursive. Let us first use this to prove that the mapping

$$n \mapsto \lambda_n$$

is order-preserving from $\leq_{f(x)}$ into the ordinals. This holds because

$$\begin{aligned} n <_{f(x)} m &\Rightarrow f(x)(\langle n, m\rangle) = 1 \ \& \ f(x)(\langle m, n\rangle) \neq 1 \\ &\Rightarrow \text{for all large } i,\ f(x_i)(\langle n, m\rangle) = 1 \ \& \ f(x_i)(\langle m, n\rangle) \neq 1 \\ &\quad\ \text{(by the continuity of } f) \\ &\Rightarrow \text{for all large } i,\ n <_{f(x_i)} m \\ &\Rightarrow \text{for all large } i,\ |\leq_{f(x_i)}\restriction n| < |\leq_{f(x_i)}\restriction m| \\ &\Rightarrow \lambda_n < \lambda_m, \end{aligned}$$

where the last implication is justified since for all large i, $|\leq_{f(x_i)}\restriction n| = \lambda_n$.

Since $n \mapsto \lambda_n$ is order-preserving, $\leq_{f(x)}$ is a wellordering, i.e. $f(x) \in WO$ and we know $P(x)$. The same fact implies that for every n,

$$|\leq_{f(x)}\restriction n| \leq \lambda_n$$

since $|\leq_{f(x)} \restriction n|$ is the rank of n in $\geq_{f(x)}$ and every order-preserving map dominates the rank function by 2G.7. Similarly,

$$|\leq_{f(x)}| \leq \lambda,$$

because

$$\begin{aligned} |\leq_{f(x)}| &= \mathit{supremum}\{|\leq_{f(x)}\restriction n| : n \in \omega\} \\ &\leq \mathit{supremum}\{\lambda_n : n \in \omega\} \leq \lambda, \end{aligned}$$

the last inequality following from the fact that for each n and all large i,

$$\lambda_n = |\leq_{f(x_i)}\restriction n| \leq |\leq_{f(x_i)}| = \lambda.$$

Thus

$$\varphi_n(x) = \langle |\leq_{f(x)}|, |\leq_{f(x)}\restriction n| \rangle \leq \langle \lambda, \lambda_n \rangle$$

and $\bar{\varphi}$ is a scale on P.

To show that $\bar{\varphi}$ is a Π^1_1-scale, find (easily) a recursive g such that for $\alpha \in LO$,

$$\leq_{g(\alpha,n)} = \leq_\alpha\restriction n,$$

and put

$$\begin{aligned} S_\Pi(n, x, y) &\Leftrightarrow f(x) \leq_\Pi f(y) \ \& \ [\neg(f(y) \leq_\Sigma f(x)) \vee g(f(x), n) \leq_\Pi g(f(y), n)], \\ S_\Sigma(n, x, y) &\Leftrightarrow f(x) \leq_\Sigma f(y) \ \& \ [\neg(f(y) \leq_\Pi f(x)) \vee g(f(x), n) \leq_\Sigma g(f(y), n)], \end{aligned}$$

where $\leq_\Pi$, $\leq_\Sigma$ are from 4A.2. ⊣

As with semiscales in the proof of the Kunen–Martin Theorem, here too we often need scales with very special properties. A scale $\bar{\varphi} = \{\varphi_n\}_{n\in\omega}$ on $P \subseteq \mathfrak{X}$ is *very good* if the following two conditions hold:

(1) If $x_0, x_1, \ldots$ are in P and for each n and all large i, $\varphi_n(x_i) = \lambda_n$, then there exists some $x \in P$ such that $limit_{i\to\infty} x_i = x$ (and hence for each n, $\varphi_n(x) \leq \lambda_n$).

(2) If x, y are in P and $\varphi_n(x) \leq \varphi_n(y)$, then for each $i \leq n$, $\varphi_i(x) \leq \varphi_i(y)$.

Condition (1) implies that $\bar{\varphi}$ is a *good semiscale* in the sense of 2G.

4E.2. LEMMA. *Let Γ be an adequate pointclass. If a pointset P of type 1 admits a Γ-scale, then P admits a very good Γ-scale.*

PROOF. Assume at first that $P \subseteq \mathcal{N}$ is a set of irrationals and let $\bar{\psi} = \{\psi_n\}_{n\in\omega}$ be a Γ-scale on P. Choose $\lambda \geq \omega$ and large enough so that all the norms ψ_n are into λ. For each n, wellorder the sequences of length $2n$ of the form $(\xi_0, k_0, \xi_1, k_1, \ldots, \xi_n, k_n)$ $(\xi_i < \lambda,\ k_i \in \omega)$ lexicographically,

$$
\begin{aligned}
(\xi_0, k_0, \ldots, \xi_n, k_n) \leq (\eta_0, l_0, \ldots, \eta_n, l_n)
&\Leftrightarrow \xi_0 < \eta_0 \\
&\quad \vee [\xi_0 = \eta_0 \ \&\ k_0 < l_0] \\
&\quad \vee [\xi_0 = \eta_0 \ \&\ k_0 = l_0 \ \&\ \xi_1 < \eta_1] \\
&\quad \vee \cdots \\
&\quad \vee [\xi_0 = \eta_0 \ \&\ \cdots \ \&\ \xi_n = \eta_n \ \&\ k_n \leq l_n]
\end{aligned}
$$

and let

$$(\xi_0, k_0, \ldots, \xi_n, k_n) \mapsto \langle \xi_0, k_0, \ldots, \xi_n, k_n \rangle$$

be an order-preserving map of this ordering into the ordinals. Finally put

$$\varphi_n(\alpha) = \langle \psi_0(\alpha), \alpha(0), \psi_1(\alpha), \alpha(1), \ldots, \psi_n(\alpha), \alpha(n) \rangle.$$

We will show that $\bar{\varphi} = \{\varphi_n\}_{n\in\omega}$ is a very good Γ-scale on P.

Suppose first that $\alpha_0, \alpha_1, \ldots$ are in P and for each n and all large i, $\varphi_n(\alpha_i)$ is constant,

$$
\begin{aligned}
\varphi_n(\alpha_i) &= \langle \psi_0(\alpha_i), \alpha_i(0), \ldots, \psi_n(\alpha_i), \alpha_i(n) \rangle \\
&= \langle \xi_0^n, k_0^n, \ldots, \xi_n^n, k_n^n \rangle.
\end{aligned}
$$

Since by the definition

$$k_j^n = \alpha_i(j) \qquad (j \leq n, \text{ all large } i),$$

it follows that

$$k_j^n = k_j$$

is independent of n and

$$limit_{i\to\infty} \alpha_i = \alpha = (k_0, k_1, \ldots).$$

Similarly,

$$\xi_j^n = \psi_j(\alpha_i) \quad (j \leq n, \text{ all large } i),$$

so that

$$\xi_j^n = \xi_j$$

is independent of n and for all large i,

$$\psi_j(\alpha_i) = \xi_j.$$

Since $\bar{\psi}$ is a scale, we thus have $\alpha \in P$ and for each j, $\psi_j(\alpha) \leq \xi_j$; from this follows immediately that for each n,

$$\varphi_n(\alpha) \leq \langle \xi_0, k_0, \xi_1, k_1, \ldots, \xi_n, k_n \rangle.$$

It is also immediate from the definition that for x, y in P,

$$\varphi_n(x) \leq \varphi_n(y) \Rightarrow \text{for each } i \leq n,\ \varphi_i(x) \leq \varphi_i(y),$$

so that $\bar{\varphi}$ is a very good scale.

To prove that $\bar{\varphi}$ is a Γ-scale, let

$$\alpha \sim_{\psi_i} \beta \Leftrightarrow \alpha \leq^*_{\psi_i} \beta \ \&\ \beta \leq^*_{\psi_i} \alpha$$

and put

$$\begin{aligned} R(n, \alpha, \beta) \Leftrightarrow\ & \alpha <^*_{\psi_0} \beta \\ & \vee [\alpha \sim_{\psi_0} \beta \ \&\ \alpha(0) < \beta(0)] \\ & \vee \cdots \\ & \vee [\alpha \sim_{\psi_0} \beta \ \&\ \alpha(0) = \beta(0) \ \&\ \cdots \ \&\ \alpha \sim_{\psi_n} \beta \ \&\ \alpha(n) \leq \beta(n)] \\ \Leftrightarrow\ & (\exists i \leq n)\{(\forall j < i)[\alpha \sim_{\psi_j} \beta \ \&\ \alpha(j) = \beta(j)] \\ & \quad \&\ [\alpha <^*_{\psi_i} \beta \vee [\alpha \sim_{\psi_i} \beta \ \&\ \alpha(i) < \beta(i)] \\ & \quad \vee [i = n \ \&\ \alpha \sim_{\psi_i} \beta \ \&\ \alpha(i) \leq \beta(i)]]\}. \end{aligned}$$

Clearly R is in Γ and

$$\alpha \leq^*_{\varphi_n} \beta \Leftrightarrow R(n, \alpha, \beta),$$

so $\bar{\varphi}$ is a Γ-scale, since the argument for $<^*_{\varphi_n}$ is similar.

Finally, if $Q \subseteq \mathfrak{X}$ is of type 1 with $\mathfrak{X} \neq \mathcal{N}$, let

$$\pi : \mathcal{N} \rightarrowtail\!\!\!\!\twoheadrightarrow \mathfrak{X}$$

be a recursive isomorphism, let

$$P = \pi^{-1}[Q]$$

and verify easily the following two propositions: if $\bar{\psi}$ is a Γ-scale of Q, then the sequence

$$\psi^*_n(\alpha) = \psi_n(\pi\alpha)$$

is a Γ-scale on P and if $\bar{\varphi}$ is a very good Γ-scale on P, then the sequence

$$\varphi^*_n(x) = \varphi_n(\pi^{-1}x)$$

is a very good Γ-scale on Q. ⊣

There are many interesting results about scales and we will look at some of them in the exercises and again in Chapter 6. Here we concentrate on the relation between scales and uniformization.

4E.3. THE UNIFORMIZATION LEMMA. *If Γ is adequate, $\mathfrak{X}$ is of type 0 or 1 and $P \subseteq \mathfrak{X} \times \mathcal{N}$ admits a Γ-scale, then P can be uniformized by some P^* in $\forall^{\mathcal{N}}\Gamma$.*

PROOF. By 4E.2, let $\bar{\varphi} = \{\varphi_n\}_{n\in\omega}$ be a very good Γ-scale on P, let

$$R(n, x, \alpha) \Leftrightarrow (\forall\beta)[(x, \alpha) \leq^*_{\varphi_n} (x, \beta)]$$

and put

$$P^*(x, \alpha) \Leftrightarrow (\forall n)R(n, x, \alpha).$$

It is sufficient to show that P^* uniformizes P, since R is obviously in $\forall^{\mathcal{N}}\Gamma$ and hence P^* is in $\forall^{\mathcal{N}}\Gamma$.

To begin with, clearly

$$P^*(x, \alpha) \Rightarrow P(x, \alpha),$$

since

$$\begin{aligned} P^*(x, \alpha) &\Rightarrow (x, \alpha) \leq^*_{\varphi_0} (x, \alpha) \\ &\Rightarrow P(x, \alpha). \end{aligned}$$

Assume now that for some fixed x, $(\exists\alpha)P(x, \alpha)$; we must show that in this case there is *exactly one* α such that $P^*(x, \alpha)$.

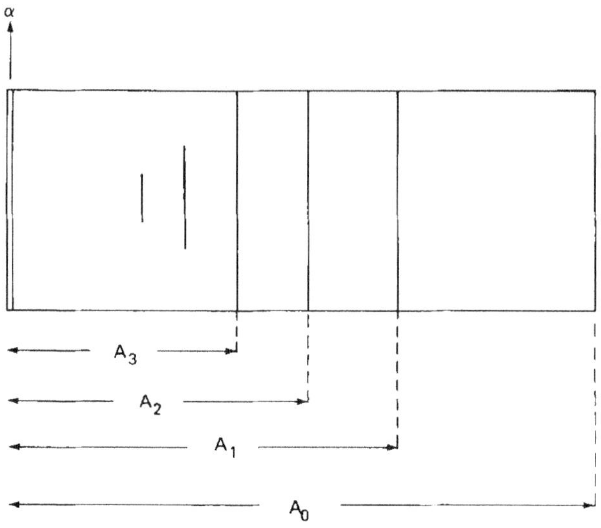

Figure 4E.1. Uniformizing via a very good scale.

Keeping x fixed, put

$$\lambda_n = \textit{infimum}\{\varphi_n(x, \alpha): P(x, \alpha)\}$$

and let (see Figure 4E.1)

$$\begin{aligned} A_0 &= \{\alpha: P(x, \alpha)\} \\ A_{n+1} &= \{\alpha: P(x, \alpha) \mathbin{\&} \varphi_n(x, \alpha) = \lambda_n\} \\ &= \{\alpha: (\forall\beta)[(x, \alpha) \leq_{\varphi_n} (x, \beta)]\} \\ &= \{\alpha: R(n, x, \alpha)\}. \end{aligned}$$

Clearly each A_n is non-empty and

$$P^*(x, \alpha) \Leftrightarrow \bigwedge_n [\alpha \in A_n],$$

so it is enough to prove that $\bigcap_n A_n$ is a singleton.

Notice that

$$A_0 \supseteq A_1$$

and by the second condition on a very good scale,

$$(\forall\beta)[(x, \alpha) \leq_{\varphi_{n+1}} (x, \beta)] \Rightarrow (\forall\beta)[(x, \alpha) \leq_{\varphi_n} (x, \beta)],$$

so that in fact

$$A_0 \supseteq A_1 \supseteq A_2 \supseteq \ldots .$$

Choose now some $\alpha_i \in A_i$, one for each i. We then have $\varphi_n(x, \alpha_i) = \lambda_n$ for each $i > n$, so by the first condition on a very good scale, there is some α such that

$$\alpha = \mathit{limit}_{i \to \infty} \alpha_i,$$

$P(x, \alpha)$, and for each n,

$$\varphi_n(x, \alpha) \leq \lambda_n;$$

by the definition of λ_n then,

$$\varphi_n(x, \alpha) = \lambda_n,$$

so $\alpha \in \bigcap_n A_n$ and this intersection is non-empty. Moreover, if also $\beta \in \bigcap_n A_n$, then the sequence

$$\alpha_0, \beta, \alpha_1, \beta, \alpha_2, \beta, \ldots = \gamma_0, \gamma_1, \gamma_2, \gamma_3, \ldots$$

has the property that for each n and all large i, $\varphi_n(x, \gamma_i)$ is constant, so that $\mathit{limit}_{i \to \infty} \gamma_i$ must converge, presumably both to β and to $\alpha = \mathit{limit}_{i \to \infty} \alpha_i$, so that $\beta = \alpha$. Hence $\bigcap_n A_n$ is the singleton $\{\alpha\}$, which is what we needed to show. ⊣

A pointclass Γ has the *uniformization property* if every $P \subseteq \mathfrak{X} \times \mathfrak{Y}$ in Γ can be uniformized by some P^* in Γ.

4E.4. THE NOVIKOV–KONDO–ADDISON UNIFORMIZATION THEOREM. (Kondo [1938].(19–22)) *The pointclasses Π^1_1, $\mathbf{\Pi}^1_1$, Σ^1_2, $\mathbf{\Sigma}^1_2$ all have the uniformization property.*

PROOF. Suppose first that $P \subseteq \mathfrak{X} \times \mathfrak{Y}$ is in Π^1_1. If $\mathfrak{Y}$ is of type 0, the result follows from 4B.4. If not, let

$$\pi : \mathcal{N} \rightarrowtail\!\!\!\!\rightarrow \mathfrak{Y}$$

be a Δ^1_1 isomorphism of $\mathcal{N}$ with $\mathfrak{Y}$, let

$$\sigma : \mathfrak{X}^* \rightarrowtail\!\!\!\!\rightarrow \mathfrak{X}$$

be a Δ^1_1 isomorphism of $\mathfrak{X}$ with some space $\mathfrak{X}^*$ of type 0 or 1 and define

$Q \subseteq \mathfrak{X}^* \times \mathcal{N}$ by

$$Q(x, \alpha) \Leftrightarrow P(\sigma(x), \pi(\alpha)).$$

Now Q is Π_1^1 and by 4E.3 we can find a Π_1^1 set $Q^* \subseteq \mathfrak{X}^* \times \mathcal{N}$ which uniformizes Q. It is immediate that the Π_1^1 set

$$P^*(x, y) \Leftrightarrow Q^*(\sigma^{-1}(x), \pi^{-1}(y))$$

uniformizes P.

The argument for $\mathbf{\Pi}_1^1$ is identical.

If $P \subseteq \mathfrak{X} \times \mathcal{Y}$ is in Σ_2^1, then

$$P(x, y) \Leftrightarrow (\exists \alpha) Q(x, y, \alpha)$$

with Q in Π_1^1. Applying the result about Π_1^1 to $Q \subseteq \mathfrak{X} \times (\mathcal{Y} \times \mathcal{N})$, we get a Π_1^1 set $Q^* \subseteq \mathfrak{X} \times (\mathcal{Y} \times \mathcal{N})$ which uniformizes Q. Then

$$P^*(x, y) \Leftrightarrow (\exists \alpha) Q^*(x, y, \alpha)$$

is easily seen to uniformize P. ⊣

We will see in Chapter 5 that this is just about the strongest uniformization theorem which can be proved in Zermelo–Fraenkel set theory; it is consistent with the axioms of Zermelo–Fraenkel (including choice) that there exists a Π_2^1 set which cannot be uniformized by any "definable" set—in particular, it cannot be uniformized by any projective set.

Among the many important consequences of the uniformization theorem, perhaps the most significant is the *basis* result for $\mathbf{\Sigma}_2^1$ which we now explain.

A set of points (in various spaces) $\mathfrak{B}$ is called *a basis* for a pointclass Γ, if every non-empty set in Γ has a member in $\mathfrak{B}$, i.e. for $P \subseteq \mathfrak{X}$ in Γ,

$$(\exists x) P(x) \Leftrightarrow (\exists x \in \mathfrak{B}) P(x).$$

We also say that a pointclass Λ is *a basis* for Γ if the set of Λ-recursive points is a basis for Γ, i.e. for P in Γ,

$$(\exists x) P(x) \Leftrightarrow (\exists x \in \Lambda) P(x).$$

In 4D10 we proved that Δ_1^1 is not a basis for Π_1^0, and hence it is not a basis for Σ_1^1 or Π_1^1.

4E.5. THE BASIS THEOREM FOR Σ_2^1. *The pointclass Δ_2^1 is a basis for Σ_2^1 and more generally, for each x, $\Delta_2^1(x)$ is a basis for $\Sigma_2^1(x)$. Thus, if $P \subseteq \mathfrak{X} \times \mathcal{Y}$ is* *in* Σ^1_2, *then*

$$(\exists y)P(x, y) \Leftrightarrow (\exists y \in \Delta^1_2(x))P(x, y).$$

PROOF. The second assertion immediately implies the first. To prove it, given $P \subseteq \mathfrak{X} \times \mathfrak{Y}$, choose $P^* \subseteq \mathfrak{X} \times \mathfrak{Y}$ in Σ^1_2 which uniformizes P. If $(\exists y)P(x, y)$, then there exists exactly one y which satisfies $P^*(x, y)$, call it y^*; clearly

$$\begin{aligned} y^* \in N_s &\Leftrightarrow (\exists y)[P^*(x, y) \;\&\; y \in N_s] \\ &\Leftrightarrow (\forall y)[P^*(x, y) \Rightarrow y \in N_s], \end{aligned}$$

so y^* is $\Delta^1_2(x)$-recursive. ⊣

Again this result is best possible in Zermelo–Fraenkel set theory, i.e. we cannot prove in this theory that every non-empty Π^1_2 set must contain a "definable" element.

EXERCISES

4E.6. Suppose Γ is an adequate pointclass closed under substitution of Δ^1_1 functions. Prove that if every pointset of type 1 in Γ admits a Γ-scale, then every pointset in Γ admits a very good Γ-scale.

In particular, Π^1_1 and $\mathbf{\Pi}^1_1$ are scaled.

Hint. Suppose $P \subseteq \mathfrak{X}$ is given, P in Γ. Using 3E.6, let

$$\pi : \mathcal{N} \twoheadrightarrow \mathfrak{X}$$

be a recursive surjection of $\mathcal{N}$ onto $\mathfrak{X}$ such that for some Π^0_1 set $A \subseteq \mathcal{N}$, $\pi[A] = \mathfrak{X}$ and $\pi f(x) = x$ for every $x \in \mathfrak{X}$, with a Δ^1_1 function such that $f[\mathfrak{X}] = A$. Put

$$Q(\alpha) \Leftrightarrow \alpha \in A \;\&\; P(\pi(\alpha)),$$

so Q is in Γ and by hypothesis and 4E.2, Q admits a very good Γ-scale $\bar{\psi} = \{\psi_n\}_{n \in \omega}$. On P set

$$\varphi_n(x) = \psi_n(f(x))$$

and show that $\bar{\varphi}$ is a very good Γ-scale. The key point is the continuity of π; this implies that if $x_0, x_1, \ldots$ are in P and $\alpha_i = f(x_i)$, then $limit_{i \to \infty}\, x_i = limit_{i \to \infty} \cdot \pi(\alpha_i) = \pi(\alpha)$ with $\pi(\alpha) \in P$. ⊣

The analog of 4B.3 also holds for scales, i.e. if Γ is scaled, adequate and closed under $\forall^{\mathcal{N}}$, then $\exists^{\mathcal{N}}\Gamma$ is also scaled. There is a bit of computation to this and we will postpone it until Chapter 6 when we will need it.

The next result is implicit in the proof of 4E.4, but we put it down for the record.

4E.7. Prove that if Γ is adequate and closed under substitution of Δ^1_1 functions and $\forall^{\mathcal{N}}$ and if every pointset of type 1 in Γ admits a Γ-scale, then both Γ and $\exists^{\mathcal{N}}\Gamma$ have the uniformization property. ⊣

Every non-empty Σ^1_1 set has a Δ^1_2 member by 4E.5 but need not have a Δ^1_1 member by 4D.10. The correct basis for Σ^1_1 is a small part of Δ^1_2, by the next result.

With each relation $P \subseteq \mathfrak{X} = \omega^k$ on a space of type 0 we associate its *contracted characteristic function* α_P,

$$\alpha_P(n) = \begin{cases} 1 & \text{if } P((n)_1, \ldots, (n)_k), \\ 0 & \text{if } \neg P((n)_1, \ldots, (n)_k). \end{cases}$$

We call a set, function or point *recursive in P* if it is recursive in α_P. *Notice that we only define these notions here for P of type* 0—the correct concept of recursion relative to an arbitrary pointset is quite complicated and we will not go into it now.

4E.8. Prove that there is a fixed Σ^1_1 set P of type 0 such that $\{x : x$ is recursive in $P\}$ is a basis for Σ^1_1. (*Kleene's Basis Theorem*, Kleene [1955b].[(23)])

Hint. Suppose

$$Q(\alpha) \leftrightarrow (\exists\beta)(\forall t)R(\bar{\alpha}(t), \bar{\beta}(t))$$

is a typical Σ^1_1 set of irrationals with R recursive. As usually, we can think of Q as the projection

$$Q = \mathrm{p}[T]$$

of the tree T on $\omega \times \omega$,

$$T = \{(a_0, b_0, a_1, b_1, \ldots, a_{t-1}, b_{t-1}) : (\forall i < t)R(\langle a_0, \ldots, a_i\rangle, \langle b_0, \ldots, b_i\rangle)\}.$$

Any infinite branch of this tree will determine an element of Q, so our aim is to find a *definable* infinite branch. The basic idea of the proof is that the *leftmost infinite branch* (see Figure 4E.2) of T is recursive in some Σ^1_1 set P of type 0.

Recall the function $u * v$ from *18 of 3A.6 and suppose that we can find integers a_0, b_0, a_1, b_1,... such that for every n,

$$(*) \qquad (\exists\alpha)(\exists\beta)(\forall t)R(\langle a_0, \ldots, a_{n-1}\rangle * \bar{\alpha}(t)), \langle b_0, \ldots, b_{n-1}\rangle * \bar{\beta}(t));$$

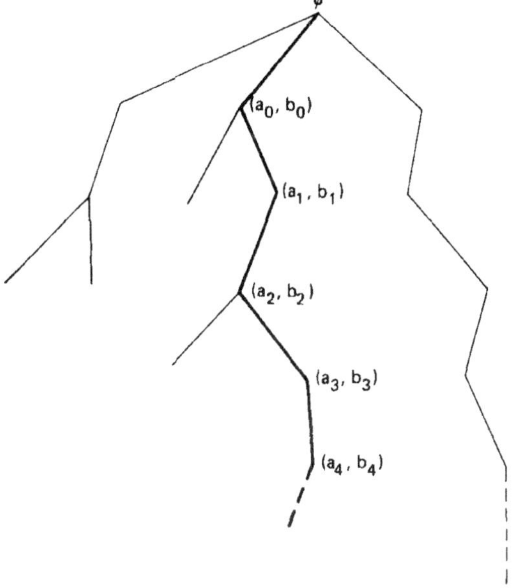

Figure 4E.2. The leftmost infinite branch.

choosing α, β to witness this, and taking $t=0$, we have in particular

$$(\forall n)R(\langle a_0,\ldots, a_{n-1}\rangle, \langle b_0,\ldots, b_{n-1}\rangle),$$

i.e. with $\alpha(n)=a_n$, $\beta(n)=b_n$ now, we have

$$(\forall n)R(\bar{\alpha}(n), \bar{\beta}(n)),$$

so $\alpha \in Q$.

It is clear that $(*)$ simply asserts that an infinite branch of T starts with the finite sequence $(a_0, b_0,\ldots, a_{n-1}, b_{n-1})$. We now choose for each n the leftmost finite sequence which is the beginning of some infinite branch. To be precise, put

$$P(u, v) \Leftrightarrow (\exists\alpha)(\exists\beta)(\forall t)R(u*\bar{\alpha}(t), v*\bar{\beta}(t));$$

clearly P is Σ_1^1. It is easy to verify that

$$P(u, v) \Rightarrow (\exists n)(\exists m)P(u*\langle n\rangle, v*\langle m\rangle).$$

Thus we can define $\alpha=(a_0, a_1,\ldots)$, $\beta=(b_0, b_1,\ldots)$ as above, recursive in P, by the simple recursion

$$\alpha(t) = (\mu s P(\bar{\alpha}(t) * \langle (s)_0 \rangle, \bar{\beta}(t) * \langle (s)_1 \rangle))_0,$$
$$\beta(t) = (\mu s P(\bar{\alpha}(t) * \langle (s)_0 \rangle, \bar{\beta}(t) * \langle (s)_1 \rangle))_1.$$

This shows how to assign to each Σ_1^1 set $Q \subseteq \mathcal{N}$ a Σ_1^1 set P of type 0 such that $\{x: x \text{ is recursive in } P\}$ is a basis for the single pointset Q. To get a single P so that $\{x: x \text{ is recursive in } P\}$ is a basis for Σ_1^1 subsets of $\mathcal{N}$, apply this procedure to some $Q \subseteq \omega \times \mathcal{N}$ which is universal for $\Sigma_1^1 \upharpoonright \mathcal{N}$. Moreover, to see that this yields a basis for Σ_1^1, use the fact that for every $\mathfrak{X}$ there is a recursive surjection $\pi: \mathcal{N} \twoheadrightarrow \mathfrak{X}$ and that if α is recursive in P and π is recursive, then $\pi(\alpha)$ is recursive in P. ⊣

It should be quite obvious by now that every basis result implies some uniformization result, at least implicity, as a corollary of its proof. The uniformization theorem that comes out of the preceding exercise is a bit messy, but it is worth putting down because it implies that we can always find *measurable* uniformizations for $\mathbf{\Sigma}_1^1$ sets.

4E.9. Prove that every $\mathbf{\Sigma}_1^1$ set $Q \subseteq \mathfrak{X} \times \mathfrak{Y}$ can be uniformized by some $Q^* \subseteq Q$ which can be constructed from $\mathbf{\Sigma}_1^1$ and $\mathbf{\Pi}_1^1$ sets using the operations &, $\vee$, $\exists^\omega$, $\forall^\omega$.

Infer that if $Q \subseteq \mathfrak{X} \times \mathfrak{Y}$ is $\mathbf{\Sigma}_1^1$ and $D = \exists^{\mathfrak{Y}} Q$ is the projection of Q on $\mathfrak{X}$, then we can find a function $f: \mathfrak{X} \to \mathfrak{Y}$ which is Baire-measurable, absolutely measurable and such that $(\forall x \in D) Q(x, f(x))$. (*The von Neumann Selection Theorem*, von Neumann [1949].[23])

Hint. It is enough to prove the result for $\mathfrak{X} \times \mathcal{N}$ with $\mathfrak{X}$ of type 1, since the smallest pointclass containing $\mathbf{\Sigma}_1^1$ and $\mathbf{\Pi}_1^1$ and closed under &, $\vee$, $\exists^\omega$, $\forall^\omega$ is easily invariant under Borel isomorphisms.

Suppose then that

$$Q(x, \alpha) \Leftrightarrow (\exists \beta)(\forall t) R(x, \bar{\alpha}(t), \bar{\beta}(t))$$

with R clopen and put

$$P(x, u, v) \Leftrightarrow (\exists \alpha)(\exists \beta)(\forall t) R(x, u * \bar{\alpha}(t), v * \bar{\beta}(t)),$$

$$\begin{aligned} P^*(x, t, u, v) \Leftrightarrow\ & Seq(u) \,\&\, Seq(v) \,\&\, lh(u) = lh(v) = t \\ & \& \ P(x, u, v) \\ & \& \ (\forall u')(\forall v')\{[Seq(u') \,\&\, Seq(v') \,\&\, lh(u') = lh(v') = t \\ & \& \ \langle u', v' \rangle < \langle u, v \rangle] \Rightarrow \neg P(x, u, v)\}. \end{aligned}$$

It is clear from the proof of 4E.8 that the relation

$$Q^*(x, \alpha) \Leftrightarrow (\forall t)(\exists v)P^*(x, t, \bar{\alpha}(t), v)$$

uniformizes Q.

For the second assertion, assume first $Q \subseteq \mathfrak{X} \times \mathcal{N}$ and define P, P^*, Q^* as above, choose a fixed $\alpha_0 \in \mathcal{N}$ and put

$$f(x) = \begin{cases} \alpha_0 & \text{if } (\forall \alpha) \neg Q(x, \alpha) \\ \alpha & \text{if } (\exists \alpha) Q(x, \alpha) \text{ and } Q^*(x, \alpha). \end{cases}$$

For any closed $F \subseteq \mathcal{N}$ we have

$$f(x) \in F \Leftrightarrow [\alpha_0 \in F \,\&\, (\forall \alpha) \neg Q(x, \alpha)] \vee (\exists \alpha)[\alpha \in F \,\&\, Q^*(x, \alpha)]$$

and since $\mathbf{\Pi}^1_1$ sets have the property of Baire and are μ-measurable for each σ-finite Borel measure μ, it is enough to prove that the set $B = \{x: (\exists \alpha)[\alpha \in F \,\&\, Q^*(x, \alpha)]\}$ has the same properties. Computing,

$$\begin{aligned} x \in B &\Leftrightarrow (\exists \alpha)(\forall t)\{(\exists v)P^*(x, t, \bar{\alpha}(t), v) \\ &\qquad \& \,(\exists \beta)[\bar{\beta}(t) = \bar{\alpha}(t) \,\&\, \beta \in F]\} \\ &\Leftrightarrow (\exists \alpha)(\forall t)[x \in S_{\bar{\alpha}(t)}], \end{aligned}$$

where

$$S_{\bar{\alpha}(t)} = \{x: (\exists v)P^*(x, t, \bar{\alpha}(t), v) \,\&\, (\exists \beta)(\bar{\beta}(t) = \bar{\alpha}(t) \,\&\, \beta \in F]\}.$$

Now each S_u is absolutely measurable and has the property of Baire by 2H.8 and 2H.5 and

$$B = \mathcal{A}_u S_u,$$

so by the same results, B is absolutely measurable and has the property of Baire.

In the general case, if $P \subseteq \mathfrak{X} \times \mathfrak{Y}$ with $\mathfrak{Y}$ perfect, let $\pi: \mathcal{N} \twoheadrightarrow \mathfrak{Y}$ be a Borel isomorphism, let $Q \subseteq \mathfrak{X} \times \mathcal{N}$ be defined by

$$Q(x, \alpha) \Leftrightarrow P(x, \pi(\alpha))$$

and choose $f: \mathfrak{X} \to \mathcal{N}$ as above. Take

$$g(x) = \pi^{-1}(f(x))$$

and verify easily that g has the required properties, since for any Borel $A \subseteq \mathfrak{Y}$, $g^{-1}[A] = \{x: f(x) \in \pi[A]\}$. ⊣

This is the strongest result we can prove in Zermelo–Fraenkel set theory in this direction. We will see in Chapter 5 that it is consistent with this theory that there exists a function $f:\mathcal{R}\to\mathcal{R}$ whose graph is $\mathbf{\Pi}_1^1$ and which is not Lebesgue-measurable; the graph of f, then, is a $\mathbf{\Pi}_1^1$ set in the plane which cannot be uniformized by the graph of a Lebesgue measurable function.

By 4E.5, every non-empty Π_1^1 set has a Δ_2^1 member. The next exercise gives another basic result for Π_1^1 which is stronger, at least superficially.

For any pointclass Γ, a point x is a *Γ-singleton* if the set $\{x\}$ is in Γ.

4E.10. Prove that the collection of Π_1^1-singletons is a basis for Π_1^1.

Hint. Given $P\subseteq\mathcal{X}$ in Π_1^1, let

$$Q(n,x)\Leftrightarrow P(x)$$

and let $Q^*\subseteq Q$ uniformize Q in Π_1^1. The unique x such that $Q^*(0,x)$ is a Π_1^1-singleton in P. ⊣

On the other hand, if we impose the weakest natural closure property on a basis for Π_1^1, then this basis must include all of Δ_2^1.

4E.11. Suppose $\mathcal{B}$ is a set of points which is a basis for Π_1^1 and which is closed under Turing reducibility $\le_T$, i.e.

$$y\in\mathcal{B} \text{ and } x \text{ is recursive in } y \Rightarrow x\in\mathcal{B}.$$

Prove that $\mathcal{B}$ contains every Δ_2^1 point.

Hint. If α is a Δ_2^1 irrational, then the set $P=\{\beta:\beta=\alpha\}$ is easily Σ_2^1,

$$P(\beta)\Leftrightarrow(\exists\gamma)\{(\forall s)[\gamma\in N_s\Rightarrow\alpha\in N_s]\ \&\ \gamma=\beta\}.$$

Let

$$P(\beta)\Leftrightarrow(\exists\gamma)Q(\beta,\gamma)$$

with Q in Π_1^1 and let $Q^*\subseteq Q$ be in Π_1^1 and uniformize Q. Now Q^* is non-empty, so it must contain a point of $\mathcal{B}$, which must be (α,γ) for some γ. Since α is recursive in (α,γ), $\alpha\in\mathcal{B}$. It follows easily that $\mathcal{B}$ contains the Δ_2^1 points in all spaces. ⊣

4E.12. Prove that for each perfect $\mathcal{X}$, the collection of Π_1^1-singletons in $\mathcal{X}$ is a Π_1^1 pointset—and hence a proper subset of $\Delta_2^1\cap\mathcal{X}$ by 4D.16.

Hint. Choose a universal Π_1^1 pointset $G\subseteq\omega\times\mathcal{X}$, let G^* uniformize G and notice that

$$x \text{ is a } \Pi_1^1\text{-singleton} \Leftrightarrow (\exists e)G^*(e,x). \quad ⊣$$

4F. Additional results about Π_1^1

Most of the results in this chapter have been about a general Spector pointclass Γ, perhaps with an additional hypothesis that Γ is closed under $\forall^{\mathcal{N}}$ or that it has the scale property. Here we will look at some very specific properties of Π_1^1 which do not follow easily from neat, axiomatic assumptions. These results too will be extended to some of the higher Kleene pointclasses in Chapter 6, using strong set-theoretic hypotheses, but we will need new proofs from them.

First, an effective version of the Perfect Set Theorem 2C.2.

4F.1. THE EFFECTIVE PERFECT SET THEOREM (Harrison [1967]; this proof due to Mansfield [1970]). *If $P \subseteq \mathfrak{X}$ is a Σ_1^1 pointset which has at least one member not in Δ_1^1, then P has a non-empty perfect subset.*

Similarly, if P is $\Sigma_1^1(z)$ with some member not in $\Delta_1^1(z)$, *then P has a non-empty perfect subset.*

In particular, if $P \subseteq \mathfrak{X}$ is $\Sigma_1^1(z)$ and countable, then $P \subseteq \Delta_1^1(z) \cap \mathfrak{X}$.

PROOF. The argument for $\Sigma_1^1(z)$ is identical with that for Σ_1^1, so we only prove the absolute version.

We may assume that P has no Δ_1^1 members, since $\{x \in P: x \notin \Delta_1^1\}$ is also Σ_1^1 by 4D.14. Suppose $\mathfrak{X} = \mathcal{N}$, to begin with, choose a recursive R such that

$$P(\alpha) \Leftrightarrow (\exists\beta)(\forall t)R(\bar{\alpha}(t), \bar{\beta}(t)),$$

and let

$$T = \{(a_0, b_0, \dots, a_{n-1}, b_{n-1}): (\forall i < n)R(\langle a_0, \dots, a_i\rangle, \langle b_0, \dots, b_i\rangle)$$
$$\& (\exists\alpha')(\exists\beta)(\forall t)R(\langle a_0, \dots, a_{n-1}\rangle * \bar{\alpha}'(t), \langle b_0, \dots, b_{n-1}\rangle * \bar{\beta}(t))\};$$

clearly T is a tree on $\omega \times \omega$ and in the notation of 2C,

$$P = \mathfrak{p}[T].$$

Look up the proof of the Perfect Set Theorem 2C.2. We claim that in the notation used there,

$$T = S;$$

because if not, then there is some $u = (a_0, b_0, \dots, a_{n-1}, b_{n-1}) \in T$ with $\mathfrak{p}[T_u]$

a singleton $\{\alpha\}$ and

$$\begin{aligned}\alpha(n)=m \Leftrightarrow (\exists\alpha')(\exists\beta)\{(\forall t)R(\langle a_0,\ldots,a_{n-1}\rangle * \bar{\alpha}'(t),\\ \langle b_0,\ldots,b_{n-1}\rangle * \bar{\beta}(t))\\ \& \ \alpha'(n)=m\},\end{aligned}$$

so easily α is Δ^1_1.

Now $\mathfrak{p}[S]=\mathfrak{p}[T]\neq\emptyset$, so $P=\mathfrak{p}[T]$ has a perfect non-empty subset as in the proof of 2C.2.

The result follows for arbitrary $\mathfrak{X}$ as in 2C.2, using 3E.6. ⊣

This theorem implies in particular that Borel sets with countable sections can be uniformized by Borel sets, see 4F.6.

The next result is a converse to 4D.3 for the case $\Gamma=\Pi^1_1$. First a lemma.

4F.2. Lemma. *There is a Π^0_1 relation $S(\alpha,\beta,\gamma)$, such that whenever $\beta\in WO$ and $\alpha\in LO$,*

$$\begin{aligned}\alpha\in WO \ \& \ |\alpha|\leq|\beta| &\Leftrightarrow (\exists\gamma)S(\alpha,\beta,\gamma)\\ &\Leftrightarrow (\exists\gamma\in\Delta^1_1(\alpha,\beta))S(\alpha,\beta,\gamma).\end{aligned}$$

Proof. The notation is that of 4A. As in the proof of 4A.7, put

$$\begin{aligned}Q(\alpha,\beta,\gamma)\Leftrightarrow \gamma \text{ maps} \leq_\alpha \text{onto an initial segment of}\\ \leq_\beta \text{in an order-preserving fashion}\\ \text{and } \gamma=0 \text{ outside the field of } \leq_\alpha,\end{aligned}$$

where we allow "initial segment" to include all of $\leq_\alpha$. As in that exercise, Q is easily Π^0_2, say

$$Q(\alpha,\beta,\gamma)\Leftrightarrow(\forall n)(\exists m)R(\alpha,\beta,\gamma,n,m)$$

with R recursive. Put further,

$$Q^*(\alpha,\beta,\gamma,\delta)\Leftrightarrow(\forall n)R(\alpha,\beta,\gamma,n,\delta(n))$$

and notice that

$$\begin{aligned}Q(\alpha,\beta,\gamma)&\Leftrightarrow(\exists\delta)Q^*(\alpha,\beta,\gamma,\delta)\\ &\Leftrightarrow(\exists\delta\in\Delta^1_1(\alpha,\beta,\gamma))Q^*(\alpha,\beta,\gamma,\delta),\end{aligned}$$

since if $(\exists\delta)Q^*(\alpha,\beta,\gamma,\delta)$, we can choose

$$\delta(n)=\text{least } mR(\alpha,\beta,\gamma,n,m),$$

where this δ is clearly in $\Delta^1_1(\alpha,\beta,\gamma)$.

Moreover, it is immediate that if $\beta \in WO$ and $\alpha \in LO$, then

$$\begin{aligned}\alpha \in WO \;\&\; |\alpha| \leq |\beta| &\Rightarrow \text{there is a unique } \gamma \text{ such that } Q(\alpha, \beta, \gamma)\\ &\Rightarrow (\exists \gamma \in \Delta_1^1(\alpha, \beta))Q(\alpha, \beta, \gamma),\end{aligned}$$

since the unique γ such that $Q(\alpha, \beta, \gamma)$ is surely in $\Delta_1^1(\alpha, \beta)$. Thus we have, for $\beta \in WO$ and $\alpha \in LO$,

$$\begin{aligned}\alpha \in WO \;\&\; |\alpha| \leq |\beta| &\Leftrightarrow (\exists \gamma)(\exists \delta)Q^*(\alpha, \beta, \gamma, \delta)\\ &\Leftrightarrow (\exists \gamma \in \Delta_1^1(\alpha, \beta))(\exists \delta \in \Delta_1^1(\alpha, \beta, \gamma))Q^*(\alpha, \beta, \gamma, \delta)\end{aligned}$$

with Q^* in Π_1^0, so by 3E.17

$$\begin{aligned}\alpha \in WO \;\&\; |\alpha| \leq |\beta| &\Leftrightarrow (\exists \gamma)(\exists \delta)Q^*(\alpha, \beta, \gamma, \delta)\\ &\Rightarrow (\exists \gamma \in \Delta_1^1(\alpha, \beta))(\exists \delta \in \Delta_1^1(\alpha, \beta))Q^*(\alpha, \beta, \gamma, \delta).\end{aligned}$$

Finally, take

$$S(\alpha, \beta, \gamma) \Leftrightarrow Q^*(\alpha, \beta, (\gamma)_0, (\gamma)_1)$$

and verify easily that the lemma holds with this S. ⊣

4F.3. THE SPECTOR–GANDY THEOREM (Spector [1960], also Gandy [1960]). *For every Π_1^1 set $P \subseteq \mathfrak{X}$ there is a Π_1^0 set $R \subseteq \mathfrak{X} \times \mathcal{N}$ such that*

$$P(x) \Leftrightarrow (\exists \alpha \in \Delta_1^1(x))R(x, \alpha).$$

PROOF. Suppose first that P is Δ_1^1. By 4A.7 there is a Π_1^0 set $A \subseteq \mathcal{N}$ and a recursive $\pi : \mathcal{N} \to \mathfrak{X}$ which is injective on A and $\pi[A] = P$. Hence,

$$\begin{aligned}P(x) &\Leftrightarrow (\exists \alpha)[\alpha \in A \;\&\; \pi(\alpha) = x]\\ &\Leftrightarrow (\exists \alpha \in \Delta_1^1(x))[\alpha \in A \;\&\; \pi(\alpha) = x],\end{aligned}$$

where the second equivalence holds because if $\pi(\alpha) = x$ and $\alpha \in A$, then α is *the unique* irrational satisfying these conditions and it is easily $\Delta_1^1(x)$. Thus for a Δ_1^1 set P we have the stronger representation:

$$\begin{aligned}P(x) &\Leftrightarrow (\exists \alpha)R(x, \alpha)\\ &\Leftrightarrow (\exists \text{ a unique } \alpha)R(x, \alpha)\\ &\Leftrightarrow (\exists \alpha \in \Delta_1^1(x))R(x, \alpha),\end{aligned}$$

where R is some Π_1^0 set.

Towards proving the result for Π_1^1 pointsets of type 0 or 1, recall first 4D.14 according to which $\{(\alpha, x) : \alpha \in \Delta_1^1(x)\}$ is Π_1^1. Hence, for $\mathfrak{X}$ of type 0 or 1, there is a recursive function $g:\mathcal{N}\times\mathfrak{X}\to\mathcal{N}$ such that for each α, x, $g(\alpha, x)\in LO$ and

$$\alpha\in\Delta_1^1(x)\Leftrightarrow g(\alpha, x)\in WO.$$

For each x, let

$$\omega_1^x = \textit{supremum}\{|\beta|: \beta \text{ is recursive in } x, \beta\in WO\};$$

by the relativized version of 4A.4, easily, for each x

(1) $$\textit{supremum}\{|g(\alpha, x)|: \alpha\in\Delta_1^1(x)\} = \omega_1^x,$$

or else $\{\alpha: \alpha\in\Delta_1^1(x)\}$ would be $\Delta_1^1(x)$, contradicting 4D.16.

Suppose now $P\subseteq\mathfrak{X}$ is Π_1^1, with $\mathfrak{X}$ of type 0 or 1, so there is a recursive $f:\mathfrak{X}\to\mathcal{N}$ such that for each x, $f(x)\in LO$ and

$$P(x)\Leftrightarrow f(x)\in WO.$$

From (1), we get immediately

$$P(x)\Leftrightarrow(\exists\alpha\in\Delta_1^1(x))[f(x)\in WO\ \&\ |f(x)|\leq|g(\alpha, x)|],$$

since for each x, $f(x)$ is recursive in x. We claim

(2) $$P(x)\Leftrightarrow(\exists\alpha\in\Delta_1^1(x))(\exists\gamma\in\Delta_1^1(x))S(f(x), g(\alpha, x), \gamma),$$

where S is the Π_1^0 set of the lemma.

To prove direction ($\Rightarrow$) of (2), assume $P(x)$; then $f(x)\in WO$ and $f(x)$ is recursive in x, so by (1) there is some $\alpha\in\Delta_1^1(x)$ such that $|f(x)|\leq|g(\alpha, x)|$. By the lemma then, there is some $\gamma\in\Delta_1^1(f(x), g(\alpha, x))$ such that $S(f(x), g(\alpha, x), \gamma)$; but clearly, $\gamma\in\Delta_1^1(x)$ by 3E.17 since $f(x)$ is recursive in x and $g(\alpha, x)$ is recursive in (α, x) and hence $\Delta_1^1(x)$.

To prove direction ($\Leftarrow$) of (2), suppose there is an $\alpha\in\Delta_1^1(x)$ and some γ such that $S(f(x), g(\alpha, x), \gamma)$. Now $g(\alpha, x)\in WO$ and $f(x)\in LO$, so by the lemma we have $f(x)\in WO$, i.e. $P(x)$.

This completes the proof of (2). From (2) we get the theorem for any pointset of type 0 or 1 by a trivial contraction of quantifiers.

Finally, suppose $P\subseteq\mathfrak{X}$ where $\mathfrak{X}$ is not of type 0 or 1, so there is a Δ_1^1 isomorphism

$$\pi:\mathcal{N}\rightarrowtail\!\!\!\!\rightarrow\mathfrak{X}.$$

If P is Π_1^1, then the inverse image

$$Q(\alpha)\Leftrightarrow P(\pi(\alpha))$$

is Π_1^1, so by the theorem for spaces of type 1,

$$Q(\alpha) \Leftrightarrow (\exists \beta \in \Delta_1^1(\alpha)) R(\alpha, \beta)$$

with R in Π_1^0. Hence

$$\begin{aligned} P(x) \Leftrightarrow Q(\pi^{-1}(x)) &\Leftrightarrow (\exists \beta \in \Delta_1^1(\pi^{-1}(x))) R(\pi^{-1}(x), \beta) \\ &\Leftrightarrow (\exists \beta \in \Delta_1^1(x)) R(\pi^{-1}(x), \beta), \end{aligned}$$

where $\Delta_1^1(x) = \Delta_1^1(\pi^{-1}(x))$ holds because both π and π^{-1} are Δ_1^1. Continuing the computation, we have

$$P(x) \Leftrightarrow (\exists \beta \in \Delta_1^1(x))(\exists \gamma \in \Delta_1^1(x))[\gamma = \pi^{-1}(x) \ \& \ R(\gamma, \beta)]$$

from which the result follows easily using the same kind of arguments and the fact that $\{(\gamma, x) : \gamma = \pi^{-1}(x)\}$ is Δ_1^1. ⊣

The Spector–Gandy theorem does not have many applications but it is undoubtedly one of the jewels of the effective theory. It gives a very elegant characterization of Π_1^1 in terms of a (restricted) existential quantifier which is particularly significant in the case of relations on ω: *$P \subseteq \omega$ is Π_1^1 if and only if there is a Π_1^0 set $R \subseteq \omega \times \mathcal{N}$ such that*

$$P(n) \Leftrightarrow (\exists \alpha \in \Delta_1^1) R(n, \alpha).$$

This corollary says in effect that the collection of Δ_1^1 irrationals somehow "determines" the collection of Π_1^1 relations on ω.

The third main result of this section is also peculiar to the effective theory, like the Spector–Gandy theorem. It differs from it in that it says something most significant about perfect product spaces.

A set $P \subseteq \mathfrak{X}$ is *thin* if P has no perfect subsets other than $\emptyset$. Countable sets are thin and by the Perfect Set Theorem, every $\mathbf{\Sigma}_1^1$ thin set is in fact countable. As we will see in the next chapter, it is consistent with the axioms of Zermelo–Fraenkel Set theory that there exist uncountable, thin Π_1^1 sets.

4F.4. THE LARGEST THIN Π_1^1 SET THEOREM (Guaspari [19??], Kechris [1975], Sacks [1976].). *For each perfect product space $\mathfrak{X}$, there is a thin, Π_1^1 set $C_1 = C_1(\mathfrak{X}) \subseteq \mathfrak{X}$ which contains every thin, Π_1^1 subset of $\mathfrak{X}$.*

PROOF. Fix $\mathfrak{X}$ and let $G \subseteq \omega \times \mathfrak{X}$ be universal for the Π_1^1 subsets of $\mathfrak{X}$, let

$$\varphi : G \to \mathit{Ordinals}$$

be a Π^1_1-norm on G. Put

(1) $R(n, x) \Leftrightarrow G(n, x) \,\&\, [\{y: G(n, y) \,\&\, \varphi(n, y) \le \varphi(n, x)\}$ is countable].

We claim:

(2) $R(n, x) \Leftrightarrow G(n, x) \,\&\, (\forall y)\{[G(n, y) \,\&\, \varphi(n, y) \le \varphi(n, x)] \Rightarrow y \in \Delta^1_1(x)\}$.

To prove direction ($\Rightarrow$) of (2), notice that if $R(n, x)$, then the set

$$\begin{aligned} A &= \{y: G(n, y) \,\&\, \varphi(n, y) \le \varphi(n, x)\} \\ &= \{y: (n, y) \le^*_\varphi (n, x)\} \\ &= \{y: \neg((n, x) <^*_\varphi (n, y))\} \end{aligned}$$

is $\Delta^1_1(x)$, so if A is countable, we must have $A \subseteq \Delta^1_1(x) \cap \mathfrak{X}$ by 4F.1. Conversely, assuming the right hand side of (2) we immediately infer that A is countable, since $\Delta^1_1(x) \cap \mathfrak{X}$ is countable.

Now (2) implies that R is Π^1_1, since it yields

$$R(n, x) \Leftrightarrow G(n, x) \,\&\, (\forall y)[(n, x) <^*_\varphi (n, y) \vee y \in \Delta^1_1(x)].$$

We define $C_1 = C_1(\mathfrak{X})$ by

$$C_1(x) \Leftrightarrow (\exists n) R(n, x).$$

Clearly C_1 is Π^1_1, so it remains to show that C_1 is thin and that it contains every thin, Π^1_1 subset of $\mathfrak{X}$.

Assume first that $P \subseteq \mathfrak{X}$ is thin and Π^1_1, so that for some fixed n_0,

$$P(x) \Leftrightarrow G(n_0, x).$$

For each x in P, $\{y: G(n_0, y) \,\&\, \varphi(n_0, y) \le \varphi(n_0, x)\}$ is $\Delta^1_1(x)$ as above; in particular it is Borel, so it must be countable, since it is a subset of P and cannot have a perfect subset. Hence

$$P(x) \Rightarrow G(n_0, x) \Rightarrow R(n_0, x) \Rightarrow C_1(x)$$

and $P \subseteq C_1$.

Suppose now, towards a contradiction, that $F \ne \emptyset$, F is perfect, $F \subseteq C_1$, put

$$Q(x, n) \Leftrightarrow F(x) \,\&\, R(n, x).$$

The relation Q is $\mathbf{\Pi}^1_1$ and $(\forall x \in F)(\exists n) Q(x, n)$, so by the Δ-selection principle 4B.5, there is a Borel function $g: \mathfrak{X} \to \omega$ such that

$$(\forall x \in F) R(g(x), x).$$

The map

$$x \mapsto (g(x), x)$$

is also Borel and maps F into G. Now G is not in $\boldsymbol{\Sigma}_1^1$ by 3E.9 (otherwise every Π_1^1 subset of $\mathfrak{X}$ would be in $\boldsymbol{\Sigma}_1^1$), hence by the Covering Lemma, 4C.11 there exists an ordinal $\lambda < |\varphi|$, such that

$$x \in F \Longrightarrow \varphi(g(x), x) \leq \lambda.$$

The ordinal λ is countable, since $|\varphi| \leq \boldsymbol{\delta}_1^1 = \aleph_1$. Letting

$$A_{n,\xi} = \{x \colon R(n, x) \mathbin{\&} \varphi(n, x) = \xi\},$$

this means that

$$F \subseteq \bigcup_{n,\xi \leq \lambda} A_{n,\xi}.$$

However, each $A_{n,\xi}$ is countable, since

$$A_{n,\xi} \subseteq \{y \colon G(n, y) \mathbin{\&} \varphi(n, y) \leq \varphi(n, x)\}$$

with x any point such that $R(n, x)$ and $\varphi(n, x) = \xi$, so F is countable, contradicting the assumption that it is perfect. ⊣

This theorem has led to an interesting theory of the structure of countable and thin Π_1^1 sets which we will not pursue here beyond 4F.7 and 4F.8. See Kechris [1975], [1973].

EXERCISES

4F.5. Prove that if $P \subseteq \mathcal{N}$ is a countable Σ_1^1 set of irrationals, then there exists a Δ_1^1 irrational ε such that

$$P \subseteq \{(\varepsilon)_0, (\varepsilon)_1, (\varepsilon)_2, \ldots\}.$$

Hint. By 4F.1, $P \subseteq \Delta_1^1 \cap \mathcal{N}$ and by 4D.14 the set $\Delta_1^1 \cap \mathcal{N}$ is Π_1^1. It follows from the Separation Theorem, 4B.11 that there exists a Δ_1^1 set Q,

$$P \subseteq Q \subseteq \Delta_1^1 \cap \mathcal{N}.$$

Let $\mathbf{c} \colon \mathcal{N} \rightharpoonup \omega$ be the Π_1^1-recursive partial function of Exercise 4D.15 and notice that

$$\begin{aligned} A &= \{i \colon (\exists \alpha)[\alpha \in Q \mathbin{\&} \mathbf{c}(\alpha) = i]\} \\ &= \{i \colon (\exists \alpha \in \Delta_1^1)[\alpha \in Q \mathbin{\&} \mathbf{c}(\alpha) = i]\} \end{aligned}$$

is easily Δ_1^1 and the parametrizing partial function $\mathbf{d} \colon \omega \rightharpoonup \mathcal{N}$ is defined on

A. Put

$$\varepsilon(\langle i, j\rangle) = \mathbf{d}(i)(j) \quad \text{if } i \in A,$$
$$\varepsilon(k) = 0 \quad \text{if } k \neq \langle i, j\rangle \quad \text{for all } i \in A. \qquad \dashv$$

4F.6. Prove that if $P \subseteq \mathcal{X} \times \mathcal{Y}$ is a Borel set such that each section $P_x = \{y : P(x, y)\}$ is countable, then the projection $\exists^{\mathcal{Y}} P$ is Borel and P can be uniformized by a Borel set P^*. (Lusin [1930a], Novikov [1931].[(19–21)])

Hint. Suppose P is $\Delta_1^1(\varepsilon)$. Each section P_x is easily in $\Delta_1^1(\varepsilon, x)$, so

$$P_x \neq \emptyset \Rightarrow (\exists y \in \Delta_1^1(\varepsilon, x))[y \in P_x]$$

by the Effective Perfect Set Theorem 4F.1. Now apply the Δ-Uniformization Criterion 4D.4, taking $\Gamma = \Pi_1^1(\varepsilon)$. $\dashv$

4F.7. Prove that a set P is thin if and only if every Borel subset of P is countable. Infer that the notion of being thin is preserved by Borel isomorphisms.

Hint. Use the Cantor–Bendixson Theorem. $\dashv$

The result implies that if $\pi : \mathcal{X} \rightarrowtail\!\!\!\!\twoheadrightarrow \mathcal{Y}$ is a Δ_1^1 isomorphism, then

$$C_1(\mathcal{Y}) = \pi[C_1(\mathcal{X})].$$

Hence all the sets $C_1(\mathcal{X})$ for perfect $\mathcal{X}$ are determined by the set

$$C_1 = C_1(\mathcal{N}),$$

the largest thin set of irrationals.

4F.8. Let $C \subseteq \mathcal{X}$ be a thin Π_1^1 set and on C define

$$x \leq y \Leftrightarrow x \text{ is } \Delta_1^1(y).$$

Prove that $\leq$ is a prewellordering, so C ramifies into a wellordered sequence of sets of points which are Δ_1^1-equivalent. Prove that the length of $\leq$ is no more than $\aleph_1$. (Kechris [1975].)

Hint. Let $\varphi : C \to$ *Ordinals* be a Π_1^1-norm on C. Suppose x, y are in C and $\varphi(x) \leq \varphi(y)$. Since the set

$$A = \{z : z \in C \ \&\ \varphi(z) \leq \varphi(y)\}$$

is easily $\Delta_1^1(y)$ and has no perfect subsets, $A \subseteq \Delta_1^1(y) \cap \mathcal{X}$ by 4F.1; hence x is $\Delta_1^1(y)$. This proves comparability and transitivity is already known from

3E.17. To prove that $\leq$ is wellfounded, assume that x, y are in C and y is not $\Delta^1_1(x)$ and prove as above that $\varphi(x) \leq \varphi(y)$. ⊣

In the remaining exercises we outline the proofs of several uniformization theorems for Borel sets.

Let us first recall a simple fact about trees which will be needed below.

4F.9. Let T be a tree on a set X which is *finitely splitting*, i.e. every sequence $u = (x_0, \ldots, x_{n-1})$ in T has at most finitely many one-point extensions in T, $(x_0, \ldots, x_{n-1}, y_1), \ldots, (x_0, \ldots, x_{n-1}, y_k)$. Prove that T is infinite if and only if it has an infinite branch. (König's Lemma.)

Hint. If $(x_0, x_1, \ldots)$ is an infinite branch, then for each n, $(x_0, \ldots, x_{n-1}) \in T$, so T is infinite. If T is infinite, then for some x_0 the subtree $T_{(x_0)}$ must be infinite, since $T = T_{(y_1)} \cup \cdots \cup T_{(y_k)}$ for some $y_1, \ldots, y_k$. Again, for some x_1 the subtree $T_{(x_0, x_1)}$ must be infinite, so inductively we get an infinite branch $(x_0, x_1, \ldots)$. ⊣

Fix a product space $\mathfrak{X}$ and for simplicity of notation let

$$N(s) = N(\mathfrak{X}, s)$$

be the s'th basic nbhd of $\mathfrak{X}$. A *finitely splitting tree of* nbhds *of* $\mathfrak{X}$ is a tree T on ω which is finitely splitting and such that if $(s_0, \ldots, s_{n-1}) \in T$, then for each $i = 0, \ldots, n-1$ $radius(N_{s_i}) \leq 2^{-i}$. For simplicity we will call these *nbhd fans* (see Figure 4F.1).

With each nbhd fan T we associate the subset of $\mathfrak{X}$

$$K = K(T) = \{x : (\exists \alpha)(\forall n)[(\alpha(0), \ldots, \alpha(n-1)) \in T \,\&\, x \in \bar{N}(\alpha(n-1))]\}.$$

It is not hard to verify that each $K(T)$ *is a compact* subset of $\mathfrak{X}$ and each compact set K is $K(T)$ with a suitable nbhd fan T. In the next result we get an effective version of this.

Let us say that a nbhd fan T *is in a pointclass* Λ if the set of codes of the sequences in T is in Λ, i.e. if

$$T^c = \{\langle s_0, \ldots, s_{n-1} \rangle : (s_0, \ldots, s_{n-1}) \in T\}$$

is in Λ.

4F.10. Let Γ be a Spector pointclass closed under $\forall^{\mathcal{N}}$ and $\mathfrak{X}$ a fixed perfect product space. Prove that a set $K \subseteq \mathfrak{X}$ is compact and in Δ if and only if there exists a nbhd fan T in Δ such that $K = K(T)$.[24]

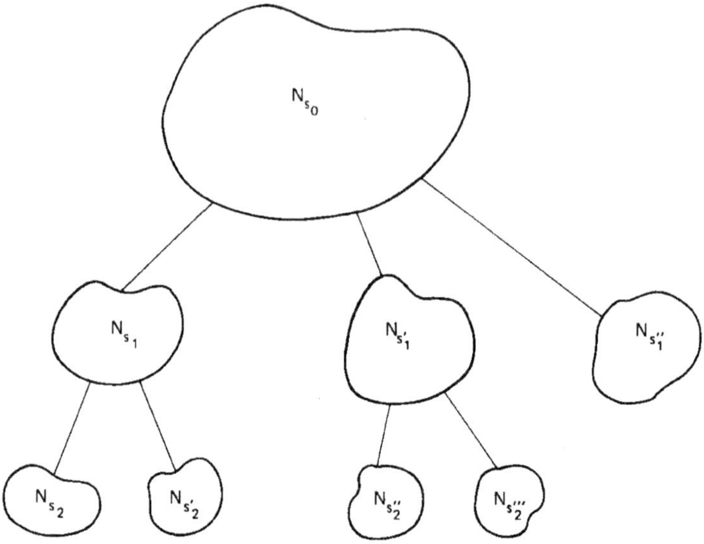

Figure 4F.1. Neighborhood fan.

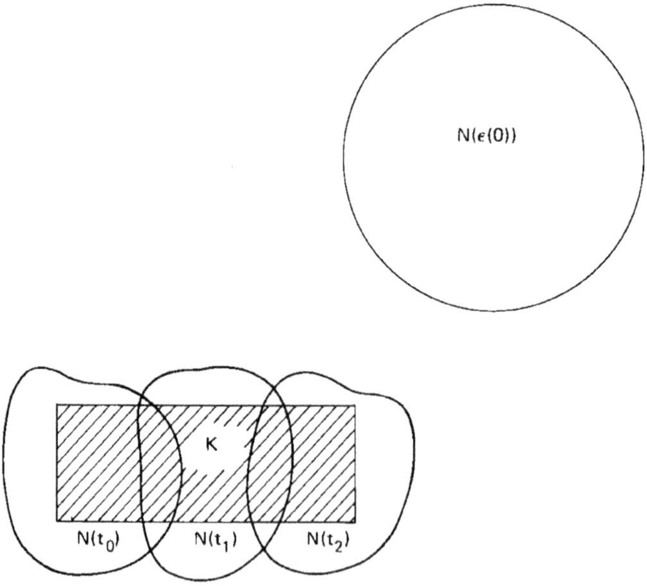

Figure 4F.2.

Hint. If T is in Δ, then

$$x \in K(T) \leftrightarrow (\forall n)(\exists u)[u \in T^c \& lh(u) = n \& (\forall i < n)[x \in \bar{N}((u)_i)]];$$

the implication $(\rightarrow)$ is immediate and the converse implication follows easily using König's Lemma. Thus $K(T)$ is in Δ and its compactness can be proved by a simple topological argument.

Conversely, suppose K is compact and in Δ. Recall from 4C.13 that there is an irrational $\varepsilon \in \Delta$ such that

$$\mathfrak{X} - K = N(\varepsilon(0)) \cup N(\varepsilon(1)) \cup \cdots$$

and for each n,

$$\bar{N}(\varepsilon(n)) \subseteq \mathfrak{X} - K.$$

To construct a nbhd fan T such that $K = K(T)$, intuitively we first find $t_0, \ldots, t_k$ as in Figure 4F.2, such that

$$K \subseteq N(t_0) \cup \cdots \cup N(t_k), \qquad \{\bar{N}(t_0) \cup \cdots \cup \bar{N}(t_k)\} \cap N(\varepsilon(0)) = \emptyset$$

and each $N(t_i)$ has radius ≤ 1. Then for each i, we find $s_0, \ldots, s_n$ such that

$$K \cap \bar{N}(t_i) \subseteq N(s_0) \cup \cdots \cup N(s_n),$$
$$\{\bar{N}(s_0) \cup \cdots \cup \bar{N}(s_n)\} \cap \{N(\varepsilon(0)) \cup N(\varepsilon(1))\} = \emptyset$$

and each $N(s_j)$ has radius $\leq \frac{1}{2}$, etc. The key to proving that this can be done in Γ is the Δ-Selection Principle 4B.5.

Put

$$\begin{aligned} P(n, s, u) \Leftrightarrow Seq(u) & \& (\forall i < lh(u))[radius(N((u)_i)) \leq 2^{-n}] \\ & \& (\forall x)[x \in K \cap \bar{N}(s) \Rightarrow (\exists i < lh(u))[x \in N((u)_i)]] \\ & \& (\forall x)[x \in N(\varepsilon(n)) \Rightarrow (\forall i < lh(u))[x \notin \bar{N}((u)_i)]]. \end{aligned}$$

Since $K \cap \bar{N}(s)$ is compact and disjoint from $\bar{N}(\varepsilon(n))$, easily

$$(\forall n)(\forall s)(\exists u)P(n, s, u).$$

Moreover, P is easily in Γ, so by 4B.5 there is a Δ-recursive function $f(n, s)$ such that $(\forall n)(\forall s)P(n, s, f(n, s))$. Choose once and for all $t_0, \ldots, t_k$ such that each $N(t_i)$ had *radius* ≤ 1,

$$K \subseteq N(t_0) \cup \cdots \cup N(t_k), \qquad \{\bar{N}(t_0) \cup \cdots \cup \bar{N}(t_k)\} \cap N(\varepsilon(0)) = \emptyset$$

and put

$$T = \{(s_0, \ldots, s_{n-1}) : s_0 \text{ is one of } t_0, \ldots, t_k$$
$$\& \, (\forall i < n-1)(\exists j < lh(f(i+1, s_i)))[s_{i+1} = (f(i+1, s_i))_j]\}.$$

Clearly T^c is in Δ and it is simple to check that $K(T) = K$. ⊣

This representation of compact sets in Δ allows us to prove that each of them (if $\neq \emptyset$) must have a member in Δ.

4F.11. Let Γ be a Spector pointclass closed under $\forall^{\mathcal{N}}$. Prove that if K is a compact, non-empty set in Δ, then K has a member in Δ.(24)

Hint. Choose T in Δ such that $K = K(T)$, let

$$R(u) \Leftrightarrow (\forall n)(\exists v)[Seq(v) \,\&\, lh(v) = n \,\&\, u * v \in T^c],$$
$$R(u, v) \Leftrightarrow \neg R(u)$$
$$\vee [R(u) \,\&\, R(v) \,\&\, v \text{ is a one-point-extension of } u]$$

and by 4C.12 let $f: \omega \to \omega$ be in Δ and such that

$$f(0) = 1 = \text{code of the empty sequence},$$
$$(\forall n) R(f(n), f(n+1)).$$

It is now easy to check that $\bigcap_n \bar{N}(f(n))$ contains a single point in K which is clearly in Δ. ⊣

4F.12. Let Γ be a Spector pointclass closed under $\forall^{\mathcal{N}}$, suppose $P \subseteq \mathfrak{X} \times \mathcal{Y}$ is in Δ and for each x, the section P_x is $\emptyset$ or contains a compact set in $\Delta(x)$. Prove that the projection $\exists^{\mathcal{Y}} P$ of P is in Δ and P can be uniformized by some P^* in Δ. Verify that the hypothesis holds if each section P_x is compact.(24)

Hint. Use 4F.11 and the Δ-Uniformization Criterion 4D.4. ⊣

This result implies immediately that Borel sets with compact sections can be uniformized by Borel sets. We proceed to show that for this it is enough to assume that the sections are *σ-compact*.

First a purely topological lemma.

4F.13. Suppose $A \subseteq \mathcal{N}$ is closed, $F: \mathcal{N} \to \mathfrak{X}$ is continuous and

$$\emptyset \neq F[A] \subseteq \bigcup_n K_n$$

with each K_n closed. Prove that for some n and for some basic nbhd N_s in $\mathcal{N}$,

$$\emptyset \neq F[A \cap N_s] \subseteq K_n.$$

(Kunugui's Lemma.)

Hint. Towards a contradiction, suppose no $F[A \cap N_s]$ is contained in some K_n. In particular, there is some $x = F(\alpha) \notin K_0$, so there is a nbhd M of x such that $\bar{M} \cap K_0 = \emptyset$. We can then find a basic nbhd N^0 of α such that $F[A \cap N^0] \subseteq M$, in particular

$$\overline{F[A \cap N^0]} \cap K_0 = \emptyset$$

as in Figure 4F.3. Assume now that $F[A \cap N^0]$ is not a subset of K_1, so there is some $x = F(\alpha)$ with $\alpha \in A \cap N^0$, $x \notin K_1$ and repeat the argument. Thus we get a sequence of basic nbhds

$$N^0 \supseteq N^1 \supseteq \cdots$$

in $\mathcal{N}$ such that each $F[A \cap N^i]$ is non-empty and $\overline{F[A \cap N^i]} \cap K_i = \emptyset$. If we also make sure that $N^i \supseteq \bar{N}^{i+1}$ and $radius(N^i) \to 0$ as $i \to \infty$, we find a point $\alpha \in A \cap N^i$ for each i, so that $F(\alpha) \notin K_i$, all i, which contradicts $F(\alpha) \in \bigcup_n K_n$. ⊣

The next lemma isolates part of the construction that we need for the main result here.

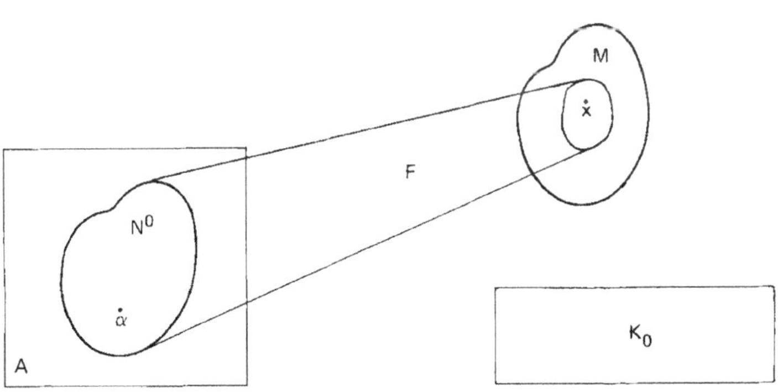

Figure 4F.3.

4F.14. Let Γ be a Spector pointclass closed under $\forall^{\mathcal{N}}$, let $\mathfrak{X}$ be a fixed product space, suppose

$$A \subseteq K \subseteq B,$$

where A is in $\neg\Gamma$, K is compact and B is in Γ. Prove that there is a nbhd fan T in Δ such that

$$A \subseteq K(T) \subseteq B.^{(24)}$$

Hint. Notice first that the closure $\bar{A}$ of A is also in $\neg\Gamma$ since

$$x \in \bar{A} \Leftrightarrow (\forall s)[x \in N_s \Rightarrow (\exists y)[y \in A \;\&\; y \in N_s]]$$

and of course $\bar{A} \subseteq K$, so $\bar{A}$ is compact.

By the Separation Theorem for $\neg\Gamma$ (4B.11) choose a Δ-set C such that

$$\bar{A} \subseteq C \subseteq B.$$

Following the method of proof of 4C.15, put

$$P(x, s) \Leftrightarrow x \notin C \;\&\; x \in N_s \;\&\; \bar{N}_s \cap \bar{A} = \emptyset \quad \text{(Figure 4F.4);}$$

P is clearly in Γ and $(\forall x \notin C)(\exists s)P(x, s)$ since $\bar{A}$ is a closed subset of C. By the Δ-Selection Principle then, there is a Δ-recursive function f such that $(\forall x \notin C)P(x, f(x))$. The set

$$\{s : (\exists x \notin C)[s = f(x)]\}$$

is in $\neg\Gamma$ and it is a subset of the Γ-set

$$\{s : \bar{N}_s \cap \bar{A} = \emptyset\};$$

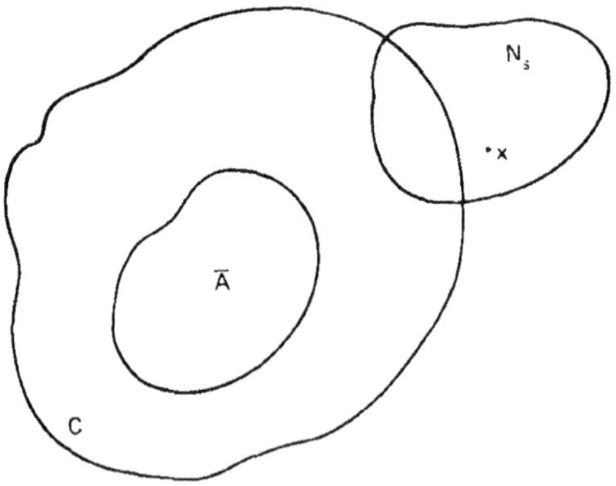

Figure 4F.4.

the separation theorem gives us a Δ set I between these two such that the open set

$$G = \bigcup \{N_s : s \in I\}$$

clearly satisfies

$$\bar{A} \cap G = \emptyset$$

$$x \notin C \Rightarrow x \in G$$

so in particular,

$$x \notin B \Rightarrow x \in G,$$

see Figure 4F.5.

We now imitate the construction of 4F.10 above to get a nbhd fan T such that the associated set $K(T)$ satisfies

$$\bar{A} \subseteq K(T)$$

$$K(T) \cap G = \emptyset.$$

This will complete the proof, since we evidently have

$$\bar{A} \subseteq K(T) \subseteq B.$$

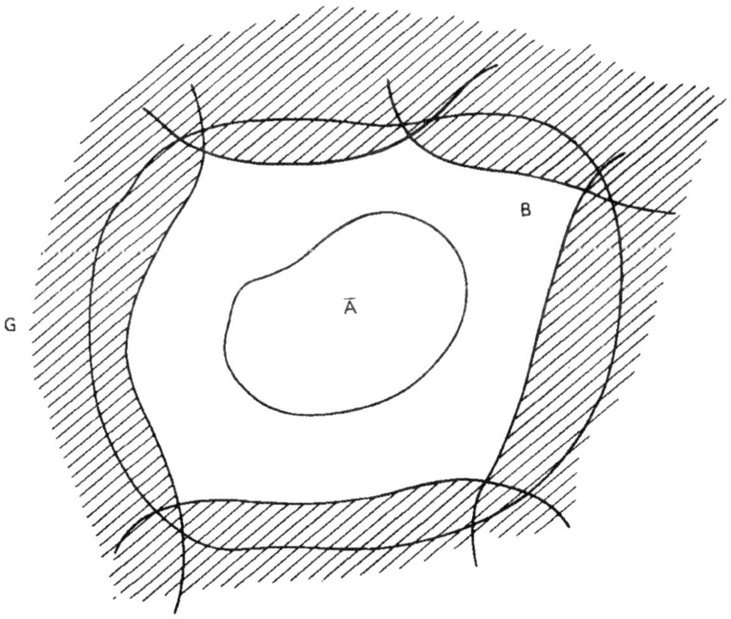

Figure 4F.5.

Briefly, we first write

$$G = N(\varepsilon(0)) \cup N(\varepsilon(1)) \cup \cdots$$

with some ε in Δ and $\bar{N}(\varepsilon(m)) \subseteq G$, all m and then we find $t_0, \ldots, t_k$ such that

$$\bar{A} \subseteq N(t_0) \cup \cdots \cup N(t_k),$$
$$\{\bar{N}(t_0) \cup \cdots \cup \bar{N}(t_k)\} \cap N(\varepsilon(0)) = \emptyset,$$

and each $N(t_i)$ has *radius* ≤ 1. Then for each i, we find $s_0, \ldots, s_n$ such that

$$\bar{A} \cap \bar{N}(t_i) \subseteq N(s_0) \cup \cdots \cup N(s_n),$$
$$\{\bar{N}(s_0) \cup \cdots \cup \bar{N}(s_n)\} \cap \{N(\varepsilon(0)) \cup N(\varepsilon(1))\} = \emptyset$$

and each $N(s_j)$ has *radius* $\leq \frac{1}{2}$, etc. The proof that this procedure determines a Δ, compact set $K(T)$ is via the Δ-Selection Principle as in 4F.10 and we omit it. ⊣

A set is *σ-compact* if it is a countable union of compact sets. (In $\mathcal{R}^n$ every $\mathbf{\Sigma}^0_2$ set is σ-compact.)

4F.15. Suppose $L \subseteq \mathfrak{X}$ is a nonempty, Δ^1_1 set which is σ-compact. Prove that L has a nonempty Δ^1_1 and compact subset; infer that L has a Δ^1_1 member.

Similarly with $\Delta^1_1(x)$ substituted for Δ^1_1 throughout.[24]

Hint. The argument for $\Delta^1_1(x)$ is identical with that for Δ^1_1.

Suppose $L = \bigcup_n K_n$ where each K_n is compact. By 4A.7, there is a Π^0_1 set $A \subseteq \mathcal{N}$ and a recursive $F: \mathcal{N} \to \mathfrak{X}$, injective on A, such that $F[A] = L$. By 4F.13 then, for some s and some n, the set

$$B = F[A \cap N_s]$$

is contained in some K_n. Now B is Δ^1_1, by 4D.7, $\bar{B}$ is compact and

$$\emptyset \neq B \subseteq \bar{B} \subseteq L;$$

by the preceding exercise then, there is a compact set K in Δ^1_1 such that

$$\emptyset \neq \bar{B} \subseteq K \subseteq L$$

and then L has a Δ^1_1 member by 4F.11 above. ⊣

We put down for the record the uniformization theorem that follows from this exercise.

4F.16. Prove that if $P \subseteq \mathfrak{X} \times \mathcal{Y}$ is $\Delta^1_1(z)$ and every section $P_x = \{y : P(x, y)\}$

is σ-compact, then the projection $\exists^{\mathcal{Y}} P$ is $\Delta_1^1(z)$ and P can be uniformized by some P^* in $\Delta_1^1(z)$.

Similarly, if $P \subseteq \mathcal{X} \times \mathcal{Y}$ is Borel and every section P_x is σ-compact, then $\exists^{\mathcal{Y}} P$ is Borel and P can be uniformized by some Borel P^*. (Arsenin [1940], Kunugui [1940]; see also Larman [1972].)

Hint. Use 4F.15 and the Δ-Uniformization Criterion 4D.4. If P is Borel, use the fact that P must be $\Delta_1^1(\varepsilon)$ in some ε, 3E.4. ⊣

The uniformization theorems of Lusin–Novikov (4F.6) and Arsenin–Kunugui can be turned into interesting structure theorems about Borel sets with "small" sections which we now proceed to show.

4F.17. Suppose $P \subseteq \mathcal{X} \times \mathcal{Y}$ is in $\Delta_1^1(z)$ and every section $P_x = \{y : P(x, y)\}$ is countable. Prove that there exists a set $P^* \subseteq \omega \times \mathcal{X} \times \mathcal{Y}$ in $\Delta_1^1(z)$ such that

$$P(x, y) \Leftrightarrow (\exists n) P^*(n, x, y)$$

and such that for each $n \in \omega$, the set

$$P_n^* = \{(x, y) : P^*(n, x, y)\} \subseteq P$$

uniformizes P.

In particular, if $P \subseteq \mathcal{X} \times \mathcal{Y}$ is Borel and each section P_x is countable, then

$$P = \bigcup_n P_n^*$$

where each P_n^* is Borel and uniformizes P. (Lusin [1930a], Novikov [1931].)

Hint. Let $P^* \subseteq P$ uniformize P in $\Delta_1^1(z)$ by 4F.6, and put

$$Q(x, y, n) \Leftrightarrow \mathbf{d}(n, z, x)\downarrow \ \& \ \mathbf{d}(n, z, x) = y$$

where $\mathbf{d}$ is the partial function which parametrizes $\Delta_1^1(z, x) \cap \mathcal{Y}$ by 4D.2. Now $(\forall (x, y) \in P)(\exists n) Q(x, y, n)$, so by the Δ-selection Theorem 4B.5 there is a $\Delta_1^1(z)$-recursive $f : \mathcal{X} \times \mathcal{Y} \to \omega$ so that

$$(\forall x)(\forall y)[P(x, y) \Rightarrow y = \mathbf{d}(f(x, y), z, x)].$$

Put then

$$P^{**}(n, x, y) \Leftrightarrow [f(x, y) = n \ \& \ P(x, y)] \vee [f(x, y) \neq n \ \& \ P^*(x, y)]$$

and check that this P^{**} satisfies the conclusion. ⊣

4F.18. Prove that if $P \subseteq \mathcal{X}$ is $\Delta_1^1(z)$ and σ-compact, then $P(x) \Leftrightarrow (\exists K)[K$ is $\Delta_1^1(z)$ and compact, $K \subseteq P$ and $x \in K]$; moreover P is $\Sigma_2^0(\alpha)$ for some

$\alpha \in \Delta_1^1(z)$, in fact P satisfies an equivalence

$$(*) \qquad P(x) \Leftrightarrow (\forall n)P^*(n, x)$$

where P^* is in $\Pi_1^0(\alpha)$ for some $\alpha \in \Delta_1^1(z)$ and such that each section $P_n^* = \{x: P^*(n, x)\}$ is compact. (Louveau.)[24]

Similarly, if $P \subseteq \mathfrak{X} \times \mathcal{Y}$ is in $\Delta_1^1(z)$ and each section P_x is σ-compact, then there is some $P^* \subseteq \omega \times \mathfrak{X} \times \mathcal{Y}$ in $\Delta_1^1(z)$ such that

$$P(x, y) \Leftrightarrow (\forall n)P^*(n, x, y)$$

and such that each section $P_{n,x}^*$ is compact.

In particular, if $P \subseteq \mathfrak{X} \times \mathcal{Y}$ is Borel and each section P_x is σ-compact, then there exist Borel sets P_n^* such that

$$P = \bigcup_n P_n^*$$

and each P_n^* is Borel with compact sections. (Saint Raymond.)[24]

Hint. To simplify notation suppose P is Δ_1^1 so that by 4A.7,

$$P = F[A]$$

for some Π_1^0 set $A \subseteq \mathcal{N}$ and a recursive $F: \mathcal{N} \to \mathfrak{X}$, with F injective on A. Put

$$\alpha \in A^* \Leftrightarrow (\exists s)\{\alpha \in N_s \;\&\; (\forall \beta)[\beta \in N_s \cap A$$
$$\Rightarrow (\exists K)[K \text{ is compact in } \Delta_1^1,\ K \subseteq P \text{ and } F(\beta) \in K]]\}.$$

It is easy to verify that A^* is a Π_1^1 set, using the representation of compact Δ_1^1 sets via nbhd fans and it is obvious that A^* is open. We will prove that $A \subseteq A^*$.

Assume towards a contradiction that

$$B = A - A^*$$

is nonempty and notice that B is Σ_1^1 and closed. Since

$$F[B] \subseteq F[A] = \bigcup_n K_n$$

with each K_n compact, 4F.13 above implies that for some s, n

$$\emptyset \neq F[B \cap N_s] \subseteq K_n,$$

so that for some s, $\overline{F[B \cap N_s]}$ is a nonempty compact subset of P. Since $\overline{F[B \cap N_s]}$ is in Σ_1^1, 4F.14 then guarantees that there is a Δ_1^1, compact set K such that

$$F[B \cap N_s] \subseteq K \subseteq P.$$

Fix $\alpha \in B \cap N_s$ and suppose $\beta \in N_s \cap A$. If $\beta \in A^*$, then $F(\beta)$ is a member of some Δ_1^1, compact set by definition; if $\beta \notin A^*$, then $\beta \in A - A^* = B$, so $F(\beta) \in K$, so again $F(\beta)$ is a member of some Δ_1^1, compact set. This establishes that $\alpha \in A^*$, *contradicting* $\alpha \in B$. Thus we have shown

$$A \subseteq A^*.$$

It then follows immediately that

$$x \in P \Rightarrow \text{for some } \Delta_1^1, \text{ compact set } K \subseteq P,\ x \in K.$$

Call $\alpha \in \mathcal{N}$ *a code of the nbhd fan* T if

$$(s_0, \ldots, s_{n-1}) \in T \Leftrightarrow \alpha(\langle s_0, \ldots, s_{n-1} \rangle) = 1$$

and put

$$Q(x, i) \Leftrightarrow x \in P \;\&\; \mathbf{d}(i)\downarrow \;\&\; \mathbf{d}(i) \text{ codes a nbhd fan } T \text{ in } \Delta_1^1$$
$$\text{such that } K(T) \subseteq P \text{ and } x \in K(T),$$

where $\mathbf{d}$ parametrizes $\Delta_1^1 \cap \mathcal{N}$ by 4D.2. Easily Q is Π_1^1 and $(\forall x \in P)(\exists i)Q(x, i)$, so we can find a Δ_1^1 function $f: P \to \omega$ such that $(\forall x \in P)Q(x, f(x))$. Put

$$R_1(i) \Leftrightarrow (\exists x \in P)[f(x) = i].$$

Put also

$$R_2(i) \Leftrightarrow \mathbf{d}(i)\downarrow \text{ and } \mathbf{d}(i) \text{ codes a } \Delta_1^1, \text{ compact subset of P}$$

and notice that R_1 is Σ_1^1, R_2 is Π_1^1 and $R_1 \subseteq R_2$, so we can find some Δ_1^1 set R such that

$$R_1 \subseteq R \subseteq R_2.$$

It is now obvious that

$$i \in R \Rightarrow \mathbf{d}(i)\downarrow \text{ and } \mathbf{d}(i) \text{ codes some } \Delta_1^1, \text{ compact set } K_i$$

and

$$P = \bigcup \{K_i : i \in R\}$$

so letting

$$P^*(n, x) \Leftrightarrow R(n) \;\&\; x \in K_n$$

we have $(*)$ in the theorem with P^* in Δ_1^1 and such that each section P_n^* is compact.

The same argument relativized to a fixed but arbitrary x gives the second assertion and then the third assertion follows trivially.

To get the full strength of the first assertion in the theorem choose $\beta \in \Delta_1^1$ so that

$$\{(\beta)_n : n = 0, 1, \ldots\} = \{\mathbf{d}(n) : R(n)\},$$

choose $\gamma \in \Delta^1_1$ so that for each n and m

$$(\gamma)_n(m) = \text{largest } u \text{ so that } Seq(u) \;\&\; lh(u) = m \;\&\; (\beta)_n(u) = 1$$

and let

$$\begin{aligned} P^*(n, x) &\Leftrightarrow x \text{ is in } K(T_n) \text{ where } T_n \text{ is the nbhd fan coded by } (\beta)_n \\ &\Leftrightarrow (\forall m)(\exists u \leq (\gamma)_n(m))[(\beta)_n(u) = 1 \;\&\; lh(u) = m \;\&\; (\forall i < m) \\ &\qquad [x \in \bar{N}((u)_i)]]. \end{aligned}$$

Again we have (*) with this P^* and this pointset is clearly $\Pi^0_1(\alpha)$ with $\alpha = \langle \beta, \gamma \rangle \in \Delta^1_1$. ⊣

The basic idea in these uniformization results about Borel sets is that we can find a Borel uniformization when the sections of the given set are "topologically small," i.e. σ-compact. We now proceed to show that we can also find Borel uniformizations when the sections are "topologically large," i.e. not meager.

The key to this type of result is a basic computation of the category of $\mathbf{\Sigma}^0_n$ and $\mathbf{\Sigma}^1_1$ sets.

4F.19. For each set $P \subseteq \mathfrak{X} \times \mathcal{Y}$, put

$$Q(x) \Leftrightarrow P_x = \{y : P(x, y)\} \text{ is not meager.}$$

Prove that if P is Σ^0_n, then Q is also Σ^0_n, if P is Σ^1_1, then Q is Σ^1_1 and if P is Π^1_1, then Q is also Π^1_1. (Kechris [1973].)

Hint. If P is Σ^0_1 and hence open, then by the Baire category Theorem 2H.2

$$P_x \text{ is not meager} \Leftrightarrow P_x \neq \emptyset \Leftrightarrow (\exists i) P(x, r_i)$$

where $\{r_0, r_1, \ldots\}$ is the recursive presentation of $\mathcal{Y}$, so Q is Σ^0_1. Now if R is Π^0_{n-1},

$$\begin{aligned} &\{y : (\exists m) R(x, y, m)\} \text{ is not meager} \\ &\quad \Leftrightarrow (\exists m)[\{y : R(x, y, m)\} \text{ is not meager}] \\ &\quad \Leftrightarrow (\exists m)(\exists s)[N_s - \{y : R(x, y, m)\} \text{ is meager}] \end{aligned}$$

and by induction hypothesis the relation in the brackets is Π^0_{n-1}, so we are done.

Suppose now

$$P(x, y) \Leftrightarrow (\exists \alpha) F(x, y, \alpha)$$

with F in Π_1^0 and fix x for the discussion. By the von Neumann Selection Theorem 4E.9 we can find a Baire-measurable $f: \mathcal{Y} \to \mathcal{N}$ which uniformizes F_x and then by 2H.10 we can find a comeager G_δ set $A \subseteq \mathcal{Y}$ such that the restriction $f \restriction A$ of f to A is continuous as in Figure 4F.6. Choose $\varepsilon \in \mathcal{N}$ so that

$$(*) \qquad \varepsilon(\langle s, t\rangle) = 1 \Leftrightarrow f[N_s \cap A] \subseteq N_t \;\&\; N_s \cap A \neq \emptyset$$

and check first that

$$(1) \qquad \varepsilon(\langle s, t\rangle) = 1 \Rightarrow F_x \cap (N_s \times N_t) \neq \emptyset,$$

since if $y \in N_s \cap A$, then $(y, f(y)) \in F_x \cap (N_s \times N_t)$. Finally put

$$(2) \qquad B_\varepsilon = \{y \in \mathcal{Y} : (\exists s)(\exists t)[\varepsilon(\langle s, t\rangle) = 1 \;\&\; y \in N_s]$$
$$\& \; (\forall s)(\forall t)(\forall k)\Big\{[\varepsilon(\langle s, t\rangle) = 1 \;\&\; y \in N_s]$$
$$\Rightarrow (\exists s')(\exists t')\Big[\varepsilon(\langle s', t'\rangle) = 1 \;\&\; y \in N_{s'}$$
$$\& \; N_{s'} \subseteq N_s \;\&\; N_{t'} \subseteq N_t$$
$$\& \; radius(N_{s'}) < \frac{1}{k+1} \;\&\; radius(N_{t'}) < \frac{1}{k+1}\Big]\Big\}$$

and check easily that

$$(3) \qquad A \cap P_x \subseteq B_\varepsilon.$$

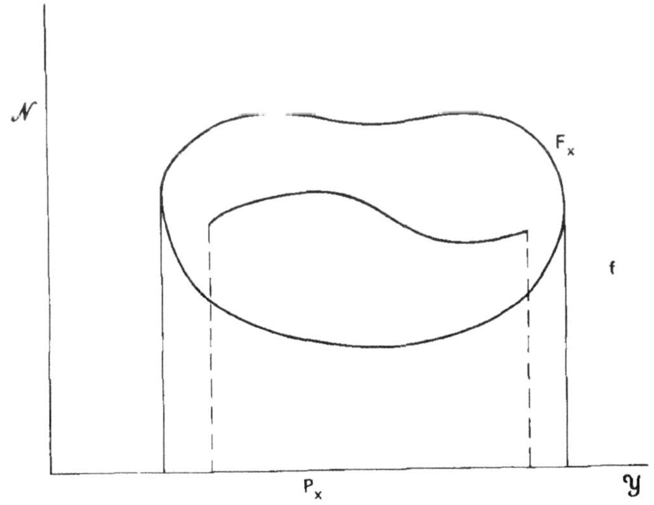

Figure 4F.6.

If P_x is not meager, then (3) implies that B_ε is not meager. Thus we have shown that

$$P_x \text{ is not meager} \Longrightarrow (\exists\varepsilon)\{\varepsilon \text{ satisfies (1) and the set } B_\varepsilon \text{ defined by (2) is not meager}\}.$$

On the other hand, the definition of B_ε makes sense for arbitrary $\varepsilon \in \mathcal{N}$, in fact the relation

$$B(\varepsilon, y) \Longleftrightarrow y \in B_\varepsilon$$

is easily Π^0_3, hence Σ^0_4. Moreover, if for some fixed x and ε the implication (1) holds, then

$$\begin{aligned} y \in B_\varepsilon \Longrightarrow\ & \text{there are sequences } s_0, s_1, \ldots, t_0, t_1, \ldots \\ & \quad \text{such that } N_{s_0} \supseteq N_{s_1} \supseteq \cdots,\ N_{t_0} \supseteq N_{t_1} \supseteq \cdots, \text{ and} \\ & \quad \text{for each } n,\ (N_{s_n} \times N_{t_n}) \cap F_x \neq \emptyset, \\ & \quad \mathit{radius}(N_{s_n}) < 1/(n+1),\ \mathit{radius}(N_{t_n}) < 1/(n+1) \text{ and } y \in N_{s_n} \\ \Longrightarrow\ & \text{there is a sequence of points } (y_n, \alpha_n) \text{ in } F_x \\ & \quad \text{such that } \mathit{limit}_{n\to\infty}\, y_n = y \text{ and } \mathit{limit}_{n\to\infty}\, \alpha_n = \alpha \text{ exists} \\ \Longrightarrow\ & (\exists\alpha)(y, \alpha) \in F_x \text{ (since } F_x \text{ is closed)} \\ \Longrightarrow\ & y \in P_x. \end{aligned}$$

Thus

$$P_x \text{ is not meager} \Longleftrightarrow (\exists\varepsilon)\{\varepsilon \text{ satisfies (1) and the set } B_\varepsilon \text{ defined by (2) is not meager}\}$$

and this relation is immediately Σ^1_1.

The claim for Π^1_1 sets follows from the remark which we have already used, that for any set P with the property of Baire,

$$P \text{ is not meager} \Longleftrightarrow (\exists s)[N_s - P \text{ is meager}]. \qquad \dashv$$

4F.20. Prove that if $P \subseteq \mathfrak{X}$ is $\Sigma^1_1(z)$ and not meager, then P has a member in $\Delta^1_1(z)$. (Thomason [1967], Hinman [1969]; see Kechris [1973].)

Infer that if $P \subseteq \mathfrak{X} \times \mathfrak{Y}$ is in $\Delta^1_1(z)$ and each section P_x is not meager in $\mathfrak{Y}$, then P can be uniformized by a set in $\Delta^1_1(z)$; similarly, every Borel set $P \subseteq \mathfrak{X} \times \mathfrak{Y}$ with non-meager sections can be uniformized by a Borel subset.

Hint. Assume P is $\Pi^1_1(z)$ and not meager and suppose $\varphi : \twoheadrightarrow \varkappa$ is a regular $\Pi_1^1(z)$-norm on P. Now

$$P = \bigcup_{\xi<\varkappa} \{x \in P: \varphi(x) = \xi\},$$

so if $\varkappa < \aleph_1$, then by the countable additivity of the collection of meager sets we get

$$(*) \qquad \text{for some } \lambda < \kappa, \quad \{x \in P: \varphi(x) = \lambda\} \text{ is not meager.}$$

If $\kappa = \aleph_1$, then P is not Borel by 4C.10. Choose a G_δ, nonmeager set $A \subseteq P$ by 2H.4 (applied to $\mathscr{X} - P$) and then use the Covering Lemma 4C.11 to infer that for some $\xi < \varkappa$, $A \subseteq \bigcup_{\zeta<\xi} \{x \in P: \varphi(x) = \zeta\}$ so that again $(*)$ holds. We have thus shown that for any regular, $\Pi_1^1(z)$-norm on P, $(*)$ holds.

Fix now a very good $\Pi_1^1(z)$-scale $\bar{\varphi} = \{\varphi_n\}$ on P, with all the norms regular and put for each n

$$\lambda_n = \text{least } \lambda \text{ such that } \{x \in P: \varphi_n(x) = \lambda\} \text{ is not meager,}$$

$$P_n = \{x \in P: \varphi_n(x) = \lambda_n\}.$$

Putting this another way,

$$\begin{aligned} x \in P_n \Leftrightarrow & \{y \in P: \varphi_n(y) = \varphi_n(x)\} \text{ is not meager} \\ & \& (\forall w)[\varphi_n(w) < \varphi_n(x) \Rightarrow \{y \in P: \varphi_n(y) = \varphi_n(w)\} \text{ is meager}] \\ \Leftrightarrow & \{y \in P: \varphi_n(y) \leq \varphi_n(x)\} \text{ is not meager} \\ & \& \{y \in P: \varphi_n(y) < \varphi_n(x)\} \text{ is meager,} \end{aligned}$$

where in the second equivalence we have used again the countable additivity of the ideal of meager sets and the fact that $\varphi_n(x) < \aleph_1$.

It is immediate that each P_n is nonempty. Notice also that by the key property of a very good scale given in 4E, easily

$$\varphi_{n+1}(y) \leq \varphi_{n+1}(x) \Rightarrow \varphi_n(y) \leq \varphi_n(x)$$

and

$$\varphi_n(y) < \varphi_n(x) \Rightarrow \varphi_{n+1}(y) < \varphi_{n+1}(x),$$

so that immediately

$$P_0 \supseteq P_1 \supseteq P_2 \ldots .$$

If $x_0, x_1, \ldots$ is any sequence of points with $x_n \in P_n$, then by the definition of a good scale again, there exists some x such that

$$\mathit{limit}_{n\to\infty} x_n = x \quad \text{and} \quad P(x).$$

Moreover, if $y_0, y_1, \ldots$ is another such sequence converging to some $y \in P$, the sequence

$$x_0, y_0, x_1, y_1, \dots$$

would have to converge to a unique point, so that $x = y$. Hence: there is a unique point x^* which is the limit of some sequence $x_0, x_1, x_2, \dots$ with $x_n \in P_n$ and this $x^* \in P$. Since obviously

$$x = x^* \Leftrightarrow (\forall s)\{x \in N_s \Rightarrow (\forall k)(\exists n \geq k)(\exists y)[P(n, y) \;\&\; y \in N_s]\}$$

it will be enough to prove that the relation

$$P^*(n, y) \Leftrightarrow y \in P_n$$

is in $\Sigma^1_1(z)$, since this would show that x^* is $\Sigma^1_1(z)$-recursive and hence (easily) in $\Delta^1_1(z)$. Finally this follows easily from the preceding exercise and the equivalence

$$\begin{aligned} x \in P_n &\Leftrightarrow \{y : \neg(x <^*_{\varphi_n} y)\} \text{ is not meager} \\ &\qquad \& \; \{y : y <^*_{\varphi_n} x\} \text{ is meager} \\ &\Leftrightarrow \{y : \neg(x <^*_{\varphi_n} y)\} \text{ is not meager} \\ &\qquad \& \; (\forall s)[(N_s - \{y : y <^*_{\varphi_n} x\}) \text{ is not meager} \end{aligned}$$

which is easy to verify.

This completes the proof of the first assertion and the rest follows from the Δ-Uniformization Criterion 4D.4. ⊣

There is a similar result for measure which we will not prove here—if $P \subseteq \mathfrak{X} \times \mathfrak{Y}$ is Borel and all sections P_x have μ-measure >0 for some σ-finite Borel measure μ on $\mathfrak{Y}$, then P can be uniformized by a Borel set. (The basic lemma for this is due to Tanaka [1968] and Sacks [1969].) Kechris [1973] has an excellent discussion of these and related results as well as additional references.

We will end this section with a negative uniformization result—an obstruction to improving the von Neumann Selection Theorem. First a computation.

Recall from 4F that

$$\omega_1^x = \textit{supremum}\{|\alpha| : \alpha \in WO \;\&\; \alpha \leq_T x\}$$

and for any two irrationals α, β let $\langle \alpha, \beta \rangle$ be the irrational coding their pair as in 1E,

$$\begin{aligned} &\langle \alpha, \beta \rangle(\langle 0, n \rangle) = \alpha(n), \\ &\langle \alpha, \beta \rangle(\langle 1, n \rangle) = \beta(n), \\ &\qquad \langle \alpha, \beta \rangle(t) = 0 \text{ if } t \text{ is not of the form } \langle 0, n \rangle \text{ or } \langle 1, n \rangle. \end{aligned}$$

4F.21. Prove that the relation

$$P(\alpha, \beta) \Leftrightarrow \omega_1^{\langle\alpha,\beta\rangle} = \omega_1^\alpha$$

is Σ^1_1 and that for each α, there is a perfect nonempty set C such that

$$\beta \in C \Rightarrow \omega_1^{\langle\alpha,\beta\rangle} = \omega_1^\alpha.$$

Hint. By a direct relativization of 4A.5, for each x,

$$\omega_1^x = \delta^1_1(x) = \textit{supremum}\{|\alpha|: \alpha \in WO \ \&\ \Delta^1_1(x)\}.$$

Compute:

$$\begin{aligned}\omega_1^{\langle\alpha,\beta\rangle} > \omega_1^\alpha \Leftrightarrow [\exists\gamma \in \Delta^1_1(\alpha, \beta)]\{&\gamma \in WO \\ &\&\ (\forall\delta)[(\delta \in LO \ \&\ \delta \leq_T \alpha) \Rightarrow (\leq_\gamma \text{cannot} \\ &\text{be mapped in an order preserving way onto } \leq_\delta)]\}.\end{aligned}$$

This implies immediately that $\{(\alpha, \beta): \omega_1^{\langle\alpha,\beta\rangle} > \omega_1^\alpha\}$ is Π^1_1 and hence P is Σ^1_1.

Since $\beta \in \Delta^1_1(\alpha) \Rightarrow P(\alpha, \beta)$, if the converse implication held we would have that $\Delta^1_1(\alpha) \cap \mathcal{N}$ is $\Sigma^1_1(\alpha)$ for each α, contradicting 4D.16; thus for each α, the $\Sigma^1_1(\alpha)$ set $\{\beta: P(\alpha, \beta)\}$ has members not in $\Delta^1_1(\alpha)$ and hence it contains a perfect set by the Effective Perfect Set Theorem 4F.1. ⊣

Lusin [1930a] claimed that every $\boldsymbol{\Sigma}^1_1$ set can be uniformized by a set which is the difference of two $\boldsymbol{\Sigma}^1_1$ sets. This is not true.

4F.22. Prove that there exists a Σ^1_1 set which cannot be uniformized by the difference of two $\boldsymbol{\Sigma}^1_1$ sets. (J. Steel, D. A. Martin.)

Hint. Take

$$P(\alpha, \beta) \Leftrightarrow \omega_1^{\langle\alpha,\beta\rangle} = \omega_1^\alpha \ \&\ \beta \notin \Delta^1_1(\alpha)$$

which is Σ^1_1 by the preceding exercise and 4D.14 and suppose that P is uniformized by

$$P^*(\alpha, \beta) \Leftrightarrow Q(\alpha, \beta) \ \&\ \neg R(\alpha, \beta)$$

where (equivalently with the hypothesis that P^* is the difference of two $\boldsymbol{\Sigma}^1_1$ sets) we assume that Q and R are $\Pi^1_1(\alpha^*)$ for some fixed α^*. Let β^* be the unique β such that

$$Q(\alpha^*, \beta^*) \ \&\ \neg R(\alpha^*, \beta^*)$$

holds.

By 4A.3, there is a recursive function f such that

$$Q(\alpha, \beta) \Leftrightarrow f(\alpha, \beta) \in WO;$$

since $Q(\alpha^*, \beta^*)$ holds, $f(\alpha^*, \beta^*) \in WO$ so that for some ξ,

$$|f(\alpha^*, \beta^*)| \leq \xi < \omega_1^{\langle \alpha^*, \beta^* \rangle} = \omega_1^{\alpha^*},$$

since $P(\alpha^*, \beta^*)$ holds and hence $\omega_1^{\langle \alpha^*, \beta^* \rangle} = \omega_1^{\alpha^*}$. The relation

$$S(\beta) \Leftrightarrow f(\alpha^*, \beta) \in WO \;\&\; |f(\alpha^*, \beta)| \leq \xi \;\&\; \neg R(\alpha^*, \beta)$$

is easily in $\Sigma_1^1(\alpha^*)$ and obviously $S(\beta^*)$ holds; but $\beta^* \notin \Delta_1^1(\alpha^*)$, so by the Effective Perfect Set Theorem 4F.1, S contains a perfect set of irrationals. This contradicts the inclusion

$$S \subseteq \{\beta : P^*(\alpha^*, \beta)\} = \{\beta^*\}. \qquad \dashv$$

4G. Historical remarks

[1] The results of this chapter are the hardest to credit, partly because we have presented them in a modern form which is the end product of the work of many researchers. In addition to this, there has been considerable duplication, and rediscovery of ideas, as the recursion theorists often did their work in ignorance of the classical theory. Since the writing of a detailed and documented history of the subject would be a formidable (though fascinating) task, I have confined myself below to a few remarks which indicate the origin of the main ideas (when this is clear from the literature) and point to the most significant papers.

[2] Let us begin with a brief summary (in somewhat modernized terminology) of the results of Lusin–Sierpinski [1923], surely one of the most important early contributions to the theory of analytic sets. Lusin *sieves* (cribles) were introduced here and they were used to obtain a representation theorem for $\mathbf{\Pi}_1^1$ sets quite similar to our 4A.3.

[3] A *sieve* is a map $r \mapsto F_r$ which assigns to each rational number r a subset F_r of a space $\mathfrak{X}$. The set *sifted* by the sieve is defined by

$$x \in \text{Sieve}_r F_r \Leftrightarrow \{r : x \in F_r\} \text{ is not well ordered},$$

where we use the customary ordering on the rationals. The basic result of Lusin–Sierpinski [1923] is that the sets of the form $\text{Sieve}_r F_r$ with each F_r closed are precisely the analytic ($\mathbf{\Sigma}_1^1$) sets; thus every coanalytic ($\mathbf{\Pi}_1^1$) set P satisfies an equivalence

$$P(x) \Leftrightarrow \{r : x \in F_r\} \text{ is well ordered}$$

with a sieve of closed sets—a representation very similar to that of 4A.3.

[4] Lusin and Sierpinski [1923] used this characterization of $\mathbf{\Pi}_1^1$ sets to give a new proof of the Suslin Theorem ($\mathbf{\Delta}_1^1 = \text{Borel}$) and also to prove

that $\mathbf{\Sigma}^1_1$ sets are both the union and the intersection of $\aleph_1$ Borel sets, our 2F.2. (Half of this result was first shown in Lusin–Sierpinski [1918] which anticipated somewhat this later joint paper.) They also established the Boundedness Theorem 4A.4 for $\mathbf{\Pi}^1_1$ sets with the natural ordinal assignment that comes from their representation.

[5] Fix an enumeration $r_0, r_1, r_2, \ldots$ of the rationals and define the set $WO^* \subseteq \mathcal{C}$ of binary infinite sequences by

$$\alpha \in WO^* = \{r_n : \alpha(n) = 1\} \text{ is wellfounded};$$

this is (essentially) the set of codes for ordinals introduced in Lusin–Sierpinski [1923] and used extensively in the classical development of the theory. Lusin and Sierpinski showed that WO^* was $\mathbf{\Pi}^1_1$ but not $\mathbf{\Sigma}^1_1$ and Kuratowski [1966] §38 Lemmas 2, 5 gives the essential content of 4A.2.

[6] The effective version of the Basic Representation Theorem 4A.3 (for $\mathfrak{X} = \omega$) was proved in Kleene [1955a], one of the most significant contributions to the effective theory. Kleene's result asserted that each Π^1_1 subset of ω satisfies

$$P(x) \Leftrightarrow f(x) \in O,$$

where f is recursive and O is a set of (integer) notations for the so-called *constructive ordinals*. (Incidentally, these had been introduced by Church and Kleene and their supremum is ω_1^{CK}, sometimes read "Church–Kleene ω_1.") Kleene's main motivation was the study of these ordinals rather than Π^1_1 and Δ^1_1; but he used his representation theorem in his [1955b] and [1955c] to study Π^1_1 and Δ^1_1 and in the second of these papers he established the effective version of the Suslin Theorem (for $\mathfrak{X} = \omega$). We will prove this result in Chapter 7, where we will also cover in some detail the very fundamental method of *definition by effective transfinite recursion* introduced in Kleene [1955a].

[7] A representation theorem for Π^1_1 (with $\mathfrak{X} = \omega$) which is much more similar to 4A.3 was established in Spector [1955], another basic source of ideas for the effective theory. Spector used integer codes of *recursive wellorderings* of ω (for ordinals below ω_1^{CK}), but other than that his basic notions were quite close to ours. He also proved 4A.2 (essentially) and 4A.4 (for $\mathfrak{X} = \omega$) as well as 4A.5.

[8] Kleene and Spector worked in almost complete ignorance of the classical theory and there is no apparent lead from the classical work to theirs—except (possibly) for one slender thread.

[9] The ordering on finite sequences of integers which we introduced in the proof of 4A.3 was first defined in Lusin–Sierpinski [1923], where it

was used in almost exactly the same way in which we used it. Kleene [1955a] used the same ordering (essentially for the same purpose) and credits Brouwer [1924] for the definition and some of its basic properties—this is Brouwer's famous intuitionistic proof that every (constructive, totally defined) real function must be uniformly continuous on closed intervals. Now, Brouwer has no list of references in his paper, but he might have seen Lusin–Sierpinski [1923]; the publication dates make this barely possible. In any case, Brouwer's background in topology makes it quite likely that he knew the early papers in descriptive set theory (including Lusin–Sierpinski [1918]) and he might have been led to the ordering along the same path followed by Lusin and Sierpinski.

[10] Recursion theorists are apt to refer to the *Kleene–Brouwer* ordering, while someone versed in the classical theory would naturally call this the *Lusin–Sierpinski* ordering.

[11] As we remarked in the introduction, the relationship between classical descriptive set theory and Kleene's theory of the arithmetical and the analytical pointclasses on ω was first perceived as a list of *analogies* between the two theories, to begin with by Mostowski and later (and more accurately) by Addison in his Thesis [1954] and later in his [1959a]. (Addison and Spector were graduate students of Kleene during the same general period 1951–1954; it is interesting that Addison's deepening interest in and knowledge of the classical theory at that time was not effectively transmitted to Kleene and Spector.) The general, unified theory which we are studying in this book evolved slowly in the years since 1955 from these analogies.

[12] The prewellordering property was first isolated explicitly by Moschovakis in 1964, in an effort to find common proofs for theorems about Π^1_1 and Σ^1_2 (on ω); see Rogers [1967]. (The original version was somewhat more complicated and this present definition is due to Kechris.) On the other hand, arguments which involve ordinal assignments to points (like the *index* in Lusin–Sierpinski [1918]) pervade the classical literature both in descriptive set theory and in recursion theory, so many of the results in 4B–4D are best viewed as elegant and strengthened versions of their classical, concrete special cases. The credits given in the text refer to these special cases.

[13] In particular, Novikov [1935] assigned ordinals to the points of a $\mathbf{\Sigma}^1_2$ set precisely as we did in 4B.3, starting with an ordinal assignment from a sieve on the given $\mathbf{\Pi}^1_1$ matrix. Novikov [1935] used 4B.3 to settle the problems of separation and non-separation for $\mathbf{\Pi}^1_1$, $\mathbf{\Sigma}^1_2$ and $\mathbf{\Pi}^1_2$—the

separation theorem for $\mathbf{\Sigma}_1^1$ is already in Lusin [1927]. Kuratowski [1936] inferred the separation property for $\mathbf{\Pi}_2^1$ from the reduction property for $\mathbf{\Sigma}_2^1$ which he introduced and established. Finally, Addison [1959a] put down the lightface results in 4B.10, 4B.11 and 4B.12, following both the classical work and Kleene [1950], where the failure of separation for Σ_1^0 was proved.

[14] The further step of using the prewellordering property as the key tool in studying the structure theory of collections of relations was taken in generalized recursion theory, particularly in recursion in higher types and inductive definability; Moschovakis [1967], [1969], [1970], [1974a], [1974b] and the present work are successive stages in the development of what is sometimes called *prewellordering theory*.

[15] The present notion of a *Spector pointclass* is the natural generalization to the context of Polish spaces of the *Spector classes* of Moschovakis [1974a]. Theorems 6B.3 and 9A.2 in that monograph correspond to the substitution property and 4C.2 here.

[16] The study of collections of relations with arguments in several spaces (and specifically Σ-pointclasses and Spector pointclasses) as opposed to studying collections of subsets of a fixed space (often σ-fields) is one of the chief methodological differences between our approach to descriptive set theory here and the classical work. We are forced to look at relations since the effective pointclasses are not closed under countable unions but they are closed under projection along ω, to take an example. At the same time, the use of relations makes the logical computations of complexity (which were also used in the classical theory) much simpler, so that there is an advantage, even if one is only interested in the projective pointclasses.

[17] The parametrization theorem for Δ, 4D.2 is an abstract version of the various "hierarchies" for the hyperarithmetical sets, for example the sets H_a in Kleene [1955b] or the sets W_γ in Spector [1955]. Similar abstract parametrizations were constructed directly from the prewellordering property in Moschovakis [1967], [1969] and [1974a] whose Theorem 5D.4 is the basic model for 4D.2 here.

[18] Kleene [1959b] has the basic version of 4D.3, for Π_1^1, with a proof based on the hierarchy of the H_a-sets. The upper classification of Δ (4D.14) is a trivial consequence of this. As for the lower classification of Δ (4D.16), it has been rediscovered by several people at various times, with the proof for Δ_2^1 usually depending on the Uniformization Theorem 4E.4. The simple argument for Δ_2^1 that we gave is due to Kechris.

[19] Lusin [1930a] introduced the fundamental problem of uniformization and announced four results. I. *Every* $\boldsymbol{\Sigma}_1^1$ *set can be uniformized by the difference of two* $\boldsymbol{\Sigma}_1^1$ *sets.* (This is actually false, see 4F.22.) II. *There is a* $\boldsymbol{\Sigma}_1^1$ *set which cannot be uniformized by a* $\boldsymbol{\Pi}_1^1$ *set.* III. *Every Borel set can be uniformized by a* $\boldsymbol{\Pi}_1^1$ *set* (joint result with Sierpinski). IV. *If* $P \subseteq \mathcal{X} \times \mathcal{Y}$ *is Borel and every section* P_x *is countable, then P is the union of countably many Borel sets* P_n^*, *each of which uniformizes P.* (There was a similar result for analytic P.)

[20] Theorem II is equivalent to 4D.11, which we obtained as an immediate corollary of Kleene's 4D.10. Novikov [1931] also gave a proof of this result, as well as a proof of a weak version of IV, that for Borel P with countable sections the projection $\exists^{\mathcal{Y}} P$ is Borel and there exists a Borel uniformization, our 4F.6. The complete IV is 4F.17 here.

[21] Sierpinski [1930] established III and asked whether every $\boldsymbol{\Pi}_1^1$ set can be uniformized by some projective set. It was a bold question, because Lusin had published an example which purported to show that one could not "effectively" (in what he called "realistic mathematics") uniformize $\boldsymbol{\Pi}_1^1$ sets. This uniformization problem was soon recognized as the outstanding problem of descriptive set theory, until Kondo [1938] solved it using the basic idea introduced by Novikov and published in Lusin–Novikov [1935]. (Kondo gives additional credit to another Novikov paper where apparently $\boldsymbol{\Pi}_1^1$ sets with finite sections were uniformized.) The lightface version was worked out by Addison in the late fifties.

[22] Kondo's solution of the uniformization problem was in many ways harder than the problem—his proof appeared to be so complicated that few people ever read it. But the difficulty is only a matter of style, as there is basically only one natural proof of this result. The present treatment in 4E.1–4E.4 via scales was worked out in 1971 by Moschovakis who was attempting to generalize the result using strong axioms. We will look at this generalization in Chapter 6.

[23] For the applications of descriptive set theory to analysis, the most important uniformization result is von Neumann's Selection Theorem 4E.9. This was proved before the war, despite the late publication of von Neumann [1949]. Here we obtained it as a direct corollary of the Kleene Basis Theorem for Σ_1^1, 4E.8.

[24] There are many results in the literature on Borel uniformizations of Borel (or even $\boldsymbol{\Sigma}_1^1$ and $\boldsymbol{\Pi}_1^1$) sets with special properties, many of them set in wider contexts than the category of Polish spaces. We have concentrated on just the basic theorems here which illustrate the applicability of

the effective theory to this kind of problem—and particularly the usefulness of the Δ-uniformization criterion 4D.4. These "effective proofs of boldface results" have been part of the folklore of the subject for a long time and there is nothing basically new in our treatment here; the final versions of 4F.9–4F.18 owe much to the seminar notes of some lectures given by Louveau after he had seen a preliminary version of this chapter.

[25] Louveau has obtained recently a very beautiful extension of 4F.18 which in particular implies the following: if $P \subseteq \mathcal{X}$ is Δ^1_1 and $\mathbf{\Sigma}^0_n$ for some n, then P is $\Sigma^0_n(\alpha)$ for some $\alpha \in \Delta^1_1$.

CHAPTER 5

THE CONSTRUCTIBLE UNIVERSE

We have already referred to several *consistency* and *independence* results, e.g. that the continuum hypothesis cannot be settled in Zermelo–Fraenkel set theory and that one cannot prove in this theory that $\mathbf{\Delta}^1_2$ sets of reals are Lebesgue measurable or that uncountable $\mathbf{\Pi}^1_1$ pointsets have perfect subsets. To prove rigorously theorems of this type, one needs the powerful metamathematical methods of modern logic; we will study these quite carefully in Chapter 8. Our main purpose here is to consider briefly a property of pointsets which holds in Godel's *universe of constructible sets* and use it to establish the independence of many important propositions of descriptive set theory.

Gödel's aim was to prove the consistency of *the axiom of choice* and the *generalized continuum hypothesis* with the classical axioms of Zermelo–Fraenkel set theory (without choice). To do this, he defined a collection L of sets with very special properties, *the constructible sets* and showed (first) that if we reinterpret "set" to mean "constructible set," then all the axioms of Zermelo–Fraenkel set theory become true. In other words, all these assertions about sets hold in the constructible universe L; it follows that every logical consequence of these axioms also holds in L. Now Gödel went on to establish that the axiom of choice and the generalized continuum hypothesis also hold in L, because of the special nature of constructible sets; it follows that the negations of these statements do not hold in L, so they cannot be logical consequences of the axioms of Zermelo–Fraenkel set theory. In other words, the axiom of choice and the generalized continuum hypothesis are *consistent* with Zermelo–Fraenkel set theory, they cannot be disproved in that theory.

Gödel's work implies that a very strong form of the continuum hypothesis holds in L—the set $\mathcal{N}$ of irrationals admits a wellordering of rank $\aleph_1$ which is "Δ^1_2-good" in a technical sense. We will make this proposition precise in 5A and we will abbreviate it by "$\mathcal{N} \subseteq L$," since it is in fact equivalent to the assertion that all points in Baire space (as sets of ordered pairs of integers) are constructible. Thus the hypothesis $\mathcal{N} \subseteq L$ is also consistent with the axioms of Zermelo–Fraenkel set theory and neither it nor any of its logical consequences can be disproved from these axioms.

We will show that the hypothesis $\mathcal{N} \subseteq L$ yields a complete structure theory for the Lusin and Kleene pointclasses, e.g. it implies that all Σ^1_n $(n \geq 2)$ are Spector pointclasses with the uniformization property. It also implies that there are $\mathbf{\Delta}^1_2$ sets of reals which are not Lebesgue measurable, that there are uncountable $\mathbf{\Pi}^1_1$ sets which have no perfect subset, etc. The consistency of $\mathcal{N} \subseteq L$ implies then that all these propositions are also consistent with the axioms of Zermelo–Fraenkel set theory (with the axiom of choice), e.g. we cannot prove in this theory that some $\mathbf{\Sigma}^1_3$ set cannot be uniformized by a $\mathbf{\Sigma}^1_3$ set of that every $\mathbf{\Delta}^1_2$ set of reals is Lebesgue measurable.

It turns out that the proofs of these results from the hypothesis $\mathcal{N} \subseteq L$ are quite easy. The more difficult metamathematical proof of the consistency of $\mathcal{N} \subseteq L$ will be given in full in Chapter 8.

At the end of this chapter we will also give a brief discussion (without proofs) of various other consistency and independence results which illustrate the limitations of classical, axiomatic set theory. These are proved by Cohen's method of *forcing* which we will not cover in this book.

Although the statements of theorems and the proofs in this chapter will be completely rigorous, the discussion will be necessarily somewhat vague since we will not say exactly what we mean by "statements about sets" or "logical consequences." These are among the basic notions of logic and they will be defined with complete precision in Chapter 8. The reader who feels uneasy about using these terms intuitively may turn to that chapter now and peruse at least the first few sections. *Almost all of Chapter* 8 *can be read at this point with no knowledge of the intervening Chapters* 5, 6 *and* 7. However, only a rudimentary, intuitive understanding of these metamathematical notions is needed to read the material here and it is perhaps best to continue in our development of descriptive set theory before we turn to look seriously at logical matters.

5A. Descriptive set theory in $L^{(1)}$.

Suppose $\leq$ is a wellordering of some product space $\mathfrak{X}$ and Γ is a pointclass. We say that $\leq$ is Γ-*good* if for every $P \subseteq \mathfrak{Z} \times \mathfrak{X}$ in Γ the relations

$$Q(z, x) \Leftrightarrow (\exists y \leq x) P(z, y),$$

$$R(z, x) \Leftrightarrow (\forall y \leq x) P(z, y),$$

are also in Γ, i.e. if Γ is closed under $\leq$*-bounded quantification.* For example, the natural ordering on ω is Γ-good for every adequate pointclass Γ.

Notice that if Γ is adequate and the identity relation $=$ on $\mathfrak{X}$ is in Δ, then every Γ-good wellordering of $\mathfrak{X}$ is in Δ, since

$$z \leq x \Leftrightarrow (\exists y \leq x)[z = y],$$

$$\neg z \leq x \Leftrightarrow z \neq x \;\&\; x \leq z.$$

We now introduce the abbreviation

$$\mathcal{N} \subseteq L \Leftrightarrow \mathcal{N} \textit{ admits a } \Sigma^1_2\textit{-good wellordering of order type (rank) } \aleph_1;$$

put another way, $\mathcal{N} \subseteq L$ asserts that there is a bijection

$$\rho : \mathcal{N} \rightarrowtail\!\!\!\!\twoheadrightarrow \aleph_1$$

of Baire space with the set of countable ordinals such that the relation

$$\alpha \leq_L \beta \Leftrightarrow \rho(\alpha) \leq \rho(\beta)$$

is a Σ^1_2-good wellordering of $\mathcal{N}$. In particular, $\leq_L$ is a Δ^1_2 pointset which wellorders $\mathcal{N}$.[2,3]

As we mentioned in the introduction, $\mathcal{N}$ admits a Σ^1_2-good wellordering of rank $\aleph_1$ exactly when every point of Baire space is in the collection L of constructible sets. This motivates our choice of notation.

The hypothesis $\mathcal{N} \subseteq L$ is almost certainly false on the basis of our intuitive understanding of the universe of sets. Not only does it imply the continuum hypothesis $2^{\aleph_0} = \aleph_1$ which is itself dubious, it further asserts the existence of a *definable*, Δ^1_2 wellordering of $\mathcal{N}$, for which we have no evidence at all. It is well known that all proofs of Zermelo's theorem that $\mathcal{N}$ can be wellordered depend heavily on the axiom of choice and fail to produce an explicit, definable wellordering.

The proof that the set of *constructible* irrationals admits a Σ^1_2-good wellordering of rank $\aleph_1$ utilizes the very special properties of constructible sets.

Notice that by Corollary 2G.3 to the Kunen–Martin Theorem, $\mathcal{N}$ does not admit a $\mathbf{\Delta}^1_1$ wellordering. Thus the existence of a Δ^1_2 wellordering is the strongest hypothesis of this type which can be consistent with the axioms of Zermelo–Fraenkel set theory.

EXERCISES

We start with a few simple facts about good analytical wellorderings. If $\leq$ is a wellordering of $\mathcal{N}$ of rank $\aleph_1$, put

$$IS(\alpha, \beta) \Leftrightarrow \{(\alpha)_i : i \in \omega\} = \{\gamma : \gamma \leq \beta\};$$

we read this "α codes the initial segment of $\leq$ with top β."

5A.1. Let $\leq$ be a wellordering of $\mathcal{N}$ of rank $\aleph_1$. Prove that $\leq$ is Σ^1_n-good ($n \geq 2$) if and only if the associated relation $IS(\alpha, \beta)$ is Δ^1_n.

Hint. Assume first that $\leq$ is Σ^1_n-good. Compute:

$$\begin{aligned} IS(\alpha, \beta) &\Leftrightarrow (\forall i)[(\alpha)_i \leq \beta] \\ &\qquad \& \ (\forall \gamma)[\gamma \leq \beta \Rightarrow (\exists i)[\gamma = (\alpha)_i]] \\ &\Leftrightarrow (\forall i)[(\alpha)_i \leq \beta] \\ &\qquad \& \ (\forall \gamma \leq \beta)(\exists i)[(\alpha)_i = \gamma], \end{aligned}$$

which implies that $IS(\alpha, \beta)$ is Δ^1_n, since $\leq$ is Σ^1_n-good and in Δ^1_n. Conversely, if $IS(\alpha, \beta)$ is Δ^1_n and $P(\delta, \gamma)$ is in Σ^1_n, then

$$\begin{aligned} (\exists \gamma \leq \beta) P(\delta, \gamma) &\Leftrightarrow (\exists \alpha)\{IS(\alpha, \beta) \ \& \ (\exists i) P(\delta, (\alpha)_i)\}, \\ (\forall \gamma \leq \beta) P(\delta, \gamma) &\Leftrightarrow (\exists \alpha)\{IS(\alpha, \beta) \ \& \ (\forall i) P(\delta, (\alpha)_i)\}. \end{aligned}$$ ⊣

5A.2. Prove that if $\mathcal{N}$ admits a Σ^1_n-good wellordering of rank $\aleph_1$, then every perfect product space $\mathfrak{X}$ admits a wellordering of rank $\aleph_1$ which is Σ^1_k-good, Π^1_k-good and Δ^1_k-good for every $k \geq n$.

Hint. Suppose $\leq$ is Σ^1_n-good of rank $\aleph_1$ on $\mathcal{N}$. The equivalences

$$\begin{aligned} (\exists \gamma \leq \beta) P(\delta, \gamma) &\Leftrightarrow (\exists \alpha)\{IS(\alpha, \beta) \ \& \ (\exists i) P(\delta, (\alpha)_i)\} \\ &\Leftrightarrow (\forall \alpha)\{IS(\alpha, \beta) \Rightarrow (\exists i) P(\delta, (\alpha)_i)\} \end{aligned}$$

and their duals show easily that $\leq$ is Σ^1_k-good, Π^1_k-good and Δ^1_k-good for each $k \geq n$. If $\mathfrak{X}$ is any perfect product space, let

$$h : \mathfrak{X} \rightarrowtail\!\!\!\!\twoheadrightarrow \mathcal{N}$$

be a Δ^1_1 isomorphism of $\mathfrak{X}$ with $\mathcal{N}$, put

$$x \leq' y \Leftrightarrow h(x) \leq h(y)$$

on $\mathfrak{X}$ and verify easily that $\leq'$ is Σ^1_k-good, Π^1_k-good and Δ^1_k-good on $\mathfrak{X}$ for each $k \geq n$. ⊣

After these preliminary results, we proceed to list the most significant facts about descriptive set theory in L.

5A.3. Prove that if $\mathcal{N} \subseteq L$, then for each $n \geq 2$, Σ^1_n is a Spector pointclass.

Hint. By the remarks following the definition of Spector pointclasses in 4C, it is enough to show that each Σ^1_n $(n \geq 2)$ has the prewellordering property. Suppose $P \subseteq \mathfrak{X}$ is Σ^1_n, so that for some Π^1_{n-1} set $Q \subseteq \mathfrak{X} \times \mathcal{N}$.

$$P(x) \Leftrightarrow (\exists\alpha)Q(x, \alpha);$$

let $\leq_L$ be a Σ^1_2-good wellordering on $\mathcal{N}$ induced by a rank function

$$\rho : \mathcal{N} \twoheadrightarrow \aleph_1$$

and put

$$\varphi(x) = inf\{\rho(\alpha) : Q(x, \alpha)\} \quad (x \in P).$$

Now

$$x \leq^*_\varphi y \Leftrightarrow (\exists\alpha)\{Q(x, \alpha) \,\&\, (\forall\beta \leq_L \alpha)[\beta = \alpha \vee \neg Q(y, \beta)]\},$$
$$x <^*_\varphi y \Leftrightarrow (\exists\alpha)\{Q(x, \alpha) \,\&\, (\forall\beta \leq_L \alpha)\neg Q(y, \beta)\},$$

so that by 5A.2, both $\leq^*_\varphi$ and $<^*_\varphi$ are Σ^1_n and φ is a Σ^1_n-norm. ⊣

Thus in L the analytical pointclasses which are normed are exactly those circled in Diagram 5A.1.

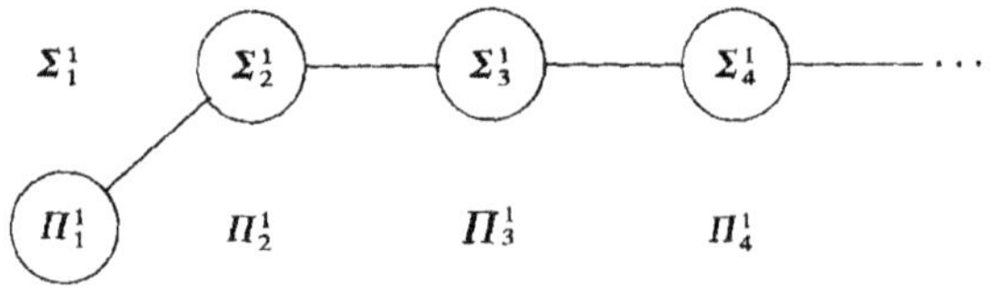

Diagram 5A.1. The normed Kleene pointclasses in L.

The diagram for the boldface pointclasses is identical by 4B.7.

5A.4. Prove that if $\mathcal{N} \subseteq L$ and $n \geq 2$, then Σ^1_n has the uniformization property and Δ^1_n is a basis for Σ^1_n. (Addison [1959b].[3])

Hint. Suppose first that $Q \subseteq \mathfrak{X} \times \mathfrak{Y}$ is Π^1_{n-1}, let $\leq$ be a Σ^1_n-good wellordering of rank $\aleph_1$ on $\mathfrak{Y}$ (by 5A.2) and put

$$Q^*(x, y) \Leftrightarrow Q(x, y) \,\&\, (\forall z < y)\neg Q(x, z);$$

clearly Q^* is Σ^1_n and it uniformizes Q. Now show that every Σ^1_n relation can be uniformized by a Σ^1_n relation as in the proof of 4E.4.

The second assertion follows by the argument we used to prove 4E.5. ⊣

5A.5. Prove that if $\mathcal{N} \subseteq L$, then every Σ^1_n $(n \geq 2)$ has the scale property.

Hint. Given P in Σ^1_n, suppose

$$P(x) \leftrightarrow (\exists\alpha)Q(x, \alpha)$$

with Q in Π^1_{n-1}, let $\leq_1$ be a Σ^1_n-good wellordering of $\mathcal{N}$ and let $\leq_2$ be a Σ^1_n-good wellordering of $\mathfrak{X}$. Define the *anti-lexicographic wellordering* $\leq$ on $\mathfrak{X} \times \mathcal{N}$ (the product of $\mathfrak{X}$ and $\mathcal{N}$) by

$$\langle x, \alpha\rangle \leq \langle y, \beta\rangle \leftrightarrow \alpha <_1 \beta \vee [\alpha = \beta \;\&\; x \leq_2 y]$$

and let

$$\pi : \mathfrak{X} \times \mathcal{N} \to \mathit{Ordinals}$$

be an order-preserving mapping of $\leq$ into the ordinals. Now put

$$\varphi(x) = \pi(\langle x, \leq_1\text{-least } \alpha \text{ such that } Q(x, \alpha),\rangle);$$

easily

$$\begin{aligned} x \leq^*_\varphi y \leftrightarrow\; & (\exists\alpha)\{Q(x, \alpha) \;\&\; (\forall\beta \leq_1 \alpha)\neg Q(y, \beta)\} \\ & \vee (\exists\alpha)\{Q(x, \alpha) \;\&\; Q(y, \alpha) \\ & \qquad \& \; (\forall\beta <_1 \alpha)[\neg Q(x, \beta) \;\&\; \neg Q(y, \beta)] \;\&\; x \leq_2 y\} \end{aligned}$$

and similarly for $<^*_\varphi$, so φ is a Σ^1_n-norm. Since φ is actually an injection, the sequence $\varphi, \varphi, \varphi, \ldots$ is a Σ^1_n-scale on P. ⊣

We now consider the regularity properties of projective sets in L. The key construction is embodied in the following simple fact.

5A.6. Assume $\mathcal{N} \subseteq L$ and let $\mathfrak{X}$, $\mathfrak{Y}$ be any two perfect product spaces. Prove that there exists a function $f : \mathfrak{X} \to \mathfrak{Y}$ whose graph

$$\mathit{Graph}(f) = \{(x, y) : f(x) = y\}$$

is Π^1_1 and thin.

Hint. Let $\leq_L$ be a Σ^1_2-good wellordering of $\mathcal{N}$ of rank $\aleph_1$ and put

$$P(\alpha, \beta) \leftrightarrow \alpha \leq_L \beta \;\&\; \beta \in \mathit{WO} \;\&\; (\forall\gamma <_L \beta)\neg\{\gamma \in \mathit{WO} \;\&\; |\gamma| = |\beta|\},$$

where WO is the set of ordinal codes of 4A. Clearly P is Σ_2^1, so let

$$P(\alpha, \beta) \Leftrightarrow (\exists\gamma)Q(\alpha, \beta, \gamma)$$

with Q in Π_1^1; considering Q as a subset of $\mathcal{N}\times(\mathcal{N}\times\mathcal{N})$, let Q^* uniformize Q in Π_1^1, so that for each α,

$$(\exists\beta)(\exists\gamma)Q(\alpha, \beta, \gamma) \Leftrightarrow (\exists\beta)(\exists\gamma)Q^*(\alpha, \beta, \gamma),$$

$$Q^*(\alpha, \beta, \gamma) \,\&\, Q^*(\alpha, \beta', \gamma') \Rightarrow \beta = \beta' \,\&\, \gamma = \gamma'.$$

Use 4A.6 to show that Q^* has no nonempty perfect subsets—the key observation is that any uncountable subset of Q^* involves uncountably many α's and hence uncountably many distinct ordinals $|\beta|$ which form an unbounded subset of $\aleph_1$.

Since Q^* is obviously the graph of a function

$$f:\mathcal{N}\to\mathcal{N}\times\mathcal{N},$$

this proves the result for $\mathfrak{X}=\mathcal{N}$, $\mathcal{Y}=\mathcal{N}\times\mathcal{N}$, from which the general fact follows by taking Δ_1^1 isomorphisms and using 4F.7. ⊣

In 8G.12 we will establish the converse of this result—the existence of a function $f:\mathcal{N}\to\mathcal{N}$ with thin, Π_1^1 graph in fact implies that $\mathcal{N}\subseteq L$.

Recall from 2H that a function $f:\mathfrak{X}\to\mathcal{Y}$ is Baire-measurable if for each open $G\subseteq\mathcal{Y}$, the inverse image $f^{-1}[G]$ has the property of Baire. Similarly $f:\mathfrak{X}\to\mathcal{Y}$ is *μ-measurable* (where μ is a σ-finite Borel measure on $\mathfrak{X}$) if $f^{-1}[G]$ is μ-measurable, for each open $G\subseteq\mathcal{Y}$.

A measure μ on $\mathfrak{X}$ is *regular* if $\mu(\mathfrak{X})>0$ and $\mu(\{x\})=0$ for all $x\in\mathfrak{X}$.

5A.7. Suppose $f:\mathfrak{X}\to\mathcal{Y}$ and *Graph*(f) is thin; prove that f is not Baire-measurable or μ-measurable for any regular σ-finite Borel measure on $\mathfrak{X}$.

Thus, if $\mathcal{N}\subseteq L$, then there are functions $f:\mathbb{R}\to\mathbb{R}$ with Π_1^1 graphs which are neither Baire-measurable nor Lebesgue-measurable.

Hint. By 2H.10, if f is Baire-measurable then there is a comeager G_δ set P such that $f\restriction P$ is continuous. Now P is uncountable, so the injective image

$$P^*=\{(x, f(x)) : x\in P\}$$

is also uncountable and a $\boldsymbol{\Sigma}_1^1$ subset of *Graph*(f); but then P^* must have a perfect subset by 2C.3, contradicting the hypothesis.

The argument for measure is similar. ⊣

5A.8. Prove that if $\mathcal{N} \subseteq L$, then each perfect product space $\mathfrak{X}$ has uncountable, thin, Π^1_1 subsets, it has Δ^1_2 subsets without the property of Baire and for each regular, σ-finite measure μ, it has Δ^1_2 subsets which are not μ-measurable. (Gödel, Addison [1959b],[(1-3)])

Hint. Take $f: \mathfrak{X} \to \mathcal{R}$ as in 5A.6.
For the first assertion use a Δ^1_1 isomorphism of $\mathfrak{X}$ with $\mathfrak{X} \times \mathcal{R}$. For the second and third assertions argue that the set $\{x: p < f(x) < q\}$ is Δ^1_2 for each pair of rationals p, q and that not all these sets can have the property of Baire—or be μ-measurable. ⊣

There is another, simpler way of obtaining sets in L which are not measurable and do not have the property of Baire, which depends on the classical theorems of Fubini and Kuratowski–Ulam.

The *Fubini Theorem* asserts (in part) that if μ is a Borel measure on $\mathfrak{X}$ and if $A \subseteq \mathfrak{X} \times \mathfrak{X}$ is measurable in the product measure $\mu \times \mu$, then

(i) the section

$$A^y = \{x: A(x, y)\}$$

is μ-measurable, for almost all $y \in \mathfrak{X}$,

(ii) if $\mu(A^y) = 0$ for almost all $y \in \mathfrak{X}$, then A has measure 0 in the product measure and the section

$$A_x = \{y: A(x, y)\}$$

has measure 0, for almost all $x \in \mathfrak{X}$.

We will not prove this here.

The corresponding result for category is not as well-known and it is worth putting it down for the record.

5A.9. Prove that if $A \subseteq \mathfrak{X} \times \mathcal{Y}$ has the property of Baire, then the section $A^y = \{x: A(x, y)\}$ has the property of Baire for a comeager set of y's and

$$\begin{aligned} A \text{ is meager} &\Leftrightarrow A^y \text{ is meager for a comeager set of } y\text{'s} \\ &\Leftrightarrow A_x = \{y: A(x, y)\} \text{ is meager for a comeager set of } x\text{'s}. \end{aligned}$$

(The Kuratowski–Ulam Theorem, see Oxtoby [1971].)

Hint. Suppose first that $A \subseteq \mathfrak{X} \times \mathcal{Y}$ is closed and nowhere dense, let $G = \mathfrak{X} \times \mathcal{Y} - A$, so G is dense and open. For each basic nbhd $N_s \subseteq \mathfrak{X}$, put

$$G_s = \{y: (\exists x \in N_s) G(x, y)\}$$

and check that each G_s is dense, open in $\mathcal{Y}$; hence by the Baire Category

Theorem 2H.2, easily,

$$H = \bigcap_s G_s \subseteq \mathcal{Y}$$

is comeager. Now if $y \in H$, then easily $G^y = \{x: G(x, y)\}$ is dense and hence $A^y = \mathcal{X} - G^y$ is nowhere dense.

Thus whenever A is closed and nowhere dense, the section A^y is nowhere dense for a comeager set of y's. This implies immediately the same result for meager A, which yields direction ($\Rightarrow$) of the second assertion.

To prove the first assertion choose an open G and a meager set P such that $A \triangle G = P$ and notice that for each y, $A^y \triangle G^y = P^y$; since each G^y is open, this proves that A^y has the property of Baire whenever P^y is meager, which is true for a comeager set of y's.

Finally, to prove direction ($\Leftarrow$) of the second assertion, suppose $A \triangle G$ is meager with G open and A is not meager, so that $G \neq \emptyset$; now $A^y \triangle G^y$ is meager for a comeager set of y's and if the same were true for A^y, it would follow that G^y is meager for a comeager set of y's. But G contains a basic nbhd of the form $N(\mathcal{X}, s) \times N(\mathcal{Y}, t)$ since it is nonempty and G^y is not meager for $y \in N(\mathcal{Y}, t)$, which implies that $N(\mathcal{Y}, t)$ is contained in a meager set, contradicting the Baire Category Theorem. ⊣

5A.10. Suppose $\rho: \mathcal{X} \to$ *Ordinals* maps a perfect product space $\mathcal{X}$ into the ordinals so that for each ξ, the set $\rho^{-1}(\xi) = \{x: \rho(x) = \xi\}$ is meager; prove that the prewellordering $\leq$ induced by ρ does not have the property of Baire (as a subset of $\mathcal{X} \times \mathcal{X}$).

Similarly, if each $\rho^{-1}(\xi)$ has μ-measure 0 for some regular Borel measure Γ, then $\leq$ is not measurable in the product measure $\mu \times \mu$.

In particular, both conclusions hold if each $\rho^{-1}(\xi)$ is countable or a singleton (i.e. if $\leq$ is a wellordering).

Hint. Let $A = \{(x, y): \rho(x) \leq \rho(y)\}$ and suppose first that $\{y: A^y$ is meager$\}$ is not comeager; choose then a least y_0 such that A^{y_0} has the property of Baire but is not meager, put $B = \{(x, y): \rho(x) \leq \rho(y) < \rho(y_0)\}$ and verify easily that B has the property of Baire. Now B^y is meager for a comeager set of y's by the choice of y_0, hence B is meager. On the other hand, for each $x \leq y_0$, $B_x = \{y: \rho(x) \leq \rho(y) < \rho(y_0)\}$ so that $A^{y_0} \subseteq B^x \cup B_x \cup \{y: \rho(y) = \rho(y_0)\}$ and hence B_x cannot be meager or else A^{y_0} would be meager.

The argument is a bit simpler if $\{y: A^y$ is meager$\}$ is comeager and the whole proof goes through word-for-word for the case of measure. ⊣

We will end this set of exercises with an interesting alternative version of 5A.6 which brings in the largest, thin Π_1^1 set of 4F.4.

5A.11. Let C_1 be the largest, thin, Π_1^1 subset of $\mathcal{N}$. Prove that if $\mathcal{N} \subseteq L$, then every irrational α is recursive in some $\beta \in C_1$.

Hint. Let $f: \mathcal{N} \to \mathcal{N}$ have thin, Π_1^1 graph and notice that (trivially) each α is recursive in the pair $(\alpha, f(\alpha))$. Now let $\pi: \mathcal{N} \times \mathcal{N} \to \mathcal{N}$ be a recursive homeomorphism and observe that $\pi[Graph(f)]$ is a thin, Π_1^1 subset of $\mathcal{N}$ and α is recursive in $\pi((\alpha, f(\alpha)))$. ⊣

5B. Independence results obtained by the method of forcing[4]

If we can prove a certain statement θ about sets from the hypothesis $\mathcal{N} \subseteq L$, then we know that θ is consistent with the axioms of Zermelo–Fraenkel set theory (with choice)—it cannot be disproved from these axioms. This is a particularly nice consistency proof for θ, as it establishes that θ actually holds in a very natural model of set theory, namely L.

The ingenious method of *forcing* was invented by Cohen in order to prove the independence of the axiom of choice from the remaining Zermelo–Fraenkel axioms as well as the independence of the continuum hypothesis in Zermelo–Fraenkel set theory with choice. This involves constructing models of set theory which are more complicated (and less natural) than L. We will not attempt to explain forcing here, but we will simply list a few of the results which are proved using it and which are relevant to descriptive set theory.

By *ZFC* we understand the classical Zermelo–Fraenkel set theory, with the axiom of choice. (The specific axioms are listed in Chapter 8.) In stating consistency results it is natural to assume that *ZFC* itself is consistent—this is surely true since all its axioms hold in its intended model, the universe of sets. We will not bother to make this hypothesis explicit.

5B.1. THEOREM. *We cannot prove in ZFC that $\mathcal{N}$ admits a projective wellordering; in particular, we cannot prove in ZFC that $\mathcal{N} \subseteq L$ (Cohen, Levy).* ⊣

An extension of this result asserts that the uniformization problem for the higher Lusin pointclasses is hopeless in *ZFC*.

5B.2. THEOREM. *We cannot prove in ZFC that every Π^1_2 set in $\mathcal{N}\times\mathcal{N}$ can be uniformized by some projective set* (*Levy* [1965]). ⊣

Even weaker structure properties cannot be proved.

5B.3. THEOREM. *We cannot prove in ZFC that either $\mathbf{\Sigma}^1_3$ or $\mathbf{\Pi}^1_3$ has the separation property; hence, we cannot prove in ZFC that either $\mathbf{\Pi}^1_3$ or $\mathbf{\Sigma}^1_3$ has the prewellordering property* (*Harrington* [198?]). ⊣

Martin's Theorem 2G.4 is also best possible in *ZFC*—it cannot be extended to the higher Lusin pointclasses. Sample result of this type:

5B.4. THEOREM. *We may assume consistently with ZFC that $2^{\aleph_0}=\aleph_{17}$ and that there are Π^1_2 wellfounded relations on $\mathcal{N}$ of rank $\aleph_{17}$* (*Harrington* [1977]). ⊣

Similarly, Sierpinski's Theorem 2F.3 cannot be extended:

5B.5. THEOREM. *We may assume consistently with ZFC that $2^{\aleph_0}=\aleph_{17}$ and that there is a Π^1_2 set A which cannot be written as the union of fewer than $\aleph_{17}$ Borel sets* (*Solovay*). ⊣

The best result about regularity properties of projective sets needs the additional hypothesis that the existence of inaccessible cardinals is consistent—see Section 6G for a definition of these. In the present context this is surely a reasonable assumption.

5B.6. THEOREM. *If the theory ZFC+(there exist inaccessible cardinals) is consistent, then the following statements (taken together) are consistent with ZFC.*

(i) *Every uncountable projective set has a perfect subset.*

(ii) *Every projective set has the property of Baire.*

(iii) *Every projective set is μ-measurable, for every σ-finite Borel measure μ.*

(iv) *There is no projective wellordering of the continuum.*

Moreover, one may consistently assume (i)–(iv) *together with either the continuum hypothesis or its negation.* (*Solovay* [1970].) ⊣

We will say something about the proofs of these results in Chapter 8, after we have studied the metamathematical method. Suffice it to say here that they are deep and intricate arguments which involve a detailed analysis of both set theory and the "axiomatic method."

5C. Historical remarks

[1] The results in 5A are all due basically to Gödel, except for some of the refinements and generalizations.

[2] Gödel [1938] proved that the collection L of constructible sets is a model of Zermelo–Fraenkel set theory which further satisfies the axiom of choice and the generalized continuum hypothesis. He also announced there without proof that in L there are non-Lebesgue measurable $\boldsymbol{\Delta}^1_2$ sets and uncountable $\boldsymbol{\Pi}^1_1$ sets with no perfect subsets. In the later second printing of the monograph Gödel [1940] (which apparently appeared in 1951) he also added a note announcing that in L, the set $\mathcal{N}$ admits a $\boldsymbol{\Sigma}^1_2$ wellordering.

[3] Addison [1959b] formulated the notion of a $\boldsymbol{\Sigma}^1_2$-good wellordering of rank $\aleph_1$ and gave the first published proof that $L \cap \mathcal{N}$ admits such a wellordering. He also derived most of the corollaries of this proposition that we have listed here, including the uniformization and basis results 5A.4 as well as 5A.8. According to his introduction in [1959b], Addison was building on earlier papers of Kuratowski and Novikov and on some very early unpublished work of Mostowski.

[4] For the results using forcing which we will not cover here in any detail, we refer the reader to Jech [1971].

CHAPTER 6

THE PLAYFUL UNIVERSE

The results of Chapter 5 make clear the basic inadequacy of the Zermelo–Fraenkel axioms for descriptive set theory. It is simply impossible to extend the classical results about Π_1^1 and Σ_2^1 to the higher Lusin and Kleene pointclasses on the basis of *ZFC*.

One way to go, at this point, would be to adopt the hypothesis $\mathcal{N} \subseteq L$ as an additional axiom of set theory. As we saw, this gives (almost trivially) a complete structure theory for projective sets. The defect of this approach is that most set theorists do not see much evidence of the truth of the hypothesis $\mathcal{N} \subseteq L$—in fact many tend to believe that it is false.

Another possibility is to give up developing a theory for the higher Lusin pointclasses and concentrate on consistency and independence results. Many logicians do this, but it is not our approach here.

Instead, we will study hypotheses which go beyond *ZFC*, which yield a rich theory of projective sets and which appear to be (at least) plausible. We should caution the reader that this "plausibility" will not be obvious on first reading; evidence for it will flow (we claim) precisely from the results which we will prove in this chapter.

Solovay [1969] was first to use strong set theoretic hypotheses (unprovable in *ZFC* and inconsistent with $\mathcal{N} \subseteq L$) to solve problems in descriptive set theory. Specifically, he assumed that there *exist measurable cardinals* (*MC*); granting this, he proved that every $\mathbf{\Sigma}_2^1$ pointset has the property of Baire, is measurable relative to every σ-finite Borel measure and is either countable or has a perfect subset.

We will define measurable cardinals and prove Solovay's results in Section 6G. Before this, however, we will study another strong hypothesis which yields a rich structure theory for projective sets. Beside its power, this hypothesis is preferable to *MC* because it is simpler to state and easier to use.

This is where *games* come in: the hypothesis of *projective determinacy* (*PD*) asserts that in certain *two-person, infinite games of perfect information* one of the two players must have a winning strategy. We will give a precise definition of these notions in Section 6A.

There is no doubt that this introduction of powerful and unfamiliar hypotheses poses serious foundational questions. Our discussion of these problems here will be somewhat vague and tentative; we will come back to them (better equipped) in Chapter 8. In the meantime, the reader who wants to go beyond the classical theory of the first four chapters should put aside his doubts about our approach, open himself to new ideas and plunge into the mathematics of the subject. If he can do this, he will be amply rewarded.

6A. Infinite games of perfect information[1,2]

Let X be a fixed nonempty set. With each set $A \subseteq {}^{\omega}X$ of infinite sequences from X, we associate a two-person game $G = G_X(A)$ as follows. Players I and II alternatively choose members of X ad infinitum, as in Diagram 6A.1, so that a sequence

$$f = (a_0, a_1, a_2, \ldots) \in {}^{\omega}X$$

is specified; I wins if $f \in A$, otherwise II wins.

It is understood here that before I chooses a_n (for n even) he is allowed to see $a_0, a_1, \ldots, a_{n-1}$, and similarly with II. This is why we call these *games of perfect information.*

We have described a *run* of the game G which resulted in a particular *play* f. The set A is the *payoff* for $G_X(A)$, although we often identify A with $G_X(A)$ and talk of *the game* A.

A *strategy for player* I is any function σ with domain all finite sequences from X of even length (including the empty sequence) and values in X. We say that I *follows* (or *plays*) σ in a run of the game G, if the resulting play

$$f = (a_0, a_1, a_2, \ldots)$$

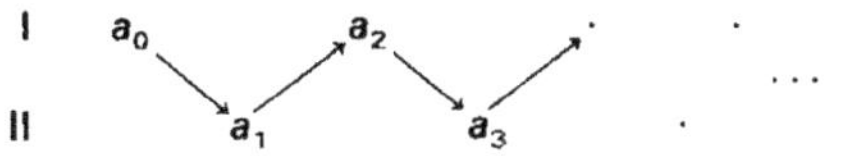

Diagram 6A.1. Playing the game.

satisfies

$$a_0 = \sigma(\emptyset),$$
$$a_2 = \sigma(a_0, a_1),$$
$$\dots$$
$$a_n = \sigma(a_0, a_1, \dots, a_{n-1}) \quad (n \text{ even}).$$

Dually, a *strategy for player* II is any function τ on the finite sequences from X of odd length, with values in X.

When I plays σ against II's τ, the resulting play

$$\sigma * \tau = (a_0, a_1, a_2, \dots)$$

is completely specified,

$$a_0 = \sigma(\emptyset) \quad a_1 = \tau(a_0),$$
$$a_n = \sigma(a_0, \dots, a_{n-1}) \quad (n \text{ even}),$$
$$a_n = \tau(a_0, \dots, a_{n-1}) \quad (n \text{ odd}).$$

We naturally call σ a *winning strategy for* I if for every τ, $\sigma * \tau \in A$—i.e. if I always wins when he plays σ, no matter what II plays; dually, τ is *winning for* II if for every σ, $\sigma * \tau \notin A$.

Finally, *the game* $G = G_X(A)$ (or *the set* A) is *determined* if either I or II has a winning strategy—or *wins the game*, as we will say.

Not all games are determined, but simple ones are. The first result of this type is fundamental for the subject.

First a definition: if $A \subseteq {}^\omega X$ and $u = (a_0, \dots a_{n-1})$ is a sequence of even length, *the subgame of* A *at* u is

$$A(u) = \{f \in {}^\omega X \colon (a_0, a_1, \dots, a_{n-1}, f(0), f(1), \dots) \in A\}.$$

6A.1. Lemma. *Let* $A \subseteq {}^\omega X$ *and suppose* $u = (a_0, a_1, \dots, a_{n-1})$ *is a finite sequence from* X *of even length. If* II *does not win the game* $A(u)$, *then there is some* a *such that for all* b, II *does not win* $A(u^\frown(a, b))$.

Proof. Towards a contradiction, suppose II does not win $A(u)$, but that for each a, there is some b and a strategy τ which is winning for II in $A(u^\frown(a, b))$. Using the axiom of choice, let

$$a \mapsto (b^a, \tau^a)$$

be a function which assigns to each a some b^a and τ^a with these properties. Now II can win $A(u)$ by answering I's first more a_0 by b^{a_0} and then following τ^{a_0}, as if he were playing in $A(u^\frown(a^0, b^{a_0}))$.

More formally, define τ by

$$\tau(a_0) = b^{a_0},$$
$$\tau(a_0,\ldots, a_{n-1}) = \tau^{a_0}(a_2,\ldots, a_{n-1}) \quad (\text{odd } n)$$

and suppose $f = (a_0, a_1,\ldots)$ is played in $A(u)$ with II following τ. Then $(a_2, a_3,\ldots) \notin A(u^\frown(a_0, a_1))$, since II has been following τ^{a_0} after the first two moves hence $(a_0, a_1, a_2,\ldots) \notin A(u)$, i.e. II has won this run of $A(u)$. ⊣

We will customarily describe in an informal way how I or II can play to win a certain game, as in the first paragraph of this proof, without bothering to define formally a winning strategy.

6A.2. THE GALE–STEWART THEOREM (Gale–Stewart [1953]). *For each $X \neq \emptyset$, every closed subset of $^\omega X$ is determined.*

PROOF. Of course we use the product topology on $^\omega X$ (with X discrete) as in Chapter 2.

Suppose then that $A \subseteq {^\omega X}$ and II does not have a winning strategy in A. We describe how I can play to win.

By the lemma, there is some a_0 such that for every b, II cannot win $A(a_0, b)$; let I start the game by playing some a_0 with this property and suppose II answers by some a_1. Now II cannot win $A(a_0, a_1)$.

By the lemma again, there is some a_2 such that for every b, II cannot win $A(a_0, a_1, a_2, b)$; let I play one such a_2 and continue in the same fashion.

At the end of this run of the game, we have a play

$$f = (a_0, a_1, a_2,\ldots)$$

and for every even n, II cannot win the subgame $A(a_0,\ldots, a_{n-1})$. This implies that there is some $f_n \in {^\omega X}$ with

$$f_n(0) = a_0,\ldots, f_n(n-1) = a_{n-1},$$
$$f_n \in A,$$

otherwise II could play at random in $A(a_0,\ldots, a_{n-1})$ and win. Clearly $limit_{n\to\infty} f_n = f$ and hence $f \in A$, since A is closed, so that I has won. ⊣

Proofs of determinacy can become very complicated, but their basic idea is always the same: to reduce in some way the problem of winning a given game to winning various associated closed games. We give here one more proof of this type.

First extend to all spaces ${}^{\omega}X$ ($X \neq \emptyset$) the basic definitions for the Borel subsets of ${}^{\omega}\omega = \mathcal{N}$:

$$\mathbf{\Sigma}^0_1 = \text{all open subsets of } {}^{\omega}X,$$

$$\mathbf{\Pi}^0_\xi = \{{}^{\omega}X - A : A \in \mathbf{\Sigma}^0_\xi\},$$

and for $\xi > 1$,

$$A \in \mathbf{\Sigma}^0_\xi \Leftrightarrow A = \bigcup_{i \in \omega} A_i \text{ for suitable } A_0, A_1, \ldots,$$
$$\text{where each } A_i \text{ is in } \mathbf{\Pi}^0_\eta \text{ for some } \eta < \xi.$$

The *Borel* subsets of ${}^{\omega}X$ are the sets which occur in some $\mathbf{\Sigma}^0_\xi$ or $\mathbf{\Pi}^0_\xi$—they obviously form the smallest collection of subsets of ${}^{\omega}X$ which contains the open sets and is closed under complementation and countable union.

Most of the trivial results about the Borel subsets of $\mathcal{N}$ extend to the Borel subsets of ${}^{\omega}X$, but one must be careful; if X is uncountable, then ${}^{\omega}X$ is not separable and theorems which depend on the separability of $\mathcal{N}$ may fail. (For example, it is not always the case that open sets are countable unions of closed sets.)

6A.3. THEOREM (Wolfe [1955]). *For each $X \neq \emptyset$, every $\mathbf{\Sigma}^0_2$ subset of ${}^{\omega}X$ is determined.*

PROOF. Suppose

$$A = \bigcup_{i \in \omega} F_i$$

with each F_i closed and by 2C.1 choose trees T^i on X such that

$$F_i = [T^i] = \{f \in {}^{\omega}X : (\forall k)(f(0), \ldots, f(k-1)) \in T^i\}.$$

The idea of the proof is to define a set of *sure winning positions* for I in A, i.e. a set W of sequences from X such that I wins $A(u)$ in a particularly obvious way, if $u \in W$. We will subsequently show that if $\emptyset \notin W$, then in fact II wins A, thus establishing the determinacy of A.

Put first

$$u \in W^0 \Leftrightarrow (\exists i)[\text{I wins } F^i(u)];$$

if $u \in W^0$, then I wins $A(u)$ almost trivially, by playing to get into a specific closed set F_i.

Suppose now that W^η has been defined for each $\eta<\xi$ and for $i\in\omega$ put

$$f\in H^{\xi,i} \Leftrightarrow (\forall \text{ even } k)[(f(0),\ldots,f(k-1))\in \bigcup_{\eta<\xi} W^\eta \cup T^i];$$

clearly $H^{\xi,i}$ is a closed set. Let

$$u\in W^\xi \Leftrightarrow (\exists i)[\text{I wins the game } H^{\xi,i}(u)]$$

and

$$W=\bigcup_\xi W^\xi.$$

We now prove by induction on ξ that

$$(*)\qquad u\in W^\xi \Rightarrow \text{I wins } A(u).$$

Granting $(*)$ for all $\eta<\xi$ and assuming that $u\in W^\xi$, choose i so that I wins $H^{\xi,i}(u)$ and let I play in $A(u)$ following his winning strategy in $H^{\xi,i}(u)$. As the game progresses, if $c_0,\ldots,c_{k-1}$ have been played after k steps, we know that

$$u^\frown(c_0,\ldots,c_{k-1})\in \bigcup_{\eta<\xi} W^\eta \cup T^i.$$

Case 1. For some k, we actually have

$$u^\frown(c_0,\ldots,c_{k-1})\in \bigcup_{\eta<\xi} W^\eta.$$

By the induction hypothesis then I can switch to a strategy which will produce from then on some $f\in A(u^\frown(c_0,\ldots,c_{k-1}))$, so that the whole play $(c_0,\ldots,c_{k-1},f(0),f(1),\ldots)$ is in $A(u)$ and I has won.

Case 2. For every k, $u^\frown(c_0,\ldots,c_{k-1})\notin \bigcup_{\eta<\xi} W^\eta$. Now the play $f=(c_0,c_1,\ldots)$ satisfies

$$(\forall k)[u^\frown(c_0,c_1,\ldots,c_{k-1})\in T^i]$$

so that $u^\frown(f(0),f(1),\ldots)\in[T^i]=F^i$ and $f\in F^i(u)\subseteq A(u)$, so that again I has won $A(u)$.

In particular, $(*)$ implies that

$$\emptyset\in W \Rightarrow \text{I wins } A.$$

To show the second implication we need, notice first that for each i,

$$\eta\le\xi \Rightarrow H^{\eta,i}\subseteq H^{\xi,i}$$

and hence trivially

$$\eta\le\xi \Rightarrow W^\eta\subseteq W^\xi.$$

Since the sequence W^ξ (ξ an ordinal) cannot increase forever, there is some ordinal $\varkappa$ such that

$$W^{\varkappa+1} = W^{\varkappa} = W.$$

Suppose now that $\emptyset \notin W^{\varkappa+1}$; we describe how II can play to win A.

By the definition of $W^{\varkappa+1}$ and the determinacy of each closed game $H^{\varkappa+1,i}$, II can actually win every $H^{\varkappa,i}$. Let him start by playing to win $H^{\varkappa,0}$; after a while then, a finite sequence $(c_0,\ldots, c_{k-1})$ has been played and

$$(c_0,\ldots, c_{k-1}) \notin W^{\varkappa} \ \&\ (c_0,\ldots, c_{k-1}) \notin T^0;$$

no matter how the game continues, we know at this stage that the final play will not be in F^0.

Let k_0 be the first k at which this happens and using $W^{\varkappa} = W^{\varkappa+1}$, let II switch to a strategy so he can win $H^{\varkappa+1,1}(c_0,\ldots, c_{k_0-1})$; again, some $k > k_0$ is reached so that

$$(c_0,\ldots, c_{k_0-1}, c_{k_0},\ldots, c_{k-1}) \notin W^{\varkappa} \ \&\ (c_0,\ldots, c_{k-1}) \notin T^1.$$

At this point we have insured that the final play will not be in F^1.

Clearly II can continue to play in this manner and guarantee that the final play will not be in any of the sets F^0, F^1, $F^2,\ldots$, thereby winning A. ⊣

If Λ is a collection of sets, put

$$Det_X(\Lambda) \Leftrightarrow \text{for every set } A \subseteq {}^{\omega}X \text{ in } \Lambda, \text{ the game } G_X(A) \text{ is determined.}$$

We will be particularly interested in the hypotheses $Det_{\omega}(\Lambda)$ and $Det_2(\Lambda)$, with Λ one of the pointclasses we have been studying.

Thus far we have shown $Det_X(\mathbf{\Sigma}^0_2)$, for every X. The determinacy of the dual class follows from the following trivial result.

6A.4. THEOREM. *Suppose Λ is a collection of subsets of some ${}^{\omega}X$ which is closed under continuous substitution; then*

$$Det_X(\Lambda) \Leftrightarrow Det_X(\neg\Lambda).$$

If $X = \omega$ or $X = 2$, then closure under recursive substitution is a sufficient hypothesis.

PROOF. Given $A \subseteq {}^{\omega}X$ in Λ, let

$$B = \{(x, f(0), f(1),\ldots) : f \in A\}$$

and verify easily that

$$\text{I wins } B \rightarrow \text{II wins } {}^{\omega}X - A,$$
$$\text{II wins } B \rightarrow \text{I wins } {}^{\omega}X - A,$$

so that if B is determined, A must be determined too. ⊣

6A.5. COROLLARY. *For each X, $Det_X(\mathbf{\Pi}^0_2)$.* ⊣

Martin has shown that all Borel games are determined—we will prove this for games in $\mathbf{\Sigma}^0_n$ $(n \in \omega)$ in Section 6F. For the moment we consider the significance for descriptive set theory of some specific types of games. The main content of the exercises below is that for adequate Λ closed under Borel substitution, $Det_\omega(\Lambda)$ implies that all sets in Λ have the property of Baire, they are absolutely measurable and if uncountable, they have perfect subsets.

EXERCISES

Let us first put down for the record that not every set is determined.

6A.6. Prove that there is a set $A \subseteq {}^{\omega}2$ which is not determined. (Gale–Stewart [1953].)

Hint. Notice that there are $2^{\aleph_0}$ possible strategies for player I (and also player II) on $2 = \{0, 1\}$ and choose wellorderings $\{\alpha_\xi\}$, $\{\sigma_\xi\}$, $\{\tau_\xi\}$ of rank $2^{\aleph_0}$ for the set $\mathcal{C}$ of binary sequences and the sets of strategies for I and II respectively. Now define by induction sets A_ξ, B_ξ such that for $\xi < 2^{\aleph_0}$,

$$A_\xi \cap B_\xi = \emptyset$$
$$card(A_\xi) < 2^{\aleph_0}, \quad card(B_\xi) < 2^{\aleph_0}$$
$$(\exists \tau)[\sigma_\xi * \tau \in B_\xi]$$
$$(\exists \sigma)[\sigma * \tau_\xi \in A_\xi]$$

and take $A = \bigcup_\xi A_\xi$. ⊣

This proof of course depends on a blatant application of the axiom of choice. No one has been able to prove without using the axiom of choice that there exist non determined games on ω or 2; neither has anyone defined a specific set $A \subseteq {}^{\omega}2$ or $A \subseteq {}^{\omega}\omega$ and then proved (in *ZFC*) that A is not determined.

It is often easier to study games on 2 instead of games on ω, but there is little difference in the results.

6A.7. Prove that if Λ is a pointclass closed under recursive substitution, then

$$Det_\omega(\Lambda) \Rightarrow Det_2(\Lambda).$$

Hint. Given $A \subseteq {}^\omega 2$ in Λ, define $g : \mathcal{N} \to \mathcal{C}$ by

$$g(\alpha)(n) = \begin{cases} 0 & \text{if } \alpha(n) = 0 \\ 1 & \text{if } \alpha(n) > 0, \end{cases}$$

let

$$\alpha \in B \Leftrightarrow g(\alpha) \in A$$

and show that the player who wins B (playing on ω) also wins A (playing on 2). ⊣

The converse implication is a bit awkward to state but it makes the point.

6A.8. Prove that there is an operation

$$A \mapsto H(A)$$

which takes subsets of ${}^\omega\omega$ into subsets of ${}^\omega 2$ such that the following hold.

(i) If Λ is closed under H, then

$$Det_2(\Lambda) \Rightarrow Det_\omega(\Lambda).$$

(ii) Σ^0_n, Π^0_n, Δ^0_n and the corresponding boldface pointclasses are closed under H, if $n \geq 3$.

(iii) If Λ is adequate, $\Lambda \supseteq \Pi^0_2$ and Λ is closed under Δ^1_1 substitution, then Λ is closed under H; in particular this holds if Λ is Σ^1_n, Π^1_n, Δ^1_n, etc.

Hint. The idea is to simulate games on ω by games on 2.

Think of a sequence α as the play in some game on ω, where $\alpha(0), \alpha(2), \alpha(4), \ldots$ are contributed by I and $\alpha(1), \alpha(3), \alpha(5), \ldots$ are contributed by II; we will code this by a play $h(\alpha)$ of an associated game on 2 which looks like this:

$$h(\alpha) = (\underbrace{1, \text{II}, 1, \text{II}, \ldots, 1, \text{II}}_{\alpha(0)\text{ 1's}}, 0, \underbrace{1, \text{I}, 1, \text{I}, \ldots, 1, \text{I}}_{\alpha(1)\text{ 1's}}, 0, \underbrace{1, \text{II}, 1, \text{II}, \ldots, 1, \text{II}}_{\alpha(2)\text{ 1's}}, 0, \ldots).$$

(Here I stands for arbitrary digits played by I and similarly for II.) Call sequences of this form *good* and give a precise Π^0_2 definition of goodness. Notice also that if β fails to be good, this is because one of the players *first* gives infinitely many 1's when it is his turn to code an integer by playing finitely many 1's and then a 0. For each $A \subseteq {}^\omega\omega$ and $\alpha \in {}^\omega 2$, put then

$$\alpha \in H(A) \Leftrightarrow \alpha \text{ is not good on account of II}$$
$$\text{or } \alpha \text{ is good and } h^{-1}(\alpha) \in A.$$

It is easy to verify that whichever player wins $H(A)$ also wins A, so we have proved (i).

(iii) is immediate, since h is recursive and hence the inverse function h^{-1} $(= t \mapsto 0$ on bad arguments) is Δ^1_1.

To prove (ii), check first that there is a recursive relation $Q(u, \alpha, t)$ such that whenever α is good,

$$Q(u, \alpha, t) \Leftrightarrow (\exists\beta)\{h(\beta) = \alpha \;\&\; \bar{\beta}(t) = u\}.$$

Suppose now that A is in Π^0_3, so that

$$\alpha \in A \Leftrightarrow (\forall n)(\exists m)(\forall t)R(\bar{\alpha}(t), n, m),$$

with R recursive, by 4A.1. Hence,

$$\alpha \in H(A) \Leftrightarrow \alpha \text{ is not good on account of II or}$$
$$\alpha \text{ is good} \;\&\; (\forall n)(\exists m)(\forall t)(\forall u)\{Q(u, \alpha, t) \rightarrow R(u, n, m)\}.$$

The argument is similar for the other pointclasses involved. ⊣

We now consider some special games which have topological or measure-theoretic significance.

Given $A \subseteq {}^\omega X$, the game $G^*_X(A)$ is played as follows (see Diagram 6A.2): I chooses a finite (non-empty) sequence from X, then II chooses a single member from X, then I chooses a finite (non-empty) sequence from X, etc. ad infinitum. I wins if the play

$$f = (a_0, a_1, \ldots)$$

is in A, otherwise II wins.[5]

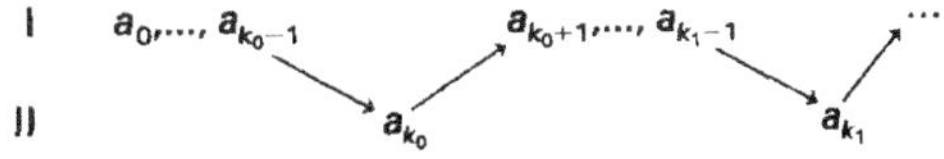

Diagram 6A.2. Playing the *-game.

In this game I is favored, since he is allowed to play more than one point from X if he wishes; in particular, if I wins $G_X(A)$, he obviously wins $G^*_X(A)$ too.

Strategies and winning strategies for these games are defined in the obvious way. We put

$$Det^*_X(\Lambda) \Leftrightarrow \text{for each } A \subseteq {}^\omega X \text{ in } \Lambda, \text{ either I or II wins the game } G^*_X(A).$$

Let us first notice the obvious.

6A.9. Prove that if Λ is a pointclass closed under recursive substitution, then

$$Det_\omega(\Lambda) \Rightarrow Det^*_\omega(\Lambda) \Rightarrow Det^*_2(\Lambda).$$

Hint. For the first implication, associate with each $A \subseteq {}^\omega\omega$ the set

$$B = \{\alpha\colon \text{for every even } n,\ Seq(\alpha(n)) \text{ and } \alpha(0)^\frown(\alpha(1))^\frown\alpha(2)^\frown(\alpha(3))^\frown\cdots \in A\}$$

and check that the player who wins B also wins A.

The second implication is proved by the method of 6A.7. ⊣

The topological significance of the *-game is evidenced in the next two results.

6A.10. Prove that I has a winning strategy for $G^*_2(A)$ if and only if $A \subseteq {}^\omega 2$ has a perfect subset. (Davis [1964].)

Hint. If σ is a winning strategy for I, then the set

$$B = \{\alpha \in {}^\omega 2\colon \alpha \text{ is the play in some run of } G^*(A), \text{ where I plays by } \sigma\}$$

is easily a perfect subset of A. Conversely, if C is a perfect subset of A, choose a tree T on 2 such that $C = [T] = \{\alpha\colon (\forall n)(\alpha(0),\ldots,\alpha(n-1)) \in T\}$ and have I start playing in $G^*_2(A)$ by moving some $(a_0,\ldots,a_{n-1}) \in T$ such that both $(a_0,\ldots,a_{n-1},0)$ and $(a_0,\ldots,a_{n-1},1)$ are in T; such a sequence exists, otherwise C would be a singleton. No matter what II moves, have I play $a_{n+1},\ldots,a_{k-1}$ such that both $(a_0,\ldots,a_{n-1},a_n,a_{n+1},\ldots,a_{k-1},0)$ and $(a_0,\ldots,a_{n-1},a_n,a_{n+1},\ldots,a_{k-1},1)$ are in T—argue that such a move is possible and proceed in the same manner. ⊣

6A.11. Prove that II has a winning strategy in $G_2^*(A)$ if and only if A is countable. (Davis [1964].)

Hint. If A is countable, then II has an obvious winning strategy—he simply plays in his n'th turn to make the play different from α_n, where $A = \{\alpha_0, \alpha_1, ...,\}$.

Suppose now that II wins via τ and let α be a fixed binary sequence. Call a sequence

$$s_0, k_0, s_1, k_1, ..., s_{l-1}\ k_{l-1}$$

good (for τ and α) if each s_i is a non-empty, finite binary sequence each k_i is 0 or 1, the sequence

$$w = s_0{}^\frown(k_0)^\frown s_1{}^\frown(k_1)^\frown\cdots{}^\frown s_{l-1}(k_{l-1})$$

is an initial segment of α and $s_0, k_0, ..., s_{l-1}k_{l-1}$ is the beginning of a run of $G_2^*(A)$ played according to τ, i.e. for $j < l$,

$$k_j = \tau(s_0, k_0, ..., s_j).$$

If every good sequence has a good proper extension, then α is the play in a run of $G_2^*(A)$ where II has followed τ and hence $\alpha \notin A$; thus if $\alpha \in A$, there must exist some $s_0, k_0, ..., s_{l-1}, k_{l-1}$ which is good for τ and α and maximal ($l = 0$ is allowed). If

$$s_0{}^\frown(k_0)^\frown\cdots{}^\frown(s_{l-1})^\frown(k_{l-1}) = \alpha(0), \alpha(1), ..., \alpha(n-1),$$

then easily for $i > n$ we must have

$$\alpha(i) = 1 - \tau(s_0, k_0, ..., s_{l-1}, k_{l-1}, (\alpha(n), ..., \alpha(n-i)))$$

so that α is completely determined by the maximal initial good part $s_0, k_0, ..., s_{l-1}, k_{l-1}$ and the value $\alpha(n)$. Since there are only countably many possible good sequences, A must be countable. ⊣

From these two simple facts we obtain the first connection between determinacy hypothesis and structural properties of pointsets.

6A.12. Prove that if Λ is an adequate pointclass closed under Δ_1^1 substitution, then

$$Det_2^*(\Lambda) \Leftrightarrow \text{every uncountable pointset in } \Lambda \text{ has a nonempty perfect subset}$$

and hence

$$Det_\omega(\Lambda) \rightarrow \text{every uncountable pointset in } \Lambda \text{ has a nonempty perfect subset.}^{(6,7)}$$

Infer $Det_2^*(\Sigma_1^1)$. Infer also that $Det_2^*(\Pi_1^1)$ and $Det_\omega(\Sigma_1^1)$ cannot be proved in *ZFC*.

Hint. If $A \subseteq \mathfrak{X}$ is uncountable in Λ, let $\pi : \mathfrak{X} \rightarrowtail\!\!\!\twoheadrightarrow {}^\omega 2$ be a Δ_1^1 isomorphism, infer that $\pi[A]$ has a perfect subset $\pi[C]$ and argue that the uncountable Borel set $C \subseteq A$ has a perfect subset.

The other assertions follow immediately using 6A.9, 5A.8 and 6A.4. ⊣

The next game we will study can be used to prove that determinacy implies the Baire property. It is easier here to work directly on arbitrary pointsets rather than prove the result for subsets of ${}^\omega 2$ and then transfer it.

Given $A \subseteq \mathfrak{X}$, the ***-game* (or *Banach–Mazur game*) $G^{**} = G^{**}(A)$ is played as follows: I chooses an integer s_0, II chooses s_1, I chooses s_2, and so on. If $N(0), N(1),...$ is the standard enumeration of a nbhd basis for $\mathfrak{X}$, then each player must move some s_i such that (see Figure 6A.3)

$$\bar{N}(s_{i-1}) \supseteq \bar{N}(s_i),$$
$$radius(N(s_i)) \leq \tfrac{1}{2}\, radius(N(s_{i-1}))$$

—otherwise the first player who does not follow this restriction loses. If they both follow the restriction, at the end they have defined a point x, the unique point in all the $\bar{N}(s_i)$; now I wins if $x \in A$, otherwise II wins.[1]

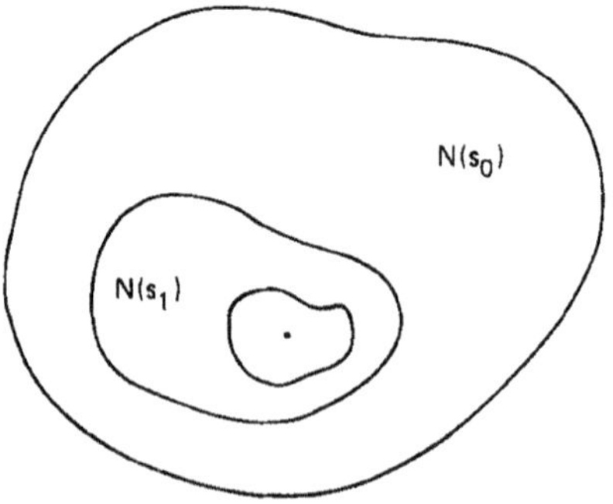

Figure 6A.3. The Banach–Mazur game.

6A.13. Prove that if Λ is adequate and closed under Δ_1^1 substitution, then

$$Det_\omega(\Lambda) \Longrightarrow \text{for each } A \in \Lambda,\ G^{**}(A) \text{ is determined.}$$

Hint. The payoff set $A^{**} \subseteq {}^\omega\omega$ for $G^{**}(A)$ is easily in Λ. ⊣

6A.14. Prove that for a fixed $A \subseteq \mathfrak{X}$,

(i) II wins $G^{**}(A) \Leftrightarrow A$ is meager,

(ii) I wins $G^{**}(A) \Leftrightarrow$ for some s, $\bar{N}(s) - A$ is meager. (Banach, Oxtoby [1957].[(1)])

Hint. (i) If A is meager, then $A \subseteq \bigcup_n F_n$ with each F_n closed and having no interior. If I plays s_0, have II play s_1 such that the restrictions are satisfied and $\bar{N}(s_1) \cap F_0 = \emptyset$, and in general let II play so that $\bar{N}(s_{2n+1}) \cap F_n = \emptyset$; then the point x determined at the end will not be in A.

Suppose now that II wins $G^*(A)$ via some strategy σ and let $x \in \mathfrak{X}$. Call a sequence $s_0, \ldots, s_n$ of even length *good* if it is the initial part of some play in $G^{**}(A)$, where the restrictions have been followed, II plays by σ and $x \in \bar{N}(s_n)$—the empty sequence is good by definition. If every good sequence has a good extension, then (easily) x is the point determined by some play where II plays σ, hence $x \notin A$; thus

$$x \in A \Longrightarrow \text{some } s_0, \ldots, s_n \text{ is maximal good (for } x).$$

Now if $s_0, \ldots, s_n$ is any even sequence, put

$$B(s_0, \ldots, s_n) = \bigcap \{\bar{N}(s_n) - N(\sigma(s_0, \ldots, s_n, s)) : \bar{N}(s) \subseteq \bar{N}(s_n)$$
$$\& \ radius(N(s)) \leq \tfrac{1}{2}\, radius(N(s_n))\};$$

each $B(s_0, \ldots, s_n)$ is easily closed and nowhere dense and we have shown

$$x \in A \Longrightarrow \text{for some } s_0, \ldots, s_n,\ x \in B(s_0, \ldots, s_n)$$

which establishes that A is meager.

(ii) Suppose I wins $G^{**}(A)$ via a strategy σ whose first move is $s = s_0$; it is now easy to check that II wins the game $G^{**}(\bar{N}(s_0) - A)$ by following σ, so that $\bar{N}(s_0) - A$ is meager by (i). Conversely, if $\bar{N}(s) - A$ is meager for some s, then I can easily win $G^{**}(A)$ by playing s to begin with and then staying out of $\bar{N}(s) - A$ as in the first part of the argument above. ⊣

6A.15. Suppose Λ is adequate $\mathbf{\Pi}_1^0 \subseteq \Lambda$ and for each A in Λ, either A is meager or there is a nbhd $N(s)$ such that $\bar{N}(s) - A$ is meager. Prove that every pointset in Λ has the property of Baire.

Hint. Given A in Λ, let

$$A^* = \bigcup \{N(s): \bar{N}(s) - A \text{ is meager}\};$$

A^* is open so it is enough to show that $A \triangle A^*$ is meager.

To begin with, $A^* - A \subseteq \bigcup \{\bar{N}(s) - A: \bar{N}(s) - A \text{ is meager}\}$, so $A^* - A$ is meager.

If $A - A^*$ is not meager, since $A - A^* \in \Lambda$ by the hypotheses, there must be some s such that $\bar{N}(s) - (A - A^*)$ is meager. Clearly then the smaller set $\bar{N}(s) - A$ is also meager so that $N(s) \subseteq A^*$ by the definition of A^* and easily $\bar{N}(s) - (A - A^*) \supseteq N(s)$; thus $N(s)$ is meager, contradicting the Baire Category Theorem. ⊣

6A.16. Prove that if Λ is adequate and closed under Borel substitution, then the following two conditions are equivalent:

(i) For each $A \in \Lambda$, $G^{**}(A)$ is determined.

(ii) Every pointset in Λ has the property of Baire.

Hence for such Λ,

$$Det_\omega(\Lambda) \Rightarrow \text{every pointset in } \Lambda \text{ has the property of Baire.}^{(6,7)}$$

Hint. One direction is immediate from 6A.14 and 6A.15. For the other direction, check easily that if A has the property of Baire, then in fact either A is meager or some $\bar{N}(s) - A$ is meager. ⊣

Let us now go to σ-finite Borel measures or simply *measures* for this discussion. The definitions are given in 2H.

First recall a few simple facts.

If μ is a measure on the Borel subsets of $\mathfrak{X}$, then for each Borel $P \subseteq \mathfrak{X}$,

$$\mu(P) = \textit{infimum}\{\mu(G): G \text{ is open, } P \subseteq G\}.$$

This is immediate for open P and follows for closed P because closed sets are countable intersections of open sets. Inductively, if $P = \bigcup_n P_n$ and for each n, $P_n \subseteq G_n$ with $\mu(G_n - P_n) < \varepsilon/2^n$, then $P \subseteq G = \bigcup_n G_n$ and $\mu(G - P) = \mu(\bigcup_n G_n - \bigcup_n P_n) \leq \Sigma_n \mu(G_n - P_n) \leq \varepsilon$. The argument is even simpler when $P = \bigcap_n P_n$ with P_n of smaller Borel order.

It follows that $\mu(P) = 0$ precisely when we can find for each $\varepsilon > 0$ an open set $G \supseteq P$ with $\mu(G) \leq \varepsilon$.

We now describe the *covering game* $G^\mu(A, \varepsilon)$ associated with the measure μ on the space ${}^\omega 2$ and each set $A \subseteq {}^\omega 2$. This is a game on ω, invented by Harrington.

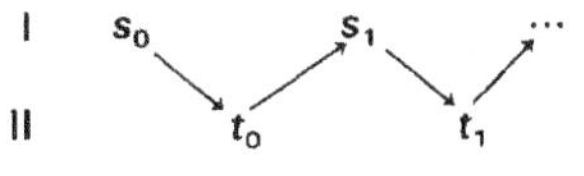

Diagram 6A.4.

Player I plays integers $s_0, s_1, s_2,\ldots$, with each $s_i = 0$ or $s_i = 1$. Thus he determines at the end a binary sequence $\alpha \in {}^\omega 2$.

Player II plays integers $t_0, t_1,\ldots$ where each t_n codes in some canonical way a finite union of basic open sets G_n, such that

$$\mu(G_n) \leq \varepsilon/2^{2n}.$$

For example, we may insist that $Seq(t_n)$ and

$$G_n = N((t_n)_0) \cup \cdots \cup N((t_n)_{lh(t_n)}).$$

Here $\varepsilon > 0$ is fixed and if II does not play the right kind of t_n he loses. The moves are made in the obvious order, as in Diagram 6A.4.

If II follows the rules, at the end he defines an open set

$$G = \bigcup_n G_n.$$

Now

$$\text{I wins} \Leftrightarrow \alpha \in A - G.$$

6A.17. Suppose μ is a σ-finite Borel measure on ${}^\omega 2$, $A \subseteq {}^\omega 2$ has no Borel subsets of μ-measure >0 and for each $\varepsilon > 0$ the game $G^\mu(A, \varepsilon)$ is determined. Prove that $\mu(A) = 0$.

Hint. Suppose first I wins $G^\mu(A, \varepsilon)$ via σ and let

$$B = \{\sigma * \tau : \tau \text{ is a strategy for II}\}.$$

Now B is a $\boldsymbol{\Sigma}^1_1$ subset of A, so it is μ-measurable by 2H.8 and then easily $\mu(B) = 0$, since $B \subseteq A$. We can easily find $G_0, G_1, G_2,\ldots$ all finite unions of basic nbhds with $\mu(G_n) \leq \varepsilon/2^{2n}$ and $B \subseteq \bigcup_n G_n$, which determines a strategy for II that beats σ contrary to hypothesis. Thus I cannot win $G^\mu(A, \varepsilon)$.

It follows that II wins, say by τ. Put

$$\begin{aligned} G = \bigcup\{G(s_0,\ldots, s_{n-1}) : {} & (s_0,\ldots, s_{n-1}) \text{ is a finite binary} \\ & \text{sequence and } G(s_0,\ldots, s_{n-1}) \text{ is the finite} \\ & \text{union of basic nbhds coded by II's move } t_n \\ & (\text{playing by } \tau) \text{ when I plays } s_0,\ldots, s_{n-1}\}. \end{aligned}$$

It is immediate that $A \subseteq G$ and G is open. But

$$\begin{aligned}\mu(G) &\leq \sum \{\mu(G(s_0, \ldots, s_{n-1})) : (s_0, \ldots, s_{n-1}) \text{ a binary sequence}\} \\ &= \sum_n \sum \{\mu(G(s_0, \ldots, s_{n-1})) : (s_0, \ldots, s_{n-1}) \text{ a binary sequence} \\ &\qquad\qquad\qquad\qquad\qquad\qquad \text{of length } n\} \\ &\leq \sum_n (\sum_{k<2^n} \varepsilon/2^{2n}) \\ &= \sum_n 2^n \cdot \varepsilon/2^{2n} = \varepsilon. \end{aligned}$$

⊣

6A.18. Suppose Λ is an adequate pointclass closed under Borel substitution and let μ be a σ-finite Borel measure on some $\mathfrak{X}$. Prove that

$$Det_\omega(\Lambda) \Rightarrow \text{every set } A \subseteq \mathfrak{X} \text{ in } \Lambda \text{ is } \mu\text{-measurable}.$$

(Mycielski–Swierczkowski [1964].[(6,7)])

Hint. Suppose first $\mathfrak{X} = {}^\omega 2$, let $A \subseteq \mathfrak{X}$. By 2H.7 there is a Borel set $\tilde{A}$ such that $A \subseteq \tilde{A}$ and $\tilde{A} - A$ contains no Borel set of μ-measure >0. Let $B = \tilde{A} - A$; clearly B is in the dual class $\neg\Lambda$. The game $G^\mu(B, \varepsilon)$ is easily in $\neg\Lambda$, hence it is determined by 6A.4; by 6A.17, $\mu(B) = 0$, so A is μ-measurable.

Every perfect space $\mathfrak{X}$ is Borel isomorphic with ${}^\omega 2$ and we can establish the result for $\mathfrak{X}$ by carrying to ${}^\omega 2$ any given measure on $\mathfrak{X}$. ⊣

6B. The first periodicity theorem

We saw in 6A that if Λ is a reasonable pointclass, then the hypothesis $Det_\omega(\Lambda)$ implies that all the sets in Λ are "nice": they have the property of Baire, they are absolutely measurable and they are uncountable precisely when they have perfect subsets. Put

$$PD \Leftrightarrow \textit{every projective set } A \subseteq \mathcal{N} \textit{ is determined};$$

this hypothesis of *projective determinacy* implies then that all projective sets are nice in this sense.

In this section we will show that if *PD* holds, then the prewellordering property oscillates between the Σ and the Π sides of the analytical hierarchy i.e. the circled pointclasses in the Diagram 6B.1 are all normed. These are Spector pointclasses then and the structure theory of Chapter 4 applies to them.

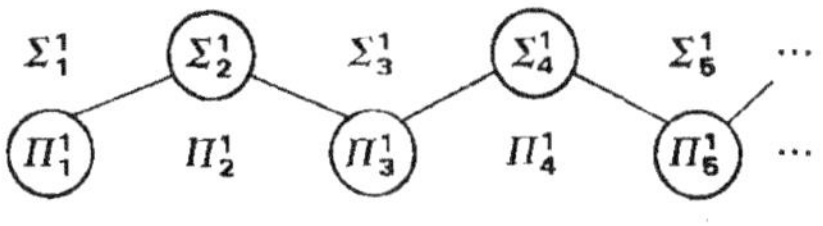

Diagram 6B.1. The normed Kleene pointclasses under *PD*.

Since we will be playing games on ω almost exclusively from now on, it is convenient to skip the subscript and abbreviate

$$Det(\Lambda) \Leftrightarrow Det_\omega(\Lambda)$$
$$\Leftrightarrow \text{every set } A \subseteq \mathcal{N} \text{ in } \Lambda \text{ is determined.}$$

In describing games on ω, it is often convenient to think of I and II as playing distinct sequences α, β, as we did in some of the exercises of the preceding section, see Diagram 6B.2. The *play* then is the sequence

$$a_0, b_0, a_1, b_1, \ldots,$$

but we often say that I *plays* α and II *plays* β in this run. We also describe games in this way, e.g. we may say that if I plays α and II plays β, then

$$\text{I wins} \Leftrightarrow P(\alpha, \beta),$$

where $P \subseteq \mathcal{N} \times \mathcal{N}$; this is obviously the game $G_\omega(A)$, where

$$A = \{(\alpha(0), \beta(0), \alpha(1), \beta(1), \ldots) : P(\alpha, \beta)\}.$$

Clearly A and P can be obtained by recursive substitutions from each other.

We identify a strategy σ for I in a game with the irrational σ defined by

$$\sigma(u) = \begin{cases} \sigma((u)_0, \ldots, (u)_{n-1}) & \text{if } Seq(u) \,\&\, lh(u) = n \text{ is even,} \\ 0 & \text{otherwise} \end{cases}$$

and similarly for strategies for II. If σ instructs I to play α when II plays β, we write

$$\alpha = \sigma * [\beta];$$

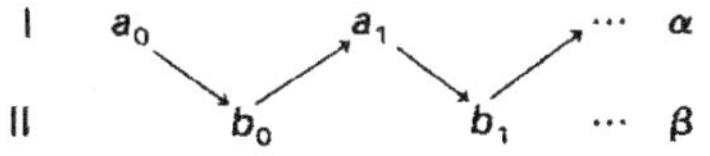

Diagram 6B.2.

similarly, if τ instructs II to play β when I plays α we write

$$\beta = [\alpha] * \tau.$$

Clearly both functions

$$(\sigma, \beta) \mapsto \sigma * [\beta], \qquad (\alpha, \tau) \mapsto [\alpha] * \tau$$

are recursive on $\mathcal{N} \times \mathcal{N}$ to $\mathcal{N}$.

The key result is the following theorem, dual to 4B.3. (For the definitions see 4B and the exercises of 4C.)

6B.1. THE FIRST PERIODICITY THEOREM (Martin, Moschovakis[9]). *Assume that Γ is adequate and Det ($\boldsymbol{\Delta}$) holds; if $P \subseteq \mathfrak{X} \times \mathcal{N}$ is in Γ and admits a Γ-norm, then the set $\forall^{\mathcal{N}} P$ admits a $\forall^{\mathcal{N}} \exists^{\mathcal{N}}$ Γ-norm.*

PROOF. Assume the hypotheses and let φ be a Γ-norm on P and

$$Q(x) \Leftrightarrow (\forall \alpha) P(x, \alpha).$$

We will define a prewellordering $\leq$ on Q and then take ψ to be the associated norm, such that $\leq_\psi = \leq$.

Given $x, y \in \mathfrak{X}$, consider the game $G(x, y)$

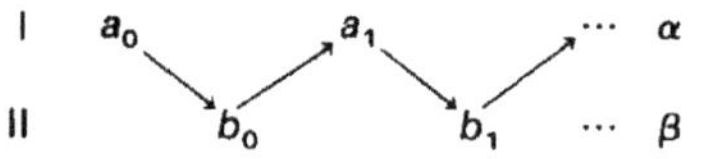

on ω, where I plays α, II plays β and

$$\text{I wins} \Leftrightarrow (y, \beta) <^*_\varphi (x, \alpha);$$

equivalently,

$$\begin{aligned} \text{II wins} &\Leftrightarrow \neg (y, \beta) <^*_\varphi (x, \alpha) \\ &\Leftrightarrow \neg P(y, \beta) \vee (x, \alpha) \leq^*_\varphi (y, \beta). \end{aligned}$$

Put

$$(*) \qquad x \leq y \Leftrightarrow \text{II has a winning strategy in } G(x, y).$$

We will prove that the restriction of $\leq$ to Q is a prewellordering with the desired properties.

The motivation for the proof comes from the natural attempt to define a norm ψ on Q by

$$\psi(x) = \textit{supremum}\{\varphi(x, \alpha) : \alpha \in \mathcal{N}\}.$$

This is a norm of course, but more often than not it is a trivial norm—e.g. if φ is an $\aleph_1$-norm, then we are likely to have $\psi(x) = \aleph_1$ for almost all

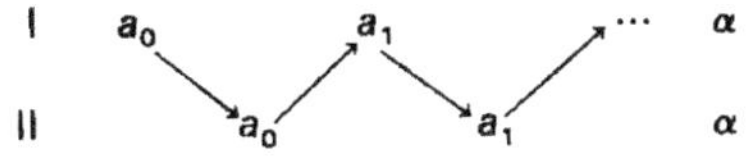

Diagram 6B.3.

$x \in Q$. Definition (*) can be interpreted as saying that $x \leq y$ holds when *supremum*$\{\varphi(x, \alpha): \alpha \in \mathcal{N}\}$ is *effectively* $\leq$ *supremum*$\{\varphi(y, \beta): \beta \in \mathcal{N}\}$, for $y \in Q$, in the sense that we have a strategy τ which correlates with each α (given bit by bit) some β such that $\varphi(x, \alpha) \leq \varphi(y, \beta)$. Because of this picture, we call $G(x, y)$ *the sup game* for the norm φ.

We now establish the properties of $\leq$ in a sequence of lemmas.

Lemma 1. For every $x \in Q$, $x \leq x$.

Proof. Have II play in $G(x, x)$ simply by copying the moves of I, as in Diagram 6B.3. Since $x \in Q$ we have $P(x, \alpha)$, hence $(x, \alpha) \leq_{\varphi}^{*} (x, \alpha)$, and II wins.

Lemma 2. If x, y, z are all in Q, then

$$(x \leq y \;\&\; y \leq z) \rightarrow x \leq z.$$

Proof. We are assuming that II has winning strategies in both $G(x, y)$ and $G(y, z)$ and we must describe how II can play to win in $G(x, z)$. Consider Diagram 6B.4.

Suppose I plays a_0 in $G(x, z)$. Then I copies a_0 in $G(x, y)$ and II answers b_0 in that game by his winning strategy. Then I copies this b_0 in

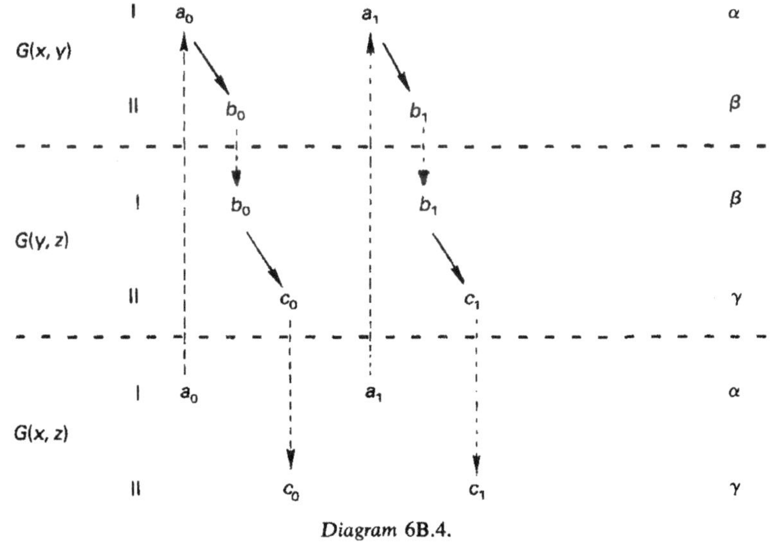

Diagram 6B.4.

$G(y, z)$ and II answers c_0 in that game by his winning strategy. Now II responds with this c_0 to a_0 in $G(x, z)$.

Now I plays a_1 in $G(x, z)$ and II responds by c_1 which is determined in the same way, as indicated in the diagram, and similarly for the moves after that. In effect II plays in $G(y, z)$ by *simulating* runs of $G(x, y)$ and $G(y, z)$ on the side and watching the moves of the second players in these games, which follow winning strategies.

It is a simple matter to give a formal definition of this strategy for II in $G(x, z)$ in terms of the given strategies for II in $G(x, y)$ and $G(y, z)$ and we will not bother.

At the end of the run, sequences α, β, γ have been played as in the diagram and we know $P(x, \alpha)$, $P(y, \beta)$, $P(z, \gamma)$ (since x, y, z are all in Q) and also $\varphi(x, \alpha) \leq \varphi(y, \beta)$, $\varphi(y, \beta) \leq \varphi(z, \gamma)$, since II wins $G(x, y)$ and $G(y, z)$. Hence $\varphi(x, \alpha) \leq \varphi(z, \gamma)$ and II has also won $G(x, z)$.

This describes a winning strategy for II in $G(x, z)$, hence $x \leq z$.

Lemma 3. For all x, y, $G(x, y)$ is determined.

Proof. If $y \notin Q$, then II can win by playing any β such that $\neg P(y, \beta)$. If $y \in Q$, then $P(y, \beta)$ holds for each β, so that

$$\begin{aligned} \text{I wins} &\Leftrightarrow (y, \beta) <^*_\varphi (x, \alpha) \\ &\Leftrightarrow \neg((x, \alpha) \leq^*_\varphi (y, \beta)) \end{aligned}$$

and the payoff set is in $\mathbf{\Delta}$, hence the game is determined by hypothesis.

Put

$$x < y \Leftrightarrow x \leq y \;\&\; \neg(y \leq x).$$

Lemma 4. If x, y are in Q, then

$$x < y \Leftrightarrow \text{I wins } G(y, x);$$

thus for $x, y \in Q$,

$$x \leq y \vee y \leq x.$$

Proof. If $x < y$, then $\neg(y \leq x)$, so II does not win $G(y, x)$ by definition, hence I wins $G(y, x)$ by Lemma 3.

Conversely, suppose I wins $G(y, x)$; then certainly II does not win $G(y, x)$, so to establish $x < y$ it is enough to show that II wins $G(x, y)$.

Fix a winning strategy for I in $G(y, x)$ and consider Diagram 6B.5.

We describe a strategy for II in $G(x, y)$ as follows. Suppose I plays a_0 in $G(x, y)$; II disregards the value of a_0 and answers by b_0, the first

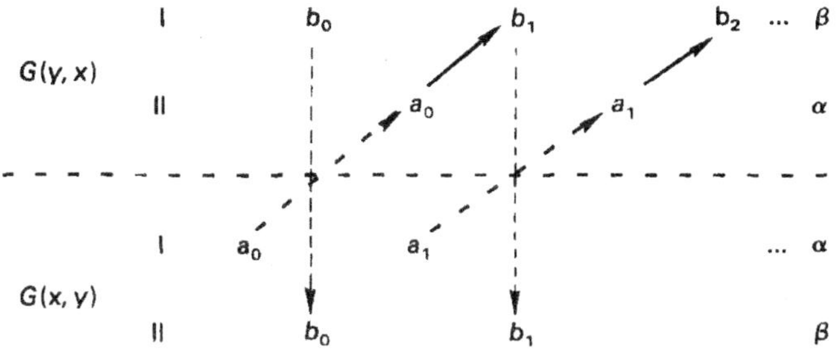

Diagram 6B.5.

winning move of I in $G(y, x)$. He then copies a_0 in $G(y, x)$ and observes that I answers this move of II by b_1 in that game. Suppose now I plays a_1 in $G(x, y)$; again II disregards the value of a_1 and responds b_1, but then copies a_1 in $G(y, x)$ and observes the response b_2, etc.

At the end of the game sequences α, β have been played and I has won $G(y, x)$, i.e. $(x, \alpha) <^*_\varphi (y, \beta)$; hence surely $(x, \alpha) \leq^*_\varphi (y, \beta)$, so II has won $G(x, y)$.

Lemma 5. The relation $\leq$ is wellfounded.

Proof. We must show that there are no infinite descending chains, so assume towards a contradiction that

$$x_0 > x_1 > x_2 > \dots,$$

i.e. by Lemma 4, I wins $G(x_i, x_{i+1})$ for every i. Fix winning strategies for I in all these games and consider the following Diagram 6B.6. Here player I follows the fixed winning strategies in all the games and the moves of II are filled in by copying along the dotted arrows. At the end of the run sequences $\alpha^0, \alpha^1, \alpha^2, \dots$ have been played and since I wins all these games we have

$$\varphi(x_0, \alpha^0) > \varphi(x_1, \alpha^1) > \varphi(x_2, \alpha^2) > \cdots$$

which is absurd.

At this point we have established that $\leq$ is a prewellordering on Q, so let $\psi : Q \to$ *Ordinals* be the regular norm associated with it, i.e.

$$x \leq y \Leftrightarrow \psi(x) \leq \psi(y) \quad (x, y \in Q).$$

Lemma 6. The norm ψ is a $\forall^{\mathcal{N}}\exists^{\mathcal{N}}\,\Gamma$-norm on Q.

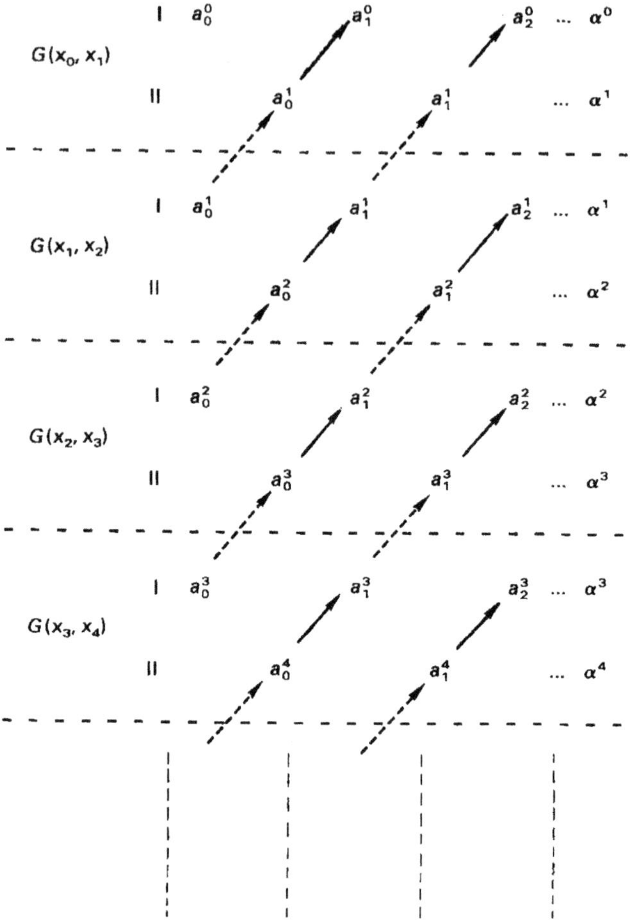

Diagram 6B.6.

Proof. From the definition,

$$x \leq^*_\psi y \Leftrightarrow Q(x) \;\&\; \text{II wins } G(x, y)$$
$$\Leftrightarrow Q(x) \;\&\; \text{I does not win } G(x, y)$$
$$\Leftrightarrow Q(x) \;\&\; (\forall\sigma)(\exists\beta)[(x, \sigma * [\beta]) \leq^*_\varphi (y, \beta)].$$

Similarly, using Lemma 4,

$$x <^*_\psi y \Leftrightarrow Q(x) \;\&\; \text{I wins } G(y, x)$$

$$\Leftrightarrow Q(x) \,\&\, \text{II does not win } G(y, x)$$

$$\Leftrightarrow Q(x) \,\&\, (\forall \tau)(\exists \alpha)[(x, \alpha) <^*_\varphi (y, [\alpha] * \tau)]. \qquad \dashv$$

6B.2. COROLLARY. *If Γ is adequate, normed and $\exists^{\mathcal{N}}\Gamma \subseteq \Gamma$ and if $Det(\boldsymbol{\Delta})$ holds, then $\forall^{\mathcal{N}}\Gamma$ is normed.*

In particular, PD implies that Π^1_1, Σ^1_2, Π^1_3, $\Sigma^1_4, \ldots, \Pi^1_n$ (n odd), Σ^1_k (k even) are all Spector pointclasses.

PROOF is immediate from 6B.1 and 4B.3. $\dashv$

We often refer to this corollary instead of 6B.1 as the (first) periodicity theorem for the obvious reason. Recall from 4B.13 that Σ^1_n and Π^1_n cannot both be normed, so that the oscillating picture of the analytical pointclasses we get from *PD* is in fact totally different from the picture when we assume $\mathcal{N} \subseteq L$.

At this point, one should go back to sections 4B–4D and recall the structure theory for Spector pointclasses developed there: these results have now been established for all Π^1_n (n odd) and Σ^1_k (k even), under the hypothesis of projective determinacy.

We should also point out here that the first periodicity theorem gives a new and interesting proof of the prewellordering theorem for Π^1_1, 4B.2, as follows.

By 4B.8, every pointset of type 1 in Σ^0_1 admits a Σ^0_1-norm. Now 3C.14 implies immediately that

$$\forall^{\mathcal{N}}\Sigma^0_1 = \forall^{\mathcal{N}}\exists^{\mathcal{N}}\Sigma^0_1 = \Pi^0_1;$$

hence by 6B.1 every Π^1_1 pointset of type 1 admits a Π^1_1-norm, using the determinacy of $\boldsymbol{\Delta}^0_1$ (clopen) sets. This yields immediately that Π^1_1 is normed, by 4B.9.

This is one of the characteristic features of the game-theoretic proofs that we will construct in this chapter: when we apply them to pointclasses whose determinacy is known (like $\boldsymbol{\Delta}^0_1$, $\boldsymbol{\Sigma}^0_1$ or $\boldsymbol{\Sigma}^0_2$), we obtain new proofs of classical results about Π^1_1 and Σ^1_2.

It is convenient to introduce the notations

$$\Sigma^1_0 = \Sigma^0_1;\ \Pi^1_0 = \Pi^0_1;\ \Delta^1_0 = \Delta^0_1$$

together with their boldface companions, so that e.g. $\boldsymbol{\Delta}^1_0$ is the class of all clopen sets. In many of the results below we will use the hypothesis $Det(\boldsymbol{\Delta}^1_{2n})$; now this makes sense even when $n = 0$, in which case it is simply true, by the Gale–Stewart Theorem.

6C. The second periodicity theorem; uniformization

The next obvious question is whether the hypothesis of projective determinacy settles the uniformization problem: must each projective set be uniformizable by some projective set? We show here that it does, but in a precise way which differs from the situation in L and which reveals further the periodicity phenomenon we uncovered in 6B.

We will in fact show that if *PD* holds, then the scale property oscillates between the Σ and Π sides of the Kleene hierarchy together with the prewellordering property. Consider first the analog for scales of 4B.3 which we did not establish in 4E—we had no use for it then.

6C.1. LEMMA (Moschovakis [1971a]). *Suppose Γ is adequate, $\mathfrak{X}$ is a space of type* 1, *$P \subseteq \mathfrak{X} \times \mathcal{N}$ is in Γ and P admits a Γ-scale; then $\exists^{\mathcal{N}} P$ admits a $\exists^{\mathcal{N}}\forall^{\mathcal{N}}\, \Gamma$-scale.*

PROOF. By 4E.2, let $\bar{\varphi} = \{\varphi_n\}$ be a very good Γ-scale on P and put

$$P^*(x, \alpha) \Leftrightarrow (\forall n)(\forall \beta)[(x, \alpha) \leq^*_{\varphi_n} (x, \beta)];$$

we showed in 4E.3 that P^* uniformizes P. Let

$$\begin{aligned} Q(x) &\Leftrightarrow (\exists \alpha) P(x, \alpha) \\ &\Leftrightarrow (\exists \alpha) P^*(x, \alpha) \end{aligned}$$

and define $\bar{\psi} = \{\psi_n\}$ on Q by

$$\psi_n(x) = \varphi_n(x, \alpha) \text{ for the unique } \alpha \text{ such that } P^*(x, \alpha).$$

We now verify that $\bar{\psi}$ is a scale on Q.

If $x_0, x_1, \ldots$ are all in Q and $\psi_n(x_i) = \lambda_n$ for each n and all large i, choose $\alpha_0, \alpha_1, \ldots$ such that $P^*(x_i, \alpha_i)$, so that by definition

$$\varphi_n(x_i, \alpha_i) = \psi_n(x_i).$$

Thus $\varphi_n(x_i, \alpha_i) = \lambda_n$ for each n and all large i and since $\bar{\varphi}$ is a very good scale, we have $x_i \to x$, $\alpha_i \to \alpha$ and

$$\varphi_n(x, \alpha) \leq \lambda_n, \text{ all } n.$$

In particular of course, $Q(x)$. Now choose α^* such that $P^*(x, \alpha^*)$ and notice that for each n.

$$\psi_n(x) = \varphi_n(x, \alpha^*) \leq \varphi_n(x, \alpha) \leq \lambda_n,$$

where the inequality $\varphi_n(x, \alpha^*) \le \varphi_n(x, \alpha)$ follows from the definition of P^*. Thus $\bar{\psi}$ is a scale.

That $\bar{\psi}$ is a $\exists^{\mathcal{N}}\forall^{\mathcal{N}}$ Γ-scale follows from the easy equivalence

$$x \le^*_{\psi_n} y \Leftrightarrow (\exists\alpha)(\forall\beta)[P^*(x, \alpha) \,\&\, (x, \alpha) \le^*_{\varphi_n} (y, \beta)]$$

and the similar one for $<^*_{\psi_n}$. ⊣

6C.2. THEOREM (Moschovakis [1971a]). *If $\Sigma^0_1 \subseteq \Gamma$ and Γ is adequate, closed under $\forall^{\mathcal{N}}$ and scaled, then $\exists^{\mathcal{N}}\Gamma$ is also scaled. In particular, Σ^1_2, $\boldsymbol{\Sigma}^1_2$ are scaled.*

PROOF is immediate from 6C.1. ⊣

There is no immediate use at this point for the fact that $\boldsymbol{\Sigma}^1_2$ is scaled, but we will apply it later. For now, 6C.2 will serve only as an induction loading device, in conjunction with the next result.

6C.3. THE SECOND PERIODICITY THEOREM (Moschovakis [1971a]). *Suppose Γ is adequate and Det($\boldsymbol{\Delta}$) holds. If $P \subseteq \mathfrak{X} \times \mathcal{N}$ is in Γ of type 1 and admits a Γ-scale, then the set $\forall^{\mathcal{N}}P$ admits a $\forall^{\mathcal{N}}\exists^{\mathcal{N}}$ Γ-scale.*

PROOF. Let $\bar{\varphi} = \{\varphi_n\}$ be a fixed very good Γ-scale on P and put

$$Q(x) \Leftrightarrow (\forall\alpha)P(x, \alpha).$$

It will be convenient to have an effective enumeration of all finite sequences of integers, so put

$u(0)$ = the empty sequence,

$u(i)$ = the sequence coded by the i'th number v such that $Seq(v)$.

It is immediate from this definition that if $u(i)$ is an initial segment of $u(j)$, then $i < j$.

If u is a finite sequence and $\alpha \in \mathcal{N}$, let

$$u < \alpha \Leftrightarrow u \text{ is an initial segment of } \alpha$$

and put

$$x \in Q_n \Leftrightarrow (\forall\alpha > u(n))P(x, \alpha).$$

Clearly $Q_0 = Q$ and $Q \subseteq Q_n$ for every n. We will define a norm ψ_n on each Q_n by considering a game $G_n(x, y)$, very much as in the proof of the First Periodicity Theorem 6B.1.

Suppose I and II play sequences α', β' in the usual fashion. We let

$$\alpha = u(n)^\frown\alpha' = u(n)^\frown(a_0, a_1, \ldots)$$
$$\beta = u(n)^\frown\beta' = u(n)^\frown(b_0, b_1, \ldots)$$

and we put

$$\text{I wins } G_n(x, y) \Leftrightarrow (y, \beta) <^*_{\varphi_n} (x, \alpha)$$

or equivalently,

$$\text{II wins } G_n(x, y) \Leftrightarrow \neg P(y, \beta) \vee (x, \alpha) \leq^*_{\varphi_n} (y, \beta).$$

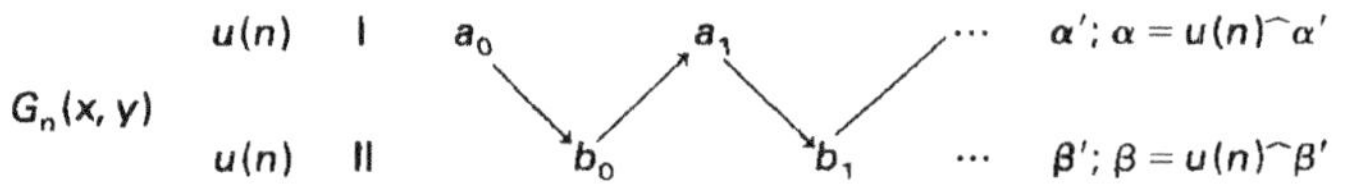

Diagram 6C.1. The sup game on $u(n)$.

We can think of $G_n(x, y)$ as a subgame of the sup game for the norm φ_n, as we defined this in 6B.1, where both players have been saddled with the same first few moves—those in the sequence $u(n)$. It will be useful to think of α and β as the plays in $G_n(x, y)$ (instead of α' and β').

Put now

$$x \leq_n y \Leftrightarrow \text{II has a winning strategy in } G_n(x, y).$$

Using the arguments in 6B.1, one easily checks that each $\leq_n$ is a prewellordering on Q_n and that the associated norm ψ_n on Q_n is a $\forall^{\mathcal{N}}\exists^{\mathcal{N}}\Gamma$-norm. Moreover, the relations

$$R(n, x, y) \Leftrightarrow x \in Q_n \;\&\; x \leq^*_{\psi_n} y,$$
$$S(n, x, y) \Leftrightarrow x \in Q_n \;\&\; x <^*_{\psi_n} y$$

are also in $\forall^{\mathcal{N}}\exists^{\mathcal{N}}\Gamma$.

It will be very easy to turn the sequence $\bar{\psi} = \{\psi_n\}$ into a $\forall^{\mathcal{N}}\exists^{\mathcal{N}}\Gamma$-scale on Q, after we prove the following key fact.

Lemma. Suppose $x_0, x_1, \ldots$ are all in Q, $limit_{i\to\infty}\, x_i = x$ and for each n and all large i, $\psi_n(x_i) = \lambda_n$; then $x \in Q$ and for each n, $\psi_n(x) \leq \lambda_n$.

Proof. By passing to a subsequence if necessary, we may assume that

$$i \geq n \Rightarrow \psi_n(x_i) = \lambda_n.$$

To show first that $x \in Q$, we must verify that for each fixed α, $P(x, \alpha)$

holds. Choose n_i so that

$$u(n_i) = (\alpha(0),\ldots, \alpha(i-1))$$

and consider the subsequence

$$x_{n_0}, x_{n_1}, x_{n_2},\ldots.$$

Now $n_i < n_{i+1}$ and hence

$$\psi_{n_i}(x_{n_i}) = \psi_{n_i}(x_{n_{i+1}}) = \lambda_{n_i},$$

so that

$$x_{n_{i+1}} \leq_{n_i} x_{n_i}$$

and II has a winning strategy in all the games $G_{n_i}(x_{n_{i+1}}, x_{n_i})$. Fix winning strategies for II in all these games and consider Diagram 6C.2 which is constructed by the following rules.

To begin with, I plays $\alpha(i)$ in $G_{n_i}(x_{n_{i+1}}, x_{n_i})$. After II responds to this (by his winning strategy) with a move that we have labeled $\alpha_i(i)$, I plays by copying $\alpha_i(i)$ into the preceding game $G_{n_{i-1}}(x_{n_i}, x_{n_{i-1}})$, as we have indicated by the dotted lines. Now II plays again to win and I responds by copying, etc. ad infinitum.

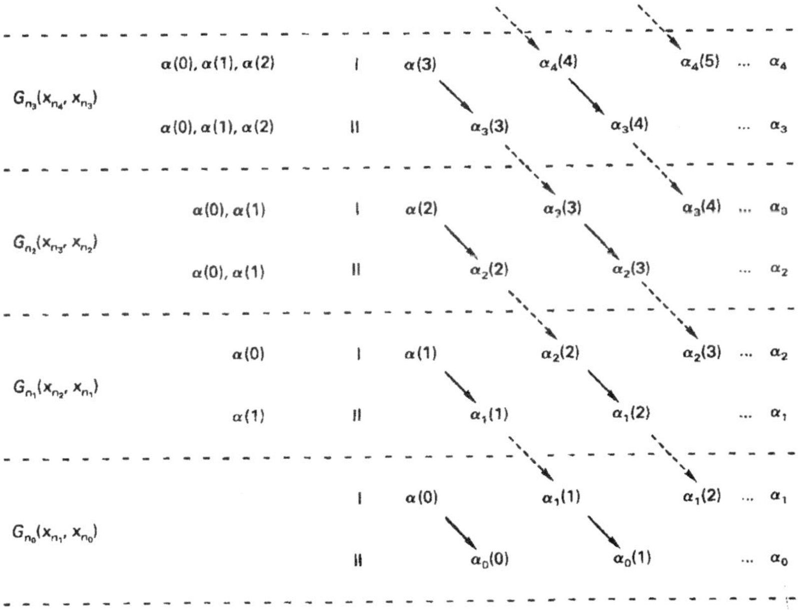

Diagram 6C.2.

At the end, plays $\alpha_0, \alpha_1, \alpha_2, \ldots$ have been determined and it is obvious that

$$limit_{i\to\infty}\, \alpha_i = \alpha,$$

since in fact

$$\alpha_i(j) = \alpha(j) \quad \text{for } j < i.$$

Moreover II wins all these runs, so we have for every i,

$$\varphi_{n_i}(x_{n_{i+1}}, \alpha_{i+1}) \leq \varphi_{n_i}(x_{n_i}, \alpha_i).$$

Since the scale $\bar{\varphi}$ is very good, this implies that for each fixed k and all i which are large enough (so that $k \leq n_i$),

$$\varphi_k(x_{n_{i+1}}, \alpha_{i+1}) \leq \varphi_k(x_{n_i}, \alpha_i),$$

so that in fact there are ordinals μ_k and

$$\varphi_k(x_{n_i}, \alpha_i) = \mu_k$$

for all large i; hence $P(x, \alpha)$, since $\bar{\varphi}$ is a scale on P. Thus $P(x, \alpha)$ holds for every α and $x \in Q$.

We now prove that for each n,

$$\psi_n(x) \leq \lambda_n.$$

If $k \leq m$, then $\psi_k(x_k) = \psi_k(x_m)$, hence $x_m \leq_k x_k$ and II has a winning strategy in each of the games $G_k(x_m, x_k)$ $(k \leq m)$. Fix winning strategies for II in each of these games and fix the number n. We will describe how II can play to win $G_n(x, x_n)$, thus showing $x \leq_n x_n$, i.e. $\psi_n(x) \leq \psi_n(x_n) = \lambda_n$.

Player II will win by utilizing many of the strategies in the games $G_k(x_m, x_k)$. In fact, he will construct on the side a diagram of games much like the one above and his moves in $G_n(x, x_n)$ will be copied from the appropriate places in that diagram. The only additional complication in this argument is that II does not know ahead of time which of the games $G_k(x_m, x_k)$ he wants to play on the side; these will depend on the moves that I makes in $G_n(x, x_n)$.

Consider then Diagram 6C.3 which is constructed as follows.

Let $n_0 = n$ and suppose that the sequence $u(n)$ has length l—these are useful notation conventions.

Suppose that I starts by playing a_l in $G_n(x, x_n)$. Choose n_1 so that $u(n_1) = u(n_0)^\frown(a_l)$, so that $n_0 < n_1$ and start the game $G_{n_0}(x_{n_1}, x_{n_0})$ with I playing a_l in it. Have II respond by his winning strategy by some $\alpha_0(l)$ (α_0

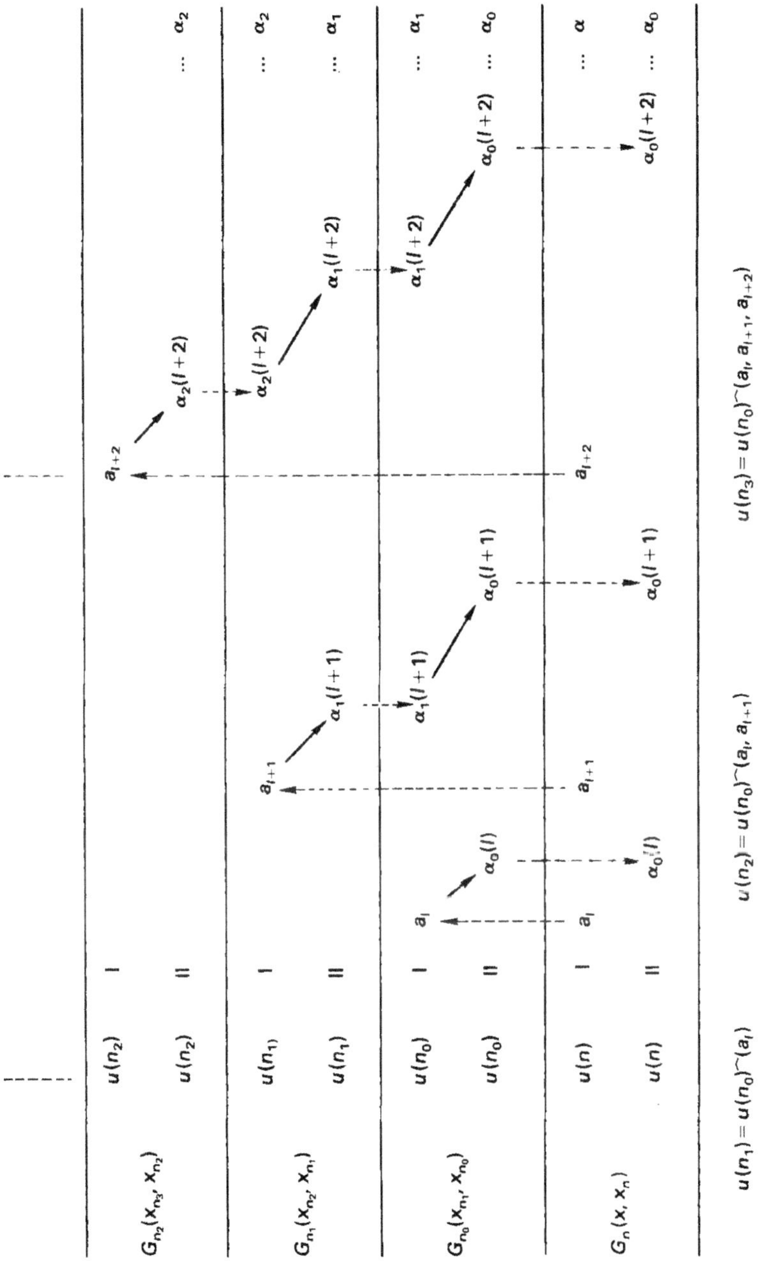

G_{n_2}(x_{n_3}, x_{n_2})
G_{n_1}(x_{n_2}, x_{n_1})
G_{n_0}(x_{n_1}, x_{n_0})
G_n(x, x_n)
u(n_2) | a_{l+2} α_2(l+2) ... α_2
u(n_1) | a_{l+1} α_1(l+1) α_2(l+2) α_1(l+2) ... α_1
u(n_0) | a_l α_0(l) α_1(l+1) α_0(l+1) α_1(l+2) α_0(l+2) ... α_0
u(n) | a_l a_{l+1} a_{l+2} ... α
u(n) || α_0(l) α_0(l+1) α_0(l+2) ... α_0
u(n_1) = u(n_0)⌢(a_l)
u(n_2) = u(n_0)⌢(a_l, a_{l+1})
u(n_3) = u(n_0)⌢(a_l, a_{l+1}, a_{l+2})

Diagram 6C.3.

will be his eventual play in this game) and have II play the same $\alpha_0(l)$ in $G_n(x, x_n)$.

Suppose now that I plays a_{l+1} in $G_n(x, x_n)$. Let $u(n_2) = u(n_0)^\frown(a_l, a_{l+1})$ so that $n_1 < n_2$ and start the game $G_{n_1}(x_{n_2}, x_{n_1})$ with I playing a_{l+1}; II responds to win by $\alpha_1(l+1)$, we copy this move in $G_{n_0}(x_{n_1}, x_{n_0})$, II responds by $\alpha_0(l+1)$ and finally II plays this $\alpha_0(l+1)$ in $G_n(x, x_n)$.

Continuing in this fashion as in the diagram, we determine successively games $G_{n_i}(x_{n_{i+1}}, x_{n_i})$ and plays α_i, so that if α is the play of I in $G_n(x, x_n)$, then

$$limit_{i\to\infty}\, \alpha_i = \alpha,$$

since in fact

$$j < l + i \Rightarrow a_i = \alpha(j) = \alpha_i(j).$$

Moreover, II wins all these games, so that

$$\varphi_{n_i}(x_{n_{i+1}}, \alpha_{i+1}) \leq \varphi_{n_i}(x_{n_i}, \alpha_i).$$

We now argue very much as in the first part of this proof: since $\bar{\varphi}$ is a very good scale, we have

$$(*) \qquad \varphi_k(x_{n_{i+1}}, \alpha_{i+1}) \leq \varphi_k(x_{n_i}, \alpha_i)$$

for all i large enough so that $k \leq n_i$, hence all the norms $\varphi_k(x_{n_i}, \alpha_i)$ are eventually constant, and hence we have $P(x, \alpha)$ and for each k,

$$(**) \qquad \varphi_k(x, \alpha) \leq limit_{i\to\infty}\, \varphi_k(x_{n_i}, \alpha_i).$$

Taking $k = n = n_0$ in $(*)$ for $i = 0, 1, 2, \ldots$, we have

$$\varphi_{n_0}(x_{n_0}, \alpha_0) \geq \varphi_{n_0}(x_{n_1}, \alpha_1) \geq \varphi_{n_0}(x_{n_2}, \alpha_2) \geq \cdots \geq limit_{i\to\infty}\, \varphi_{n_0}(x_{n_i}, \alpha_i),$$

so that by $(**)$

$$\varphi_{n_0}(x_{n_0}, \alpha_0) \geq \varphi_{n_0}(x, \alpha)$$

and II wins the game $G_n(x, x_n)$.

This completes the proof of the lemma.

Going back to the proof of the theorem, suppose all the norms ψ_n are into the ordinal $\varkappa$, let

$$(\xi, n) \mapsto \langle \xi, n \rangle$$

be an order-preserving map of $\varkappa \times \varkappa$ (ordered lexicographically) into the ordinals and put

$$\psi'_n(x) = \langle \psi_0(x), \psi_n(x) \rangle.$$

It is easy to check (as in the proof of 4E.1) that $\bar{\psi}' = \{\psi'_n\}$ is a $\forall^{\mathcal{N}}\exists^{\mathcal{N}}$ Γ-scale on Q. ⊣

6C.4. Corollary. *If $\Pi^0_1 \subseteq \Gamma$ and Γ is adequate, scaled and $\exists^{\mathcal{N}}\Gamma \subseteq \Gamma$ and if $Det(\mathbf{\Delta})$ holds, then $\forall^{\mathcal{N}}\Gamma$ is also scaled.*

In particular, PD implies that Π^1_1, Σ^1_2, Π^1_3, $\Sigma^1_4, \ldots$, Π^1_n (n odd), Σ^1_k (k even) are all scaled. ⊣

6C.5. The Uniformization Theorem. *If PD holds, then every projective set can be uniformized by a projective set and every analytical set can be uniformized by an analytical set.*

More specifically, $Det(\mathbf{\Delta}^1_{2n})$ implies that Π^1_{2n+1}, Σ^1_{2n+2}, $\mathbf{\Pi}^1_{2n+1}$, $\mathbf{\Sigma}^1_{2n+2}$ all have the uniformization property. (*Moschovakis* 1971a].)

Proof is immediate from 6C.3 and 4E.7. ⊣

6C.6. The Basis Theorem. *If PD holds, then every nonempty analytical pointset contains an analytical point.*

More specifically, $Det(\mathbf{\Delta}^1_{2n})$ implies that Δ^1_{2n+2} is a basis for Σ^1_{2n+2} and $\Delta^1_{2n+2}(x)$ is a basis for $\Sigma^1_{2n+2}(x)$. (*Moschovakis* [1971a].)

Proof is immediate as in 4E.5. ⊣

These two results are the most obvious and significant consequences of the second periodicity theorem, but there are others. We will consider some of them in the exercises here and in the next two sections.

Recall the notational convention

$$\mathbf{\Sigma}^1_0 = \mathbf{\Sigma}^0_1$$

which we introduced at the end of 6B. Many of the results in the exercises depend on the hypothesis $Det(\mathbf{\Delta}^1_{2n})$ which is true when $n = 0$.

Exercises

Let us take up first a few facts about bases which complement 6C.6.

6C.7. Prove that if $Det(\mathbf{\Delta}^1_{2n})$ holds, then Δ^1_{2n+1} is not a basis for Π^1_{2n}—i.e. there is a nonempty Π^1_{2n} set $A \subseteq \mathcal{N}$ which has no Δ^1_{2n+1}-recursive member.

Hint. See 4D.10. ⊣

6C.8. Prove that if $Det(\mathbf{\Delta}^1_{2n})$ holds, then there exists a Π^1_{2n} set $P \subseteq \mathcal{N} \times \mathcal{N}$ which cannot be uniformized by any $\mathbf{\Sigma}^1_{2n+1}$ set.

Hint. See 4D.11. ⊣

Kleene's basis theorem for Σ^1_1 (4E.8) does not extend to all Σ^1_{2n+1}, but Martin and Solovay have found a better basis for this pointclass than Δ^1_{2n+2}. In our presentation of their results here we will use the important notion of the (Δ^1_k)-*hull* of a pointset introduced in Kechris [1975],

$$Hull_k(A) = \{\alpha \in \mathcal{N}: (\forall x)[x \in A \rightarrow \alpha \in \Delta^1_k(x)].$$

Recall the definition of α_P (for P a pointset of type 0) from 4E.

6C.9. Assume $Det(\mathbf{\Delta}^1_{2n})(n \geq 1)$ and prove the following three properties of hulls.

(i) If A is Σ^1_{2n+1}, then $Hull_{2n+1}(A)$ is Π^1_{2n+1}.

(ii) If $A \neq \emptyset$ and A is Σ^1_{2n+1}, then there exists a Π^1_{2n} set $B \neq \emptyset$, $B \subseteq \mathcal{N}$ such that $B \cap Hull_{2n+1}(A) = \emptyset$.

(ii) If P is a Σ^1_{2n+1} pointset of type 0 and α_P is its contracted characteristic function, then there is a nonempty Σ^1_{2n+1} set A such that $\alpha_P \in Hull_{2n+1}(A)$.

Infer that for any Σ^1_{2n+1} set P of type 0, $\{x: x \in \Delta^1_{2n+1}(\alpha_P)\}$ is not a basis for Π^1_{2n}. (Martin–Solovay.)

Hint. (i) is a trivial computation using 4D.14.

To prove (ii) check first that it is enough to find a Σ^1_{2n+1} set $B \subseteq \mathcal{N}$ such that $B \neq \emptyset$ but $B \cap Hull_{2n+1}(A) = \emptyset$ and then (assuming for simplicity that $A \subseteq \mathcal{N}$) take

$$B = \{\langle \alpha, \beta \rangle: \alpha \in A \;\&\; \beta \notin \Delta^1_{2n+1}(\alpha)\}.$$

Clearly $B \neq \emptyset$ if $A \neq \emptyset$ and if $\langle \alpha, \beta \rangle \in B \cap Hull_{2n+1}(A)$, then $\alpha \in A$ and hence $\langle \alpha, \beta \rangle \in \Delta^1_{2n+1}(\alpha)$ which implies $\beta \in \Delta^1_{2n+1}(\alpha)$ contradicting $\langle \alpha, \beta \rangle \in B$.

For (iii), assume for simplicity that $P \subseteq \omega$, let $\varphi: Q \rightarrow$ *Ordinals* be a Π^1_{2n+1}-*norm on* $Q = \omega - P$ and put

$$\begin{aligned} \alpha \in A \leftrightarrow & \{(n, m): \alpha(\langle n, m \rangle) = 1 \text{ is a prewellordering}\} \\ & \& (\forall m)[Q(m) \rightarrow (\forall n)[(Q(n) \;\&\; \varphi(n) \leq \varphi(m)) \leftrightarrow \alpha(\langle n, m \rangle) = 1]]. \end{aligned}$$

If $\alpha \in A$, then obviously $\{(n, m): \alpha(\langle n, m \rangle) = 1\}$ is a prewellordering which extends the prewellordering $\leq^\varphi$ induced by φ; thus

$$\alpha \in A \Rightarrow \{(n, m): Q(n) \,\&\, Q(m) \,\&\, \varphi(n) \leq \varphi(m)\} \text{ is recursive in } \alpha$$
$$\Rightarrow \alpha_Q \text{ is recursive in } \alpha \Rightarrow \alpha_P \text{ is recursive in } \alpha$$

so that $\alpha_P \in Hull_{2n+1}(A)$.

The last assertion follows immediately from (ii) and (iii). ⊣

6C.10. Assume $Det(\boldsymbol{\Delta}^1_{2n})$ and suppose α_0 is a Π^1_{2n+1} singleton but $\alpha_0 \notin \Delta^1_{2n+1}$; prove that $\{x: x \in \Delta^1_{2n+1}(\alpha_0)\}$ is a basis for Σ^1_{2n+1}. (Martin–Solovay.)

Hint. It is of course enough to prove that if A is Π^1_{2n} and $\emptyset \neq A \subseteq \mathcal{N}$, then there is some $\alpha \in A \cap \Delta^1_{2n+1}(\alpha_0)$. Given such an A, let

$$B = \{\alpha: \alpha \in A \,\&\, \alpha_0 \notin \Delta^1_{2n+1}(\alpha)\}$$

and check first that $B \neq \emptyset$; because if $B = \emptyset$, then this means precisely that $\alpha_0 \in Hull(A)$ and then

$$\widehat{\alpha_0}(s) = t \Leftrightarrow (\forall \beta)[\beta \in A \Rightarrow (\exists \alpha \in \Delta^1_{2n+1}(\beta))[\alpha = \alpha_0 \,\&\, \alpha(s) = t]]$$

which implies directly that α_0 has Π^1_{2n+1} graph and hence is in Δ^1_{2n+1} (recall that $\{\alpha: \alpha = \alpha_0\}$ is Π^1_{2n+1} by hypothesis). Check also that B is Σ^1_{2n+1} since

$$\alpha \in B \Leftrightarrow \alpha \in A \,\&\, [\forall \beta \in \Delta^1_{2n+1}(\alpha)[\beta \neq \alpha_0].$$

Fix a very good Π^1_{2n+1}-scale $\bar{\varphi}$ on A and check that it is in fact a Δ^1_{2n+1}-scale since A is Π^1_{2n}. The idea is to pick some α_B in A by choosing the leftmost branch on the tree determined by $\bar{\varphi}$ on B; it will not in general be true that $\alpha_B \in B$, but of course we only need some $\alpha_B \in A$.

As in the proof of 4F.20 then, put

$$\lambda_s = \text{least } \lambda \text{ such that } (\exists \beta)[\beta \in B \,\&\, \varphi_s(\beta) = \lambda],$$
$$B_s = \{\beta \in B: \varphi_s(\beta) = \lambda_s\}$$

and check by a simple very-good-scale argument that each $B_s \neq \emptyset$ and that there is a unique $\alpha_B \in A$ such that if $\alpha_0 \in B_0, \alpha_1 \in B_1, \ldots$, then $limit_{s\to\infty} \alpha_s = \alpha_B$. It remains to show that $\alpha_B \in \Delta^1_{2n+1}(\alpha_0)$.

Computing,

$$\alpha_B(s) = t \Leftrightarrow (\exists \beta)\{(\forall i)[(\beta)_i \in B_i] \,\&\, (\forall i)(\exists j \geq i)[(\beta)_j(s) = t]\}$$

so that it is enough to check that the relation

$$P(s, \beta) \Leftrightarrow \beta \in B_s$$

is in $\Sigma^1_{2n+1}(\alpha_0)$.

Put

$$Q(s, \beta) \Leftrightarrow (\forall \gamma)[\gamma \in B \rightarrow \beta \leq^*_{\varphi_s} \gamma]$$

so that

$$\beta \in B_s \Leftrightarrow \beta \in B \ \&\ Q(s, \beta)$$

and Q is obviously Π^1_{2n+1}; so is $\{(0, \alpha): \alpha = \alpha_0\}$ by hypothesis, so let $G \subseteq \omega \times (\omega \times \mathcal{N})$ be universal in Π^1_{2n+1}, let ψ be a Π^1_{2n+1}-norm on G and choose k_0, l_0, so that

$$\alpha = \alpha_0 \Leftrightarrow G(k_0, 0, \alpha)$$
$$Q(s, \beta) \Leftrightarrow G(l_0, s, \beta).$$

Suppose that there is some $\beta \in B$ so that $Q(s, \beta)$ and $\psi(k_0, 0, \alpha_0) < \psi(l_0, s, \beta)$ for some s; then

$$\alpha_0(m) = w \Leftrightarrow (\forall \alpha)[(l_0, s, \beta) \leq^*_\psi (k_0, 0, \alpha) \vee \alpha(m) = w]$$

and hence $\alpha_0 \in \Delta^1_{2n+1}(\beta)$ which contradicts $\beta \in B$. Hence for each s and each $\beta \in B$,

$$Q(s, \beta) \Leftrightarrow G(l_0, s, \beta) \ \&\ \psi(l_0, s, \beta) \leq \psi(k_0, 0, \alpha_0)$$
$$\Leftrightarrow \neg(k_0, 0, \alpha_0) <^*_\psi (l_0, s, \beta)$$

so that finally

$$\beta \in B_s \Leftrightarrow \beta \in B \ \&\ \neg(k_0, 0, \alpha_0) <^*_\psi (l_0, s, \beta). \qquad \dashv$$

This result takes a more interesting form if we add to it a simple observation.

6C.11. Assume $Det(\mathbf{\Delta}^1_{2n})$ and prove that the collection of Π^1_{2n+1}-singletons in $\mathcal{N}$ is prewellordered by the relation

$$\alpha \leq_{2n+1} \beta \Leftrightarrow \alpha \in \Delta^1_{2n+1}(\beta).$$

Thus if α_0 is $\leq_{2n+1}$-minimal among the non-Δ^1_{2n+1} singletons in Π^1_{2n+1}, then $\{x: x \in \Delta^1_{2n+1}(\alpha_0)\}$ is a basis for Σ^1_{2n+1}. (Martin–Solovay.)

Hint. Let $G(e, \alpha)$ be Π^1_{2n+1}-universal, suppose β_0, γ_0 are Π^1_{2n+1}-singletons and choose m, n so that

$$\beta = \beta_0 \leftrightarrow G(m, \beta),$$
$$\gamma = \gamma_0 \leftrightarrow G(n, \gamma).$$

Let φ be a Π^1_{2n+1}-norm on G and suppose that $\varphi(m, \beta_0) \leq \varphi(n, \gamma_0)$; then

$$\beta_0(s) = t \leftrightarrow (\exists \beta)[\neg(n, \gamma_0) <^*_\varphi (m, \beta) \ \&\ \beta(s) = t],$$

so that $\beta_0 \in \Delta^1_{2n+1}(\gamma_0)$. ⊣

In 7C.7 we will show that the Σ^1_1 set P of type 0 in Kleene's Basis Theorem 4E.8 can be chosen so that α_P is a Π^1_1-singleton; in this sense, the Martin–Solovay theorem above gives a natural generalization to all odd n of the result of Kleene.

Recall the definition of the ordinals $\boldsymbol{\delta}^1_n$ in the exercises of 4C. The next few results are easy, but they are interesting as they reveal the nature of the second periodicity theorem as a *structure theorem* for projective pointsets.

6C.12. Assume $Det(\boldsymbol{\Delta}^1_{2n})$; prove that every pointset in $\boldsymbol{\Sigma}^1_{2n+2}$ is $\boldsymbol{\delta}^1_{2n+1}$-Suslin and every pointset in $\boldsymbol{\Sigma}^1_{2n+1}$ is λ-Suslin for some $\lambda < \boldsymbol{\delta}^1_{2n+1}$. (Moschovakis, Kechris.)

Hint. For the first assertion, it is enough to show that $\boldsymbol{\Pi}^1_{2n+1}$ sets are $\boldsymbol{\delta}^1_{2n+1}$-Suslin (by 2B.2) and this follows immediately from 6C.4, 4C.9 and 2B.1.

For the second assertion, again it is enough to show that $\boldsymbol{\Pi}^1_{2n}$ sets are λ-Suslin for some $\lambda < \boldsymbol{\delta}^1_{2n+1}$. If A is in $\boldsymbol{\Pi}^1_{2n}$, then it admits a $\boldsymbol{\Pi}^1_{2n+1}$-scale $\bar{\varphi} = \{\varphi_i\}$ by 6C.4. Putting together all the prewellorderings $\leq^{\varphi_i}$ into one, we easily see that for some λ = order type of a $\boldsymbol{\Delta}^1_{2n+1}$ prewellordering $< \boldsymbol{\delta}^1_{2n+1}$ and each i, $|\varphi_i| \leq \lambda$; hence A is λ-Suslin. ⊣

6C.13. Assume $Det(\boldsymbol{\Delta}^1_{2n})$; prove that every $\boldsymbol{\Delta}^1_{2n+1}$ set is $\boldsymbol{\delta}^1_{2n+1}$-Borel and every $\boldsymbol{\Sigma}^1_{2n+2}$ set is the union of $\boldsymbol{\delta}^1_{2n+1}$ sets each of which is $\boldsymbol{\delta}^1_{2n+1}$-Borel. (Moschovakis [1971*a*].)

Hint. The first assertion is immediate from 6C.12 and the Suslin Theorem 2E.2. The second follows easily as in 2F.2 and 2F.3. ⊣

This last exercise generalizes part of the Suslin Theorem 2E.2 and the Sierpinski Theorem 2F.3 to all the odd levels of the hierarchy. How good this generalization is depends on how large the ordinals $\boldsymbol{\delta}^1_n$ are; this turns out to be a very difficult problem and we will come back to it in the next two chapters.

6D. The game quantifier $\Game$

With each pointset $P \subseteq \mathfrak{X} \times \mathcal{N}$ we associate the set $\Game P$,

$$x \in \Game P \Leftrightarrow (\Game \alpha) P(x, \alpha)$$
$$\Leftrightarrow \text{I wins the game } \{\alpha : P(x, \alpha)\};$$

$\Game$ is a set operation, a quantifier like $\exists^{\mathcal{N}}$ and $\forall^{\mathcal{N}}$. We read $\Game\alpha$ as "game α" or "gee α."

Our main result here is that under reasonable closure and determinacy hypotheses, the prewellordering property transfers from a pointclass Γ to

$$\Game\Gamma = \{\Game P : P \subseteq \mathfrak{X} \times \mathcal{N}, P \in \Gamma\}.$$

We will also show that if Γ is adequate and $Det(\boldsymbol{\Gamma})$ holds, then

$$\forall^{\mathcal{N}}\Gamma \subseteq \Gamma \Rightarrow \Game\Gamma = \exists^{\mathcal{N}}\Gamma,$$
$$\exists^{\mathcal{N}}\Gamma \subseteq \Gamma \Rightarrow \Game\Gamma = \forall^{\mathcal{N}}\Gamma,$$

so that

$$\Game\Sigma^0_1 = \Pi^1_1, \quad \Game\Pi^1_1 = \Sigma^1_2, \quad \Game\Sigma^1_2 = \Pi^1_3, \quad \Game\Pi^1_3 = \Sigma^1_4, \ldots.$$

Thus the transfer theorem gives an "explanation" of the periodicity phenomenon. It will also have several concrete applications in the next section.

It is often very useful to think of $\Game\alpha$ as an infinite string of alternating quantifiers.

$$(*) \qquad (\Game\alpha)P(x, \alpha) \Leftrightarrow \{(\exists a_0)(\forall a_1)(\exists a_2)(\forall a_3)\ldots\}P(x, (a_0, a_1, a_2, \ldots));$$

intuitively, I wins $\{\alpha : P(x, \alpha)\}$ if there is a beginning move a_0 for I such that whatever move a_1 II makes, there is a next move a_2 for I, such that ... etc ... eventually, $P(x, (a_0, a_1, a_2, \ldots))$ is true. Formally, $(*)$ is definition of the expression on the right in terms of $\Game$, for which we have a perfectly precise definition via strategies:

$$(\Game\alpha)P(x, \alpha) \Leftrightarrow (\exists\sigma)(\forall\tau)P(x, \sigma * \tau).$$

More generally, suppose

$$\bar{Q} = (Q_0, Q_1, Q_2, \ldots)$$

is an infinite string where for each i,

$$Q_i = \exists \quad \text{or} \quad Q_i = \forall$$

and let $A \subseteq \mathcal{N}$. We associate with $\bar{Q}$ and A a game G with two players, call them $\exists$ and $\forall$; a run of G consists of their choosing an infinite sequence $a_0, a_1, a_2, \ldots$ with $\exists$ choosing a_i if $Q_i = \exists$ and $\forall$ choosing a_i if $Q_i = \forall$. When the play $\alpha = (a_0, a_1, a_2, \ldots)$ is determined, we put

$$\exists \text{ wins} \Leftrightarrow \alpha \in A.$$

The notions of strategy, winning strategy, etc. are defined for these more general games in the obvious way. Of course the game $G_\omega(A)$ which we defined in 6A corresponds to the infinite alternating string

$$\exists, \forall, \exists, \forall, \ldots.$$

Now each such infinite string defines in a natural way a set operation,

$$\{(Q_0 a_0)(Q_1 a_1)(Q_2 a_2)\cdots\} P(x, (a_0, a_1, a_2, \ldots))$$
$$\Leftrightarrow \exists \text{ has a winning strategy in the game } G.$$

Let us call the string $\bar{Q} = (Q_0, Q_1, Q_2, \ldots)$ *recursive* if the function

$$f(i) = \begin{cases} 0 & \text{if } Q_i = \exists \\ 1 & \text{if } Q_i = \forall \end{cases}$$

is recursive.

6D.1. Lemma. *Let Γ be a pointclass closed under recursive substitution, let $\bar{Q} = Q_0, Q_1, Q_2, \ldots$ be a recursive infinite string of quantifiers and let $P \subseteq \mathfrak{X} \times \mathcal{N}$ be in Γ. Then the relation*

$$R(x) \Leftrightarrow \{(Q_0 a_0)(Q_1 a_1)(Q_2 a_2)\cdots\} P(x, (a_0, a_1, \ldots))$$

is in $\Game\Gamma$.

Moreover, if Det(Γ) holds, then for each x, the game G determined by $\bar{Q}$ and $\{\alpha : P(x, \alpha)\}$ is determined.

Proof. Define $g : \omega \to \omega$ by

$$g(i) = \begin{cases} 2i & \text{if } Q_i = \exists \\ 2i+1 & \text{if } Q_i = \forall, \end{cases}$$

so that g is recursive and (easily)

$$\{(Q_0 a_0)(Q_1 a_1)\cdots\}P(x, (a_0, a_1, \ldots))$$
$$\Leftrightarrow \{(\exists a_0)(\forall a_1)(\exists a_2)(\forall a_3)\cdots\}P(x, (a_{g(0)}, a_{g(1)}, \ldots)$$
$$\Leftrightarrow (\Game \alpha)P(x, i \mapsto \alpha(g(i))). \quad -$$

This simple lemma implies directly all the closure properties of the pointclass $\Game\Gamma$.

6D.2. THEOREM. *If Γ is an adequate pointclass, then the following hold*

(i) *$\Game\Gamma$ is adequate and closed under $\exists^\omega$ and $\forall^\omega$.*

(ii) $\exists^{\mathcal{N}}\Gamma \subseteq \Game\Gamma$; $\forall^{\mathcal{N}}\Gamma \subseteq \Game\Gamma$.

(iii) $\Game\Gamma \subseteq \exists^{\mathcal{N}}\forall^{\mathcal{N}}\Gamma$.

(iv) *If Det($\mathbf{\Gamma}$) holds, then $\Game\Gamma \subseteq \forall^{\mathcal{N}}\exists^{\mathcal{N}}\Gamma$.*

(v) $\forall^{\mathcal{N}}\Gamma \subseteq \Gamma \Rightarrow \Game\Gamma = \exists^{\mathcal{N}}\Gamma$.

(vi) *If Det($\mathbf{\Gamma}$) holds and $\exists^{\mathcal{N}}\Gamma \subseteq \Gamma$, then $\Game\Gamma = \forall^{\mathcal{N}}\Gamma$.*

(vii) *If Γ is $\mathcal{Y}$-parametrized, then so is $\Game\Gamma$.*

PROOF. For (i) we use the lemma and the obvious equivalences

$$(\exists t)(\Game\alpha)P(x, t, \alpha) \Leftrightarrow \{(\exists t)(\exists a_0)(\forall a_1)(\exists a_2)\cdots\}P(x, t, (a_0, a_1, \ldots)),$$
$$(\forall t)(\Game\alpha)P(x, t, \alpha) \Leftrightarrow \{(\forall t)(\exists a_0)(\forall a_1)(\exists a_2)\cdots\}P(x, t, (a_0, a_1, \ldots)).$$

For (ii):

$$(\exists\beta)P(x, \beta) \Leftrightarrow \{(\exists b_0)(\exists b_1)(\exists b_2)\cdots\}P(x, (b_0, b_1, b_2, \ldots)),$$
$$(\forall\beta)P(x, \beta) \Leftrightarrow \{(\forall b_0)(\forall b_1)(\forall b_2)\cdots\}P(x, (b_0, b_1, b_1, \ldots)).$$

For (iii) and (iv) we use the coding of strategies by irrationals,

$$(\Game\alpha)P(x, \alpha) \Leftrightarrow (\exists\sigma)(\forall\tau)P(x, \sigma * \tau)$$
$$\Leftrightarrow (\forall\tau)(\exists\sigma)P(x, \sigma * \tau),$$

where the second equivalence depends on the determinacy of $\{\alpha : P(x, \alpha)\}$.

Finally, (v) and (vi) follow immediately from (ii) and (iii) and (vii) is trivial. $-$

We now come to the main result of this section. This is stated in a strong and detailed form because it will have applications later beyond the transfer of the prewellordering property from Γ to $\Game\Gamma$ which concerns us here.

6D.3. THE NORM-TRANSFER THEOREM FOR $\Game$ (Moschovakis). *Suppose Γ is an adequate pointclass, $Det(\mathbf{\Gamma})$ holds, $P \subseteq \mathfrak{X} \times \mathcal{N}$ is in Γ and*

$$Q(x) \Leftrightarrow \{(\forall a_0)(\exists a_1)(\forall a_2)\cdots\} P(x, (a_0, a_1, \ldots)).$$

If φ is a Γ-norm on P, then there exists a $\Game\Gamma$-norm ψ on Q such that

$$x \leq^*_\psi y \Leftrightarrow \{(\forall a_0)(\exists b_0)(\forall b_1)(\exists a_1)(\forall a_2)(\exists b_2)(\forall b_3)(\exists a_3)\cdots\} \\ (x, (a_0, a_1, \ldots)) \leq^*_\varphi (y, (b_0, b_1, \ldots)),$$

$$x <^*_\psi y \Leftrightarrow \{(\exists b_0)(\forall a_0)(\exists a_1)(\forall b_1)(\exists b_2)(\forall a_2)(\exists a_3)(\forall b_3)\cdots\} \\ (x, (a_0, a_1, \ldots)) <^*_\varphi (y, (b_0, b_1, \ldots)).$$

In particular, if Γ is adequate and normed and if $Det(\mathbf{\Gamma})$ holds, then $\Game\Gamma$ is also normed.

PROOF. Assume the hypotheses and for each x, y define the game $H(x, y)$ which is played as in Diagram 6D.1. There are two players, as usual, whom we have named F (first) and S (second). We have also indicated in the diagram which player makes each move. At the end of the game, plays α and β have been determined and

$$S \text{ wins the run} \Leftrightarrow (x, \alpha) \leq^*_\varphi (y, \beta)$$

i.e.

$$F \text{ wins the run} \Leftrightarrow \neg P(x, \alpha) \vee (y, \beta) <^*_\varphi (x, \alpha).$$

Put

$$x \leq^* y \Leftrightarrow S \text{ wins } H(x, y) \\ \Leftrightarrow \{(\forall a_0)(\exists b_0)(\forall b_1)(\exists a_1)\cdots\}[(x, (a_0, a_1, \ldots)) \leq^*_\varphi (y, (b_0, b_1, \ldots))].$$

By Lemma 6D.1 each $H(x, y)$ is determined and the relation $\leq^*$ is in $\Game\Gamma$.

In $H(x, y)$ we are (in effect) playing simultaneously two games, the one corresponding to the assertion

(1) $$Q(x) \Leftrightarrow \{(\forall a_0)(\exists a_1)(\forall a_2)\cdots\} P(x, (a_0, a_1, \ldots))$$

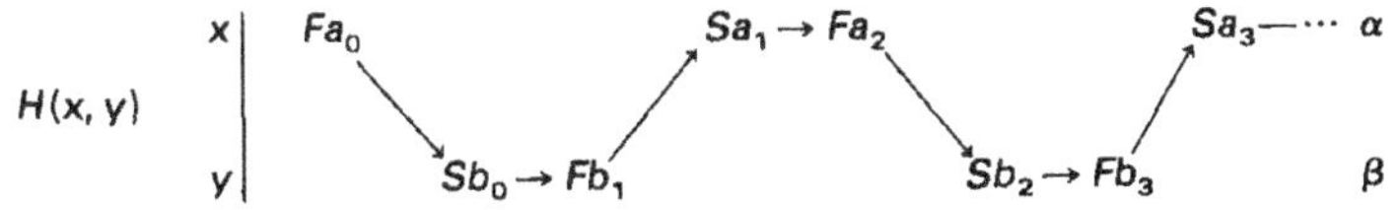

Diagram 6D.1. The game $H(x, y)$.

on the top board and on the bottom board the game associated with the assertion

(2) $$Q(y) \leftrightarrow \{(\forall b_0)(\exists b_1)(\forall b_2)\cdots\}P(y, (b_0, b_1, \ldots)).$$

Player S makes the moves of $\exists$ on the top board and the moves of $\forall$ on the bottom board; to win he must win on the top board, producing some α such that $P(x, \alpha)$ and either win also on the bottom board so that $\neg P(y, \beta)$ or at least insure $\varphi(x, \alpha) \leq \varphi(y, \beta)$.

The sequence of moves by which we have interweaved these two games in defining $H(x, y)$ is important for the argument.

We now verify in a sequence of lemmas that there is a norm ψ on Q such that

$$\leq^* = \leq^*_\psi$$

and such that $<^*_\psi$ satisfies the equivalence in the statement of the theorem.

Lemma 1. The relation $\leq^*$ is transitive.

Proof. Assume $x \leq^* y$ and $y \leq^* z$ and consider Diagram 6D.2 which describes a strategy of S in $H(x, z)$, given winning strategies for S in

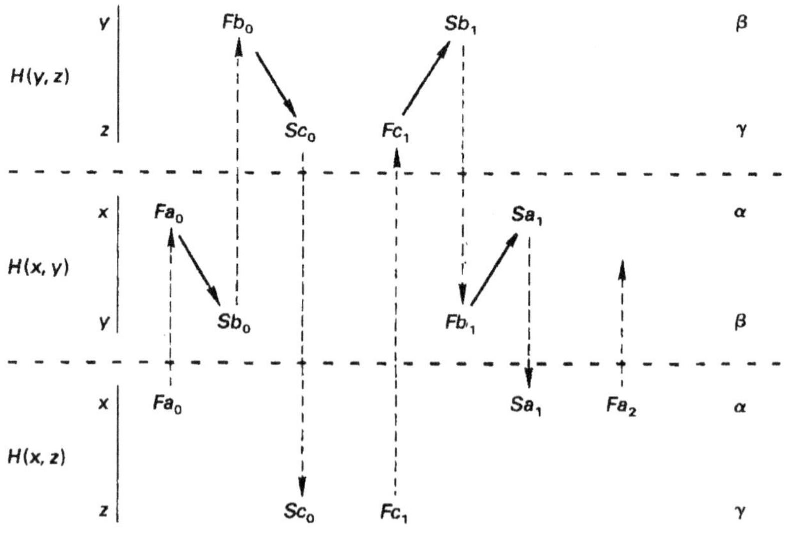

Diagram 6D.2.

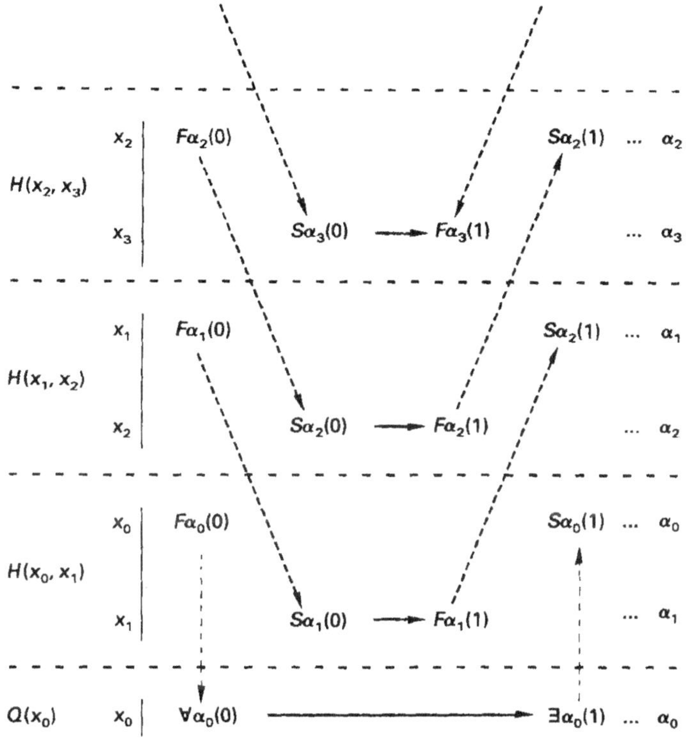

Diagram 6D.3.

$H(x, y)$ and $H(y, z)$. As usual, broken arrows indicate copies of moves and solid arrows show responses by the fixed winning strategies.

It is clear that this strategy in winning for S in $H(x, z)$ since at the end of the run we have plays α, β, γ and

$$(x, \alpha) \leq^*_\varphi (y, \beta); \qquad (y, \beta) \leq^*_\varphi (z, \gamma).$$

Lemma 2. There is no infinite sequence of points $x_0, x_1, x_2, \ldots$ such that $Q(x_0)$ and for every i, F wins $H(x_i, x_{i+1})$.

Proof. Assume towards a contradiction that there were such a sequence and fix winning strategies for F in all the games $H(x_i, x_{i+1})$. Fix also a winning strategy for $\exists$ in the game that verifies the assertion

$$Q(x_0) \leftrightarrow \{(\forall a_0)(\exists a_1)(\forall a_2)\cdots\} P(x_0, (a_0, a_1, \ldots))$$

and consider Diagram 6D.3; as usually, the moves of S (and $\forall$ in the game for $Q(x_0)$) are obtained by copying along the broken arrows and the moves for F and $\exists$ are by the fixed winning strategies.

At the end of the games plays $\alpha_0, \alpha_1, \alpha_2, \ldots$ have been determined and $\exists$ wins the game on the bottom line, so that we have $P(x, \alpha_0)$; however, F wins each $H(x_i, x_{i+1})$ so that we have $\neg[(x_0, \alpha_0) \leq_\varphi^* (x_1, \alpha_1)]$, $\neg[(x_1, \alpha_1) \leq_\varphi^* (x_2, \alpha_2)], \ldots$ and successively $P(x_1, \alpha_1)$, $P(x_2, \alpha_2), \ldots$ and

$$\varphi(x_0, \alpha_0) > \varphi(x_1, \alpha_1) > \varphi(x_2, \alpha_2) > \cdots$$

which is absurd.

Lemma 3. The restriction of $\leq^*$ to Q is a prewellordering.

Proof. We already know that $\leq^*$ is transitive. If $x, y \in Q$ and we do not have $x \leq^* y$, we have that F wins $H(x, y)$; if F also won $H(y, x)$, then the infinite sequence $x, y, x, y, \ldots$ would violate Lemma 7, so that S wins $H(y, x)$ and $y \leq^* x$. The assertion $x \leq^* x$ $(x \in Q)$ is proved similarly and then the lemma follows immediately.

Let ψ be the regular norm on Q associated with $\leq^*$, i.e.

$$x \leq^* y \Leftrightarrow \psi(x) \leq \psi(y) \quad (x, y \in Q).$$

Lemma 4. For every x, y,

$$x \leq_\psi^* y \Leftrightarrow x \leq^* y.$$

Proof. Assume first $x \leq_\psi^* y$, so that in particular $x \in Q$. If also $y \in Q$, then $x \leq^* y$ since on Q the relations $\leq_\psi^*$ and $\leq^*$ coincide by definition. If $y \notin Q$, have S play in $H(x, y)$ to insure $P(x, \alpha)$ on the top board and $\neg P(y, \beta)$ on the bottom board.

Conversely, assume $x \leq^* y$. If $x \in Q$, then immediately $x \leq_\psi^* y$, taking cases on $y \in Q$ or $y \notin Q$. But $x \leq^* y$ easily implies that $x \in Q$, since S's winning strategy in $H(x, y)$ restricted to the top board gives a winning strategy for $\exists$ in the game verifying $Q(x)$.

To prove that $<_\psi^*$ satisfies the formula in the statement of the theorem let $H'(x, y)$ be the game corresponding to this formula which is played as in Diagram 6D.4. The payoff is given by

$$F \text{ wins} \Leftrightarrow (x, \alpha) <_\varphi^* (y, \beta)$$

and we must show:

Lemma 5. For each x, y,

$$x <_\psi^* y \Leftrightarrow F \text{ wins } H'(x, y).$$

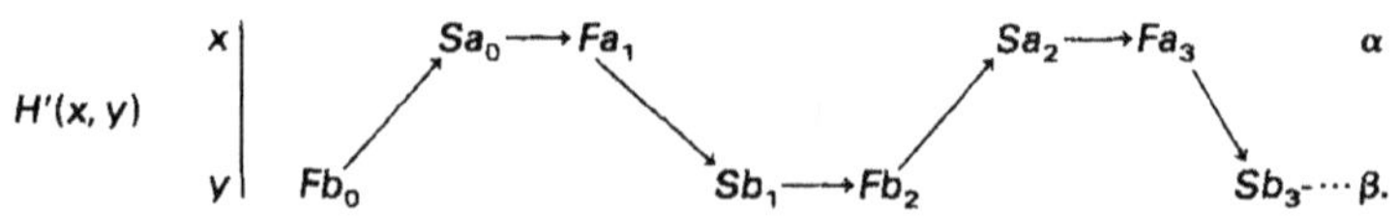

Diagram 6D.4. The game $H'(x, y)$.

Proof. Assume first $x <^*_\psi y$ and $x \in Q$ but $y \notin Q$. In this case F can easily win $H'(x, y)$ by playing on the top board to insure $P(x, \alpha)$ while playing on the bottom board to insure $\neg P(y, \beta)$.

If $x <^*_\psi y$ and both $x, y \in Q$, then by Lemma 4 we must have that $\neg(y \leq^* x)$ so that F wins the game $H(y, x)$. Assume also (towards a contradiction) that S wins $H'(x, y)$, fix winning strategies for these two games and fix also a strategy for $\exists$ in the game verifying that $y \in Q$. Now play these three games against each other as in Diagram 6D.5, where we indicate copied moves by broken arrows and moves by the winning strategies by solid arrows in the usual way.

After all the games have been played we have determined plays β_0, α_1, β_1, α_2, β_2,... and the following relations hold:

$P(y, \beta_0)$, since $\exists$ wins the game on the top row,

$$\neg(y, \beta_0) \leq^*_\varphi (x, \alpha_1), \quad \text{hence} \quad (x, \alpha_1) <^*_\varphi (y, \beta_0),$$
$$\text{since } F \text{ wins } H(y, x),$$
$$\neg(x, \alpha_1) <^*_\varphi (y, \beta_1), \quad \text{hence} \quad (y, \beta_1) \leq^*_\varphi (x, \alpha_1),$$
$$\text{since } S \text{ wins } H'(x, y)$$

etc. But then we obviously have

$$\varphi(y, \beta_0) > \varphi(x, \alpha_1) \geq \varphi(y, \beta_1) > \varphi(x, \alpha_2) > \cdots$$

which is absurd.

Finally suppose F wins $H'(x, y)$ but $\neg(x <^*_\psi y)$. Since F's winning strategy in $H'(x, y)$ restricted to the top board implies immediately that $x \in Q$, we must then have that $y \leq^*_\psi x$ so that S wins $H(y, x)$. Fix then winning strategies for F in $H'(x, y)$ and S in $H(y, x)$ and play them against each other as in Diagram 6D.6. We obtain plays α, β such that

$$(x, \alpha) <^*_\varphi (y, \beta) \;\&\; (y, \beta) \leq^*_\varphi (x, \alpha)$$

which is absurd.

To prove the second assertion of the theorem notice that

$$\begin{aligned}(\Game\alpha)P(x, \alpha) &\Leftrightarrow \{(\exists a_0)(\forall a_1)(\exists a_2)\cdots\}P(x, (a_0, a_1, \ldots))\\ &\Leftrightarrow \{(\forall b_0)(\exists a_0)(\forall a_1)(\exists a_2)\cdots\}P(x, (a_0, a_1, a_2, \ldots))\\ &\Leftrightarrow \{(\forall a_0)(\exists a_1)(\forall a_2)\cdots\}P^*(x, (a_0, a_1, a_2, \ldots))\end{aligned}$$

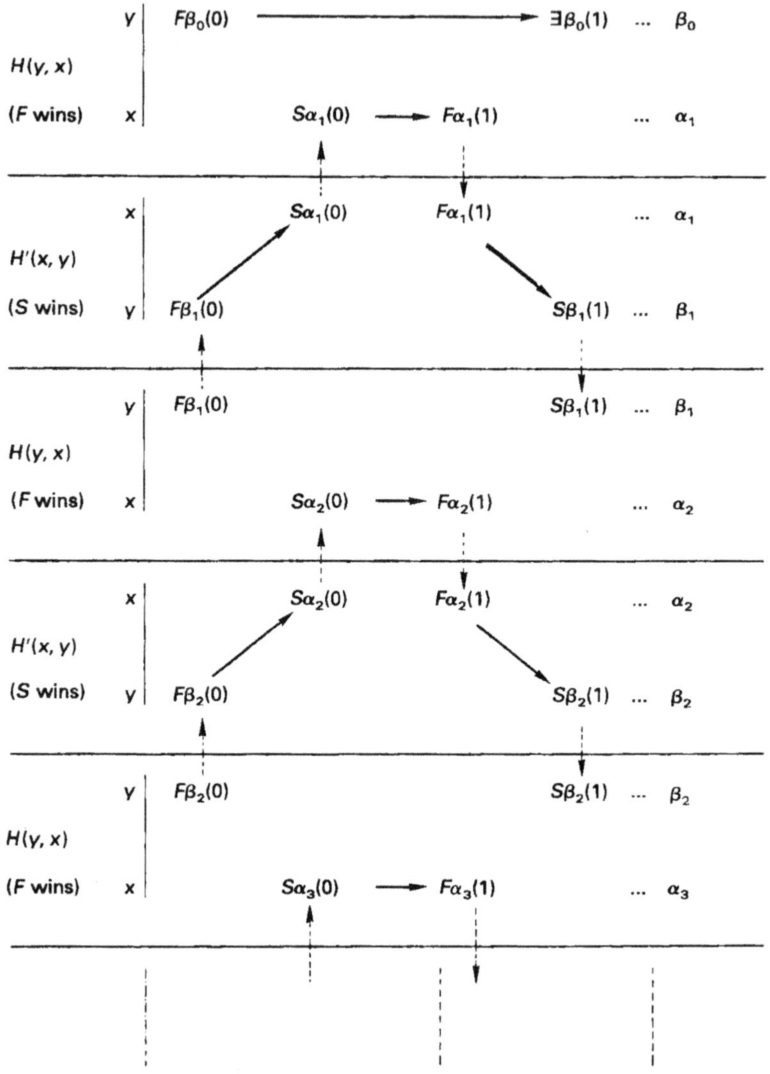

Diagram 6D.5.

with

$$P^*(x, \alpha) \leftrightarrow P(x, \alpha^\star),$$

so by the first part, if $P \in \Gamma$, then $\mathfrak{D}P$ admits a $\mathfrak{D}\Gamma$-norm. ⊣

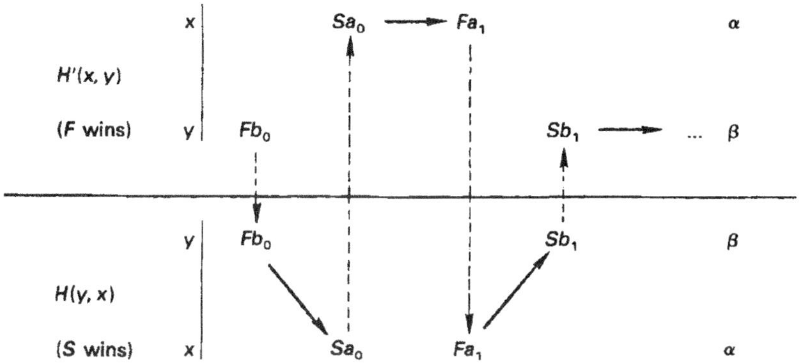

Diagram 6D.6.

This result combines with 6D.2 to give us a collection of new and interesting Spector pointclasses.

6D.4. THEOREM (i) *If $\Pi^0_1 \subseteq \Gamma$ and Γ is adequate, closed under $\exists^\omega$, normed and ω-parametrized and if Det(Γ) holds, then $\Game\Gamma$ is a Spector pointclass.*

(ii) *$\Game\Sigma^0_1 = \Pi^1_1$ and $\Game\Sigma^0_2$ are Spector pointclasses and so is each $\Game\Sigma^0_n(n \geq 3)$ granting Det($\mathbf{\Sigma}^0_n$).* (Kechris–Moschovakis.)

PROOF. Because of 6D.1 and 6D.2 we need only check the *substitution property* for $\Game\Gamma$, as this was explained in 4C.

Suppose then that

$$Q(y) \leftrightarrow (\Game\alpha)Q^*(y, \alpha)$$

and

$$P(x, s) \leftrightarrow (\Game\beta)P^*(x, s, \beta),$$

where P computes some partial function $f: \mathcal{X} \to \mathcal{Y}$ on its domain. We must find some $R \subseteq \mathcal{X}$ in $\Game\Gamma$ such that

$$f(x)\downarrow \rightarrow [R(x) \leftrightarrow Q(f(x))].$$

Fix a recursive surjection

$$\pi : \mathcal{N} \twoheadrightarrow \mathcal{Y}$$

and consider the following game $G(x)$ in which the two players I and II define sequences α, β, γ, δ as indicated in Diagram 6D.7. At the end of

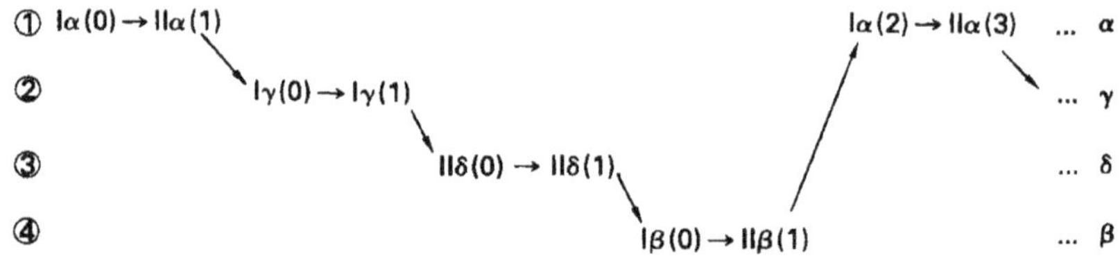

Diagram 6D.7.

the run,

$$\begin{aligned}\text{I wins} \leftrightarrow\ & Q^*(\pi(\gamma), \alpha)\\ & \& \{(\forall j)[\delta(j)=0 \rightarrow j \text{ is odd}]\\ & \quad \vee (\exists j)[\delta(j)>0 \ \&\ j \text{ is even} \ \&\ (\forall i<j)[i \text{ even} \rightarrow \delta(i)=0]\\ & \quad\quad \& [\pi(\gamma)\notin N(\delta(j)-1) \vee P^*(x, \delta(j)-1, t \mapsto \beta(j+t))]]\}.\end{aligned}$$

Intuitively, I is attempting to define some $y=\pi(\gamma)$ by giving γ and then win the game $\{\alpha: Q^*(y, \alpha)\}$ so as to guarantee $(\Game\alpha)Q^*(y, \alpha)$; he must give the correct y however, so that

$$(*) \qquad (\forall s)[y \in N_s \leftrightarrow (\Game\beta)P^*(x, s, \beta)]$$

and $y=f(x)$. To insure this, II is allowed to give $\delta(0)$, $\delta(1)$,... which may all be 0, but at any even j he may play

$$\delta(j)=s+1,$$

at which point either $y=\pi(\gamma)\notin N_s$ or I must win the game $\{\beta: P^*(x, s, \beta)\}$ insuring $(\Game\beta)P^*(x, s, \beta)$.

We claim that if $f(x)=y$ so that $(*)$ holds, then

$$Q(y) \rightarrow \text{I wins } G(x);$$

simply have I play γ so that $\pi(\gamma)=y$, play on board ① so that ultimately $Q^*(y, \alpha)$ and if and when II plays some $\delta(j)=s+1$ (with j even and $i<j \rightarrow \delta(i)=0$) and with $y \in N_s$, have I play on board ④ to define some $\beta'=t \mapsto \beta(j+t)$ so that $P^*(x, s, \beta')$.

Conversely, if $f(x)=y$ and $(*)$ holds,

$$\text{I wins } G(x) \rightarrow Q(y).$$

To check this consider Diagram 6D.8 where I plays in $G(x)$ by his winning strategy. We claim that $\pi(\gamma)=y$ and $Q^*(y, \alpha)$, so that this defines a winning strategy for I in $\{\alpha: Q^*(y, \alpha)\}$ insuring $Q(y)$.

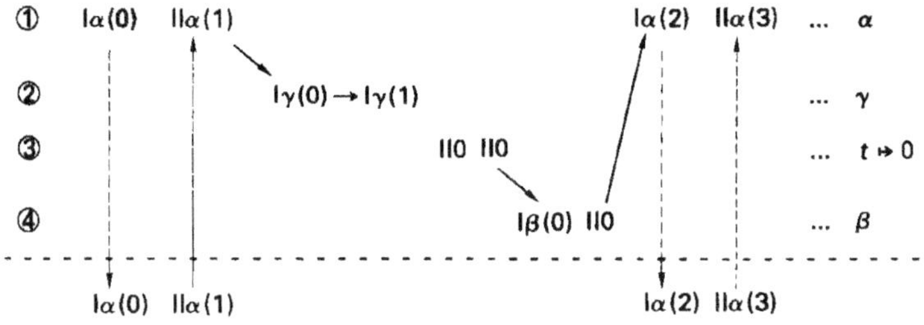

Diagram 6D.8.

To verify this, notice that if $\pi(\gamma) \neq y$, then for some s, $y \notin N_s$ but $\pi(\gamma) \in N_s$. Choose j large enough and even so that

$$\bar{\gamma}(j) = \bar{\gamma}'(j) \Longrightarrow \pi(\gamma') \in N_s$$

and have II play against I (who is using his fixed winning strategy) by giving

$$\delta(i) = 0 \quad \text{for} \quad i < j, \qquad \delta(j) = s + 1$$

and then play on board ④ to insure $\neg P^*(x, s, \beta')$, which he can do since $y \notin N_s$. No matter what γ' is played by I, we have $\pi(\gamma') \in N_s$, so I loses the run, contradicting the assumption that he is following a winning strategy.

Once we know that $\pi(\gamma) = y$ and $Q^*(\pi(\gamma), \alpha)$ (since I wins $G(x)$) we have $Q^*(y, \alpha)$ as required.

It follows from these claims that if $f(x)\downarrow$ and $f(x) = y$, then

$$Q(y) \Leftrightarrow \text{I wins } G(x)$$

and we can take

$$R(x) \Leftrightarrow \text{I wins } G(x);$$

this is in $⅁\Gamma$ since $G(x)$ is a game defined by an infinite recursive string of quantifiers and payoff in Γ, by 6D.1.

The second assertion of the theorem follows immediately, except for the part $⅁\Sigma^0_1 \subseteq \Pi^1_1$; for this we express $⅁P$ with P in Σ^0_1 using strategies

$$(⅁\alpha)P(x, \alpha) \Leftrightarrow (\forall\tau)(\exists\sigma)P(x, \sigma * \tau)$$

and then we use closure of Σ^0_1 under $\exists^{\mathcal{N}}$, 3C.14. ⊣

We will prove $Det(\mathbf{\Sigma}^0_n)$ in 6*F*, so no determinacy hypotheses are needed to insure that each $⅁\Sigma^0_n$ $(n \geq 2)$ is a Spector pointclass; in any case, we know this now for $⅁\Sigma^0_2$ by 6A.3.

These pointclasses are quite interesting and we will come back to them in the exercises of the next section. See also 7C.10 for an important characterization of $\mathfrak{D}\Sigma_2^0$ due to Solovay.

EXERCISES

We stated 6D.3 directly for relations of the form

$$\{(\forall a_0)(\exists a_1)(\forall a_2)(\exists a_3)\cdots\}P(x,(a_0,a_1,\ldots))$$

rather than

$$(\mathfrak{D}\alpha)P(x,\alpha)\Leftrightarrow\{(\exists a_0)(\forall a_1)(\exists a_2)(\forall a_3)\cdots\}P(x,(a_0,a_1,\ldots)).$$

This was because we will need the explicit formulas of 6D.3 in the next section, but of course there are similar formulas for $\mathfrak{D}P$.

6D.5. Suppose Γ is an adequate pointclass, $Det(\mathbf{\Gamma})$ holds, $P\subseteq\mathfrak{X}\times\mathcal{N}$ is in Γ and

$$Q(x)\Leftrightarrow(\mathfrak{D}\alpha)P(x,\alpha).$$

If φ is a Γ-norm on P, show that there exists a $\mathfrak{D}\Gamma$-norm ψ on Q such that

$$x\leq_\psi^* y\Leftrightarrow\{(\exists a_0)(\forall a_1)(\forall b_0)(\exists b_1)(\exists a_2)(\forall a_3)(\forall b_2)(\exists b_3)\cdots\}$$
$$(x,(a_0,a_1,\ldots))\leq_\psi^*(y,(b_0,b_1,\ldots)),$$

$$x<_\psi^* y\Leftrightarrow\{(\forall b_0)(\exists b_1)(\exists a_0)(\forall a_1)(\forall b_2)(\exists b_3)(\exists a_2)(\forall a_3)\cdots\}$$
$$(x_1,(a_0,a_1,\ldots))<_\varphi^*(y,(b_0,b_1,\ldots)).\quad ⊣$$

In the next section we will show that the scale property also transfers from Γ to $\mathfrak{D}\Gamma$. Here we confine ourselves to a restatement of the second periodicity theorem in terms of $\mathfrak{D}$.

6D.6. Assume that Γ is adequate and $P\subseteq\mathfrak{X}\times\mathcal{N}$ is in Γ of type 1 and admits a Γ-scale. Prove that $\exists^{\mathcal{N}}P$ admits a $\mathfrak{D}\Gamma$-scale; prove also that if $Det(\mathbf{\Delta})$ holds, then $\forall^{\mathcal{N}}P$ admits a $\mathfrak{D}\Gamma$-scale.

Hint. Look up the proofs of 6C.1, 6C.3 and 6D.2. ⊣

6E. The third periodicity theorem; definable winning strategies

Suppose $A\subseteq\mathcal{N}$ is a Σ_2^1 set and player I has a winning strategy in the game A. Now the set W of strategies winning for I is Π_3^1,

$$\sigma\in W\Leftrightarrow(\forall\beta)(\sigma*[\beta]\in A),$$

hence it has a Δ^1_4 member (if $Det(\mathbf{\Delta}^1_2)$ holds) by the Basis Theorem 6C.6. We will show here that in fact, if $Det(\mathbf{\Sigma}^1_2)$ holds, then I has a Δ^1_3 winning strategy. In its proper, general context, this is the last basic result we need in order to extend most of the structure theory of Π^1_1 and Σ^1_2 to all the higher levels—and to many other Spector pointclasses besides.

For the first time here we will use the existence of scales as a hypothesis to obtain results other than uniformization. Actually semiscales will suffice.

A *Γ-semiscale* on a pointset P is a sequence $\bar{\varphi} = \{\varphi_n\}$ of norms on P which is a semiscale in the sense of 2B and such that the relations

$$R(n, x, y) \Leftrightarrow x \leq^*_{\varphi_n} y,$$
$$S(n, x, y) \Leftrightarrow x <^*_{\varphi_n} y,$$

are in Γ. As with scales (which have the additional lower semicontinuity property), we call $\bar{\varphi}$ *very good* if the following two conditions hold:

(1) If $x_0, x_1, \ldots$ are in P and if for each n and all large i, $\varphi_n(x_i) = \lambda_n$, then there exists some $x \in P$ such that $limit_{i \to \infty} x_i = x$.

(2) If x, y are in P and $\varphi_n(x) \leq \varphi_n(y)$, then for each $i \leq n$, $\varphi_i(x) \leq \varphi_i(y)$.

It is very easy to check (as in 4E.2) that if a pointset P of type 1 in an adequate pointclass Γ admits a Γ-semiscale, then P admits a very good Γ-semiscale.

6E.1. THE THIRD PERIODICITY THEOREM (Moschovakis [1973]). *Suppose Γ is adequate, $Det(\mathbf{\Gamma})$ holds and $A \subseteq \mathcal{N}$ is in Γ and admits a Γ-semiscale; if player I wins the game A, then I has a $\mathfrak{D}\Gamma$-recursive winning strategy.*

PROOF. Fix a very good Γ-semiscale $\bar{\varphi}$ on A and for each even integer k put

$$\begin{aligned} u \in W_k \Leftrightarrow\ & Seq(u)\ \&\ lh(u) = k+1 \\ & \&\ \{(\forall a_{k+1})(\exists a_{k+2})(\forall a_{k+3})(\exists a_{k+4})\cdots\} \\ & \qquad ((u)_0, (u)_1, \ldots, (u)_k, a_{k+1}, a_{k+2}, \ldots) \in A \end{aligned}$$

so that W_k consists of all the *winning positions for* I in the game A—when it is next II's turn to play. Clearly each W_k is in $\mathfrak{D}\Gamma$.

If $u = \langle a_0, \ldots, a_k \rangle$, $v = \langle b_0, \ldots, b_k \rangle$, let $H_k(u, v)$ be the game played as in Diagram 6E.1. At the end of each run plays

$$\alpha = u^\frown \alpha^* = (a_0, a_1, \ldots, a_k, a_{k+1}, a_{k+2}, \ldots),$$
$$\beta = v^\frown \beta^* = (b_0, b_1, \ldots, b_k, b_{k+1}, b_{k+2}, \ldots),$$

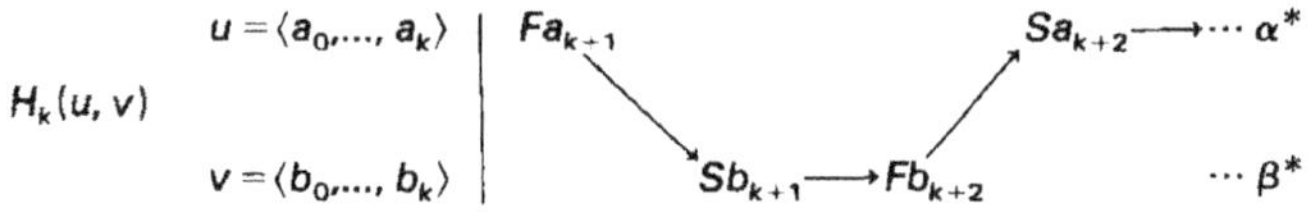

Diagram 6E.1.

have been constructed and

$$S \text{ wins the run} \Leftrightarrow \alpha \leq^*_{\varphi_k} \beta$$

i.e.

$$F \text{ wins the run} \Leftrightarrow \alpha \notin A \vee \beta <^*_{\varphi_k} \alpha.$$

If we rewrite the definition of W_k in the form

$$u \in W_k \Leftrightarrow \{(\forall a_{k+1})(\exists a_{k+2}) \cdots\}$$
$$[Seq(u) \;\&\; lh(u) = k+1 \;\&\; ((u)_0, \ldots, (u)_k, a_{k+1} \cdots) \in A],$$

it becomes completely obvious that this is a special case of the construction in 6D.3 with

$$P(u, \alpha) \Leftrightarrow Seq(u) \;\&\; lh(u) = k+1 \;\&\; ((u)_0, \ldots, (u)_k, \alpha(0), \ldots) \in A.$$

Thus we know that there is a $\Game\Gamma$-norm ψ_k on W_k such that for each u, v,

$$u \leq^*_{\psi_k} v \Leftrightarrow S \text{ wins the game } H_k(u, v).$$

It is worth for the motivation here to recall the meaning of the game $H_k(u, v)$.

In $H_k(u, v)$, we are in effect playing simultaneously two runs of the game A. On the top board we are given the starting position $a_0, \ldots, a_k$ and S makes the moves of I while F makes the moves of II; on the bottom board we start from $b_0, \ldots, b_k$ and the roles are reversed, with F making the moves for I and S making the moves for II. Now S wins $H_k(u, v)$ if he wins A (as I) on the top board and either he also wins A (as II) on the bottom board or at least he does not lose there with an ordinal $\varphi_k(\beta)$ less than the ordinal $\varphi_k(\alpha)$ assigned to his winning play on the top board.

It is obvious that the relations

$$R(k, u, v) \Leftrightarrow Seq(u) \;\&\; Seq(v) \;\&\; lh(u) = lh(v) = k+1 \;\&\; u \leq^*_{\psi_k} v,$$
$$S(k, u, v) \Leftrightarrow Seq(u) \;\&\; Seq(v) \;\&\; lh(u) = lh(v) = k+1 \;\&\; u <^*_{\psi_k} v,$$

are both in $\Game\Gamma$. To simplify notation we will write

$$u \leq^*_k v \Leftrightarrow u \leq^*_{\psi_k} v; \qquad u <^*_k v \Leftrightarrow u <^*_{\psi_k} v.$$

Call an odd sequence code $u = \langle a_0, \dots, a_{k-1}, a_k \rangle$ *minimal* if for every b,

$$\langle a_0, \dots, a_{k-1}, a_k \rangle \leq_k^* \langle a_0, \dots, a_{k-1}, b \rangle.$$

The next lemma is the crucial argument in the proof of this theorem.

Lemma. Suppose $\alpha = (a_0, a_1, a_2, \dots)$ is such that for every even k, the initial segment $\langle a_0, \dots, a_k \rangle$ is minimal; then $\alpha \in A$.

Proof. The argument is quite similar to the key lemma in the proof of the Second Periodicity Theorem 6C.3, but a bit more elaborate. We will construct a master diagram of games $H_k(u, v)$, one for each even k, which will determine plays $\alpha_0, \alpha_2, \alpha_4, \dots$ in A such that $limit_{i\to\infty} \alpha_{2i} = \alpha$ and all norms $\varphi_k(\alpha_{2i})$ are eventually fixed. This will imply that $\alpha \in A$, since $\bar{\varphi}$ is a semiscale.

For each even k, we will have $u = \langle a_0, \dots, a_k \rangle$ in the game $H_k(u, v)$ which we will play; but we will take $v = \langle a_0, \dots, a_{k-1}, b \rangle$ with a certain b which will depend on the various moves which are made as the construction of the diagram progresses.

To begin with, fix a winning strategy for S in every game of the form $H_k(\langle a_0, \dots, a_k \rangle, \langle a_0, \dots, a_{k-1}, b \rangle)$. Fix also some winning strategy for I in A.

Suppose I's winning strategy in A starts with a move $\alpha_0(0)$. Take $u_0 = \langle a_0 \rangle$, $v_0 = \langle \alpha_0(0) \rangle$ and start the game $H_0 = H_0(u_0, v_0)$ as in Diagram 6E.2, with F playing a_1.

It is obvious how this *Stage* 0 of the construction is built up. The play $\alpha_2(2)$ determined by S's winning strategy in H_0 is important, as it initiates *Stage* 2 of the construction. Put

$$u_2 = \langle a_0, a_1, a_2 \rangle, \qquad v_2 = \langle a_0, a_1, a_2(2) \rangle$$
$$H_2 = H_2(u_2, v_2)$$

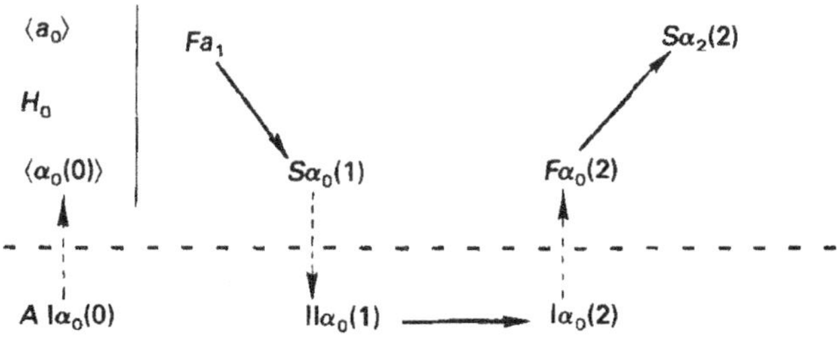

Diagram 6E.2. *Stage* 0.

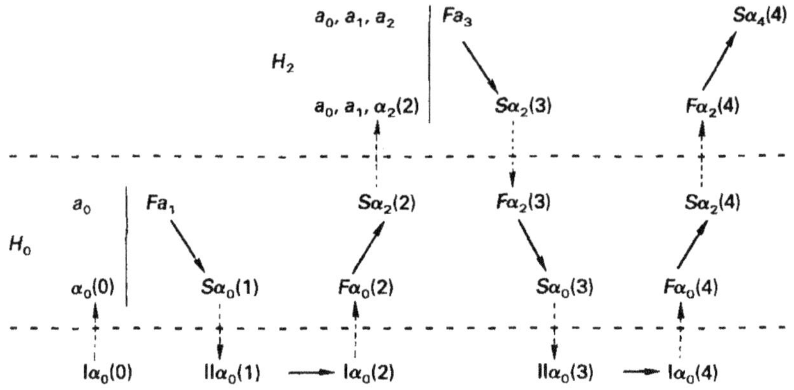

Diagram 6E.3. *Stage* 2.

and start H_2 with F playing a_3. The other moves in this second stage are filled in by copying and using the fixed winning strategies in the obvious way; see Diagram 6E.3.

Now the key move is the last one by S in H_2, $\alpha_4(4)$. Put

$$u_4 = \langle a_0, a_1, a_2, a_3, a_4 \rangle, \qquad v_4 = \langle a_0, a_1, a_2, a_3\ \alpha_4(4) \rangle$$
$$H_4 = H_4(u_4, v_4)$$

and start H_4 with F playing a_5. This will be *Stage* 4 of the construction.

It is clear how we can continue this construction successively with stages numbered by the even integers 0, 2, 4, 6,... . At stage $2n$ we determine the values $\alpha_0(i), \alpha_2(i), \dots, \alpha_{2n+2}(i)$ for all $i \le 2n+2$ and using $\alpha_{2n+2}(2n+2)$ we can start the next stage. At the end plays $\alpha_0, \alpha_2, \alpha_4, \dots$ are determined and we have established that I wins A, so that $\alpha_0 \in A$ and S wins every H_{2n}, so that all α_{2n} are in A and

$$\varphi_{2n+2}(\alpha_{2n+2}) \le \varphi_{2n+2}(\alpha_{2n}).$$

Using the fact that $\bar{\varphi}$ is a very good semiscale, it is easy to check (as in the proof of 6C.3) that all the norms $\varphi_i(\alpha_{2n})$ are ultimately constant, as $n \to \infty$. It follows that $\alpha = limit_{n\to\infty}\, \alpha_{2n} \in A$, completing the proof of the lemma.

The import of the lemma is that I can win A by playing each time so that the successive initial pieces of the run

$$\langle a_0 \rangle, \langle a_0, a_1, a_2 \rangle, \langle a_0, a_1, a_2, a_3, a_4 \rangle, \dots$$

are minimal. We will complete the proof of the theorem by verifying that he can do this following a $\mathfrak{D}\Gamma$-recursive strategy.

Let u, v vary over sequence codes (integers) and put

$$\begin{aligned} Min(u) \Leftrightarrow\ & Seq(u)\ \&\ lh(u) \text{ is odd } \&\ u \in W_{lh(u)\dot{-}1} \\ & \&\ (\forall v)\{[Seq(v)\ \&\ lh(v) = lh(u) \\ & \qquad \&\ (\forall i < lh(u) \dot{-} 1)[(u)_i = (v)_i]] \\ & \qquad \Rightarrow S \text{ wins } H_{lh(u)\dot{-}1}(u, v)\}. \end{aligned}$$

Using 6D.1, the relation $Min(u)$ is easily in $\mathfrak{D}\Gamma$.

Call a sequence code $\langle a_0, \ldots, a_{k-1}, a_k\rangle$ *best* (k *even*) *if it is minimal and* if in addition there is no $b < a_k$ so that $\langle a_0, \ldots, a_{k-1}, b\rangle$ is minimal. Thus

$$\begin{aligned} & Best(\langle a_0, \ldots, a_{k-1}, a_k\rangle) \\ & \quad \Leftrightarrow Min(\langle a_0, \ldots, a_{k-1}, a_k\rangle) \\ & \qquad \&\ (\forall b < a_k)[(a_0, \ldots, a_{k-1}, a_k) <^*_k (a_0, \ldots, a_{k-1}, b)]. \end{aligned}$$

Since the relation $<^*_k$ is in $\mathfrak{D}\Gamma$, so is the relation $Best(u)$.

Finally we get a $\mathfrak{D}\Gamma$-recursive winning strategy for I by putting

$$\begin{aligned} & \sigma(a_0, c_0, \ldots, a_{k-1}, c_{k-1}) = a_k \\ & \quad \Leftrightarrow (\exists a'_0) \cdots (\exists a'_k)[(\forall j < k) Best(\langle a'_0, c_0, \ldots, a'_j\rangle) \\ & \qquad \&\ Best(\langle a'_0, c_0, \ldots, a'_{k-1}, c_{k-1}, a_k\rangle)]. \end{aligned}$$

⊣

There are many applications of this theorem which we will pursue in the exercises. For some of them we will need to go into the proof of 6E.1 and use specifically the notions of a *minimal* or a *best* strategy. It is important to notice that these are defined for a given game $A \subseteq \mathcal{N}$ (which I can win) and a given semiscale $\bar{\varphi}$ on A independently of any definability hypotheses; σ is *minimal* (or *best*) if each odd initial segment $\langle a_0, \ldots, a_k\rangle$ of a play following σ is minimal (or best).

Let us just put down here the main corollary of 6E.1 for the Kleene pointclasses.

6E.2. COROLLARY (Moschovakis [1973]). *If* $Det(\mathbf{\Sigma}^1_{2n})$ *holds and if* I *wins a* $\Sigma^1_{2n}(x)$ *game* A, *then* I *has a winning strategy in* $\Delta^1_{2n+1}(x)$.

Similarly, if $Det(\mathbf{\Delta}^1_{2n})$ *holds and* I *wins a* $\Pi^1_{2n+1}(x)$ *game* A, *then* I *has a winning strategy in* $\Delta^1_{2n+2}(x)$.

In particular, granting PD, for each Δ^1_n *game* A *either* I *or* II *has a winning* Δ^1_{n+1} *winning strategy, and similarly with* $\Delta^1_n(x)$, $\Delta^1_{n+1}(x)$.

PROOF. The first assertion comes directly from 6E.1, taking $\Gamma = \Sigma^1_{2n}(x)$ so that $\mathfrak{D}\Gamma = \mathfrak{D}\Sigma^1_{2n}(x) = \Pi^1_{2n+1}(x)$ by 6D.2 and using the fact that if σ is $\Pi^1_{2n+1}(x)$-recursive then surely σ is in $\Delta^1_{2n+1}(x)$.

The second assertion is a trivial consequence of the Basis Theorem 6C.6. If I wins a $\Pi^1_{2n+1}(x)$ set A, then the set

$$P = \{\sigma : (\forall\tau)A(\sigma * \tau)\}$$

is non-empty and in $\Pi^1_{2n+1}(x)$, so it has a member in $\Delta^1_{2n+2}(x)$. ⊣

Taking $n = 0$ in this Corollary, we get in particular that if I wins a Σ^1_0(i.e. a Σ^0_1) game, then I has a Δ^1_1 winning strategy. It is not too hard to see this directly, without the elaborate analysis of games of 6E.1.

EXERCISES

First we put down two simple results which are needed for completeness.

6E.3. Prove that if Γ is adequate and a pointset P of type 1 in Γ admits a Γ-semiscale, then P admits a very good Γ-semiscale.

Hint. See the proof of 4E.2. ⊣

6E.4. Suppose Γ is adequate, $Det(\boldsymbol{\Gamma})$ holds and $A \subseteq \mathcal{N}$ is in Γ and admits a Γ-semiscale, let

$$\bar{Q} = (Q_0, Q_1, Q_2, \ldots)$$

be a recursive infinite string of quantifiers and let G be the game determined by A and $\bar{Q}$; prove that if $\exists$ wins, then $\exists$ has a $\mathfrak{D}\Gamma$-recursive winning strategy.

Hint. See the proof of 6D.1 and apply the Third Periodicity Theorem 6E.1. ⊣

As a first application of the third periodicity theorem, let us show that Harrison's result 4F.1 generalizes to all odd levels.

6E.5. Assume $Det(\boldsymbol{\Sigma}^1_{2n})$ and suppose $P \subseteq \mathfrak{X}$ is in Σ^1_{2n+1} and has at least one member not in Δ^1_{2n+1}; prove that P has a perfect subset.

Similarly, if P is $\Sigma^1_{2n+1}(z)$ with some member not in $\Delta^1_{2n+1}(z)$, then P has a perfect subset.

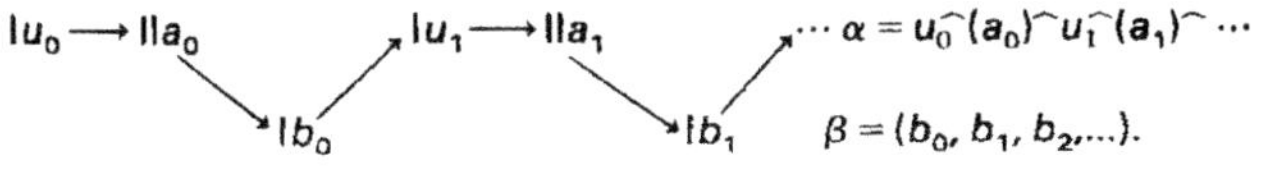

Diagram 6E.4.

In particular, if $P \subseteq \mathfrak{X}$ is $\Sigma^1_{2n+1}(z)$ and countable, then $P \subseteq \Delta^1_{2n+1}(z) \cap \mathfrak{X}$.

(The Effective Perfect Set Theorem for the odd levels, Martin.)

Hint. If we know the result for subsets of $\mathcal{C} = {}^{\omega}2$ and $P \subseteq \mathfrak{X}$, let $\pi : \mathcal{C} \rightarrowtail\!\!\!\!\rightarrow \mathfrak{X}$ be a Δ^1_1 isomorphism, take $Q = \pi^{-1}[P]$ and apply the result to Q to get a perfect subset K—now use the fact that $\pi[K]$ is an uncountable Borel subset of P and hence has a perfect subset.

If $P \subseteq \mathcal{C}$ and

$$P(\alpha) \Leftrightarrow (\exists\beta)Q(\alpha, \beta)$$

with Q in Π^1_{2n}, consider the game G played as in Diagram 6E.4. Here each u_i is a finite (non-empty binary) sequence

$$u_i = c^i_0, \ldots, c^i_{k_i},$$

a_j is 0 or 1 and $b_j \in \omega$, so that in effect I and II define an infinite binary sequence

$$\alpha = (c^0_0, \ldots, c^0_{k_0}, a_0, c^1_0, \ldots, c^1_{k_1}, a_1, \ldots)$$

and an irrational

$$\beta = (b_0, b_1, \ldots).$$

At the end of the game,

$$\text{I wins} \Leftrightarrow Q(\alpha, \beta).$$

Argue that this game is determined since it is essentially in Π^1_{2n}, then argue that if I wins, then P has a perfect subset.

To complete the proof, we must show that if II wins, then $P \subseteq \Delta^1_{2n+1}$. Suppose then that τ is winning for II and $\alpha \in P$, fix β so that $Q(\alpha, \beta)$ and call an initial part of the game

$$(*) \qquad u_0, a_0, b_0, u_1, a_1, b_1, \ldots, u_n, a_n, b_n$$

good, if

$$u_0\frown(a_0)\frown\cdots\frown u_n\frown(a_n) \subseteq \alpha$$

and the part is played by τ, i.e.

$$a_0 = \tau(u_0), \quad a_1 = \tau(u_0, a_0, b_0, u_1), \ldots, \quad a_n = \tau(u_0, \ldots, u_n).$$

If every good part had a good extension, then $\neg Q(\alpha, \beta)$ hence some good part has no good extension, say the one in $(*)$ above. It follows then that if $u_0\frown(a_0)\frown\cdots\frown u_n\frown(a_n) = (\alpha(0), \alpha(1),\ldots, \alpha(t))$, then for all $k \geq 2$

$$\alpha(t+k) = 1 \dot{-} \tau(u_0, a_0, b_0,\ldots, u_n, a_n, b_n, (\alpha(t+1),\ldots, \alpha(t+k-1))),$$

so that α is recursive in τ.

By the third periodicity theorem (and specifically 6E.4) we may assume that τ is in Δ^1_{2n+1}, so that $P \subseteq \Delta^1_{2n+1}$. ⊣

This result is both interesting in its own right and very useful. Here is one immediate consequence.

6E.6. Assume $Det(\mathbf{\Sigma}^1_{2n})$; prove that if $P \subseteq \mathfrak{X} \times \mathcal{Y}$ is $\mathbf{\Delta}^1_{2n+1}$ and all sections $P_x = \{y: P(x, y)\}$ are countable, then P can be uniformized by some $P^* \subseteq P$ in $\mathbf{\Delta}^1_{2n+1}$.

Hint. See 4F.6. ⊣

A set $P \subseteq \mathfrak{X} \times \mathcal{Y}$ in $\mathbf{\Delta}^1_{2n+1}$ with compact sections is uniformizable in $\mathbf{\Delta}^1_{2n+1}$ by 4F.12 (granting $Det(\mathbf{\Delta}^1_{2n})$), but the corresponding generalization of the Arsenin–Kunugui Theorem 4F.16 is still open. (Kechris and Martin have recently proved this for $\mathbf{\Delta}^1_3$, but they use methods that are quite deep and do not generalize immediately to arbitrary $\mathbf{\Delta}^1_{2n+1}$.) On the other hand, 4F.19 and the resulting uniformization theorems for $\mathbf{\Delta}^1_1$ sets with large sections (4F.20) generalize to all odd levels of the hierarchy, essentially by the same arguments we gave in 4F.19, 4F.20. See Kechris [1973].

We now aim towards a generalization of the Spector–Gandy Theorem 4F.3 to all odd levels. The proof is new even for Π^1_1 and it is in some ways simpler than our original proof in 4F.3.

6E.7. Assume $Det(\mathbf{\Delta}^1_{2n})$; prove that for every Π^1_{2n+1} set $P \subseteq \mathfrak{X}$ there is some Π^1_{2n} set $R \subseteq \mathfrak{X} \times \mathcal{Y}$ such that

$$P(x) \Leftrightarrow (\exists \alpha \in \Delta^1_{2n+1}(x)) R(x, \alpha).$$

(The Spector–Gandy Theorem for the odd levels, Moschovakis [1973].)

Hint. Notice first that taking $\Gamma = \Delta^1_{2n}(x)$ in 6E.1, $Det(\mathbf{\Delta}^1_{2n})$ implies that every $\Delta^1_{2n}(x)$ game admits a $\Delta^1_{2n+1}(x)$ winning strategy for one of the players. Check also as in 4F.3 that it is enough to prove the result for $P \subseteq \mathcal{N}$.

Let

$$G(e, \alpha) \Leftrightarrow (\forall\beta)R(e, \alpha, \beta)$$

be universal in Π^1_{2n+1} and choose e_0 such that

$$P(\alpha) \Leftrightarrow G(e_0, \alpha).$$

Put on G the canonical Π^1_{2n+1}-norm ψ which we defined in the proof of the First Periodicity Theorem 6B.1. Using 4D.14, choose also some fixed k so that

$$\beta \in \Delta^1_{2n+1}(\alpha) \Leftrightarrow G(k, \langle\beta, \alpha\rangle).$$

We claim that

$$P(\alpha) \Leftrightarrow (\exists\beta \in \Delta^1_{2n+1}(\alpha))[(e_0, \alpha) \leq^*_\psi (k, \langle\beta, \alpha\rangle)];$$

if this failed for some fixed $\alpha \in P$, we would have

$$\beta \in \Delta^1_{2n+1}(\alpha) \Leftrightarrow (k, \langle\beta, \alpha\rangle) <^*_\psi (e_0, \alpha)$$

which implies that $\Delta^1_{2n+1}(\alpha) \cap \mathcal{N}$ is in $\Delta^1_{2n+1}(\alpha)$ contradicting 4D.16.

By the construction in the proof of the First Periodicity Theorem 6B.1, there is a fixed Π^1_{2n} relation $S(\alpha, \beta, \gamma)$ such that whenever $G(k, \langle\beta, \alpha\rangle)$,

$$\begin{aligned}(e_0, \alpha) \leq^*_\psi (k, \langle\beta, \alpha\rangle) &\Leftrightarrow \text{II wins the game } \{\gamma: S(\alpha, \beta, \gamma)\}\\ &\Leftrightarrow (\exists\tau)(\forall\sigma)S(\alpha, \beta, \sigma * \tau)\\ &\Leftrightarrow (\exists\tau \in \Delta^1_{2n+1}(\alpha, \beta))(\forall\sigma)S(\alpha, \beta, \sigma * \tau),\end{aligned}$$

where for the last equivalence we have used the fact that for $(k, \langle\beta, \alpha\rangle) \in G$, the set $\{\gamma: S(\alpha, \beta, \gamma)\}$ is actually $\Delta^1_{2n}(\alpha, \beta)$ and of course we have also used the third periodicity theorem 6E.1.

We now have

$$P(\alpha) \Leftrightarrow (\exists\beta \in \Delta^1_{2n+1}(\alpha))(\exists\tau \in \Delta^1_{2n+1}(\alpha, \beta))(\forall\sigma)S(\alpha, \beta, \sigma * \tau)$$

which implies the result easily by contraction of quanitfiers. ⊣

There is a simple but interesting converse to 6E.2.

6E.8. Assume $Det(\mathbf{\Delta}^1_{2n})$; prove that for each $\alpha \in \Delta^1_{2n+1}$, there is a Δ^1_{2n} set $A \subseteq \mathcal{N}$ such that II wins the game (with payoff) A and α is recursive in every winning strategy for II in A.

Thus $\Delta^1_{2n+1} \cap \mathcal{N}$ is the smallest set which is closed under "recursive in" and contains a winning strategy (for one of the players) for each Δ^1_{2n} game.

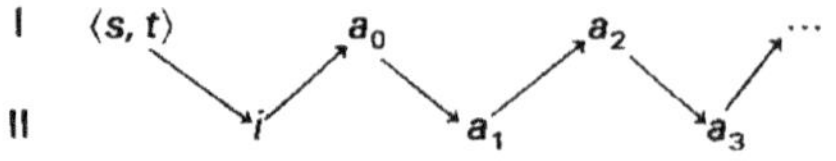

Diagram 6E.5.

Hint. Let $H \subseteq \omega \times (\omega \times \omega)$ be universal in Π^1_{2n+1}, let ψ be the canonical Π^1_{2n+1}-norm that is assigned to H by the First Periodicity Theorem 6B.1 and choose some k_0 such that

$$\alpha(s) = t \Leftrightarrow H(k_0, s, t).$$

By the Covering Lemma 4C.11, there are fixed integers l_0, l_1, l_2 such that $H(l_0, l_1, l_2)$ and

$$\alpha(s) = t \Leftrightarrow \psi(k_0, s, t) \leq \psi(l_0, l_1, l_2).$$

It is now obvious from the proof of 6B.1 that (with the fixed k_0, l_0, l_1, l_2) there are Δ^1_{2n} sets $P(s, t, \alpha)$ and $Q(s, t, \alpha)$ such that

$$(k_0, s, t) \leq^*_\psi (l_0, l_1, l_2) \Leftrightarrow \text{II wins } \{\alpha : P(s, t, \alpha)\}$$
$$(l_0, l_1, l_2) <^*_\psi (k_0, s, t) \Leftrightarrow \text{II wins } \{\alpha : Q(s, t, \alpha)\}.$$

Define the game A played as in Diagram 6E.5 where

$$\text{II wins} \Leftrightarrow i = 0 \ \& \ P(s, t, (a_0, a_1, \ldots)) \\ \vee i > 0 \ \& \ Q(s, t, (a_0, a_1, \ldots))$$

and check that A has the required properties. ⊣

Taking $n = 0$ here, we have a new game-theoretic characterization of $\Delta^1_1 \cap \mathcal{N}$ as the smallest set which is closed under "recursive in" and contains a winning strategy for every Δ^0_1 game.

We now turn to the generalization of the largest thin Π^1_1 set theorem 4F.4.

6E.9. Assume $Det(\mathbf{\Delta}^1_{2n+1})$; prove that for each perfect space $\mathfrak{X}$, there is a thin Π^1_{2n+1} set $C_{2n+1}(\mathfrak{X}) \subseteq \mathfrak{X}$ which contains every thin, Π^1_{2n+1} subset of $\mathfrak{X}$.

In particular, if *PD* holds, then for each n, there is a largest, countable Π^1_{2n+1} subset of $\mathfrak{X}$. (*The Largest Countable* Π^1_{2n+1} *Set Theorem*, Kechris [1975].)

Hint. Follow the proof of 4F.4 until the point where we have produced a perfect set F such that

$$(\forall x \in F)[R(g(x), x) \,\&\, \varphi(g(x), x) \leq \lambda],$$

where $\lambda < \boldsymbol{\delta}^1_{2n+1}$. The last part of the argument in 4F.4 depends on the fact that $\boldsymbol{\delta}^1_1 = \aleph_1$ and we must replace it by something more sophisticated when $n > 0$.

Since F is uncountable and $g: F \to \omega$, there must exist a fixed k such that $g(x) = k$ for uncountably many x's; since the set $\{x : x \in F \,\&\, R(k, x) \,\&\, \varphi(k, x) \leq \lambda\}$ is $\boldsymbol{\Delta}^1_{2k+1}$ and uncountable, it must have a perfect subset, by the hypothesis $Det(\boldsymbol{\Delta}^1_{2k+1})$ and 6A.12. Calling this new perfect set F again, we have

$$(\forall x \in F)[R(k, x) \,\&\, \varphi(k, x) \leq \lambda].$$

On F we have an obvious $\boldsymbol{\Delta}^1_{2n+1}$ prewellordering,

$$x \leq y \Leftrightarrow \varphi(k, x) \leq \varphi(k, y)$$

such that every initial segment $\{y : y \leq x\}$ is countable, since

$$\begin{aligned} y \leq x &\Rightarrow \varphi(k, y) \leq \varphi(k, x) \\ &\Rightarrow y \in \Delta^1_{2n+1}(x) \end{aligned}$$

by the definition of $R(k, x)$.

We can consider F as a perfect Polish space with the topology induced on it by $\mathfrak{X}$, so by 1A.3 there is a continuous injection

$$\pi : \mathcal{C} \rightarrowtail F;$$

this carries the prewellordering $\leq$ on F to a $\boldsymbol{\Delta}^1_{2n+1}$ prewellordering on $\mathcal{C}$ whose initial segments are again countable and which does not have the property of Baire (as a subset of $\mathcal{C} \times \mathcal{C}$) by 5A.10. Our assumption $Det(\boldsymbol{\Delta}^1_{2n+1})$ and 6A.16 do not allow such sets, so we have reached a contradiction and completed the proof of the first assertion.

The second assertion follows by 6A.12, since under *PD* every thin projective set is countable. ⊣

The hypothesis $Det(\boldsymbol{\Sigma}^1_{2n})$ is sufficient for this result, see 6G.10 and 6G.11.

Granting *PD*, we can also find largest countable sets at the even levels, but on the Σ side.

6E.10. Assume $Det(\mathbf{\Sigma}^1_{2n+1})$; prove that for each perfect product space $\mathfrak{X}$, there is a largest countable Σ^1_{2n+2} subset of $\mathfrak{X}$. (Kechris–Moschovakis [1972].)

Hint. Let $C = C_{2n+1}(\mathfrak{X} \times \mathcal{N})$ be the largest countable Π^1_{2n+1} subset of $\mathfrak{X} \times \mathcal{N}$ and put

$$x \in D \Leftrightarrow (\exists \alpha) C(x, \alpha).$$

Clearly D is Σ^1_{2n+2} and countable, since the map $(x, \alpha) \mapsto x$ is a surjection of C onto D.

If $P \subseteq \mathfrak{X}$ is countable and Σ^1_{2n+2}, choose some Π^1_{2n+1} set $Q \subseteq \mathfrak{X} \times \mathcal{N}$ such that

$$P(x) \Leftrightarrow (\exists \alpha) Q(x, \alpha),$$

let $Q^* \subseteq Q$ uniformize Q in Π^1_{2n+1} by 6C.5 and notice that Q^* must be countable, so that $Q^* \subseteq C$; hence $P \subseteq D$. ⊣

Assuming PD, put

$$C_n = C_n(\mathcal{N}) = \text{the largest countable } \Pi^1_n \text{ subset of } \mathcal{N} \text{ (if } n \text{ is odd)},$$

$$C_k = C_k(\mathcal{N}) = \text{the largest countable } \Sigma^1_k \text{ subset of } \mathcal{N} \text{ (if } k \text{ even } >0).$$

Since the property of being countable and Π^1_n or Σ^1_k is obviously preserved under Δ^1_1 isomorphisms, these sets $C_1, C_2, \ldots$ determine the largest countable Π^1_n (n odd) and Σ^1_k (k even >0) sets in all the perfect product spaces.

6E.11. Assume *PD*; prove that there is no largest countable Σ^1_{2n+1} subset of $\mathcal{N}$ and there is no largest countable Π^1_{2n+2} subset of $\mathcal{N}$. (Kechris [1975].)

Hint. If $A \subseteq \mathcal{N}$ is countable and Σ^1_{2n+1}, then $A \subseteq \Delta^1_{2n+1}$ by 6E.5 above; on the other hand, if A were the largest countable Σ^1_{2n+1} set, then $\Delta^1_{2n+1} \cap \mathcal{N} \subseteq A$, since for each $\alpha \in \Delta^1_{2n+1}$ the singleton $\{\alpha\}$ is obviously Σ^1_{2n+1}. Thus the largest countable Σ^1_{2n+1} set would have to be $\Delta^1_{2n+1} \cap \mathcal{N}$—and this set is not in Σ^1_{2n+1} by 4D.16.

The even case is easier: if $A \subseteq \mathcal{N}$ is countable and Π^1_{2n+2}, then $\mathcal{N} - A$ is Σ^1_{2n+2} and nonempty, hence $\mathcal{N} - A$ has a member in Δ^1_{2n+2} by the Basis Theorem 6C.6, hence we cannot have $\Delta^1_{2n+2} \cap \mathcal{N} \subseteq A$—which the largest countable Π^1_{2n+2} set would have to satisfy as above. ⊣

According to these last two exercises, the property of possessing a largest countable set of irrationals oscillates between the Π and the Σ side of the Kleene hierarchy together with the prewellordering property.

The sets $C_1, C_2, \ldots$ have a very interesting structure which we will not pursue here—see Kechris [1975].

A result which is somewhat related to the Spector–Gandy Theorem but which is really much deeper is the characterization of Δ^1_1 sets as precisely the injective, recursive images of Π^0_1 sets; similarly, the Borel sets are precisely the injective, continuous images of closed sets, see 1G.5, 2E.7, 2E.8, 4A.7 and 4D.9. Before going into the extension of this to all odd levels (with *PD*), let us look at a related and basic theorem about the quantifier $\Game$.

6E.12. Assume that Γ is adequate and ω-parametrized, that every pointset in Γ admits a Γ-semiscale and that $Det(\Gamma)$ holds. Prove that every pointset $Q \subseteq \mathcal{X}$ in $\Game\Gamma$ satisfies a triple equivalence of the form

$$\begin{aligned} Q(x) &\Leftrightarrow (\exists\sigma)(\forall\alpha)P(x, \sigma, \alpha) \\ &\Leftrightarrow (\exists!\sigma)(\forall\alpha)P(x, \sigma, \alpha) \\ &\Leftrightarrow (\exists\sigma)[\sigma \text{ is } \Game\Gamma(x)\text{-recursive } \& \ (\forall\alpha)P(x, \sigma, \alpha)], \end{aligned}$$

with P in Γ. (Moschovakis.)

Hint. Choose a good parametrization of Γ as in 3H—we will denote by $G \subseteq \mathcal{N} \times \mathcal{Y}$ (ambiguously) all good universal sets, no matter which $\mathcal{Y}$ is involved. Define "best strategy" for a game A as in the proof of 6E.1 so that if I can win A, then there is *exactly one* $\sigma \in \mathcal{N}$ which is I's best strategy. Put

$$\begin{aligned} R(\varepsilon, \sigma, x) \Leftrightarrow \ & \text{I wins } \{\alpha : G(\varepsilon, x, \alpha)\} \\ & \& \ \sigma \text{ is I's best strategy for } \{\alpha : G(\varepsilon, x, \alpha)\}; \end{aligned}$$

this relation is in $\Game\Gamma$ by 6E.1, so there is a fixed recursive $\varepsilon^* \in \mathcal{N}$ so that

$$\begin{aligned} R(\varepsilon, \sigma, x) &\Leftrightarrow (\Game\beta)G(\varepsilon^*, \varepsilon, \sigma, x, \beta) \\ &\Leftrightarrow (\Game\beta)G(S(\varepsilon^*, \varepsilon, \sigma), x, \beta), \end{aligned}$$

where S is the recursive function associated with the good parametrizations. Now for any ε_0, compute

$$\begin{aligned} (\Game\alpha)G(\varepsilon_0, x, \alpha) &\Leftrightarrow \text{I wins } \{\alpha : G(\varepsilon_0, x, \alpha)\} \\ &\Leftrightarrow (\exists\sigma_0)[(\forall\beta)G(\varepsilon_0, x, \sigma_0 * [\beta]) \ \& \ R(\varepsilon_0, \sigma_0, x)] \end{aligned}$$

$$\Leftrightarrow (\exists\sigma_0)[(\forall\beta)G(\varepsilon_0, x, \sigma_0 * [\beta])$$
$$\& \text{ I wins } \{\alpha : G(S(\varepsilon^*, \varepsilon_0, \sigma_0), x, \alpha)\};$$

by repeating the computation on the last conjunct of this equivalent we get

$$\begin{aligned}(\mathfrak{D}\alpha)G(\varepsilon_0, x, \alpha) \Leftrightarrow\ & \{(\exists\sigma_0)(\exists\sigma_1)(\exists\sigma_2)\cdots\} \\ & [(\forall\beta)G(\varepsilon_0, x, \sigma_0 * [\beta]) \\ & \& (\forall\beta)G(S(\varepsilon^*, \varepsilon_0, \sigma_0), x, \sigma_1 * [\beta]) \\ & \& (\forall\beta)G(S(\varepsilon^*, S(\varepsilon^*, \varepsilon_0, \sigma_0), \sigma_1), \sigma_2 * [\beta]) \\ & \& \cdots] \\ \Leftrightarrow\ & \{(\exists\sigma_0)(\exists\sigma_1)(\exists\sigma_2)\cdots\} \\ & (\exists\gamma)\{[(\gamma)_0 = \varepsilon_0 \ \& \ (\forall i)[(\gamma)_{i+1} = S(\varepsilon^*, (\gamma)_i, \sigma_i)]] \\ \rightarrow\ & (\forall i)(\forall\beta)G((\gamma)_i, x, \sigma_i * [\beta])\},\end{aligned}$$

where in fact if the left-hand-side holds, then by 6E.1 there are *unique* $\sigma_0, \sigma_1, \ldots$ which satisfy the right-hand-side and they are all $\mathfrak{D}\Gamma(\varepsilon_0, x)$-recursive. Now choose an arithmetical function $\pi : \omega \times \mathcal{N} \rightarrow \mathcal{N}$ such that the map $\sigma \mapsto (\pi(0, \sigma), \pi(1, \sigma), \ldots)$ is a bijection of $\mathcal{N}$ with ${}^{\omega}\mathcal{N}$, replace in this formula each σ_i by $\pi(i, \sigma)$ and the infinite string $(\exists\sigma_0)(\exists\sigma_1)\cdots$ by $(\exists\sigma)$ and prove by a standard prewellordering argument that if there exist σ_0, $\sigma_1, \ldots$ which are $\mathfrak{D}\Gamma(\varepsilon_0, x)$-recursive and satisfy the $\mathfrak{D}\Gamma$ matrix above, then (the unique) σ which codes all the $\sigma_i = \pi(i, \sigma)$ is also $\mathfrak{D}\Gamma(\varepsilon_0, x)$-recursive. This yields equivalences of the form

$$\begin{aligned}(\mathfrak{D}\alpha)G(\varepsilon_0, x, \alpha) & \Leftrightarrow (\exists\sigma)(\forall\beta)P(\varepsilon_0, x, \sigma, \beta) \\ & \Leftrightarrow (\exists!\sigma)(\forall\beta)P(\varepsilon_0, x, \sigma, \beta) \\ & \Leftrightarrow (\exists\sigma)[\sigma \text{ is } \mathfrak{D}\Gamma(\varepsilon_0, x)\text{-recursive} \ \& \ (\forall\beta)P(\varepsilon_0, x, \sigma, \beta)]\end{aligned}$$

which are what we need to complete the proof. ⊣

The representation of sets in Δ^1_{2n+1} as recursive, injective images of Π^1_{2n} sets can be shown by a minor variation of this proof. First a simple lemma of some independent interest.

6E.13. Assume $Det(\mathbf{\Delta}^1_{2n})$. Show that

$$\Delta^1_{2n+1} = \mathfrak{D}\Delta^1_{2n},$$

i.e. $Q \subseteq \mathfrak{X}$ is in Δ^1_{2n+1} if and only if there is some $P \subseteq \mathfrak{X} \times \mathcal{N}$ in Δ^1_{2n} such that

$$Q(x) \leftrightarrow (\Game\alpha)P(x, \alpha).$$

Hint. It is clear that $\Game\Delta^1_{2n} \subseteq \Delta^1_{2n+1}$. To prove the converse inclusion, choose $A \subseteq \omega$ in $\Pi^1_{2n+1} - \Delta^1_{2n+1}$ and such that $0 \notin A$ and for any Q in Δ^1_{2n+1} put

$$R(k, x) \leftrightarrow k \in A \vee [k = 0 \,\&\, Q(x)].$$

Now R is Π^1_{2n+1} so let φ be any Π^1_{2n+1}-norm on R; an easy argument by contradiction shows that there is some fixed $k^* \in A$ so that

$$(\forall x)R(k^*, x),$$
$$Q(x) \leftrightarrow (0, x) \leq^*_\varphi (k^*, x),$$

(otherwise A is Σ^1_{2n+1}).

Suppose now that

$$R(k, x) \leftrightarrow (\forall\alpha)P(k, x, \alpha)$$

for some P in Σ^1_{2n} and φ comes from a Σ^1_{2n}-norm ψ on P by the construction in the First Periodicity Theorem 6B.1. In this case,

$$\begin{aligned} Q(x) &\leftrightarrow (0, x) \leq^*_\varphi (k^*, x) \\ &\leftrightarrow \{(\forall a_0)(\exists b_0)(\forall a_1)(\exists b_1)\cdots\}[(0, x, (a_0, a_1, \ldots)) \leq^*_\psi (k^*, x, (b_0, b_1, \ldots))]. \end{aligned}$$

The result follows easily from 6D.1 and the fact that $(\forall x)(\forall\beta)P(k^*, x, \beta)$ which implies that

$$\{(\alpha, \beta, x) : (0, x, \alpha) \leq^*_\psi (k^*, x, \beta)\} \text{ is in } \Delta^1_{2n}. \qquad \dashv$$

6E.14. Assume $Det(\mathbf{\Sigma}^1_{2n})$. Prove that a set $Q \subseteq \mathfrak{X}$ is in Δ^1_{2n+1} if and only if Q is the recursive, injective image of some Π^1_{2n} set $P \subseteq \mathcal{N}$.

Similarly, Q is in $\Delta^1_{2n+1}(x)$ if and only if it is the recursive, injective image of some $\Pi^1_{2n}(x)$ set $P \subseteq \mathcal{N}$ and Q is in $\mathbf{\Delta}^1_{2n+1}$ if and only if it is the continuous, injective image of some $P \subseteq \mathcal{N}$ in $\mathbf{\Pi}^1_{2n}$. (Moschovakis [1973].)

Hint. We work with Δ^1_{2n+1}, the relativized case following similarly and then implying immediately the boldface result.

By 3E.6 we may assume that $\mathfrak{X} = \mathcal{N}$ so that $\mathfrak{X} \times \mathcal{N}$ is recursively homeomorphic with $\mathcal{N}$ and it is enough to produce a Π^1_{2n} set $P^* \subseteq \mathfrak{X} \times \mathcal{N}$ so that

$$\begin{aligned} Q(x) &\leftrightarrow (\exists\sigma)P^*(x, \sigma) \\ &\leftrightarrow (\exists!\sigma)P^*(x, \sigma). \end{aligned}$$

Let $\Gamma = \Sigma^1_{2n}$. We will modify the argument of 6E.12 to work with $\Delta = \Delta^1_{2n}$ instead of Γ so that ultimately we will have

$$Q(x) \Leftrightarrow (\exists\sigma)(\forall\alpha)P(x, \sigma, \alpha)$$
$$\Leftrightarrow (\exists!\sigma)(\forall\alpha)P(x, \sigma, \alpha)$$

with P in Δ and then we can take

$$P^*(x, \sigma) \Leftrightarrow (\forall\alpha)P(x, \sigma, \alpha)$$

so that P^* is in $\forall^{\mathcal{N}}\Delta = \Pi^1_{2n}$ as required. Since Δ is not parametrized, we must work with *codes* of sets rather than universal sets.

Choose then a good parametrization for $\Gamma = \Sigma^1_{2n}$ and for any $A \subseteq \mathcal{Y}$, call ε a $\mathbf{\Delta}$-code of A if

$$y \in A \Leftrightarrow G((\varepsilon)_0, y)$$
$$\Leftrightarrow \neg G((\varepsilon)_1, y),$$

where $G \subseteq \mathcal{N} \times \mathcal{Y}$ is a good universal set.

If Q is in Δ^1_{2n+1}, then by 6E.13 there is a Δ^1_{2n} set P so that $Q = \mathfrak{D}P$ and if P has recursive code ε^*, we have

$$Q(x) \Leftrightarrow (\mathfrak{D}\alpha)G((\varepsilon^*)_0, x, \alpha) \Leftrightarrow (\mathfrak{D}\alpha)G(S((\varepsilon^*)_0, x), \alpha)$$
$$\Leftrightarrow (\mathfrak{D}\alpha) \neg G((\varepsilon^*)_1, x, \alpha) \Leftrightarrow (\mathfrak{D}\alpha) \neg G(S((\varepsilon^*)_1, x), \alpha)$$

where S is the recursive function associated with the good parametrization and we have used the fact that $\mathfrak{X} = \mathcal{N}$ is of type 1. Taking

$$u(x) = \langle S((\varepsilon^*)_0, x), S((\varepsilon^*)_1, x)\rangle,$$

we have a recursive u such that for each x, $u(x)$ codes a $\mathbf{\Delta}^1_{2n}$ set $A_x \subseteq \mathcal{N}$ and

(1) $\qquad Q(x) \Leftrightarrow$ I *wins the game* A_x with $\mathbf{\Delta}^1_{2n}$-code $u(x)$.

Next we need a uniformity result.

LEMMA. *There is a recursive function* $v: \mathcal{N} \times \mathcal{N} \to \mathcal{N}$ *such that whenever* ε *is a* $\mathbf{\Delta}^1_{2n}$-*code of some set* $A(\varepsilon)$, *then for each* σ, $v(\varepsilon, \sigma)$ *is a* $\mathbf{\Delta}^1_{2n}$-*code of some set* $B(\varepsilon, \sigma)$ *such that*

$$\text{I wins } B(\varepsilon, \sigma) \Leftrightarrow \text{I wins } A(\varepsilon)$$
$$\& \ \sigma \text{ is the best winning strategy}$$
$$\text{for I in } A(\varepsilon).$$

This can be checked easily by going through the proof of the Third Periodicity Theorem 6E.1 and using the fact that we can pass uniformly from a $\mathbf{\Delta}^1_{2n}$-code of A to a $\mathbf{\Delta}^1_{2n}$-code of some scale on A (using a fixed Σ^1_{2n}-scale on a good universal Σ^1_{2n} set).

Finally as in 6E.12,

$$
\begin{aligned}
Q(x) \Leftrightarrow \{(\exists\sigma_0)(\exists\sigma_1)(\exists\sigma_2)\cdots\}\\
&[\sigma_0 \text{ wins } A^0_x \text{ with } \mathbf{\Delta}^1_{2n}\text{-code } \mathsf{u}(x)\\
&\ \&\ \sigma_1 \text{ wins } A^1_x \text{ with } \mathbf{\Delta}^1_{2n}\text{-code } \mathsf{v}(\mathsf{u}(x), \sigma_0)\\
&\ \&\ \sigma_2 \text{ wins } A^2_x \text{ with } \mathbf{\Delta}^1_{2n}\text{-code } \mathsf{v}(\mathsf{v}(\mathsf{u}(x), \sigma_0), \sigma_1)\\
&\ \&\ \cdots]\\
\Leftrightarrow \{(\exists\sigma_0)(\exists\sigma_1)(\exists\sigma_2)\cdots\}\\
&(\forall\gamma)\{[(\gamma)_0 = \mathsf{u}(x)\ \&\ (\forall i)[(\gamma)_{i+1} = \mathsf{v}((\gamma)_i, \sigma_i)]]\\
&\Rightarrow (\forall i)[\sigma_i \text{ wins the game with } \mathbf{\Delta}^1_{2n}\text{-code } (\gamma)_i]\}.
\end{aligned}
$$

The result follows as in 6E.12, noticing that for γ that satisfy the hypotheses above,

$$\sigma_i \text{ wins the game with } \mathbf{\Delta}^1_{2n}\text{-code } (\gamma)_i \Leftrightarrow (\forall\beta)\ \neg G(((\gamma)_i)_1, \sigma_i * [\beta]). \quad \dashv$$

We now come to the relationship between the operation $\mathfrak{D}$ and scales.

6E.15. Suppose Γ is adequate, $Det(\mathbf{\Gamma})$ holds and $P \subseteq \mathfrak{X} \times \mathcal{N}$ is in Γ and admits a Γ-scale; show that $\mathfrak{D}P$ admits a $\mathfrak{D}\Gamma$-scale. (Moschovakis.)

Hint. The argument is an elaboration of the proofs of the second and third periodicity theorems.

Suppose

$$Q(x) \Leftrightarrow (\mathfrak{D}\alpha)P(x, \alpha)$$

with P in Γ and put for each even m,

$$
\begin{aligned}
Q^*_m(x, u) \Leftrightarrow (\exists a_0) \cdots (\exists a_m)[u = \langle a_0, \dots, a_m\rangle\\
\&\ \{(\forall a_{m+1})(\exists a_{m+2})(\forall a_{m+3})\cdots\}P(x, (a_0, a_1, \dots, a_m, a_{m+1}, \dots))].
\end{aligned}
$$

Given a sequence of norms $\varphi_0, \varphi_1, \dots$ on P, we can define a norm Ψ^*_m on Q^*_m for each even m using φ_m, by the construction in 6D.3. To recall this and set up notation for the proof, consider the game $H_m(x, u, y, v)$ for each even m, x, $u = \langle a_0, \dots, a_m\rangle$, y and $v = \langle b_0, \dots, b_m\rangle$ which is played as in Diagram 6E.6. At

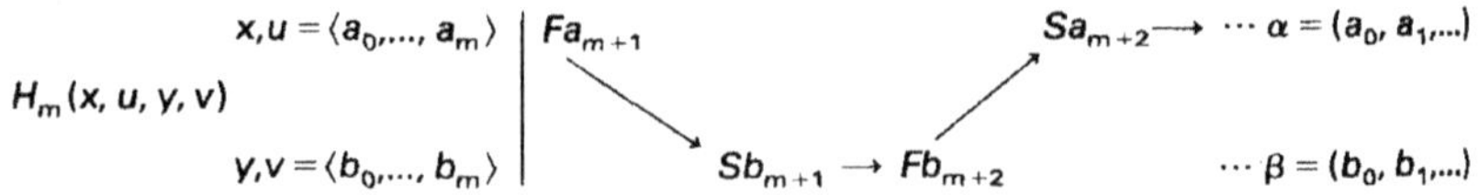

Diagram 6E.6.

the end of the game,

$$S \text{ wins the run} \Leftrightarrow (x, \alpha) \leq^*_{\varphi_m} (y, \beta)$$

and by 6D.3, the norms ψ^*_m satisfy

$$(x, u) \leq^*_{\psi^*_m} (y, v) \Leftrightarrow (x, u) \leq^*_m (y, v) \Leftrightarrow S \text{ wins } H_m(x, u, y, v).$$

Moreover from the formulas of 6D.3, if $\bar{\varphi} = \{\varphi_n\}$ is a Γ-semiscale on P, then each ψ^*_m is a $\Game\Gamma$-norm on Q^*_m and in fact the relations

$$R(x, u, y, v) \Leftrightarrow Q^*_m(x, u) \;\&\; (x, u) \leq^*_m (y, v) \qquad (m = lh(u) \dot{-} 1)$$
$$S(x, u, y, v) \Leftrightarrow Q^*(x, u) \;\&\; (x, u) <^*_m (y, v)$$

are in $\Game\Gamma$.

We now assume that P is of type 1 and that $\bar{\varphi}$ is a very good Γ-scale on P. We will use the ψ^*_m to construct a $\Game\Gamma$-scale on Q and then the result will follow by 4E.6.

Let $u(i)$ be the i'th sequence code (essentially) as in the proof of 6C.3, with $u(0) = 1 =$ the code of the empty sequence and so that if $u(i)$ is an initial segment of $u(j)$, we must have $i < j$. For each n and each $x \in Q$, let $w(x, n)$ *be the code of the initial segment of length* $2lh(u(n)) + 1$ *in the game establishing* $Q(x)$, *where* I *follows his best winning strategy and* II *plays following* $u(n)$. To clear this up, suppose

$$u(n) = \langle b_0, \ldots, b_{k-1} \rangle;$$

let $a_0, a_1, \ldots, a_k$ be the first $k+1$ moves of I in the game $\{\alpha : P(x, \alpha)\}$ where I plays his best (winning) strategy as we defined it in 6E.1 and set

$$w(x, n) = \langle a_0, b_0, \ldots, a_{k-1}, b_{k-1}, a_k \rangle.$$

It goes without saying that in defining the best strategy for I we use the scale $\bar{\varphi}$. By the construction in 6E.1 then, the best strategy is minimal so in particular for every a'_k, if $w(x, n)$ is as above,

$$(x, \langle a_0, b_0, \ldots, a_{k-1}, b_{k-1}, a_k \rangle) \leq^*_{2k} (x, \langle a_0, b_0, \ldots, a_{k-1}, b_{k-1}, a'_k \rangle).$$

Finally, if

$$u(n) = \langle b_0, \ldots, b_{k-1} \rangle,$$

choose $n_0, \ldots, n_{k-1} = n$ such that

$$u(n_i) = \langle b_0, \ldots, b_i \rangle \quad (i < k)$$

and for $x \in Q$ put

$$\psi_n(x) = \langle \psi_0^*(x, w(x, 0)), w(x, 0), \psi_2^*(x, w(x, n_0)), w(x, n_0),$$
$$\ldots \quad \ldots$$
$$\psi_{2k}^*(x, w(x, n_{k-1})), w(x, n_{k-1}) \rangle$$

where (as always) we use an order-preserving map $\langle \cdots \rangle$ from the alphabetic ordering of tuples of ordinals into the ordinals. We now proceed to show that $\bar{\psi} = \langle \psi_n \rangle$ is a $\Game\Gamma$-scale on Q.

The definability part is quite easy and we will omit it.

It remains to check that $\bar{\psi}$ is a scale on Q, so suppose $x_0, x_1, \ldots$ are all in Q, $limit_{i \to \infty} x_i = x$ and all $\psi_n(x_i)$ are ultimately fixed for large i; we must show that $x \in Q$ and for each n,

$$\psi_n(x) \le limit_{i \to \infty} \psi_n(x_i).$$

The hypothesis means in particular that $limit_{i \to \infty} w(x_i, n) = w_n$ for each n, i.e. the best strategy for establishing $Q(x_i)$ converges as $i \to \infty$. We will call this *the limiting best strategy* σ^*. In particular

$$\sigma^*(\emptyset) = a_0 = w(x_i, 0) \quad \text{for all large } i.$$

In addition, all $\psi_{2k}^*(x_i, w_n)$ are eventually constant. We will assume without loss of generality that both $a_0 = w(x_i, 0)$ and $\psi_0^*(x_i, \langle a_0 \rangle)$ are fixed for all $i \ge 0$.

We will show that

$$(x, \langle a_0 \rangle) \le_0^* (x_0, \langle a_0 \rangle);$$

this will establish in particular that $x \in Q$ and that

$$\psi_0^*(x, \langle a_0 \rangle) \le \psi_0^*(x_0, \langle a_0 \rangle)$$

and hence (easily) $\psi_0(x) \le \psi_0(x_i)$ for all i. A slight modification of the argument shows that for given $b_0, \ldots, b_{k-1}$, if $a_0, \ldots, a_k$ are the first $k+1$

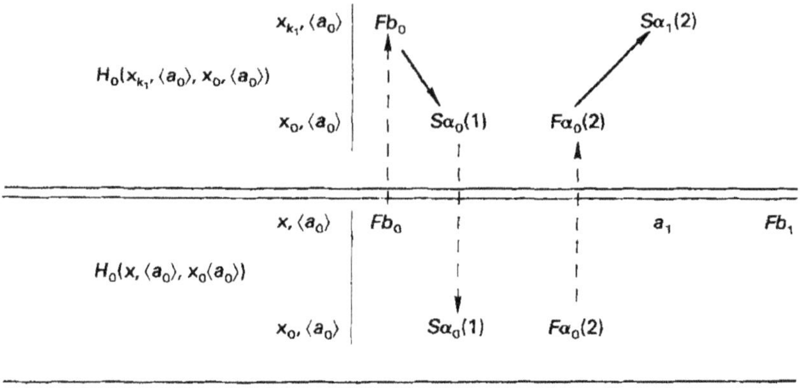

Diagram 6E.7. *Stage* 1.

moves of the limiting best strategy for the x_i, then for all large i,

$$(x, \langle a_0, b_0, \ldots, a_{k-1}, b_{k-1}, a_k\rangle) \leq^*_{2k} (x_i, \langle a_0, b_0, \ldots, a_{k-1}, b_{k-1}, a_k\rangle)$$

from which the result follows easily.

Suppose then that in the game $H_0(x, \langle a_0\rangle, x_0, \langle a_0\rangle)$ F starts with a move b_0. Suppose

$$\sigma^*(a_0, b_0) = a_1$$

and choose $k_1 > 0$ large enough so that if $u(j) = \langle b_0\rangle$, then $\psi_j(x_i)$ is constant for all $i \geq k_1$. Since also

$$(x_{k_1}, \langle a_0\rangle) \leq^*_0 (x_0, \langle a_0\rangle),$$

fix a winning strategy for S in $H_0(x_{k_1}, \langle a_0\rangle, x_0, \langle a_0\rangle)$ and construct Diagram 6E.7 in the usual way.

In this first stage of the construction we see how S can play in his first two moves of the "*master game*" $H_0(x, \langle a_0\rangle, x_0, \langle a_0\rangle)$, by copying $\alpha_0(1)$ and then playing a_1. The key moves that allow us to start the next stage are the numbers $\alpha_1(2)$ (by the winning strategy of S in $H_0(x_{k_1}, \langle a_0\rangle, x_0, \langle a_0\rangle)$) and b_1, which is F's next move in the master game.

To begin with, by the choice of k_1 we know that

$$(x_{k_1}, \langle a_0, b_0, a_1\rangle) \leq^*_2 (x_{k_1}, \langle a_0, b_0, \alpha_1(2)\rangle)$$

so fix a strategy for S in the game witnessing this. Also let

$$\sigma^*(a_0, b_0, a_1, b_1) = a_2$$

and choose k_2 so large that if $u(j)=\langle b_0, b_1\rangle$, then $\psi_j(x_i)$ is constant for all $i \geq k_2$. By the choice of k_1, we have

$$(x_{k_2}, \langle a_0, b_0, a_1\rangle) \leq_2^* (x_{k_1}, \langle a_0, b_0, a_1\rangle)$$

so we can fix a winning strategy for S in the game witnessing this and construct the second stage by starting with Fb_1 at the top; see Diagram 6E.8.

In this second stage we obtained the moves $\alpha_0(3)$ and a_2 for S in the master game. The new key moves that start the third stage are $\alpha_2(4)$ and b_3 and from then on we proceed in the obvious fashion.

It is clear that at the end we will have plays $\alpha_0, \alpha_1, \beta_1, \alpha_2, \beta_2, \dots$ and that the following will hold:

(1) $P(x_{k_1}, \alpha_1), P(x_{k_1}, \beta_1), P(x_{k_2}, \alpha_2), P(x_{k_2}, \beta_2), \dots,$

(2) for a suitable increasing sequence of integers

$$0 = j_0 < j_1 < j_2, \dots,$$

$$(x_{k_1}, \alpha_1) \leq^*_{\varphi_{j_0}} (x_0, \alpha_0),$$

$$\varphi_{j_1}(x_{k_1}, \beta_1) \leq \varphi_{j_1}(x_{k_1}, \alpha_1),$$

$$\varphi_{j_1}(x_{k_2}, \alpha_2) \leq \varphi_{j_1}(x_{k_1}, \beta_1),$$

$$\varphi_{j_2}(x_{k_2}, \beta_2) \leq \varphi_{j_2}(x_{k_2}, \alpha_2),$$

$$\dots \qquad \dots$$

(3) $limit_{i\to\infty}(x_{k_i}, \alpha_i) = (x, (a_0, b_0, a_1, b_1, \dots))$.

Now the fact that $\bar{\varphi}$ is a very good scale on P implies that $(x, (a_0, b_0, a_1, b_1, \dots)) \in P$ and

$$\varphi_0(x, (a_0, b_0, a_1, b_1, \dots)) \leq \varphi_0(x_0, \alpha_0),$$

(unless $\neg P(x_0, \alpha_0)$) so that S wins the master game. ⊣

The results in this section make it clear that almost everything we have proved about the analytical pointclasses Σ^1_k (k even) Π^1_n (n odd) can be extended to an arbitrary

$$\Gamma = \Game\Gamma_1,$$

where $\Pi^0_1 \subseteq \Gamma_1$, Γ_1 is adequate, ω-parametrized and scaled. Only occasionally we need the additional hypothesis that Γ is closed under $\forall^{\mathcal{N}}$ or $\exists^{\mathcal{N}}$.

We end this section with a simple result which implies that all the pointclasses $\Game\Sigma^0_n$, $\Game\Game\Sigma^0_n$, etc. are scaled.

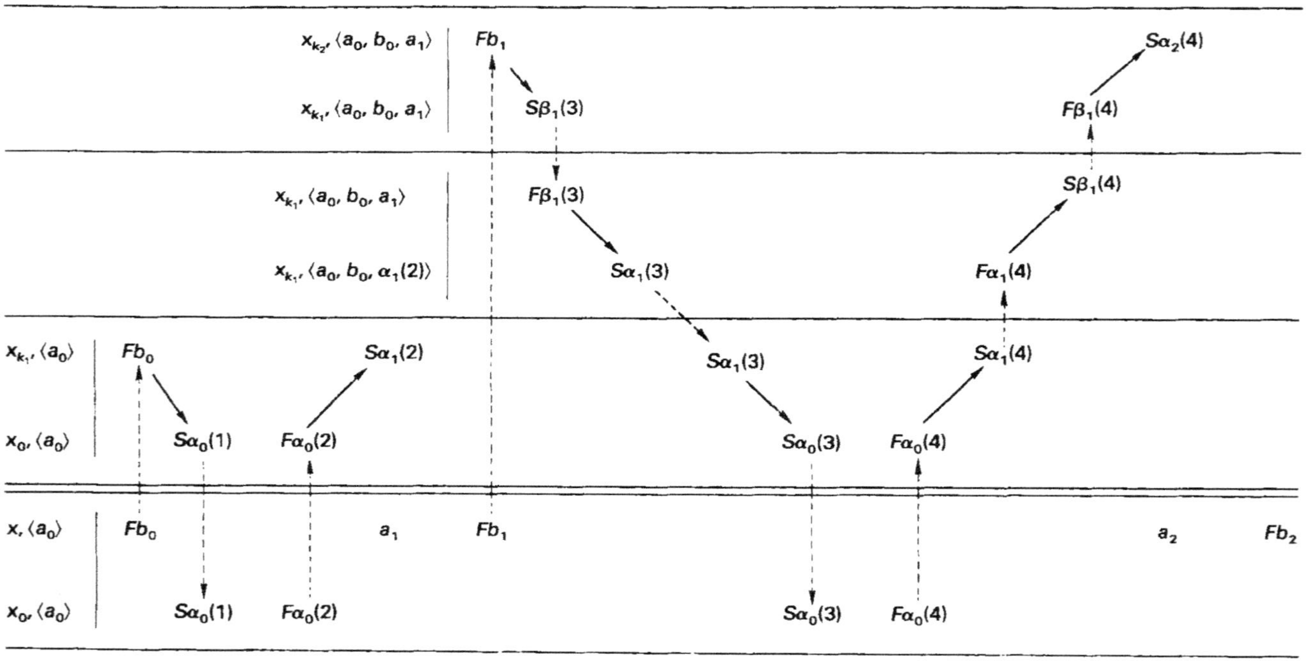

Diagram 6E.8. *Second Stage.*

6E.16. Show that for each $n \geq 1$, each Σ^0_n pointset of type 1 admits a Σ^0_n-scale (Kechris [1973]).

Infer that $\mathfrak{D}\Sigma^0_n$, $\mathfrak{D}\mathfrak{D}\Sigma^0_n$, etc. all have the scale property, granting the appropriate determinacy hypotheses.

Hint. Call $\bar{\varphi} = \{\varphi_n\}$ a *weak-*Π^0_k*-scale* on P if it is a scale on P and if the relations

$$R(n, x, y) \Leftrightarrow P(x) \,\&\, P(y) \,\&\, \varphi_n(x) \leq \varphi_n(y),$$

$$S(n, x, y) \Leftrightarrow P(x) \,\&\, P(y) \,\&\, \varphi_n(x) < \varphi_n(y),$$

are both in Π^0_k. Prove by induction on $k \geq 1$ that each Π^0_k set of type 1 admits a weak-Π^0_k-scale and each Σ^0_k set of type 1 admits a Σ^0_k-scale. The basis case $k = 1$ is trivial. If

$$Q(x) \Leftrightarrow (\forall m) P(x, m)$$

and $\bar{\varphi}$ is a Σ^0_k-scale on P, put

$$\begin{aligned}\psi_n(x) = \langle &\varphi_0(x, 0),\\ &\varphi_1(x, 0), \varphi_0(x, 1), \varphi_1(x, 1),\\ &\varphi_2(x, 0), \varphi_2(x, 1), \varphi_0(x, 2), \varphi_1(x, 2), \varphi_2(x, 2),\\ &\ldots\\ &\varphi_n(x, 0), \ldots, \varphi_n(x, n-1), \varphi_0(x, n), \varphi_1(x, n), \ldots, \varphi_n(x, n)\rangle\end{aligned}$$

and check that $\bar{\psi}$ is a weak-Π^0_{k+1}-scale on Q. If

$$Q(x) \Leftrightarrow (\exists m) P(x, m)$$

and $\bar{\varphi}$ is a weak-Π^0_k-scale on P, put

$$\psi_0(x) = \text{least } m, P(x, m),$$

$$\psi_{n+1}(x) = \langle \psi_0(x), \varphi_n(x, \psi_0(x))\rangle$$

and check that $\bar{\psi}$ is a Σ^0_{k+1}-scale on Q. (Here the tuples of ordinals are ordered lexicographically.) ⊣

6F. The determinacy of Borel sets[2,3]

By 6A.12 we know that $Det(\mathbf{\Sigma}^1_1)$ cannot be established in *ZFC*, but the next best result can be proved in this theory: all Borel sets are determined. This important result of Martin [1975] answered a long-standing question and lent considerable respectability to the practice of adopting determinacy hypotheses.

Here we will prove only $Det(\mathbf{\Sigma}^0_n)$ $(n<\omega)$ for which the argument is relatively simple and elegant. The complete result is established in Martin [1975] (in somewhat abbreviated form) and in the forthcoming Martin [19?c].

Let us first introduce a couple of simple notions which will be useful in the proof.

It is often convenient to describe a game G on X by giving a payoff set $A \subseteq {}^{\omega}X$ and *a set of rules*, i.e. a tree T on X. The game then must proceed along some branch of T, otherwise the first player who gets outside T loses. Formally, if

$$f=(x_0, x_1, x_2,...)$$

is the play, then

$$\text{I wins } G \leftrightarrow (\exists n)\{(x_0,..., x_{2n}) \in T \mathbin{\&} (x_0,..., x_{2n}, x_{2n+1}) \notin T\}$$
$$\vee (x_0, x_1,...) \in A.$$

We will often describe rules for a game informally, by putting down restrictions on the choice of x_n (by the appropriate player) which depend on the preceding moves $x_0,..., x_{n-1}$. One obtains a tree T from such restrictions in the obvious way,

$$(x_0,..., x_n) \in T \leftrightarrow \text{for each } i \leq n,\ x_i \text{ is allowed by the restrictions.}$$

If A is in $\mathbf{\Sigma}^0_\xi$ or $\mathbf{\Pi}^0_\xi$ with $\xi \geq 2$, then the game G defined by A with any set of rules T is obviously in the same class of Borel sets.

Suppose $(x_0,..., x_n)$ is an initial piece of a play. A I-*imposed subgame on* $(x_0,..., x_n)$ is a set Σ of finite sequences from X such that

(i) every sequence in Σ extends $(x_0,..., x_n)$,

(ii) if m is even and $(x_0,..., x_m) \in \Sigma$, then for every y, $(x_0,..., x_m, y) \in \Sigma$.

Intuitively, a I-imposed subgame is a set of conditions which restricts the possible moves of I but does not restrict at all II's moves. II-*imposed subgames* are defined similarly.

6F.1. THEOREM (Martin [1975]). *For each $X \neq \emptyset$ and each $n \geq 1$, every $\mathbf{\Sigma}^0_n$ game on X is determined.*

PROOF is by induction on n, but we will put it down only for $n=3$, to simplify notation—it will be clear that the argument is perfectly general.

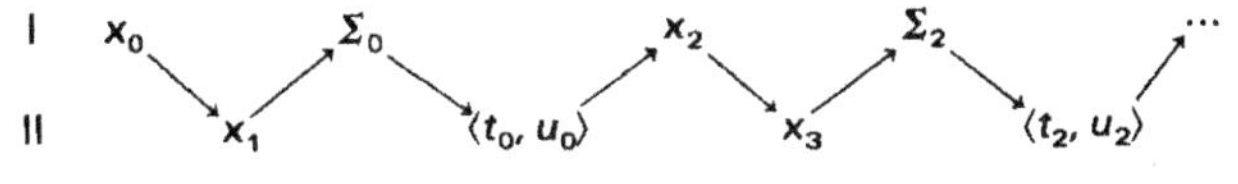

Diagram 6F.1.

The general idea is to assign to each $\mathbf{\Sigma}_3^0$ game A on X some $\mathbf{\Sigma}_2^0$ game A^* (on some other set X^*) and then prove that whichever player wins A^* also wins A.

Suppose then that

$$f \in A \Leftrightarrow (\exists i)(\forall j)[f \notin F_{i,j}]$$

where each $F_{i,j} \subseteq {}^\omega X$ is closed and let

$$F^0, F^1, F^2, \ldots$$

be a fixed enumeration of the countably many sets $F_{i,j}$. For each n, choose a tree T_n on X such that $F^n = [T_n]$ (by 2C.1) and notice that

$$(x_0, x_1, \ldots) \notin F^n \Leftrightarrow \text{for some } j, (x_0, \ldots, x_j) \notin T_n.$$

The players in A^* make moves in X and they also make various *auxiliary moves*, as in Diagram 6F.1. The following conditions are among *the rules* for A^*:

(1) $x_0, x_1, x_2, \ldots$ are chosen from X.

(2) Σ_0 is a I-imposed subgame on (x_0, x_1), Σ_2 is a I-imposed subgame on (x_0, x_1, x_2, x_3), etc.

(3) Each t_{2n} is 0 or 1 and each u_{2n} is a finite sequence from X of odd length which is compatible with $(x_0, x_1, \ldots, x_{2n+1})$.

There is one more restriction on the rules of the players which is the most significant. To motivate it as well as (1)–(3), consider the following "interpretation" of the game A^*.

The moves $x_0, x_1, \ldots$ are clearly meant to be moves in A—in effect the players are playing the game A, but they are also making side moves which reveal part of their intended strategy for winning in A.

When I plays Σ_0, he is offering his opponent to restrict his further moves in A according to Σ_0. Now II can respond to this in one of two ways.

(i) If II plays $t_0 = 0$, he is signifying that he accepts the offered restriction of I and in return he promises to see to it that at the end of the run

$$(x_0, x_1, x_2, \ldots) \in F^0.$$

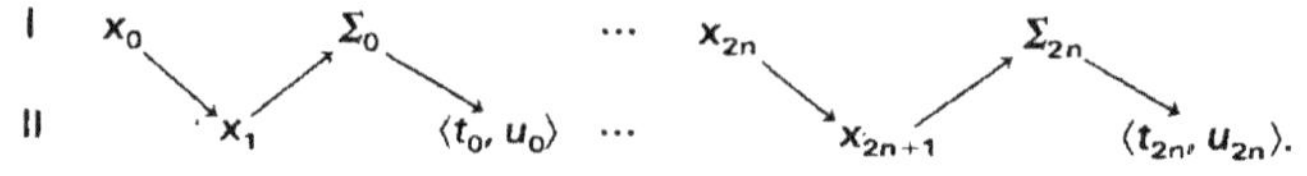

Diagram 6F.2.

From now on, I must play in Σ_0 and II must see to it that for every n,

$$(x_0, x_1, \ldots, x_{2n+1}) \in T_0,$$

or else he will lose the game. (In this case the move u_0 of II is irrelevant and might as well be omitted.)

(ii) If II plays $t_0 = 1$, he is signifying that he rejects the offered restriction of I. Instead he plays a sequence

$$u_0 = (x_0, x_1, \ldots, x_{2m}) \in \Sigma_0 - T_0$$

which is compatible with (x_0, x_1), lies in Σ_0 and avoids T_0. From now on both players are restricted to play consistently with u_0, so that at the end of the run,

$$(x_0, x_1, \ldots) \notin F^0.$$

The later auxiliary moves Σ_{2n}, $\langle t_{2n}, u_{2n}\rangle$ are interpreted in a similar way. To be precise, we define for each n a tree S_n on X which depends on the initial piece of the game in Diagram 6F.2.

(i) S_{-1} = all finite sequences from X.

(ii) If $t_{2n} = 0$, then $S_n = S_{n-1} \cap \Sigma_{2n}$.

(iii) If $t_{2n} = 1$, then

$$S_n = \{u \in S_{n-1} : u \text{ is compatible with } u_{2n}\}.$$

Now the last part of the rules for A^* reads:

(4) For each n, x_{2n}, x_{2n+1} must be such that the sequences $(x_0, \ldots, x_{2n})$, $(x_0, \ldots, x_{2n}, x_{2n+1})$ are in S_{n-1} and if $t_{2n} = 1$, then u_{2n} must be in $\Sigma_{2n} - T_n$.

Finally we define the payoff for A^* on runs which are played according to the rules:

$$(x_0, x_1, \Sigma_0, \langle t_0, u_0\rangle, \ldots) \in A^* \leftrightarrow \text{for some } n,\ t_{2n} = 0 \ \&\ (x_0, x_1, x_2, \ldots) \notin F^n$$
$$\text{or } (\exists i)(\forall j)(\forall n)[F_{i,j} = F^n \rightarrow t_{2n} = 1].$$

It is clear that the game A^* is in $\boldsymbol{\Sigma}^0_2$. To see the key connection between A and A^*, let us say that II *has kept all his commitments* in a run $(x_0, x_1,$

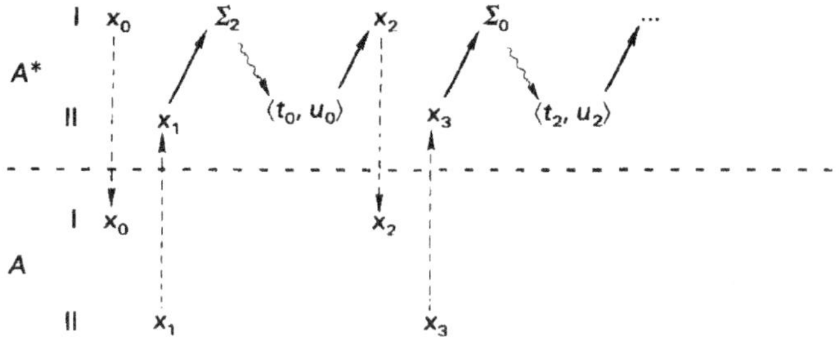

Diagram 6F.3.

$\Sigma_0, \langle t_0, u_0\rangle,\ldots)$ of A^* if for every n,

$$(x_0, x_1, x_2,\ldots) \in F^n \Leftrightarrow t_{2n} = 0.$$

For a run of this type then, we have

$$\begin{aligned}(x_0, x_1, \Sigma_0, \langle t_0, u_0\rangle,\ldots) \in A^* &\Leftrightarrow (\exists i)(\forall j)[F_{i,j} = F^n \rightarrow (x_0, x_1,\ldots) \notin F^n] \\ &\Leftrightarrow (x_0, x_1,\ldots) \in A.\end{aligned}$$

We now show that whichever player wins A^* also wins A.

Case 1. I *wins* A^*. Fix a winning strategy σ^* for I in A^* and have I play in A by "guessing" or "simulating" II's auxiliary moves in A^* according to Diagram 6F.3. As usual, we have indicated copying moves by broken arrows and responses according to the fixed strategy σ^* by solid arrows; we use wavy arrows $\rightsquigarrow$ to indicate the moves of II in A^* which must be simulated.

To begin with, I simulates $t_0 = 0$, thus commiting II in A^* to insure that the play $(x_0, x_1,\ldots)$ will be in F^0. Suppose we have reached stage n in this fashion, with the initial simulations t_{2i} $(i < n)$ always being 0 and consider Diagram 6F.4. Suppose also that for a fixed $i < n$ it is now realized that

$$(x_0, x_1,\ldots, x_{2i}, x_{2i+1},\ldots, x_{2n}) \notin T_i,$$

so that II will not be able to keep his commitment in A^* and insure that the ultimate play will be in F^i. At this point, I goes back to stage i and changes the simulated move $\langle 0, u_{2i}\rangle$ to $\langle 1, (x_0, x_1,\ldots, x_{2n})\rangle$; this of course changes some of the moves following that in A^*, as we always follow the instructions of σ^* in that game-consider Diagram 6F.5. Notice, however, that the moves $x_{2i+2}, x_{2i+3},\ldots, x_{2n}$, cannot be different, since the restriction $u'_{2i} = (x_0, x_1,\ldots, x_{2n})$ forces I to play along u'_{2i} until stage n.

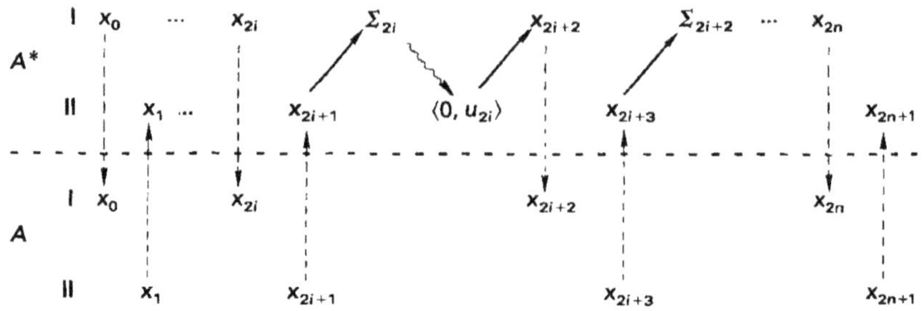

Diagram 6F.4.

It is clear that I can play in A in this manner by simulating moves in A^* and occasionally changing his earlier simulations, *so that no move in A^* is changed more than a finite number of times;* this is because a simulated move $\langle 0, u_{2i}\rangle$ can change at most once to $\langle 1, u'_{2i}\rangle$ and moves Σ_{2j} can change only when some $\langle 0, u_{2i}\rangle$ is changed with $i<j$. Thus at the end we obtain a run of A^* in which both players have followed the rules and in which II has kept all his commitments. For such runs, we know that by the definitions,

$$(x_0, x_1, \Sigma_0, \langle t_0, u_0\rangle, \ldots) \in A^* \Leftrightarrow (x_0, x_1, x_2, \ldots) \in A.$$

But in this specific run I has played according to the winning strategy σ^*, so that the left-hand-side above is true and $(x_0, x_1, x_2, \ldots) \in A$, i.e. I has won A.

Case 2. II *wins* A^*. Again fix a winning strategy τ^* for II in A^* and have II play in A by simulating I's auxiliary moves in A^* according to Diagram 6F.6.

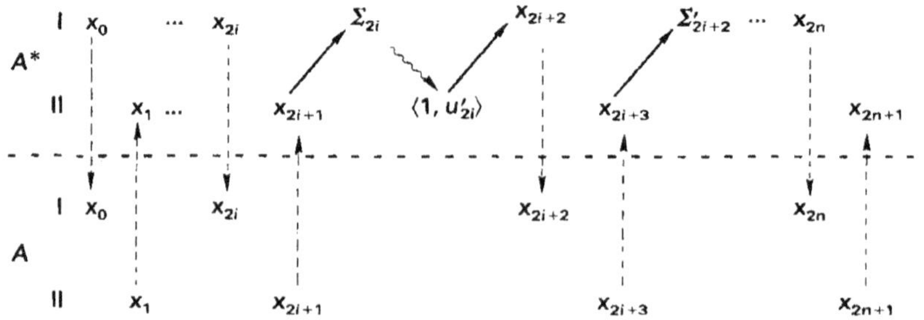

Diagram 6F.5.

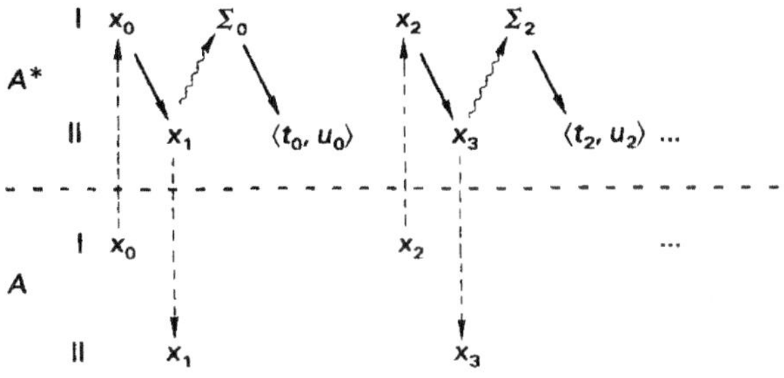

Diagram 6F.6.

To begin the simulation, take

$$\Sigma_0 = \{u : u \text{ extends } \langle x_0, x_1\rangle \text{ and for every } \Sigma, \langle 1, u\rangle \neq \tau^*(x_0, x_1, \Sigma)\}.$$

Since Σ_0 contains all sequences of even length, it is a I-imposed subgame on (x_0, x_1) and hence a legitimate move for I in A^*. Notice that if $\langle t_0, u_0\rangle = \tau^*(x_0, x_1, \Sigma_0)$, then we must have $t_0 = 0$; because if $t_0 = 1$, then by the rules of A^* we should have $u_0 \in \Sigma_0$ which is clearly impossible by the definition of Σ_0. Thus u_0 introduces no restrictions on the further moves of I.

Suppose we have reached stage n in this fashion, with the initial simulation Σ_{2j} defined by

$$\Sigma_{2j} =$$

$$\{u : \text{for every } \Sigma, \langle 1, u\rangle \neq \tau^*(x_0, x_1, \Sigma_0, \ldots, x_{2j-2}, x_{2j-1}, \Sigma_{2j-2}, x_{2j}, x_{2j+1}, \Sigma)\}$$

consider Diagram 6F.7. Suppose also that for a fixed $i < n$ it now happens that

$$(x_0, x_1, \ldots, x_{2n}) \notin \Sigma_{2i}$$

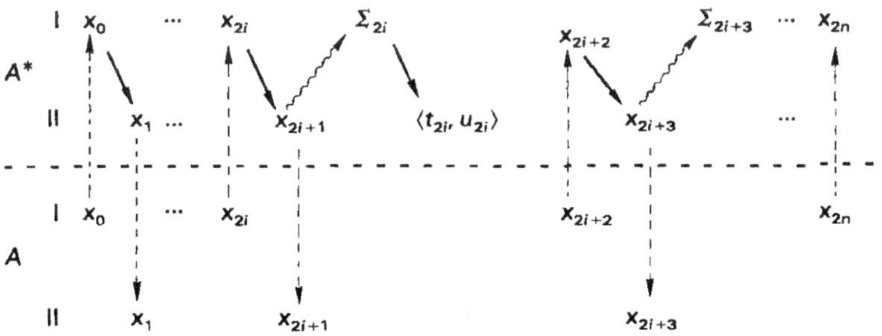

Diagram 6F.7.

so that I is not keeping with the restriction we imposed on him by having him play Σ_{2i}. This means that there is some Σ'_{2i} such that

$$\tau^*(x_0, x_1, \ldots, x_{2i}, x_{2i+1}, \Sigma'_{2i}) = \langle 1, (x_0, x_1, \ldots, x_{2n})\rangle,$$

by the definition of Σ_{2i}, so we can go back and change the simulated move Σ_{2i} to Σ'_{2i}; this of course will change some of the further moves in A^* as we proceed to follow the instructions of τ^* in this game and resimulate, but of course none of the moves $(x_0, x_1, \ldots, x_{2n})$ will change as II has now restricted himself to play along this sequence.

As in the first case, it is now clear that II can continue to play in A in this fashion and simulate moves in A^* so that he constructs a run of A^* played according to τ^* and in which I has played according to the rules; again the key fact is that no move in A^* is changed more than a finite number of times. At the end II wins in A^*, and this implies immediately as before that he also wins in A. ⊣

6G. Measurable cardinals[10]

Before proving the determinacy of Borel sets, Martin [1970] showed that $Det(\mathbf{\Sigma}^1_1)$ follows from a "large cardinal hypothesis," the existence of measurable cardinals. Our main purpose in this section is to define measurable cardinals and prove this result.

Recall that a *filter* on a set X is a collection $\mathfrak{F}$ of subsets of X such that

(i) $\emptyset \notin \mathfrak{F}$, $X \in \mathfrak{F}$,

(ii) if $A \in \mathfrak{F}$ and $A \subseteq B$, then $B \in \mathfrak{F}$,

(iii) if A, $B \in \mathfrak{F}$, then $A \cap B \in \mathfrak{F}$.

For any cardinal number $\varkappa$, a filter $\mathfrak{F}$ is *$\varkappa$-complete* if whenever $\lambda < \varkappa$ and $\{A_\xi\}_{\xi<\lambda}$ is a family of λ subsets of X, then

$$(\forall \xi < \lambda) A_\xi \in \mathfrak{F} \Rightarrow \bigcap_{\xi<\lambda} A_\xi \in \mathfrak{F}.$$

Finally, $\mathfrak{F}$ is an *ultrafilter* (maximal filter) if for each $A \subseteq X$, either $A \in \mathfrak{F}$ or $X - A \in \mathfrak{F}$.

Each point $x_0 \in X$ determines a *principal ultrafilter*

$$\mathfrak{U}(x_0) = \{A \subseteq X : x_0 \in A\}$$

which is obviously κ-complete for every κ. To get non-principal ultrafilters one generally needs the axiom of choice; with it one can prove in fact that every filter $\mathfrak{F}$ on X is contained in some ultrafilter (which must be non-principal if $\mathfrak{F}$ contains all complements of singletons, e.g. if $\mathfrak{F} = \{A \subseteq X : X - A \text{ is finite}\}$.

A cardinal number $\varkappa$ is *measurable* if $\varkappa > \omega$ and some set X of cardinality $\varkappa$ carries a $\varkappa$-complete non-principal ultrafilter. By the usual conventions of set theory, $\varkappa$ itself is a specific set of cardinality $\varkappa$ (the set of ordinals preceding it) and it follows easily that $\varkappa$ is measurable exactly when it carries a $\varkappa$-complete non-principal ultrafilter.

Let us abbreviate the hypothesis that we will be using:

$MC \Leftrightarrow$ *there exists at least one measurable cardinal.*

It is easy to check that MC is equivalent to the assumption that some set carries an $\aleph_1$-complete, non-principal ultrafilter (6G.8).

We cannot hope to prove MC in Zermelo–Fraenkel set theory because, as we will see, it fails in the constructible universe L. On the other hand, although one could theoretically refute MC, this does not appear likely on the basis of the presently available evidence. We will discuss the plausibility of MC and similar hypotheses in Chapter 8.

With each ultrafilter $\mathcal{U}$ on $\varkappa$ we associate the (two-valued) *measure* $\mu = \mu_{\mathcal{U}}$ on the power of $\varkappa$,

$$\mu(A) = \begin{cases} 1 & \text{if } A \in \mathcal{U} \\ 0 & \text{if } \kappa - A \in \mathcal{U}. \end{cases}$$

If $\mathcal{U}$ is $\varkappa$-complete, then easily μ is *$\varkappa$-additive*, i.e. whenever $\lambda < \varkappa$ and $\{A_\xi\}_{\xi<\lambda}$ is a λ-sequence of *pairwise disjoint* subsets of $\varkappa$, then

$$\mu(\bigcup_{\xi<\lambda} A_\xi) = \sum_{\xi<\lambda} \mu(A_\xi).$$

(To check this take cases on whether some $A_\xi \in \mathcal{U}$ or not.) Conversely, if μ is a $\varkappa$-additive two-valued measure on the power of $\varkappa$ such that $\mu(\varkappa) = 1$ and for each ξ, $\mu(\{\xi\}) = 0$, then $\mathcal{U}_\mu = \{A : \mu(A) = 1\}$ is a $\varkappa$-complete non-principal ultrafilter on $\varkappa$ and $\varkappa$ is measurable. We will speak interchangeably of μ and $\mathcal{U}$ when they are related in this way—e.g. we will often refer to members of $\mathcal{U}$ as *sets of measure* 1.

For each ultrafilter $\mathcal{U}$ on $\varkappa$ and for any two functions $f, g : \varkappa \to \varkappa$, put

$$f \leq g \Leftrightarrow \{\xi : f(\xi) \leq g(\xi)\} \in \mathcal{U}$$

6G.1. LEMMA. *If $\mathcal{U}$ is an $\aleph_1$complete ultrafilter on $\varkappa$, then the associated relation $\leq$ is a prewellordering on the set ${}^\varkappa\varkappa$ of functions on $\varkappa$ to $\varkappa$.*

PROOF. Clearly $f \leq f$ for each f and since

$$\{\xi: f(\xi) \leq h(\xi)\} \supseteq \{\xi: f(\xi) \leq g(\xi)\} \cap \{\xi: g(\xi) \leq h(\xi)\},$$

if $f \leq g$ and $g \leq h$ then $f \leq h$. Also, if $\neg(f \leq g)$, then $\{\xi: f(\xi) \leq g(\xi)\} \notin \mathcal{U}$ so that $\{\xi: g(\xi) < f(\xi)\} \in \mathcal{U}$ and hence $g < f$ which implies immediately $g \leq f$. Thus $\leq$ is a preordering. Finally, if

$$f_0 > f_1 > \cdots,$$

then each set

$$A_n = \{\xi: f_n(\xi) > f_{n+1}(\xi)\}$$

is of measure 1 and hence $\bigcap_n A_n$ has measure 1 and in particular $\bigcap_b A_n \neq \emptyset$; for any $\lambda \in \bigcap_n A_n$ then,

$$f_0(\lambda) > f_1(\lambda) > f_2(\lambda) > \cdots,$$

which is absurd. ⊣

It follows that with each $\aleph_1$-complete ultrafilter on $\varkappa$ we can associate a *rank function*

$$\rho : {}^{\varkappa}\varkappa \to \mathit{Ordinals}$$

such that

$$\rho(f) \leq \rho(g) \leftrightarrow \{\xi \in \varkappa: f(\xi) \leq g(\xi)\} \in \mathcal{U}.$$

An ultrafilter $\mathcal{U}$ (or the corresponding measure μ) on $\varkappa$ is *normal* if $\mathcal{U}$ is nonprincipal, $\varkappa$-complete and such that for each function $f: \varkappa \to \varkappa$,

$$\{\xi \in \varkappa: f(\xi) < \xi\} \in \mathcal{U} \to \text{there is a fixed } \lambda_0 < \varkappa$$
$$\text{such that } \{\xi \in \varkappa: f(\xi) = \lambda_0\} \in \mathcal{U}.$$

6G.2. LEMMA. *Every measurable cardinal carries a normal ultrafilter.*

PROOF. Let $\mathcal{U}$ be a $\varkappa$-complete, non-principal ultrafilter on $\varkappa$ with associated function ρ and for each $\lambda < \varkappa$ let C_λ be the constant function with value λ,

$$C_\lambda(\xi) = \lambda.$$

Clearly

$$\lambda < \lambda' \to \rho(C_\lambda) < \rho(C_{\lambda'})$$

so that

$$\rho(C_\lambda) \geq \lambda.$$

If *id* is the identity function,

$$id(\xi) = \xi,$$

then for each λ

$$\{\xi : C_\lambda(\xi) \leq id(\xi)\} = \varkappa - \lambda = \bigcap_{\xi<\lambda} (\varkappa - \{\xi\}) \in \mathfrak{U},$$

so $\rho(C_\lambda) \leq \rho(id)$ and hence

$$\rho(id) \geq \varkappa.$$

It follows that for some f_0 we must have

$$\rho(f_0) = \varkappa.$$

Fix such an f_0 then and define $\mathfrak{U}^*$ by

$$\begin{aligned} A \in \mathfrak{U}^* &\leftrightarrow f_0^{-1}[A] \in \mathfrak{U} \\ &\leftrightarrow \{\xi : f_0(\xi) \in A\} \in \mathfrak{U}. \end{aligned}$$

Proof that $\mathfrak{U}^*$ is $\varkappa$-complete and non-principal is routine. To check that $\mathfrak{U}^*$ is normal, suppose $\{\xi : f(\xi) < \xi\} \in \mathfrak{U}^*$, so that $\{\xi : f(f_0(\xi)) < f_0(\xi)\} \in \mathfrak{U}$. If g is the composition

$$g(\xi) = f(f_0(\xi)),$$

we have $\rho(g) < \rho(f_0) = \varkappa$ which implies easily by $\varkappa$-completeness that for some $\lambda < \varkappa$, $\{\xi : f(f_0(\xi)) = \lambda\} \in \mathfrak{U}$. By the definition of $\mathfrak{U}^*$ then, $\{\xi : f(\xi) = \lambda\} \in \mathfrak{U}^*$ which is what we needed to show. ⊣

Suppose $\{A_\xi\}_{\xi<\varkappa}$ is a $\varkappa$-sequence of subsets of $\varkappa$. The *diagonal intersection of* $\{A_\xi\}_{\xi<\varkappa}$ is defined by

$$\lambda \in A \leftrightarrow (\forall \xi < \lambda)[\lambda \in A_\xi].$$

6G.3. LEMMA. *Suppose $\mathfrak{U}$ is a normal ultrafilter on $\varkappa$ and each $A_\xi (\xi < \varkappa)$ is in $\mathfrak{U}$; then the diagonal intersection $A = \{\lambda : (\forall \xi < \lambda)[\lambda \in A_\xi]\}$ is also in $\mathfrak{U}$.*

PROOF. Assume not, so that $(\varkappa - A) \in \mathfrak{U}$ and for each $\lambda \in \varkappa - A$ choose $f(\lambda) < \lambda$ so that $\lambda \notin A_{f(\lambda)}$ (and set $f(\lambda) = 0$ for $\lambda \in A$). Now $f(\lambda) < \lambda$ on a set of measure 1, so by normality, $f(\lambda) = \lambda^*$ for a fixed λ^* and all λ in a set B of measure 1. But then $B \cap A_{\lambda^*}$ has measure 1, so it contains some $\lambda > \lambda^*$, $\lambda \in \varkappa - A$; this λ then satisfies both $\lambda \in A_{\lambda^*}$ and $\lambda \notin A_{f(\lambda)} = A_{\lambda^*}$ which is absurd. ⊣

After these preliminary results we are ready to state and prove the key *partition property* of measurable cardinals which will be our main tool.

For each $n \geq 1$ let

$$\varkappa^{[n]} = \text{all subsets of } \varkappa \text{ with exactly } n \text{ members}$$

and put

$$\begin{aligned} \varkappa^{<\omega} &= \text{all finite subsets of } \varkappa \\ &= \bigcup_n \varkappa^{[n]}. \end{aligned}$$

A map

$$F : \varkappa^{[n]} \to \lambda$$

is often called *a partition of* $\varkappa^{[n]}$ into λ parts. A subset $I \subseteq \varkappa$ is *homogeneous* for F if for all A, $B \in \varkappa^{[n]}$

$$A \subseteq I, B \subseteq I \Rightarrow F(A) = F(B),$$

i.e. if all the n-element subsets of I are put into the same bin by the partition F.

6G.4. THEOREM (Rowbottom [1971]). *Let* $\mathcal{U}$ *be a normal ultrafilter on* $\varkappa$ *and suppose* $F : \varkappa^{[n]} \to \lambda$ *is a partition of the* n*-element subsets of* $\varkappa$ *into* $\lambda < \varkappa$ *parts. Then there exists a homogeneous set* I *for* F *which is in* $\mathcal{U}$.

PROOF is by induction on n with the basis $n = 1$ being an immediate consequence of the fact that $\mathcal{U}$ is $\varkappa$-complete.

Suppose then that $F : \varkappa^{[n+1]} \to \lambda$. For each fixed $\xi < \varkappa$, define a partition

$$F_\xi : \varkappa^{[n]} \to \lambda$$

by the formulas

$$\begin{aligned} F_\xi(A) &= F(\{\xi\} \cup A) \quad &&\text{if } \xi \notin A, \\ F_\xi(A) &= 0 &&\text{if } \xi \in A, \end{aligned}$$

and by the induction hypothesis choose I_ξ in $\mathcal{U}$ to be homogeneous for F_ξ

and put

$$G(\xi) = F_\xi(A) \text{ for any } n\text{-element } A \subseteq I_\xi.$$

Since $G: \varkappa \to \lambda$, by $\varkappa$-completeness easily there is a set $J \subseteq \varkappa$ in $\mathcal{U}$ and an ordinal $\lambda_0 < \lambda$ so that

$$\xi \in J \Rightarrow G(\xi) = \lambda_0.$$

Put then

$$I^*_\xi = J \cap I_\xi$$

and let I be the diagonal intersection of the I^*_ξ's,

$$\lambda \in I \Leftrightarrow \lambda \in J \;\&\; (\forall \xi < \lambda)[\lambda \in I_\xi].$$

It remains to verify that I is homogeneous for F.

Given an $(n+1)$-element $A \subseteq I$, let ξ be its least member, let $B = A - \{\xi\}$ and notice that $B \subseteq I_\xi$; this because if $\lambda \in B$, then $\lambda \in I$ and also $\xi < \lambda$, so that $\lambda \in I_\xi$. Thus

$$\begin{aligned} F(B) &= F_\xi(A) && \text{by definition of } F_\xi \\ &= G(\xi) && \text{since } \xi \in J \text{ and } A \subseteq I_\xi \\ &= \lambda_0 \end{aligned}$$

so that F is constant on the $(n+1)$-element subsets of I. ⊣

A map

$$F: \varkappa^{<\omega} \to \lambda$$

is a *partition* of the finite subsets of $\varkappa$ into λ parts. We call $I \subseteq \varkappa$ *homogeneous* for F if for $A, B \in \varkappa^{<\omega}$

$$A \subseteq I,\ B \subseteq I \text{ and } card(A) = card(B) < \aleph_0 \Rightarrow F(A) = F(B).$$

6G.5. COROLLARY. *Let $\mathcal{U}$ be a normal ultrafilter on $\varkappa$ and for each i let $F_i: \varkappa^{<\omega} \to \lambda$ be a partition of the finite subsets of $\varkappa$ into $\lambda < \varkappa$ parts. Then there exists a set I in $\mathcal{U}$ which is simultaneously homogeneous for all the F_i.*

PROOF. Define $F_{i,n}: \varkappa^{[n]} \to \lambda$ by

$$F_{i,n}(A) = F_i(A) \quad (A \in \varkappa^{[n]}),$$

let $I_{i,n}$ be homogeneous for $F_{i,n}$ in $\mathcal{U}$ and take $I = \bigcap_{i,n} I_{i,n}$. ⊣

We will not study *partition calculus* here, but it will be useful to have around a bit of notation from this part of combinatorial set theory. For given cardinals $\varkappa' \leq \varkappa$, put

$$\varkappa \to (\varkappa') \Leftrightarrow \text{for each sequence } \{F_i\}_{i\in\omega} \text{ of partitions } F_i : \varkappa^{<\omega} \to \omega \text{ there exists a set } I \subseteq \varkappa \text{ of cardinality } \varkappa' \text{ which is homogeneous for all the } F_i.$$

We say that $\varkappa$ is *Ramsey* if $\varkappa \to (\varkappa)$, so that by 6G.5 every measurable cardinal is Ramsey. (This is equivalent to the more usual definition of Ramsey cardinals, see Drake [1974].) All the applications of measurable cardinals to descriptive set theory follow from the (weaker) consequence of 6G.5, that

$$\varkappa \text{ measurable} \Rightarrow \varkappa \to (\aleph_1);$$

this is due to Erdős, Hajnal [1958].

To simplify the proof of Martin's theorem we first reformulate the basic representation theorem for $\mathbf{\Pi}^1_1$ sets, 4A.3.

6G.6. LEMMA. *If $A \subseteq \mathcal{N}$ is a $\mathbf{\Pi}^1_1$ set of irrationals, then there exists a function D with the following properties:*

(i) *The domain of D consists of all codes u of finite sequences of even length, i.e. $\{u: Seq(u)\ \&\ lh(u) \text{ is even}\}$.*

(ii) *If $Seq(u)\ \&\ lh(u) = 2n$, then $D(u)$ is an ordering with field some set of n integers.*

(iii) *If $t < s$, then $D(\bar{\alpha}(2t))$ is a subordering of $D(\bar{\alpha}(2s))$.*

(iv) *The following equivalence holds:*

$$\alpha \in A \Leftrightarrow \bigcup_t D(\bar{\alpha}(2t)) \text{ is a wellordering.}$$

PROOF. By 4A.3 choose a continuous $f: \mathcal{N} \to \mathcal{N}$ so that for each α, $f(\alpha) \in LO$ and

$$\alpha \in A \Leftrightarrow f(\alpha) \in WO$$

and let R be such that

$$\begin{aligned} n \leq_{f(\alpha)} k &\Leftrightarrow f(\alpha)(\langle n, k\rangle) = 1 \\ &\Leftrightarrow (\exists s) R(\bar{\alpha}(s), n, k). \end{aligned}$$

If $u = \langle u_0, \ldots, u_{2t-1}\rangle$ is a sequence code with even length $2t$, put

$$C(u) = \{(2n, 2k) : (\exists s \leq 2t) R(\langle u_0, \ldots, u_{s-1}\rangle, n, k) \;\&\; n, k \leq t\}$$

so that in particular, for each α, t

$$C(\bar{\alpha}(2t)) = \{(2n, 2k) : (\exists s \leq 2t) R(\bar{\alpha}(s), n, k) \;\&\; n, k < t\}.$$

Clearly each $C(\bar{\alpha}(2t))$ is a partial ordering whose domain consists of even numbers $<2t$,

$$s \leq t \Rightarrow C(\bar{\alpha}(2s)) \subseteq C(\bar{\alpha}(2t))$$

for each α, $\bigcup_t C(\bar{\alpha}(2t))$ is a total ordering and

$$\alpha \in A \Leftrightarrow \bigcup_t C(\bar{\alpha}(2t)) \text{ is a wellordering.}$$

We can now define $D(u)$ with the required properties so that

$$D(\bar{\alpha}(2t)) = D_1 \cup D_2,$$

where D_1 is a totally ordered subrelation of $C(\bar{\alpha}(2t))$ and D_2 is a finite tail end of odd integers. In detail, take

$$D(\bar{\alpha}(0)) = D(1) = \emptyset$$

and let $D(\bar{\alpha}(2t+2))$ be the extension of $D(\bar{\alpha}(2t))$ obtained as follows. If the smallest even number in the field of $C(\bar{\alpha}(2t+2))$ which is not in the field of $D(\bar{\alpha}(2t))$ is comparable in $C(\bar{\alpha}(2t+2))$ to every even number in the field of $D(\bar{\alpha}(2t))$, then extend $D(\bar{\alpha}(2t))$ by adding this number to its field and putting in the appropriate place. If this fails to hold (or if $C(\bar{\alpha}(2t+2))$ has the same even numbers in its field as $D(\bar{\alpha}(2t))$, then extend $D(\bar{\alpha}(2t))$ by adding some unused odd number at the top.

It is now easy to check (i), (ii) and (iii) and (iv) follows from the fact that for each α, $\bigcup_t D(\bar{\alpha}(2t))$ differs from $\bigcup_t C(\bar{\alpha}(2t))$ only by (possibly) having some odd integers at its top, with their natural ordering. (To prove this check by induction on $2n \in \omega$ that if $2n$ is in the field of $\bigcup_t C(\bar{\alpha}(2t))$, then $2n$ is in the field of $\bigcup_t D(\bar{\alpha}(2t))$.) ⊣

6G.7. THEOREM. (Martin [1970]). *If there exists a cardinal number $\varkappa$ such that $\varkappa \to (\aleph_1)$, then $Det(\mathbf{\Sigma}^1_1)$ holds.*

PROOF. Given $A \subseteq \mathcal{N}$ in $\mathbf{\Sigma}^1_1$, choose D as in the lemma so in particular

$$\alpha \notin A \Leftrightarrow \bigcup_t D(\bar{\alpha}(2t)) \text{ is a wellordering.}$$

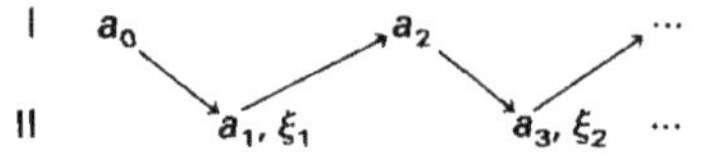

Diagram 6G.1.

We define a new game A^* where player II makes additional auxiliary moves in $\varkappa$ as in Diagram 6G.1. At the end of the run an irrational

$$\alpha = (a_0, a_1, \ldots)$$

has been played as well as an infinite sequence of ordinals $\xi_0, \xi_1, \ldots$ below $\varkappa$. For each t, let

$$\mathit{Field}(D(\bar{\alpha}(2t))) = \{x_1, \ldots, x_t\}$$

so that

$$\mathit{Field}(\bigcup_t D(\bar{\alpha}(2t))) = \{x_1, x_2, \ldots\};$$

now II wins the run if the map

$$x_t \mapsto \xi_t$$

is order-preserving from $\bigcup_t D(\bar{\alpha}(2t))$ into the natural ordering on $\varkappa$.

It is obvious that the game A^* is open, so it is determined. Also if II wins A^*, then obviously II wins A since he can play in A with the same strategy he has in A^* (disregarding his ordinal moves) and at the end he has an order-preserving map from $\bigcup_t D(\bar{\alpha}(2t))$ into $\varkappa$, so $\bigcup_t D(\bar{\alpha}(2t))$ is a wellordering and $\alpha \notin A$.

Assume then that I wins A^* by some strategy σ^*—we must show that I can also win A.

Given t distinct ordinals $\xi_1, \ldots, \xi_t$ and a sequence code $u = \langle a_0, \ldots, a_{2t-1} \rangle$, let $\xi_{n(1)}, \ldots, \xi_{n(t)}$ be some ordering of $\{\xi_1, \ldots, \xi_t\}$ and consider the sequence of moves in Diagram 6G.2 as an initial piece of a play in A^*. Clearly, there is *exactly one* ordering of $\{\xi_1, \ldots, \xi_t\}$ so that II has not already lost after these first t moves; let us denote it by

$$\xi_{n(u,1)}, \ldots, \xi_{n(u,t)}.$$

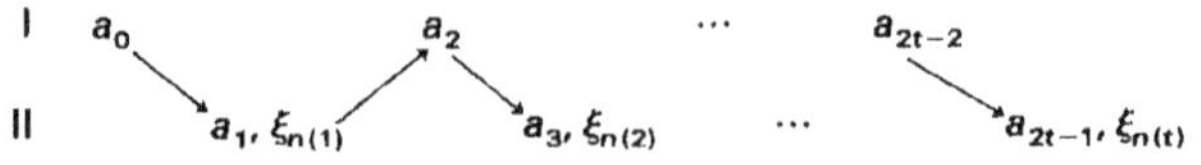

Diagram 6G.2.

Now for each sequence code $u = \langle a_0, \dots, a_{2t-1}\rangle$ of length $2t$ consider the partition

$$F_u : \varkappa^{[t]} \to \omega$$

given by

$$F_u(\{\xi_1, \dots, \xi_t\}) = \sigma^*(a_0, (a_1, \xi_{n(u,1)}), \dots, a_{2t-2}, (a_{2t-1}, \xi_{n(u,t)}))$$

and let $J \subseteq \varkappa$ be of cardinality $\aleph_1$ and homogeneous for all these partitions. Finally, put

$$\sigma(a_0, a_1, \dots, a_{2t-1}) = F_u(\{\xi_1, \dots, \xi_t\})$$

where $u = \langle a_0, a_1, \dots, a_{2t-1}\rangle$ and $\xi_1, \dots, \xi_t$ are arbitrary distinct members of J. We will show that σ is a winning strategy for I in the game A.

In effect we define σ from σ^* by simulating the ordinal moves of II in A^* in some homogeneous set J whose members give no information to I in that game.

Suppose then that I follows σ in some run of A and the play

$$\alpha = (a_0, a_1, \dots)$$

results, but $\alpha \notin A$. Then $\bigcup_t D(\bar{\alpha}(2t))$ is a wellordering of countable rank, so if $\{x_1, x_2, \dots\}$ is its field, there is some order-preserving map

$$x_t \mapsto \xi_t$$

with all the $\xi_t \in J$, since J has cardinality $\aleph_1$. It is now obvious that in the run of A^* pictured in Diagram 6G.3 player I is following his winning strategy σ^* and yet he loses, which is a contradiction. ⊣

One can extend this method to prove the determinacy of simple combinations of $\mathbf{\Sigma}^1_1$ sets (e.g. *differences*) granting that some $\varkappa \to (\aleph_1)$. In fact, Martin has established the determinacy of a reasonably large subclass of $\mathbf{\Delta}^1_2$ from the hypothesis that there exist *long* (infinite) *sequences* of measurable cardinals. On the other hand it is known that the existence of any number of measurable cardinals does not imply $Det(\mathbf{\Delta}^1_2)$ and no one has succeeded in proving $Det(\mathbf{\Delta}^1_2)$ from any (reasonable) large cardinal

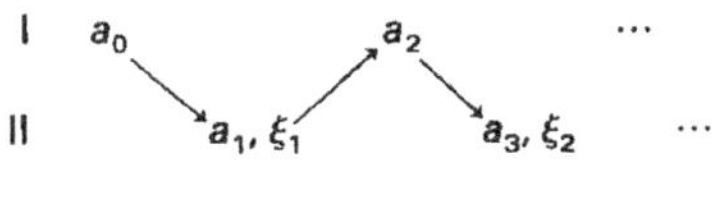

Diagram 6G.3.

hypothesis. Martin [19?c] gives a comprehensive treatment of the known results in this important problem of relating determinacy and large cardinal hypotheses.

EXERCISES

6G.8. Prove that if some cardinal carries an $\aleph_1$-complete non-principal ultrafilter, then there exists a measurable cardinal.

Hint. Let $\varkappa$ be the least cardinal which carries an $\aleph_1$-complete non-principal ultrafilter $\mathfrak{U}$ and suppose towards a contradiction that for some $\lambda < \varkappa$, there are sets A_η of measure 0 (not in $\mathfrak{U}$) such that $\bigcup_{\eta<\lambda} A_\eta \in \mathfrak{U}$. Pick the least λ for which such a sequence exists and take $B_\eta = A_\eta - \bigcup_{\zeta<\eta} A_\zeta$, so that the B_η are pairwise disjoint of measure 0 and $\bigcup_{\eta<\lambda} B_\eta \in \mathfrak{U}$. Now for $X \subseteq \lambda$, put

$$X \in \mathfrak{U}^* \Leftrightarrow \bigcup_{\eta \in X} B_\eta \in \mathfrak{U}$$

so that

$$X \notin \mathfrak{U}^* \Leftrightarrow \bigcup_{\eta \in X} B_\eta \notin \mathfrak{U}$$

and verify that $\mathfrak{U}^*$ is $\aleph_1$-complete on λ contradicting the choice of $\varkappa$.

⊣

We have been referring to "large cardinal hypotheses" but there is no hint in what we have proved that measurable cardinals are large. In fact it is consistent with the axioms of Zermelo–Fraenkel set theory (without the axiom of choice) that $\aleph_1$ is measurable—granting that the hypothesis *MC* is consistent at all, Jech [1968]. On the other hand, the axiom of choice implies that measurable cardinals are very large indeed. We will give here only a glimpse of the results that can be proved in this direction.

Recall that $\varkappa$ is *regular* if there is no function $f:\lambda \to \varkappa$ (with some $\lambda < \varkappa$) which is unbounded in $\varkappa$. Also, $\varkappa$ is *strongly inaccessible* if $\varkappa$ is regular and for each $\lambda < \varkappa$, $card(power(\lambda)) = 2^\lambda < \varkappa$.

6G.9. Prove that a measurable cardinal is regular and in fact (granting the axiom of choice) strongly inaccessible.

Hint. If $f:\lambda \to \varkappa$ is unbounded, then

$$\varkappa = \bigcup_{\eta<\lambda} \{\xi : \xi < f(\eta)\}$$

and each of the sets in this union has measure 0 by $\varkappa$-completeness.

Suppose now there is a $\lambda < \varkappa$ so that $\varkappa \le 2^\lambda$, where we have used the axiom of choice in comparing κ with power (λ). There is then an injection

$$\xi \mapsto X_\xi \subseteq \lambda,$$

i.e. such that

$$\begin{aligned}\xi \neq \eta &\Longrightarrow X_\xi \neq X_\eta\\ &\Longrightarrow (\exists \zeta < \lambda)[\zeta \in (X_\xi - X_\eta) \vee \zeta \in (X_\eta - X_\xi)].\end{aligned}$$

Choose then some function $f(\xi, \eta)$ such that

$$\xi \neq \eta \Longrightarrow f(\xi, \eta) \in (X_\xi - X_\eta) \vee f(\xi, \eta) \in (X_\eta - X_\xi).$$

Define now a partition of $\varkappa^{[2]}$ into λ parts by

$$F(\{\xi, \eta\}) = \begin{cases} f(\xi, \eta) & \text{if } \xi < \eta \\ f(\eta, \xi) & \text{if } \eta < \xi \end{cases}$$

and using 6G.5, let $I \subseteq \varkappa$ be a homogeneous set for this partition, $card(I) = \varkappa$. Now check that we cannot have $\xi_1 < \xi_2 < \xi_3$ all $\xi_1, \xi_2, \xi_3 \in I$ without an obvious contradiction. ⊣

The proof obviously shows that if every partition of $\varkappa^{[2]}$ into $\lambda < \varkappa$ parts has a homogeneous set of cardinality at least 3, then $2^\lambda < \varkappa$. Much stronger results can be proved about a measurable cardinal $\varkappa$—it cannot be the first strongly inaccessible, it must have $\varkappa$ strongly inaccessibles below it, it cannot be the least λ with λ strongly inaccessibles below it, etc.

In 6A we saw that the hypothesis $Det(\Lambda)$ implies a good deal of regularity for the pointsets in Λ. By modifying a bit those proofs, we can establish the same results for the sets in $\exists^{\mathcal{N}}\Lambda$, so in particular $Det(\mathbf{\Sigma}^1_1)$ (which is equivalent to $Det(\mathbf{\Pi}^1_1)$ by 6A.4) implies that every $\mathbf{\Sigma}^1_2$ pointset P is absolutely measurable, it has the property of Baire and it is either countable or it has a perfect subset. These regulation results for $\mathbf{\Sigma}^1_2$ then follow from the hypothesis that some $\varkappa \to (\aleph_1)$.

We now give brief outlines of these arguments.

6G.10. Suppose Λ is an adequate pointclass closed under Borel substitution and assume $Det(\Lambda)$; prove that every uncountable set P in $\exists^{\mathcal{N}}\Lambda$ has a perfect subset. (Martin.)

Infer that if there exists a cardinal $\varkappa$ such that $\varkappa \to (\aleph_1)$, then every uncountable $\mathbf{\Sigma}^1_2$ set has a perfect subset. (Solovay [1969].)

Hint. It is enough to prove the result for $P \subseteq \mathcal{N}$ as in 6A.12, so assume

$$P(\alpha) \Leftrightarrow (\exists\beta)Q(\alpha, \beta)$$

with Q in Λ and recall the game G which we associated with Q in the hint to 6E.5. Following the same hint, G is determined and if I wins G then easily P has a perfect subset; if II wins G then any winning strategy τ for II can be used to enumerate P.

The second assertion follows immediately by Martin's theorem 6G.7. ⊣

6G.11. Suppose Λ is an adequate pointclass closed under Borel substitution and assume $Det(\Lambda)$; prove that every pointset in $\exists^{\mathcal{N}}\Lambda$ has the property of Baire. (Kechris [1973].)

Infer that if there exists a cardinal $\varkappa$ such that $\varkappa \to (\aleph_1)$, then every $\boldsymbol{\Sigma}^1_2$ pointset has the property of Baire. (Solovay.)

Hint. We will prove that under the hypotheses, each A in $\exists^{\mathcal{N}}\Lambda$ is either meager or there is an s such that $\bar{N}(s) - A$ is meager, from which the result follows by 6A.15.

Suppose

$$x \in A \Leftrightarrow (\exists\alpha)Q(x, \alpha)$$

with Q in Λ and consider the following game which is a modification of the game G^{**} used in 6A.14. The players move as in Diagram 6G.4. The restrictions on the players are the same as in G^{**}; if both players follow the rules to the end, then

$$\text{I wins the run} \Leftrightarrow Q(x, \alpha)$$

where x is the unique point in all the basic nbhds $N(s_i) \subseteq \mathfrak{X}$ and $\alpha = (a_0, a_1, a_2, \ldots)$.

Clearly G is determined and it is easy to see as in 6A.14 that if A is meager then II wins G. Conversely, if II wins G via σ and $Q(x, \alpha)$ holds, call a sequence $s_0, \ldots, s_n$ of even length *good* (for x, α, σ) if s_0, s_1, $\alpha(0)$, $\alpha(1), \ldots, \alpha(n-3)$, $\alpha(n-2)$, s_{n-1}, s_n is the initial part of some play in G in which the restrictions have been obeyed, II has followed σ and $x \in \bar{N}(s_n)$.

Diagram 6G.4

If $Q(x, \alpha)$, then clearly there must be a good sequence with no good extension; hence,

$$x \in A \rightarrow \text{for some } s_0, a_1, a_0, a_1, \ldots, s_n, a_{n-1}, a_n,$$
$$x \in \bigcap \{\bar{N}(s_n) - N(\sigma(s_0, s_1, a_0, a_1, \ldots, s_n, a_{n-1}, a_n, s)) : \bar{N}(s) \subseteq \bar{N}(s_n) \;\&\; radius(N(s)) \leq \tfrac{1}{2}\, radius(N(s_n))\}$$

and the set on the right is clearly meager. Thus,

(1) $$\text{II wins } G \leftrightarrow A \text{ is meager.}$$

We also claim that

$$\text{I wins } G \rightarrow \text{for some } s, \bar{N}(s) - A \text{ is meager.}$$

To check this, let $s = s_0$ be the first move of I by a winning strategy σ and for any x call a sequence $s_0, s_1, a_0, a_1, \ldots, s_n$ (n even) *good* (for x and σ) if it is played by the rules with I following σ and $x \in \bar{N}(s_n)$. Easily, if $x \in (\bar{N}(s_0) - A)$ then there must be a maximal good sequence (which may be the one-term sequence s_0) or else we would get a play establishing that $Q(x, \alpha)$ holds for some α; thus

$$x \in (\bar{N}(s_0) - A) \rightarrow \text{for some } s_1, a_0, a_1, \ldots, s_n$$
$$x \in \bigcap \{\bar{N}(s_n) - N(\sigma_*(s_0, s_1, \ldots, s_n, s)) : \bar{N}(s) \subseteq \bar{N}(s_n) \;\&\; radius(N(s)) \leq \tfrac{1}{2}\, radius(N(s_n))\}$$

where $\sigma_*(s_0, s_1, \ldots, s_n, s) = s_{n+2}$ is the "nbhd code response" of I to II's play s. This implies immediately that $\bar{N}(s_0) - A$ is meager. ⊣

6G.12. Suppose Λ is an adequate pointclass closed under Borel substitution and assume $Det(\Lambda)$; prove that every pointset $A \subseteq \mathfrak{X}$ in $\exists^{\mathcal{N}}\Lambda$ is absolutely measurable. (Kechris.)

Infer that if there exists a $\varkappa$ such that $\varkappa \rightarrow (\aleph_1)$, then every $\boldsymbol{\Sigma}^1_2$ pointset is absolutely measurable. (Solovay.)

Hint. Suppose first $A \subseteq {}^{\omega}2$ and for some $P \subseteq {}^{\omega}2 \times {}^{\omega}2$,

(*) $$\alpha \in A \leftrightarrow (\exists \beta \in {}^{\omega}2) P(\alpha, \beta).$$

For each fixed Borel measure μ on ${}^{\omega}2$ and each $\varepsilon > 0$ consider the *modified covering game* $G^{\mu}_1(A, \varepsilon)$ defined as follows. Players I and II make moves as in Diagram 6G.5. The restrictions are exactly like those in $G^{\mu}(A, \varepsilon)$ defined in 6A.17, except that the additional moves $b_0, b_1, \ldots$ of I must also be 0 or 1 and we insist that the sets G_n produced by II satisfy

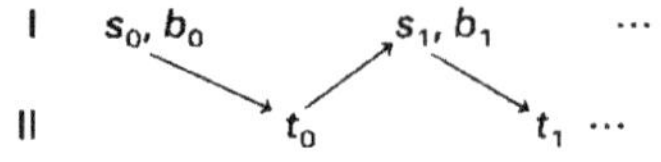

Diagram 6G.5.

$\mu(G_n) \leq \varepsilon/2^{2n+1}$. At the end of a run, binary sequences

$$\alpha = (s_0, s_1, \ldots)$$
$$\beta = (b_0, b_1, \ldots),$$

have been played by I and II has defined a sequence $\{G_n\}_{n\in\omega}$ of finite unions of basic nbhds in ${}^\omega 2$ with special properties. Set

$$G = \bigcup_n G_n$$

and put

$$\text{I wins the run} \Leftrightarrow \alpha \notin G \ \&\ P(\alpha, \beta).$$

It is now easy to mimic the proof of 6A.17 and show that if $G_1^\mu(A, \varepsilon)$ is determined for each $\varepsilon > 0$ and A has no Borel subsets of μ-measure >0, then $\mu(A) = 0$.

Assume now the hypotheses on Λ and check first that if $A \subseteq {}^\omega 2$ is in $\exists^{\mathcal{N}}\Lambda$, then A can be defined from some P as in $(*)$ above, using the Δ_1^1 isomorphism of ${}^\omega 2$ with $\mathcal{N}$. It follows that for each $A \subseteq {}^\omega 2$, if $A \in \exists^{\mathcal{N}}\Lambda$, then the game $G_1^\mu(A, \varepsilon)$ is determined and hence *if $A \in \exists^{\mathcal{N}}\Lambda$ and A has no Borel subsets of μ-measure >0, then $\mu(A) = 0$.*

Given $A \subseteq {}^\omega 2$ in $\exists^{\mathcal{N}}\Lambda$, let $C = {}^\omega 2 - A$ and choose a Borel set $\tilde{C} \supseteq C$ by 2H.7 so that $\tilde{C} - C = \tilde{C} \cap A$ contains no Borel set of μ-measure >0. Since $\tilde{C} \in \exists^{\mathcal{N}}\Lambda$, we then have that $\mu(\tilde{C} \cap A) = 0$. so that C and hence $A = {}^\omega 2 - C$ is μ-measurable.

The result holds for arbitrary product spaces $\mathfrak{X}$ because it is clearly preserved under Borel isomorphisms. ⊣

Solovay's original proofs of these regularity results for $\boldsymbol{\Sigma}_2^1$ depended heavily on metamathematical ideas. We will come back to them in Chapter 8, as the metamathematical approach illuminates these theorems from an interesting and very different point of view.

6H. Historical remarks

[1] As with so many other basic notions of our subject, infinite games were introduced into descriptive set theory by the Polish mathematicians

of the midwar period. Mazur invented the **-game (for the reals) and conjectured its connection with category, 6A.14; Banach verified the conjecture but did not publish his proof. (Later, Oxtoby [1957] proved a generalization of 6A.14 to arbitrary topological spaces.)

[2] Gale–Stewart [1953] introduced into the literature the general notion of an *infinite game of perfect information* and began a systematic study of these games. They proved that closed (and open) games are determined and that not all games are determined and they asked some basic questions, e.g. if all Borel games are determined.

[3] Wolfe [1955] proved $Det(\boldsymbol{\Sigma}^0_2)$ and some time later, Davis [1964] established $Det(\boldsymbol{\Sigma}^0_3)$ in one of the fundamental early papers in the subject. For many years this was the strongest result in the direction of establishing determinacy, until Martin [1970] proved $Det(\boldsymbol{\Sigma}^1_1)$ granting *MC* (the hypothesis that there exists at least one measurable cardinal). Using the new methods introduced by Martin and ideas of Baumgartner, Paris [1972] established $Det(\boldsymbol{\Sigma}^0_4)$ in *ZFC*. Finally Martin [1975] completed this cycle of results by proving the determinacy of all Borel sets in *ZFC*.

[4] Martin's forthcoming monograph will give a detailed exposition of these positive results as well as the known (quite intricate) connections between determinacy and large cardinal hypotheses.

[5] In addition to proving $Det(\boldsymbol{\Sigma}^0_3)$, Davis [1964] introduced the *-game (which he attributed to L. Dubins) and established the connection of this game with perfect sets, 6A.10 and 6A.11. From our present point of view, these results and the theorem of Banach relate determinacy hypotheses with regularity properties of sets.

[6] This point of view was propounded in an extreme form at about the same time by Mycielski–Steinhaus [1962] who considered the blatantly false proposition

$$\begin{aligned} AD &\Leftrightarrow Det(Power(\mathcal{N})) \\ &\Leftrightarrow \text{every subset of } \mathcal{N} \text{ is determined.} \end{aligned}$$

Mycielski and Steinhaus suggested that we replace the (true) axiom of choice by the (false) hypothesis *AD* in our development of ordinary mathematics; our results must then be interpreted in some universe of sets presumably smaller than the usual collection of all sets (where *AD* fails and *AC* holds). The alleged advantage of replacing *AC* by *AD* is the resulting regularity of pointsets—all sets are absolutely measurable, they have the property of Baire, etc.

[7] Proofs of these "regularity results" from the "axiom of determinacy" *AD* were given in the sequence of papers Mycielski [1964], Mycielski–Swierczkowski [1964] and Mycielski [1965]. In our exposition in 6A we have taken the point of view that these results relate $Det(\Lambda)$ with the regularity of the pointsets in Λ, whenever Λ is an arbitrary pointclass with certain reasonable closure properties. Now 6G.10–6G.12 appear as refinements which establish the regularity of sets in $\exists^{\mathcal{N}}\Lambda$ from $Det(\Lambda)$.

[8] Despite the dubious reasons given by Mycielski–Steinhaus [1962] for studying consequences of the false hypothesis *AD*, it has turned out that this was in fact a non-trivial and significant program. We will turn briefly to it in 7D. Let us just notice here that one of the first results that focused the attention on "mainstream" set theorists to determinacy was Solovay's proof in 1967 that *AD* implies that $\aleph_1$ is a measurable cardinal, see 7D.18.

[9] In the most important single contribution to the theory presented in this Chapter, Blackwell [1967] gave a new proof of the separation theorem for $\mathbf{\Sigma}^1_1$ sets which used the Gale–Stewart theorem. Addison and Martin instantly saw the possibilities of this approach and independently established that $Det(\mathbf{\Delta}^1_2) \rightarrow \text{Reduction}(\Pi^1_3)$; then Martin again and Moschovakis (who heard Addison lecture on his results) proved independently the First Periodicity Theorem 6B.1. These theorems appeared in Addison–Moschovakis [1968] and Martin [1968] and started the sequence of results which has led to the present substantial structure theory for the projective sets and pointclasses on the basis of the hypothesis of projective determinacy.

[10] The few results on measurable and Ramsey cardinals which we covered in 6G are well-known and we will not attempt to trace their history here; see Drake [1974].

[11] As we pointed out in the introduction to this chapter, Solovay's regularity results about $\mathbf{\Sigma}^1_2$ sets were the first applications of the hypothesis *MC* (or any strong axioms for that matter) to problems in descriptive set theory. These were established at about 1965 and they were very well known among set theorists long before their (partial) publication in Solovay [1969]; they were instrumental in creating the climate where the use of strong hypotheses in descriptive set theory became tenable. Solovay's proofs were metamathematical (he used forcing) and had a very different flavor from the game-theoretic arguments we gave in 6G.10, 6G.11 and 6G.12. We will come back to them in Chapter 8.

CHAPTER 7

THE RECURSION THEOREM

Kleene's recursion theorem is a very simple fact with remarkably broad and important consequences. Combined with techniques also pioneered by Kleene, it allows us in effect to define recursive partial functions by transfinite recursion and to obtain *uniform versions* of many results, in the sense of 3H.

After proving the recursion theorem in a wide context in 7A, we will use it in 7B to establish the *Suslin–Kleene theorem*, the central result of the effective theory. In 7C we will consider briefly the general theory of *inductive definability* (of relations) and in 7D we will look at some of the consequences of the so-called *axiom of* (full) *determinacy*.

It is perhaps an indication of the significance of the recursion theorem that this section on full determinacy come in this chapter; as it happens, one of the key lemmas in this most set theoretic part of our subject depends on the recursion theorem for its proof.

7A. Recursion in a Σ^*-pointclass

Let us call for convenience Γ a *Σ^*-pointclass* if it is a *Σ-pointclass* which is ω-parametrized and has the substitution property, as in 3G—these are the pointclasses which carry a very smooth theory of Γ-recursion.

For each space $\mathfrak{X}$, let

$$G_\Gamma^{\mathfrak{X}\times\omega} = G \subseteq \mathcal{N} \times \mathfrak{X} \times \omega$$

be the fixed (good) universal set for the $\mathbf{\Gamma}$-subsets of $\mathfrak{X}\times\omega$ and for each $\mathfrak{Y}$ define the partial function

$$U_\Gamma^{\mathfrak{X},\mathfrak{Y}} = U : \mathcal{N} \times \mathfrak{X} \to \mathfrak{Y}$$

as follows:

$$U(\varepsilon, x)\downarrow \leftrightarrow \text{there exists a unique } y \in \mathfrak{Y} \text{ such that}$$
$$(\forall s)[y \in N_s \leftrightarrow G(\varepsilon, x, s)],$$
$$U(\varepsilon, x) = \text{the unique } y \text{ such that}$$
$$(\forall s)[y \in N_s \leftrightarrow G(\varepsilon, x, s)];$$

in other words, U is *the largest* partial function on $\mathfrak{X}$ to $\mathfrak{Y}$ which is computed on its domain by G. Finally, for each $\varepsilon \in \mathcal{N}$ define the partial function

$$\{\varepsilon\}_{\Gamma}^{\mathfrak{X},\mathfrak{Y}} = \{\varepsilon\} : \mathfrak{X} \to \mathfrak{Y}$$

by

$$\{\varepsilon\}(x) = U(\varepsilon, x).$$

(Sometimes φ_ε or f_ε is used for $\{\varepsilon\}$ but Kleene's original notation is well established and really easier to use in the long run.)

We will always omit the cumbersome superscripts and subscripts $\mathfrak{X}, \mathfrak{Y}, \Gamma$, unless they are necessary for clarity.

7A.1. Theorem. *Let Γ be a fixed Σ^*-pointclass.*

(i) *For each $\mathfrak{X}, \mathfrak{Y}$, the partial function*

$$U_{\Gamma}^{\mathfrak{X},\mathfrak{Y}} = U : \mathcal{N} \times \mathfrak{X} \to \mathfrak{Y}$$

is Γ-recursive on its domain.

(ii) *A partial function $f : \mathfrak{X} \to \mathfrak{Y}$ is $\boldsymbol{\Gamma}$-recursive on its domain, if and only if there is some $\varepsilon \in \mathcal{N}$ such that $f \subseteq \{\varepsilon\}$* i.e.

$$f(x)\downarrow \rightarrow f(x) = \{\varepsilon\}(x).$$

(iii) *A partial function $f : \mathfrak{X} \to \mathfrak{Y}$ is Γ-recursive on its domain if and only if there is some recursive $\varepsilon \in \mathcal{N}$ such that $f \subseteq \{\varepsilon\}$, i.e.*

$$f(x)\downarrow \rightarrow f(x) = \{\varepsilon\}(x).$$

(iv) *For each space $\mathfrak{X}$ of type* 0 *or* 1 *and each $\mathfrak{W}, \mathfrak{Y}$ there is a recursive function*

$$S_{\Gamma}^{\mathfrak{X},\mathfrak{W},\mathfrak{Y}} = S : \mathcal{N} \times \mathfrak{X} \to \mathcal{N}$$

such that for all $\varepsilon \in \mathcal{N}$, $x \in \mathfrak{X}$,

$$\{\varepsilon\}(x, w) = \{S(\varepsilon, x)\}(w).$$

(*Kleene.*)

PROOF. (i) is immediate and (ii) and (iii) follow trivially from the properties of a good parametrization, 3H.1. To prove (iv), let

$$S : \mathcal{N} \times \mathfrak{X} \to \mathcal{N}$$

be chosen by 3H.1 so that for all ε, x, w, s,

$$G(\varepsilon, x, w, s) \leftrightarrow G(S(\varepsilon, x), w, s). \qquad \dashv$$

The recursion theorem follows from this result by a simple (if somewhat subtle) diagonalization argument.

7A.2. KLEENE'S RECURSION THEOREM.[1] *Let Γ be a Σ^*-pointclass and suppose*

$$f : \mathcal{N} \times \mathfrak{X} \to \mathfrak{Y}$$

is Γ-recursive on its domain; then there exists a fixed $\varepsilon^* \in \mathcal{N}$ such that for all $x \in \mathfrak{X}$,

$$(*) \qquad f(\varepsilon^*, x)\downarrow \to [f(\varepsilon^*, x) = \{\varepsilon^*\}(x)].$$

In fact there is a fixed recursive function $R(\alpha)$ depending only on the spaces $\mathfrak{X}, \mathfrak{Y}$, so that if α is a code of f in the sense that

$$f(\varepsilon, x)\downarrow \to f(\varepsilon, x) = \{\alpha\}(\varepsilon, x),$$

then we can take

$$\varepsilon^* = R(\alpha)$$

in $(*)$.

In particular, if f is Γ-recursive on its domain, then we can find a recursive ε^* which satisfies $(*)$.

PROOF. Given $\mathfrak{X}$ and $\mathfrak{Y}$, define

$$g : \mathcal{N} \times \mathcal{N} \times \mathfrak{X} \to \mathfrak{Y}$$

by

$$g(\beta, \alpha, x) = \{\alpha\}(S(\beta, \beta, \alpha), x),$$

where

$$S : \mathcal{N} \times (\mathcal{N} \times \mathcal{N}) \to \mathcal{N}$$

is recursive and satisfies (iv) of 7A.1. Clearly g is Γ-recursive on its domain, so by 7A.1 there is a fixed recursive ε_0 so that

$$\begin{aligned} g(\beta, \alpha, x)\downarrow \to g(\beta, \alpha, x) &= \{\varepsilon_0\}(\beta, \alpha, x) \\ &= \{S(\varepsilon_0, \beta, \alpha)\}(x) \end{aligned}$$

by the key property of the function S. Taking $\beta = \varepsilon_0$ in this implication, we obtain

$$\{\alpha\}(S(\varepsilon_0, \varepsilon_0, \alpha), x) = \{S(\varepsilon_0, \varepsilon_0, \alpha)\}(x)$$

and we can satisfy (*) by setting

$$\varepsilon^* = R(\alpha) = S(\varepsilon_0, \varepsilon_0, \alpha). \qquad \dashv$$

The recursion theorem has been described as a *fixed point theorem* for maps on the collection of Γ-recursive partial functions which are *uniform* in the coding for these objects introduced by 7A.1. This point of view is a little artificial when we consider partial functions whose domain is not in Γ, but in any case, the applications of the theorem are hard to couch in topological terms. They tend rather to exhibit a connection between this result and definition by recursion as the next theorem plainly shows.

7A.3. THEOREM. *If Γ is a Σ^*-pointclass, then the collection of Γ-recursive (total) functions is closed under primitive recursion.*

PROOF. We are given total Γ-recursive maps

$$g : \mathfrak{X} \to \mathcal{Y}$$
$$h : \mathcal{Y} \times \omega \times \mathfrak{X} \to \mathcal{Y}$$

and we define

$$f : \omega \times \mathfrak{X} \to \mathcal{Y}$$

by the recursion

$$\begin{cases} f(0, x) = g(x) \\ f(m+1, x) = h(f(m, x), m, x). \end{cases}$$

To see that f is also Γ-recursive, let

$$\varphi(\varepsilon, m, x) = \begin{cases} g(x) & \text{if} \quad m = 0, \\ h(\{\varepsilon\}(m-1, x), m-1, x) & \text{if} \quad m > 0, \end{cases}$$

where

$$\{\varepsilon\}(k, x) = U(\varepsilon, k, x)$$

is Γ-recursive on its domain as a function of ε, k, x by 7A.1. It is easy to check (using the substitution property) that φ is Γ-recursive on its domain, so by the recursion theorem there is a fixed recursive ε^* so that

$$\varphi(\varepsilon^*, m, x)\!\downarrow \;\rightarrow [\varphi(\varepsilon^*, m, x) = \{\varepsilon^*\}(m, x)].$$

For this ε^*, then we have

$$\{\varepsilon^*\}(0, x) = g(x),$$
$$\{\varepsilon^*\}(m+1, x) = h(\{\varepsilon^*\}(m, x), m, x)$$

which implies by a trivial induction on m that $\{\varepsilon^*\}(m, x)$ is always defined and for each m, x,

$$\{\varepsilon^*\}(m, x) = f(m, x).$$

Now f is Γ-recursive by 7A.1, since ε^* is recursive. ⊣

We will see in the exercises that the collection of partial functions which are Γ-recursive on their domain is also closed under primitive recursion (when Γ is a Σ^*-pointclass), but this simpler result already shows the power of the recursion theorem. Even for the simple case $\Gamma = \Sigma^0_1$ of ordinary recursion, this is the simplest known proof of 7A.3 (for functions into spaces $\mathcal{Y}$ which are not of type 0).

EXERCISES

Let us first consider a simple case of definition by effective *transfinite* recursion.

Suppose $\prec$ is a (strict) wellfounded relation with $Field(\prec) \subseteq \mathcal{X}$ and

$$f : Field(\prec) \to \mathcal{Y}$$

is defined by *recursion on* $\prec$, i.e. f satisfies the equation

$$(*) \qquad f(x) = G(\{(u, f(u)) : u \prec x\}, x)$$

which determines it uniquely on $Field(\prec)$ (by induction). Now the map G is defined on

$$Domain(G) = \{(h, x) : x \in Field(\prec) \text{ and } h : \mathcal{X} \to \mathcal{Y} \text{ is a partial function with } Domain(h) = \{u : u \prec x\}\};$$

we will say that G is *Γ-effective* (Γ a given Σ^*-pointclass) if there is a partial function

$$g : \mathcal{N} \times \mathcal{X} \to \mathcal{Y}$$

which is Γ-recursive on its domain and such that for each $\varepsilon \in \mathcal{N}$ and each $x \in Field(<)$,

$$(\forall u)[u < x \rightarrow \{\varepsilon\}(u)\downarrow] \\ \rightarrow g(\varepsilon, x)\downarrow \ \& \ g(\varepsilon, x) = G(\{(u, \{\varepsilon\}(u)) : u < x\}, x).$$

If f is defined by $(*)$ with a Γ-effective G, we say that f is defined by *Γ-effective recursion* on $<$.

7A.4. Show that if Γ is a Σ^* pointclass and $f : \mathfrak{X} \rightarrow \mathfrak{Y}$ is defined by Γ-effective recursion on some wellfounded relation $<$ with $Field(<) \subseteq \mathfrak{X}$, then f is Γ-recursive on $Field(<)$. (Kleene.[1])

Hint. Let g "compute" G as above and choose a recursive $\varepsilon^* \in \mathcal{N}$ by the recursion theorem so that

$$g(\varepsilon^*, x)\downarrow \rightarrow \{\varepsilon^*\}(x) = g(\varepsilon^*, x);$$

now show by induction on $x \in Field(<)$ that $f(x) = \{\varepsilon^*\}(x)$. ⊣

Notice that there are no effectivity hypotheses on the relation $<$ in this result and in fact we can obtain ε^* in the proof directly from g, with no knowledge of the relation $<$; this is important in more subtle applications of this method where we define ε^* before we even know that $<$ is wellfounded and then show that it has whatever properties we need *if* the relevant relation $<$ happens to be wellfounded.

7A.5. Prove that if Γ is a Σ^*-pointclass, then the collection of partial functions which are Γ-recursive on their domain is closed under both minimalization and primitive recursion.

Hint. One must be a bit careful with the definitions. For minimalization,

$$\mu i[g(x, i) = 0] = w \leftrightarrow g(x, w) = 0 \ \& \ (\forall i < w)(\exists j)[g(x, i) = j + 1]$$

and the argument uses only the fact that Γ is a Σ-pointclass. For primitive recursion we must understand the basic equations literally for partial functions g, h,

$$f(0, x) = g(x),$$
$$f(m + 1, x) = h(f(m, x), m, x),$$

so that for example if $f(m, x)\downarrow$, we must have $f(0, x), \ldots, f(m-1, x)$ all defined. The proof is the same as that of 7A.3. ⊣

It is occasionally useful (particularly in the effective theory) to give a coding in ω for the partial functions which are Γ-recursive on their domain. Fix a Σ^*-pointclass Γ then and choose $G^{\mathfrak{X}} \subseteq \omega \times \mathfrak{X}$ for each $\mathfrak{X}$ to be universal for $\Gamma \restriction \mathfrak{X}$ by 3H.3, so that the parametrization system $\{G^{\mathfrak{X}}\}$ is good. Using the same notation as in the case of parametrizations in $\mathcal{N}$ (no conflict can arise), define the partial function

$$U_\Gamma^{\mathfrak{X},\mathfrak{Y}} = U : \omega \times \mathfrak{X} \to \mathfrak{Y}$$

by

$$\begin{aligned} U(e, x)\downarrow &\Leftrightarrow \text{there exists a unique } y \in \mathfrak{Y} \text{ such} \\ &\qquad \text{that } (\forall s)[y \in N_s \Leftrightarrow G(e, x, s)], \\ U(e, x) &= \text{the unique } y \text{ such that } (\forall s)[y \in N_s \Leftrightarrow G(e, x, s)] \end{aligned}$$

and for each $e \in \omega$, define the partial function

$$\{e\}_\Gamma^{\mathfrak{X},\mathfrak{Y}} = \{e\} : \mathfrak{X} \to \mathfrak{Y}$$

by

$$\{e\}(x) = U(e, x).$$

7A.6. Let Γ be a Σ^*-pointclass and define $U_\Gamma^{\mathfrak{X},\mathfrak{Y}}$, $\{e\}_\Gamma^{\mathfrak{X},\mathfrak{Y}}$ as above.

(i) Show that each $U_\Gamma^{\mathfrak{X},\mathfrak{Y}}$ is Γ-recursive on its domain.

(ii) Show that a partial function $f : \mathfrak{X} \to \mathfrak{Y}$ is Γ-recursive on its domain if and only if there is some $e \in \omega$ so that

$$f(x)\downarrow \Rightarrow f(x) = \{e\}(x).$$

(iii) Show that for each space $\mathfrak{X}$ of type 0 and each $\mathfrak{W}$, $\mathfrak{Y}$ there is a recursive function

$$S_\Gamma^{\mathfrak{X},\mathfrak{W},\mathfrak{Y}} = S : \omega \times \mathfrak{X} \to \omega$$

such that for each $e \in \omega$, $x \in \mathfrak{X}$,

$$\{e\} = \{S(e, x)\},$$

i.e. for all $e \in \omega$, $x \in \mathfrak{X}$, $w \in \mathfrak{W}$

$$\{e\}(x, w) = \{S(e, x)\}(w).$$

(iv) Show that if $f : \omega \times \mathfrak{X} \to \mathfrak{Y}$ is Γ-recursive on its domain, then there exists some $e^* \in \omega$ so that for all $x \in \mathfrak{X}$,

$$f(e^*, x)\downarrow \Rightarrow [f(e^*, x) = \{e^*\}(x)];$$

in fact we can take $e^* = r(a)$ where r is a fixed recursive function (depending only on $\mathfrak{X}, \mathfrak{Y}$) and a is any member of ω such that

$$f(e, x)\downarrow \twoheadrightarrow f(e, x) = \{a\}(e, x).$$

Hint. Follow the proofs of 7A.1 and 7A.2. ⊣

In the simple case $\Gamma = \Sigma_1^0$ and on spaces of type 0, this result gives Kleene's original calculus of recursive partial functions.

7B. The Suslin–Kleene theorem

The key ingredient in the proof of this central result of the effective theory is the method of *definition by effective transfinite recursion* which we described first in 7A.4. In fact the "constructive proof" of the Strong Separation Theorem 2E.1 which we gave in Chapter 2 defines the separating sets by an effective recursion and all we have to do here is to recast that argument in the language of codings.

Let us first introduce a new coding of Borel sets which is somewhat easier to work with than the coding of 3H. In this definition and in the rest of this section *recursive* always means Σ_1^0*-recursive* i.e.

$$\{\varepsilon\}(x) = \{\varepsilon\}_{\Sigma_1^0}(x).$$

We define by recursion on the countable ordinal ξ the set BC_ξ of *Borel Codes for* $\mathbf{\Sigma}_\xi^0$ as follows:

$$BC_0 = \{\alpha : \alpha(0) = 0\},$$

$$BC_\xi = \{\alpha : \alpha(0) = 1 \;\&\; (\forall n)[\{\alpha^\star\}(n)\downarrow \;\&\; \{\alpha^\star\}(n) \in \textstyle\bigcup_{\eta<\xi} BC_\eta]\}$$

if $\xi > 0$, where

$$\alpha^\star(t) = \alpha(t+1)$$

and $\{\alpha^\star\} : \omega \rightharpoonup \eta$ is the partial function of 7A which is (Σ_1^0-) recursive on its domain. For each fixed space $\mathfrak{X}$ and each ξ we define the coding

$$\pi c_\xi^{\mathfrak{X}} : BC_\xi \twoheadrightarrow \mathbf{\Sigma}_\xi^0 \restriction \mathfrak{X}$$

by the recursion

$$\pi c_0^{\mathfrak{X}}(\alpha) = N(\mathfrak{X}, \alpha(1)),$$

$$\pi c_\xi^{\mathfrak{X}}(\alpha) = \textstyle\bigcup_n (\mathfrak{X} - \pi c_{\eta(n)}^{\mathfrak{X}}(\{\alpha^\star\}(n)))$$

where

$$\eta(n) = \text{least } \eta \text{ so that } \{\alpha^{\star}\}(n) \in BC_{\eta}.$$

Finally, put

$$BC = \bigcup_{\xi} BC_{\xi} = \text{the set of Borel codes,}$$
$$\pi c^{\mathfrak{X}} = \bigcup_{\xi} \pi c_{\xi}^{\mathfrak{X}}.$$

We will call $\pi c^{\mathfrak{X}}(\alpha)$ *the set with Borel code* α.

In addition to starting with the basic nbhds rather than arbitrary open sets, this coding differs from that of 3H in the way we chose to code infinite sequencies of irrationals: instead of the simple mapping

$$\alpha \mapsto (\alpha)_0, (\alpha)_1, (\alpha)_2, \ldots,$$

we took

$$\alpha \mapsto \{\alpha^{\star}\}(0), \{\alpha^{\star}\}(1), \{\alpha^{\star}\}(2), \ldots$$

which depends on the messy basic definitions of Σ_1^0-recursion and is not defined for all α. There are technical advantages to this new coding which will become clear soon—and it is equivalent to the coding of 3H as we will show in 7B.8.

Let us first prove a couple of simple lemmas about this coding.

7B.1. LEMMA. (i) $\eta \leq \xi \rightarrow BC_{\eta} \subseteq BC_{\xi}$ & $\pi_{\eta} = \pi_{\xi} \restriction BC_{\eta}$, *so that* π *is a coding of the Borel subsets of* $\mathfrak{X}$ *with* BC *the set of codes.*

(ii) *The class of Borel subsets of* $\mathfrak{X}$ *is uniformly closed under complementation, countable union and countable intersection in the following precise sense.*

(a) *There is a recursive function* $u_1(\alpha)$ *such that if* α *is a Borel code of some* $A \subseteq \mathfrak{X}$, *then* $u_1(\alpha)$ *is a Borel code of* $\mathfrak{X} - A$.

(b) *There is a recursive function* $u_2(\varepsilon)$ *such that if for each* i, $\{\varepsilon\}(i)\downarrow$ *and* $\{\varepsilon\}(i)$ *is a Borel code of some set* $A_i \subseteq \mathfrak{X}$, *then* $u_2(\varepsilon)$ *is a Borel code of* $\bigcup_i A_i$.

(c) *There is a recursive function* $u_3(\varepsilon)$ *such that if for each* i, $\{\varepsilon\}(i)\downarrow$ *and* $\{\varepsilon\}(i)$ *is a Borel code of some* $A_i \subseteq \mathfrak{X}$, *then* $u_3(\varepsilon)$ *is a Borel code of* $\bigcap_i A_i$.

(iii) *If* $P \subseteq \mathfrak{X} \times \mathcal{Y}$ *is in* Σ_n^0 *or* Π_n^0 *and* $\mathfrak{X}$ *is of type* ≤ 1, *then there is a recursive function* $u_4 : \mathfrak{X} \rightarrow \mathcal{N}$ *such that for each* x, $u_4(x)$ *is a Borel code of* $P_x = \{y : P(x, y)\}$.

PROOF. (i) is immediate by induction on ξ.

For (ii) (a) choose a recursive ε_1 so that

$$\{\varepsilon_1\}(\alpha, t) = \alpha \quad (\text{for all } t \in \omega, \alpha \in \mathcal{N})$$

and define

$$u_1(\alpha) = (1)\widehat{\ } S(\varepsilon_1, \alpha);$$

now if $u_1(\alpha) = \beta$, then $\beta(0) = 1$ and $\beta^* = S(\varepsilon_1, \alpha)$ so that if α codes A, then for all i

$$\{\beta^*\}(i) = \{S(\varepsilon_1, \alpha)\}(i) = \alpha$$

and hence β codes $\bigcup_i (\mathfrak{X} - A) = \mathfrak{X} - A$.

Similarly, for (ii) (b), choose a recursive ε_2 so that for all ε, t

$$\{\varepsilon_2\}(\varepsilon, t) = u_1(\{\varepsilon\}(t))$$

and let

$$u_2(\varepsilon) = (1)\widehat{\ } S(\varepsilon_2, \varepsilon);$$

now if ε satisfies the hypothesis, so each $\{\varepsilon\}(i)$ is a Borel code of A_i, then $\{S(\varepsilon_2, \varepsilon)\}(i) = u_1(\{\varepsilon\}(i))$ is a Borel code of $\mathfrak{X} - A_i$ for each i and hence $u_2(\varepsilon)$ is a Borel code of $\bigcup_i (\mathfrak{X} - (\mathfrak{X} - A_i)) = \bigcup_i A_i$.

The construction for part (c) is similar.

To prove (iii) by induction on n, suppose first that $P(x, y)$ is Σ^0_1 with $\mathfrak{X}$ of type 0 or 1. By 3C.5. now (considering $\{(y, x): P(x, y)\}$) easily, there is a recursive $P^* \subseteq \mathfrak{X} \times \omega^2$ so that

$$P(x, y) \Leftrightarrow (\exists t)\{y \in N(\mathcal{Y}, t) \,\&\, (\exists u)P^*(x, t, u)\}$$

and contracting quantifiers,

$$P(x, y) \Leftrightarrow (\exists s)\{y \in N(\mathcal{Y}, (s)_0) \,\&\, P^*(x, (s)_0, (s)_1)\}.$$

Put now

$$f(x, s) = (i \mapsto 0) \quad \text{if} \quad \neg P^*(x, (s)_0, (s)_1),$$

and if $P^*(x, (s)_0, (s)_1)$ holds, let

$$f(x, s) = \beta_{x,s}$$

where

$$\beta_{x,s}(0) = 0, \qquad \beta_{x,s}(i+1) = (s)_0.$$

Clearly f is recursive; if $\hat{f} \in \mathcal{N}$ is a *recursive code* for f so that

$$\begin{aligned} f(x, s) &= \{\hat{f}\}(x, s) \\ &= \{S(\hat{f}, x)\}(s), \end{aligned}$$

then easily for each s, $\{S(\hat{f}, x)\}(s)$ is a Borel code of some nbhd $N^{\mathfrak{X},s} \subseteq \mathcal{Y}$ and

$$P_x = \bigcup_s N^{\mathfrak{X},s};$$

thus we can take

$$\mathsf{u}_4(x) = \mathsf{u}_2(S(\hat{f}, x)).$$

The induction step is even easier using u_1 and u_3. ⊣

The second lemma is also quite simple but its proof illustrates the use of definition by effective transfinite recursion.

Let us call a map

$$\pi : \mathfrak{X} \to \mathcal{Y}$$

effectively Borel (or Borel in the coding) if there is a recursive function $\mathsf{v} : \omega \to \mathcal{N}$ such that for each s, $\mathsf{v}(s)$ is a Borel code of $\pi^{-1}[N[\mathcal{Y}, s)]$.

7B.2. LEMMA, (i) *For each product space $\mathfrak{X}$, there is an effectively Borel and Δ^1_1-recursive injection*

$$\pi_* : \mathfrak{X} \rightarrowtail \mathcal{N}.$$

(ii) *If $\pi : \mathfrak{X} \to \mathcal{Y}$ is effectively Borel, then there is a recursive function $\mathsf{u} : \mathcal{N} \to \mathcal{N}$ such that whenever α is a Borel code of some $A \subseteq \mathcal{Y}$, then $\mathsf{u}(\alpha)$ is a Borel code of $\pi^{-1}[A]$.*

PROOF. (i) Go back to the proof of 1G.2 and take for π_* the g defined there. Recalling that definition,

$$\pi_*(x) = \alpha,$$

where

$$\alpha(n) = \text{least } k \text{ such that } d(x, r_k) \leq 2^{-n-2}$$

so that π_* is easily an injection and if

$$B_{n,k} = \{\alpha : \alpha(n) = k\},$$

then

$$\pi_*^{-1}[B_{n,k}] = \{x : d(x, r_k) \leq 2^{-n-2}\} \cap \{x : (\forall j < k) d(x, r_j) > 2^{-j-2}\}:$$

thus the set

$$P(n, k, x) \Leftrightarrow \pi_*(x) \in B_{n,k}$$

is easily Π^0_2 and by 7B.1 there is a recursive $\mathbf{v}_1$ so that $\mathbf{v}_1(n, k)$ is a Borel code of $\pi_*^{-1}[B_{n,k}]$. From this it is easy to get a recursive $\mathbf{v}$ witnessing that π_* is effectively Borel using again 7B.1.

(ii) If $\mathbf{v}$ witnesses that π is effectively Borel, we clearly want to put

$$\mathbf{u}(\alpha) = \mathbf{v}(\alpha(1)), \quad \text{if} \quad \alpha(0) = 0,$$

but we need a transfinite recursion to define $\mathbf{u}(\alpha)$ for Borel codes α with $\alpha(0) = 1$. We will define $\mathbf{u}$ by the recursion theorem, i.e. we will put

$$\mathbf{u}(\alpha) = \{\varepsilon^*\}(\alpha)$$

where $\{\varepsilon^*\}$ is defined for ordinary (Σ^0_1-) recursion and ε^* is a recursive irrational so that for a suitable h,

$$h(\varepsilon, \alpha)\downarrow \rightarrow \{\varepsilon^*\}(\alpha) = h(\varepsilon^*, \alpha);$$

in fact h will be total recursive, so that $\mathbf{u}$ will also be total.

Having decided on this plan of the proof, we are left with a simple coding problem in defining h.

Define g by

$$g(\varepsilon, \alpha, t) = \{\varepsilon\}(\{\alpha^*\}(t))$$

and let $\hat{g}$ be a recursive irrational so that

$$\begin{aligned}\{\varepsilon\}(\{\alpha^*\}(t))\downarrow \rightarrow \{\varepsilon\}(\{\alpha^*\}(t)) &= \{\hat{g}\}(\varepsilon, \alpha, t)\\ &= \{S(\hat{g}, \varepsilon, \alpha)\}(t).\end{aligned}$$

Now put

$$h(\varepsilon, \alpha) = \begin{cases} \mathbf{v}(\alpha(1)) & \text{if} \quad \alpha(0) = 0,\\ (1)^\frown S(\hat{g}, \varepsilon, \alpha) & \text{if} \quad \alpha(0) \neq 0,\end{cases}$$

and choose a recursive ε^* so that

$$h(\varepsilon^*, \alpha) = \{\varepsilon^*\}(\alpha).$$

We claim that the function

$$\mathbf{u}(\alpha) = \{\varepsilon^*\}(\alpha)$$

has the required properties.

To check this, we establish by induction on ξ that if $\alpha \in BC_\xi$ and α codes the set $A = \pi c^{\mathcal{Y}}(\alpha)$, then $\mathbf{u}(\alpha)$ codes the set $\pi^{-1}[A] \subseteq \mathcal{Y}$.

The result is immediate for $\xi = 0$ by the choice of $\mathbf{v}$.

Asssuming $\xi > 0$, if $\alpha \in BC_\xi$, then $\alpha(0) = 1$ and α codes the set

$$A = \bigcup_n (\mathcal{Y} - A_n)$$

where for each n, $\{\alpha^*\}(n)$ codes A_n and $\{\alpha^*\}(n) \in \bigcup_{\eta<\xi} BC_\eta$. By induction hypothesis then, for each n the irrational

$$\begin{aligned}\{S(\hat{g}, \varepsilon^*, \alpha)\}(n) &= \{\varepsilon^*\}(\{\alpha^*\}(n)) \\ &= u(\{\alpha^*\}(n))\end{aligned}$$

codes $\pi^{-1}[A_n]$, so that

$$u(\alpha) = (1)^\frown S(\hat{g}, \varepsilon^*, \alpha)$$

codes

$$\begin{aligned}\bigcup_n (\mathcal{X} - \pi^{-1}[A_n]) &= \pi^{-1} \bigcup_n [\mathcal{Y} - A_n] \\ &= \pi^{-1}[A]. \end{aligned}$$

⊣

We wrote this proof up in full detail to illustrate the method—quite often in the future we will simply summarize the key ideas involved and leave out the details.

Let us now turn to the key result of this section.

It is convenient here to define a *tree* T on $\omega \times \omega$ as a set of *sequence codes* (in ω) $\langle\langle t_0, \xi_0\rangle, \ldots, \langle t_{n-1}, \xi_{n-1}\rangle\rangle$ of finite sequences of pairs of integers closed under initial segments. Let us say that an irrational $\tau \in \mathcal{N}$ *codes* the tree T if

$$\langle\langle t_0, \xi_0\rangle, \ldots, \langle t_{n-1}, \xi_{n-1}\rangle\rangle \in T \Leftrightarrow \tau(\langle\langle t_0, \xi_0\rangle, \ldots, \langle t_{n-1}, \xi_{n-1}\rangle\rangle) = 1.$$

Similary for a tree J on $\omega \times \omega \times \omega$; α codes J, if

$$\begin{aligned}&\langle\langle t_0, \xi_0, \eta_0\rangle, \ldots, \langle t_{n-1}, \xi_{n-1}, \eta_{n-1}\rangle\rangle \in J \\ &\qquad \Leftrightarrow \alpha(\langle\langle t_0, \xi_0, \eta_0\rangle, \ldots, \langle t_{n-1}, \xi_{n-1}, \eta_{n-1}\rangle\rangle) = 1.\end{aligned}$$

It is clear that any code τ (or α) determines completely the tree T (or J).

As in Chapter 2, for each tree T on $\omega \times \omega$ the projection $\mathrm{p}[T]$ is a subset of $\mathcal{N}$,

$$\mathrm{p}[T] = \{\alpha : (\exists f \in {}^\omega\omega)(\forall n)[\langle\langle \alpha(0), f(0)\rangle, \ldots, \langle \alpha(n-1), f(n-1)\rangle\rangle \in T].$$

7.B.3. THE EFFECTIVE STRONG SEPARATION THEOREM. *There is a recursive function* $u : \mathcal{N} \times \mathcal{N} \to \mathcal{N}$ *such that whenever* τ *and* σ *code respectively trees* T *and* S *with*

$$\mathrm{p}[T] \cap \mathrm{p}[S] = \emptyset,$$

then $\mathfrak{u}(\tau, \sigma)$ *is a Borel code of some set C which separates* $\mathfrak{p}[T]$ *from* $\mathfrak{p}[S]$, *i.e.*

$$\mathfrak{p}[T] \subseteq C, \qquad C \cap \mathfrak{p}[S] = \emptyset.^{(2)}$$

PROOF. Following closely the constructive proof of 2E.1, let us associate with each two trees of pairs T and S on $\omega \times \omega$ the tree of triples J,

$$\begin{aligned} \langle\langle t_0, \xi_0, \eta_0\rangle, \ldots, \langle t_{n-1}, \xi_{n-1}, \eta_{n-1}\rangle\rangle \in J \\ \Leftrightarrow \langle\langle t_0, \xi_0\rangle, \ldots, \langle t_{n-1}, \xi_{n-1}\rangle\rangle \in T \\ \& \ \langle\langle t_0, \eta_0\rangle, \ldots, \langle t_{n-1}, \eta_{n-1}\rangle\rangle \in S. \end{aligned}$$

To simplify notation, let $f(u)$, $h(u)$ be recursive functions such that if

$$u = \langle\langle t_0, \xi_0, \eta_0\rangle, \ldots, \langle t_{n-1}, \xi_{n-1}, \eta_{n-1}\rangle\rangle,$$

then

$$\begin{aligned} f(u) &= \langle\langle t_0, \xi_0\rangle, \ldots, \langle t_{n-1}, \xi_{n-1}\rangle\rangle, \\ h(u) &= \langle\langle t_0, \eta_0\rangle, \ldots, \langle t_{n-1}, \eta_{n-1}\rangle\rangle. \end{aligned}$$

(Notice that throughout this argument the variables ξ_i, η_j vary over ω, as the $\varkappa$ of 2E.1 is ω here.)

Now if τ and σ code trees T and S, a code of J is easily $j(\tau, \sigma)$ where j is a fixed recursive function such that

$$j(\tau, \sigma)(u) = 1 \Leftrightarrow \tau(f(u)) = 1 \ \& \ \sigma(h(u)) = 1.$$

As in 2E.1, let

$$A_v = \mathfrak{p}[T_v], \qquad B_v = \mathfrak{p}[S_v],$$

where v varies over sequence codes. Clearly

$$A = \mathfrak{p}[T] = \mathfrak{p}[T_1], \qquad B = \mathfrak{p}[S] = \mathfrak{p}[S_1],$$

recalling that $1 = \langle\ \rangle$ codes the empty sequence.

We aim to define a recursive function

$$\mathfrak{u}^* : \mathcal{N} \times \mathcal{N} \times \omega \to \mathcal{N}$$

such that whenever τ and σ code trees T and S with $\mathfrak{p}[T] \cap \mathfrak{p}[S] = \emptyset$ and $u = \langle\langle t_0, \xi_0, \eta_0\rangle, \ldots, \langle t_{n-1}, \xi_{n-1}, \eta_{n-1}\rangle\rangle$, then $\mathfrak{u}^*(\tau, \sigma, u)$ is a Borel code of a set

$$C_u = C(\tau, \sigma, u)$$

which separates $\mathfrak{p}[T_{f(u)}]$ from $\mathfrak{p}[S_{h(u)}]$. The proof will be completed by taking

$$\mathsf{u}(\tau, \sigma) = \mathsf{u}^*(\tau, \sigma, 1).$$

The definition of u^* will be by the recursion theorem, i.e. we will take

$$\mathsf{u}^*(\tau, \sigma, u) = \{\varepsilon^*\}(\tau, \sigma, u),$$

where $\{\varepsilon^*\}$ is defined for ordinary (Σ^0_1-) recursion and ε^* is a recursive irrational chosen so that

$$g(\varepsilon^*, \tau, \sigma, u)\downarrow \rightarrow \{\varepsilon^*\}(\tau, \sigma, u) = g(\varepsilon^*, \tau, \sigma, u);$$

in fact g will be a total recursive function, so that u^* will also be total.

In the proof of 2E.1, the sets C_u were defined by bar recursion on the tree J determined by T and S which is well-founded when $\mathfrak{p}[T] \cap \mathfrak{p}[S] = \emptyset$. Here, obviously we must choose g so that (when τ and σ code trees T and S with $\mathfrak{p}[T] \cap \mathfrak{p}[S] = \emptyset$) it proves that this bar recursion is *effective* in the sense of 7A. Of course, the definition must make sense for arbitrary τ, σ, u so that we get a total recursive function.

With this plan of attack, it is only a simple matter of manipulating the coding to define g. Let us first prove a lemma that reduces the construction of a code for $C(\tau, \sigma, u)$ to constructing codes for the sets $D_{t,\xi,s,\eta}$ in the proof of 2E.1.

Lemma. There is a recursive function $\mathsf{v}(\varepsilon)$ such that if $\{\varepsilon\}(t, \xi, s, \eta)$ is defined for each t, ξ, s, η and codes a Borel set $D_{t,\xi,s,\eta}$, then $\mathsf{v}(\varepsilon)$ is a Borel code of the set

$$\bigcup_{t,\xi} \bigcap_{s,\eta} D_{t,\xi,s,\eta}.$$

Proof of the lemma. Recall the recursive functions u_2 and u_3 of 7B.1 which construct codes for countable unions and intersections of Borel sets and define the recursive functions $\mathsf{v}_1, \mathsf{v}_2, \mathsf{v}_3$,

$$\mathsf{v}_1(\varepsilon, t, \xi, s) = \mathsf{u}_3(S(\varepsilon, t, \xi, s)),$$

$$\mathsf{v}_2(\varepsilon, t, \xi) = \mathsf{u}_3(S(\hat{\mathsf{v}}_1, \varepsilon, t, \xi),$$

$$\mathsf{v}_3(\varepsilon, t) = \mathsf{u}_2(S(\hat{\mathsf{v}}_2, \varepsilon, t)),$$

$$\mathsf{v}(\varepsilon) = \mathsf{u}_2(S(\hat{\mathsf{v}}_3, \varepsilon)),$$

where $\hat{\mathsf{v}}_1, \hat{\mathsf{v}}_2, \hat{\mathsf{v}}_3$ are recursive irrationals successively chosen so that

$$\mathsf{v}_1(\varepsilon, t, \xi, s) = \{\hat{\mathsf{v}}_1\}(\varepsilon, t, \xi, s),$$

$$\mathsf{v}_2(\varepsilon, t, \xi) = \{\hat{\mathsf{v}}_2\}(\varepsilon, t, \xi),$$

$$\mathsf{v}_3(\varepsilon, t) = \{\hat{\mathsf{v}}_3\}(\varepsilon, t).$$

Asssuming now that ε satisfies the hypotheses,

$$\{S(\varepsilon, t, \xi, s)\}(\eta) = \{\varepsilon\}(t, \xi, s, \eta)$$

codes $D_{t,\xi,s,\eta}$ for each η, hence

$$\mathbf{v}_1(\varepsilon, t, \xi, s) = \mathbf{u}_3(S(\varepsilon, t, \xi, s))$$

codes $\bigcap_\eta D_{t,\xi,s,\eta}$; but

$$\mathbf{v}_1(\varepsilon, t, \xi, s) = \{S(\hat{\mathbf{v}}_1, t, \xi)\}(s),$$

so that

$$\mathbf{v}_2(\varepsilon, t, \xi) = \mathbf{v}_3(S(\hat{\mathbf{v}}_1, t, \xi))$$

codes $\bigcap_s \bigcap_\eta D_{t,\xi,s,\eta} = \bigcap_{s,n} D_{t,\xi,s,\eta}$. Continuing the same argument we easily check that $\mathbf{v}(\varepsilon)$ codes the required set. ⊣ (lemma)

Going back to the proof of the theorem, let us define the partial function $d(\varepsilon, \tau, \sigma, u, t, \xi, s, \eta)$ (by cases) so that the following hold. (After ε is fixed by the recursion theorem, $d(\varepsilon, \tau, \sigma, u, t, \xi, s, \eta)$ will give a Borel code of the set $D_{t,\xi,s,\eta}$ in the proof of 2E.1.)

(1) If $t = s$ and $j(\tau, \sigma)(u * \langle t, \xi, \eta\rangle) = 1$, then

$$d(\varepsilon, \tau, \sigma, u, t, \xi, s, \eta) = \{\varepsilon\}(\tau, \sigma, u * \langle t, \xi, \eta\rangle).$$

(2) If $t \neq s$, then

$$d(\varepsilon, \tau, \sigma, u, t, \xi, s, \eta) = d_2(u, t)$$

where $d_2(u, t)$ is a recursive function giving a Borel code of $\{\alpha : \alpha(lh(u)) = t\}$—this is easy to get.

(3) If $t = s$ & $\tau(u * \langle t, \xi, \eta\rangle) \neq 1$, then

$$d(\varepsilon, \tau, \sigma, u, t, \xi, s, \eta) = d_0$$

where d_0 is a fixed (recursive) Borel code of $\emptyset$.

(4) If $t = s$ & $\sigma(u * \langle t, \xi, \eta\rangle) \neq 1$, then

$$d(\varepsilon, \tau, \sigma, u, t, \xi, s, \eta) = d_1$$

where d_1 is a fixed (recursive) Borel code of $\mathcal{N}$.

It is clear that d is recursive on its domain, so let $\hat{d} \in \mathcal{N}$ be recursive so that

$$\begin{aligned} &d(\varepsilon, \tau, \sigma, u, t, \xi, s, \eta)\downarrow \\ &\quad \rightarrow d(\varepsilon, \tau, \sigma, u, t, \xi, s, \eta) = \{\hat{d}\}(\varepsilon, \tau, \sigma, u, t, \xi, s, \eta) \\ &\qquad = \{S(\hat{d}, \varepsilon, \tau, \sigma, u)\}(t, \xi, s, \eta) \end{aligned}$$

and finally put

$$g(\varepsilon, \tau, \sigma, u) = \mathbf{v}(S(\hat{d}, \varepsilon, \tau, \sigma, u)).$$

Obviously g is a total recursive function.

Now following our original plan for the proof, choose a recursive ε^* by the recursion theorem so that

$$g(\varepsilon^*, \tau, \sigma, u) = \{\varepsilon^*\}(\tau, \sigma, u)$$

and let

$$\mathbf{u}^*(\tau, \sigma, u) = \{\varepsilon^*\}(\tau, \sigma, u).$$

To prove that $\mathbf{u}^*$ has the required properties, suppose τ and σ code trees T and S so that $\mathrm{p}[T] \cap \mathrm{p}[S] = \emptyset$ and let J be the tree of triples (with code $j(\tau, \sigma)$) that we associated with T and S and which is now well-founded. We check by bar recursion on J, that if $u \in J$, then $\mathbf{u}^*(\tau, \sigma, u)$ codes a Borel set C_u that separates $\mathrm{p}[T_{f(u)}]$ from $\mathrm{p}[S_{h(u)}]$ by looking over the steps in the constructive proof of 2E.1 and verifying that (for the fixed T, S), $d(\varepsilon^*, \tau, \sigma, u, t, \xi, s, \eta)$ codes the set $D_{t,\xi,s,\eta}$ and $\mathbf{u}^*(\tau, \sigma, u)$ codes the set C_u. ⊣

Recall the coding of $\mathbf{\Sigma}^1_1$ sets which we introduced in 3H: for each $\mathfrak{X}$, fix a Σ^1_1 set $G^{\mathfrak{X}} \subseteq \mathcal{N} \times \mathfrak{X}$ which is universal for $\mathbf{\Sigma}^1_1$ and call α a $\mathbf{\Sigma}^1_1$-code for $A \subseteq \mathfrak{X}$ if

$$A = G^{\mathfrak{X}}_\alpha = \{x : G^{\mathfrak{X}}(\alpha, x)\}.$$

A $\mathbf{\Delta}^1_1$-code for a set C is any α such that $(\alpha)_0$ is a $\mathbf{\Sigma}^1_1$-code for C and $(\alpha)_1$ is a $\mathbf{\Sigma}^1_1$-code for $\mathfrak{X} - C$.

We customarily assume that the system $\{G^{\mathfrak{X}}\}$ is a good parametrization (in Σ^1_1 for $\mathbf{\Sigma}^1_1$) in the sense of 3H, but this is not necessary for the Suslin–Kleene theorem.

7B.4. THE SUSLIN–KLEENE THEOREM (Kleene[(2)]). *For each product space $\mathfrak{X}$, there is a recursive function $\mathbf{v} : \mathcal{N} \times \mathcal{N} \to \mathcal{N}$ such that if α, β are $\mathbf{\Sigma}^1_1$-codes of sets $A, B \subseteq \mathfrak{X}$ respectively and $A \cap B = \emptyset$, then $\mathbf{v}(\alpha, \beta)$ is a Borel code of some set C which separates A from B, i.e.*

$$A \subseteq C, \qquad C \cap B = \emptyset.$$

In particular, for each $\mathfrak{X}$ there is a recursive function $\mathbf{v}^* : \mathcal{N} \to \mathcal{N}$ such that if α is a $\mathbf{\Delta}^1_1$-code of $A \subseteq \mathfrak{X}$, then $\mathbf{v}^*(\alpha)$ is a Borel code of A.

PROOF. Take first $\mathfrak{X}=\mathcal{N}$, let $G\subseteq\mathcal{N}\times\mathcal{N}$ be Σ^1_1 and universal for $\mathbf{\Sigma}^1_1$ and choose by 4A.1(i) a recursive Q such that

$$G(\alpha,\beta)\Leftrightarrow(\exists\gamma)(\forall t)Q(\alpha,\bar\beta(t),\bar\gamma(t)),$$
$$Q(\alpha,\bar\beta(t),\bar\gamma(t))\ \&\ s<t\Rightarrow Q(\alpha,\bar\beta(s),\bar\gamma(s)).$$

For each α, the set of sequence codes

$$T^{(\alpha)}=\{\langle\bar\beta(t),\bar\gamma(t)\rangle: t\in\omega\ \&\ \neg Q(\alpha,\bar\beta(t),\bar\gamma(t))\}$$

is clearly a tree and in fact

$$G_\alpha=\mathrm{p}[T^{(\alpha)}];$$

moreover (easily) there is a recursive function $\mathbf{u}_1$ such that for each α, $\mathbf{u}_1(\alpha)$ is a code of $T^{(\alpha)}$. Letting $\mathbf{u}:\mathcal{N}\times\mathcal{N}\to\mathcal{N}$ be the recursive function of 7B.3, we can then take for the space $\mathcal{N}$,

$$\mathbf{v}(\alpha,\beta)=\mathbf{u}(\mathbf{u}_1(\alpha),\mathbf{u}_1(\beta)).$$

If $\mathfrak{X}$ is any product space, let

$$\pi:\mathfrak{X}\rightarrowtail\mathcal{N}$$

be an effectively Borel injection by 7B.2 and notice that the image $\pi[\mathfrak{X}]$ is easily Δ^1_1. Fix a good universal set $G\subseteq\mathcal{N}\times\mathcal{N}$ in Σ^1_1 and let $G^{\mathfrak{X}}\subseteq\mathcal{N}\times\mathfrak{X}$ be any Σ^1_1 set universal for $\mathbf{\Sigma}^1_1$, relative to which we have defined the codings. Put

$$P(\alpha,\beta)\Leftrightarrow\beta\in\pi[\mathfrak{X}]\ \&\ G^{\mathfrak{X}}(\alpha,\pi^{-1}(\beta));$$

now P is Σ^1_1, so that for a fixed recursive ε_0,

$$\begin{aligned}P(\alpha,\beta)&\Leftrightarrow G(\varepsilon_0,\alpha,\beta)\\&\Leftrightarrow G(S(\varepsilon_0,\alpha),\beta)\end{aligned}$$

with a recursive S. Thus,

$$\begin{aligned}G^{\mathfrak{X}}(\alpha,x)&\Leftrightarrow P(\alpha,\pi(x))\\&\Leftrightarrow G(S(\varepsilon_0,\alpha),\pi(x)).\end{aligned}$$

From the result about $\mathcal{N}$, it follows that if $G^{\mathfrak{X}}_\alpha\cap G^{\mathfrak{X}}_\beta=\emptyset$, then $\mathbf{v}(S(\varepsilon_0,\alpha), S(\varepsilon_0,\beta))$ is a Borel code of a set that separates $\pi[G^{\mathfrak{X}}_\alpha]$ from $\pi[G^{\mathfrak{X}}_\beta]$. If we now take $\mathbf{u}:\mathcal{N}\to\mathcal{N}$ to be the recursive function of 7B.2 for π, we see that $\mathbf{u}(\mathbf{v}(S(\varepsilon_0,\alpha),S(\varepsilon_0,\beta)))$ is a Borel code of some $C\subseteq\mathfrak{X}$ which separates $G^{\mathfrak{X}}_\alpha$ from $G^{\mathfrak{X}}_\beta$.

The second assertion follows immediately. ⊣

There is a simple converse to the Suslin–Kleene Theorem.

7B.5. THEOREM. *Assume that we code the* $\mathbf{\Delta}_1^1$ *pointsets using good parametrizations for* $\mathbf{\Sigma}_1^1$ *in* Σ_1^1; *then for each space* $\mathfrak{X}$, *there is a recursive function* $u : \mathcal{N} \to \mathcal{N}$ *such that if* α *is a Borel code of some* $A \subseteq \mathfrak{X}$, *then* $u(\alpha)$ *is a* $\mathbf{\Delta}_1^1$ *code of* A.

PROOF. Suppose $G \subseteq \mathcal{N} \times \mathfrak{X}$ parametrizes $\mathbf{\Sigma}_1^1$ in Σ_1^1 so that a $\mathbf{\Delta}_1^1$-code for $A \subseteq \mathfrak{X}$ is any α such that

$$\begin{aligned} x \in A &\leftrightarrow G((\alpha)_0, x) \\ &\leftrightarrow \neg G((\alpha)_1, x). \end{aligned}$$

We will define the function u by the recursion theorem, so that

$$u(\alpha) = g(\varepsilon^*, \alpha)$$

where g will be a (total) recursive function and ε^* will be recursive and chosen so that

$$(*) \qquad g(\varepsilon^*, \alpha) = \{\varepsilon^*\}(\alpha).$$

After that we will prove by induction on ξ that if $\alpha \in BC_\xi$ and codes the set $A \subseteq \mathfrak{X}$, then $u(\alpha)$ is a $\mathbf{\Delta}_1^1$-code for A.

In defining $g(\varepsilon, \alpha)$ it is convenient to talk of the "basis," the "induction step" and the "induction hypothesis" that $\{\varepsilon\}(\beta)$ "must satisfy" when $\beta \in \bigcup_{\eta<\xi} BC_\eta$, as if we were giving an ordinary definition by transfinite recursion and we already knew that $g(\varepsilon, \alpha) = \{\varepsilon\}(\alpha)$ and $\alpha \in BC_\xi$ for some ξ. Of course, the definition must make sense for arbitrary ε, α and g must be recursive; but after g is defined and ε is fixed to be some ε^* satisfying $(*)$ above, then these informal comments in the definition of g lead easily to a proof that $u(\alpha) = \{\varepsilon^*\}(\alpha)$ has the required properties.

Basis. If $\alpha(0) = 0$, set

$$(1) \qquad g(\varepsilon, \alpha) = g_0(\alpha(1))$$

where g_0 is recursive and such that for each s, $g_0(s)$ is a $\mathbf{\Delta}_1^1$-code of the basic *nbhd* $N(\mathfrak{X}, s)$. This is easy to construct.

Induction Step. $\alpha(0) \neq 0$ (including the case $\alpha(0) = 1$ in which we are interested). We may assume here that $\{\alpha^\star\}(i)$ is defined for each i and is a Borel code of some A_i so that α is the code of

$$A = \bigcup_i (\mathfrak{X} - A_i).$$

By induction hypothesis then, $\{\varepsilon\}(\{\alpha^\star\}(i))$ is a $\mathbf{\Delta}_1^1$-code of each A_i, so that we have the following equivalences satisfied by A:

$$x \in A \leftrightarrow (\exists i)\neg G((\{\varepsilon\}(\{\alpha^\star\}(i)))_0, x),$$
$$x \in A \leftrightarrow (\exists i)G((\{\varepsilon\}(\{\alpha^\star\}(i)))_1, x).$$

Now A is obviously $\Delta^1_1(\varepsilon, \alpha)$ and it simply remains to find effectively a $\mathbf{\Delta}^1_1$-code for it.

Put

$$P_0(\varepsilon, \alpha, x) \leftrightarrow (\forall i)[\{\varepsilon\}(\{\alpha^\star\}(i))\downarrow \& \ G((\{\varepsilon\}(\{\alpha^\star\}(i)))_0, x)],$$
$$P_1(\varepsilon, \alpha, x) \leftrightarrow (\exists i)[\{\varepsilon\}(\{\alpha^\star\}(i))\downarrow \& \ G((\{\varepsilon\}(\{\alpha^\star\}(i)))_1, x)],$$

and check (easily) that both P_0 and P_1 are in Σ^1_1, so that for fixed recursive $\varepsilon_0, \varepsilon_1$,

$$P_0(\varepsilon, \alpha, x) \leftrightarrow G(\varepsilon_0, \varepsilon, \alpha, x) \leftrightarrow G(S(\varepsilon_0, \varepsilon, \alpha), x),$$
$$P_1(\varepsilon, \alpha, x) \leftrightarrow G(\varepsilon_1, \varepsilon, \alpha, x) \leftrightarrow G(S(\varepsilon_1, \varepsilon, \alpha), x).$$

If the induction hypothesis now holds, we clearly have

$$x \in A \leftrightarrow \neg P_0(\varepsilon, \alpha, x) \leftrightarrow \neg G(S(\varepsilon_0, \varepsilon, \alpha), x),$$
$$x \in A \leftrightarrow P_1(\varepsilon, \alpha, x) \leftrightarrow G(S(\varepsilon_1, \varepsilon, \alpha), x),$$

so that we can set in this case

$$g(\varepsilon, \alpha) = \langle S(\varepsilon_1, \varepsilon, \alpha), S(\varepsilon_0, \varepsilon, \alpha)\rangle. \tag{2}$$

The definition of g is now complete by the equations (1) and (2) which *do not depend on the discussion* (this is important) and determine g as a total recursive function. We fix $\varepsilon = \varepsilon^*$ so that $(*)$ above holds and we take $\mathbf{u}(\alpha) = \{\varepsilon^*\}(\alpha)$; for this ε^*, the discussion in the definition of g translates easily into a rigorous proof by induction on ξ that if $\alpha \in BC_\xi$ and codes A, then $\mathbf{u}(\alpha)$ is a $\mathbf{\Delta}^1_1$-code of A. ⊣

Since all recursive functions are continuous, the Suslin–Kleene Theorem and its converse imply weak, "classical" versions where the uniformities $\mathbf{v}, \mathbf{u}$ are only claimed to be continuous. These results are of obvious interest to the classical theory, but their proof seems to need the recursion theorem.

In addition to these results, there are many more powerful applications of Kleene's method of definition by effective transfinite recursion. Complete proofs by this method seem a bit technical and hard to read in the beginning. However, once one understands the key idea, one can skip most of the technicalities and focus on the key points of these arguments—which is that certain assertions hold uniformly in the codings in the sense of 3H.

EXERCISES

A pointset $A \subseteq \mathfrak{X}$ *is hyperarithmetical* if it is Borel and has a *recursive* Borel code α; similarly, A is hyperarithmetical in z if it has a Borel code which is recursive in z.

7B.6. Prove that a set is hyperarithmetical if and only if it is Δ^1_1.

Hint. If α is a recursive Borel code for A then the recursive irrational $\mathsf{u}(\alpha)$ is a $\mathbf{\Delta}^1_1$-code for A by 7B.5 so easily A is Δ^1_1. The converge follows similarly by the Suslin–Kleene theorem. ⊣

The hyperarithmetical sets have a direct characterization which is natural in the effective theory as an *analog* of the Borel sets. To establish this we will use the machinery of (Σ^0_1-) recursive partial functions coded in ω that we established in 7A.6, i.e. $\{e\}(x)$ below is $\{e\}^{\mathfrak{X},\mathcal{Y}}_{\Gamma}$ with $\Gamma = \Sigma^0_1$.

Recall from 3H that a *coding* of a set S is any mapping

$$\pi : D \twoheadrightarrow S$$

onto S. We say that the coding is *in* ω if *the set of codes* D is a subset of ω.

Suppose S is a collection of subsets of a space $\mathfrak{X}$. We say that S is an *effective* σ*-field* if there is a coding $\pi : D \twoheadrightarrow S$ for S in ω such that the following properties hold.

(1) Each nbhd in $\mathfrak{X}$ is uniformly in S: i.e. for some recursive $\mathsf{u}_1 : \omega \to \omega$, $\mathsf{u}_1(s)$ is a code of $N(\mathfrak{X}, s)$ (i.e. $\pi(\mathsf{u}_1(s)) = N_s$).

(2) S is uniformly closed under complementation; i.e. there is a recursive $\mathsf{u}_2 : \omega \to \omega$ such that if $\pi(a) = A \in S$, then $\pi(\mathsf{u}_2(a)) = \mathfrak{X} - A$.

(3) S is uniformly closed under recursive unions in the following sense: there is a recursive function $\mathsf{u}_3 : \omega \to \omega$ such that if $\{e\}(i)$ is defined for each $i \in \omega$ and gives a code in S of some $A_i \subseteq \mathfrak{X}$, then $\mathsf{u}_3(e)$ gives a code in S of $\bigcup_i A_i$.

The last condition is one natural way of insuring that from each recursive description of a sequence of sets in S we can effectively find a code of the union of the sequence.

7B.7. Prove that for each space $\mathfrak{X}$, the collection $\Delta^1_1 \restriction \mathfrak{X}$ of Δ^1_1 subsets of $\mathfrak{X}$ is the smallest effective σ-field on $\mathfrak{X}$ (Kleene [1955c]).[2]

Hint. Choose a good parametrization of Σ_1^1 in ω by 3H.3, code a set $A \subseteq \mathfrak{X}$ in Δ_1^1 by any $a \in \omega$ such that

$$\begin{aligned} x \in A &\leftrightarrow G((a)_0, x) \\ &\leftrightarrow \neg G((a)_1, x) \end{aligned}$$

and check that this coding witnesses that $\Delta_1^1 \upharpoonright \mathfrak{X}$ is an effective σ-field. This is a simple coding problem.

To prove that every effective σ-field on $\mathfrak{X}$ contains $\Delta_1^1 \upharpoonright \mathfrak{X}$, suppose π is a coding of S as above with recursive uniformities $\mathbf{u}_1, \mathbf{u}_2, \mathbf{u}_3$. Using the recursion theorem for recursive functions coded in ω (7A.6) prove the following result: there is a recursive function $\mathbf{v}: \omega \to \omega$ such that whenever $\{a\}(i)$ is defined for every i (with values in ω) and the resulting irrational $\{a\}$ is a Borel code of some set $A \subseteq \mathfrak{X}$, then $\mathbf{v}(a)$ is a code in S of A. (The construction is a bit messy but direct.) Now use 7B.6. ⊣

We have introduced two different codings for Borel sets and the question arises if the results we have obtained (which refer explicitly to the codings) depend on the particular coding we used. The answer is: not essentially.

Suppose $\pi: D_\pi \twoheadrightarrow \mathcal{S}$, $\rho: D_\rho \twoheadrightarrow \mathcal{S}$ are two codings of the same set $\mathcal{S}$ in the sense of 3H, where $D_\pi \subseteq \mathfrak{X}$ and $D_\rho \subseteq \mathcal{Y}$. We will say that π and ρ are (*recursively*) *equivalent* if there are partial functions $f: \mathfrak{X} \to \mathcal{Y}$, $g: \mathcal{Y} \to \mathfrak{X}$ which are (Σ_1^0-) recursive on D_π and D_ρ respectively so that Diagram 7B.1 commutes;

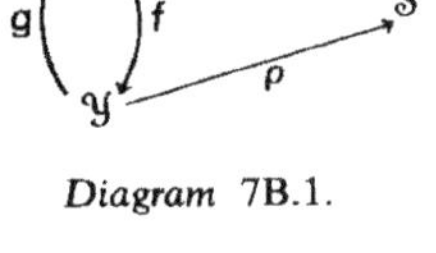

Diagram 7B.1.

thus $f(x)$ gives a code in ρ of the object coded by x in π and $g(y)$ gives a code in π of the object coded by y in ρ.

It is clear that theorems which assert the existence of Γ-recursive uniformities (Γ a Σ^*-pointclass) for a given coding π are automatically true for any equivalent coding ρ. The technical advantage of the coding for Borel sets that we used in this section is that we can get *total recursive uniformities* for it, whereas for the coding of 3H the uniformities are often partial; but the codings are equivalent.

7B.8. Prove that the coding of Borel sets introduced in 3H is recursively equivalent to the coding introduced in this section.

Hint. For both directions the recursion theorem is used. Let us consider briefly the direction in which partial functions are introduced.

We want to define $\mathfrak{u}$ by

$$\mathfrak{u}(\alpha) = g(\varepsilon^*, \alpha),$$

where g will be partial, recursive on its domain and ε^* will be recursive, chosen so that

$$g(\varepsilon^*, \alpha)\downarrow \rightarrow [g(\varepsilon^*, \alpha) = \{\varepsilon^*\}(\alpha)];$$

moreover, if $\alpha \in BC_\xi$ is a Borel code of A, then $\mathfrak{u}(\alpha)$ will be a Borel code in the sense of 3H (say 3H-Borel code for short) of the same set A.

(1) If $\alpha(0) = 0$, define $g(\varepsilon, \alpha)$ trivially to give a 3H-Borel code of $N_{\alpha(1)}$.

(2) If $\alpha(0) = 1$, then we may assume that $\{\alpha^\star\}(i)$ is defined for each i and codes some A_i, so by the induction hypothesis, $\{\varepsilon\}(\{\alpha^\star\}(i))$ gives a 3H-Borel code of A_i; define then $g(\varepsilon, \alpha)$ so that *whenever for each* i, $\{\varepsilon\}(\{\alpha^\star\}(i))\downarrow$, *then* $g(\varepsilon, \alpha) = \beta$ *for some* β *such that*

$$\beta(0) = 1,$$
$$(\beta^\star)_i = \{\varepsilon\}(\{\alpha^\star\}(i)) \quad (i \in \omega).$$

It is quite easy to do this and have g recursive on its domain.

If we now get $\mathfrak{u}$ from g as above, we have no problem showing (by induction on ξ) that for $\alpha \in BC_\xi$, $\mathfrak{u}(\alpha)$ is defined and gives the right thing. But it is not possible to get a *total* recursive function g with the properties we need. ⊣

We will end these exercises with a simple but interesting application of the Suslin–Kleene Theorem—there are may others of this type.

7B.9. Show that a function $f : \mathfrak{X} \to \mathfrak{Y}$ is effectively Borel if and only if it is Δ^1_1-recursive. ⊣

7C. Inductive definability

We will survey here briefly the theory of *inductive definability of relations* which is intimately connected with some of our main concerns. One of our aims is to get a new and interesting characterization of Π^1_1; we will also prove a theorem of Solovay about the pointclass $\mathfrak{D}\Sigma_2^0$ of 6D which relates games with inductive definability and we will introduce the pointclass of *inductive* sets, a natural extension of the class of projective sets.

Suppose

$$\Phi: Power(\mathfrak{X}) \to Power(\mathfrak{X})$$

is a pointset operation which takes subsets of $\mathfrak{X}$ into subsets of $\mathfrak{X}$ and suppose further that Φ is *monotone*, i.e.

$$A \subseteq B \Longrightarrow \Phi(A) \subseteq \Phi(B).$$

For each ordinal ξ, we define *the ξ'th iterate* Φ^ξ of Φ by the transfinite recursion

$$\Phi^\xi = \Phi(\bigcup_{\eta<\xi} \Phi^\eta)$$

and we call

$$\Phi^\infty = \bigcup_\xi \Phi^\xi$$

the fixed point of Φ or the set built up by Φ.

7C.1. Lemma. *Let Φ be a monotone operation on Power($\mathfrak{X}$).*

(i) *If $\zeta \leq \xi$, then $\Phi^\zeta \subseteq \Phi^\xi$.*

(ii) *For some ordinal $\varkappa$ of cardinality $\leq card(\mathfrak{X})$,*

$$\Phi^\infty = \Phi^\varkappa = \bigcup_{\xi<\varkappa} \Phi^\xi.$$

(iii) *The set Φ^∞ built up by Φ is the least fixed point of Φ, i.e.*

$$\Phi(\Phi^\infty) = \Phi^\infty,$$

$$\Phi^\infty = \bigcap \{A \subseteq \mathfrak{X}: \Phi(A) \subseteq A\}.$$

Proof. (i) is immediate by monotonicity,

$$\Phi^\zeta = \Phi(\bigcup_{\eta<\zeta} \Phi^\eta) \subseteq (\bigcup_{\eta<\xi} \Phi^\eta) = \Phi^\xi.$$

To prove (ii) notice that if we had

$$\bigcup_{\eta<\xi} \Phi^\eta \subsetneq \Phi^\xi$$

for each ordinal $\xi < card(\mathfrak{X})^+$, then we could choose some $x_\xi \in \Phi^\xi - \bigcup_{\eta<\xi} \Phi^\eta$ for each such ξ and then the set $\{x_\xi: \xi < card(\mathfrak{X})^+\}$ would be a subset of $\mathfrak{X}$ of cardinality $card(\mathfrak{X})^+$ which is absurd. Thus for some

$\varkappa < card(\mathfrak{X})^+$ we have

$$\Phi^\varkappa = \bigcup_{\xi<\varkappa} \Phi^\xi$$

and then easily for each $\zeta \geq \varkappa$ we have $\Phi^\zeta = \Phi^\varkappa$ so that $\Phi^\infty = \Phi^\varkappa$.

This argument also proves that Φ^∞ is a fixed point of Φ, since (choosing $\varkappa$ as in (ii)) we have

$$\Phi(\Phi^\infty) = \Phi(\Phi^\varkappa) = \Phi(\bigcup_{\xi<\varkappa} \Phi^\xi) = \Phi^\varkappa = \Phi^\infty.$$

On the other hand, if $\Phi(A) \subseteq A$, then an easy induction shows that for each ξ, $\Phi^\xi \subseteq A$ (using monotonicity), so that $\Phi^\infty \subseteq A$. ⊣

The least ordinal $\varkappa$ which satisfies (ii) in this lemma is called *the closure ordinal of* Φ—it gives us the length of the recursion determined by the operation Φ.

In order to study the more useful *definitions by recursion with parameters*, it is convenient to switch from set operations to *set relations*, i.e. relations with arguments both points and pointsets. We will denote a typical set relation by

$$\varphi(x_1, \ldots, x_n, A_1, \ldots, A_k),$$

where $x_1, \ldots, x_n$ vary over specified spaces $\mathfrak{X}_1, \ldots, \mathfrak{X}_n$ and $A_1, \ldots, A_k$ vary over the subsets of specified spaces $\mathfrak{Y}_1, \ldots, \mathfrak{Y}_k$ respectively. (Of course, we may have $n = 0$ or $k = 0$, so in particular, all pointsets are set relations.)

We call φ *monotone* if

$$\varphi(x_1, \ldots, x_n, A_1, \ldots, A_k) \,\&\, A_1 \subseteq B_1 \,\&\, \cdots \,\&\, A_k \subseteq B_k \implies \varphi(x_1, \ldots, x_n, B_1, \ldots, B_k).$$

A set relation is *operative on the space* $\mathfrak{W}$ if it is monotone and of the form

$$\varphi(w, x, A)$$

where w *varies over* $\mathfrak{W}$ *and* A *varies over the subsets of* $\mathfrak{W}$. Such a relation determines naturally for each $x \in \mathfrak{X}$ a monotone set operation on the subsets of $\mathfrak{W}$,

$$\Phi_x(A) = \{w : \varphi(w, x, A)\}$$

which we can iterate and set

$$\varphi^\xi(w, x) \iff w \in \Phi_x^\xi;$$

it is clear that each relation φ^ξ is defined directly by the recursion

$$\varphi^\xi(w, x) \leftrightarrow \varphi(w, x, \{w' : (\exists\eta < \xi)\varphi^\eta(w', x)\}).$$

To simplify this further, put

$$\varphi^{<\xi}(w, x) \leftrightarrow (\exists\eta < \xi)\varphi^\eta(w, x) \tag{1}$$

so that the determining equivalence for φ^ξ becomes

$$\varphi^\xi(w, x) \leftrightarrow \varphi(w, x, \{w' : \varphi^{<\xi}(w', x)\}). \tag{2}$$

We also put

$$\varphi^\infty(w, x) \leftrightarrow (\exists\xi)\varphi^\xi(w, x)$$

and we call φ^∞ *the fixed point of* φ *or the relation built up by* φ.

The *closure ordinal* of φ is the least $\varkappa$ such that for all $w \in \mathcal{W}$, $x \in \mathfrak{X}$

$$\varphi^\infty(w, x) \leftrightarrow \varphi^\varkappa(w, x) \leftrightarrow \varphi^{<\varkappa}(w, x);$$

this is easily the supremum of the closure ordinals of the set operations $\Phi_x (x \in \mathfrak{X})$ associated with φ.

To illustrate the notions, let us reconsider from this point of view the analysis of Π^1_1 relations that we gave in 4C in order to prove that Π^1_1 is contained in every Spector pointclass.

Suppose that $Q \subseteq \mathfrak{X} \times \omega$ is given and we put

$$P(x, w) \leftrightarrow (\forall\alpha)(\exists t)Q(x, w * \bar{\alpha}(t)). \tag{3}$$

Define the set relation $\varphi(w, x, A)$ (with w ranging over ω and A ranging over subsets of ω) by

$$\varphi(w, x, A) \leftrightarrow Q(x, w) \vee (\forall s)A(w * \langle s \rangle); \tag{4}$$

now φ is clearly operative on ω and by (1) and (2) above,

$$\begin{aligned}\varphi^\xi(w, x) &\leftrightarrow \varphi(w, x, \{w' : \varphi^{<\xi}(w', x)\})\\ &\leftrightarrow Q(x, w) \vee (\forall s)\varphi^{<\xi}(w * \langle s \rangle, x).\end{aligned}$$

In the notation that we used in the beginning of 4C then, clearly

$$\varphi^\xi(w, x) \leftrightarrow P^\xi(x, w)$$

and hence by the argument given there,

$$\begin{aligned}P(x, w) &\leftrightarrow (\exists\xi)P^\xi(x, w)\\ &\leftrightarrow (\exists\xi)\varphi^\xi(w, x)\\ &\leftrightarrow \varphi^\infty(w, x),\end{aligned}$$

i.e. except for the order of the variables, P is the fixed point of φ. Using the canonical representation of Π^1_1 sets (4A.1) and setting $w = 1$ in (3), we thus have that *every* Π^1_1 *pointset* $R \subseteq \mathfrak{X}$ *satisfies*

$$R(x) \leftrightarrow \varphi^{\infty}(1, x)$$

where φ^{∞} *is the fixed point of a set relation defined by* (4) *above with* Q *some* Σ^0_1 *set; in fact if* $\mathfrak{X}$ *is of type* 0 *or* 1, *we can take* Q *to be recursive.*

In order to define a relation R on $\mathfrak{X}$ here, we perform for each x an induction on ω and we ask at the end whether the specific constant 1 belongs to the fixed point of this induction; this key example motivates the following basic definition.

Suppose Γ is a collection of monotone set relations and $R \subseteq \mathfrak{X}$; we say that R is Γ-*inductive on* $\mathcal{W}$ if there is a set relation $\varphi(w, x, A)$ in Γ which is operative on $\mathcal{W}$ and some point $w_0 \in \mathcal{W}$, such that

(5) $$R(x) \leftrightarrow \varphi^{\infty}(w_0, x).$$

If we can choose a *recursive* w_0 in (5), we call R *absolutely* Γ-*inductive on* $\mathcal{W}$.

Let us now look at some specific examples of collections of set relations and (in particular) try to find a simple Γ so that Π^1_1 consists precisely of all pointsets which are Γ-inductive on ω. We will extend to (some) set relations the Kleene hierarchy of arithmetical and analytical pointclasses.

A collection Γ of set relations is a *monotone* Σ-*collection* if the following conditions hold.

(i) All set relations in Γ are monotone.

(ii) For each space $\mathcal{W}$, Γ contains the relation of *evaluation on* $\mathcal{W}$,

(6) $$E_{\mathcal{W}}(w, A) \leftrightarrow A(w) \quad (A \subseteq \mathcal{W});$$

this is obviously monotone.

(ii) Γ contains all the Σ^0_1 pointsets (viewed as set relations with no set arguments) and it is closed under $\&$, $\vee$, $\exists^{\leq}$, $\forall^{\leq}$, $\exists^{\omega}$ and trivial substitutions; by *trivial substitution* here we mean any map of the form

$$(x_1, \ldots, x_n, A_1, \ldots, A_k) \mapsto (x_{j_1}, \ldots, x_{j_s}, A_{l_1}, \ldots, A_{l_t})$$

where for each i, $1 \leq Rj_i \leq n$ and $1 \leq l_i \leq k$, so that we can add, permute or identify variables of any sort and stay in Γ.

A *monotone* Π-*collection* is defined in the same way, except that we replace the conditions $\Sigma^0_1 \subseteq \Gamma$ and closure under $\exists^{\omega}$ by $\Pi^0_1 \subseteq \Gamma$ and closure under $\forall^{\omega}$.

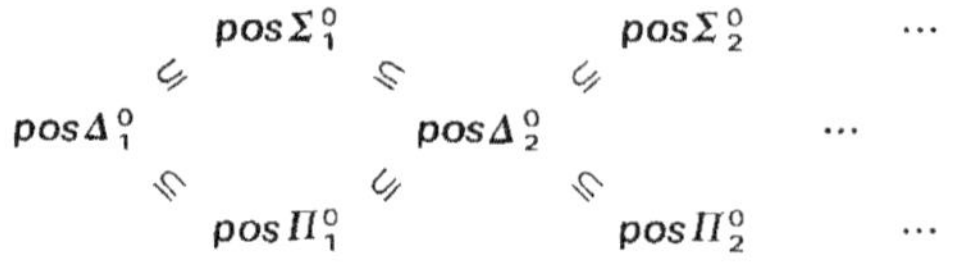

Diagram 7C.1. The positive arithmetical collections of set relations.

The collection $pos\,\Sigma^0_1$ of *positive* Σ^0_1 *set relations* is the smallest monotone Σ-collection and $pos\,\Pi^0_1$ is the smallest monotone Π collection. Proceeding inductively, put

$$pos\,\Pi^0_2 = \forall^\omega\, pos\,\Sigma^0_1$$
$$pos\,\Sigma^0_3 = \exists^\omega\, pos\,\Pi^0_2,$$

etc. and for the dual collections,

$$pos\,\Sigma^0_2 = \exists^\omega\, pos\,\Pi^0_1,$$
$$pos\,\Pi^0_3 = \forall^\omega\, pos\,\Sigma^0_2,$$

etc.

It is trivial to check that the canonical diagram of inclusions hold for these collections, see Diagram 7C.1, where of course

$$pos\,\Delta^0_n = pos\,\Sigma^0_n \cap pos\,\Pi^0_n.$$

Most of the other formal properties of the arithmetical pointsets also extend trivially to these *positive arithmetical set relations* with one obvious exception: the negation of the basic relation $E_{\mathcal{W}}$ in (6) is not positive arithmetical, so that this collection of set relations is not closed under $\neg$.

The extension of these definitions to the *positive analytical set relations* is immediate,

$$pos\,\Sigma^1_1 = \exists^{\mathcal{N}}(\textstyle\bigcup_n pos\,\Pi^0_n),$$
$$pos\,\Pi^1_1 = \forall^{\mathcal{N}}(\textstyle\bigcup_n pos\,\Sigma^0_n),$$

and inductively,

$$pos\,\Sigma^1_{n+1} = \exists^{\mathcal{N}}(pos\,\Pi^1_n),$$
$$pos\,\Pi^1_{n+1} = \forall^{\mathcal{N}}(pos\,\Sigma^1_n).$$

Again the diagram of inclusions is easy to establish.

We now have all notions we need to state the characterization of Π^1_1 in terms of inductive definability on ω.

7C.2. THEOREM. (i) *All the positive analytical set relations are monotone.*

(ii) *Every* Π^1_1 *pointset* R *is positive* Δ^0_2*-inductive on* ω; *in fact, if* R *is of type* 0 *or* 1, *then* R *is positive* Π^0_1*-inductive on* ω (*Kleene*[(3)]).

(iii) *If* R *is positive* Π^1_n*-inductive on* ω, *then* R *is* Π^1_n (*Spector* [1961]).

(iv) *If* R *is positive* Σ^1_n*-inductive on* ω *and* $n \geq 2$, *then* R *is* Σ^1_n (*Spector* [1961]).

PROOF. (i) holds for $pos\,\Sigma^0_1$ and $pos\,\Pi^0_1$ because the collection of monotone set relations is easily both a Σ-collection and a Π-collection and it contains E. For the higher collections use induction.

(ii) was proved above, since the set relation defined by (*) is trivially in $pos\,\Delta^0_2$,

$$\varphi(w, x, A) \leftrightarrow Q(x, w) \vee (\forall s)(\forall u)[u \neq w * \langle s \rangle \vee A(u)].$$

With each set relation $\varphi(x, A)$ where A varies over subsets of ω we associate the pointset $Q_\varphi \subseteq \mathfrak{X} \times \mathcal{N}$,

$$Q_\varphi(x, \alpha) \leftrightarrow \varphi(x, \{t : \alpha(t) = 1\});$$

it is trivial to check that if φ is $pos\,\Sigma^1_n$ or $pos\,\Pi^1_n$, then Q_φ is Σ^1_n or Π^1_n respectively.

To prove (iii) we use the characterization of Φ^∞ in 7C.1(iii), as the least fixed point of Φ: if $\varphi(w, x, A)$ is operative on ω, tracing the definitions,

$$\begin{aligned}
\varphi^\infty(w, x) &\leftrightarrow w \in \Phi^\infty_x \\
&\leftrightarrow (\forall A)\{\Phi_x(A) \subseteq A \rightarrow w \in A\} \\
&\leftrightarrow (\forall A)\{(\forall w')[\varphi(w', x, A) \rightarrow w' \in A] \rightarrow w \in A\} \\
&\leftrightarrow (\forall A)\{(\exists w')[\varphi(w', x, A) \;\&\; w' \notin A] \wedge w \in A\} \\
&\leftrightarrow (\forall \alpha)\{(\exists w')[Q_\varphi(w', x, \alpha) \;\&\; \alpha(w') \neq 1] \vee \alpha(w) = 1\},
\end{aligned}$$

so that if φ is $pos\,\Pi^1_n$, then Q_φ is Π^1_n and hence φ^∞ is Π^1_n.

To prove (iv) we analyze the definition of Φ^ξ by transfinite recursion. Check first, using monotonicity that

$$\begin{aligned}
\varphi^\infty(w, x) &\leftrightarrow w \in \Phi^\infty_x \\
&\leftrightarrow \text{there exists a sequence of sets } \{A_\xi\}_{\xi<\varkappa} \text{ indexed} \\
&\quad\ \text{by some countable ordinal } \varkappa \text{ such that} \\
&\quad\ (\forall \xi < \varkappa)[A_\xi \subseteq \Phi_x(\textstyle\bigcup_{\eta<\xi} A_\eta)] \;\&\; w \in \textstyle\bigcup_{\xi<\varkappa} A_\xi.
\end{aligned}$$

Using the canonical codes for ordinals and associating with each α the set $\{w: \alpha(w)=1\}$, we then have

$$\varphi^{\infty}(w, x) \Leftrightarrow (\exists\beta)(\exists\alpha)\{\beta \in WO \\ \& (\forall n)[n \leq_\beta n \rightarrow \{w': (\alpha)_n(w')=1\} \\ \subseteq \Phi_x(\bigcup\{w': (\exists m <_\beta n)[(\alpha)_m(w')=1]\})] \\ \& (\exists n)[n \leq_\beta n \& (\alpha)_n(w)=1]\}.$$

Now using monotonicity again,

$$w'' \in \Phi_x(\bigcup\{w': (\exists m <_\beta n)[(\alpha)_m(w')=1]\}) \\ \Leftrightarrow (\exists\gamma)[(\forall w')[\gamma(w')=1 \rightarrow (\exists m <_\beta n)[(\alpha)_m(w')=1]] \\ \& Q_\varphi(w'', x, \gamma)];$$

thus if φ is *pos* Σ^1_n, this last relation is Σ^1_n and then easily φ^∞ is also Σ^1_n, provided of course that $n \geq 2$ so that the Π^1_1 relation "$\beta \in WO$" is Σ^1_n. ⊣

This simple result does not handle the case of positive Σ^1_1-induction on ω—we will deal with this very interesting example in 7C.10.

Let us now turn to induction on $\mathcal{N}$ which is more directly relevant to descriptive set theory.

A pointset $R \subseteq \mathfrak{X}$ is *inductive* if it is *positive analytical inductive on* $\mathcal{N}$, i.e. if there is a set relation $\varphi(\alpha, x, A)$ which is in some *pos* Σ^1_n and operative on $\mathcal{N}$ and some fixed $\alpha_0 \in \mathcal{N}$ so that

$$R(x) \Leftrightarrow \varphi^\infty(\alpha_0, x). \tag{7}$$

We call R *hyperprojective* if both R and $\neg R$ are inductive and we denote the pointclasses of inductive and hyperprojective pointsets respectively by ***IND*** and ***HYP***.

If (7) holds with a recursive irrational α_0, then R is *absolutely inductive* and if both R and $\neg R$ are absolutely inductive, then R is *absolutely hyperprojective*. We will denote these two "lightface" pointclasses by *IND* and *HYP* respectively.[4]

The definition of inductive sets together with the characterization of Π^1_1 sets in 7C.2 suggest that *IND* is somehow a "generalization" or "second-order analog" of the pointclass Π^1_1. In fact (with suitable hypotheses) almost the entire theory of Π^1_1 can be extended to *IND*. We will look at some of these results in the exercises, but it is worth putting down here the statements of three basic facts.

7C.3. THEOREM (Moschovakis). *The class IND of all absolutely inductive sets is a Spector pointclass, in fact it is the smallest Spector pointclass which is closed under both* $\forall^{\mathcal{N}}$ *and* $\exists^{\mathcal{N}}$; *its associated boldface class is* **IND**, *i.e. a set* $R \subseteq \mathfrak{X}$ *is inductive if and only if there is some absolutely inductive* $R^* \subseteq \mathcal{N} \times \mathfrak{X}$ *and some* $\varepsilon \in \mathcal{N}$ *so that*

$$R = R^*_\varepsilon = \{x : R^*(\varepsilon, x)\}.$$

PROOF. See 7C.12–7C.15. ⊣

7C.4. THEOREM (Moschovakis [1971b]). *A pointset* $R \subseteq \mathfrak{X}$ *is absolutely inductive if and only if there is an analytical (or even* Σ^1_1*) pointset P such that*

$$R(x) \Leftrightarrow \{(\exists\alpha_0)(\forall\alpha_1)(\exists\alpha_2)(\forall\alpha_3)\cdots\}(\exists n)P(x, \langle\alpha_0,\dots,\alpha_{n-1}\rangle).$$

PROOF. See 7C.18. ⊣

The interpretation of the infinite string of quantifiers here is via games, as in 6D, except that now the games are played in $\mathcal{N}$. This representation of *IND* is obviously analogous to the representation of Π^1_1 as $\Game\Sigma^0_1$ in 6D in terms of open games on ω.

7C.5. THEOREM (Moschovakis [1978]). *If every hyperprojective set is determined, then every absolutely inductive set admits an IND-scale.*

In particular, every absolutely inductive $R \subseteq \mathfrak{X} \times \mathcal{Y}$ *can be uniformized by some absolutely inductive* $R^* \subseteq R$ *and every nonempty absolutely inductive pointset has an absolutely hyperprojective member.*

PROOF. See 7C.19. ⊣

The collection of inductive sets is the largest known pointclass on which we can put definable scales, from any hypotheses; *whether coinductive sets admit definable scales* appears to be one of the critical open problems of the subject.

EXERCISES

Let us start with a simple result which relates these new positive analytical set relations with the Kleene pointclasses.

Recall from 4E that with each pointset $P \subseteq \omega^k$ of type 0 we have associated its *contracted characteristic function*

$$\alpha_P(n) = \begin{cases} 1 & \text{if} \quad P((n)_1, \ldots, (n)_k), \\ 0 & \text{if} \quad \neg P((n)_1, \ldots, (n)_k); \end{cases}$$

conversely, for each $\alpha \in \mathcal{N}$ and fixed k, let

$$P^\alpha(t_1, \ldots, t_k) \Leftrightarrow \alpha(\langle t_1, \ldots, t_k \rangle) = 1,$$

so that

$$P^{\alpha_P} = P.$$

Finally, if $\varphi(x_1, \ldots, x_n, A_1, \ldots, A_m)$ is a set relation where each A_i varies over the subsets of some space $\mathfrak{X}_i$ of type 0, put

$$Q_\varphi(x_1, \ldots, x_n, \alpha_1, \ldots, \alpha_m) \Leftrightarrow \varphi(x_1, \ldots, x_n, P^{\alpha_1}, \ldots, P^{\alpha_m}).$$

7C.6. (i) Prove that if $\varphi(x_1, \ldots, x_n, A_1, \ldots, A_m)$ is in *pos* Σ^0_k or *pos* Π^0_k, then the associated pointset Q_φ is in Σ^0_k or Π^0_k respectively.

(ii) Prove that if each A_i ranges over the subsets of some space $\mathfrak{Y}_i$ of type 0, then $\varphi(x_1, \ldots, x_n, A_1, \ldots, A_m)$ is in *pos* Σ^1_k or *pos* Π^1_k if and only if φ is monotone and the associated Q_φ is in the corresponding pointclass Σ^1_k or Π^1_k.

Hint. (i) and half of (ii) can be verified by a simple induction on the definition of the positive analytical classes of set relations. For the other direction of (ii) notice (for example) that if $\varphi(x, A)$ is monotone and A varies over subsets of ω, then

$$\begin{aligned} \varphi(x, A) &\Leftrightarrow (\exists \beta)\{(\forall t)[\beta(t) = 1 \rightarrow A(t)] \;\&\; Q_\varphi(x, \beta)\} \\ & (\forall \beta)\{(\forall t)[A(t) \rightarrow \beta(t) = 1] \rightarrow Q_\varphi(x, \beta)\}, \end{aligned}$$

from which it follows immediately that if Q_φ is in Σ^1_k or Π^1_k, then φ is in *pos* Σ^1_k or *pos* Π^1_k. ⊣

The next result clarifies the connection between the Kleene Basis Theorem for Σ^1_1 (4E.8) and the Martin–Solovay Basis Theorem for Σ^1_{2n+1} (6C.10, 6C.11).

7C.7. (i) Prove that every Σ^1_1 pointset of type 0 is recursive in some Σ^1_1 set $P \subseteq \omega^k$ such that α_P is a Π^1_1-singleton.

(ii) Prove that if P is Σ^1_1 of type 0 and α is any Π^1_1-singleton which is not Δ^1_1, then $\alpha_P \in \Delta^1_1(\alpha)$.

Thus in the notation of 6C.11, the Kleene Basis Theorem asserts that there is a Σ_1^1 set P of type 0 such that $\{x: x \in \Delta_1^1(\alpha_P)\}$ is a basis for Σ_1^1 and such that α_P is $\leq_1$-minimal among the non-Δ_1^1 singletons in Π_1^1.

Hint. Let $G \subseteq \mathfrak{X}$ be any Π_1^1 set of type 0, choose $\varphi(w, x, A)$ in *pos* Π_1^0 by 7C.2 so that for suitable integers w^*,

$$G(x) \Leftrightarrow \varphi^\infty(w^*, x)$$

and let

$$H(w, x) \Leftrightarrow \varphi^\infty(w, x).$$

Now, easily

$$\begin{aligned}
\alpha = \alpha_H \Leftrightarrow & (\forall x)(\forall w)[\alpha(\langle w, x\rangle) \leq 1] \\
& \& \ (\forall x)(\forall w)[Q_\varphi(w, x, w' \mapsto \alpha(\langle w', x\rangle)) \Rightarrow \alpha(\langle w, x\rangle) = 1] \\
& \& \ (\forall x)(\forall \gamma)\{(\forall w)[Q_\varphi(w, x, \gamma) \Leftrightarrow \gamma(\langle w\rangle) = 1] \\
& \quad \Rightarrow (\forall w)[\gamma(\langle w\rangle) = 1 \Leftrightarrow \alpha(\langle w, x\rangle) = 1],
\end{aligned}$$

so that α_H is a Π_1^1-singleton. This establishes (i) by considering the complements of G and H.

To prove (ii), let $G \subseteq \omega \times (\omega \times \mathcal{N})$ be universal Π_1^1 and choose n_0, n_1 so that

$$\neg P(s) \Leftrightarrow G(n_0, s, \beta) \quad (\text{any } \beta),$$
$$\beta = \alpha \Leftrightarrow G(n_1, s, \beta) \quad (\text{any } s).$$

Let φ be a Π_1^1-norm on G and argue that $(\forall s)[\neg P(s) \Rightarrow \varphi(n_0, s, \alpha) \leq \varphi(n_1, s, \alpha)]$—otherwise easily α is a Σ_1^1-singleton and hence Σ_1^1; this then yields immediately that P is $\Delta_1^1(\alpha)$. ⊣

Before going into the discussion of specific pointclasses introduced by inductive definitions, it is worth putting down a soft and very general—but useful—remark about inductive definability.

Suppose Γ is a pointclass and $\varphi(x, A)$ is a set relation where A ranges over the subsets of some $\mathcal{W}$; we say that φ is *Γ on Γ* if for every $Q \subseteq Z \times \mathcal{W}$ in Γ, the relation

$$P(x, z) \Leftrightarrow \varphi(x, \{w: Q(z, w)\})$$

is also in Γ. The definition extends in the obvious way to more general set relations $\varphi(x_1, \ldots, x_n, A_1, \ldots, A_m)$.

7C.8. Suppose Γ is adequate, ω-parametrized and normed and suppose that $\varphi(w, x, A)$ is Γ on Γ and operative on $\mathcal{W}$; prove that the fixed point $\varphi^{\infty}(w, x)$ is in Γ. (Moschovakis [1974b].)

Hint. Choose good parametrizations of Γ by 3H.4, let ψ be a Γ-norm on a universal set $G \subseteq \omega \times \mathcal{W} \times Z$ and put

$$Q(m, w, x) \Leftrightarrow \varphi(w, x, \{w' : (m, w', x) <^*_{\psi} (m, w, x)\});$$

Q is easily in Γ, so by 3H.4 there is a fixed $m^* \in \omega$ so that

$$\begin{aligned} Q(m^*, w, x) &\Leftrightarrow G(m^*, w, x) \\ &\Leftrightarrow \varphi(w, x, \{w' : (m^*, w', x) <^*_{\psi} (m^*, w, x)\}). \end{aligned}$$

Now check by induction on $\psi(m^*, w, x)$ that

$$G(m^*, w, x) \Rightarrow \varphi^{\infty}(w, x)$$

and by induction on ξ that

$$\varphi^{\xi}(w, x) \Rightarrow G(m^*, w, x),$$

so that

$$\varphi^{\infty}(w, x) \Leftrightarrow G(m^*, w, x).$$

(Both arguments appeal to the monotonicity of φ.) ⊣

In order to identify the relations which are positive Σ^1_1-inductive on ω, we compute first a simple normal form for *pos* Σ^1_1.

7C.9. Suppose $\varphi(x, A)$ is a set relation, where A ranges over subsets of ω. Prove that φ is *pos* Σ^1_1 if and only if there exist Σ^1_1 pointsets $Q(x)$ and $R(x, \alpha)$ so that

$$(*) \qquad \varphi(x, A) \Leftrightarrow Q(x) \vee (\exists \alpha)\{(\forall t) A(\alpha(t)) \;\&\; R(x, \alpha)\}.$$

Hint. Use induction on the definition of *pos* Σ^1_1. The only parts of the argument where some care is required, are in checking that the class of relations defined by $(*)$ is closed under & and $\forall^{\omega}$. ⊣

7C.10. Prove that a pointset R is positive Σ^1_1-inductive on ω if and only if R is in the pointclass $\mathfrak{D}\Sigma^0_2$ defined in 6D. (Solovay.)

Hint. To check that every pointset in $\mathfrak{D}\Sigma^0_2$ is positive Σ^1_1-inductive on ω, go back and look carefully at the proof of *Det*$(\boldsymbol{\Sigma}^0_2)$ in 6A.3.

For the converse, it is enough by 7C.8 and 7C.9 to check that if $Q(x, n)$

is in $\mathfrak{D}\Sigma_2^0$ and $R(z, \alpha)$ is Σ_1^1, then the relation

(1) $$P(x, z) \Leftrightarrow (\exists\alpha)\{(\forall t)Q(x, \alpha(t)) \;\&\; R(z, \alpha)\}$$

is in $\mathfrak{D}\Sigma_2^0$, so suppose

(2) $$Q(x, n) \Leftrightarrow (\exists u_1)(\forall v_1)(\exists u_2)(\forall v_2)\cdots Q_1(x, n, (u_1, v_1, \ldots)),$$

(3) $$R(z, \alpha) \Leftrightarrow (\exists\beta)(\forall k)R_1(z, \bar{\alpha}(k), \bar{\beta}(k)),$$

where Q_1 is Σ_2^0 and R_1 is recursive. For each x, z consider the game $G(x, z)$ where I starts by playing

$$\alpha(0), \ldots, \alpha(l),$$
$$\beta(0), \ldots, \beta(l),$$

until at some step l he is told by II to stop; I loses at this step if for some $k \leq l$, $\neg R_1(z, \bar{\alpha}(k), \bar{\beta}(k))$ and wins if he is never told to stop. If II does ask I to stop at some l, then II chooses some t and then the game continues as in Diagram 7C.2. At the end of the run I wins if $(\forall k)R_1(z, \bar{\alpha}(k), \bar{\beta}(k))$ and $Q_1(x, \alpha(l), (u_1, v_1, u_2, v_2, \ldots))$. This is clearly a Σ_2^0 game (for each x, z) and it is easy to check that if (1), (2) and (3) hold, then

$$P(x, z) \Leftrightarrow \text{I } wins\ G(x, z),$$

so that P is in $\mathfrak{D}\Sigma_2^0$. ⊣

Together with the Third Periodicity Theorem 6E.1, the last exercise implies that if player I can win a Σ_2^0 game, then I in fact has a winning strategy which is positive Σ_1^1-inductive on ω. Thomas John has recently computed the complexity of winning strategies for Σ_3^0 games (in terms of recursion in higher types) but the corresponding problem for Σ_n^0 $(n \geq 4)$ is still open.

Let us now turn to an outline of the proofs of 7C.3–7C.5; it will be convenient to frame some of the lemmas in a context wider than positive

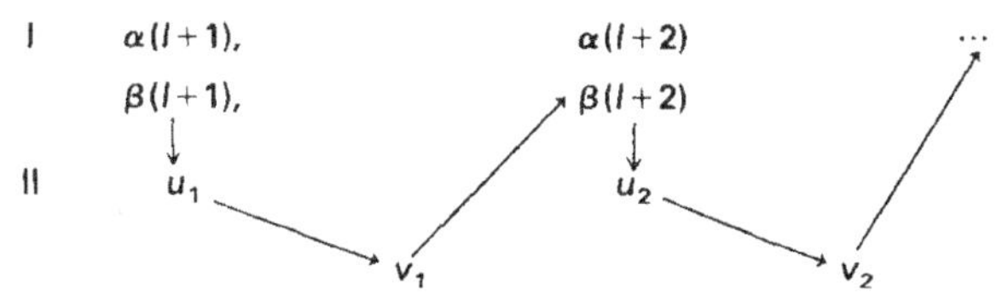

Diagram 7C.2.

analytical induction, although the proofs will make it clear that we have not aimed at the most general results.

A collection Γ of set relations is (temporarily) called "*suitable for our purposes*" if Γ is both a Σ-collection and a Π-collection (as we defined these preceding 7C.2) and if in addition Γ is closed under recursive substitution in the following sense: if $P(z, A_1, \ldots, A_k)$ is in Γ and $f(x), g_1(x, y_1), \ldots, g_k(x, y_k)$ are recursive functions, then $Q(x, B_1, \ldots, B_k)$ is also in Γ, where

$$Q(x, B_1, \ldots, B_k) \Leftrightarrow P(f(x), \{y_1: B_1(g_1(x, y_1))\}, \ldots, \{y_k: B_k(g_k(x, y_k))\}).$$

It is not hard to verify that the collections $pos\,\Sigma^1_n$, $pos\,\Pi^1_n$ are all suitable for our purposes.

For any Γ, put

$$\boldsymbol{IND}(\Gamma; \mathcal{N}) = \text{all relations } \Gamma\text{-inductive on } \mathcal{N},$$

$$IND(\Gamma; \mathcal{N}) = \text{all relations absolutely } \Gamma\text{-inductive on } \mathcal{N}.$$

7C.11. Suppose that Γ is suitable for our purposes and $\varphi_1(y_1, A_1, \ldots, A_n), \varphi_2(y_2, A_1, \ldots, A_n), \ldots, \varphi_n(y_n, A_1, \ldots, A_n)$ are in Γ, where each y_i ranges over the space $\mathcal{Y}_i$ of type 1 and each A_i ranges over the subsets of $\mathcal{Y}_i$. Define the sets $J^\xi_i \subseteq \mathcal{Y}_i$ by the simultaneous induction:

$$y_1 \in J^\xi_1 \Leftrightarrow \varphi_1(y_1, J^{<\xi}_1, J^{<\xi}_2, \ldots, J^{<\xi}_n),$$

$$y_2 \in J^\xi_2 \Leftrightarrow \varphi_2(y_2, J^{<\xi}_1, J^{<\xi}_2, \ldots, J^{<\xi}_n),$$

$$\ldots$$

$$y_n \in J^\xi_n \Leftrightarrow \varphi_n(y_n, J^{<\xi}_1, J^{<\xi}_2, \ldots, J^{<\xi}_n),$$

where of course

$$J^{<\xi}_i = \bigcup_{\eta<\xi} J^\eta_i.$$

Prove that each

$$J^\infty_i = \bigcup_\xi J^\xi_i$$

is absolutely Γ-inductive on $\mathcal{N}$.

Similarly with parameters: given $\varphi_1(y_1, x, A_1, \ldots, A_n), \ldots, \varphi_n(y_n, x, A_1, \ldots, A_n)$ in Γ with the obvious restrictions, if we set

$$J^\xi_i(y_i, x) \Leftrightarrow \varphi_i(y_1, x, \{y'_1: J^{<\xi}_1(y'_1, x)\}, \ldots, \{y'_n: J^{<\xi}_n(y'_n, x)\}),$$

then each J^∞_i is absolutely Γ-inductive on $\mathcal{N}$. In this case we allow that for some i we may have $\mathcal{Y}_i = \emptyset$, i.e. $\varphi_i(x, A_1, \ldots, A_n)$ is given and

$$J_i^\xi(x) \Leftrightarrow \varphi_i(x, \{y_1' : J_1^{<\xi}(y_1', x)\}, \ldots, \{y_n' : J_n^{<\xi}(y_n', x)\}).$$

In particular, if $\varphi(y, x, A)$ is in Γ and operative on the space $\mathcal{Y}$, then the fixed point $\varphi^\infty(y, x)$ is absolutely Γ-inductive on $\mathcal{N}$.

Hint. Take $n = 2$ without parameters, so that $\varphi_1(y_1, A_1, A_2)$, $\varphi_2(y_2, A_1, A_2)$ are given, let

$$\pi_1 : \mathcal{N} \rightarrowtail\!\!\!\twoheadrightarrow \mathcal{Y}_1, \qquad \pi_2 : \mathcal{N} \rightarrowtail\!\!\!\twoheadrightarrow \mathcal{Y}_2$$

be recursive isomorphisms and put

$$\begin{aligned}\psi(\alpha, y_1, A) \Leftrightarrow & [\alpha(0) = 1 \;\&\; \varphi_1(\pi_1(\alpha^\star), \{\pi_1(\beta) : A(1\frown\beta)\},\\ & \quad \{\pi_2(\gamma) : A(2\frown\gamma)\})]\\ & \vee [\alpha(0) = 2 \;\&\; \varphi_2(\pi_2(\alpha^\star), \{\pi_1(\beta) : A(1\frown\beta)\},\\ & \quad \{\pi_2(\gamma) : A(2\frown\gamma)\})]\\ & \vee [\alpha(0) = 3 \;\&\; (1, \pi_1^{-1}(y_1)) \in A]\end{aligned}$$

where $\alpha^\star(t) = \alpha(t+1)$ and $i\frown\delta = (i, \delta(0), \delta(1), \ldots)$. Clearly ψ is in Γ and by a simple induction on ξ,

$$\begin{aligned}\psi^\xi(1\frown\beta, y_1) &\Leftrightarrow J_1^\xi(\pi_1(\beta)),\\ \psi^\xi(2\frown\gamma, y_1) &\Leftrightarrow J_2^\xi(\pi_2(\gamma))\end{aligned}$$

so that if we substitute the recursive constant

$$\alpha_3(t) = 3,$$

we have

$$\begin{aligned}\psi^\xi(\alpha_3, y_1) &\Leftrightarrow (1, \pi_1^{-1}(y_1)) \in \psi^{<\xi}\\ &\Leftrightarrow J_1^{<\xi}(y_1);\end{aligned}$$

thus

$$J_1^\infty(y_1) \Leftrightarrow \psi^\infty(\alpha_3, y_1)$$

and J_1^∞ is absolutely Γ-inductive on $\mathcal{N}$. The argument for J_2^∞ is similar. ⊣

7C.12. Prove that if Γ is suitable for our purposes, then $IND(\Gamma; \mathcal{N})$ contains all the pointsets in Γ and is closed under $\&$, $\vee$, $\exists^\omega$, $\forall^\omega$ and recursive substitutions; if Γ is also closed under either $\exists^\mathcal{N}$ or $\forall^\mathcal{N}$, then so is $IND(\Gamma; \mathcal{N})$.

Hint. To show closure under $\exists^{\mathcal{N}}$ as an example, suppose

$$R(x, \beta) \Leftrightarrow \varphi^{\infty}(\alpha_0, x, \beta)$$

with α_0 recursive and $\varphi(\alpha, x, \beta, A)$ in Γ and operative on $\mathcal{N}$ and consider the system

$$\varphi_1(\alpha, \beta, x, A_1, A_2) \Leftrightarrow \varphi(\alpha, x, \beta, \{\alpha' : A_1(\alpha', \beta)\}),$$
$$\varphi_2(x, A_1, A_2) \Leftrightarrow (\exists\beta)A_1(\alpha_0, \beta);$$

check easily that

$$(\exists\beta)R(x, \beta) \Leftrightarrow J_2^{\infty}(x).$$

(In this case we are applying 7C.11 with $\mathcal{Y}_2 = \emptyset$, which is valid by the proof of that result outlined in the hint.)

Closure of $IND(\Gamma; \mathcal{N})$ under recursive substitutions follows from the simple observation that if $\varphi(\alpha, z, y, A)$ is operative on $\mathcal{N}$, if $f : \mathfrak{X} \to \mathbf{Z}$ is any function and if

$$\psi(\alpha, x, y, A) \Leftrightarrow \varphi(\alpha, f(x), y, A),$$

then

$$\psi^{\xi}(\alpha, x, y) \Leftrightarrow \varphi^{\xi}(\alpha, f(x), y). \qquad \dashv$$

7C.13. Prove that every absolutely inductive relation is in $IND(pos\,\Pi_2^1; \mathcal{N})$ and also in $IND(pos\,\Sigma_2^1; \mathcal{N})$.

Hint. Suppose for example that

$$\varphi(\alpha, x, A) \Leftrightarrow (\exists\beta)\psi(\alpha, x, \beta, A)$$

where ψ is in $pos\,\Pi_2^1$ and φ is operative on $\mathcal{N}$ and consider the system:

$$\varphi_1(\alpha, \beta, x, A_1, A_2) \Leftrightarrow \psi(\alpha, x, \beta, A_2)$$
$$\varphi_2(\alpha, x, A_1, A_2) \Leftrightarrow (\exists\beta)A_1(\alpha, \beta).$$

Check easily (using monotonicity) that

$$\varphi^{\infty}(\alpha, x) \Leftrightarrow J_2^{\infty}(\alpha, x);$$

now if α_0 is recursive, then the relation $\{x : \varphi^{\infty}(\alpha_0, x)\}$ which is a typical member of $IND(pos\,\Sigma_3^1; \mathcal{N})$ is in $IND(pos\,\Pi_2^1; \mathcal{N})$, by closure of this latter pointclass under recursive substitution. $\dashv$

7C.14. Prove that every set relation $\varphi(\alpha, x, A)$ in $pos\,\Sigma_1^1$ with $A \subseteq \mathcal{N}$ satisfies an equivalence of the form

$$(*) \qquad \varphi(\alpha, x, A) \Leftrightarrow Q(\alpha, x) \vee (\exists\gamma)\{(\forall n)A((\gamma)_n) \,\&\, R(\alpha, x, \gamma)\}$$

with Q and R in Σ_1^1; infer that the pointclass of absolutely inductive pointsets is ω-parametrized.

Hint. Check that the collection of relations which admits this representation has all the right closure properties. Derive a similar representation for *pos* Π_2^1 by quantifying both sides of $(*)$ and infer that $IND(pos\,\Pi_2^1; \mathcal{N})$ is ω-parametrized using the parametrization theorem for Σ_1^1 (and the closure properties of $IND(pos\,\Pi_2^1; \mathcal{N})$). ⊣

7C.15. Prove that the pointclass *IND* has the prewellordering property (Moschovakis [1974a]).

Hint. If $\varphi(\alpha, x, A)$ is positive analytical with $A \subseteq \mathcal{N}$, we have an obvious norm on the fixed point $\varphi^\infty(\alpha, x)$ given by

$$\sigma(\alpha, x) = \text{least } \xi \text{ such that } \varphi^\xi(\alpha, x);$$

we will prove that this norm is inductive, so that for any recursive α_0, the induced norm

$$\sigma_0(x) = \sigma(\alpha_0, x)$$

on $\{x: \varphi^\infty(\alpha_0, x)\}$ will also be inductive.

Let

$$\psi(\alpha, \beta, x, A) \leftrightarrow \varphi(\alpha, x, \{\alpha': \neg\varphi(\beta, x, \{\beta': \neg A(\alpha', \beta')\})\})$$

and verify that ψ is positive analytical; thus $\psi^\infty(\alpha, \beta, x)$ is inductive by 7C.11 (with $n = 1$) and it is not hard to check that

$$(\alpha, x) \leq_\sigma^* (\beta, x) \leftrightarrow \psi^\infty(\alpha, \beta, x).$$

The construction of a ψ' which defines inductively $<_\sigma^*$ is similar. ⊣

7C.16. Prove 7C.3.

Hint. From 7C.12–7C.15, we know that *IND* is a Spector pointclass with all the right closure properties and the fact that **IND** is its associated boldface class follows from closure of **IND** under the substitution of constants which is elementary.

Suppose now that Γ is any Spector pointclass which is closed under both $\exists^\mathcal{N}$ and $\forall^\mathcal{N}$; then easily, each positive analytical $\varphi(\alpha, x, A)$ is Γ on Γ, so that by 7C.8, $\varphi^\infty \in \Gamma$ and hence $IND \subseteq \Gamma$. ⊣

7C.17. Prove that for every set relation $\varphi(x, A)$ in *pos* Σ_3^1 with A ranging over subsets of $\mathcal{N}$, there is some Σ_1^1 relation $\varphi(x, \alpha, \beta)$ so that for all $A \subsetneq \mathcal{N}$,

$$\varphi(x, A) \leftrightarrow (\exists\alpha)(\forall\beta)(\exists\gamma)(\forall\delta)[Q(x, \alpha, \beta, \gamma, \delta) \vee A(\delta)].$$

Hint. Start with

$$\varphi(x, A) \leftrightarrow (\exists\alpha)(\forall\beta)\psi(x, \alpha, \beta, A)$$

where ψ is in *pos* Σ_1^1 so that by 7C.14 there are Σ_1^1 relations Q_1 and R_1 with which

$$\psi(x, \alpha, \beta, A) \leftrightarrow Q_1(x, \alpha, \beta) \vee (\exists\gamma)\{(\forall n)A((\gamma)_n) \,\&\, R_1(x, \alpha, \beta, \gamma)\}.$$

Check that for $A \subsetneq \mathcal{N}$, easily

$$\begin{aligned}(\forall n)A((\gamma)_n) \,\&\, R_1(x, \alpha, \beta, \gamma)\\ \leftrightarrow (\forall\delta)\{[(\forall n)[\delta \neq (\gamma)_n] \,\&\, R_1(x, \alpha, \beta, \gamma)] \vee A(\delta)\};\end{aligned}$$

thus with

$$Q_2(x, \alpha, \beta, \gamma, \delta) \leftrightarrow (\forall n)[\delta \neq (\gamma)_n] \,\&\, R_1(x, \alpha, \beta, \gamma)$$

we have for $A \subsetneq \mathcal{N}$

$$\psi(x, \alpha, \beta, A) \leftrightarrow (\exists\gamma)(\forall\delta)[Q_2(x, \alpha, \beta, \gamma, \delta) \vee A(\delta)]$$

from which the required representation follows immediately. ⊣

7C.18. Prove 7C.4.

Hint. By 7C.13, if $R(x)$ is absolutely inductive, then

$$R(x) \leftrightarrow \varphi^\infty(\alpha_0, x)$$

with $\varphi(\alpha, x, A)$ in *pos* Σ_3^1 and α_0 recursive and by 7C.17 there is a Σ_1^1 pointset $Q(\alpha, x, \beta, \gamma, \delta, \varepsilon)$ so that for $A \subsetneq \mathcal{N}$,

$$\varphi(\alpha, x, A) \leftrightarrow (\exists\beta)(\forall\gamma)(\exists\delta)(\forall\varepsilon)[Q(\alpha, x, \beta, \gamma, \delta, \varepsilon) \vee A(\varepsilon)].$$

We claim that

$$\begin{aligned}\varphi^\infty(\alpha, x) \leftrightarrow \{(\exists\beta_1)(\forall\gamma_1)(\exists\delta_1)(\forall\varepsilon_1)(\exists\beta_2)(\forall\gamma_2)(\exists\delta_2)(\forall\varepsilon_2)\cdots\}\\ [Q(\alpha, x, \beta_1, \gamma_1, \delta_1, \varepsilon_1) \vee Q(\varepsilon_1, x, \beta_2, \gamma_2, \delta_2, \varepsilon_2)\\ \vee Q(\varepsilon_2, x, \beta_3, \gamma_3, \delta_3, \varepsilon_3) \vee \cdots];\end{aligned}$$

to check this, show by induction on ξ that $\varphi^\xi(\alpha, x)$ implies the right-hand-side and show that if $\neg\varphi^\infty(\alpha, x)$, then II can easily win the game determining the truth of the right-hand side. Now half of 7C.4 follows immediately.

To prove the other half consider the inductive analysis of the open game on the right-hand side. ⊣

7C.19. Prove 7C.5.

Hint. It is enough to show (granting the hypotheses) that every fixed point φ^∞ of a positive analytical $\varphi(\alpha, x, A)$ admits an absolutely inductive scale. We will skip the parameter x in outlining the argument.

There is an analytical scale $\bar{\psi}^0 = \{\psi_n^0\}_{n\in\omega}$ on $\varphi^0 = \{\alpha: \varphi(\alpha, \emptyset)\}$ by 6C.2 and the Second Periodicity Theorem 6C.3 and from any hyperprojective scale $\bar{\chi}^\lambda = \{\chi_n^\lambda\}_{n\in\omega}$ on $\varphi^{<\lambda} = \bigcup_{\xi<\lambda} \varphi^\xi$ we can easily construct a scale $\bar{\psi}^\lambda$ on $\varphi^\lambda = \{\alpha: \varphi(\alpha, \varphi^{<\lambda})\}$ by the same two theorems. On the other hand, given scales $\bar{\psi}^\xi$ on φ^ξ for each $\xi < \lambda$, define on $\varphi^{<\lambda}$,

$$\chi_0^\lambda(\alpha) = \text{least } \xi \text{ such that } \alpha \in \varphi^\xi,$$
$$\chi_{n+1}^\lambda(\alpha) = \psi_n^\xi(\alpha), \quad \text{where } \xi = \chi_0^\lambda(\alpha);$$

it is easy to check that $\bar{\chi}^\lambda$ is a scale on $\varphi^{<\lambda}$. Finally define $\bar{\chi}^\varkappa$ on $\varphi^\infty = \bigcup_{\xi<\varkappa} \varphi^\xi$ in the same way, to get a scale on φ^∞.

It remains to check that each $\bar{\varphi}^\lambda$ and each $\bar{\chi}^\lambda$ are hyperprojective in a "uniform" way so that $\bar{\chi}^\varkappa$ is absolutely inductive—and in particular so that the hypotheses of determinacy needed to go from $\bar{\chi}^\lambda$ to $\bar{\psi}^\lambda$ actually hold.

Let $G \subseteq \mathcal{N} \times \mathcal{N}$ be universal absolutely inductive, call $\gamma = \langle \alpha, \beta \rangle$ a code of the hyperprojective set $A \subseteq \mathcal{N}$ if $A = \{\delta: G(\alpha, \delta)\} = \{\delta: \neg G(\beta, \delta)\}$ and argue by using 3H.2 that the pointclass ***HYP*** is uniformly closed under all the obvious operations in this coding. Code also hyperprojective scales on sets in the obvious way.

Finally, use the Recursion Theorem 7A.2 to define a recursive function

$$f(\alpha) = f(\varepsilon^*, \alpha) = \{\varepsilon^*\}(\alpha)$$

with the following property: for each $\alpha \in \varphi^\infty$, if $\lambda = $ least ξ such that $\alpha \in \varphi^\xi$, then $f(\alpha)$ is a code of the scale $\bar{\psi}^\lambda$ on φ^λ as described above; there is a bit of checking to be done, but the result follows easily from this. ⊣

7D. The completely playful universe

In the historical remarks at the end of Chapter 6 we mentioned the so-called *axiom of determinacy*,

$$AD \Leftrightarrow \text{every subset of } \mathcal{N} \text{ is determined.}$$

This is false by 6A.6, but the proof of that result depended on a blatant application of the axiom of choice and none has yet succeeded in violating

AD without some appeal to choice. Thus it remains possible that *AD* is consistent with the axioms of Zermelo–Fraenkel set theory (without choice) and one might attempt to study its consequences. We will derive a very few of them in this section; the interested reader should go to the collection of papers Kechris–Moschovakis [1978a] and to Kleinberg [1977] for a deeper study of this fascinating theory.

What is the value of proving theorems on the basis of a false assumption?

Mycielski and Steinhaus [1962] suggested that we simply replace the axiom of choice by *AD* in our thinking about sets, because of its many useful and beautiful consequences about sets of reals-they all become then Lebesgue measurable, they acquire the property of Baire, etc. However, mathematicians with a realistic approach to set theory have resisted this temptation to accept "desirable falsehoods" for the sake of utility or simplicity. Moreover, many consequences of *AD* (proved since 1962) give a picture of the universe of sets which is by no means "simple" and tends to contradict our basic intuitions about sets at least as much as the alleged "undesirable" consequences of the axiom of choice.

One might also study the consequences of *AD* in an attempt to prove it false without using the axiom of choice. Some have worked on this worthy program but without success so far.

Most mathematicians who work on *AD* are motivated by the hope that there is a natural model of Zermelo-Fraenkel set theory in which *AD* holds; by this we mean that there is a collection M of sets such that if we reinterpret "set" by "member of M," then all the classical axioms of set theory (except choice) become true and so does the proposition *AD*. One such candidate is the collection of sets

$$L(\mathcal{N}) = \text{the smallest model of Zermelo–Fraenkel set theory which contains } \mathcal{N}$$

which has the additional advantage that it contains all the real numbers. (We will give a precise definition of $L(\mathcal{N})$ in Chapter 8.) Consequences of *AD* then become true assertions about the sets in $L(\mathcal{N})$—which in turn yield theorems about *some of the sets* in our intended interpretation of set theory.

There is also an obvious practical reason for learning to think in the context of full determinacy: if you prove a result from *AD*, you can almost always rework the argument to get a weaker theorem—or a reformulation—from more restricted hypotheses of definable determinacy such as we have been assuming.

Although AD is inconsistent with the full axiom of choice, it implies a very weak *countable principle of choice for pointsets.*

7D.1. LEMMA. *Assume AD and suppose $P \subseteq \omega \times \mathfrak{X}$ where $\mathfrak{X}$ is any product space; then*

$$(\forall n)(\exists x)P(n, x) \twoheadrightarrow (\exists f : \omega \to \mathfrak{X})(\forall n)P(n, f(n)).$$

PROOF. Let $\pi : \mathcal{N} \twoheadrightarrow \mathfrak{X}$ be a canonical surjection of $\mathcal{N}$ onto $\mathfrak{X}$ and consider the game where I plays n (his future moves being irrelevant) and II plays α; II wins if $P(n, \pi(\alpha))$. Now I cannot win, since any strategy would fix some n and then II can beat it by playing some α so that $P(n, \pi(\alpha))$. By AD then, II has some winning strategy τ and we can take

$$f(n) = \pi([n] * \tau). \qquad \dashv$$

The countable principle of choice is not strong enough and we will often assume (together with AD the following very reasonable proposition.

AXIOM OF DEPENDED CHOICES (DC). *For every set of pairs $P \subseteq A \times A$ from a nonempty set A,*

$$(\forall x \in A)(\exists y \in A)P(x, y) \twoheadrightarrow (\exists f : \omega \to A)(\forall n)P(f(n), f(n+1)).$$

In effect, DC says that we can make a countable number of choices, each choice depending on the preceding choice. It is a very useful principle, for example it is the only kind of choice we need to prove that a relation with no infinite descending chains must be wellfounded (i.e. every nonempty subset of its field must have a minimal member).

The full axiom of choice implies DC easily and in Chapter 8 we will show that DC holds in $L(\mathcal{N})$, so that for our purposes, DC is an innocuous assumption. On the other hand, Solovay [1978a] has shown (from strong hypotheses) that AD does not formally imply DC.

It is very important to point out now that *in this book we have not used any choice principles other than DC without explicitly taking note of the fact.* Actually there are only a few results in whose proofs we used the full axiom of choice and we have listed them all in an appendix to this chapter, Section 7F.

Thus when we assume $AD+DC$, we can appeal to the whole theory developed so far except for these few results. In the cases where we only assume AD and we use previously proved theorems, the reader should be able to check easily that the proofs of these theorems appeal at most to the countable principle of choice for pointsets which is a consequence of AD.

Let us put down for the record the regularity results already established in Chapter 6.

7D.2. THEOREM. *Assume AD; then every uncountable pointset has a nonempty perfect subset, every pointset has the property of Baire and for every* σ-finite Borel measure μ *on a space* $\mathfrak{X}$, *every subset of* $\mathfrak{X}$ *is* μ*-measurable* (*Davis* [1964], *Mycielski* [1964], *Mycielski–Swierczkowski* [1964], *Mycielski* [1965]).

PROOF. Take $\Lambda = power(\mathcal{N})$ in 6A.12, 6A.16 and 6A.18. ⊣

To look at a regularity result of a different kind, consider the following relation between pointsets $A \subseteq \mathfrak{X}$ and $B \subseteq \mathfrak{Y}$:

$$\begin{aligned} A \leq_w B &\Leftrightarrow A \text{ is a continuous preimage of } B \\ &\Leftrightarrow \text{there is a continuous function } f:\mathfrak{X} \to \mathfrak{Y} \\ &\quad\ \text{such that for all } x, \\ &\quad\ x \in A \Leftrightarrow f(x) \in B. \end{aligned}$$

If $A \leq_w B$, we say that A is *Wadge reducible* or *continuously reducible to* B. It is obvious that $\leq_w$ is a transitive relation on the collection of all pointsets.

7D.3. WADGE'S LEMMA (Wadge [19??]). *Assume AD and suppose A and B are pointsets of type* 1; *then either* $A \leq_w B$ *or* $B \leq_w \neg A = \mathcal{N} - A$.

PROOF. If A, B are sets of irrationals, let $G(A, B)$ be the game where I plays α, II plays β and II wins if the following equivalence holds:

$$\alpha \in A \Leftrightarrow \beta \in B.$$

If II wins $G(A, B)$ with a strategy τ, then

$$\alpha \in A \Leftrightarrow [\alpha] * \tau \in B$$

and since the map $\alpha \mapsto [\alpha] * \tau$ is continuous, we have $A \leq_w B$. If I wins $G(A, B)$ with a strategy σ, then

$$\sigma * [\beta] \notin A \Leftrightarrow \beta \in B$$

and since again the map $\beta \mapsto \sigma * [\beta]$ is continuous, we have $B \leq_w \neg A$.

The result follows immediately for pointsets of type 1. ⊣

Wadge's lemma says essentially that any two subsets of Baire space are comparable in terms of the operations of continuous substitution and complementation. It has many interesting consequences and we will come back to it in the exercises.

We now turn to one of the most fascinating problems in the theory of full determinacy. How large is the continuum in a fully playful universe, a model of $AD + DC$? From one point of view it is very small indeed.

7D.4. THEOREM. *If AD holds, then every pointset which can be wellordered is countable; in particular, $\mathcal{N}$ cannot be wellordered and if*

$$\pi : \lambda \rightarrowtail \mathcal{N}$$

is any injection from an ordinal λ into $\mathcal{N}$, then λ is countable.

PROOF. If A is uncountable, then A has a perfect subset which (as a space with the induced topology) is Borel isomorphic with $\mathcal{N}$, so that any wellordering of A induces a well ordering of $\mathcal{N}$. If $\rho : \mathcal{N} \rightarrowtail Ordinals$ is the rank function of this wellordering, then $\rho^{-1}(\xi)$ is a singleton for each ξ so that 5A.10 applies and $\{(\alpha, \beta) : \alpha < \beta\}$ does not have the property of Baire, contradicting 7D.1. ⊣

Thus $\mathcal{N}$ is very small relative to the ordinals if we use injections to compare sizes. When choice is not available, however, it is more natural to use surjections for this purpose and think of an ordinal λ as *not larger than* $\mathcal{N}$ if we can map $\mathcal{N}$ *onto* λ.

Put

$$\Theta = supremum\{\lambda : \text{there exists a surjection } \pi : \mathcal{N} \twoheadrightarrow \lambda\};$$

AD implies that Θ is very large.

The key result is a coding lemma which asserts that bounded subsets of Θ and functions on ordinals below Θ are "definable."

Suppose $<$ is a (strict) wellfounded relation on some $S \subseteq \mathfrak{X}$ and let

$$\rho : S \twoheadrightarrow \lambda$$

be its rank function as in 2G defined by the recursion

$$\rho(x) = \mathit{supremum}\{\rho(y)+1 : y < x\}.$$

We can obviously think of each $x \in S$ as a *code* for the ordinal $\xi = \rho(x)$.

Let

$$f : \lambda^n \to \mathit{Power}(\mathcal{Y})$$

be a function which assigns subsets of $\mathcal{Y}$ to n-tuples from λ. A *choice set for* f (relative to a given wellfounded relation $<$ that codes λ) is any pointset

$$C \subseteq \mathcal{X}^n \times \mathcal{Y}$$

such that

(i) $(x_1, \ldots, x_n, y) \in C \Rightarrow x_1, \ldots, x_n \in S \;\&\; y \in f(\rho(x_1), \ldots, \rho(x_n))$,

(ii) $f(\xi_1, \ldots, \xi_n) \neq \emptyset \Rightarrow (\exists x_1) \cdots (\exists x_n)(\exists y)[\rho(x_1) = \xi_1$
$\& \cdots \& \; \rho(x_n) = \xi_n \;\&\; y \in f(\xi_1, \ldots, \xi_n)$
$\& \; (x_1, \ldots, x_n, y) \in C]$.

7D.5. THE CODING LEMMA (I) (Moschovakis [1970]).[6] *Assume AD, let $<$ be a strict wellfounded relation on some $S \subseteq \mathcal{X}$ with rank function*

$$\rho : S \twoheadrightarrow \lambda$$

and let Γ be any pointclass such that $< \in \Gamma$, $\Sigma^1_1 \subseteq \Gamma$ and Γ is ω-parametrized and closed under $\&$, $\vee$, $\exists^\omega$, $\forall^\omega$ and $\exists^\mathcal{N}$. Then every function

$$f : \lambda^n \to \mathit{Power}(\mathcal{Y})$$

has a choice set in the associated boldface pointclass $\boldsymbol{\Gamma}$.

PROOF. For each unary function

$$f : \lambda \to \mathit{Power}(\mathcal{Y})$$

and each $\xi \leq \lambda$, put

$$f_\xi(\eta) = \begin{cases} f(\eta) & \text{if } \eta < \xi, \\ \emptyset & \text{if } \xi \leq \eta. \end{cases}$$

We will prove the result for unary functions first by assuming that some f_ξ does not have a choice set in $\boldsymbol{\Gamma}$ and deducing a contradiction.

Suppose C_ξ is a choice set for $f_\xi (\xi < \lambda)$ in $\boldsymbol{\Gamma}$. If $f(\xi) = \emptyset$, then C_ξ is obviously a choice set for $f_{\xi+1}$ also, while if $y_0 \in f(\xi)$ and $\rho(x_0) = \xi$, then

$$C_{\xi+1} = C_\xi \cup \{(x_0, y_0)\}$$

is a choice set for $f_{\xi+1}$ and it is also in $\mathbf{\Gamma}$. Thus the ordinal

$$\varkappa = \text{least } \xi \text{ such that } f_\xi \text{ does not have a choice set in } \mathbf{\Gamma}$$

is a limit ordinal $\leq \lambda$.

Fix a good parametrization for Γ by 3H.1 and suppose $G \subseteq \mathcal{N} \times (\mathfrak{X} \times \mathfrak{Y})$ is universal. As usually,

$$G_\alpha = \{(x, y) : G(\alpha, x, y)\}.$$

We consider the following two-person game: if I plays α and II plays β, then

$$\begin{aligned} \text{II wins} \Leftrightarrow & [G_\alpha \text{ is not a choice set for any } f_\xi, \xi < \varkappa] \\ & \vee [G_\alpha \text{ is a choice set for } f_\xi \text{ and there is} \\ & \quad \text{some } \eta > \xi \text{ such that } G_\beta \text{ is a choice set for } f_\eta]. \end{aligned}$$

The game is determined by *AD*.

Case 1. I *has a* winning *strategy* σ. Now for each β there is some $\xi = \xi(\beta) < \varkappa$ such that $G_{\sigma * [\beta]}$ is a choice set for $f_{\xi(\beta)}$. Put

$$\xi = \textit{supremum}\{\xi(\beta) : \beta \in \mathcal{N}\}$$

and check easily that the set

$$C_\xi = \bigcup_\beta G_{\sigma * [\beta]}$$

is a choice set for f_ξ, so that by the choice of $\varkappa$, $\xi < \varkappa$; but since $\varkappa$ is limit, there is then some $\eta > \xi$, $\eta < \varkappa$ and II can beat σ by playing any β so that G_β is a choice set for f_η.

Case 2. II *has a winning strategy* τ. Let

$$\{\varepsilon\}(x) = U^{\mathfrak{X},\mathfrak{Y}}(\varepsilon, x)$$

be the partial function on $\mathfrak{X}$ to $\mathcal{N}$ which is universal for all the partial functions on $\mathfrak{X}$ to $\mathcal{N}$, ($\mathbf{\Sigma}^0_1$-) recursive on their domain and for each $\varepsilon \in \mathcal{N}$, $w \in \mathfrak{X}$, put

$$\begin{aligned} (x, y) \in A_{\varepsilon,w} \Leftrightarrow (\exists z)\{z < w \ \& \ & \{\varepsilon\}(z) \text{ is defined} \\ & \& \ G(\{\varepsilon\}(z), x, y)\}; \end{aligned}$$

each $A_{\varepsilon,w}$ is obviously in $\mathbf{\Gamma}$ and by 3H.2 there is a recursive function $\pi(\varepsilon, w)$ such that

$$A_{\varepsilon,w} = \{(x, y) : G(\pi(\varepsilon, w), x, y)\}.$$

Now the map

$$(\varepsilon, w) \mapsto [\pi(\varepsilon, w)] * \tau$$

is continuous and total, so by the Recursion Theorem 7A.2 we can find a fixed $\varepsilon^* \in \mathcal{N}$ such that for all $w \in \mathfrak{X}$,

$$\{\varepsilon^*\}(w) = [\pi(\varepsilon^*, w)] * \tau;$$

to simplify notation, put

$$g(w) = \{\varepsilon^*\}(w).$$

Sublemma. For each $w \in S = Field(<)$, there is some $\eta = \eta(w) > \rho(w)$, such that $G_{g(w)}$ is a choice set for $f_{\eta(w)}$.

Proof of the sublemma is by induction on $\rho(w)$. If the sublemma holds for all $x < w$, then $A_{\varepsilon^*, w}$ is clearly a choice set for f_ξ with

$$\xi = supremum\{\eta(x) \colon x < w\} \geq \rho(w)$$

and since $g(w) = [\pi(\varepsilon^*, w)] * \tau$ is II's response to I's play $\pi(\varepsilon^*, w)$, it must be that $G_{g(w)}$ is a choice set for some f_η with $\eta > \xi \geq \rho(w)$.

It follows from this sublemma that

$$supremum\{\eta(w) \colon w \in S\} = \varkappa$$

and that

$$C_\varkappa = \bigcup_{w \in S} G_{g(w)}$$

is a choice set for $f_\varkappa$ in $\mathbf{\Gamma}$, contrary to the choice of $\varkappa$.

This completes the proof of the coding lemma for unary functions. To prove it for functions of any number of variables by induction, suppose

$$f \colon \lambda^{n+1} \to Power(\mathfrak{Y})$$

is given and for $\eta < \lambda$, put

$$f^\eta(\xi_1, \ldots, \xi_n) = f(\xi_1, \ldots, \xi_n, \eta)$$

and let

$$g(\eta) = \{\alpha \colon G_\alpha \text{ is a choice set for } f^\eta\}.$$

Let $C_g \subseteq \mathfrak{X} \times \mathcal{N}$ be a choice set for g in $\mathbf{\Gamma}$ and check that

$$C(x_1, \ldots, x_n, x, y) \Leftrightarrow (\exists \alpha)[C_g(x, \alpha) \,\&\, G(\alpha, x_1, \ldots, x_n, y)]$$

is a choice set for f. ⊣

If $\leq$ is a prewellordering with field $S \subseteq \mathfrak{X}$ and

$$\rho \colon S \twoheadrightarrow \lambda$$

is the associated rank function, then for any $A \subseteq \lambda$, put

$$Code(A; \leq) = \{x \in S: \rho(x) \in A\};$$

similary, for prewellordering $\leq_1, \ldots, \leq_n$ with ranks $\rho_1 : S_1 \twoheadrightarrow \lambda_1, \ldots, \rho_n : S_n \twoheadrightarrow \lambda_n$ and $A \subseteq \lambda_1 \times \cdots \times \lambda_n$,

$$Code(A; \leq_1, \ldots, \leq_n) = \{(x_1, \ldots, x_n) : (\rho_1(x_1), \ldots, \rho_n(x_n)) \in A)\}.$$

7D.6. THE CODING LEMMA (II) (Moschovakis [1970]).[(6)] *Assume AD, let* $\leq_1, \ldots, \leq_n$ *be prewellorderings on* $S_1 \subseteq \mathfrak{X}_1, \ldots, S_n \subseteq \mathfrak{X}_n$ with rank functions

$$\rho_1 : S_1 \twoheadrightarrow \lambda_1, \quad \rho_2 : S_2 \twoheadrightarrow \lambda_2, \ldots, \qquad \rho_n : S_n \twoheadrightarrow \lambda_n$$

and let Γ *be any pointclass such that* $\leq \in \mathbf{\Delta}$, $\Sigma^1_1 \subseteq \Gamma$ and Γ is ω-*parametrized and closed under* &, $\vee, \exists^\omega, \forall^\omega$ *and* $\exists^{\mathcal{N}}$. *Then for every set* $A \subseteq \lambda_1 \times \cdots \times \lambda_n$, *the pointset Code* $(A; \leq_1, \ldots, \leq_n)$ *is in* $\mathbf{\Delta}$.

PROOF. On $\mathfrak{X} = \mathfrak{X}_1 \times \cdots \times \mathfrak{X}_n$ define

$$\begin{aligned}(x_1, \ldots, x_n) < (x'_1, \ldots, x'_n) \Leftrightarrow\ & x_1 <_1 x'_1 \\ & \vee x_1 \sim_1 x'_1 \;\&\; x_2 <_2 x'_2 \\ & \vee \cdots \\ & \vee x_1 \sim_1 x'_1 \;\&\; \cdots \;\&\; x_{n-1} \sim_{n-1} x'_{n-1} \;\&\; x_n <_n x'_n\end{aligned}$$

so that $<$ is strict, wellfounded in $\mathbf{\Delta}$ with rank

$$\rho(x_1, \ldots, x_n) = \langle \rho_1(x_1), \ldots, \rho_n(x_n) \rangle,$$

where $(\xi_1, \ldots, \xi_n) \mapsto \langle \xi_1, \ldots, \xi_n \rangle$ takes $\lambda_1 \times \cdots \times \lambda_n$ with the lexicographic ordering onto λ, in an order preserving way. Let $C \subseteq \mathfrak{X} \times \omega$ be a choice set in $\mathbf{\Gamma}$ for the function

$$f(\langle \xi_1, \ldots, \xi_n \rangle) = \begin{cases} 1 & \text{if} \quad (\xi_1, \ldots, \xi_n) \in A \\ 0 & \text{if} \quad (\xi_1, \ldots, \xi_n) \notin A \end{cases}$$

and check that

$$\begin{aligned}(x_1, \ldots, x_n) \in Code(A; \leq_1, \ldots, \leq_n) \Leftrightarrow\ & (\exists x'_1) \cdots (\exists x'_n)[x_1 \sim_1 x'_1 \;\&\; \cdots \\ & \;\&\; x_n \sim_n x'_n \;\&\; C(x'_1, \ldots, x'_n, 1)], \\ (x_1, \ldots, x_n) \in Code(A; \leq_1, \ldots, \leq_n) \Leftrightarrow\ & (\exists x'_1) \cdots (\exists x'_n)[x_1 \sim_1 x'_1 \;\&\; \cdots \\ & \;\&\; x_n \sim_n x'_n \;\&\; C(x'_1, \ldots, x'_n, 0)]. \quad \dashv\end{aligned}$$

In applying the coding lemma, we often take $\Gamma = pos\,\Sigma^1_1(Q_1, \ldots, Q_n)$,

where

$$pos\,\Sigma_1^1(Q_1,\dots,Q_n) = \text{the smallest pointclass which contains } Q_1,\dots,Q_n \text{ and all } \Sigma_1^1 \text{ pointsets and which is closed under } \&, \vee, \exists^\omega, \forall^\omega \text{ and } \exists^{\mathcal{N}}.$$

7D.7. LEMMA. *For each $Q_1,\dots,Q_n$, the pointclass $pos\,\Sigma_1^1(Q_1,\dots,Q_n)$, is closed under $\&$, $\vee$, $\exists^\omega$, $\forall^\omega$ and $\exists^{\mathcal{N}}$, it contains $Q_1,\dots,Q_n$ and all Σ_1^1 pointsets and it is ω-parametrized.*

PROOF. Assume without loss of generality that we have only one set *$Q \neq \emptyset$ and consider the collection of all pointsets which satisfy some equivalence of the form*

$$(*) \qquad P(x) \Leftrightarrow (\exists\alpha)\{(\forall n)Q(\pi((\alpha)_n))\ \&\ R(x,\alpha)\},$$

where R is Σ_1^1 and $\pi : \mathcal{N} \twoheadrightarrow \mathcal{Z}$ is a canonical recursive surjection. This contains Q and each P in Σ_1^1 (taking $R(x,\alpha) \Leftrightarrow P(x)$) and it is easily closed under $\&$, $\vee$, $\exists^\omega$, $\forall^\omega$, $\exists^{\mathcal{N}}$, e.g.,

$$\begin{aligned}(\forall t)(\exists\alpha)\{(\forall n)Q(\pi((\alpha)_n))\ \&\ R(x,t,\alpha)\}\\ \Leftrightarrow (\exists\beta)\{(\forall m)Q(\pi((\beta)_m))\\ \&\ (\forall t)(\exists\alpha)[(\forall n)(\exists m)[(\alpha)_n = (\beta)_m]\ \&\ R(x,t,\alpha)].\end{aligned}$$

Thus every P in $pos\,\Sigma_1^1(Q)$ satisfies $(*)$ with some R in Σ_1^1 and we can use universal sets in Σ_1^1 to get the universal sets we need. ⊣

Let us now put down one result which implies that Θ is large.

7D.8. THOEREM. *Assume AD and let Γ be an ω-parametrized pointclass such that $\Sigma_1^1 \subseteq \Gamma$ and Γ is closed under $\&$, $\vee$, $\exists^\omega$, $\forall^\omega$ and $\exists^{\mathcal{N}}$. Let*

$$\boldsymbol{\delta} = \text{supremum}\{|\leq| : \leq \textit{ is a prewellordering of } \mathcal{N} \textit{ in } \boldsymbol{\Delta}\},$$

$$\boldsymbol{\gamma} = \textit{supremum}\{|\prec| : \prec \textit{ is a strict wellfounded relation in } \boldsymbol{\Gamma} \textit{ with field in } \mathcal{N}\}.$$

(i) *$\boldsymbol{\delta}$ is a cardinal.*

(ii) *$\boldsymbol{\gamma}$ is a regular cardinal.*

(iii) *If Γ is also closed under $\forall^{\mathcal{N}}$, then $\boldsymbol{\delta}$ is a regular limit cardinal (i.e. weakly inaccessible).*

In particular, taking Γ = all inductive sets, the ordinal $\varkappa$ = supremum of all hyperprojective prewellordings of $\mathcal{N}$ is weakly inaccessible.[7,8]

PROOF. (i) Assume towards a contradiction that for some $\lambda < \boldsymbol{\delta}$ we have a bijection

$$g : \lambda \rightarrowtail\!\!\!\!\twoheadrightarrow \boldsymbol{\delta}$$

and let $\leq$ be a prewellordering of some $\mathfrak{X}$ in $\boldsymbol{\Delta}$ with rank function

$$\rho : \mathfrak{X} \twoheadrightarrow \lambda.$$

The relation

$$P(\xi, \eta) \Leftrightarrow g(\xi) \leq g(\eta) \quad (\xi, \eta < \lambda)$$

is wellfounded with length $\boldsymbol{\delta}$ and by the Coding Lemma 7D.6 it has its code set in $\boldsymbol{\Delta}$; but then $\boldsymbol{\delta}$ is the length of a $\boldsymbol{\Delta}$ prewellordering which is absurd.

(ii) Assume again towards a contradiction that some map

$$g : \lambda \to \boldsymbol{\gamma}$$

is cofinal with $\boldsymbol{\gamma}$, where $\lambda < \boldsymbol{\gamma}$ and λ is the length of some strict wellfounded relation $\prec$ in $\boldsymbol{\Gamma}$. Let $G \subseteq \mathcal{N} \times (\mathcal{N} \times \mathcal{N})$ be universal in Γ and put

$$f(\xi) = \{\alpha : G_\alpha \text{ is wellfounded with field in } \mathcal{N} \text{ and length } g(\xi)\};$$

by the coding lemma, choose a choice set $C \in \boldsymbol{\Gamma}$ for f and let

$$(x, \alpha, \beta) \prec' (x', \alpha', \beta') \Leftrightarrow x = x' \;\&\; \alpha = \alpha' \;\&\; C(x, \alpha) \;\&\; G(\alpha, \beta, \beta').$$

It is obvious that $\prec'$ is wellfounded in $\boldsymbol{\Gamma}$ with length $\boldsymbol{\gamma}$, and this is easily absurd since for each wellfounded relation in $\boldsymbol{\Gamma}$, we can find a longer one.

(iii) If $\leq$ is a prewellordering in $\boldsymbol{\Delta}$, take $\Gamma_0 = pos\,\Sigma^1_1(\leq, <)$ and notice by the closure properties that $\Gamma_0 \subseteq \boldsymbol{\Delta}$. By (i), the ordinal $\boldsymbol{\delta}_0$ of Γ_0 is a cardinal and of course it is bigger than the length of $\leq$ and less than $\boldsymbol{\delta}$; thus $\boldsymbol{\delta}$ is a limit cardinal.

To show that $\boldsymbol{\delta}$ is regular, suppose

$$g : \lambda \to \boldsymbol{\delta}$$

is cofinal in $\boldsymbol{\delta}$, where λ is the length of a prewellordering $\leq$ in $\boldsymbol{\Delta}$ and let $G \subseteq \mathcal{N} \times (\mathcal{N} \times \mathcal{N})$ be universal in Γ. Put

$$f(\xi) = \{(\alpha, \beta) : G_\alpha = \mathcal{N} \times \mathcal{N} - G_\beta \quad \text{and} \quad G_\alpha \text{ is a prewellordering of } \mathcal{N} \text{ of length } g(\xi)\}$$

and let C be some choice set for f in $pos\,\Sigma_1^1(<)\subseteq\mathbf{\Delta}$. For each (α, β) such that $G_\alpha = \mathcal{N}\times\mathcal{N} - G_\beta$ and G_α is a prewellordering of $\mathcal{N}$, let

$$\rho_{\alpha,\beta}(\gamma) = \rho(\alpha, \beta, \gamma)$$

be the rank function mapping $\mathcal{N}$ onto some $\xi_{\alpha,\beta}$ and finally put

$$\begin{aligned}(x, \alpha, \beta, \gamma) <^* (x', \alpha', \beta', \gamma') \leftrightarrow x < x' \\ \vee (\exists y)(\exists y')[x \sim y \sim x' \sim y' \\ \& \; C(y, \alpha, \beta) \\ \& \; C(y', \alpha', \beta') \\ \& \; \rho(\alpha, \beta, \gamma) \leq \rho(\alpha', \beta', \gamma')].\end{aligned}$$

It is clear that $<^*$ is a prewellordering of length $\boldsymbol{\delta}$, so we will have the desired contradiction if we can show that $<^* \in \mathbf{\Delta}$.

The only clause in the definition of $<^*$ which causes difficulty is the last one: we must find a $\mathbf{\Delta}$ relation $P(\alpha, \beta, \gamma, \alpha', \beta', \gamma)$ such that if $G_\alpha = \mathcal{N}\times\mathcal{N} - G_\beta$ and $G_{\alpha'} = \mathcal{N}\times\mathcal{N} - G_{\beta'}$ are both prewellorderings in $\mathbf{\Delta}$, then

$$\rho(\alpha, \beta, \gamma) \leq \rho(\alpha', \beta', \gamma') \leftrightarrow P(\alpha, \beta, \gamma, \alpha', \beta', \gamma').$$

Now

$\rho(\alpha, \beta, \gamma) \leq \rho(\alpha', \beta', \gamma') \leftrightarrow$ there is a set $A \subseteq \xi_{\alpha,\beta} \times \xi_{\alpha',\beta'}$ which is an order preserving map of the initial segment of G_α up to γ onto the initial segment of $G_{\alpha'}$ up to γ';

by the coding lemma, for every such set A the pointset $Code(A; G_\alpha, G_\beta)$ will be in $\mathbf{\Delta}$, i.e. it will be G_δ and $\mathcal{N}\times\mathcal{N} - G_\varepsilon$ for some δ, ε. The proof is completed by verifying easily that the conditions on G_δ, G_ε which guarantee that $G_\delta = \mathcal{N}\times\mathcal{N} - G_\varepsilon$ is the code set of some similarity of initial segments translate to Δ conditions on δ, ε. ⊣

Kechris has recently verified that the least non-hyperprojective ordinal is Mahlo in the context of AD, but even before that Solovay had proved that there are Mahlo cardinals below Θ. Moschovakis [1970] has some results of this type—as a matter of fact, AD implies that Θ is quite immense.

The last theorem also has some obvious consequences about the ordinals $\boldsymbol{\delta}^1_n$ that we introduced in 4C. We will come back to it after we establish a very important result of Martin.

7D.9. THE SUSLIN THEOREM FOR THE ODD LEVELS (Martin and Moschovakis). *Assume AD and let Γ be a Spector pointclass closed under $\forall^{\mathcal{N}}$, put*

$$\boldsymbol{\delta} = supremum\{|\leq| : \leq \text{ is a prewellordering of } \mathcal{N} \text{ in } \boldsymbol{\Delta}\}.$$

If $\{A_\xi\}_{\xi<\lambda}$ is a sequence of subsets of some $\mathfrak{X}$ with $\lambda < \boldsymbol{\delta}$ and each A_ξ in $\boldsymbol{\Delta}$, then $\bigcup_{\xi<\lambda} A_\xi \in \boldsymbol{\Delta}$ (Martin [19?a].)

In particular, if AD + DC holds, then for each odd n

$$\begin{aligned}\boldsymbol{\Delta}^1_n &= \boldsymbol{B}_{\boldsymbol{\delta}^1_n}\\ &= \text{the least pointclass which contains all}\\ &\quad \text{open sets and is closed under complementation}\\ &\quad \text{and unions of length} < \boldsymbol{\delta}^1_n.\end{aligned}$$

PROOF. The second assertion follows immediately from the first and 6C.12.

To prove the first assertion by contradiction, let λ be the least ordinal such that for some sequence of sets $\{A_\xi\}_{\xi<\lambda}$, each $A_\xi \subseteq \mathfrak{X}$ is in $\boldsymbol{\Delta}$ but $\bigcup_{\xi<\lambda} A_\xi \notin \boldsymbol{\Delta}$. Clearly λ is limit and there is a prewellordering $\leq$ of $\mathcal{N}$ in $\boldsymbol{\Delta}$ with rank function

$$\rho : \mathcal{N} \twoheadrightarrow \lambda.$$

Let $G \subseteq \mathcal{N} \times \mathfrak{X}$ be universal in Γ and put

$$f(\xi) = \{\beta : A_\xi = \mathfrak{X} - G_\beta\};$$

if $C \subseteq \mathcal{N} \times \mathcal{N}$ is a choice set for f in $\neg\boldsymbol{\Gamma}$ by the coding lemma, then clearly

$$x \in A_{\rho(\alpha)} \leftrightarrow (\exists\alpha')(\exists\beta)[\alpha \sim \alpha' \;\&\; C(\alpha', \beta) \;\&\; \neg G(\beta, x)]$$

so that the relation

$$(1) \qquad P(x, \alpha) \leftrightarrow x \in A_{\rho(\alpha)}$$

is in $\neg\boldsymbol{\Gamma}$.

By the choice of λ, for each $\zeta < \lambda$ the unions

$$\bigcup_{\xi<\zeta} A_\xi, \quad \bigcup_{\xi\leq\zeta} A_\xi$$

are in $\mathbf{\Delta}$, so put

$$g_1(\zeta) = \{\beta: \bigcup_{\xi<\zeta} A_\xi = G_\beta\},$$
$$g_2(\zeta) = \{\beta: \bigcup_{\xi\leq\zeta} A_\xi = G_\beta\},$$

take choice sets C_1 and C_2 for these two functions in $\neg\mathbf{\Gamma}$ and use them to show that the following two relations are in $\neg\mathbf{\Gamma}$:

$$Q(x, \alpha) \Leftrightarrow x \notin \bigcup_{\xi<\rho(\alpha)} A_\xi, \tag{2}$$

$$R(x, \alpha) \Leftrightarrow x \notin \bigcup_{\xi\leq\rho(\alpha)} A_\xi. \tag{3}$$

From the fact that P is in $\neg\mathbf{\Gamma}$, it follows immediately that the union $A = \bigcup_\xi A_\xi$ is in $\neg\mathbf{\Gamma}$. On this union we define the obvious norm:

$$\varphi(x) = \mu\xi[x \in A_\xi];$$

now

$$x \leq^*_\varphi y \Leftrightarrow (\exists\alpha)[P(x, \alpha) \,\&\, Q(y, \alpha)],$$
$$x <^*_\varphi y \Leftrightarrow (\exists\alpha)[P(x, \alpha) \,\&\, R(y, \alpha)],$$

so that φ is a $\neg\mathbf{\Gamma}$-norm.

What we have proved so far is that *the union* $A = \bigcup_{\xi<\lambda} A_\xi$ *is a set in* $\neg\mathbf{\Gamma}$ *which admits a* $\neg\mathbf{\Gamma}$*-norm.* If A were also in $\mathbf{\Gamma}$, there would be nothing to prove; if not, then Wadge's Lemma 7D.3 implies immediately that *every set B in $\neg\mathbf{\Gamma}$ is a continuous preimage* of A (since A cannot be a continuous preimage of $\neg B$) and then easily, *every set in $\neg\mathbf{\Gamma}$ admits a* $\neg\mathbf{\Gamma}$-norm. But this contradicts 4B.13—$\mathbf{\Gamma}$ and $\neg\mathbf{\Gamma}$ cannot both be normed. ⊣

The Suslin theorem for the odd levels is one of the most appealing structural consequences of *AD*—it is clear that we cannot hope for any neat characterization like this for $\mathbf{\Delta}^1_3$ in the real world. It is complemented nicely by the following simple fact.

7D.10. Theorem (Moschovakis [1970]). *Assume AD and let Γ be a pointclass such that $\Sigma^1_1 \subseteq \Gamma$ and Γ is closed under* &, $\vee, \exists^\omega, \forall^\omega$ *and* $\exists^{\mathcal{N}}$, *let*

$$\boldsymbol{\gamma} = \textit{supremum}\{|<|: < \textit{is a strict wellfounded relation in } \mathbf{\Gamma}\}.$$

If $\lambda < \boldsymbol{\gamma}$ *and each* $A_\xi \subseteq \mathfrak{X}$ *is in* $\mathbf{\Gamma}(\xi < \lambda)$, *then*

$$\bigcup_{\xi<\lambda} A_\xi \in \mathbf{\Gamma}.$$

In particular, if AD + DC holds, then for each even n,

$$A \in \mathbf{\Sigma}^1_n \Leftrightarrow A \textit{ is the union of } \boldsymbol{\delta}^1_{n-1} \textit{ sets in } \mathbf{\Delta}^1_{n-1}$$

and for $n = 2$,

$$A \in \mathbf{\Sigma}^1_2 \Leftrightarrow A \text{ is the union of } \aleph_1 \text{ Borel sets.}$$

PROOF. For the first assertion, let $\rho : Field(<) \twoheadrightarrow \lambda$ be the rank function of $<$, let $G \subseteq \mathcal{N} \times \mathfrak{X}$ be universal in Γ, put

$$f(\xi) = \{\alpha : G_\alpha = A_\xi\},$$

let $C \subseteq \mathcal{N} \times \mathfrak{X}$ be a choice set for f in $\mathbf{\Gamma}$ and check that

$$x \in \bigcup_{\xi < \lambda} A_\xi \Leftrightarrow (\exists y)(\exists \alpha)[C(y, \alpha) \;\&\; G(\alpha, x)].$$

To prove direction ($\Leftarrow$) of the second assertion, suppose

$$x \in A \Leftrightarrow (\exists \alpha) B(x, \alpha)$$

with B in $\mathbf{\Pi}^1_{n-1}$ and let B^* uniformize B by the Uniformization Theorem 6C.5. Now by 4C.9, we have

$$(x, \alpha) \in B^* \Leftrightarrow (\exists \xi < \boldsymbol{\delta}^1_{n-1})(x, \alpha) \in C_\xi$$

with suitable sets $C_\xi \in \mathbf{\Delta}^1_{n-1}$, so that

$$x \in A \Leftrightarrow (\exists \xi < \boldsymbol{\delta}^1_{n-1})(\exists \alpha) C_\xi(x, \alpha)$$

and moreover, since each $C_\xi \subseteq B^*$, we have

$$C_\xi(x, \alpha) \;\&\; C_\xi(x, \beta) \Rightarrow \alpha = \beta.$$

Thus taking

$$D_\xi = \{x : (\exists \alpha) C_\xi(x, \alpha)\},$$

we know that D_ξ is the image of C_ξ under the recursive injection $(x, \alpha) \mapsto x$, so that by 4D.8, $D_\xi \in \mathbf{\Delta}^1_{n-1}$ and of course

$$A = \bigcup_{\xi < \boldsymbol{\delta}^1_{n-1}} D_\xi. \qquad \dashv$$

The last two results make clear that under the hypotheses $AD + DC$, the projective sets are completely determined by the ordinals $\boldsymbol{\delta}^1_1, \boldsymbol{\delta}^1_2, \boldsymbol{\delta}^1_3, \ldots$ and the operations of wellordered union and complementation. The exact place of these ordinals in the sequence of the alephs is not known—finding it is one of the most intriguing and apparently difficult open problems in this part of descriptive set theory.

Let us collect in one theorem the facts about $\boldsymbol{\delta}^1_1, \boldsymbol{\delta}^1_2, \ldots$ which we can prove at this point. Following the notation established in 7D.8, we put

$$\boldsymbol{\sigma}^1_n = supremum\{|<| : < \text{is a strict wellfounded relation in } \mathbf{\Sigma}^1_n\}.$$

7D.11. THEOREM.[8] Assume $AD + DC$.

(i) *For each* n, $\boldsymbol{\delta}^1_n = \boldsymbol{\sigma}^1_n$.

(ii) *Each* $\boldsymbol{\delta}^1_n$ *is a regular cardinal.*

(iii) $\aleph_1 = \boldsymbol{\delta}^1_1 < \boldsymbol{\delta}^1_2 < \boldsymbol{\delta}^1_3 < \cdots$

(iv) *If* n *is even, then*

$$\boldsymbol{\delta}^1_n = (\delta^1_{n-1})^+;$$

in particular,

$$\boldsymbol{\delta}^1_2 = \aleph_2.$$

(v) *If* n *is odd, then there is a cardinal* $\varkappa_n$ *of cofinality* ω *such that* $\boldsymbol{\delta}^1_{n-1} < \varkappa_n < \boldsymbol{\delta}^1_n$ *and*

$$\boldsymbol{\delta}^1_n = (\varkappa_n)^+.$$

PROOF. All (i)–(v) follow from the following facts.

(a) Each $\boldsymbol{\delta}^1_n$ is a cardinal.

Proof. By 7D.8.

(b) For even n, $\boldsymbol{\delta}^1_{n-1} < \boldsymbol{\delta}^1_n$.

Proof. $\boldsymbol{\Pi}^1_{n-1}$ is a Spector pointclass by 6B.2 and hence by 4C.14, $\boldsymbol{\delta}^1_{n-1}$ is the length of some $\boldsymbol{\Pi}^1_{n-1}$ prewellordering.

(c) For even n, $\boldsymbol{\sigma}^1_n = \boldsymbol{\delta}^1_n = (\boldsymbol{\delta}^1_{n-1})^+$.

Proof. Each $\boldsymbol{\Sigma}^1_n$ relation is $\boldsymbol{\delta}^1_{n-1}$-Suslin by 6C.11, so that by the Kunen–Martin Theorem 2G.2, $\boldsymbol{\sigma}^1_n \leq (\boldsymbol{\delta}^1_{n-1})^+$; now use (b) and (a).

(d) For all n, $\boldsymbol{\delta}^1_n = \boldsymbol{\sigma}^1_n$.

Proof. For odd n, use the fact that $\boldsymbol{\Pi}^1_n$ is a Spector pointclass closed under $\forall^{\mathcal{N}}$ and 4C.14 which then implies $\boldsymbol{\sigma}^1_n \leq \boldsymbol{\delta}^1_n$.

(e) Each $\boldsymbol{\delta}^1_n$ is a regular cardinal.

Proof. By 7D.8, each $\boldsymbol{\sigma}^1_n$ is a regular cardinal and then (d) applies.

(f) For each odd n, there is a cardinal $\varkappa_n$ of cofinality ω such that $\boldsymbol{\delta}^1_n = (\varkappa_n)^+$.

Proof. By 6C.11, each $\boldsymbol{\Sigma}^1_n$ set is λ-Suslin for some $\lambda < \boldsymbol{\delta}^1_n$, so take

$$\varkappa_n = \text{least } \lambda \text{ such that some universal } \boldsymbol{\Sigma}^1_n \text{ set is } \lambda\text{-Suslin}.$$

Now $\varkappa_n$ is a cardinal by definition, the Kunen–Martin theorem implies $\boldsymbol{\sigma}^1_n \leq (\varkappa_n)^+$ and hence since $\boldsymbol{\sigma}^1_n$ is a cardinal, we have $\boldsymbol{\sigma}^1_n = (\varkappa_n)^+$.

Suppose now that $\varkappa_n$ has cofinality $> \omega$. By 2B.4 then, each $\boldsymbol{\Sigma}^1_n$ set A can be written in the form

$$A = \bigcup_{\xi < \varkappa_n} A_\xi$$

where each A_ξ is λ-Suslin for some $\lambda < \varkappa_n$. Applying 2F.2 and using the fact that $\varkappa_n$ is a cardinal so that $\lambda^+ \leq \varkappa_n$, each A_ξ is the union of $\varkappa_n$ sets which are $\varkappa_n$-Borel, so that by Martin's Theorem 7D.9, each A_ξ is in $\mathbf{\Delta}^1_n$ and then by 7D.9 again, A is in $\mathbf{\Delta}^1_n$. This contradicts the hypothesis that A was an arbitrary set in $\mathbf{\Sigma}^1_n$.

(g) For each odd n, $\boldsymbol{\delta}^1_{n-1} < \varkappa_n < \boldsymbol{\delta}^1_n$.

Proof. Assuming towards a contradiction that $\boldsymbol{\delta}^1_{n-1} = \boldsymbol{\delta}^1_n$, we get

$$(\varkappa_n)^+ = \boldsymbol{\delta}^1_n = \boldsymbol{\delta}^1_{n-1} = (\boldsymbol{\delta}^1_{n-2})^+$$

so that $\varkappa_n = \boldsymbol{\delta}^1_{n-2}$ which is absurd, because $\boldsymbol{\delta}^1_{n-2}$ is regular while $cf(\varkappa_n) = \omega$. Thus $\boldsymbol{\delta}^1_{n-1} < \boldsymbol{\delta}^1_n$. But then $\boldsymbol{\delta}^1_{n-1} \leq \varkappa_n$ and using again the regularity of $\boldsymbol{\delta}^1_{n-1}$, $\boldsymbol{\delta}^1_{n-1} < \varkappa_n$. ⊣

Martin has shown (from $AD + DC$) that

$$\boldsymbol{\delta}^1_3 = \aleph_{\omega+1}$$

and that consequently

$$\boldsymbol{\delta}^1_4 = \aleph_{\omega+2};$$

we will give his argument in Section 8H. On the other hand, the computation of $\boldsymbol{\delta}^1_5$ has resisted many valiant attempts. Notice that the smallest possible value for $\boldsymbol{\delta}^1_5$ is $\aleph_{\omega 2+1}$, but there is evidence that it may be larger than that; see Kechris [1978], Solovay [1978b].

Exercises

Recall the game $G(A, B)$ that we associated with any two sets $A, B \subseteq \mathcal{N}$ in the proof of Wadge's Lemma 7D.3: I plays α, II plays β and II wins if

$$\alpha \in A \leftrightarrow \beta \in B.$$

We say that A is *Lipschitz reducible to* B, $A \leq_l B$ if II wins $G(A, B)$. Clearly,

$$A \leq_l B \Rightarrow A \leq_w B$$

and by the proof of 7D.3, if AD holds then for each A, B, $A \leq_l B$ or $B \leq_l \mathcal{N} - A$.

Put

$$A \sim_l B \Leftrightarrow A \leq_l B \ \&\ B \leq_l A,$$

$$A <_l B \Leftrightarrow A \leq_l B \ \&\ A \not\sim_l B;$$

and similarly with $\leq_w, <_w, \sim_w$ in place of $\leq_l, <_l, \sim_l$; we define *the Wadge degree* and the *Lipschitz degree* of a set $A \subseteq \mathcal{N}$ by

$$[A]_w = \{B \subseteq \mathcal{N} : A \sim_w B\},$$

$$[A]_l = \{B \subseteq \mathcal{N} : A \sim_l B\}.$$

By the proof of 7D.3 again, for each A

$$[A]_l \subseteq [A]_w.$$

The reducibilities $\leq_w, \leq_l$ induce partial orderings on the sets of respective degrees in the obvious way; we will denote these partial orderings by the same symbols "$\leq_w$," "$\leq_l$."

7D.12. Assume *AD* and prove that for all $A, B \subseteq \mathcal{N}$,

$$A \leq_l \neg B \rightarrow \neg A \leq_l B$$

$$A <_l B \rightarrow A <_l \neg B$$

and similarly for the Wadge reducibility.

Hint. The first implication is immediate. To prove the second, assume $A <_l B$ and notice that by Wadge's lemma, if $A \leq_l \neg B$ does not hold, we have $B \leq_l A$ which contradicts $A <_l B$; thus $A \leq_l \neg B$ holds, and we cannot have $\neg B \leq_l A$ also, since this would give $\neg B \leq_l B$ by transitivity, thus $B \leq_l \neg B$ by the first assertion, thus $A <_l \neg B$ by transitivity again. ⊣

7D.13. Assume *AD* and let a, b be arbitrary Lipschitz degrees. Prove that exactly one of the following holds:

(i) $a = b$,
(ii) $a = \neg b$ (i.e. $a = [A]_l$ and $b = [\neg A]_l$),
(iii) $a <_l b$ and $a <_l \neg b$,
(iv) $b <_l a$ and $\neg b <_l a$.

Prove the same result for Wadge degrees. ⊣

Thus every *self-dual* degree ($b = \neg b$) is comparable with all degrees and if b is not self-dual, then for every $a \neq b$, $a \neq \neg b$, one of the following two patterns must hold in these partial orderings:

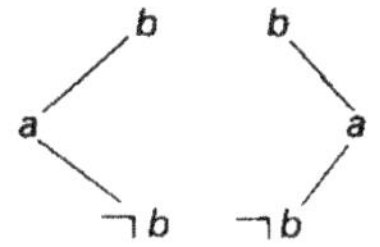

The most significant fact about Wadge and Lipschitz degrees is the following theorem of Martin which essentially asserts that AD puts a certain "hierarchy" on the power of $\mathcal{N}$.

7D.14. Assume AD and prove that the relation $<_l$ on $Power(\mathcal{N})$ is wellfounded; infer that $<_w$ is also wellfounded (Martin).

Hint. Assume towards a contradiction that

$$A_0 >_l A_1 >_l A_2 \cdots$$

so that by 7D.13 we also have

$$A_0 >_l \neg A_1,\ A_1 >_l \neg A_2, \ldots$$

etc. Rename the games,

$$G_0^n = G(A_n, A_{n+1}),$$
$$G_1^n = G(A_n, \neg A_{n+1}),$$

and by the countable axiom of choice for pointsets, let σ_0^n be a winning strategy for I in G_0^n and let σ_1^n be a winning strategy for I in G_1^n.

For each binary sequence $\alpha : \omega \to \{0, 1\}$ consider the diagram of games 7D.1. Here we read each pair of successive lines as a run of $G^n_{\alpha(n)}$ in which I plays by his winning strategy $\sigma^n_{\alpha(n+1)}$ (which is his winning strategy as player I in the next game $G^{n+1}_{\alpha(n+1)}$).

$$\begin{array}{lllllll}
G^0_{\alpha(0)} & x^0_0 & x^0_1 & x^0_2 & & \cdots & x^0(\alpha) \\
G^1_{\alpha(1)} & x^1_0 & x^1_1 & x^1_2 & & \cdots & x^1(\alpha) \\
G^2_{\alpha(2)} & x^2_0 & x^2_1 & x^2_2 & & \cdots & x^2(\alpha) \\
\vdots & & & & & & \\
G^n_{\alpha(n)} \left\{ \begin{array}{l} x^n_0 \\ x^{n+1}_0 \end{array} \right. & \begin{array}{l} \\ \end{array} & \begin{array}{l} x^n_1 \\ x^{n+1}_1 \end{array} & \begin{array}{l} x^n_2 \\ x^{n+1}_2 \end{array} & & \begin{array}{l} \cdots \\ \cdots \end{array} & \begin{array}{l} x^n(\alpha) \\ x^{n+1}(\alpha) \end{array}
\end{array}$$

Diagram 7D.1.

Plays $x^0(\alpha), x^1(\alpha), \ldots$ are produced and we clearly have

$$\alpha(n) = 0 \rightarrow x^n(\alpha) \in A_n \leftrightarrow x^{n+1}(\alpha) \notin A_{n+1},$$
$$\alpha(n) = 1 \rightarrow x^n(\alpha) \in A_n \leftrightarrow x^{n+1}(\alpha) \in A_{n+1}.$$

Notice that by the construction,

$$(\forall m \geq n)[\alpha(m) = \beta(m)] \rightarrow x^n(\alpha) = x^n(\beta)$$

and put

$$T_n = \{\alpha : \text{for some (and hence for each) binary sequence } s = s_0, \ldots, s_{n-1}, x^n(s \frown \alpha) \in A_n\}.$$

From the definition of T_n we get

(1) $$\alpha \in T_{n+1} \leftrightarrow (1) \frown \alpha \in T_n,$$

(2) $$\alpha \notin T_{n+1} \leftrightarrow (0) \frown \alpha \in T_n,$$

which imply immediately that T_n cannot be either meager or comeager, or else the whole space ${}^{\omega}2$ would be the union of two meager sets.

In particular, T_0 is not meager, so by full determinacy and 6A.14 (or using the property of Baire for T_0), there is a binary sequence $s = s_0, \ldots, s_{k-1}$ so that $\{s \frown \beta : s \frown \beta \notin T_0\}$ is meager. But it is clear from (1) and (2) above iterated k times, that either

$$\alpha \in T_k \leftrightarrow s \frown \alpha \notin T_0$$

so that T_k is meager, or

$$\alpha \notin T_k \leftrightarrow s \frown \alpha \notin T_0$$

so that T_k is comeager and in either case we have reached a contradiction. ⊣

Wadge and Libschitz degrees have been studied extensively although relatively little has been published on them; see Wadge [19??], Steel [1977], Van Wesep [1977], [197?], [1978].

There is another simple but very useful lemma of Martin, about Turing degrees. Recall from 3D that for each $\alpha \in \mathcal{N}$, *the Turing degree of* α is defined by

$$[\alpha]_T = \{\beta : \alpha \leq_T \beta \ \& \ \beta \leq_T \alpha\},$$

where $\leq_T$ stands for "recursive in." The transitive relation $\leq_T$ induces a partial ordering on the set $\mathfrak{D}_T$ of all Turing degrees.

7D.15. Assume AD and Suppose $A \subseteq \mathfrak{D}_T$ is any set of Turing degrees. Prove that there is some degree d_0 such that

$$\text{either} \quad \{d : d_0 \leq d\} \subseteq A \quad \text{or} \quad \{d : d_0 \leq d\} \subseteq \mathfrak{D}_T - A.$$

(Martin [1968].)

Hint. Consider the game G where I and II produce α and

$$\text{I wins} \Leftrightarrow [\alpha]_T \in A.$$

If I wins with a strategy σ, take $d_0 = [\sigma]_T$ and for any degree $d \geq d_0$, have II play some β with $[\beta]_T = d$; now the resulting play $\sigma * [\beta]$ easily has degree d, so that $d \in A$. The argument is similar in the case II wins the game. ⊣

This is the typical result where we assume full determinacy but where the proof makes it clear just how much determinacy is needed for each specific application. For example:

7D.16. Prove that if $A \subseteq \mathcal{N}$ is a Borel set such that

$$\alpha \in A \;\&\; \alpha \equiv_T \beta \Rightarrow \beta \in A,$$

then there is a degree d_0 such that

$$\text{either} \quad \{\alpha : d_0 \leq [\alpha]_T\} \subseteq A \quad \text{or} \quad \{\alpha : d_0 \leq [\alpha]_T\} \subseteq \mathcal{N} - A.$$

(Martin [1975].)

Hint. Appeal to the determinacy of Borel sets. ⊣

This result has interesting consequences in the theory of Turing degrees which we will not pursue here.

At the same time, 7D.15 also has some interesting and surprising consequences in the presence of full determinacy.

7D.17. Assume AD and let $\mathfrak{U}$ be the set of all subsets of the set of Turing degrees which contain cones,

$$A \in \mathfrak{U} \Leftrightarrow \text{there is some degree } d_0 \text{ such that}$$

$$\{d : d_0 \leq d\} \subseteq A.$$

Prove that $\mathfrak{U}$ is an $\aleph_1$-complete, non-principal ultrafilter on $\mathfrak{D}_T$ so that the function

$$\mu(A) = \begin{cases} 1 & \text{if} \quad A \in \mathfrak{U}, \\ 0 & \text{if} \quad A \notin \mathfrak{U}, \end{cases}$$

is an $\aleph_1$-additive measure on $\mathfrak{D}_T$. (Martin [1968].)

Hint. Use the countable axiom of choice for pointsets to show that for every sequence of degrees $d_0, d_1, d_2, \ldots$ there is some d^* above all of them—this comes up in checking $\aleph_1$-completeness. ⊣

What is perhaps even more surprising is that we can carry the Martin measure on $\mathfrak{D}_T$ to a measure on $\aleph_1$.

7D.18. Assume AD and show that $\aleph_1$ is a measurable cardinal (Solovay).

Hint. Recall from 4F that for each $\alpha \in \mathcal{N}$,

$$\omega_1^\alpha = supremum\{|\beta| : \beta \leq_T \alpha \ \& \ \beta \in WO\}$$

and for each degree d put

$$\omega_1^d = \omega_1^\alpha, \quad \text{where } \alpha \text{ is any irrational in } d.$$

Let $\mathfrak{U}$ be the Martin ultrafilter on $\mathfrak{D}_T$ and for $A \subseteq \aleph_1$, put

$$A \in \mathfrak{U}^* \leftrightarrow \{d : \omega_1^d \in A\} \in \mathfrak{U}. \qquad ⊣$$

As we mentioned in the historical remarks at the end of Chapter 6, this early result of Solovay was instrumental in focusing the attention of set theorists to determinacy. It came before Blackwell [1967] and the subsequent development of the structure theory from *PD*.

The proof of 7D.18 sketched above is obviously due to Martin. Solovay's orginal proof was a bit more complicated but also more amenable to generalization: Solovay extended it to show from AD that $\aleph_2$ is also measurable and then Kunen proved further that all the $\boldsymbol{\delta}_n^1$ are in fact measurable cardinals. (It was already know that in the absence of the axiom of choice measurable cardinals need not be large, see Jech [1968].)

For these and other results relating determinacy with measurability see Kechris [1978] (and the references given there) as well as the forthcoming monograph Martin [19?c].

The next exercise is implicit in the proof of 7D.8 but it is worth point to it.

7D.19. Assume AD and prove that if $\lambda < \Theta$, then there is a surjection $\pi : \mathcal{N} \twoheadrightarrow Power(\lambda)$ (Moschovakis [1970]); infer that $\lambda < \Theta \rightarrow \lambda^+ < \Theta$ (Friedman). ⊣

Finally, the last exercise gives an interesting corollary of the coding lemma.

7D.20. Let Γ be a Spector pointclass closed under $\forall^{\mathcal{N}}$, let

$$\varphi : P \twoheadrightarrow \lambda$$

be a regular Γ-norm on some set $P \in \Gamma - \mathbf{\Delta}$ and for each set $A \subseteq \lambda$, put

$$Code(A;\varphi) = \{x : \varphi(x) \in A\}.$$

Assume AD and prove that for each set A, $Code(A;\varphi)$ is in $\mathbf{\Gamma}$. (Solovay for $\mathbf{\Gamma} = \mathbf{\Pi}^1_1$, Moschovakis [1970] in general.)

Hint. Assume $P \subseteq \mathcal{N}$ for simplicity and consider the game where I plays α, II plays β and

$$\text{II wins} \Leftrightarrow \alpha \notin P$$
$$\vee \{\alpha \in P \,\&\, (\forall \gamma)[\gamma \leq^*_\varphi \alpha \rightarrow [\varphi(\gamma) \in A \leftrightarrow G(\beta, \gamma)]]\},$$

where $G \subseteq \mathcal{N} \times \mathcal{N}$ is universal in Γ.

If I wins with some strategy σ, then by the Covering Lemma 4C.11 there is some $\xi < \lambda$ such that

$$(\forall \beta)[\sigma * [\beta] \in P \,\&\, \varphi(\sigma * [\beta]) < \xi]$$

and then the coding lemma implies easily that II can beat this σ; thus II wins with some τ and

$$\alpha \in Code(A;\varphi) \leftrightarrow G([\alpha] * \tau, \alpha). \qquad \dashv$$

These games where we insure that player II wins by forcing I to play ordinal codes are called *Solovay games;* they were used by Solovay in one of his original proofs (from AD) of the measurability of $\aleph_1$.

7E. Historical remarks

[1] The recursion theorem was first proved in Kleene [1938] in connection with the theory of *constructive ordinals.* Kleene used it erroneously in his [1944] to claim (in effect) that every relation on ω which is positive Π^0_1-inductive on ω is Π^0_2. This false claim in one of the basic first papers of the effective theory is amusingly reminiscent of the similar false claim in Lebesgue [1905], but there is little to connect the actual mathematical

mistakes in the two papers. In any case, Kleene was his own Suslin: he corrected the mistake himself in his fundamental paper [1955a] where he characterized (essentially) the Π^1_1 pointsets in terms of positive Π^0_1-induction on ω (7C.2) and where he laid the foundations of the technique of *effective transfinite recursion.*

[2] Kleene proved his version of what we have called here the *Suslin–Kleene theorem* in his [1955c], after a good deal of preliminary work in his [1955b]. He established that *a relation on ω is Δ^1_1 exactly when it is hyperarithmetical* but with a definition of "hyperarithmetical" which was bound up with the notion of constructive ordinal. The essential content of his theorem is given in 7B.7. On the other hand, his proof (though technically quite complicated) was certainly sufficient to yield the full Suslin–Kleene Theorem 7B.4, at least for pointsets of type 0 and 1.

[3] As is obvious from the remarks above Kleene established the basic properties of positive induction on ω (especially in his [1955a]) although he did not introduce the notions explicitly; this was done by Spector [1961] who also proved (iii) and (iv) of 7C.2.

[4] Moschovakis [1974a] gives a detailed development of positive elementary induction on abstract structures, which includes (and has as one of its most interesting examples) what we have called here positive analytical induction on $\mathcal{N}$. Before that, Moschovakis [1969], [1971b] introduced and established the basic properties of inductive and hyper-projective pointsets (of type 1), using a different, recursion-theoretic approach.

[5] We have tried to give the credits for results from AD in the main text of 7D, but a few clarifications must be made.

[6] The Coding Lemma 7D.5, 7D.6 in its present form was proved by Moschovakis [1970], but it owes much to earlier (unpublished) results of Friedman and Solovay.

[7] Theorem 7D.8 gives a collection of results by several persons which are hard to untangle. It is a bit easier to give specific credits for the facts about the ordinals $\boldsymbol{\delta}^1_n$ in 7D.11 since these were the preoccupation of early research in the area.

[8] Moschovakis [1970] introduced the ordinals $\boldsymbol{\delta}^1_n$ and proved (from AD of course) that they are all cardinals and that if n is odd, then $\boldsymbol{\sigma}^1_n = \boldsymbol{\delta}^1_n$, $\boldsymbol{\delta}^1_n$ is regular and $\boldsymbol{\delta}^1_{n-1} < \boldsymbol{\delta}^1_n$. Martin and Kunen showed (independently) in 1971 that for even n, $\boldsymbol{\sigma}^1_n = \boldsymbol{\delta}^1_n$, $\boldsymbol{\delta}^1_n$ is regular and $\boldsymbol{\delta}^1_n = (\boldsymbol{\delta}^1_{n-1})^+$. Finally Kechris [1974] proved that for odd n, $\boldsymbol{\delta}^1_n = \varkappa^+_n$ for a cardinal $\varkappa_n$ of cofinality ω and that for even n, $\boldsymbol{\delta}^1_{n-1} < \boldsymbol{\delta}^1_n$.

7F. Appendix; a list of results which depend on the axiom of choice

A. Results which depend essentially on the full axiom of choice

2C.4. There is an uncountable set of reals which has no non-empty perfect subset.

2H.6. There is a set of reals without the property of Baire.

2H.9. There is a set of reals which is not Lebesgue measurable.

6A.1. If II does not win the game $A(u)$, then there is some a such that for all b, II does not win $A(u^\frown(a, b))$.

6A.6. There is a set $A \subseteq {}^\omega 2$ which is not determined.

6G.9. Every measurable cardinal is strongly inaccessible.

It is not hard to see that the innocuous sounding 6A.1 is in fact equivalent to the axiom of choice. Given $R \subseteq A \times B$ such that $(\forall x \in A)(\exists y \in B)R(x, y)$, consider the two-move game A where I plays x, II plays y and II wins if $R(x, y)$; now if there is no choice function $f: A \to B$ such that $(\forall x \in A)R(x, f(x))$, then II does not win the game, while for each x there is obviously some y such that II wins $A(x, y)$—which is finished before it starts.

B. Results which can be reformulated so that they do not depend on the axiom of choice

2F.4. If $P \subseteq \mathfrak{X}$ is $\aleph_n$-Suslin ($n \geq 1$), then

$$P = \bigcup_{\xi < \aleph_n} P_\xi$$

where each P_ξ is Borel.

Reformulation (for $n = 2$, as an example). If P is $\aleph_2$-Suslin, then there is a sequence of sets $\{P_\xi : \xi < \aleph_2\}$ such that

$$P = \bigcup_{\xi < \aleph_2} P_\xi$$

and such that for each $\xi < \aleph_2$, there is a sequence of sets $\{Q_\eta : \eta < \aleph_1\}$ such that each Q_η is Borel and

$$P_\xi = \bigcup_{\eta < \aleph_1} Q_\eta.$$

To prove 2F.4 for $n = 2$ from this we must choose for each $\xi < \aleph_2$ a sequence $\{Q_\eta : \eta < \aleph_1\}$ with the requisite property.

6A.2. Every closed subset of ${}^\omega X$ is determined.

6A.3. Every $\boldsymbol{\Sigma}^0_2$ subset of ${}^\omega X$ is determined.

6F.1. Every $\boldsymbol{\Sigma}^0_n$ subset of ${}^\omega X$ is determined.

The problem in the proofs of these basic theorems is that we need to make choices from X in order to define the required strategies—naively understood, 6A.2 in fact is equivalent to the full axiom of choice.

First reformulation. Assume that the space X in which the games are played is countable, e.g. ω. Now the proof of 6A.1 uses only the countable axiom of choice and 6A.2, 6A.3 follows from 6A.1 without any use of choice.

Second reformulation. A *multiple-valued strategy* or *quasistrategy* for player I in the space X is any set Σ of finite sequences from X such that:

(i) For some x_0, $(x_0) \in \Sigma$.

(ii) If $(x_0, \ldots, x_{2n}) \in \Sigma$, then for every $y \in X$,

$$(x_0, \ldots, x_{2n}, y) \in \Sigma.$$

(iii) If $(x_0, \ldots, x_{2n+1}) \in \Sigma$, then for some y,

$$(x_0, \ldots, x_{2n+1}, y) \in \Sigma.$$

We say that I *follows* Σ in a run $(x_0, x_1, x_2, \ldots)$ of a game G on X if for each n,

$$(x_0, \ldots, x_n) \in \Sigma$$

and we call Σ *winning for* I *in* G if whenever I follows Σ in $(x_0, x_1, x_2, \ldots)$, we have

$$(x_0, x_1, x_2, \ldots) \in G.$$

Quasistrategies for II are defined in the obvious way and we say that G is *weakly determined* if either I or II has a winning quasistrategy.

It is clear that if the set X admits a wellordering, then every winning quasistrategy for I or II determines a winning strategy for I or II, so that a game G on a wellorderable space X is determined exactly when it is weakly determined.

Reformulation of 6A.2, 6A.3 and 6F.1: replace "determined" by "weakly determined" in the statements of these results.

Proof of the reformulation of 6A.2. Given a closed $F \subseteq {}^{\omega}X$ with open complement G, define the sets W^{ξ} of sequences of even length by induction on ξ (as in the proof of 6A.3),

$$u \in W^0 \Leftrightarrow \text{if } f \in {}^{\omega}X \text{ is an extension of } u, \text{ then } f \in G,$$

$$u \in W^{\xi} \Leftrightarrow (\forall a)(\exists b)(\exists \eta < \xi)[u^\frown(a, b) \in W^{\eta}].$$

If $\emptyset \notin W^{\infty} = \bigcup_{\xi} W^{\xi}$, then

$$\Sigma = \{(x_0, \ldots, x_{n-1}) : \text{for each } i < n,\ (x_0, \ldots, x_i) \notin W^{\infty}\}$$

is easily a winning quasistrategy for I. If $\emptyset \in W^\infty$, let

$$|u| = \text{least } \xi \text{ such that } u \in W^\xi \ (u \in W^\infty)$$

and check easily that

$$\Sigma = \{(x_0, \ldots, x_{n-1}) : \text{for each odd } i < n, (x_0, \ldots, x_i) \in W^\infty \\ \text{and if } i+2 < n, |x_0, \ldots, x_i| > |x_0, \ldots, x_i, x_{i+1}, x_{i+2}|\}$$

is a winning quasistrategy for II.

The proofs we gave in the text for 6A.3 and 6F.1 are easily modified to show without appealing to the full axiom of choice that $\boldsymbol{\Sigma}^0_n$ games on an arbitrary $X \neq \emptyset$ are weakly determined—and as a corollary, that on the wellorderable space ω, $\boldsymbol{\Sigma}^0_n$ games are determined. (These proofs use *DC*.)

7C.4. A pointset $R \subseteq \mathfrak{X}$ is absolutely inductive if and only if there is a Σ^1_1 set P such that

$$R(x) \Leftrightarrow \{(\exists \alpha_0)(\forall \alpha_1)(\exists \alpha_2)(\forall \alpha_3) \cdots\}(\exists n) P(x, \langle \alpha_0, \ldots, \alpha_{n-1} \rangle).$$

Here too we must understand the infinite alternating string on the right in terms of winning quasistrategies in order to prove 7C.4 without appealing to the full axiom of choice.

7C.1. If Φ is a monotone operator on *Power*$(\mathfrak{X})$, then for some ordinal $\varkappa$ of cardinality $\leq card(\mathfrak{X})$,

$$(*) \qquad \Phi^\infty = \Phi^\varkappa.$$

If $\mathfrak{X}$ is wellorderable so that $card(\mathfrak{X})$ is an ordinal, then there is no problem with this. In the case where $\mathfrak{X}$ is not wellorderable, then we may only assert that $(*)$ holds for some ordinal $\varkappa$ such that there is a surjection

$$\pi : \mathfrak{X} \twoheadrightarrow \varkappa.$$

C. *Results whose proofs must be modified*

In addition to the determinacy of $\boldsymbol{\Sigma}^0_n$ games on ω, there are two other significant theorems which can be proved without appeal to the full axiom of choice but where we chose to use choice in the proofs we gave.

2F.3. Every $\boldsymbol{\Sigma}^1_2$ set is the union of $\aleph_1$ Borel sets.

6C.12. If $Det(\boldsymbol{\Delta}^1_{2n})$ holds, then every $\boldsymbol{\Sigma}^1_{2n+2}$ set is the union of $\boldsymbol{\delta}^1_{2n+1}$ sets each of which is $\boldsymbol{\delta}^1_{2n+1}$-Borel.

A proof of these two results which does not use the axiom of choice is part of the proof of 7D.10.

CHAPTER 8

METAMATHEMATICS

In this last chapter we will study briefly the *metamathematical method*, the key tool for establishing consistency and independence results. Here too we presuppose no knowledge of formal logic—we will develop in some detail all the preliminary material that we need. We are, however, assuming a good understanding of (informal, axiomatic) set theory, as we have been using it in the first seven chapters of this book.

The chief aim of mathematics is to study certain concrete *mathematical structures*, e.g. *the semiring* ω *of integers, the field* $\mathcal{R}$ *of real numbers* or, in set theory, the *universe* V *of sets*. What we do in actual fact is to consider various *propositions* about these structures and attempt to determine their truth or falsity. We often use the *axiomatic method* for precision and elegance: certain fairly obvious propositions are designated *axioms* and whatever assertions we make after this are supposed to follow from the axioms *by logic alone*.

The essence of the metamathematical method consists in identifying and making precise the *language* $\mathcal{L}$ in which we make assertions about a particular structure $\mathfrak{A}$. Typically we take $\mathcal{L}$ to be the *first order language* associated with $\mathfrak{A}$—this is simple but sufficiently expressive so we can formulate in it most of the propositions about $\mathfrak{A}$ we care to consider. We will look at these languages in 8A.

Suppose θ is a particular proposition of $\mathcal{L}$ which may be true or false in $\mathfrak{A}$—perhaps we have not been able to determine this yet. Suppose we can find an *alternative interpretation* of all the propositions of $\mathcal{L}$, such that all the axioms are true in this interpretation but θ fails; this clearly establishes that θ cannot follow from the axioms *by logic alone*, assuming at least that truth in our alternative interpretation is preserved under logical deduction. In these circumstances we say that θ is *independent* of the axioms or that *the negation of* θ is *consistent* with the axioms.

One powerful method for constructing alternative interpretations of the *language of set theory* is to specify a collection V' of sets with very special properties and reinterpret "set" to mean "set in V'." Some of these *inner models* of set theory are interesting mathematical objects in their own right, particularly Gödel's universe L of *constructible sets* whose properties we considered briefly in Chapter 5. Cohen's *method of forcing* introduces more complicated reinterpretations of the language.

A fascinating thing about the metamathematical method is that it can be used to establish positive results about the universe V, our intended interpretation of the language of set theory. We will look at some of these applications of metamathematics, as we are naturally more interested here in facts about sets rather than theorems about systems of axioms.

It may be useful to review at this point our intuitive conception of the standard model for set theory, the universe V of sets. This *does not* contain all "arbitrary collections of objects"—it is wellknown that this naive approach leads to contradictions. Instead, we admit as "sets" only those collections which occur in the *complete (transfinite) cumulative sequence of types*—the hierarchy obtained by starting with the empty set and iterating "indefinitely" the "power operation."

To be more precise, suppose we are given an operation P on sets which assigns to each set x a collection $P(x)$ of subsets of x

$$y \in P(x) \Rightarrow y \subseteq x. \tag{1}$$

Suppose we are also given a collection $\mathcal{S}$ of *stages*, wellordered by a relation $\leq$, i.e. for ζ, η, ξ in $\mathcal{S}$,

$$\zeta \leq \zeta,\ (\zeta \leq \eta \ \&\ \eta \leq \xi) \Rightarrow \zeta \leq \xi,\ (\zeta \leq \eta \ \&\ \eta \leq \zeta) \Rightarrow \zeta = \eta \tag{2}$$

$$\zeta \leq \eta \quad \text{or} \quad \eta \leq \zeta,$$

(3) if $A \subseteq \mathcal{S}$ is any collection of stages, $A \neq \emptyset$, then there is some $\xi \in A$ such that for every $\eta \in A$, $\xi \leq \eta$.

Call *the least stage* 0 and for $\xi \in \mathcal{S}$, let $\xi + 1$ be *the next stage*—the least stage which is greater than ξ. If λ is a stage $\neq 0$ and $\neq \xi + 1$ for every ξ, we call it a *limit stage*.

For fixed P, $\mathcal{S}$, $\leq$ satisfying (1)–(3) we define the hierarchy

$$V_\xi = V_\xi(P, \mathcal{S}, \leq) \quad (\xi \in \mathcal{S})$$

by induction on $\xi \in \mathcal{S}$:

$$V_0 = \emptyset,$$
$$V_{\xi+1} = V_\xi \cup P(V_\xi),$$
$$V_\lambda = \bigcup_{\xi<\lambda} V_\xi \quad \text{if } \lambda \text{ is a limit stage.}$$

The collection of sets

$$V = V(P, \mathcal{S}, \leq) = \bigcup_{\xi \in \mathcal{S}} V_\xi$$

is *the universe generated with P as the power operation, on the stages* $\mathcal{S}$. It is very easy to check that

$$\xi \leq \eta \Rightarrow V_\xi \subseteq V_\eta$$

and that each V_ξ is a *transitive set*, i.e.

$$(x \in V_\xi \ \&\ y \in x) \Rightarrow y \in V_\xi.$$

For example, suppose we take

$$P(x) = Power(x) = \{y : y \subseteq x\}$$

and

$$\mathcal{S} = \{0, 1, 2, \ldots, \omega, \omega+1, \omega+2, \ldots\},$$

where the stages $0, 1, 2, \ldots, \omega, \omega+1, \omega+2, \ldots$ are all assumed distinct and ordered as we have enumerated them. In this case we obtain *the universe of Zermelo*,

$$V^Z = V_0 \cup V_1 \cup V_2 \cup \cdots \cup V_\omega \cup V_{\omega+1} \cup \ldots.$$

It is well-known that all the familiar structures of classical mathematics have *isomorphic copies* within V^Z—we can locate in V^Z the integers, the reals, all functions on the reals to the reals, etc.

For a very different universe of sets, we might choose a small power operation, e.g.

$$P(x) = \{y : y \subseteq x \text{ and } y \text{ is definable}\}.$$

This appears vague, but there are many mathematicians who will argue that the notion of a *definable* subset of x is at least as clear as that of an *arbitrary* subset of x. In any case, assuming that this operation P is meaningful, we can iterate it on any collection of stages $\mathcal{S}$ and define a universe of sets. We may want to take $\mathcal{S}$ quite long this time, say

$$\mathcal{S} = \{0, 1, 2, \ldots, \omega, \omega+1, \ldots, \omega 2, \omega 2+1, \ldots, \omega n, \omega n+1, \ldots, \ldots\}.$$

It is clear that the universe $V(P, \mathcal{S}, \leq)$ does not depend on the particular objects that we have chosen to call *stages* but only on the *length* (the *order type*) of the ordering $\leq$; i.e. if we have a one-to-one correspondence of $\mathcal{S}$ with $\mathcal{S}'$ which takes the ordering $\leq$ onto $\leq'$, then

$$V(P, \mathcal{S}, \leq) = V(P, \mathcal{S}', \leq').$$

Most mathematicians accept that there is *a largest meaningful* operation P satisfying (1) above, *the true power operation* which takes x to the collection of *all* subsets of x. This is one of the cardinal assumptions of *realistic* (meaningful, not formal) set theory. Similarly, it is not unreasonable to assume that there is a *longest* collection of stages $\mathcal{S}$ along which we can *meaningfully* iterate the power operation. *We take then our standard universe of sets to be $V(P, \mathcal{S}, \leq)$, where P is the true power operation and $\mathcal{S}$, $\leq$ is the longest meaningful collection of stages.*

This definition of the universe V is admittedly vague. It is clear that we cannot expect to give a precise, mathematical definition of the basic notions of set theory, unless we use notions of some richer theory which in turn would require interpretation. We claim only that the intuitive description of V given above is sufficiently clear so we can formulate meaningful propositions about sets and argue rationally about their truth or falsity.

One last remark about the axiom of choice. Although we take it as evident (throughout this book) that the axiom of choice is true (in the standard universe of sets), it is obten useful for technical, metamathematical reasons to keep track of its (rare) uses. *In this chapter we will include among the hypotheses of our theorems whichever special case of the axiom of choice we need for the proof.*

8A. Structures and languages

Here we will explain briefly the basic notions of logic and model theory. The reader who is knowledgeable in these matters will want to skip through this section very quickly.

Let us consider first some important examples of mathematical "structures."

Example 1. $\mathbf{A}^1 = (\omega, +, \cdot, 0, 1)$, the structure of (first order) *arithmetic*.

We think of $\mathbf{A}^1$ as an algebraic system with domain ω, two binary relations, + (addition) and · (multiplication) and two specified constants, 0 and 1.

Example 2. $\mathsf{A}^2 = (\omega, \mathcal{N}, +, \cdot, ap, 0, 1)$, the structure of *second order arithmetic.*

Now we have two domains, ω and $\mathcal{N} = {}^{\omega}\omega$, the same two binary operations + and · defined on ω and again 0 and 1. We also have the binary operation of *application*

$$ap : \mathcal{N} \times \omega \to \omega,$$

where of course

$$ap(\alpha, n) = \alpha(n).$$

Example 3. $\mathsf{R} = (\mathcal{R}, +, \cdot, 0, 1, \leq, Z)$, the structure of *the real numbers* or *analysis.*

Just one domain this time, the reals $\mathcal{R}$, together with the operations +, ·, the constants 0, 1 and the ordering $\leq$ which turns $\mathcal{R}$ into a *complete ordered field.* We also have the distinguished subset Z of rational integers,

$$Z = \{\cdots -2, -1, 0, 1, 2, \ldots\}$$

which we take as an additional relation on $\mathcal{R}$,

$$Z(x) \Leftrightarrow x \text{ is a rational integer.}$$

Example 4. $\mathsf{V} = (V, \in)$, the structure of *set theory.*

Again just one domain, the universe V of sets in the cummulative sequence of types, together with the membership relation $\in$ on V,

$$x \in y \Leftrightarrow x \text{ is a member of } y.$$

In this structure the domain is a *collection of objects* which is not a set.

In general, a *structure* is determined by certain *domains* (collections of objects) and certain *functions, relations* and *distinguished elements* of these domains. Allowing for the possibility of infinitely many objects in each of these categories, a structure $\mathfrak{A}$ is given by

$$\mathfrak{A} = (\{A_i\}_{i \in I}, \{f_j\}_{j \in J}, \{R_k\}_{k \in K}, \{c_l\}_{l \in L}),$$

where the following hold.

(1) The index set I is nonempty and each A_i is a nonempty collection of objects.

(2) Each f_j is a function

$$f_j : A_{i_1} \times \cdots \times A_{i_m} \to A_n$$

with domain some cartesian product $A_{i_1} \times \cdots \times A_{i_m}$ and range some A_n.

(3) Each R_k is a relation

$$R_k \subseteq A_{i_1} \times \cdots \times A_{i_m}.$$

on some cartesian product $A_{i_1} \times \cdots \times A_{i_m}$

(4) Each c_l is an element of some A_i.

We allow the possibility that some of the index sets J, K, L and empty, so the simplest possible structure is of the form

$$\mathfrak{A} = (A),$$

with A some nonempty collection. Most of the time we have finitely many domains, functions, relations and constants and we exhibit the structures without indexings, as in the examples.

One can think of many natural structures that come up in mathematics, particularly in algebra—groups, rings, fields, etc.

With each structure

$$\mathfrak{A} = (\{A_i\}_{i \in I}, \{f_j\}_{j \in J}, \{R_k\}_{k \in K}, \{c_l\}_{l \in L})$$

we now associate a *formal language* $\mathcal{L}^{\mathfrak{A}}$, the *first order* or *elementary* language of $\mathfrak{A}$. Like a natural language, $\mathcal{L}^{\mathfrak{A}}$ will have an *alphabet* (a set of symbols) and a *grammar*, a system of rules which determine which combinations of symbols in the alphabet are *meaningful*. There will be two kinds of meaningful expressions, *terms* (or *nouns*) which will name elements in the domains of $\mathfrak{A}$ and *formulas* which will express propositions about $\mathfrak{A}$.

One difference between $\mathcal{L}^{\mathfrak{A}}$ and natural languages is that the grammar of $\mathcal{L}^{\mathfrak{A}}$ will be very simple and completely precise; there will be no exceptions to its rules and no ambiguities with variant spellings, double meanings, etc.

It will be good to keep in mind while going through the formal details below, that every object of $\mathcal{L}^{\mathfrak{A}}$ has a natural translation into English—the symbols, the terms and the formulas. What we are doing is to isolate and make precise a small part of the natural language in which we can make reasonably complicated assertions about the structure $\mathfrak{A}$.

The *alphabet* of $\mathcal{L}^{\mathfrak{A}}$ consists of the following (distinct) symbols.

Variables: We have an infinite sequence of variables

$$\mathbf{v}_0^i, \mathbf{v}_1^i, \mathbf{v}_2^i, \ldots$$

for each domain A_i.

We will use the variables $\mathbf{v}_j^i$ to name unspecified objects of A_i. For example, in the case of arithmetic $\mathbf{A}^1$ we have variables $\mathbf{v}_0$, $\mathbf{v}_1$, $\mathbf{v}_2$,... over ω and in the case of second order arithmetic $\mathbf{A}^2$ we have variables $\mathbf{v}_0$, $\mathbf{v}_1$, $\mathbf{v}_2$,... over ω and $\boldsymbol{\alpha}_0$, $\boldsymbol{\alpha}_1$, $\boldsymbol{\alpha}_2$,... over $\mathcal{N}$.

Function symbols: a function symbol $\mathbf{f}_j$ for each $j \in J$.

Relation symbols: a relation symbol $\mathbf{R}_k$ for each $k \in K$.

Constant symbols: a constant symbol $\mathbf{c}_l$ for each $l \in L$.

Logical symbols: the usual symbols of logic $\neg$, &, $\exists$, to be read "not", "and", "there exists".

Identity: the symbol

$$=$$

for equality.

Punctuation marks: the *parentheses*

$$(,)$$

and the *comma*

$$, .$$

Before we go on to define the grammar of $\mathfrak{L}^{\mathfrak{A}}$, we can get an idea of its expressive power by glancing at its alphabet. We can refer to the functions, relations and constants of $\mathfrak{A}$, we can assert that two objects in some domain A_i are equal, we can say "*and*" and "*not*" or "*it is not the case.*" More significantly, using the variables we can say "*there exists an object in* A_i *such that* ⋯" but we cannot say "*there exists a subset of* A_i *such that* ...", because we have no variables over subsets of the basic domains. This is why $\mathfrak{L}^{\mathfrak{A}}$ is called a *first order* language.

Here are the precise rules of the *grammar* of $\mathfrak{L}^{\mathfrak{A}}$.

Terms. For each domain A_i there are terms of *type* A_i which will *name* objects of A_i. We define all these simultaneously by the following induction.

(1) For each A_i, every variable $\mathbf{v}_j^i$ over A_i is a term of type A_i.

(2) If the distinguished constant c_l belongs to A_i, then the constant symbol $\mathbf{c}_l$ is a term of type A_i.

(3) If $f_j: A_{i_1} \times \cdots \times A_{i_m} \to A_n$ is one of the functions in $\mathfrak{A}$ with corresponding function symbol $\mathbf{f}_j$ and if $\mathbf{t}_1, \ldots, \mathbf{t}_m$ are terms of types $A_{i_1}, \ldots, A_{i_m}$ respectively, then the finite sequence

$$\mathbf{f}_j(\mathbf{t}_1, \ldots, \mathbf{t}_m)$$

is a term of type A_n. By $\mathbf{f}_j(\mathbf{t}_1,\ldots,\mathbf{t}_m)$ we mean the finite sequence obtained by stringing along $\mathbf{f}_j$, then (, then the sequence $\mathbf{t}_1$, then the comma , , then the sequence $\mathbf{t}_2$, etc.

These clauses determine by induction on the *length* of a given finite sequence of symbols $\mathbf{t}$ whether or not $\mathbf{t}$ is a term of type A_i. It is easy to prove by induction on the length of a given term $\mathbf{t}$ that $\mathbf{t}$ is a variable or a constant symbol or else $\mathbf{t}$ is $\mathbf{f}_j(\mathbf{t}_1,\ldots,\mathbf{t}_m)$ with uniquely determined $\mathbf{f}_j$, $\mathbf{t}_1,\ldots,\mathbf{t}_m$. Thus we can define a function F on terms by specifying outright the values $F(\mathbf{t})$ for *prime* $\mathbf{t}$ (those given by clauses (1) and (2) above) and then giving instructions for computing F at $\mathbf{f}_j(\mathbf{t}_1,\ldots,\mathbf{t}_m)$ in terms of the value of F at $\mathbf{t}_1,\ldots,\mathbf{t}_m$. This process of *definition by induction on the length of terms* is very useful.

In the case of $\mathsf{V}=(V,\in)$ where we have no functions and no distinguished constants, the only terms are the variables.

Formulas. These too are finite sequences of symbols defined inductively by the following clauses.

(1) If $R_k\subseteq A_{i_1}\times\cdots\times A_{i_m}$ is a relation of $\mathfrak{A}$ with corresponding relation symbol $\mathbf{R}_k$ and if $\mathbf{t}_1,\ldots,\mathbf{t}_m$ are terms of type $A_{i_1},\ldots,A_{i_m}$ respectively, then the finite sequence

$$\mathbf{R}_k(\mathbf{t}_1,\ldots,\mathbf{t}_m)$$

is a formula.

(2) If $\mathbf{t}$ and $\mathbf{s}$ are terms of the same type, then

$$\mathbf{t}=\mathbf{s}$$

is a formula.

(3) If φ is a formula, then $\neg(\varphi)$ is also a formula.

(4) If φ and ψ are formulas, then $(\varphi)\,\&\,(\psi)$ is also a formula.

(5) If φ is a formula, then for each variable $\mathbf{v}^i_j$, $(\exists\mathbf{v}^i_j)(\varphi)$ is also a formula.

Again, one can show by induction on the length of a given formula χ, that χ is of *exactly one* of the forms

$$\mathbf{R}_k(\mathbf{t}_1,\ldots,\mathbf{t}_m),\ \mathbf{t}=\mathbf{s},\ \neg(\varphi),\ (\varphi)\,\&\,(\psi),\ (\exists\mathbf{v}^i_j)(\varphi)$$

with uniquely determined $\mathbf{R}_k$, $\mathbf{t}_1,\ldots,\mathbf{t}_m$, $\mathbf{t}$, $\mathbf{s}$, φ, ψ, $\mathbf{v}^i_j$, see 8A.2. This justifies *definition by induction on the length of formulas*, where we define $F(\chi)$ by specifying the values outright for χ *prime* (given by clauses (1) or (2) above) and then defining F at $\neg(\varphi)$, $(\varphi)\,\&\,(\psi)$, $(\exists v^i_j)(\varphi)$ in terms of the values of F at φ and ψ.

The natural interpretation of the terms and formulas of $\mathfrak{L}^{\mathfrak{A}}$ in $\mathfrak{A}$ is completely obvious from the way in which we read the formal symbols. Notice that some formulas express propositions which are outright true or false, e.g. $(\exists \mathbf{v}_0^i)(\mathbf{v}_0^i = \mathbf{v}_0^i)$ asserts the existence of some object in A_i which is equal to itself and is obviously true, since we are assuming $A_i \neq \emptyset$. Other formulas impose conditions on unspecified members in the domains of $\mathfrak{A}$ named by the variables, e.g. $\mathbf{R}_k(\mathbf{v}_0^i)$ is true just in case the relation R_k holds at the object of A_i we are (for the moment) calling $\mathbf{v}_0^i$. Similarly, some terms name specific objects, e.g. $\mathbf{f}_j(\mathbf{c}_l, \mathbf{c}_{l'})$ names $f_j(c_l, c_{l'})$, while the value of others (like $\mathbf{f}_j(\mathbf{v}_0^i, \mathbf{c}_l)$) depends on our interpretation of the variables.

An *assignment* of values to the variables is a function

$$\bar{x} = \{x_j^i : i \in I, j = 0, 1, 2, \ldots\}$$

which assigns to each domain A_i and each variable $\mathbf{v}_j^i$ over A_i a member x_j^i of A_i. With each assignment $\bar{x}$ and each term $\mathbf{t}$ we associate *the value* $\mathbf{t}^{\mathfrak{A}}[\bar{x}] = \mathbf{t}[\bar{x}]$ *of* $\mathbf{t}$ *at* $\bar{x}$ by the obvious induction

(1) $$\mathbf{v}_j^i[\bar{x}] = x_j^i,$$

(2) $$\mathbf{c}_l[\bar{x}] = c_l,$$

(3) $$\mathbf{f}_j(\mathbf{t}_1, \ldots, \mathbf{t}_m)[\bar{x}] = f_j(\mathbf{t}_1[\bar{x}], \ldots, \mathbf{t}_m[\bar{x}]).$$

Clearly $\mathbf{t}[\bar{x}] \in A_i$, if $\mathbf{t}$ is a term of type A_i.

If $\bar{x} = \{x_j^i : i \in I, j \in \omega\}$ is an assignment and χ is a formula of $\mathfrak{L}^{\mathfrak{A}}$, let us say

$$\bar{x} \text{ satisfies } \chi \text{ in } \mathfrak{A}$$

and write

$$\mathfrak{A}, \bar{x} \models \chi$$

if χ is true when we interpret each $\mathbf{v}_j^i$ by x_j^i. This relation of *satisfaction* between structures, assignments and formulas has the following obvious properties:

(1) $\mathfrak{A}, \bar{x} \models \mathbf{R}_k(\mathbf{t}_1, \ldots, \mathbf{t}_m) \Leftrightarrow R_k(\mathbf{t}_1[\bar{x}], \ldots, \mathbf{t}_m[\bar{x}])$,

(2) $\mathfrak{A}, \bar{x} \models \mathbf{t} = \mathbf{s} \Leftrightarrow \mathbf{t}[\bar{x}] = \mathbf{s}[\bar{x}]$,

(3) $\mathfrak{A}, \bar{x} \models \neg(\varphi) \Leftrightarrow$ it is not the case that $\mathfrak{A}, \bar{x} \models \varphi$,

(4) $\mathfrak{A}, \bar{x} \models (\varphi) \,\&\, (\psi) \Leftrightarrow \mathfrak{A}, \bar{x} \models \varphi$ and $\mathfrak{A}, \bar{x} \models \psi$,

(5) $\mathfrak{A}, \bar{x} \models (\exists \mathbf{v}_j^i)(\varphi) \Leftrightarrow$ there exists some assignment $\bar{y} = \{y_n^k : k \in I, n \in \omega\}$ such that $\mathfrak{A}, \bar{y} \models \varphi$ and if $k \neq i$ or $n \neq j$, then $y_n^k = x_n^k$.

These equivalences codify the natural translation of the formal language $\mathcal{L}^{\mathfrak{A}}$ into English—only (5) needs a bit of reflection to verify this. Alternatively, we may construe (1)–(5) as a precise definition of satisfaction by induction on the length of formulas which does not appeal directly to the meaning of the formulas of $\mathcal{L}^{\mathfrak{A}}$.

The truth or falsity of $(\exists \mathbf{v}_1^0)\mathbf{R}_k(\mathbf{v}_1^0))$ does not depend on our interpretation of the variable $\mathbf{v}_1^0$—it is true or false accordingly as there exists or not some x_1^0 in A_0 such that $R_k(x_1^0)$ holds. On the other hand, whether $\bar{x}$ satisfies $\mathbf{R}_k(\mathbf{v}_1^0)$ depends on what x_1^0 is. To make this important distinction precise, we associate with each term $\mathbf{t}$ and each formula χ the set $FV(\mathbf{t})$ or $FV(\chi)$ of *free variables* of $\mathbf{t}$ or χ by the following inductions:

(1) $$FV(\mathbf{v}_j^i) = \{\mathbf{v}_j^i\},$$

(2) $$FV(\mathbf{c}_l) = \emptyset,$$

(3) $$FV(\mathbf{f}_j(\mathbf{t}_1, \ldots, \mathbf{t}_m)) = FV(\mathbf{t}_1) \cup \cdots \cup FV(\mathbf{t}_m),$$

(1) $$FV(\mathbf{R}_k(\mathbf{t}_1, \ldots, \mathbf{t}_m)) = FV(\mathbf{t}_1) \cup \cdots \cup FV(\mathbf{t}_m),$$

(2) $$FV(\mathbf{t} = \mathbf{s}) = FV(\mathbf{t}) \cup FV(\mathbf{s}),$$

(3) $$FV(\neg(\varphi)) = FV(\varphi),$$

(4) $$FV((\varphi) \mathbin{\&} (\psi)) = FV(\varphi) \cup FV(\psi),$$

(5) $$FV((\exists \mathbf{v}_j^i)(\varphi)) = FV(\varphi) - \{\mathbf{v}_j^i\}.$$

Clearly $FV(\mathbf{t})$, $FV(\chi)$ is a *finite* set of variables for each $\mathbf{t}$ and each χ.

It is clear that *the free variables of a term* $\mathbf{t}$ (the members of $FV(\mathbf{t})$) are exactly all the variables which actually *occur* in $\mathbf{t}$. On the other hand, a variable $\mathbf{v}_j^i$ is free in some formula χ only if $\mathbf{v}_j^i$ occurs some place in χ not *within the scope* of a quantifier $\exists \mathbf{v}_j^i$, because of the last clause in the definition of $FV(\chi)$.

It is also clear that to compute the value $\mathbf{t}[\bar{x}]$ we need only know x_j^i for i, j such that $\mathbf{v}_j^i$ is free in t. Similarly, whether $\mathfrak{A}, \bar{x} \models \chi$ holds or not depends only on the values x_j^i for $\mathbf{v}_j^i$ free in χ. This is made precise in the following proposition which can be proved by a simple induction on the length of terms and formulas (see 8A.2): if $\bar{x} = \{x_j^i : i \in I, j \in \omega\}$, $\bar{y} = \{y_j^i : i \in I, j \in \omega\}$ are assignments in $\mathfrak{A}$ and $x_j^i = y_j^i$ whenever $\mathbf{v}_j^i$ is free in a formula χ, then

$$\mathfrak{A}, \bar{x} \models \chi \Leftrightarrow \mathfrak{A}, \bar{y} \models \chi.$$

Formulas with no free variables are called *sentences*. These are obviously satisfied by some assignment if and only if they are satisfied by all assignments, so if φ is a sentence, we put

$$\mathfrak{A} \models \varphi \Leftrightarrow \text{for some } \bar{x}, \mathfrak{A}, \bar{x} \models \varphi$$
$$\Leftrightarrow \text{for all } \bar{x}, \mathfrak{A}, \bar{x} \models \varphi.$$

We say that φ is *true* in $\mathfrak{A}$, if $\mathfrak{A} \models \varphi$, otherwise φ is *false* in $\mathfrak{A}$.

We did not include in the vocabulary of $\mathfrak{L}^{\mathfrak{A}}$ a special symbol $\forall$ to express "for all." However, in the intended interpretation of the symbols, if φ is a formula then clearly

$$(\forall \mathbf{v}_j^i)(\varphi) \Leftrightarrow \neg((\exists \mathbf{v}_j^i)(\neg(\varphi))).$$

We will consider the expression $(\forall \mathbf{v}_j^i)(\varphi)$ (which is not a formula, strictly speaking) as an *abbreviation* of the formula $\neg((\exists \mathbf{v}_j^i)(\neg(\varphi)))$. Here are three more very useful abbreviations:

$(\varphi) \vee (\psi) \Leftrightarrow \neg((\neg(\varphi)) \& (\neg(\psi)))$ (read: φ *or* ψ),

$(\varphi) \rightarrow (\psi) \Leftrightarrow (\neg(\varphi)) \vee (\psi)$ (read: φ *implies* ψ),

$(\varphi) \leftrightarrow (\psi) \Leftrightarrow ((\varphi) \rightarrow (\psi)) \& ((\psi) \rightarrow (\varphi))$ (read: φ *if and only if* ψ).

In addition to using these simple abbreviations we will also be very sloppy in *spelling* correctly the formulas of $\mathfrak{L}^{\mathfrak{A}}$. We will freely omit parentheses when it is obvious where they should be inserted, we will use brackets [,] or {,} instead of parentheses when they improve readability and we will use variables $\mathbf{k}$, $\mathbf{n}$, $\mathbf{m}$, $\mathbf{x}$, $\mathbf{y}$, $\mathbf{z}$, $\boldsymbol{\alpha}$, $\boldsymbol{\beta}$, $\boldsymbol{\gamma}$,... etc. instead of the doubly subscripted formula $\mathbf{v}_j^i$ to bring the formal language closer to our informal mathematical notation. If certain functions or relations have customary notations in some familiar structures, we will use these rather than formal functions and relation symbols. For example, in the case of second order arithmetic we will write

$$\boldsymbol{\alpha}(\mathbf{n})$$

instead of

$$ap(\mathbf{v}_0^1, \mathbf{v}_0^0),$$

in the case of analysis we will write

$$\mathbf{x} + \mathbf{y} \leq \mathbf{z}$$

instead of something like

$$\leq (+(\mathbf{v}_0, \mathbf{v}_1), \mathbf{v}_2),$$

and in the case of set theory we will certainly write

$$\mathbf{x} \in \mathbf{y}$$

instead of

$$\in(\mathbf{v}_0, \mathbf{v}_1).$$

In fact, only rarely will we put down a correct formula of a formal language $\mathfrak{L}^{\mathfrak{A}}$. The usual practice will be to write fairly simple expressions in "symbolized English" which can be translated into correct formulas of $\mathfrak{L}^{\mathfrak{A}}$ (in principle at least) by any competent student. To give one more example of this, the expression

$$(\exists \mathbf{x})(\forall \mathbf{y})[\mathbf{y} \notin \mathbf{x}]$$

clearly asserts the existence of the empty set in the language of set theory. It is a misspelling of the horrendous

$$(\exists \mathbf{v}_1^0)(\neg((\exists \mathbf{v}_2^0)(\neg(\neg(\in(\mathbf{v}_2^0, \mathbf{v}_1^0)))))).$$

In the exercises of this section we will outline, the proofs of a few useful results from *model theory*, the study of structures and their languages. Particularly significant (for us) are the *Skolem–Löwenheim Theorem*, 8A.4, and the computation of the *satisfaction relation* for countable structures, 8A.6.

EXERCISES

8A.1. Prove that every formula χ in the language $\mathfrak{L}^{\mathfrak{A}}$ of a structure $\mathfrak{A} = (\{A_i\}_{i\in I}, \{f_j\}_{j\in J}, \{R_k\}_{k\in K}, \{c_l\}_{l\in L})$ is of exactly one of the forms

$$\mathbf{R}_k(\mathbf{t}_1,\dots,\mathbf{t}_m), \quad \mathbf{t}=\mathbf{s}, \quad \neg(\varphi), \quad (\varphi)\,\&\,(\psi), \quad (\exists \mathbf{v}_j^i)(\varphi)$$

with uniquely determined $\mathbf{R}_k$, $\mathbf{t}_1,\dots,\mathbf{t}_m$, $\mathbf{t}$, $\mathbf{s}$, φ, ψ, $\mathbf{v}_j^i$.

Hint. Use induction on the length of χ. Show first that the number of left parentheses (in a formula χ is equal to the number of right parentheses) in χ and that if σ is an initial segment of χ (thought of as a sequence of symbols), then the number of left parentheses in σ is greater than or equal to the number of right parentheses in σ. ⊣

8A.2. Prove that if $\bar{x} = \{x_j^i : i \in I, j = 0, 1,\dots\}$, $\bar{y} = \{y_j^i, i \in I, j = 0, 1,\dots\}$ are assignments in a structure $\mathfrak{A}$ and $x_j^i = y_j^i$ for each i, j such that $\mathbf{v}_j^i$ is free in a formula χ, then

$$\mathfrak{A}, \bar{x} \models \chi \Leftrightarrow \mathfrak{A}, \bar{y} \models \chi.$$

Hint. Show first by induction on the length of terms that if $x_j^i = y_j^i$ for each i, j such that $\mathbf{v}_j^i$ is free in $\mathbf{t}$, then $\mathbf{t}[\bar{x}] = \mathbf{t}[\bar{y}]$, then use induction on the length of χ. ⊣

After these preliminary trivial facts, we turn to some simple but important model-theoretic results. To simplify matters, let us restrict ourselves to structures with one domain, finitely many relations and no functions or distinguished elements,

$$\mathfrak{A} = (A, R_1, \ldots, R_K).$$

The *characteristic* or *similarity type* of such a structure is the sequence code which describes the number and arity of the relations $R_1, \ldots, R_K$,

$$u = ch(\mathfrak{A}) = \langle n_1, \ldots, n_K \rangle,$$

where for $i = 1, \ldots, K$, $R_i \subseteq A^{n_i}$. For example, the characteristic of the structure $\mathsf{V} = (V, \in)$ of set theory is given by

$$ch(\mathsf{V}) = \langle 2 \rangle = 2^{2+1} = 8.$$

It is clear that the language $\mathfrak{L}^{\mathfrak{A}}$ of a structure $\mathfrak{A} = (A, R_1, \ldots, R_K)$ is completely determined by the characteristic $u = ch(\mathfrak{A})$, since all we need to define $\mathfrak{L}^{\mathfrak{A}}$ is to choose an n_i-ary relation symbol $\mathbf{R}_i$ for $i = 1, \ldots, k$. Fix once and for all a *language* $\mathfrak{L}^u$ *for each characteristic* u; now the formulas of $\mathfrak{L}^u$ can be interpreted in every structure of characteristic u.

Consider first the very simple notion of *isomorphic structures*. Suppose $\mathfrak{A} = (A, R_1, \ldots, R_K)$, $\mathfrak{B} = (B, P_1, \ldots, P_K)$ have the same characteristic $u = \langle n_1, \ldots, n_K \rangle$. An *isomorphism* of $\mathfrak{A}$ with $\mathfrak{B}$ is any bijection

$$\pi : A \rightarrowtail\!\!\!\!\twoheadrightarrow B$$

such that for all $x_1, \ldots, x_{n_i} \in A$ $(i = 1, \ldots, K)$,

$$R_i(x_1, \ldots, x_{n_i}) \Leftrightarrow P(\pi(x_1), \ldots, \pi(x_{n_i})).$$

8A.3. Suppose $\pi : A \rightarrowtail\!\!\!\!\twoheadrightarrow B$ is an isomorphism of $\mathfrak{A} = (A, R_1, \ldots, R_K)$ with $\mathfrak{B} = (B, P_1, \ldots, P_K)$, both of the same characteristic u. Prove that for every formula φ of the common language $\mathfrak{L}^u$ and every assignment $\bar{x} = \{x_j : j = 0, 1, \ldots\}$ into A,

$$\mathfrak{A}, x_0, x_1, x_2, \ldots \vDash \varphi \Leftrightarrow \mathfrak{B}, \pi(x_0), \pi(x_1), \pi(x_2) \cdots \vDash \varphi.$$

Infer that $\mathfrak{A}$ and $\mathfrak{B}$ satisfy the same sentences of $\mathfrak{L}^u$.

Hint. Use induction on the length of formulas. ⊣

Structures of the same characteristic u which satisfy the same sentences of $\mathfrak{L}^u$ are called *elementarily equivalent*; thus 8A.3 asserts (partly) that isomorphic structures are elementarily equivalent.

Suppose again that $\mathfrak{A} = (A, R_1, \ldots, R_K)$ and $\mathfrak{B} = (B, P_1, \ldots, P_K)$ are structures of the same characteristic. We say that $\mathfrak{A}$ *is a substructure of* $\mathfrak{B}$ and write

$$\mathfrak{A} \subseteq \mathfrak{B}$$

if $A \subseteq B$ and every R_i is the restriction of the corresponding P_i to A, i.e.

$$x_1, \ldots, x_{n_i} \in A \Longrightarrow [R_i(x_1, \ldots, x_{n_i}) \Longleftrightarrow P_i(x_1, \ldots, x_{n_i})].$$

We say that $\mathfrak{A}$ is an *elementary substructure of* $\mathfrak{B}$ and write

$$\mathfrak{A} \preceq \mathfrak{B}$$

if for every formula φ of the common language $\mathfrak{L}^u$ and every assignment $\bar{x} = \{x_j : j = 0, 1, \ldots\}$ into A,

$$\mathfrak{A}, \bar{x} \models \varphi \Longleftrightarrow \mathfrak{B}, \bar{x} \models \varphi.$$

It is immediate that if $\mathfrak{A} \preceq \mathfrak{B}$, then $\mathfrak{A}$ and $\mathfrak{B}$ are elementarily equivalent.

8A.4. Suppose $\mathfrak{B} = (B, P_1, \ldots, P_K)$ is a structure of characteristic $u = \langle n_1, \ldots, n_K \rangle$ and assume that B is an infinite set which admits a well-ordering, let $A \subseteq B$ be any subset of B. Prove that there exists an elementary substructure of $\mathfrak{B}$

$$\mathfrak{A}^* = (A^*, R_1, \ldots, R_K) \preceq \mathfrak{B}$$

where $A \subseteq A^*$ and

$$card(A^*) = card(A) + \aleph_0;$$

in particular, if A is infinite, then $card(A^*) = card(A)$. Infer that every infinite wellorderable structure has a countable elementary substructure. (The *Skolem–Löwenheim* Theorem.)

Hint. Suppose $\mathfrak{C} = (C, R_1, \ldots, R_K) \subseteq \mathfrak{B}$ is a substructure of $\mathfrak{B}$ and φ is a formula of $\mathfrak{L}^u$. We say that φ is *absolute for* $\mathfrak{C}$ if for every assignment $\bar{x} = \{x_j : j = 0, 1, \ldots\}$ into C,

$$\mathfrak{C}, \bar{x} \models \varphi \Longleftrightarrow \mathfrak{B}, \bar{x} \models \varphi.$$

Thus $\mathfrak{C}$ is an elementary substructure of $\mathfrak{B}$ if every formula of $\mathfrak{L}^u$ is absolute for $\mathfrak{C}$.

The key notion of the proof is that of a *Skolem set* (of functions) for a formula. If $\mathcal{S}$ is a set of functions on B to B (of any number of variables) and $\mathfrak{C} = (C, R_1, \ldots, R_K)$ is a substructure of $\mathfrak{B}$, we say that $\mathfrak{C}$ *is closed under* $\mathcal{S}$ if for every n-ary $f : B^n \to B$ in $\mathcal{S}$,

$$x_1, \ldots, x_n \in C \Longrightarrow f(x_1, \ldots, x_n) \in C.$$

A set of functions $\mathcal{S}$ is a *Skolem set* for a formula φ if φ is absolute for every substructure of $\mathfrak{B}$ which is closed under $\mathcal{S}$.

Show by induction on the length of formulas that *every formula* χ *has a finite Skolem set* $\mathcal{S}(\chi)$. This is the heart of the proof. The most interesting case is when χ is $(\exists \mathbf{v}_j)(\varphi)$, where (say) the free variables of φ are among $\mathbf{v}_0, \mathbf{v}_1, \ldots, \mathbf{v}_m$ and (for simplicity) assume $j = m$. Here we can take

$$\mathcal{S}(\chi) = \mathcal{S}(\phi) \cup \{f\}$$

where $f : B^m \to B$ has the property that for all $x_0, \ldots, x_{m-1} \in B$,

$$\text{there exists some } x_m \text{ such that } \mathfrak{B}, x_0, \ldots, x_{m-1}, x_m \models \varphi$$
$$\Longrightarrow \mathfrak{B}, x_0, \ldots, x_{m-1}, f(x_0, \ldots, x_{m-1}) \models \varphi$$

where we have put down only the part of an assignment which is relevant to the satisfaction of φ. (We need a wellordering of B or the axiom of choice, to define f.)

Since there are only countably many formulas, the union

$$\mathcal{S} = \bigcup_\varphi \mathcal{S}(\varphi)$$

is a countable set and it is obviously a Skolem set for every formula of $\mathcal{L}^u$. Use a simple set-theoretic argument to construct some $A^* \subseteq B$ such that $A \subseteq A^*$, A^* is closed under $\mathcal{S}$ and $card(A^*) = card(A) + \aleph_0$. The required structure is $\mathfrak{A}^* = (A^*, R_1, \ldots, R_K)$, where each R_i is the restriction of P_i to A^*. ⊣

The Skolem–Löwenheim theorem tells us in particular that there are many interesting countable structures.

In the remaining exercises of this section we will compute the *complexity* of the satisfaction relation on countable structures, coded by irrationals.

For each characteristic $u = \langle n_1, \ldots, n_K \rangle$ and each irrational α, define the structure

$$\mathfrak{A}(u, \alpha) = (A, R_1, \ldots, R_K)$$

by

$$A = \{n \in \omega : (\alpha)_0(n) = 1\},$$

$$R_i(x_1, \ldots, x_{n_i}) \Longleftrightarrow x_1, \ldots, x_{n_i} \in A \; \& \; (\alpha)_i(\langle x_1, \ldots, x_{n_i} \rangle) = 1.$$

Of course, in order to get a structure we must have $A \neq \emptyset$, i.e. $(\exists n)[(\alpha)_0(n)=1]$.

We assign a *code* $[\chi]^u \in \omega$ to each formula of $\mathcal{L}^u$ by the following induction on the length of formulas:

$$[\mathbf{R}_i(\mathbf{v}_{j_1},\ldots,\mathbf{v}_{j_{n_i}})]^u = \langle 0, i, j_1,\ldots, j_{n_i}\rangle, \quad (1 \le i \le K)$$
$$[\mathbf{v}_{j_1} = \mathbf{v}_{j_2}]^u = \langle 0, 0, j_1, j_2\rangle,$$
$$[\neg(\varphi)]^u = \langle 1, [\varphi]^u\rangle,$$
$$[(\varphi)\,\&\,(\psi)]^u = \langle 2, [\varphi]^u, [\psi]^u\rangle,$$
$$[(\exists \mathbf{v}_j)(\varphi)]^u = \langle 3, j, [\varphi]^u\rangle.$$

If m is the code of a formula χ, i.e. $[\chi]^u = m$, we put

$$\chi = \chi_m.$$

The next exercise is routine by the methods of Chapter 3.

8A.5. Prove that the following relations on ω are semirecursive:

$Char(u) \Leftrightarrow u$ is the characteristic of some structure

$$\mathfrak{A} = (A, R_1,\ldots, R_K),$$

$$Str(u,\alpha) \Leftrightarrow Char(u)\,\&\,[\alpha \text{ codes a structure of characteristic } u]$$
$$\Leftrightarrow Char(u)\,\&\,(\exists n)[(\alpha)_0(n)=1],$$
$$Fmla(u,m) \Leftrightarrow Char(u)\,\&\,[m=[\chi]^u \text{ for some formula } \chi \text{ of } \mathcal{L}^u],$$
$$Free(u,m,j) \Leftrightarrow Fmla(u,m)\,\&\,[\mathbf{v}_j \text{ is free in } \chi_m],$$
$$Assgn(u,\alpha,m,x) \Leftrightarrow Str(u,\alpha)\,\&\,Fmla(u,m)$$
$$\&\,(\forall j)[Free(u,m,j) \to (\alpha)_0((x)_j)=1]. \quad \dashv$$

The last relation holds if the mapping $j \mapsto (x)_j$ is an assignment to the domain of the structure $\mathfrak{A}(u,\alpha)$, at least as far as the free variables of χ_m are concerned.

Put

$$Sat(u,\alpha,m,x) \Leftrightarrow Assgn(u,\alpha,m,x)\,\&\,\mathfrak{A}(u,\alpha), \bar{x} \models \chi_m,$$

where $\bar{x} = \{(x)_j : j=0,1,\ldots\}$. This is the coding of the *satisfaction relation*. We will have many occasions to use the basic computational estimate in the next exercise.

8A.6. Prove that the relation *Sat* is Δ^1_1.

Hint. Put

$$P(u, \alpha, \beta) \Leftrightarrow (\forall m)(\forall x)[Sat(u, \alpha, m, x) \Leftrightarrow \beta(\langle m, x\rangle) = 1]$$

and notice that

$$\begin{aligned} Sat(u, \alpha, m, x) &\Leftrightarrow Assgn(u, \alpha, m, x) \,\&\, (\exists\beta)[P(u, \alpha, \beta) \,\&\, \beta(\langle m, x\rangle) = 1] \\ &\Leftrightarrow Assgn(u, \alpha, m, x) \,\&\, (\forall\beta)[P(u, \alpha, \beta) \Rightarrow \beta(\langle m, x\rangle) = 1], \end{aligned}$$

so that it suffices to prove that P is arithmetical. This is easy to verify using 8A.5, the inductive definition of the codings above and the satisfaction relation: in order to have $P(u, \alpha, \beta)$, $\beta(\langle m, x\rangle)$ must give the correct value when m codes a prime formula and for more complicated formulas the correct value of $\beta(\langle m, x\rangle)$ can be computed in terms of $\beta(\langle s, y\rangle)$ for codes s of shorter formulas. ⊣

It is clear that the results in 8A.4–8A.6 can be extended easily to structures of the form

$$\mathfrak{A} = (A_1, \ldots, A_I, f_1, \ldots, f_J, R_1, \ldots, R_K, C_1, \ldots, C_L)$$

with finitely many domains, functions, relations and constants. The computations are a bit messier.

8B. Elementary definability

Let us introduce a very useful notational convention: *if the free variables of a formula are among* $\mathbf{x}_1, \ldots, \mathbf{x}_n$, *we will use a symbol like*

$$\varphi(\mathbf{x}_1, \ldots, \mathbf{x}_n)$$

to name that formula. For example, we might denote the formula

$$(\mathbf{R}(\mathbf{v}_0^i)) \,\&\, ((\exists\mathbf{v}_1^2)(\mathbf{v}_1^2 = \mathbf{v}_0^i))$$

by the symbols

$$\varphi(\mathbf{v}_0^i) \quad \text{or} \quad \psi(\mathbf{v}_0^i, \mathbf{v}_1^3)$$

or any $\chi(\mathbf{x}_1, \ldots, \mathbf{x}_n)$ where $\mathbf{x}_1, \ldots, \mathbf{x}_n$ are variables and one of them is $\mathbf{v}_0^i$, the only free variable of the formula above.

This is similar to the algebraic practice of referring to a polynomial

$f(x, y, z)$ in three indeterminates over some ring, where $f(x, y, z)$ might by $x \cdot y$ or $x \cdot z + x$ or $x + y + z$.

As in algebra, the advantage of this convention is that it allows a very compact notation for the operation of *evaluation*. If $\varphi(\mathbf{x}_1, \ldots, \mathbf{x}_n)$ is a formula (whose free variables are among $\mathbf{x}_1, \ldots, \mathbf{x}_n$) and $x_1, \ldots, x_n$ are members of the domains of $\mathfrak{A}$, with x_i in A_i whenever $\mathbf{x}_i$ is of type A_i, put

$$\begin{aligned}\mathfrak{A} \models \varphi(x_1, \ldots, x_n) &\Leftrightarrow \text{there is some assignment } \bar{x} \text{ which} \\ &\quad\ \text{assigns } x_i \text{ to } \mathbf{x}_i,\ i = 1, \ldots, n \text{ and} \\ &\quad\ \mathfrak{A}, \bar{x} \models \varphi(\mathbf{x}_1, \ldots, \mathbf{x}_n) \\ &\Leftrightarrow \text{for every assignment } \bar{x} \text{ which assigns } x_i \\ &\quad\ \text{to } \mathbf{x}_i,\ \mathfrak{A}, \bar{x} \models \varphi(\mathbf{x}_1, \ldots, \mathbf{x}_n).\end{aligned}$$

For example, if R_i is one of the basic n-ary relations of a structure $\mathfrak{A}$ denoted by the formal symbol $\mathbf{R}_i$, then for $x_1, \ldots, x_n$ in the appropriate domains of $\mathfrak{A}$,

$$R_i(x_1, \ldots, x_n) \Leftrightarrow \mathfrak{A} \models \mathbf{R}_i(x_1, \ldots, x_n).$$

Here we are applying the convention above to the formula $\mathbf{R}_i(\mathbf{x}_1, \ldots, \mathbf{x}_n)$, where $\mathbf{x}_1, \ldots, \mathbf{x}_n$ are any n distinct variables of the proper types.

Similarly, if $\varphi(\mathbf{x}_1, \ldots, \mathbf{x}_n)$, $\psi(\mathbf{x}_1, \ldots, \mathbf{x}_n)$ are formulas of some language and we abbreviate their conjunction by $\chi(\mathbf{x}_1, \ldots, \mathbf{x}_n)$ in some context,

$$\chi(\mathbf{x}_1, \ldots, \mathbf{x}_n) \Leftrightarrow \varphi(\mathbf{x}_1, \ldots, \mathbf{x}_n) \mathbin{\&} \psi(\mathbf{x}_1, \ldots, \mathbf{x}_n),$$

then for $x_1, \ldots, x_n$ in the appropriate domains of a structure $\mathfrak{A}$,

$$\begin{aligned}\mathfrak{A} \models \chi(x_1, \ldots, x_n) &\Leftrightarrow \mathfrak{A} \models \varphi(x_1, \ldots, x_n) \mathbin{\&} \psi(x_1, \ldots, x_n) \\ &\Leftrightarrow \mathfrak{A} \models \varphi(x_1, \ldots, x_n) \text{ and } \mathfrak{A} \models \psi(x_1, \ldots, x_n).\end{aligned}$$

According to this convention, from now on we reserve the symbols φ, χ, ψ etc. for *sentences*, i.e. formulas without free variables.

We now come to the basic notion of *first order* or *elementary* definability in a language.

Let

$$\mathfrak{A} = (\{A_i\}_{i \in I}, \{f_j\}_{j \in J}, \{R_k\}_{k \in K}, \{c_l\}_{l \in L})$$

be a structure with associated language $\mathfrak{L}^{\mathfrak{A}}$. A relation

$$R \subseteq A_{i_1} \times \cdots \times A_{i_n}$$

is *first order definable* or *elementary* in $\mathfrak{A}$ (or $\mathfrak{L}^{\mathfrak{A}}$) if there is a formula

$\varphi(\mathbf{x}_1,\ldots,\mathbf{x}_n)$ of $\mathcal{L}^{\mathfrak{A}}$ where each $\mathbf{x}_j$ is a variable of type A_{i_j} and

$$(1)\qquad R(x_1,\ldots,x_n) \Leftrightarrow \mathfrak{A} \models \varphi(x_1,\ldots,x_n);$$

a subset

$$B \subseteq A_i$$

of one of the domains of $\mathfrak{A}$ is elementary, if there is a formula $\varphi(\mathbf{x})$ of $\mathcal{L}^{\mathfrak{A}}$, with just one free variable such that

$$(2)\qquad x \in B \Leftrightarrow \mathfrak{A} \models \varphi(x);$$

a function

$$F\colon A_{i_1} \times \cdots \times A_{i_n} \to A_j$$

is elementary if for some $\varphi(\mathbf{x}_1,\ldots,\mathbf{x}_n,\mathbf{y})$,

$$(3)\qquad F(x_1,\ldots,x_n) = y \Leftrightarrow \mathfrak{A} \models \varphi(x_1,\ldots,x_n,y);$$

finally a member a of some A_i is elementary if for some $\varphi(\mathbf{x})$,

$$(4)\qquad x = a \Leftrightarrow \mathfrak{A} \models \varphi(x).$$

We will say that a formula which satisfies (1), (2), (3) or (4) *defines* R, B, F or a respectively in the structure $\mathfrak{A}$.

Let us collect in a simple result the basic properties of these notions.

8B.1. THEOREM. (i) *The collection of elementary relations in a structure* $\mathfrak{A}$ *contains all the relations* R_k *of* $\mathfrak{A}$ *and* $=$ *and is closed under* $\neg$, &, $\vee$, $\exists^{A_i}$ *and* $\forall^{A_i}$ *for each* $i \in I$ *as well as substitution of elementary functions.*

(ii) *A set* $B \subseteq A_i$ *is elementary in a structure* $\mathfrak{A}$, *if and only if its representing relation*

$$R_B(x) \Leftrightarrow x \in B$$

is elementary in $\mathfrak{A}$.

(iii) *The collection of elementary functions in a structure* $\mathfrak{A}$ *contains all the functions* f_j *of* $\mathfrak{A}$, *the projection functions*

$$(x_1,\ldots,x_n) \mapsto x_j$$

and the constants

$$(x_1,\ldots,x_n) \mapsto c_l,$$

where c_l *is a distinguished element of* $\mathfrak{A}$; *this collection is closed under addition and permutation of variables, definition by cases (determined by elementary conditions) and composition.*

(iv) *An element* $a \in A_i$ *is elementary in a structure* $\mathfrak{A}$ *if and only if every constant function*

$$(x_i, \ldots, x_n) \mapsto a$$

is elementary in $\mathfrak{A}$.

PROOF is trivial. ⊣

It is clear that a relation, function or element is elementary in $\mathfrak{A}$ if we can put down a definition for it in the small part of English which is formalized in $\mathfrak{L}^{\mathfrak{A}}$.

There are four main results which exhibit the connection between elementary definability and the analytical pointclasses. We will state them here and outline their proofs in the exercises.

Recall that a pointset of type 0 is any $R \subseteq \omega^k$ and a pointset of type 1 is any $R \subseteq X_1 \times \cdots \times X_k$ with each $X_i = \omega$ or $\mathcal{N}$ and some $X_j = \mathcal{N}$.

8B.2. THEOREM. *A pointset of type* 0 *is arithmetical if and only if* R *is elementary in the structure of arithmetic* $\mathsf{A}^1 = (\omega, +, \cdot, 0, 1)$. ⊣

8B.3. THEOREM. *A pointset* R *of type* 0 *or* 1 *is analytical if and only if* R *is elementary in the structure of second order number theory* $\mathsf{A}^2 = (\omega, \mathcal{N}, +, \cdot, ap, 0, 1)$. ⊣

8B.4. THEOREM. *An* n-ary *relation on the real numbers* $R \subseteq \mathbb{R}^n$ *is analytical if and only if* R *is elementary in the structure of analysis* $\mathsf{R} = (\mathbb{R}, +, \cdot, 0, 1, \leq, Z)$. ⊣

Sometimes we can obtain finer results if we restrict attention to definability by formulas with special properties.

8B.5. THEOREM. *A pointset* R *of type* 0 *or* 1 *is arithmetical if and only if* R *is definable in the structure of second order arithmetic* $\mathsf{A}^2 = (\omega, \mathcal{N}, +, \cdot, ap, 0, 1)$ *by some formula* $\varphi(\mathbf{x}_1, \ldots, \mathbf{x}_n)$ *which has no quantifiers over* $\mathcal{N}$. ⊣

If we use variables $\mathbf{v}_0, \mathbf{v}_1, \mathbf{v}_2, \ldots$ over ω and $\boldsymbol{\alpha}_0, \boldsymbol{\alpha}_1, \boldsymbol{\alpha}_2, \ldots$ over $\mathcal{N}$ in the language of A^2, this means that no part of $\varphi(\mathbf{x}_1, \ldots, \mathbf{x}_n)$ in 8A.4 looks like $(\exists \boldsymbol{\alpha}_j)$.

In addition to the proofs of these results, we will also formulate in the exercises several more characterizations of various pointclasses in terms of elementary definability.

EXERCISES

8B.6. Prove that every relation on ω which is elementary in the structure A^1 of arithmetic is arithmetical.

Hint. Show by induction on the length of a formula $\varphi(\mathbf{x}_1, \ldots, \mathbf{x}_n)$ that the corresponding relation

$$R(x_1, \ldots, x_n) \Leftrightarrow \mathsf{A}^1 \models \varphi(x_1, \ldots, x_n)$$

is arithmetical. It will help to notice first that if the free variables of a term $\mathbf{t}$ are among $\mathbf{x}_1, \ldots, \mathbf{x}_m$ and $f : \omega^m \to \omega$ is defined by

$$f(x_1, \ldots, x_m) = \mathbf{t}[x_1, \ldots, x_m],$$

then f is recursive. ⊣

The same method also establishes with no problem the easy directions of 8B.2–8B.4.

8B.7. Prove that if a pointset R is elementary in the structure A^2 of second order arithmetic, then R is analytical. ⊣

8B.8. Prove that if a relation $R \subseteq \mathcal{R}^n$ on the reals is elementary in the structure R of analysis, then R is analytical.

Hint. By 3D.8 and 3D.4, all terms of the language of R define recursive functions. By 3C.6, $x = y$ is Π^0_1 and by 3C.11, $x \leq y$ is Π^0_2. Show that Z is also Π^0_1 and then use induction on the length of formulas. ⊣

8B.9. Prove that if a pointset R of type 0 or 1 is definable in A^2 by a formula with no quantifiers over $\mathcal{N}$, then R is arithmetical. ⊣

Proof of the converse implications in 8B.6–8B.9 is a bit more interesting. We break it down in several steps.

8B.10. Prove that the function

$$\beta(s, t, i) = rm(s, 1 + (i+1)t)$$

is elementary in A^1 and that for every finite sequence $w_0, \dots, w_{n-1}$ of integers, there exists some s and some t such that

$$\beta(s, t, 0) = w_0, \quad \beta(s, t, 1) = w_1, \dots, \quad \beta(s, t, n-1) = w_{n-1}.$$

(*Gödel's lemma on the* β *function.*)

Hint. That β is elementary on A^1 is easy. To prove that it has the required property, notice that if $d_0, d_1, \dots, d_{n-1}$ is an n-tuple of relatively prime numbers, then the function

$$s \mapsto (rm(s, d_0), rm(s, d_1), \dots, rm(s, d_{n-1}))$$

is one-to-one from $\{s : s < d_0 \cdot d_1 \cdot \cdots \cdot d_{n-1}\}$ into $\{(w_0, \dots, w_{n-1}) : w_0 < d_0 \;\&\; w_1 < d_1 \;\&\; \cdots \;\&\; w_{n-1} < d_{n-1}\}$; this is an easy divisibility argument. Since these two sets have the same finite cardinality, it follows that *for relatively prime* $d_0, d_1, \dots, d_{n-1}$ *and every* $w_0 < d_0$, $w_1 < d_1, \dots, w_{n-1} < d_{n-1}$, *there exists some* $s < d_0 \cdot d_1 \cdot \cdots \cdot d_{n-1}$ *with*

$$w_0 = rm(s, d_0), \quad w_1 = rm(s, d_1), \dots, \quad w_{n-1} = rm(s, d_{n-1})$$

(*the Chinese remainder theorem*). Now given $w_0, w_1, \dots, w_{n-1}$, let

$$m = max(w_0, w_1, \dots, w_{n-1}, n)$$

and take

$$t = m!.$$

Another easy divisibility argument shows that the numbers

$$d_0 = 1 + t, \quad d_1 = 1 + 2t, \dots, \quad d_{n-1} = 1 + nt$$

are relatively prime, hence by the Chinese remainder theorem there is some s such that

$$\beta(s, t, i) = rm(s, 1 + (i+1)t) = w_i \quad (i < n). \qquad \dashv$$

8B.10 allows us to code tuples of arbitrary length by pairs using operations which are elementary in A^1.

8B.11. Prove Theorem 8B.2, that a relation on ω is arithmetical if and only if it is elementary in A^1.

Hint. By 8B.6 and 8B.1, it is enough to prove that every recursive relation on ω is elementary on A^1. For this again it is enough to show that every recursive $f : \omega^k \to \omega$ is elementary on A^1, and for this we need only show that the collection of functions which are elementary in A^1 contains the functions S, C^k_w, P^k_i of 3A and is closed under minimalization, composition and primitive recursion. The only non-trivial case is that of primitive recursion,

$$\begin{cases} f(0,x) = g(x), \\ f(n+1,x) = h(f(n,x),n,x) \end{cases}$$

which we analyze as follows.

$$\begin{aligned} f(n,x) = m \Leftrightarrow\ & \text{there exists a sequence } w_0, w_1, \ldots, w_n \\ & \text{such that } g(x) = w_0 \text{ and for every } i < n, \\ & h(w_i, i, x) = w_{i+1} \text{ and } w_n = m \\ \Leftrightarrow\ & (\exists s)(\exists t)\{\beta(s,t,0) = g(x) \\ & \qquad \& \ (\forall i < n)[\beta(s,t,i+1) = h(\beta(s,t,i),i,x)] \\ & \qquad \& \ \beta(s,t,n) = m\}. \end{aligned}$$ ⊣

8B.12. Prove Theorem 8B.3, that a pointset of type 0 or 1 is analytical if and only if it is elementary in A^2.

Hint. Use the preceding exercise, the closure of the pointsets which are elementary in A^1 under $\exists^{\mathcal{N}}$, $\forall^{\mathcal{N}}$ and the representation of Σ^0_1 pointsets given in 4A.1. ⊣

8B.13. Prove Theorem 8B.4, that a pointset $R \subseteq \mathcal{R}^n$ is analytical if and only if it is elementary in R.

Hint. We can think of ω as imbedded in the integers of $\mathcal{R}$,

$$\omega = \{0, 1, 2, \ldots\} \subseteq Z = \{\cdots -2, -1, 0, 1, 2, \ldots\};$$

as a subset of $\mathcal{R}$, ω is clearly elementary in R. Every formula $\chi(\mathbf{x}_1, \ldots, \mathbf{x}_n)$ in the language of A^1 has a natural translation $\chi(\mathbf{x}_1, \ldots, \mathbf{x}_n)^*$ in the language of R, where $\chi(\mathbf{x}_1, \ldots, \mathbf{x}_n)^*$ is $\chi(\mathbf{x}_1, \ldots, \mathbf{x}_n)$ if $\chi(\mathbf{x}_1, \ldots, \mathbf{x}_n)$ is prime, $\neg(\varphi(\mathbf{x}_1, \ldots, \mathbf{x}_n))^*$ is $\neg(\varphi(\mathbf{x}_1, \ldots, \mathbf{x}_n)^*)$, $(\varphi(\mathbf{x}_1, \ldots, \mathbf{x}_n))$ & $(\psi(\mathbf{x}_1, \ldots, \mathbf{x}_n))^*$ is $(\varphi(\mathbf{x}_1, \ldots, \mathbf{x}_n)^*)$ & $(\psi(\mathbf{x}_1, \ldots, \mathbf{x}_n)^*)$ and in the significant case, the translation of $(\exists \mathbf{y})(\varphi(\mathbf{x}_1, \ldots, \mathbf{x}_n, \mathbf{y}))$ is $(\exists \mathbf{y})[Z(\mathbf{y})\ \&\ 0 \leq \mathbf{y}\ \&\ \varphi(\mathbf{x}_1, \ldots, \mathbf{x}_n, \mathbf{y})^*]$. It is clear that for values of the variables in ω, the translation of $\chi(\mathbf{x}_1, \ldots, \mathbf{x}_n)$ has the same truth value as $\chi(\mathbf{x}_1, \ldots, \mathbf{x}_n)$.

Use 8B.11 to prove that every recursive relation on ω is elementary in R and then use 3C.5 to show easily that every arithmetical $R \subseteq \mathcal{R}^n$ is elementary in R.

To prove that analytical pointsets are elementary in R we must reduce quantification over $\mathcal{N}$ to quantification over $\mathcal{R}$. Let $f:\mathcal{X} \rightarrowtail \mathcal{N}$ be the injection defined in the proof of 1G.2 with $\mathcal{X} = \mathcal{R}$ and show that the relation

$$S(x, n, k) \Leftrightarrow f(x)(n) = k$$

is elementary in **R**. Now use the representation of analytical pointsets given in 4A.1; for example, if

$$P(x) \Leftrightarrow (\exists\alpha)(\forall t)Q(x, \bar{\alpha}(t)),$$

then

$$P(x) \Leftrightarrow (\exists y)(\forall t)(\exists u)[Seq(u) \mathbin{\&} lh(u) = t \mathbin{\&} (\forall i < t)S(y, i, (u)_i) \mathbin{\&} Q(x, u)]. \qquad \dashv$$

8B.14. Prove Theorem 8B.5, that a pointset of type 0 or 1 is arithmetical if and only if it is definable in $\mathbf{A}^2$ by a formula which has no quantifiers over $\mathcal{N}$. $\dashv$

A formula $\chi(\mathbf{x}_1,\ldots,\mathbf{x}_n)$ in the language of $\mathbf{A}^2$ is Σ_1^1 if it is of the form

$$(\exists\boldsymbol{\alpha})\varphi(\mathbf{x}_1,\ldots,\mathbf{x}_n,\boldsymbol{\alpha})$$

where $\varphi(\mathbf{x}_1,\ldots,\mathbf{x}_n,\boldsymbol{\alpha})$ has no quantifiers over $\mathcal{N}$. Proceeding inductively, $\chi(\mathbf{x}_1,\ldots,\mathbf{x}_n)$ is Σ_{n+1}^1 if it is of the form $(\exists\boldsymbol{\alpha})\neg\varphi(\mathbf{x}_1,\ldots,\mathbf{x}_n,\boldsymbol{\alpha})$ with $\varphi(\mathbf{x}_1,\ldots,\mathbf{x}_n,\boldsymbol{\alpha})$ some Σ_n^1 formula and $\chi(\mathbf{x}_1,\ldots,\mathbf{x}_n)$ is Π_n^1 if $\neg\chi(\mathbf{x}_1,\ldots,\mathbf{x}_n)$ is Σ_n^1. The next result is an immediate corollary of 8B.14.

8B.15. Prove that a pointset of type 0 or 1 is Σ_n^1 $(n \geq 1)$ if and only if it is definable by a Σ_n^1 formula in the language of $\mathbf{A}^2$; similarly for Π_n^1. $\dashv$

There is an obvious way to enrich the languages of the structures we have been considering by adding names for all the objects in their domains. That is equivalent to *expanding* the structures $\mathbf{A}^2$, **R**, **V** by adding all their members as distinguished elements,

$$\mathbf{A}^2 = (\omega, \mathcal{N}, +, \cdot, 0, 1, ap, \{\alpha\}_{\alpha\in\mathcal{N}}),$$

$$\mathbf{R} = (\mathcal{R}, +, \cdot, \leq, Z, \{x\}_{x\in\mathcal{R}}),$$

$$\mathbf{V} = (V, \in, \{x\}_{x\in V}).$$

Notice that the distinguished elements of **V** do not form a set, but this will cause no problem. Notice also that we do not need to add names for the members of ω in $\mathbf{A}^2$ or $\mathbf{A}^1$ since they are all named by the terms 0, 1, 1+1, 1+1+1,... .

In general, for each structure

$$\mathfrak{A} = (\{A_i\}_{i \in I}, \{f_j\}_{j \in J}, \{R_k\}_{k \in K}, \{c_l\}_{l \in L}),$$

let

$$\boldsymbol{\mathfrak{A}} = (\{A_i\}_{i \in I}, \{f_j\}_{j \in J}, \{R_k\}_{k \in K}, \{x : x \in \bigcup_{i \in I} A_i\}).$$

8B.16. Let $\mathfrak{A}$, $\boldsymbol{\mathfrak{A}}$ be structures as above, let $R \subseteq A_{i_1} \times \cdots \times A_{i_m}$ be a relation. Prove that R is elementary in $\boldsymbol{\mathfrak{A}}$ if and only if there exists some $P \subseteq A_{i_1} \times \cdots \times A_{i_m} \times A_{j_1} \times \cdots \times A_{j_n}$ and elements $a_1, \dots, a_n$ of $A_{j_1}, \dots, A_{j_n}$ respectively, such that

$$R(x_1, \dots, x_m) \Leftrightarrow P(x_1, \dots, x_m, a_1, \dots, a_n)$$

and P is elementary in $\mathfrak{A}$.

Hint. Show that every formula of $\mathfrak{L}^{\boldsymbol{\mathfrak{A}}}$ can be obtained by substituting constants for some of the free variables in a formula of $\mathfrak{L}^{\mathfrak{A}}$. ⊣

8B.17. Prove that a pointset of type 1 is projective if and only if it is elementary in $\mathbf{A}^2$ and that a relation $R \subseteq \mathcal{R}^n$ on the reals is projective if and only if it is elementary in $\mathbf{R}$. ⊣

There is an obvious combination of the methods in the last three exercises which gives characterizations of the $\mathbf{\Sigma}^1_n$ and $\mathbf{\Pi}^1_n$ pointsets in $\mathcal{N}$ and $\mathcal{R}$ in terms of the $\mathbf{\Sigma}^1_n$ and $\mathbf{\Pi}^1_n$ formulas of $\mathbf{A}^2$ and $\mathbf{R}$, naturally defined.

We should also point out that these characterizations of the pointclasses we have been studying in terms of definability by formulas can be extended easily to any product space by choosing an appropriate structure to represent the space and taking formulas in its language. We will not bother to do this here.

8C. Definability in the universe of sets

We now turn to the study of elementary definability in the structures

$$\mathbf{V} = (V, \in), \qquad \boldsymbol{\mathsf{V}} = (V, \in, \{x\}_{x \in V})$$

associated with the universe of sets. For simplicity in notation, we will denote the languages of these structures by "$\mathcal{L}^{\in,}$" and "$\mathcal{L}^{\in}$" respectively—in the terminology of 8A, $\mathcal{L}^{\in}$ is the language of characteristic $\langle 2\rangle = 8$.

To avoid confusion with the ordinary functions of mathematics which are often taken to be members of V (sets of ordered pairs), we will call functions

$$F : V^n \to V$$

on sets to sets, *operations*—e.g. the power operation assigns to each x its powerset $\{u : u \subseteq x\}$.

In the first result here we compile a list of relations and operations on sets which are definable in $\mathcal{L}^{\in}$. The proof of 8C.1 is trivial, but it will be useful to have this catalogue of definitions and equivalences for future reference. We will use the common abbreviations,

$$(\exists x \in z)R(x, y_1,\ldots, y_l) \Leftrightarrow (\exists x)[x \in z \;\&\; R(x, y_1,\ldots, y_l)],$$
$$(\forall x \in z)R(x, y_1,\ldots, y_l) \Leftrightarrow (\forall x)[x \in z \Rightarrow R(x, y_1,\ldots, y_l)].$$

8C.1. THEOREM. *The following relations and operations on V and objects in V are definable in $\mathcal{L}^{\in}$.*

\# 1. $x \in y \Leftrightarrow x$ *is a member of* y.

\# 2. $x \subseteq y \Leftrightarrow (\forall t \in x)[t \in y]$.

\# 3. $x = y \Leftrightarrow x$ *is equal to* y.

\# 4. $\{x, y\} =$ *the unordered pair of* x *and* y;
$\{x, y\} = w \Leftrightarrow x \in w \;\&\; y \in w \;\&\; (\forall t \in w)[t = x \vee t = y]$.

\# 5. $\bigcup x = \{t : (\exists s \in x)[t \in s]\}$;
$\bigcup x = w \Leftrightarrow (\forall s \in x)(\forall t \in s)[t \in w] \;\&\; (\forall t \in w)(\exists s \in x)[t \in s]$.

\# 6. $\langle x, y\rangle = \{\{x\}, \{x, y\}\}$,
$\langle x_1,\ldots, x_{n+1}\rangle = \langle\langle x_1,\ldots, x_n\rangle, x_{n+1}\rangle$.

(*We are using boldface angles to avoid confusion with the codes $\langle k_1,\ldots, k_n\rangle$ for tuples of integers. Notice that for each x, y,*

$$\langle x, y\rangle \in r \Rightarrow x, y \in \bigcup \bigcup r.)$$

\# 7. $u \times v = \{\langle x, y\rangle : x \in u \;\&\; y \in v\}$,
$u_1 \times \cdots \times u_{n+1} = (u_1 \times \cdots \times u_n) \times u_{n+1}$.

\# 8. *OrdPair*$(w) \Leftrightarrow w$ *is an ordered pair*

$$\Leftrightarrow (\exists x \in \bigcup w)(\exists y \in \bigcup w)[w = \langle x, y\rangle].$$

\# 9. $\mathit{Relation}(r) \Leftrightarrow r$ *is a set of ordered pairs*

$$\Leftrightarrow (\forall w \in r)\, \mathit{OrdPair}(w).$$

\# 10. $\mathit{Domain}(r) = \{x \in \bigcup \bigcup r : (\exists y \in \bigcup \bigcup r)[\langle x, y\rangle \in r]\}$,

$$\mathit{Domain}(r) = w \Leftrightarrow (\forall x \in \bigcup \bigcup r)(\forall y \in \bigcup \bigcup r)[\langle x, y\rangle \in r \Rightarrow x \in w]$$
$$\&\ (\forall x \in w)(\exists y \in \bigcup \bigcup r)[\langle x, y\rangle \in r].$$

\# 11. $\mathit{Image}(r) = \{y \in \bigcup \bigcup r : (\exists x \in \bigcup \bigcup r)[\langle x, y\rangle \in r]\}$;

$$\mathit{Image}(r) = w \Leftrightarrow (\forall x \in \bigcup \bigcup r)(\forall y \in \bigcup \bigcup r)[\langle x, y\rangle \in r \Rightarrow y \in w]$$
$$\&\ (\forall y \in w)(\exists x \in \bigcup \bigcup r)[\langle x, y\rangle \in r].$$

\# 12. $x \cup y =$ *the union of* x *and* y

$$= \bigcup \{x, y\}.$$

\# 13. $\mathit{Field}(r) = \mathit{Domain}(r) \cup \mathit{Image}(r)$.

\# 14. $\mathit{Function}(f) \Leftrightarrow f$ *is a function (as a set of ordered pairs)*

$$\Leftrightarrow \mathit{Relation}(f)$$
$$\&\ (\forall x \in \mathit{Domain}(f))(\forall y \in \mathit{Image}(f))(\forall y' \in \mathit{Image}(f))$$
$$[[\langle x, y\rangle \in f\ \&\ \langle x, y'\rangle \in f] \Rightarrow y = y'].$$

If f *is a function, we put*

$$f(x) = y \Leftrightarrow \langle x, y\rangle \in f.$$

\# 15. $r \restriction u = \{\langle x, y\rangle \in r : x \in u\}$;

$$r \restriction u = w \Leftrightarrow w \subseteq r\ \&\ \mathit{Relation}(w)$$
$$\&\ (\forall x \in \mathit{Domain}(r))(\forall y \in \mathit{Image}(r))$$
$$[\langle x, y\rangle \in w \Leftrightarrow x \in u].$$

\# 16. $\emptyset =$ *the empty set*;

$$\emptyset = w \Leftrightarrow (\forall t)[t \notin w].$$

\# 17. $\mathit{Transitive}(x) \Leftrightarrow x$ *is a transitive set*

$$\Leftrightarrow (\forall s \in x)(\forall t \in s)[t \in x].$$

\# 18. $\mathit{Ordinal}(\xi) \Leftrightarrow \xi$ *is an ordinal*

$$\Leftrightarrow \mathit{Transitive}(\xi)\ \&\ (\forall x \in \xi)(\forall y \in \xi)[x \in y \vee y \in x \vee x = y]].$$

\# 19. $x' = x \cup \{x\}$.

\# 20. $\omega =$ *the least infinite ordinal*;

$$\omega = w \Leftrightarrow \mathit{Ordinal}(w)\ \&\ (\forall x \in w)(\exists y \in w)[y = x']$$
$$\&\ (\forall x \in w)[x \neq \emptyset \Rightarrow (\exists y \in w)[x = y']].$$

PROOF is immediate by 8B.1. ⊣

For reasons which will become clear later, we have omitted the *power operation* from this list,

$$Power(x) = \{u : u \subseteq x\};$$

this too is definable in $\mathcal{L}^\in$,

$$Power(u) = w \Leftrightarrow (\forall t)[t \in w \Leftrightarrow t \subseteq u].$$

Similarly, the binary operation

$$(x, y) \mapsto {}^y x$$

is definable,

$${}^y x = w \Leftrightarrow (\forall t)[t \in w \Leftrightarrow [Function(t) \;\&\; Domain(t) = y \;\&\; Image(t) \subseteq x]].$$

In particular, Baire space is a definable set, $\mathcal{N} = {}^\omega\omega$.

We could go on and give formal definitions in $\mathcal{L}^\in$ of the rationals, the reals and all the familiar mathematical objects we have been studying. As usual, we take these to be sets—objects in V-constructed successively, starting with the set ω of integers and using operations which are easily definable in $\mathcal{L}^\in$. There is no point in doing this in detail, as it should be obvious by now that all reasonable sets and relations and operations on sets are definable in $\mathcal{L}^\in$.

By the same token, all ordinary mathematical assertions about sets are expressible by sentences of $\mathcal{L}^\in$ and in particular, the axioms of Zermelo–Fraenkel set theory can be so expressed. We list them here for reference, indicating briefly with some "symbolized English" how each can be expressed in $\mathcal{L}^\in$.

There are seven basic axioms in *ZF*, but we need infinitely many sentences of $\mathcal{L}^\in$ to express the fifth and most significant of these, the axiom of replacement. We take up first the simpler axioms 1–4, 6 and 7.

AXIOM 1 (Extensionality). *Two sets are equal if they have the same members*,

$$(\forall x)(\forall y)[(\forall t)[t \in x \Leftrightarrow t \in y] \Rightarrow x = y].$$

In a simpler form, skipping the trivial initial quantifiers,

$$(\forall t)[t \in x \Leftrightarrow t \in y] \Rightarrow x = y.$$

We will also omit the initial quantifiers in our abbreviations of the remaining axioms.

AXIOM 2 (Pairing). *If x, y are sets, so is the unordered pair* $\{x, y\}$,

$$(\exists z)(\forall t)[t \in z \leftrightarrow [t = x \vee t = y]].$$

AXIOM 3 (Union). *For each set x, the union*

$$\bigcup x = \{t: \textit{for some } s \in x, t \in s\}$$

is also a set,

$$(\exists z)(\forall t)[t \in z \leftrightarrow (\exists s)[t \in s \;\&\; s \in x]].$$

AXIOM 4 (Power). *For each set x, the power set* $\{t : t \subseteq x\}$ *is also a set,*

$$(\exists z)(\forall t)[t \in z \leftrightarrow t \subseteq x].$$

AXIOM 6 (Infinity). There exists a set z such that $\emptyset \in z$ *and for every x, if* $x \in z$, *then* $x \cup \{x\} \in z$;

$$(\exists z)\{\emptyset \in z \;\&\; (\forall x)[x \in z \rightarrow x \cup \{x\} \in z]\}.$$

AXIOM 7 (Foundation or Regularity). *The membership relation is well-founded, i.e. every nonempty set has an* $\in$*-minimal member*;

$$x \neq \emptyset \rightarrow (\exists y)[y \in x \;\&\; (\forall t \in y)[t \notin x]].$$

Consider now the classical axiom of replacement.

AXIOM 5 (Replacement). *For each set x and for each operation*

$$F : V \rightarrow V$$

which is definable in $\mathcal{L}^{\in}$, *the image*

$$F[x] = \{F(t): t \in x\}$$

is also a set.

We cannot express this directly in $\mathcal{L}^{\in}$, because it is not possible in this simple language to quantify over all operations definable in $\mathcal{L}^{\in}$. Instead, we must assert *separately* for each formula of $\mathcal{L}^{\in}$, that if it defines a set operation (in terms of given parameters), then the image of any set by that operation is also a set. Using the abbreviation

$$(\exists! \mathbf{t})\varphi(\mathbf{x}_1, \ldots, \mathbf{x}_n, \mathbf{t}) \leftrightarrow (\exists \mathbf{y})(\forall \mathbf{t})[\varphi(\mathbf{x}_1, \ldots, \mathbf{x}_n, \mathbf{t}) \leftrightarrow \mathbf{t} = \mathbf{y}],$$

we then take as Axiom 7 the collection of all sentences of the form

$$(\forall \mathbf{y}_1) \cdots (\forall \mathbf{y}_n)(\forall \mathbf{x})\{(\forall \mathbf{s})(\exists ! \mathbf{t})\psi(\mathbf{y}_1,\dots,\mathbf{y}_n,\mathbf{s},\mathbf{t}) \rightarrow (\exists \mathbf{z})(\forall \mathbf{t})[\mathbf{t} \in \mathbf{z} \leftrightarrow (\exists \mathbf{s})[\mathbf{s} \in \mathbf{x} \;\&\; \psi(\mathbf{y}_1,\dots,\mathbf{y}_n,\mathbf{s},\mathbf{t})]]\},$$

one for each formula $\psi(\mathbf{y}_1,\dots,\mathbf{y}_n,\mathbf{s},\mathbf{t})$ of $\mathcal{L}^{\in}$.

Since it takes infinitely many sentences of $\mathcal{L}^{\in}$ to express Axiom 7, we often talk of *the axiom scheme of replacement.*

From now on, by ZF we will mean this infinite list of sentences of $\mathcal{L}^{\in}$ *which express the axioms of Zermelo–Fraenkel set theory.* It is very important that *ZF* is an infinite set of sentences. The finite subsets of *ZF* determine axiomatic set theories which *approximate* Zermelo–Fraenkel set theory—the more axioms we have, the better the approximation.

The theory of *Zermelo–Fraenkel with choice ZFC* is obtained by adding to *ZF* the *axiom of choice* (*AC*) in any of its well-known equivalent forms. All of them are easily expressible by sentences of $\mathcal{L}^{\in}$.

According to usual mathematical practice, when we prove a theorem about sets, *we customarily assume without explicit mention that the structure* $\mathsf{V} = (V, \in)$ *satisfies all the axioms of ZFC,* but we are careful to list among the hypotheses of our theorems any additional assumptions about sets—like the continuum hypothesis or determinacy hypotheses. We have followed this practice scrupulously in this book. In fact, as we pointed out in 7D, we have identified all blatant uses of the axiom of choice which cannot be justified on the basis of the weaker axiom of dependent choices and in this chapter we are including whatever choice we need among the hypotheses of the theorems.

Since we have been emphasizing the fact that all "ordinary mathematical assertions" about sets can be expressed by sentences of $\mathcal{L}^{\in}$, it is perhaps worth pointing out that in metamathematics we often consider assertions about sets which are not immediately or naturally expressible in $\mathcal{L}^{\in}$—or which may not be expressible in $\mathcal{L}^{\in}$ at all. For example, Theorem 8C.1 is not easy to translate into $\mathcal{L}^{\in}$ because it refers to *relations* and *operations on V* while $\mathcal{L}^{\in}$ can only speak directly about *members* of *V*. Sometimes we will seek *indirect* ways of expressing the meaning of a certain proposition by a sentence of $\mathcal{L}^{\in}$ because this will be important in an argument but of course most of the time we do not care whether or not our theorems are stated in the small part of English formalized in $\mathcal{L}^{\in}$.

We now return to our study of elementary definability in the structure V.

It is useful to call a collection of sets

$$M \subseteq V$$

a *class* if membership in M is *definable in* $\mathfrak{L}^{\in}$, i.e. if there is some formula $\varphi(\mathbf{s}, \mathbf{x}_1, \ldots, \mathbf{x}_n)$ of $\mathfrak{L}^{\in}$ and sets $x_1, \ldots, x_n \in V$ such that

$$s \in M \Leftrightarrow \mathsf{V} \vDash \varphi(s, x_1, \ldots, x_n).$$

If we can find a formula $\varphi(\mathbf{s})$ in $\mathfrak{L}^{\in}$ (without parameters) so that

$$s \in M \Leftrightarrow \mathsf{V} \vDash \varphi(s),$$

we then call M a *definable class*. For example, the class of ordinals

$$ON = \{\xi : Ordinal(\xi)\}$$

is definable.

If $M_1, \ldots, M_n$ are classes, then an operation

$$F : M_1 \times \cdots \times M_n \to V$$

is any n-ary operation on sets such that

$$x_1 \notin M_1 \vee \cdots \vee x_n \notin M_n \Rightarrow F(x_1, \ldots, x_n) = 0.$$

Such an operation is then determined by its values $F(x_1, \ldots, x_n)$ when $x_1 \in M_1, \ldots, x_n \in M_n$.

8C.2. THEOREM. *Let $G : V^{n+1} \to V$ be an $(n+1)$-ary operation which is definable in $\mathfrak{L}^{\in}$. There exists a unique operation*

$$F : ON \times V^n \to V$$

which is also definable in $\mathfrak{L}^{\in}$ and which satisfies the following equation for each ordinal ξ and $x_1, \ldots, x_n \in V$:

$$F(\xi, x_1, \ldots, x_n) = G(\{\langle \eta, F(\eta, x_1, \ldots, x_n)\rangle : \eta < \xi\}, x_1, \ldots, x_n).$$

(*Definition by recursion on the ordinals.*)

PROOF. Put

$$\begin{aligned} P(h, x_1, \ldots, x_n) \Leftrightarrow {} & Function(h) \,\&\, (\forall y \in Domain(h))[Ordinal(y)] \\ & \&\, (\forall \xi \in Domain(h))(\forall \eta \in \xi)[\eta \in Domain(h)] \\ & \&\, (\forall \xi \in Domain(h))[h(\xi) = G(\{\langle \eta, h(\eta)\rangle : \eta \in \xi\}, x_1, \ldots, x_n) \end{aligned}$$

and notice that by 8C.1 and 8B.1, P is definable in $\mathfrak{L}^{\in}$. An easy induction on the ordinals shows that

$$\begin{aligned} P(h, x_1, \ldots, x_n) \,\&\, P(h', x_1, \ldots, x_n) \,\&\, \xi \in Domain(h) \cap Domain(h') \\ \Rightarrow h(\xi) = h'(\xi), \end{aligned}$$

so put

$$F(\xi, x_1,\ldots, x_n) = \begin{cases} y & \text{if } (\exists h)[P(h, x_1,\ldots, x_n) \ \&\ \xi \in Domain(h) \ \&\ h(\xi) = y], \\ \emptyset & \text{if } (\forall h)[P(h, x_1,\ldots, x_n) \Rightarrow \xi \notin Domain(h)]. \end{cases}$$

To complete the proof, it is enough to show that for each $x_1,\ldots, x_n$,

$$(*) \qquad (\forall \xi)(\exists h)[P(h, x_1,\ldots, x_n) \ \&\ \xi \in Domain(h)].$$

Assuming that $(*)$ fails for some ξ, let λ be the least ordinal ξ such that $(\forall h)[P(h, x_1,\ldots, x_n) \Rightarrow \xi \notin Domain(h)]$ and define

$$H(\xi) = \begin{cases} \langle \xi, F(\xi, x_1,\ldots, x_n)\rangle & \text{if } \xi < \lambda, \\ \emptyset & \text{if } \lambda \leq \xi; \end{cases}$$

now H is definable in $\mathcal{L}^{\in}$, so by the replacement axiom, the image

$$h = H[\lambda] = \{\langle \xi, F(\xi, x_1,\ldots, x_n)\rangle : \xi < \lambda\}$$

is a set and hence

$$h' = h \cup \{\langle \lambda, G(h, x_1,\ldots, x_n)\rangle\}$$

is also a set. But easily

$$P(h', x_1,\ldots, x_n) \ \&\ \lambda \in Domain(h'),$$

contradicting the choice of λ. ⊣

We allow $n = 0$ in this theorem, in which case

$$G : V \to V$$

defines uniquely the operation $F : ON \to V$ which satisfies

$$F(\xi) = G(F \restriction \xi).$$

Using this basic result we can define the transfinite sequence $\{V_\xi : \xi \in ON\}$ of partial universes by the recursion

$$\begin{aligned} V_0 &= \emptyset, \\ V_{\xi+1} &= Power(V_\xi), \\ V_\lambda &= \textstyle\bigcup_{\xi<\lambda} V_\xi, \text{ if } \lambda \text{ is a limit ordinal} \end{aligned}$$

(see 8C.8). It is easy to verify (by induction on ξ) that each V_ξ is a transitive set,

$$V_0 \subsetneq \cdots \subseteq V_\xi \subsetneq V_{\xi+1} \subsetneq \cdots,$$
$$V = \textstyle\bigcup_\xi V_\xi,$$

and

$$\xi \in V_{\xi+1} - V_\xi.$$

This hierarchy of partial universes gives a precise version of the intuitive construction for the universe of sets which we discussed in the introduction to this chapter, where *for stages we take the ordinals.* It also suggests drawing the universe V in the form of an inverted conve, growing upwards, with the ordinals plotted along the main axis, see Figure 8C.1.

Recall that we are taking cardinals to be *initial ordinals,*

$$\begin{aligned} Cardinal(\varkappa) \leftrightarrow\ & Ordinal(\varkappa) \\ & \& (\forall \xi \in \varkappa)(\forall f)\{[Function(f) \,\&\, Domain(f) = \xi] \\ & \qquad \Rightarrow (\exists \eta \in \varkappa)[\eta \notin Image(f)]\}. \end{aligned}$$

The relation $Cardinal(\varkappa)$ is clearly definable in $\mathcal{L}^{\in}$.

The familiar indexing of cardinals by ordinals is also defined by recursion,

$$\begin{aligned} \aleph_0 &= \omega, \\ \aleph_{\xi+1} &= \aleph_\xi^+ = \text{least cardinal } \varkappa > \aleph_\xi, \\ \aleph_\lambda &= supremum\{\aleph_\xi : \xi < \lambda\}, \text{ if } \lambda \text{ is a limit ordinal.} \end{aligned}$$

Intuitively, one would think that when ξ is a "typical" very large ordinal, then the partial universe V_ξ should "look very much like" the completed universe of sets V. We next prove a very important precise version of this idea.

First a lemma about closed, unbounded classes of ordinals.

A class K or ordinals in *unbounded* if

$$(\forall \xi)(\exists \eta > \xi)[\eta \in K];$$

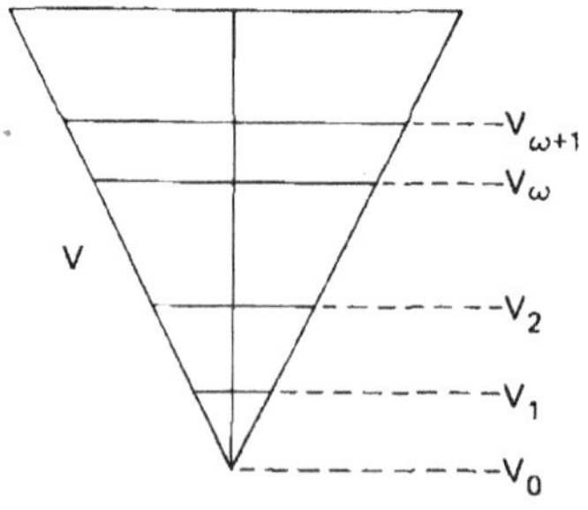

Figure 8C.1.

K is *closed* if for every limit ordinal ξ,

$$(\forall\eta<\xi)(\exists\zeta)[\eta<\zeta<\xi \;\&\; \zeta\in K] \Rightarrow \xi\in K,$$

i.e. if K is closed in the natural order topology on ON.

8C.3. LEMMA. (i) *If K_1 and K_2 are closed, unbounded classes of ordinals, then $K_1\cap K_2$ is also closed and unbounded.*

(ii) *If*

$$F: ON \to ON$$

is an operation on ordinals which is definable in $\mathcal{L}^{\in}$, then the class

$$K^* = \{\xi : (\forall\eta<\xi)[F(\eta)<\xi]\}$$

is closed and unbounded.

PROOF. (i) $K_1\cap K_2$ is obviously closed. To see that it is unbounded, given ξ, choose successively ξ_0, ξ_1, ξ_2,... so that

$$\begin{aligned} &\xi<\xi_0 \quad\text{and}\quad \xi_0\in K_1,\\ &\xi_0<\xi_1 \quad\text{and}\quad \xi_1\in K_2,\\ &\xi_1<\xi_2 \quad\text{and}\quad \xi_2\in K_1,\\ &\qquad\text{etc.}\end{aligned}$$

and check that $\xi^* = limit_n\, \xi_n \in K_1\cap K_2$ because both K_1, K_2 are closed.

(ii) Again, K^* is obviously closed. Given ξ, define ξ_n by the recursion on ω

$$\xi_0 = \xi,$$
$$\xi_{n+1} = \xi_n + 1 + supremum\{f(\zeta) : \zeta < \xi_n\},$$

where the supremum exists by replacement and verify that $\eta = \xi_0 < \xi_1 < \cdots$ and $limit_{n\to\infty}\, \xi_n \in K^*$. ⊣

For each class M (which may be a set), let $(M, \in)$ be the structure obtained by restricting the membership relation to M, i.e.

$$(M, \in) = (M, E^M)$$

where

$$E^M(x, y) \Leftrightarrow x\in M \;\&\; y\in M \;\&\; x\in y.$$

This notational convention will simplify many formulas and cannot cause any confusion.

We will prove the reflection theorem in a general context because it has many applications, but in a first reading one way as well take $C_\xi = V_\xi$.

8C.4. THE REFLECTION THEOREM. *Let $\xi \mapsto C_\xi$ be an operation on ordinals to sets which is definable in $\mathfrak{L}^\epsilon$ and satisfies the following two conditions:*

(i) $\zeta \leq \xi \Rightarrow C_\zeta \subseteq C_\xi$.

(ii) *If λ is a limit ordinal, then*

$$C_\lambda = \bigcup_{\xi<\lambda} C_\xi.$$

Let

$$C = \bigcup_\xi C_\xi.$$

Then for each formula $\varphi(\mathbf{x}_1,\ldots,\mathbf{x}_n)$ of $\mathfrak{L}^\epsilon$, there is a closed, unbounded class of ordinals K such that for $\xi \in K$ and $x_1,\ldots, x_n \in C_\xi$,

$$(C, \in) \models \varphi(x_1,\ldots, x_n) \Leftrightarrow (C_\xi, \in) \models \varphi(x_1,\ldots, x_n).$$

In particular, if φ is any sentence of $\mathfrak{L}^\epsilon$, then

$$(C, \in) \models \varphi \Rightarrow \textit{for some } \xi,\ (C_\xi, \in) \models \varphi.$$

PROOF. We use induction on $\varphi(\mathbf{x}_1,\ldots,\mathbf{x}_n)$, the result being trivial for prime formulas and following easily from the induction hypothesis for negations and conjunctions.

Suppose $(\exists \mathbf{y})\varphi(\mathbf{y}, \mathbf{x}_1,\ldots,\mathbf{x}_n)$ is given and assume that K satisfies the result for $\varphi(\mathbf{y}, \mathbf{x}_1,\ldots,\mathbf{x}_n)$. Let

$$G(x_1,\ldots, x_n) = \begin{cases} \text{least } \xi \text{ such that } (\exists y \in C_\xi)[(C, \in) \models \varphi(y, x_1,\ldots, x_n)] \\ \qquad \text{if one such } \xi \text{ exists,} \\ 0 \quad \text{otherwise} \end{cases}$$

and take

$$F(\xi) = \textit{supremum}\{G(x_1,\ldots, x_n) \colon x_1,\ldots, x_n \in C_\xi\}$$

by replacement. By the lemma then, the class of ordinals

$$K \cap \{\xi : (\forall \eta < \xi)[F(\eta) < \xi]\} \cap \{\xi : \xi \text{ is limit}\}$$

is closed and unbounded and it is easy to verify that it satisfies the theorem for the formula $(\exists \mathbf{y})\varphi(\mathbf{y}, \mathbf{x}_1,\ldots,\mathbf{x}_n)$. ⊣

EXERCISES

We take up first a small improvement of 8C.1 which will prove useful in the next section.

Let Σ_0 be the smallest collection of formulas in the language $\mathcal{L}^{\in}$ such that all prime formulas

$$\mathbf{v}_i \in \mathbf{v}_j$$

are in Σ_0 and such that if φ and ψ are in Σ_0, then the formulas

$$\neg(\varphi), \quad (\varphi) \,\&\, (\psi), \quad (\exists \mathbf{v}_i)[\mathbf{v}_i \in \mathbf{v}_j \,\&\, \varphi]$$

are also in Σ_0.

8C.5. Prove that the relations #1, #2, #8, #9, #14, #17 and #18 of 8C.1 are definable by Σ_0 formulas. ⊣

Consider next a trivial consequence of 8C.2 which is however worth putting down.

8C.6. Let $G_1 : V^n \to V$ and $G_2 : V^{n+2} \to V$ be definable in $\mathcal{L}^{\in}$. Show that there exists a unique operation $F : \omega \times V^n \to V$ which is definable in $\mathcal{L}^{\in}$ and satisfies

$$F(0, x_1, \ldots, x_n) = G_1(x_1, \ldots, x_n),$$
$$F(k+1, x_1, \ldots, x_n) = G_2(F(k, x_1, \ldots, x_n), k, x_1, \ldots, x_n).$$

It follows that every recursive function on ω (extended to be $=0$ for arguments not in ω) is definable in $\mathcal{L}^{\in}$ and that every recursive relation on ω is definable in $\mathcal{L}^{\in}$ (as a set of tuples).

Hint. The first assertion follows easily from 8C.2 and the second is proved by induction on the definition of recursive function. ⊣

There is also a generalization of 8C.2 which is useful. (Recall from Chapter 2 that $x <_R y \Leftrightarrow R(x, y) \,\&\, \neg R(y, x)$.)

8C.7. Let $G : V \to V$ be an operation definable in $\mathcal{L}^{\in}$, let S be a class and suppose $R \subseteq S \times S$ is a relation which is definable in $\mathcal{L}^{\in}$ and which is wellfounded, i.e.

$$x \subseteq S \,\&\, x \neq \emptyset \Rightarrow (\exists y \in x)(\forall z \in x)\, \neg z <_R y.$$

Assume further that each initial segment of R is a set, i.e. $(\forall x \in S)(\exists z)(\forall y)[y \in z \Leftrightarrow y <_R x]$. Prove that there is a unique operation

$$F : S \to V$$

which is definable in $\mathcal{L}^{\in}$ and satisfies for each $x \in S$ the equation

$$F(x) = G(\{\langle y, F(y)\rangle : y <_R x\}).$$

Show moreover that if G, S and R are definable in $\mathcal{L}^{\in}$, then so is F.
Hint. Imitate the proof of 8C.2. ⊣

8C.8. Prove that the operation

$$\xi \mapsto V_\xi$$

is definable in $\mathcal{L}^{\in}$.
Hint. Apply 8C.2 with

$$G(h) = \begin{cases} \emptyset \quad \text{if } (\forall \xi \in ON)[\xi \notin Domain(h)], \\ Power(h(\xi)) \text{ where } \xi \\ \qquad\qquad = infimum\{\eta \in Domain(h) : \eta + 1 \notin Domain(h)\}, \\ \quad \text{if } [\exists \eta \in Domain(h)][\eta + 1 \notin Domain(h)], \\ \bigcup \{h(\xi) : \xi \in Domain(h)\}, \quad \text{otherwise.} \end{cases}$$ ⊣

A class M is *transitive* if

$$x \in M \;\&\; y \in x \Longrightarrow y \in M.$$

In seeking models of Zermelo–Fraenkel set theory we will concentrate on structure of the form $(M, \in)$ with transitive M, partly because of the following basic fact.

8C.9. Show that if M is a transitive class, then the structure $(M, \in)$ satisfies the axiom of extensionality.

Conversely, suppose M is a class, $E \subseteq M \times M$ is a binary strict well-founded relation on M which is definable in $\mathcal{L}^{\in}$ and such that for each $x \in M$, $\{y : E(y, x)\}$ is a set. Assume that (M, E) satisfies the axiom of extensionality; prove that there is a unique transitive class $\bar{M}$ and a unique bijection

$$\pi : M \rightarrowtail\!\!\!\!\rightarrow \bar{M}$$

which is an isomorphism of (M, E) with $(\bar{M}, \in)$. Moreover, if there is a transitive set $y \subseteq M$ such that $x \in y$ and E agrees with $\in$ on y, then $\pi(x) = x$.

Show also that if M, E are definable in $\mathcal{L}^{\in}$, then so are $\bar{M}$, π. (*The Mostowski Collapsing Lemma.*)

Hint. The first assertion is quite easy and will follow from the more general 8D.3, but the inexperienced reader will do well to check it out.

For the second assertion define $\pi : M \to V$ by the recursion

$$\pi(x) = \{\pi(y) : E(y, x)\}$$

and take $\bar{M} = \pi[M] = \{\pi(x) : x \in M\}$.

If y is transitive and $y \subseteq M$, show by $\in$-induction that

$$t \in y \Rightarrow \pi(t) = t. \qquad \dashv$$

(By $\in$-induction in this hint we mean the method of proof where you show

$$(\forall x)[(\forall t \in x)R(t) \Rightarrow R(x)]$$

and you infer

$$(\forall x)R(x);$$

this is easy to justify using the axiom of foundation.)

The Mostowski collapsing lemma is a very useful fact to which we will appeal often. In the typical case we will be applying it to a structure of the form $(M, \in)$, where M will not be necessarily transitive. In this case of course we will only need to check that $(M, \in)$ satisfies extensionality, since the relation in $(M, \in)$

$$E^M(x, y) \Leftrightarrow x \in M \;\&\; y \in M \;\&\; x \in y$$

is automatically wellfounded; if it does, then we have an canonical isomorphism

$$\pi : M \rightarrowtail\!\!\!\!\twoheadrightarrow \bar{M}$$

of M with the transitive class $\bar{M}$, the so-called *Mostowski collapsing map of M*. (Of course, if M is a set, then so is $\bar{M}$.)

If $\mathfrak{A}$ is a structure and T is a set of sentences (a *theory*) in the language of $\mathfrak{A}$, we write

$$\mathfrak{A} \models T$$

and we call $\mathfrak{A}$ *a model of T* just in case

$$\text{for all } \varphi \in T, \mathfrak{A} \models \varphi.$$

8C.10. Prove that (granting the axiom of choice), for every finite set T^0 of true sentences of $\mathfrak{L}^{\in}$, there is a countable, transitive set A such that $(A, \in) \models T^0$. (*The countable reflection theorem.*)

Hint. *Choose* V_ξ such that $(V_\xi, \in) \models T^0$ by the Reflection Theorem 8C.4, find a countable $B \subseteq V_\xi$ such that $(B, \in)$ is an elementary submodel of $(V_\xi, \in)$ by the Skolem–Löwenheim Theorem 8A.4 and take $A = \bar{B} =$ the transitive collapse of B. (Notice that 8C.9 applies because $(V_\xi, \in)$ and hence $(B, \in)$ satisfies extensionality; notice also that you are using the axiom of choice in this proof because the Skolem–Löwenheim theorem needs V_ξ to be wellorderable.) ⊣

What we would really like to have is a countable transitive set A such that $(A, \in) \models ZF$, but we cannot prove that such a set exists assuming just the axioms of *ZFC* for V. The next result gives us these countable, transitive models of *ZFC*, granting a strongly inaccessible cardinal.

8C.11. Show that if the axiom of choice holds and $\varkappa$ is strongly inaccessible, then $(V_\varkappa, \in) \models ZFC$; infer that if there exists a strongly inaccessible cardinal, then there exists a countable, transitive model of *ZFC*.

Hint. All the axioms of *ZFC* except perhaps replacement hold in every $(V_\varkappa, \in)$ with limit $\varkappa$—and replacement is easy to check using the strong inaccessibility. The second assertion is proved as in 8C.10. ⊣

This result appears a bit paradoxical at first sight, since a transitive model $(A, \in)$ of *ZFC* satisfies the formal sentence of $\mathcal{L}^\in$

"there exists an uncountable set",

so for some $x \in A$,

$$(A, \in) \models \text{"}x \text{ is uncountable"}.$$

If A is countable, surely x is countable, so in the real world (the universe V) there exists a bijection $f : \omega \rightarrowtail\!\!\!\!\twoheadrightarrow x$; the explanation of the "paradox" is that no such bijection can be a member of A.

8D. Gödel's model of constructible sets

We will define here the class L of constructible sets and we will prove that the structure $(L, \in)$ satisfies all the axioms of *ZF*.

The key idea is to imitate the definition of the partial universes V_ξ but to replace the power operation (about which we know very little) by the much more tractable operation on sets

$$\boldsymbol{Def}(A) = \{x \subseteq A : x \text{ is elementary in the structure } (A, \in, \{s : s \in A\})\}.$$

We must first show that ***Def*** is definable in $\mathcal{L}^{\in}$.

Recall that in 8A we assigned *integer codes* to the formulas of a language $\mathcal{L}^{u}$ of a given characteristic. In the next result we use these codes to refer (indirectly) to formulas and definability notions in general within the language $\mathcal{L}^{\in}$. We continue the numbering of Theorem 8C.1.

8D.1. THEOREM. *The following relations and operations on V and objects in V are definable in* $\mathcal{L}^{\in}$. *(When we use variables m, n, k, it is understood that the relations in question are false and the operations in question are set* $= \emptyset$, *unless* $m, n, k \in \omega$.)

#21. $TC(x) = $ *the transitive closure of* x

$= $ *the smallest transitive set* y *such that* $x \in y$.

#22. ${}^{n}A = \{h : h$ *is a function with domain*

$n = \{0, 1, \ldots, n-1\}$ *and values in* $A\}$.

#23. $Formula(m, n) \Leftrightarrow m$ *is the code of some formula*

$\varphi(\mathbf{v}_0, \ldots, \mathbf{v}_{n-1})$ *of the language* $\mathcal{L}^{\in}$ *whose free variables are among* $\mathbf{v}_0, \ldots, \mathbf{v}_{n-1}$.

#24. $Sat(m, n, x, A, e) \Leftrightarrow Formula(m, n)$

$\& \; x \in {}^{n}A$

$\& \; e \subseteq A \times A$

$\&$ [*if* $\varphi(\mathbf{v}_0, \ldots, \mathbf{v}_{n-1})$ *is the formula with code* m, *then*

$(A, e) \models \varphi(x(0), \ldots, x(n-1))$].

#25. $\boldsymbol{Def}_1(m, n, x, A, e) = \{s \in A : Sat(m, n+1, x \cup \{\langle n, s\rangle\}, A, e)\}$.

#26. $\boldsymbol{Def}(A) = \{Def_1(m, n, x, A, \{\langle u, v\rangle : u \in v \;\&\; u \in A \;\&\; v \in A\}$

$: m \in \omega \;\&\; n \in \omega \;\&\; x \in {}^{n}A\}$.

PROOF. #21. $TC(x) = \bigcup \{TC(n, x) : n \in \omega\}$, where $TC(n, x)$ is defined by the recursion

$$TC(0, x) = \{x\},$$
$$TC(n+1, x) = \bigcup TC(n, x).$$

#22. Use recursion again,

$$
\begin{aligned}
{}^{0}a &= \emptyset,\\
{}^{(n+1)}A &= \{y \cup \{\langle n, t\rangle\} : y \in {}^{n}A \;\&\; t \in A\}.
\end{aligned}
$$

#23 is immediate since *Formula*(m, n) is recursive.

#25 and #26 will follow immediately, once we prove that the satisfaction relation is definable.

To prove the latter, define

$$
F_1(m, n, x, A, e) = \begin{cases} 1 & \text{if } m \text{ is the code of some formula} \\ & \varphi(\mathbf{v}_0, \ldots, \mathbf{v}_{n-1}) \text{ with free variables among} \\ & \text{the } \mathbf{v}_0, \ldots, \mathbf{v}_{n-1} \text{ and } x \in {}^{n}A \text{ and} \\ & e \subseteq A \times A \text{ and } (A, e) \models \varphi(x(0), \ldots, x(n-1)), \\ 0 & \text{otherwise} \end{cases}
$$

and put

$$
F(m, A, e) = \{\langle i, n, x, F_1(i, n, x, A, e)\rangle : n \in \omega \;\&\; i < m \in \omega \;\&\; x \in {}^{n}A \;\&\; e \subseteq A \times A\};
$$

it is enough to show that F is definable in $\mathcal{L}^{\in}$, since

$$
Sat(m, n, x, A, e) \Leftrightarrow \langle m, n, x, 1\rangle \in F(m + 1, A, e).
$$

To show that F is definable by applying 8C.6, we need definable operations G_1, G_2 such that

$$
\begin{aligned}
F(0, A, e) &= G_1(A, e),\\
F(m+1, A, e) &= G_2(F(m, A, e), m, A, e).
\end{aligned}
$$

The first of these is trivial, since

$$
F(0, A, e) = \emptyset.
$$

On the other hand,

$$
F(m+1, A, e) = F(m, A, e) \cup G_3(m, A, e)
$$

where $G_3(m, A, e) = \emptyset$, unless m is the code of some formula $\varphi(\mathbf{v}_0, \ldots, \mathbf{v}_{n-1})$; and if m is the code of some such formula, then we can easily compute $G_3(m, A, e)$ from $F(m, A, e)$ because of the inductive nature of the definition of satisfaction—and the fact that in our coding formulas are assigned bigger codes than their subformulas. For example, if

$$m = \langle 3, j, k \rangle$$

so that m is the code of some formula

$$(\exists \mathbf{v}_j)(\varphi)$$

where k is the code of φ, then

$$\begin{aligned} G_3(m, A, e) = \{\langle m, n, x, 1\rangle &: n \in \omega \ \&\ x \in {}^nA \ \&\ e \subseteq A \times A \\ &\& (\exists l)(\exists y \in {}^lA)[j < l \ \&\ n \leq l \ \&\ (\forall i < n)[i \neq j \Rightarrow x(i) = y(i) \\ &\& \langle k, l, y, 1\rangle \in F(m, A, e)]\} \\ &\cup \{\langle m, n, x, 0\rangle : n \in \omega \ \&\ x \in {}^nA \ \&\ e \subseteq A \times A \\ &\& (\forall l)(\forall y \in {}^lA)[[j < l \ \&\ n \leq l \ \&\ (\forall i < n)[i \neq j \Rightarrow x(i) = y(i)] \\ &\Rightarrow \langle k, l, y, 0\rangle \in F(m, A, e)]\}. \end{aligned}$$

Similar expressions for $G_3(m, A, e)$ can be found for the other cases where m codes a formula and then we can easily put these together to define $G_2(w, m, A, e)$. We will omit the details. ⊣

It is obvious from the definition of the operation ***Def*** that

$$\begin{aligned} x \in \mathbf{Def}(A) \Leftrightarrow x \subseteq A &\text{ and there is a formula } \varphi(\mathbf{v}_0, \ldots, \mathbf{v}_{n-1}, \mathbf{v}_n) \\ &\text{in the language } \mathcal{L}^{\in} \text{ and members } x_0, \ldots, x_{n-1} \text{ of} \\ &A, \text{ such that for all } s \in A, \\ &s \in x \Leftrightarrow (A, \in) \models \varphi(x_0, \ldots, x_{n-1}, s). \end{aligned}$$

We now define the *constructible hierarchy* $\{L_\xi : \xi \in ON\}$ by the recursion

$$\begin{aligned} L_0 &= \emptyset, \\ L_{\xi+1} &= \mathbf{Def}(L_\xi), \\ L_\lambda &= \textstyle\bigcup_{\xi<\lambda} L_\xi, \quad \text{if } \lambda \text{ is a limit ordinal} \end{aligned}$$

and we let

$$L = \textstyle\bigcup_\xi L_\xi$$

be *the class of constructible sets*. More generally, for any set A, put

$$\begin{aligned} L_0(A) &= TC(A), \\ L_{\xi+1}(A) &= \mathbf{Def}(L_\xi(A)), \\ L_\lambda(A) &= \textstyle\bigcup_{\xi<\lambda} L_\xi(A), \quad \text{if } \lambda \text{ is a limit ordinal} \end{aligned}$$

and

$$L(A) = \bigcup_\xi L_\xi(A).$$

8D.2. THEOREM. (i) *The operation* $\xi \mapsto L_\xi$ *is definable in* $\mathcal{L}^\in$ *and the class* L *is a definable class.*

(ii) $\eta \leq \xi \Rightarrow L_\eta \subseteq L_\xi$.

(iii) *Each* L_ξ *is a transitive set and* L *is a transitive class.*

Similarly:

(iv) *The operation* $(\xi, A) \mapsto L_\xi(A)$ *is definable in* $\mathcal{L}^\in$ *and if* A *is a definable set, then* $L(A)$ *is a definable class.*

(v) $\eta \leq \xi \Rightarrow L_\eta(A) \subseteq L_\xi(A)$.

(vi) *Each* $L_\xi(A)$ *is a transitive set and* $L(A)$ *is a transitive class.*

PROOF. (i) follows from 8C.2 since it is easy to find a definable $G: V \to V$ such that

$$L_\xi = G(\{\langle \eta, L_\eta \rangle : \eta < \xi\}),$$

see 8C.8.

To prove (ii) and (iii) we show simultaneously by induction that for each ξ,

$$L_\xi \text{ is transitive and } \eta < \xi \Rightarrow L_\eta \subseteq L_\xi.$$

This is trivial for $\xi = 0$ or limit ordinals ξ.

If $\xi = \zeta + 1$, suppose first that $\eta < \xi$ and $x \in L_\eta$; by induction hypothesis then $x \in L_\zeta$ and $x \subseteq L_\zeta$, so that $x \in L_{\zeta+1}$, since we can obviously define x as a subset of L_ζ using x as a parameter,

$$s \in x \Leftrightarrow (L_\zeta, \in) \models s \in x.$$

In particular, $L_\zeta \subseteq L_{\zeta+1}$ and hence for any $x \in L_{\zeta+1}$ and $y \in x$, we have $y \in L_\zeta$ and hence $y \in L_{\zeta+1}$, so $L_{\zeta+1}$ is also transitive.

Now L is easily transitive as the union of transitive sets and (iv)–(vi) are proved similarly. ⊣

The transitivity of L was well worth pointing out because of the following general fact about transitive classes. (Σ_0 formulas are defined in the Exercises of 8C.)

8D.3. LEMMA. *Let* M *be a transitive class.*

(i) *If* $\varphi(\mathbf{x}_1,\dots,\mathbf{x}_n)$ *is a formula in* Σ_0 *with the indicated free variables and* $x_1, \ldots, x_n \in M$, *then*

$$(V, \in) \models \varphi(x_1, \ldots, x_n) \Leftrightarrow (M, \in) \models \varphi(x_1, \ldots, x_n).$$

(ii) *The structure* $(M, \in)$ *satisfies the axioms of extensionality and foundation in the list ZF.*

(iii) *If* M *is closed under pairing and union, then* $(M, \in)$ *satisfies the axioms of pairing and union.*

(iv) *If some infinite ordinal* $\lambda \in M$, *then* $(M, \in)$ *satisfies the axiom of infinity.*

PROOF. (i) Reverting to the notation of 8A which is more appropriate here, we must show that if φ is a formula in Σ_0 and $\bar{x} = \{x_j\}$ is any assignment into M, then

$$(V, \in), \bar{x} \models \varphi \Leftrightarrow (M, \in), \bar{x} \models \varphi.$$

This is immediate for prime formulas,

$$\begin{aligned}(V, \in), \bar{x} \models \mathbf{v}_i \in \mathbf{v}_j &\Leftrightarrow x_i \in x_j \\ &\Leftrightarrow (M, \in), \bar{x} \models \mathbf{v}_i \in \mathbf{v}_j\end{aligned}$$

and if it holds for φ and ψ, it obviously holds for $\neg(\varphi)$ and for $(\varphi) \,\&\, (\psi)$. By induction on the length of formulas then, in the nontrivial case,

$$(V, \in), \bar{x} \models (\exists \mathbf{v}_i)[\mathbf{v}_i \in \mathbf{v}_j \,\&\, \varphi] \Leftrightarrow \text{for some } z \in x_j, (V, \in), \bar{x}^z \models \varphi$$

where

$$x_k^z = \begin{cases} x_k & \text{if } k \neq i, \\ z & \text{if } k = i; \end{cases}$$

but since M is transitive and $x_j \in M$, we have $x_j \subseteq M$ and hence for every $z \in x_j$ the assignment $\bar{x}^z$ is into M, so that by the induction hypothesis,

$$\begin{aligned}\text{for some } z \in x_j, (V, \in), \bar{x}^z \models \varphi &\Leftrightarrow \text{for some } z \in x_j, (M, \in), \bar{x}^z \models \varphi \\ &\Leftrightarrow (M, \in), \bar{x} \models (\exists \mathbf{v}_i \in \mathbf{v}_j \,\&\, \varphi].\end{aligned}$$

(ii) It is easy to check that both of these axioms are expressed in $\mathcal{L}^{\in}$ by formulas of the form

$$(\forall \mathbf{x}_1) \cdots (\forall \mathbf{x}_n) \varphi(\mathbf{x}_1, \ldots, \mathbf{x}_n)$$

where $\varphi(\mathbf{x}_1,\ldots,\mathbf{x}_n)$ is in Σ_0. Hence

$$\begin{aligned}(V,\in)\models(\forall\mathbf{x}_1)\cdots(\forall\mathbf{x}_n)\varphi(\mathbf{x}_1,\ldots,\mathbf{x}_n)\\ &\Rightarrow \text{for all } x_1,\ldots,x_n,\ (V,\in)\models\varphi(x_1,\ldots,x_n)\\ &\Rightarrow \text{for all } x_1,\ldots,x_n\in M,\ (M,\in)\models\varphi(x_1,\ldots,x_n) \text{ using (i))}\\ &\Rightarrow (M,\in)\models(\forall\mathbf{x}_1)\cdots(\forall\mathbf{x}_n)\varphi(\mathbf{x}_1,\ldots,\mathbf{x}_n)\end{aligned}$$

and since these axioms hold in $(V,\in)$, they must also hold in $(M,\in)$.

(iii) Again, it is easy to find a formula $\varphi(\mathbf{x},\mathbf{y},\mathbf{z})$ in Σ_0 such that

$$z=\{x,y\}\Leftrightarrow(V,\in)\models\varphi(x,y,z).$$

To show that $(M,\in)$ satisfies the pairing axiom then, we must verify that for each $x\in M$, $y\in M$, there is some $z\in M$ such that $(M,\in)\models\varphi(x,y,z)$; of course, we take $z=\{x,y\}$ and we use (i).

The argument for the union axiom is similar.

(iv) If $\lambda\in M$ and λ is infinite, then either $\omega=\lambda$ or $\omega\in\lambda$ and in either case, by the transitivity of M, $\omega\in M$. Checking #20 and then #19 and #18 of 8C.1, we can construct a Σ_0 formula $\varphi(\mathbf{x})$ such that

$$x=\omega\Leftrightarrow(V,\in)\models\varphi(x);$$

in part $\varphi(\mathbf{x})$ asserts tha $\mathbf{x}$ is the z required to exist by the axiom of infinity. Clearly $(V,\in)\models\varphi(\omega)$ and then by (i), $(M,\in)\models\varphi(\omega)$ so that $(M,\in)$ satisfies the axiom of infinity. ⊣

The lemma implies immediately that $(L,\in)$ satisfies all the axioms of *ZF* except perhaps for the power and replacement axioms. The key to deriving these for $(L,\in)$ is the Reflection Theorem 8C.4.

It is worth putting down a general result.

8D.4. THOEREM. *Let $\xi\mapsto C_\xi$ be an operation on ordinals to sets which is definable in $\mathfrak{L}^\in$ and satisfies the following four conditions.*

(i) *Each C_ξ is a transitive set.*

(ii) $\zeta\leq\xi\Rightarrow C_\zeta\subseteq C_\xi$.

(iii) *If λ is a limit ordinal, then $C_\lambda=\bigcup_{\xi<\lambda}C_\xi$.*

(iv) *For each ξ, if $x\subseteq C_\xi$ is elementary in the structure*

$$\mathfrak{C}_\xi=(C_\xi,\in,\{s: s\in C_\xi\}),$$

then there is some ζ such that $x\in C_\zeta$. Let

$$C=\bigcup_\xi C_\xi;$$

then the structure

$$\mathfrak{C} = (C, \in)$$

is a model of ZF which furthermore contains all the ordinals.

In particular, $(L, \in)$ *and each* $(L(A), \in)$ *are models of ZF which contains all the ordinals.*

PROOF. To begin with, we know from 8D.3 that $\mathfrak{C}$ satisfies extensionality, foundation, pairing and union, since condition (iv) in the hypothesis implies easily that C is closed under pairing and union.

We argue that C must contain all ordinals; if not, let λ be the least ordinal not in C and choose ξ large enough so that $\lambda \subseteq C_\xi$. Let $\varphi(\mathbf{x})$ be a formula in Σ_0 such that

$$Ordinal(x) \Leftrightarrow (V, \in) \models \varphi(x);$$

this is easy to construct from the expression given in #18 of 8C.1. Since no $ordinal \geq \lambda$ can be in C_ξ (by transitivity), we have

$$\{x \in C_\xi : \mathfrak{C}_\xi \models \varphi(x)\} = \lambda;$$

hence by condition (iv), $\lambda \in C$, which is a contradiction.

It follows in particular that $\omega \in C$, so that $\mathfrak{C}$ satisfies the axiom of infinity by 8D.3.

Verification of the power axiom. It is enough to show that for each $x \in C$, there is some $z \in C$ such that z has as members precisely all the members of C which are subsets of x—from this we can infer that $\mathfrak{C}$ satisfies the power axiom by arguments familiar from the proof of 8D.3 and above.

Consider the operation

$$F(u) = \begin{cases} \text{least } \xi \text{ such that } u \in C_\xi, & \text{if} \quad u \in C \;\&\; u \subseteq x, \\ 0 & \text{otherwise;} \end{cases}$$

this is obviously definable in $\mathfrak{L}^\in$, so by the replacement axiom, the image $F[Power(x)]$ is a set. (We are using the fact that x has a power set in V.) Now $F[Power(x)]$ is a set of ordinals, so there must be some ordinal λ above all of them and we have: if $u \in C$ and $u \subseteq x$, then $u \in C_\lambda$. Thus

$$z = \{u \in C_\lambda : u \subseteq x\}$$

has as members precisely the subsets of x which are in C and since z is clearly definable in $\mathfrak{C}_\lambda$, it is a member of C by (iv).

Verification of the axiom scheme of replacement. Suppose $x \in C$ and $F: C \to C$ is an operation which is definable (with parameters) on $\mathfrak{C}$, i.e. for some formula $\psi(\mathbf{v}_1,\ldots,\mathbf{v}_n,\mathbf{s},\mathbf{t})$ and fixed $y_1,\ldots,y_n \in C$,

$$F(s)=t \Leftrightarrow \mathfrak{C} \models \psi(y_1,\ldots,y_n,s,t) \quad (s,t \in C);$$

as above, it is enough to show that the image

$$F[x]=\{F(s): s \in x\}$$

is also a member of C.

Using the reflection theorem, choose λ so that $x, y_1,\ldots,y_n \in C_\lambda$ and for $s, t \in C_\lambda$,

$$\mathfrak{C} \models \psi(y_1,\ldots,y_n,s,t) \Leftrightarrow (C_\lambda, \in) \models \psi(y_1,\ldots,y_n,s,t)$$

and make sure as in the argument above that $F[x] \subseteq C_\lambda$; clearly

$$F[x]=\{t \in C_\lambda : \mathfrak{C}_\lambda \models (\exists \mathbf{s})[\mathbf{s} \in x \ \& \ \psi(y_1,\ldots,y_n,\mathbf{s},t)\}$$

and hence $F[x]$ is elementary in $\mathfrak{C}_\lambda$ and must be in C by (iv).

This concludes the proof of the main part of the theorem and the fact that L and $L(A)$ satisfy the hypotheses follows easily from their definitions. ⊣

EXERCISES

Let us take up first a few simple exercises which will help clarify the definability notions we have been using.

8D.5. Show that if $R(x_1,\ldots,x_n)$ is definable by a Σ_0 formula, then the relation

$$R^*(k_1,\ldots,k_n) \Leftrightarrow k_1 \in \omega \ \& \cdots \& \ k_n \in \omega \ \& \ R(k_1,\ldots,k_n)$$

is recursive. ⊣

A little thinking is needed for the next one.

8D.6. Prove that the relation of satisfaction in #24 of 8C.4 is not definable by a Σ_0 formula. ⊣

A formula of $\mathcal{L}^\epsilon$ is in Σ_1 if it is of the form

$$(\exists \mathbf{y}_1)\cdots(\exists \mathbf{y}_n)\varphi(\mathbf{y}_1,\ldots,\mathbf{y}_n,\mathbf{x}_1,\ldots,\mathbf{x}_m),$$

where $\varphi(\mathbf{y}_1,\ldots,\mathbf{y}_n,\mathbf{x}_1,\ldots,\mathbf{x}_m)$ is Σ_0. The next result is not entirely trivial and we will not appeal to it later, but it is useful to understand.

8D.7. Prove that every relation, operation and object in #1–#26 of 8C.1 and 8D.1 is definable by a Σ_1 formula. (A weak form of the axiom of choice is needed.)

Hint. The key fact is to check that if a relation $R(y, x_1,\ldots, x_n)$ is definable by a Σ_1 formula, then so is the relation

$$Q(z, x_1,\ldots, x_n) \Leftrightarrow (\forall y \in z)R(y, x_1,\ldots, x_n). \qquad \dashv$$

8D.8. Prove that if $\varphi(\mathbf{x}_1,\ldots,\mathbf{x}_n)$ is a Σ_1 formula, M is a transitive class, $x_1,\ldots, x_n \in M$ and $(M, \in)\models\varphi(x_1,\ldots, x_n)$, then $(V, \in)\models\varphi(x_1,\ldots, x_n)$. $\dashv$

8D.9. Suppose $F: V \to V$ is an operation definable by a Σ_1 formula $\varphi(\mathbf{x}, \mathbf{y})$ and suppose that we can prove in ZF that $(\forall\mathbf{x})(\exists!\mathbf{y})\varphi(\mathbf{x}, \mathbf{y})$; prove that for every transitive model $(M, \in)$ of ZF,

$$x \in M \Rightarrow F(x) \in M.$$

Hint. Use the fact that $(M, \in)\models(\forall\mathbf{x})(\exists!\mathbf{y})\varphi(\mathbf{x}, \mathbf{y})$, which follows from the hypotheses that $(M, \in)\models ZF$ and that $(\forall\mathbf{x})(\exists!\mathbf{y})\varphi(\mathbf{x}, \mathbf{y})$ is a theorem of ZF. $\dashv$

The rank of a set x is the least ordinal ξ such that $x \in V_{\xi+1}$. It is easy to check that this can be defined also by the recursion on $\in$

$$rank(x) = supremum\{rank(y)+1: y \in x\}$$

(where $supremum(\emptyset) = 0$) and that for each ordinal ξ,

$$rank(\xi) = \xi.$$

8D.10. Show that for each ξ, $L_\xi \subseteq V_\xi$ so that

$$x \in L_\xi \Rightarrow rank(x) \leq \xi.$$

Show also that for each ξ,

$$\xi \in L_{\xi+1}.$$

Hint. $L_\xi \subseteq V_\xi$ is easy by induction. The second assertion follows from

$$\xi = \{x \in L_\xi: Ordinal(x)\},$$

8D.3 and the fact that $Ordinal(x)$ can be defined by a Σ_0 formula. $\dashv$

8D.11. Show that for each ordinal $\xi \geq \omega$,

$$card(L_\xi) = card(\xi).$$

Hint. Use induction on ξ and a bit of cardinal arithmetic; you need to check (without using the axiom of choice) that if $\varkappa$ is a cardinal, then $\bigcup\{{}^n\varkappa : n \in \omega\}$ and $\varkappa \times \varkappa$ do not have bigger cardinality than $\varkappa$. ⊣

The nature of the constructible hierarchy $\{L_\xi : \xi \in ON\}$ makes it possible to define explicitly a wellordering of L. We will outline a proof of this result in some detail in the next exercise, as it is the key to our showing in the next section that the axiom of choice holds in $(L, \in)$.

8D.12. Prove that there is a binary relation $x \leq_L y$ which is definable in $\mathcal{L}^\in$ and wellorders L and such that

$$x \leq_L y \ \&\ y \in L_\xi \Rightarrow x \in L_\xi.$$

In particular, if $V = L$ (i.e. if every set is in L), then the axiom of choice holds.

Hint. The idea is to construct an operation

$$F : ON \to V$$

which is definable in $\mathcal{L}^\in$ and such that for each ξ, $F(\xi)$ is a wellordering of L_ξ—i.e. $F(\xi) \subseteq L_\xi \times L_\xi$ and the relation

$$u \leq_\xi v \Leftrightarrow \langle u, v\rangle \in F(\xi)$$

wellorders L_ξ.

We will build up F step-by-step.

1. There is an operation $F_1 : \omega \times V \times V \to V$ which is definable in $\mathcal{L}^\in$ and such that if w wellorders A, then $F_1(n, w, A)$ wellorders ${}^nA$.

Hint. Order the n-tuples from A lexicographically, using w.

2. There is an operation $F_2 : V^2 \to V$ which is definable in $\mathcal{L}^\in$ and such that if w wellorders A, then $F_2(w, A)$ wellorders $\bigcup\{{}^nA : n \in \omega\}$.

Hint. For x, x' in $\bigcup\{{}^nA : n \in \omega\}$ put

$$\begin{aligned}\langle x, x'\rangle \in F(w, A) \Leftrightarrow\ & Domain(x) < Domain(x') \\ & \vee (\exists n)[Domain(x) = Domain(x') = n \\ & \quad \& \ \langle x, x'\rangle \in F_1(n, w, A)].\end{aligned}$$

3. There is an operation $F_3 : V^2 \to V$ which is definable in $\mathcal{L}^\in$ and such that if w wellorders A, then $F_3(w, A)$ wellorders $\textbf{\textit{Def}}(A)$.

Hint. Let

$$G_1(m, n, x, A) = \mathbf{Def}_1(m, n, A, \{\langle u, v\rangle : u \in A \ \&\ v \in A \ \&\ u \in v\})$$

and for $y \in \mathbf{Def}(A)$ define successively:

$$G_2(y, w, A) = \text{least } m \text{ such that } (\exists n)(\exists x \in {}^nA)[y = G_1(m, n, x, A)],$$

$$G_3(y, w, A) = \text{least } n \text{ such that } (\exists x \in {}^nA)[y = G_1(G_2(y, w, A), n, x, A)],$$

$$G_4(y, w, A) = \text{least } x \text{ in the ordering } F_2(w, A) \text{ such that}$$
$$y = G_1(G_2(y, w, A), G_3(y, w, A), x, A).$$

Now each $y \in \mathbf{Def}(A)$ is completely determined by the triple

$$(G_2(y, w, A), G_3(y, w, A), G_4(y, w, A))$$

and we can order these triples lexicographically, using the wellordering $F_2(w, A)$ in the last component.

4. There is an operation $F : ON \to V$ which is definable in L and such that for each ξ, $F(\xi)$ is a wellordering of L_ξ.

Hint. We define F by recursion on the ordinals (8C.2), in the form where we take cases on 0, successors and limits as in 8C.8.

Clearly $F(0) = \emptyset$ and at the successor step,

$$F(\xi + 1) = F_3(F(\xi), L_\xi).$$

If λ is limit, define first $G : L \to ON$ by

$$G(x) = \text{least } \xi \text{ such that } x \in L_\xi$$

and put

$$F(\lambda) = \{\langle x, y\rangle \in L_\lambda \times L_\lambda : G(x) < G(y)$$
$$\vee [G(x) = G(y) \ \&\ \langle x, y\rangle \in F(G(x))]\}.$$

The result follows from this by setting again

$$x \leq_L y \Leftrightarrow G(x) < G(y) \vee [G(x) = G(y) \ \&\ \langle x, y\rangle \in F(G(x))].$$

Notice that initial segments are sets, since

$$x \leq_L y \Rightarrow x \in L_{G(y)}.$$

All the usual forms of the axiom of choice follow trivially from the first assertion of the result. ⊣

8E. Absoluteness

In the last result of the preceding section (8D.12) we just about proved that the structure $(L, \in)$ satisfies the axiom of choice. In fact we showed that there is a certain definable relation $x \leq_L y$ such that

(1) *if every set is in L, then* $\{(x, y): x \leq_L y\}$ *wellorders the collection of all sets;*

since in the structure $(L, \in)$ every set is obviously in L, the structure $(L, \in)$ satisfies the hypothesis of (1) and therefore it must satisfy the conclusion, which is a very strong "global" form of the axiom of choice.

To see that matters are not quite as simple as that, let us try to express (1) in the language $\mathcal{L}^{\in}$. Choose first a formula $\varphi_L(\mathbf{x}, \boldsymbol{\xi})$ of $\mathcal{L}^{\in}$ by 8D.2 so that

(2) $$x \in L_\xi \Leftrightarrow (V, \in) \models \varphi_L(x, \xi)$$

and let

(3) $$V = L \Leftrightarrow (\forall \mathbf{x})(\exists \boldsymbol{\xi})\varphi_L(\mathbf{x}, \boldsymbol{\xi}),$$

so that this formal sentence "$V = L$" clearly expresses in $\mathcal{L}^{\in}$ the proposition that *every set is constructible.* Choose then another formula $\psi_L(\mathbf{x}, \mathbf{y})$ of $\mathcal{L}^{\in}$ by 8D.12 such that

$$x \leq_L y \Leftrightarrow (V, \in) \models \psi_L(x, y)$$

and take

$$\psi^* \Leftrightarrow \text{"}\{(\mathbf{x}, \mathbf{y}): \psi_L(\mathbf{x}, \mathbf{y})\} \text{ is a wellordering of } V\text{"},$$

where it is easy to turn the symbolized English in quotes into a formal sentence of $\mathcal{L}^{\in}$. Now (1) is expressed by the formal sentence of $\mathcal{L}^{\in}$

(4) $$V = L \rightarrow \psi^*$$

and what we would like to prove is that

(5) $$(L, \in) \models \psi^*.$$

It is important here that 8D.12 was proved on the basis of the axioms in *ZF* without appeal to the axiom of choice. Since $(L, \in)$ is a model of *ZF* by 8D.4, it must also satisfy all the consequences of *ZF* and certainly

(6) $$(L, \in) \models V = L \rightarrow \psi^*.$$

Now the hitch is that in order to infer (5) from (6), we must prove

$$(L, \in) \models V = L; \tag{7}$$

this is what we took as "obvious" in the first paragraph above, after expressing it sloppily in English by "every set in the structure $(L, \in)$ is in L." But is (7) obvious?

By the definition of satisfaction and the construction of the sentence $V = L$ above, (7) is equivalent to

(8) for each $x \in L$, there exists $\xi \in L$ such that $(L, \in) \models \varphi_L(x, \xi)$

while what we know is

(9) for each $x \in L$, there exists $\xi \in L$ such that $(V, \in) \models \varphi_L(x, \xi)$.

Thus, to complete the proof of (7) and verify that $(L, \in)$ satisfies the axiom of choice, we must prove that we can choose the formula $\varphi_L(\mathbf{x}, \boldsymbol{\xi})$ so that in addition to (2), it also satisfies

$$(V, \in) \models \varphi_L(x, \xi) \Leftrightarrow (L, \in) \models \varphi_L(x, \xi), \tag{10}$$

when $x \in L$. In other words, we must show that *the basic relation of constructibility can be defined in* $\mathcal{L}^\in$ *so that the model* $(L, \in)$ *recognizes that each of its members is constructible.*

The theory of *absoluteness* which we will develop to do this is the key to many other results, including the fact that $V = L$ implies the *generalized continuum hypothesis*. We will study here the basic facts about absoluteness and then we will derive the consequences about L in 8F.

Since we will be considering only structures of the form $(M, \in)$ let us simplify notation and write

$$M \models \varphi(x_1, \ldots, x_n) \Leftrightarrow (M, \in) \models \varphi(x_1, \ldots, x_n);$$

similarly, for sets of sentences of $\mathcal{L}^\in$,

$$M \models T \Leftrightarrow (M, \in) \models T$$
$$\Leftrightarrow \text{for each } \varphi \in T, (M, \in) \models \varphi.$$

We will call M a *standard model* of a set of sentences T in $\mathcal{L}^\in$ (*a set theory*) if M is a transitive class (perhaps a set) and $M \models T$; if in addition M contains all the ordinals, we will call M an *inner model* of T—so that by 8D.4, *L and each $L(A)$ are inner models of ZF.*

Let $\mathfrak{D}$ be a collection of transitive classes and let R be an n-ary relation on V. We say that R is *definable in* $\mathcal{L}^{\epsilon}$ *absolutely for* $\mathfrak{D}$ or simply *absolute for (classes in)* $\mathfrak{D}$ if there exists a formula $\varphi(\mathbf{x}_1,\ldots,\mathbf{x}_n)$ of $\mathcal{L}^{\epsilon}$ such that for every M in $\mathfrak{D}$ and $x_1,\ldots,x_n$ in M,

$$R(x_1,\ldots,x_n) \Leftrightarrow M \models \varphi(x_1,\ldots,x_n).$$

This notion is the key metamathematical tool for the study of models of set theory.

Notice that if $\varphi(\mathbf{x}_1,\ldots,\mathbf{x}_n)$ *defines* R *absolutely for* $\mathfrak{D}$, then in particular for M, N in $\mathfrak{D}$, if $M \subseteq N$ and $x_1,\ldots,x_n \in M$, then

$$M \models \varphi(x_1,\ldots,x_n) \Leftrightarrow N \models \varphi(x_1,\ldots,x_n).$$

In all the cases we will consider, the universe V will be in $\mathfrak{D}$; then for each M in $\mathfrak{D}$ and $x_1,\ldots,x_n \in M$, we have

$$\begin{aligned} R(x_1,\ldots,x_n) &\Leftrightarrow V \models \varphi(x_1,\ldots,x_n) \\ &\Leftrightarrow M \models \varphi(x_1,\ldots,x_n). \end{aligned}$$

Following the same idea, an operation $F: C_1 \times \cdots \times C_n \to V$ (where $C_1,\ldots,C_n$ are given classes) is *definable in* $\mathcal{L}^{\epsilon}$ *absolutely for* $\mathfrak{D}$ or just *absolute for* $\mathfrak{D}$, if three things hold.

(1) The classes $C_1, C_2,\ldots, C_n$ are absolute for $\mathfrak{D}$—i.e. each membership relation $x \in C_i$ is absolute for $\mathfrak{D}$.

(2) If $M \in \mathfrak{D}$ and $x_1 \in C_1 \cap M,\ldots, x_n \in C_n \cap M$, then

$$F(x_1,\ldots,x_n) \in M.$$

(3) There is a formula $\varphi(\mathbf{x}_1,\ldots,\mathbf{x}_n,\mathbf{y})$ of $\mathcal{L}^{\epsilon}$ such that for each $M \in \mathfrak{D}$ and $x_1 \in C_1 \cap M,\ldots, x_n \in C_n \cap M$,

$$F(x_1,\ldots,x_n) = y \Leftrightarrow M \models \varphi(x_1,\ldots,x_n,y).$$

An object c is *absolute for* $\mathfrak{D}$ if for each M in $\mathfrak{D}$,

$$c \in M$$

and the relation

$$R_c(x) \Leftrightarrow x = c$$

is absolute for $\mathfrak{D}$.

It is common to also call *absolute for* $\mathfrak{D}$ the relevant formula $\varphi(\mathbf{x}_1,\ldots,\mathbf{x}_n)$ of $\mathcal{L}^{\epsilon}$ which defines a given relation, operation or constant as above.

If $\mathfrak{D}$ consists of just two classes, V and some M, we will say *absolute for M* instead of *absolute for $\mathfrak{D}$*.

We now come to the important metamathematical concept of *ZF*-absoluteness.

Let us collectively call *notions* the relations and operations on V as well as the members of V. A notion N is *ZF-absolute* if there exists a finite set $T^0 \subseteq ZF$ of axioms in ZF such that N is absolute for the collection $\mathfrak{D}$ of all standard models of T^0,

$$M \in \mathfrak{D} \Leftrightarrow M \models T^0.$$

Intuitively, a notion N is *ZF*-absolute if there is a formula of $\mathcal{L}^{\in}$ which defines N in all sufficiently good approximations to standard models of *ZF*.

We will need to know that a good many notions are *ZF*-absolute. Before embarking on this let us establish the closure properties of the collection of *ZF*-absolute notions in the next simple but basic theorem.

8E.1. THEOREM. (i) *The collection of ZF-absolute relations contains* $\in$ *and is closed under the propositional operations* $\neg, \&, \vee, \Rightarrow, \Leftrightarrow$.

(ii) *The collection of ZF-absolute operations is closed under addition and permutation of variables and under composition; each n-ary projection operation*

$$F(x_1,\ldots, x_n) = x_i$$

is ZF-absolute.

(iii) *An object $c \in V$ is ZF-absolute if and only if each n-ary constant operation*

$$F(x_1,\ldots, x_n) = c$$

is ZF-absolute.

(iv) *If $R \subseteq V^m$ and $F_1\colon C_1 \times \cdots \times C_n \to V,\ldots, F_m\colon C_1 \times \cdots \times C_n \to V$ are all ZF-absolute and*

$$P(x_1,\ldots, x_n) \Leftrightarrow x_1 \in C_1 \,\&\, \cdots \,\&\, x_n \in C_n \,\&\, R(F_1(x_1,\ldots, x_n),\ldots, F_m(x_1,\ldots, x_n)),$$

then P is also ZF-absolute.

(v) *If $R \subseteq V^{n+1}$ is ZF-absolute and*

$$P(x_1,\ldots, x_n, z) \Leftrightarrow (\exists y \in z) R(x_1,\ldots, x_n, y),$$

$$Q(x_1,\ldots, x_n, z) \Leftrightarrow (\forall y \in z) R(x_1,\ldots, x_n, y),$$

then P and Q are also ZF-absolute.

(vi) *Suppose* $P \subseteq V^{n+1}$ *is ZF-absolute,* $Q \subseteq V^{n+1}$ *and there exists a finite* $T^0 \subseteq ZF$ *such that for each transitive* $M \models T^0$ *and* $x_1, \ldots, x_n \in M$,

$$(\exists y \in M)P(x_1, \ldots, x_n, y) \Leftrightarrow (\forall y \in M)Q(x_1, \ldots, x_n, y);$$

then the relation $R \subseteq V^n$ *defined by*

$$R(x_1, \ldots, x_n) \Leftrightarrow (\exists y)P(x_1, \ldots, x_n, y)$$

is ZF-absolute.

(vii) *If* $G: V^{n+1} \to V$ *is ZF-absolute, then so is* $F: V^{n+1} \to V$ *defined by*

$$F(x_1, \ldots, x_n, w) = \{G(x_1, \ldots, x_n, t): t \in w\};$$

similarly with more variables, if $G: V^{n+m} \to V$ *is ZF-absolute, so is*

$$F(x_1, \ldots, x_n, w_1, \ldots, w_m) = \{G(x_1, \ldots, x_n, t_1, \ldots, t_m): t_1 \in w_1 \,\&\, \cdots \,\&\, t_m \in w_m\}.$$

(viii) *If* $R \subseteq V^{n+1}$ *is ZF-absolute, then so is the operation*

$$F(x_1, \ldots, x_n, w) = \{t \in w: R(x_1, \ldots, x_n, t)\}.$$

PROOF. Parts (i)–(iv) are very easy, using the basic properties of the language $\mathcal{L}^{\in}$.

For example if

$$R(x_1, \ldots, x_n) \Leftrightarrow P(x_1, \ldots, x_n) \,\&\, Q(x_1, \ldots, x_n)$$

with P and Q given ZF-absolute relations, choose finite $T^0 \subseteq ZF$, $T^1 \subseteq ZF$ and formulas $\varphi(\mathbf{x}_1, \ldots, \mathbf{x}_n)$, $\psi(\mathbf{x}_1, \ldots, \mathbf{x}_n)$ of $\mathcal{L}^{\in}$ such that for $M \models T^0$, $x_1, \ldots, x_n \in M$,

$$P(x_1, \ldots, x_n) \Leftrightarrow M \models \varphi(x_1, \ldots, x_n)$$

and for $M \models T^1$, $x_1, \ldots, x_n \in M$,

$$Q(x_1, \ldots, x_n) \Leftrightarrow M \models \psi(x_1, \ldots, x_n).$$

It is clear that if $M \models T^0 \cup T^1$ and $x_1, \ldots, x_n \in M$, then

$$R(x_1, \ldots, x_n) \Leftrightarrow M \models \varphi(x_1, \ldots, x_n) \,\&\, \psi(x_1, \ldots, x_n),$$

so the formula $\varphi(\mathbf{x}_1, \ldots, \mathbf{x}_n) \,\&\, \psi(\mathbf{x}_1, \ldots, \mathbf{x}_n)$ defines R absolutely on all standard models of $T^0 \cup T^1$.

Suppose again that

$$F(x) = G(H_1(x), H_2(x))$$

where G, H_1, H_2 are ZF-absolute and we have chosen on binary and two unary operations to simplify notation. Choose finite subsets T^G, T^1, T^2 of ZF and formulas $\psi(\mathbf{u}, \mathbf{v}, \mathbf{z})$, $\varphi_1(\mathbf{x}, \mathbf{u})$, $\varphi_2(\mathbf{x}, \mathbf{v})$ of $\mathcal{L}^\in$ such that for $M \models T^G$ and $u, v, z \in M$, we have $G(u, v) \in M$ and

$$G(u, v) = z \Leftrightarrow M \models \psi(u, v, z)$$

and similarly with H_1, T^1 and $\varphi_1(\mathbf{x}, \mathbf{u})$, H_2, T^2 and $\varphi_2(\mathbf{x}, \mathbf{v})$. (It is easy to arrange that the free variables in these formulas are as indicated.) Now it is clear that if

$$M \models T^G \cup T^1 \cup T^2,$$

then

$$x \in M \Rightarrow F(x) \in M$$

and for $x, z \in M$,

$$F(x) = z \Leftrightarrow M \models \chi(x, z)$$

where

$$\chi(\mathbf{x}, \mathbf{z}) \Leftrightarrow (\exists \mathbf{u})(\exists \mathbf{v})[\varphi_1(\mathbf{x}, \mathbf{u}) \;\&\; \varphi_2(\mathbf{x}, \mathbf{v}) \;\&\; \psi(\mathbf{u}, \mathbf{v}, \mathbf{z})].$$

Proof of (iv) is very similar to this.

(v) The argument is very similar to the proof of (i) in 8D.3 and we will omit it—the transitivity of M is essential here.

(vi) Choose a formula $\varphi(\mathbf{x}_1, \ldots, \mathbf{x}_n, \mathbf{y})$ and a finite $T^P \subseteq ZF$ such that for all $M \models T^P$ and $x_1, \ldots, x_n \in M$,

$$P(x_1, \ldots, x_n, y) \Leftrightarrow M \models \varphi(x_1, \ldots, x_n, y)$$

and take

$$\chi(\mathbf{x}_1, \ldots, \mathbf{x}_n) \Leftrightarrow (\exists \mathbf{y})\varphi(\mathbf{x}_1, \ldots, \mathbf{x}_n, \mathbf{y}).$$

If $M \models T^P \cup T^0$ and $x_1, \ldots, x_n \in M$, then

$$\begin{aligned}
R(x_1, \ldots, x_n) &\Rightarrow (\exists y)P(x_1, \ldots, x_n, y) \\
&\Rightarrow (\forall y)Q(x_1, \ldots, x_n, y) && \text{(since } V \models T^0\text{)} \\
&\Rightarrow (\forall y \in M)Q(x_1, \ldots, x_n, y) && \text{(obviously)} \\
&\Rightarrow (\exists y \in M)P(x_1, \ldots, x_n, y) && \text{(since } M \models T^0\text{)} \\
&\Rightarrow \text{for some } y \in M,\ M \models \varphi(x_1, \ldots, x_n, y) && \text{(since } M \models T^P\text{)} \\
&\Rightarrow M \models (\exists \mathbf{y})\varphi(x_1, \ldots, x_n, \mathbf{y});
\end{aligned}$$

Conversely,

$$\begin{aligned}
M \models (\exists \mathbf{y})\varphi(x_1, \ldots, x_n, \mathbf{y}) &\Rightarrow (\exists y \in M)P(x_1, \ldots, x_n, y) \\
&\Rightarrow (\exists y)P(x_1, \ldots, x_n, y) \\
&\Rightarrow R(x_1, \ldots, x_n),
\end{aligned}$$

so $\chi(\mathbf{x}_1,\ldots,\mathbf{x}_n,\mathbf{y})$ defines R on all models of $T^P \cup T^0$ and hence R is ZF-absolute.

(viii) Suppose that if $M \models T^0$, then

$$x_1,\ldots,x_n, t \in M \Rightarrow G(x_1,\ldots,x_n,t) \in M$$

and

$$G(x_1,\ldots,x_n,t) = s \Leftrightarrow M \models \varphi(x_1,\ldots,x_n,s,t).$$

Let ψ be the instance of the replacement axiom scheme which concerns $\varphi(\mathbf{x}_1,\ldots,\mathbf{x}_n,\mathbf{t},\mathbf{s})$,

$$\psi \Leftrightarrow (\forall \mathbf{x}_1)\cdots(\forall \mathbf{x}_n)(\forall \mathbf{w})\{(\forall \mathbf{t})(\exists ! \mathbf{s})\varphi(\mathbf{x}_1,\ldots,\mathbf{x}_n,\mathbf{t},\mathbf{s}) \\ \rightarrow (\exists \mathbf{z})(\forall \mathbf{s})[\mathbf{s} \in \mathbf{z} \leftrightarrow (\exists \mathbf{t})[\mathbf{t} \in \mathbf{w} \mathbin{\&} \varphi(\mathbf{x}_1,\ldots,\mathbf{x}_n,\mathbf{t},\mathbf{s})]]\}$$

and take

$$T^1 = T^0 \cup \{\psi\}.$$

If $M \models T^1$ and $x_1,\ldots,x_n$, $w \in M$, this means easily that there is some $z \in M$ so that for all $s \in M$,

$$\begin{aligned} s \in z &\Leftrightarrow \text{for some } t \in w,\ M \models \varphi(x_1,\ldots,x_n,t,s) \\ &\Leftrightarrow (\exists t \in w)[G(x_1,\ldots,x_n,t) = s]. \end{aligned}$$

Since $M \models T^0$ and hence M is closed under G, this implies that in fact

$$\begin{aligned} z &= \{G(x_1,\ldots,x_n,t) : t \in w\} \\ &= F(x_1,\ldots,x_n,w), \end{aligned}$$

hence M is closed under F. Moreover, taking

$$\chi(\mathbf{x}_1,\ldots,\mathbf{x}_n,\mathbf{w},\mathbf{z}) \Leftrightarrow (\forall \mathbf{s})[\mathbf{s} \in \mathbf{z} \leftrightarrow (\exists \mathbf{t})[\mathbf{t} \in \mathbf{w} \mathbin{\&} \varphi(\mathbf{x}_1,\ldots,\mathbf{x}_n,\mathbf{t},\mathbf{s})]],$$

it is clear that

$$F(x_1,\ldots,x_n,w) = z \Leftrightarrow M \models \chi(x_1,\ldots,x_n,w,z),$$

so F is ZF-absolute.

The argument with $m > 1$ is similar.

(viii) Let

$$G(x_1,\ldots,x_n,z,t) = \begin{cases} t & \text{if} \quad R(x_1,\ldots,x_n,t), \\ z & \text{if} \quad \neg R(x_1,\ldots,x_n,t), \end{cases}$$

so that G is ZF-absolute and by (vii), the operation

$$F_1(x_1,\ldots,x_n,z,w) = \{G(x_1,\ldots,x_n,z,t) : t \in w\} \cap w$$

is also (easily) ZF-absolute. Clearly

$$s \in F_1(x_1,\dots, x_n, z, w) \Leftrightarrow s \in w \;\&\; R(x_1,\dots, x_n, s)$$
$$\vee[s = z \;\&\; z \in w \;\&\; (\exists t)\neg R(x_1,\dots, x_n, t)];$$

since ZF implies that we cannot have $w \in w$, we then have

$$s \in F_1(x_1,\dots, x_n, w, w) \Leftrightarrow s \in w \;\&\; R(x_1,\dots, x_n, s)$$

and we can take

$$F(x_1,\dots, x_n, w) = F_1(x_1,\dots, x_n, w, w). \qquad \dashv$$

We will now apply this basic theorem to show that many natural notions are ZF-absolute.

8E.2. THEOREM. *The notions #1–#20 of Theorem* 8C.1 *are all ZF-absolute.*

PROOF. It is enough to establish the ZF-absoluteness of #1–#5, since the notions following #5 are easily proved absolute using 8E.1. Now $\in$, $\subseteq$ and $=$ are definable by Σ_0 formulas, so they are ZF-absolute by 8D.3; we will outline the proof for pairing, that for union being similar.

Let ψ be the formal sentence of $\mathcal{L}^{\in}$ which expresses the axiom of pairing and suppose M is a transitive class which satisfies ψ. This means that for each x, y in M there is some w in M such that

$$(*) \qquad M \vDash (\forall \mathbf{t})[\mathbf{t} \in w \Leftrightarrow (\mathbf{t} = x \vee \mathbf{t} = y)].$$

We claim that in fact $w = \{x, y\}$; this is because $(*)$ simply means that (in V)

$$(\forall t \in M)[t \in w \Leftrightarrow (t = x \vee t = y)]$$

and since M is transitive we have $w \subseteq M$, so easily

$$(\forall t)[t \in w \Leftrightarrow (t = x \vee t = y)],$$

i.e. $w = \{x, y\}$.

This means that every transitive M which satisfies ψ is closed under the pairing operation. Since the relation $\{x, y\} = w$ is obviously definable by a Σ_0 formula which is absolute for all transitive classes, the operation $(x, y) \mapsto \{x, y\}$ is ZF-absolute. $\dashv$

Before proceeding to show the ZF-absoluteness of several other notions, it will be instructive to notice that *many natural and useful notions are not ZF-absolute.* Roughly speaking, no notion related to *cardinality* is *ZF*-absolute. The key for these proofs is the Countable Reflection Theorem 8C.10.

8E.3. THEOREM. *None of the following notions is ZF-absolute:* $\mathcal{N}$, $\mathcal{R}$, *Cardinal*$(\varkappa)$, $x \mapsto$ *Power*(x), $x \mapsto$ *Card*(x), (*Card*(x) = *least ordinal equinumerous with* x).

PROOF. If $\mathcal{N}$ were *ZF*-absolute, then $\mathcal{N}$ would be a member of every standard model A of some finite $T^0 \subseteq ZF$; but $\mathcal{N}$ is uncountable and T^0 has countable standard models by 8C.10.

To take another example, suppose the relation *Cardinal*$(\varkappa)$ were *ZF*-absolute, so that there exists a formula $\varphi(\boldsymbol{\varkappa})$ of $\mathcal{L}^{\in}$ and a finite $T^0 \subseteq ZF$ such that for every transitive $M \models T^0$ and for $\varkappa \in M$,

$$Cardinal(\varkappa) \Leftrightarrow M \models \varphi(\varkappa).$$

Now it is true in V that

$$(\exists x)(\exists \varkappa)[x = \omega \ \&\ Cardinal(\varkappa) \ \&\ \omega \in \varkappa]$$

and of course we can express this proposition by a formal sentence ψ, using $\varphi(\mathbf{x})$ to express *Cardinal*$(\varkappa)$ and expressing $x = \omega$ by some $\chi(\mathbf{x})$ which is absolute for all standard models of some $T^1 \subseteq ZF$. By the *ZF*-absoluteness of ω, we can also make sure that ω belongs to all standard models of T^1.

By the Countable Reflection Theorem 8C.10, choose a countable, transitive M such that $M \models T^0 \cup T^1 \cup \{\psi\}$. Clearly $\omega \in M$ and for some $\varkappa \in M$,

$$M \models \varphi(\varkappa) \ \&\ \omega \in \varkappa$$

so by the alleged absoluteness property of $\varphi(\boldsymbol{\varkappa})$,

$$Cardinal(\varkappa) \ \&\ \omega \in \varkappa.$$

However this is absurd since $\varkappa$ is a countable set (it belongs to a transitive countable set) and $\varkappa \neq \omega$. ⊣

(The proof we gave used the axiom of choice, but the result does not depend on this axiom by 8F.14.)

The next result is fundamental.

8E.4. Mostowski's Theorem. *The relation*

$$WF(r) \Leftrightarrow r \text{ is a wellfounded relation}$$

is ZF-absolute.

Proof. Put

$$\begin{aligned} P(r, x) \Leftrightarrow\ & r \text{ is a relation} \\ & \& \{\text{either } x \text{ is not a subset of } Field(r) \\ & \quad \text{or } x = \emptyset \text{ or } x \text{ has an } r\text{-minimal member}\} \\ \Leftrightarrow\ & Relation(r) \\ & \& \{(\exists t \in x)[t \notin Field(r) \\ & \quad \vee x = \emptyset \\ & \quad \vee (\exists u \in x)(\forall v \in x)[\langle v, u\rangle \in r \Rightarrow \langle u, v\rangle \in r]\}, \end{aligned}$$

using the notation of 8C.1. Clearly P is ZF-absolute and

$$WF(r) \Leftrightarrow (\forall x)P(r, x).$$

Similarly, let

$$\begin{aligned} Q(r, f) \Leftrightarrow\ & r \text{ is a relation} \\ & \& \{f \text{ is a function which maps } Field(r) \text{ into the} \\ & \quad \text{ordinals in an order-preserving fashion}\} \\ \Leftrightarrow\ & Relation(r) \ \&\ Function(f) \\ & \& Domain(f) = Field(r) \\ & \& (\forall \xi \in Image(f))Ordinal(\xi) \\ & \& (\forall x \in Field(r))(\forall y \in Field(r))\{[\langle x, y\rangle \in r \ \&\ \langle y, x\rangle \notin r \\ & \quad \Rightarrow f(x) < f(y)\}. \end{aligned}$$

Again Q is ZF-absolute and obviously

$$WF(r) \Leftrightarrow (\exists f)Q(r, f).$$

Hence

$$(*) \qquad (\forall r)\{(\forall x)P(x, r) \Leftrightarrow (\exists f)Q(r, f)\}.$$

Now we come to a subtle point in the argument. How did we recognize that $(*)$ is true? The answer is that *we proved* $(*)$ *from the axioms of ZFC*, the only assumptions we make about sets without explicit notice. If the reader takes the time to actually write down a proof of (∗), he will realize that *in that proof he does not use the axiom of choice and he appeals to only finitely many axioms of ZF.*

Let T^0, T^1 be finite subsets of ZF such that P and Q are absolute for standard models of T^0 and T^1 respectively and let T^* be the finite subset of ZF we needed to establish (∗). A moment's reflection shows that if M is a standard model of $T^0 \cup T^1 \cup T^*$, then M satisfies the formal sentence of $\mathcal{L}^{\in}$ which expresses (∗); in other words, for $r \in M$,

$$(**) \qquad (\forall x \in M)P(x, r) \leftrightarrow (\exists f \in M)Q(r, f).$$

Now part (vi) of 8E.1 implies immediately that $WF(r)$ is ZF-absolute. ⊣

The argument in this proof is typically metamathematical and will undoubtedly cause some uneasiness to those without a good background in logic.

One simple fact we used was that if $M \models T$ and φ is a logical consequence of T, then $M \models \varphi$. This is a basic property of mathematical proofs which has nothing to do with set theory—any logical consequence of group theory will hold in all groups, any property of Banach spaces whose proof does not appeal to the completeness of the norm will in fact hold for all normed linear spaces, etc.

The observation that in proving (∗) we only used a finite number of axioms from ZF (and that therefore (∗) holds in all models of these finitely many axioms) is a bit more subtle and it would be a good idea for the novice in metamathematics to actually put down a proof of (∗) and list all the axioms of ZF that are needed. (Assuming $WF(r)$ define $f: Field(r) \to ON$ by the recursion

$$f(x) = \bigcup \{f(y) \cup \{f(y)\}: y <_r x\},$$

see 8C.2.)

In fact, *any theorem of Zermelo–Fraenkel set theory which is expressible by a sentence of $\mathcal{L}^{\in}$ can be proved using only finitely many axioms of ZF*; this is because a "proof" in Zermelo–Fraenkel set theory is nothing but a finite sequence of propositions, all of them expressible in $\mathcal{L}^{\in}$ and each being either an axiom of ZF or a "purely logical consequence" of propositions preceding it. It follows that *the formal sentence of $\mathcal{L}^{\in}$ expressing some theorem of ZF holds in all standard models M of some sufficiently large, finite $T^0 \subseteq ZF$.*

We will often appeal to these metamathematical observations to save ourselves from having to put down long, complicated proofs. In principle, the reader could always supply the finite $T^0 \subseteq ZF$ needed, by working out in full detail a proof of the relevant theorem.

The same kind of metamathematical argument is needed in the proof of the next result.

8E.5. THEOREM. *Suppose $G: V^{n+1} \to V$ is a ZF-absolute operation and*

$$F: ON \times V^n \to V$$

is the unique operation satisfying

$$F(\xi, x_1,\ldots, x_n) = G(\{\langle \eta, F(\eta, x_1,\ldots, x_n)\rangle : \eta < \xi\}, x_1,\ldots, x_n);$$

then F is also ZF-absolute. (*ZF*-absoluteness of definition by recursion on the ordinals.)

PROOF. Assume G is absolute for all transitive models of $T^0 \subseteq ZF$. Go back to the proof of 8C.2 to recall that F is defined by an expression of the form

$$F(\xi, x_1,\ldots, x_n) = w \Leftrightarrow (\exists h)\{P(\xi, x_1,\ldots, x_h, h) \,\&\, Function(h) \\ \&\, \xi \in Domain(h) \,\&\, h(\xi) = w\},$$

where P is easily absolute for all models of T^0. Moreover, we can prove

$$(\forall \xi, x_1,\ldots, x_n)(\exists h)P(\xi, x_1,\ldots, x_n, h)$$

using only finitely many additional instances of the axiom scheme of replacement, say $T^1 \subseteq ZF$. Thus for every standard model M of $T^0 \cup T^1$ and $\xi, x_1,\ldots, x_n$ in M we have $(\exists h \in M)P(\xi, x_1,\ldots, x_n, h)$, which implies immediately that M is closed under F.

We can also prove easily in ZF (using only some finite $T^2 \subseteq ZF$) that

$$\begin{aligned}(\forall \xi, x_1,\ldots, x_n, w)\{(\exists h)[&P(\xi, x_1,\ldots, x_n, h) \,\&\, Function(h) \\ &\&\, \xi \in Domain(h) \,\&\, h(\xi) = w] \\ &\Leftrightarrow (\forall h)[[P(\xi, x_1,\ldots, x_n, h) \,\&\, Function(h) \\ &\&\, \xi \in Domain(h)] \Rightarrow h(\xi) = w]\};\end{aligned}$$

thus by part (vi) of 8E.1 the relation

$$R(\xi, x_1,\ldots, x_n, w) \Leftrightarrow F(\xi, x_1,\ldots, x_n) = w$$

is ZF-absolute and then easily F is ZF-absolute.

A special case of definition by recursion on ON is simple recursion on ω.

8E.6. THEOREM. *Suppose $F(k, x_1, \ldots, x_n)$ satisfies the recursion*

$$F(0, x_1, \ldots, x_n) = G_1(x_1, \ldots, x_n),$$
$$F(k+1, x_1, \ldots, x_n) = G_2(F(k, x_1, \ldots, x_n), k, x_1, \ldots, x_n)$$

where G_1 and G_2 are ZF-absolute. Then F is also ZF-absolute.

PROOF. Define

$$G(f, k, x_1, \ldots, x_n) = \begin{cases} G_1(x_1, \ldots, x_n) & \text{if } m = 0, \\ G_2(f(k-1, x_1, \ldots, x_n), k-1, x_1, \ldots, x_n) & \text{if } k \in \omega, k \neq 0, \\ 0 & \text{otherwise} \end{cases}$$

and verify easily that G is ZF-absolute and F is definable from G as in 8E.5. ⊣

8E.7. COROLLARY. *The relations and operations* #21–#26 *of* 8D.1 *are ZF-absolute.*

PROOF. Go back and reread the proof of 8D.1, keeping in mind the results of this section. The key part is the ZF-absoluteness of the satisfaction relation which comes directly from 8E.6. ⊣

EXERCISES

A very natural question to ask at this point is: *which analytical pointsets are ZF-absolute?* We have to be careful here, because the simplest pointset of type 1, $\mathcal{N}$ itself is not ZF-absolute as an object by 8E.3—simply because it is uncountable. On the other hand, *membership in $\mathcal{N}$* is easily ZF-absolute and it will turn out that all Σ^1_1 pointsets have this property.

To make the notions precise, call a set A absolute for (a collection of classes) $\mathfrak{D}$ *as a relation* or *ZF-absolute as a relation* if the corresponding membership relation in A,

$$R_A(x) \Leftrightarrow x \in A$$

is absolute in the relevant sense.

Similarly, a function

$$f: A \to B$$

is *absolute for* $\mathfrak{D}$ *as an operation* or *ZF-absolute as an operation*, if the operation

$$F(x) = \begin{cases} f(x) & \text{if} \quad x \in A, \\ 0 & \text{if} \quad x \notin A \end{cases}$$

is absolute in the relevant sense. For example, the identity function on $\mathcal{N}$ which sends α to α is ZF-absolute as an operation, but not as a set of ordered pairs (which is uncountable).

8E.8. Show that all arithmetical pointsets of type 0 or 1 are ZF-absolute as relations.

Hint. For type 0 the result follows easily from 8C.1 and the closure properties of ZF-absolute notions, 8E.1.

For type 1, first compute

$$\alpha \in \mathcal{N} \Leftrightarrow \mathit{Function}(\alpha) \;\&\; \mathit{Domain}(\alpha) = \omega \;\&\; \mathit{Image}(\alpha) \subseteq \omega;$$

so the relation

$$P(\alpha) \Leftrightarrow \alpha \in \mathcal{N}$$

is ZF-absolute. Again,

$$\alpha(n) = m \Leftrightarrow \alpha \in \mathcal{N} \;\&\; n \in \omega \;\&\; m \in \omega \;\&\; \langle n, m \rangle \in \alpha,$$

so the relation

$$Q(\alpha, n, m) \Leftrightarrow \alpha(n) = m$$

is ZF-absolute. The rest follows by 8E.1. ⊣

8E.9. Show that every Σ^1_1 pointset of type 0 or 1 is ZF-absolute as a relation; similarly, every Δ^1_1 function

$$f: \mathfrak{X} \to \mathfrak{Y}$$

with $\mathfrak{X}, \mathfrak{Y}$ of type 0 or 1 is ZF-absolute as an operation.

Hint. Given $Q \subseteq X_1 \times \cdots \times X_n$ with each X_i either ω or $\mathcal{N}$ and such that

$$Q(x_1, \ldots, x_n) \Leftrightarrow (\exists \alpha)(\forall t) P(x_1, \ldots, x_n, \bar{\alpha}(t))$$

with P recursive as in 4A.1, put

$$F(x_1,\ldots,x_n)=\{\langle u,v\rangle\in\omega\times\omega:\ Seq(u)\ \&\ Seq(v)$$
$$\&\ v \text{ codes an initial segment of the sequence coded by } u$$
$$\&\ P(x_1,\ldots,x_n,u)\}.$$

Clearly F is ZF-absolute as an operation by 8E.1 and 8E.8 and

$$Q(x_1,\ldots,x_n)\Leftrightarrow\neg WF(F(x_1,\ldots,x_n)),$$

so Q is ZF-absolute, by Mostowski's Theorem 8E.4.

To prove the second assertion, suppose

$$f:\mathfrak{X}\to\mathcal{N}$$

is Δ_1^1 and put

$$Q(x,\alpha,t,s)\Leftrightarrow f(x)(t)=s;$$

now Q is Δ_1^1 and hence ZF-absolute as a relation and clearly

$$f(x)=\{\langle t,x\rangle:\ t,s\in\omega\ \&\ Q(x,\alpha,t,s)\},$$

so that f is ZF-absolute as an operation by 8E.1.

The result for $f:\mathfrak{X}\to Y_1\times\cdots\times Y_k$ with each $Y_i=\omega$ or $\mathcal{N}$ follows easily. ⊣

In trying to extend this result to arbitrary product spaces we meet a problem: for some basic space X, the recursive function

$$i\mapsto r_i$$

which enumerates the fixed recursive presentation of X may already fail to be ZF-absolute. Suppose, for example, that we have carelessly adopted the definition of real numbers as equivalence classes of Cauchy sequences of rationals. Now each real x is an *uncountable* object and cannot be ZF-absolute by the argument of 8E.3.

In the case of R we can easily correct this situation by adopting some other definition of real numbers, e.g. in terms of Dedekind cuts, but for the general case we need the following lemma.

8E.10. Show that each basic space (X,d) (which admits a recursive presentation) is isomorphic with a space (X^*,d^*) where $X^*\subseteq\mathcal{N}$ and

(X^*, d^*) admits a recursive presentation $\{r_0^*, r_1^*, \ldots\}$ for which the relations

$$P(x) \Leftrightarrow x \in X^*,$$

$$Q(x, y, m, k) \Leftrightarrow d^*(x, y) < \frac{m}{k+1}$$

and the operation

$$i \mapsto r_i^*$$

are all ZF-absolute.

Hint. Given X with metric d and recursive presentation $\{r_0, r_1, \ldots\}$, choose by 4A.7 a Π_1^0 set $A \subseteq \mathcal{N}$ and a recursive $\pi : \mathcal{N} \to X$, such that π is injective on A and $\pi[A] = X$. Take $X^* = A$ and on X^* define

$$d^*(\alpha, \beta) = d(\pi(\alpha), \pi(\beta)).$$

Put also

$$r_i^* = f(r_i),$$

where $f : X \to \mathcal{N}$ is a Δ_1^1 inverse of π.

It is immediate from 8E.8 that the relations P and Q above are ZF-absolute, since they are arithmetical relations on spaces of type 1. It is also clear that $\{r_0^*, r_1^*, \ldots\}$ is a recursive presentation of X^*. Moreover, the map

$$i \mapsto r_i^*$$

is easily Δ_1^1 on ω to $\mathcal{N}$, so it is ZF-absolute by 8E.9. ⊣

From now on we assume that we have replaced all basic spaces by isomorphic copies, if necessary, so that the conditions of 8E.10 *hold.* This implies immediately that the same conditions hold for every product space. We might as well put this down as part of the stronger result that we need.

8E.11. Show that all Σ_1^1 pointsets are ZF-absolute as relations and all Δ_1^1 functions between product spaces are ZF-absolute as operations.

Hint. Check first that for each $\mathfrak{X} = X_1 \times \cdots \times X_k$ the membership relation

$$\begin{aligned} P^{\mathfrak{X}}(x) &\Leftrightarrow x \in \mathfrak{X} \\ &\Leftrightarrow (\exists x_1) \cdots (\exists x_k)[x_1 \in X_1 \;\&\; \cdots \;\&\; x_k \in X_k \\ &\qquad \& \; x = \langle x_1, \ldots, x_k \rangle] \end{aligned}$$

is ZF-absolute. (For transitive M, if $\langle x_1,\ldots,x_k\rangle \in M$, then $x_1,\ldots,x_k \in M$.) Similarly, the function $i \mapsto r_i$ and the basic nbhd relation $\{(x,s): x \in N(\mathfrak{X}, s)\}$ are easily ZF-absolute, from the same results about the basic spaces. Now every Σ^0_1 pointset is of the form

$$P(x) \Leftrightarrow (\exists s)\{x \in N(\mathfrak{X}, s) \,\&\, P^*(s)\}$$

with $P^* \subseteq \omega$, P^* semirecursive, so P is ZF-absolute. Finally, if P is Σ^1_1, then

$$P(x) \Leftrightarrow (\exists \alpha)(\forall t) R(x, \bar{\alpha}(t))$$

with a semirecursive R and the argument that this is ZF-absolute as a relation is exactly as in 8E.9 by appealing to Mostowski's Theorem 8E.4.

For Δ^1_1 functions again the argument is the same as in 8F.3, using the fact that for each basic space (X^*, d^*) we have $X^* \subseteq \mathcal{N}$ by our convention. ⊣

8F. The basic facts about L

Let us start by collecting in one theorem the basic absoluteness facts about the constructible hierarchy that follow from the results of 8E.

8F.1. THEOREM. (i) *The operation*

$$\xi \mapsto L_\xi$$

and the relation

$$P(x, \xi) \Leftrightarrow x \in L_\xi$$

are both ZF-absolute.

(ii) *There is a canonical wellordering of L, $x \leq_L y$ which is ZF-absolute and such that*

$$y \in L_\xi \,\&\, x \leq_L y \Rightarrow x \in L_\xi.$$

(iii) *The operation*

$$(\xi, A) \mapsto L_\xi(A)$$

and the relation

$$P'(x, \xi, A) \Leftrightarrow x \in L_\xi(A)$$

are both ZF-absolute.

(iv) *The relations*

$$Q(x) \Leftrightarrow x \in L,$$
$$Q'(x, A) \Leftrightarrow x \in L(A)$$

are both absolute for inner models of ZF.

PROOF. (i) and (iii) follows immediately from the definitions, 8E.7, 8E.5 and of course, the basic closure properties of ZF-absoluteness listed in 8E.1. Part (ii) also follows easily by examining the proof of 8D.12.

To prove (iv), let $\varphi_L(\mathbf{x}, \boldsymbol{\xi})$ be a formula of $\mathcal{L}^\in$ by (i) such that for some finite $T^0 \subseteq ZF$, whenever M is transitive and $M \models T^0$,

$$Ordinal(\xi) \;\&\; x \in L_\xi \Leftrightarrow M \models \varphi_L(x, \xi)$$

and let $\psi(\mathbf{x})$ be the formula

$$\psi(\mathbf{x}) \Leftrightarrow (\exists \boldsymbol{\xi})\varphi_L(\mathbf{x}, \boldsymbol{\xi}).$$

If M is an inner model of ZF so that $M \models ZF$ and M contains all the ordinals, then for $x \in M$, obviously

$$\begin{aligned} x \in L &\Leftrightarrow \text{for some } \xi,\, x \in L_\xi \\ &\Leftrightarrow \text{for some } \xi \in M,\, M \models \varphi_L(x, \xi) \\ &\Leftrightarrow M \models (\exists \boldsymbol{\xi})\varphi_L(x, \boldsymbol{\xi}) \end{aligned}$$

The argument for $x \in L(A)$ is similar. ⊣

We are now in a position to prove what we claimed in the beginning of 8E.

Fix once and for all a formula $\varphi_L(\mathbf{x}, \boldsymbol{\xi})$ such that for all transitive models M of some finite $T^0 \subseteq ZF$ and x, ξ in M,

$$x \in L_\xi \Leftrightarrow M \models \varphi_L(x, \xi)$$

and let "$V = L$" abbreviate the formal sentence of $\mathcal{L}^\in$ which says that every set is constructible:

$$V = L \Leftrightarrow (\forall \mathbf{x})(\exists \boldsymbol{\xi})\varphi_L(\mathbf{x}, \boldsymbol{\xi}).$$

We also construct a similar formula

$$V = L(\mathbf{A})$$

with a free variable $\mathbf{A}$ which says that "every set is constructible from $\mathbf{A}$".

8F.2. THEOREM. (i) $L \models V = L$.

(ii) *For each set* A,

$$L(A) \models V = L(A).$$

PROOF. Compute:

$$\begin{aligned} L \models V = L &\Leftrightarrow L \models (\forall \mathbf{x})(\exists \boldsymbol{\xi})\varphi_L(\mathbf{x}, \boldsymbol{\xi}) \\ &\Leftrightarrow \text{for each } x \in L, \text{ there exists } \xi \in L, L \models \varphi_L(x, \xi) \\ &\Leftrightarrow \text{for each } x \in L, \text{ there exists } \xi \in L, x \in L_\xi, \end{aligned}$$

and the last assertion is true by the definition of L and the fact that it contains all the ordinals. ⊣

This is a very basic result about L. One of its applications is that it allows us to prove theorems about L without constant appeal to metamathematical results and methods: we simply assume $V = L$ in addition to the axioms of ZF and any consequence of these assumptions must hold in L.

We have already argued about the axiom of choice in the beginning of 8E, but we should put the result down for the record.

8F.3. THEOREM (Gödel [1938], [1940]). *There is a formula* $\psi_L(x, y)$ *of* $\mathcal{L}^\in$ *such that*

$$L \models \text{“}\{(x, y) : \psi_L(x, y)\} \textit{ is a wellordering of } V\text{”}.$$

In particular, L satisfies the axiom of choice.

PROOF. If ψ^* is the formal sentence of $\mathcal{L}^\in$ expressing the symbolized English in quotes, then by 8D.12 and the fact that $L \models ZF$,

$$L \models V = L \rightarrow \psi^*$$

while by 8F.2 we have $L \models V = L$. ⊣

For the generalized continuum hypothesis we need a basic fact about L, a version of what is sometimes called *the condensation lemma.*

8F.4. THEOREM. *There is a finite set of sentences* T^L *of* $\mathcal{L}^\in$ *such that the following hold.*

(i) $L \models T^L$.

(ii) If A *is a transitive set and* $A \models T^L$, *then* $A = L_\lambda$ *for some limit ordinal* λ.

(iii) *For every infinite ordinal ξ and every set $x \in L$ such that $x \subseteq L_\xi$, there is some ordinal λ such that*

$$\xi \leq \lambda < \xi^+,$$

$L_\lambda \models T^L$ *and* $x \in L_\lambda$.

PROOF. Take

$$T^L = T^0 \cup \{V = L\}$$

where both operations

$$\xi \mapsto \xi + 1,$$
$$\xi \mapsto L_\xi,$$

are absolute for the standard models of T^0 and the relation $x \in L_\xi$ is defined on all standard models of T^0 by the specific formula $\varphi_L(\mathbf{x}, \boldsymbol{\xi})$ which we used to construct the sentence $V = L$.

Clearly $L \models T^L$.

If A is transitive and $A \models T^L$, let

$$\lambda = \text{least ordinal not in } A$$

and notice that λ is a limit ordinal, since A is closed under the successor operation. Now

$$\xi < \lambda \Longrightarrow L_\xi \in A,$$

by the absoluteness of $\xi \mapsto L_\xi$, so

$$L_\lambda = \bigcup_{\xi < \lambda} L_\xi \subseteq A.$$

On the other hand, $A \models V = L$, so that

$$\text{for each } x \in A, \text{ there exists } \xi \in A, A \models \varphi_L(x, \xi)$$

i.e. (by the absoluteness of $\varphi_L(x, \xi)$), $A \subseteq L_\lambda$.

To prove (iii) suppose $x \subseteq L_\xi$ and $x \in L_\zeta$. Using the Reflection Theorem 8C.4 on the hierarchy $\{L_\xi : \xi \in ON\}$ and the fact that $L \models T^L$, choose $\mu > \zeta$, $\mu > \xi$ such that $L_\mu \models T^L$—so now $x \in L_\mu$ and $L_\mu \models T^L$.

By the Skolem–Löwenheim Theorem 8A.4 applied to the (wellorderable) structure $(L_\mu, \in)$, we can find an elementary substructure

$$(M, \in) \preceq (L_\mu, \in)$$

such that $L_\xi \subseteq M$, $x \in M$ and $card(M) = card(L_\xi) = card(\xi)$ by 8D.11.

Since $(M, \in)$ is elementarily equivalent with $(L_\mu, \in)$, it satisfies in particular the extensionality axiom, so by the Mostowski Collapsing Lemma 8C.9, there is a transitive set $\bar{M}$ and an isomorphism

$$\pi : M \to \bar{M}$$

of $(M, \in)$ with $(\bar{M}, \in)$. Moreover, since the transitive set

$$y = L_\xi \cup \{x\} \subseteq M,$$

we have

$$\pi(x) = x$$

and hence $x \in \bar{M}$.

Now $(L_\mu, \in) \models T^L$ and therefore the elementarily equivalent structure $(M, \in) \models T^L$, so that the isomorphic structure $(\bar{M}, \in) \models T^L$; by (ii) then,

$$\bar{M} = L_\lambda$$

for some λ and of course, $\lambda < \xi^+$, since $card(\bar{M}) = card(\xi)$. ⊣

From this key theorem we get immediately the generalized continuum hypothesis for L.

8F.5. THEOREM (Gödel [1938], [1940]). *If $V = L$, then for each cardinal λ, $2^\lambda = \lambda^+$.*

PROOF. By 8F.4, $Power(\lambda) \subseteq L_{\lambda^+}$ and hence $card(Power(\lambda)) \leq card(L_{\lambda^+}) = \lambda^+$. ⊣

We should point out that the models $L(A)$ need not satisfy either the axiom of choice or the continuum hypothesis. For example, if in V truly $2^{\aleph_0} > \aleph_1$, then there is some surjection

$$\pi : \mathcal{N} \twoheadrightarrow \aleph_2$$

and obviously

$$L(\{\langle \alpha, \pi(\alpha) \rangle : \alpha \in \mathcal{N}\}) \models 2^{\aleph_0} \geq \aleph_2.$$

As another application of the basic Theorem 8F.1, we obtain intrinsic characterizations of the models L, $L(A)$.

8F.6. THEOREM. *L is the smallest inner model of ZF and for each set A, $L(A)$ is the smallest inner model of ZF which contains A.*

PROOF. Suppose M is an inner model of ZF and $A_0 \in M$. Since the operation

$$(\xi, A) \mapsto L_\xi(A)$$

is ZF-absolute, M is closed under this operation; since $A_0 \in M$ and every ordinal $\xi \in M$, we have $(\forall \xi)[L_\xi(A_0) \in M]$ so that $L(A_0) \subseteq M$. ⊣

One of the consequences of this result is that for every perfect product space $\mathcal{X}$,

$$L(\mathcal{X}) = L(\mathcal{N}),$$

simply because there is a Δ^1_1 isomorphism between $\mathcal{X}$ and $\mathcal{N}$ which is ZF-absolute as an operation by 8E.11. In particular,

$$\begin{aligned} L(\mathcal{N}) &= L(\mathbb{R}) \\ &= \text{the smallest inner model which} \\ &\quad\ \text{contains all real numbers.} \end{aligned}$$

At this point we can deliver on our promise in Chapter 5 and show that if $\mathcal{N} \subseteq L$, then $\mathcal{N}$ admits a Σ^1_2-good wellordering.

Recall that $\leq_L$ is the canonical wellordering of L which by 8F.1 satisfies

$$y \in L_\xi \ \&\ x \leq_L y \Rightarrow x \in L_\xi.$$

8F.7. THEOREM (Gödel [1940], Addison [1959b]). (i) *The pointset $\mathcal{N} \cap L$ of constructible irrationals is Σ^1_2.*

(ii) *The restriction of $\leq_L$ to $\mathcal{N}$ is a Σ^1_2-good wellordering of $\mathcal{N} \cap L$; i.e. if $P \subseteq \mathcal{N} \times \mathcal{X}$ is in Σ^1_2, then so are the relations*

$$\begin{aligned} Q(\alpha, x) &\Leftrightarrow (\exists \beta \leq_L \alpha) P(\beta, x), \\ R(\alpha, x) &\Leftrightarrow \alpha \in L \ \&\ (\forall \beta \leq_L \alpha) P(\beta, x). \end{aligned}$$

(iii) *If $\mathcal{N} \subseteq L$, then $\mathcal{N}$ admits a Σ^1_2-good wellordering of rank $\aleph_1$.*

PROOF. (i) is any easy consequence of (ii), but it is more instructive to show (i) first.

First of all, we claim that if T^L is the finite set of sentences of the basic Lemma 8F.4, then

$$(1) \qquad \begin{aligned} \alpha \in L \Leftrightarrow\ &\text{there exists a countable, transitive set } A \\ &\text{such that } (A, \in) \models T^L \text{ and } \alpha \in A. \end{aligned}$$

The implication ($\Leftarrow$) in (1) is immediate, because by 8F.4, if $(A, \in) \models T^L$, then $A = L_\lambda$ for some ordinal λ. For the other direction, notice that (as a set of pairs of integers), each α is a subset of L_ω, so by (iii) of 8F.4

$$\alpha \in L \Rightarrow \text{for some countable } \lambda$$
$$\alpha \in L_\lambda \quad \text{and} \quad L_\lambda \models T^L.$$

The key idea of the proof is that the structures of the form $(A, \in)$ with countable transitive A can be characterized *up to isomorphism* by the Mostowski Collapsing Lemma 8C.9. In fact, if (M, E) is any structure with countable M and $E \subseteq M \times M$, then by 8C.9, immediately

$$(M, E) \text{ is isomorphic with some } (A, \in) \text{ where } A \text{ is transitive}$$
$$\Leftrightarrow E \text{ is wellfounded and } (M, E) \models \text{"axiom of extensionality"}; \tag{2}$$

thus

$$\begin{aligned}\alpha \in L \Leftrightarrow\ & \text{there exists a countable, wellfounded structure} \\ & (M, E) \text{ such that } (M, E) \models \text{"axiom of extensionality"}, \\ & (M, E) \models T^L \text{ and} \\ & \alpha \in \bar{M} = \text{the unique transitive set such that} \\ & \qquad (M, E) \text{ is isomorphic with } (\bar{M}, \in).\end{aligned} \tag{3}$$

To see how to express the last condition in a model-theoretic way, recall that the relation "$\alpha \in \mathcal{N}$" is ZF-absolute by 8E.8 and choose some $\varphi_0(\boldsymbol{\alpha})$ such that for all transitive models M of some $T_0 \subseteq ZF$,

$$\alpha \in \mathcal{N} \Leftrightarrow M \models \varphi_0(\alpha).$$

Next define for each integer n a formula $\psi_n(\mathbf{x})$ which asserts that $\mathbf{x} = n$, by the recursion

$$\psi_0(\mathbf{x}) \Leftrightarrow \mathbf{x} = 0,$$
$$\psi_{n+1}(\mathbf{x}) \Leftrightarrow (\exists \mathbf{y})[\psi_n(\mathbf{y})\ \&\ \mathbf{x} = \mathbf{y} \cup \{\mathbf{y}\}]$$

and for each n, m, let

$$\psi_{n,m}(\boldsymbol{\alpha}) \Leftrightarrow (\exists \mathbf{x})(\exists \mathbf{y})[\psi_n(\mathbf{x})\ \&\ \psi_m(\mathbf{y})\ \&\ \langle \mathbf{x}, \mathbf{y} \rangle \in \boldsymbol{\alpha}].$$

It is now obvious that

$$
\begin{aligned}
&\alpha \in L \Leftrightarrow \text{there exists a countable, wellfounded structure}\\
(4)\qquad &\qquad (M, E) \text{ such that } (M, E) \models \text{“axiom of extensionality”},\\
&\qquad (M, E) \models T^L \text{ and for some } a \in M, (M, E) \models \varphi_0(a) \text{ and}\\
&\qquad \text{for all } n, m,\ \alpha(n) = m \Leftrightarrow (M, E) \models \psi_{n,m}(a).
\end{aligned}
$$

The point of this model-theoretic computation is that we can code countable structures by irrationals, as we did in 8A and then the satisfaction relation is Δ_1^1 by 8A.6. Recall that in the notation established for 8A.5 and 8A.6, we associated with each characteristic u and each irrational β a structure $\mathfrak{A}(u, \beta)$, so that in the case $u = 8$ which corresponds to the language of set theory $\mathfrak{L}^\epsilon$,

$$
\begin{aligned}
\mathfrak{A}(8, \beta) = (\{t : (\beta)_0(t) = 1\}, \{(t, s) : (\beta)_0(t) = (\beta)_0(s) = 1\\
\& \ (\beta)_1(\langle t, s\rangle) = 1\});
\end{aligned}
$$

moreover

$$
\begin{aligned}
Sat(8, \beta, m, x) \Leftrightarrow\ & \mathfrak{A}(8, \beta) \text{ is a structure (i.e. it has a nonempty domain)}\\
& \ \& \ m \text{ is the code of a formula } \chi_m \text{ of } \mathfrak{L}^\epsilon\\
& \ \& \ x \text{ is an assignment in } \mathfrak{A}(8, \beta) \text{ to the free}\\
& \qquad \text{variables of } \chi_m \text{ (i.e. whenever } \mathbf{v}_i \text{ is}\\
& \qquad \text{free in } \chi_m, \text{ then } (\beta)_0((x)_i) = 1)\\
& \ \& \ \mathfrak{A}(8, \beta), (x)_0, (x)_1, \ldots \models \chi_m
\end{aligned}
$$

and this relation Sat is Δ_1^1.

Let

$$f(m, n) = \text{the code of the formula } \psi_{m,n}(\boldsymbol{\alpha}),$$

so that f is obviously a recursive function. Let also k_0 be the code of the conjunction of the sentences in T^L and the axiom of extensionality and let k_1 be the code of the formula $\varphi_0(\boldsymbol{\alpha})$ which defines $\alpha \in \mathcal{N}$; we are assuming that both in $\psi_{m,n}(\boldsymbol{\alpha})$ and in $\varphi_0(\boldsymbol{\alpha})$, the free variable $\boldsymbol{\alpha}$ is actually the first variable $\mathbf{v}_0$. It is now clear that

$$
\begin{aligned}
\alpha \in L \Leftrightarrow (\exists \beta)\{& Sat(8, \beta, k_0, 1)\\
& \& \ \{(t, s) : (\beta)_0(t) = (\beta)_0(s) = 1 \ \& \ (\beta)_1(\langle t, s\rangle) = 1\}\\
& \quad \text{is wellfounded}\\
& \& \ (\exists a)[Sat(8, \beta, k_1, \langle a\rangle)\\
& \& \ (\forall n)(\forall m)[\alpha(n) = m \leftrightarrow Sat(8, \beta, f(n, m), \langle a\rangle)]]\},
\end{aligned}
$$

which implies directly that $L\cap\mathcal{N}$ is Σ^1_2 (using the fact that wellfoundedness is Π^1_1).

To prove (ii), let $\psi_L(\mathbf{v}_0, \mathbf{v}_1)$ be a formula which defines the canonical wellordering of L absolutely on all models of some finite $T^L_1 \subseteq ZF$ (by (ii) of 8F.1) and let $S^L \subseteq ZF$ be finite and large enough to include T^L_1, T^L, the axiom of extensionality and the set T_0 of part (i), chosen so that $\varphi_0(\boldsymbol{\alpha})$ defines $\boldsymbol{\alpha}\in\mathcal{N}$ on all transitive models of T_0. Using the key fact

$$\alpha\in L_\xi \;\&\; \beta\leq_L\alpha \Longrightarrow \beta\in L_\xi$$

and Mostowski collapsing as above, we can verify directly that for $\alpha\in L$ and arbitrary $P\subseteq\mathcal{N}\times\mathfrak{X}$,

$$\begin{aligned}(\forall\beta\leq_L\alpha)P(\beta,x) \Longleftrightarrow\ & \text{there exists a countable, wellfounded}\\ & \text{structure } (M,E)\models S^L \text{ and some}\\ & a\in M \text{ such that } (M,E)\models\varphi_0(a) \text{ and}\\ & (\forall n)(\forall m)[\alpha(n)=m \Longleftrightarrow (M,E)\models\psi_{n,m}(a)]\\ & \&\ (\forall b)\{(M,E)\models\varphi_0(b)\ \&\ \psi_L(b,a)\\ & \Longrightarrow (\exists\beta)[(\forall n)(\forall m)[\beta(n)=m \Longleftrightarrow (M,E)\models\psi_{n,m}(b)]\\ & \&\ P(\beta,x)]\}.\end{aligned}$$

If P is Σ^1_2, then it is easy to see that this whole expression on the right leads to a Σ^1_2 relation by coding the structures (M,E) by irrationals as above—the key being that the universal quantifier $\forall\beta$ has been turned to the number quantifier $\forall b$. ⊣

We put down the argument for (i) in considerable detail, because it illustrates a very useful technique for making analytical computations of relations defined by set-theoretic constructions.

For the next result we will do the opposite, i.e. we will give a set-theoretic construction for $\boldsymbol{\Sigma}^1_2$ pointsets which will establish that they are all absolute as relations for L. It is useful to derive this fundamental result of Shoenfield from a strong representation theorem for Σ^1_2 subsets of $\mathcal{N}$, which is nothing more but the metamathematical content of the proof of 2D.3—that $\boldsymbol{\Sigma}^1_2$ sets are $\aleph_1$-Suslin.

8F.8. Shoenfield's Lemma. *Suppose $A\subseteq\mathcal{N}$ is a Σ^1_2 set of irrationals. Then there exists a ZF-absolute operation*

$$\xi\mapsto T^\xi$$

which assigns to each ordinal $\xi \geq \omega$ a tree T^ξ on $\omega \times \xi$ such that the following holds, where λ is any uncountable ordinal:

$$\begin{aligned} \alpha \in A &\Leftrightarrow (\exists \xi \geq \omega)[T^\xi(\alpha) \text{ is not wellfounded}] \\ &\Leftrightarrow (\exists \xi \geq \omega)[\xi < \aleph_1 \ \&\ T^\xi(\alpha) \text{ is not wellfounded}] \\ &\Leftrightarrow T^\lambda(\alpha) \text{ is not wellfounded.} \end{aligned}$$

PROOF. One can derive this quite easily by re-examining the proof of 2D.3 and applying the absoluteness theory at the key points, but it is also simple enough to put a complete proof down.

Choose a recursive R so that

$$\alpha \in A \Leftrightarrow (\exists \beta)(\forall \gamma)(\exists t) R(\bar{\alpha}(t), \bar{\beta}(t), \bar{\gamma}(t))$$

by 4A.1 so that for each α, β, the set of sequences

$$S^{\alpha,\beta} = \{(c_0,\dots, c_{s-1}) : (\forall t < s) \neg R(\bar{\alpha}(t), \bar{\beta}(t), \langle c_0,\dots, c_{t-1}\rangle)\}$$

is a tree and easily

$$\begin{aligned} \alpha \in A &\Leftrightarrow (\exists \beta)\{S^{\alpha,\beta} \text{ is wellfounded}\} \\ &\Leftrightarrow (\exists \beta)(\exists f: S^{\alpha,\beta} \to \aleph_1)\{\text{if } (c_0,\dots, c_{s-1}) \in S^{\alpha,\beta} \text{ and} \\ &\qquad t < s, \text{ then} \\ &\qquad f(c_0,\dots, c_{t-1}) > f(c_0,\dots, c_{s-1})\}. \end{aligned} \tag{1}$$

In the computation below we will represent $S^{\alpha,\beta}$ by the set of sequence codes $\langle c_0,\dots, c_{s-1}\rangle$ of members $(c_0,\dots, c_{s-1}) \in S^{\alpha,\beta}$.

For each $\xi \geq \omega$, define first a tree S^ξ on $\omega \times (\omega \times \xi)$ as follows:

$$\begin{aligned} &((a_0, b_0, \xi_0),\dots, (a_{n-1}, b_{n-1}, \xi_{n-1})) \in S^\xi \\ &\qquad \Leftrightarrow \xi_0,\dots, \xi_{n-1} < \xi \\ &\qquad \&\ \{\text{if } i = \langle c_0,\dots, c_{s-1}\rangle \text{ and } j = \langle c_0,\dots, c_{t-1}\rangle \\ &\qquad\quad \text{where } j < i < n \text{ and for each } m \leq s, \\ &\qquad\quad \neg R(\langle a_0,\dots, a_{m-1}\rangle, \langle b_0,\dots, b_{m-1}\rangle, \langle c_0,\dots, c_{m-1}\rangle), \\ &\qquad\quad \text{then } \xi_j > \xi_i\}. \end{aligned}$$

Notice that the operation

$$\xi \mapsto S^\xi$$

is clearly ZF-absolute and

$$\xi \leq \eta \Rightarrow S^\xi \subseteq S^\eta.$$

In the notation established in Chapter 2, for each ξ, α,

$$S^{\xi}(\alpha)=\{((b_0,\xi_0),\dots,(b_{n-1},\xi_{n-1})):$$

$$((\alpha(0),b_0,\xi_0),\dots,(\alpha(n-1),b_{n-1},\xi_{n-1}))\in S^{\xi}\}$$

is a tree on $\omega\times\xi$ and it is almost immediate that

$$\alpha\in A\Rightarrow(\exists\xi<\aleph_1)[S^{\xi}(\alpha)\text{ is not wellfounded}],\tag{2}$$

$$S^{\xi}(\alpha)\text{ is not wellfounded}\Rightarrow\alpha\in A.\tag{3}$$

(To prove (2) choose $\beta=(b_0,b_1,\dots)$ such that $S^{\alpha,\beta}$ is wellfounded, choose $f:S^{\alpha,\beta}\to\aleph_1$ as in (1) above and for $i=\langle c_0,\dots,c_{s-1}\rangle$ with $(c_0,\dots,c_{s-1})\in S^{\alpha,\beta}$, take $\xi_i=f(c_0,\dots,c_{s-1})$—for i not of this form take $\xi_i=0$. To prove (3), choose an infinite branch (b_0,ξ_0), $(b_1,\xi_1),\dots$ in $S^{\xi}(\alpha)$, take $\beta=(b_0,b_1,\dots)$ and define $f:S^{\alpha,\beta}\to\xi$ by

$$f(c_0,\dots,c_{s-1})=\xi_i\Leftrightarrow i=\langle c_0,\dots,c_{s-1}\rangle$$

so that it immediately satisfies the condition in (1).)

Now (2) and (3) imply directly the assertions in the theorem taking $T^{\xi}=S^{\xi}$, except that S^{ξ} is a tree on $\omega\times(\omega\times\xi)$ rather than a tree on $\omega\times\xi$. To complete the proof, put

$$\begin{aligned}T^{\xi}=&\text{ all initial segments of sequences of the form}\\&((a_0,b_0),(a_1,\xi_0),(a_2,b_1),(a_3,\xi_1),\dots,(a_{2n},b_n),(a_{2n+1},\xi_n))\\&\text{such that }((a_0,b_0,\xi_0),(a_1,b_1,\xi_1),\dots,(a_n,b_n,\xi_n))\in S^{\xi}\end{aligned}$$

so that T^{ξ} is a tree on $\omega\times\xi$ (because $\omega\subseteq\xi$) and easily, for any α,

$$T^{\xi}(\alpha)\text{ is not wellfounded}\Leftrightarrow S^{\xi}(\alpha)\text{ is not wellfounded.}\qquad\dashv$$

8F.9. SHOENFIELD'S THEOREM (I) (Shoenfield [1961]). *Each Σ^1_2 pointset is absolute as a relation for all standard models M of some finite $T_*\subseteq ZF$ such that $\aleph_1\subseteq M$.*

In particular, if $A\subseteq\omega^n$ is $\Sigma^1_2(\alpha_0)$ and $\alpha_0\in L$, then $A\in L$; similarily, if $\beta\in\Delta^1_2(\alpha_0)$ and $\alpha_0\in L$, then $\beta\in L$.

PROOF. Take first $A\subseteq\mathcal{N}$ *and let* $\varphi(\xi,\mathbf{T})$ be a formula of $\mathfrak{L}^{\in}$ by the lemma such that for all standard models M of some finite $T_1\subseteq ZF$,

$$\xi\in M\Rightarrow T^{\xi}\in M,$$

$$T=T^{\xi}\Leftrightarrow M\vDash\varphi(\xi,T).$$

Notice also that the operation

$$(\alpha, T) \mapsto T(\alpha)$$

is easily ZF-absolute, so choose $\psi(\boldsymbol{\alpha}, \mathbf{S}, \mathbf{T})$ so that for all standard models M of some finite $T_2 \subseteq ZF$,

$$\alpha, T \in M \Rightarrow T(\alpha) \in M,$$
$$S = T(\alpha) \Leftrightarrow M \models \psi(\alpha, S, T).$$

Finally use Mostowski's Theorem 8E.4 to construct a formula $\chi(\mathbf{S})$ of $\mathcal{L}^{\varepsilon}$ such that for all standard models M of some finite $T_3 \subseteq ZF$ and $S \in M$,

$$S \text{ is wellfounded} \Leftrightarrow M \models \chi(S).$$

Now if M is any standard model of

$$T_* = T_1 \cup T_2 \cup T_3$$

such that $\aleph_1 \subseteq M$, then by the lemma, for $\alpha \in M$

$$\begin{aligned} \alpha \in A &\Leftrightarrow \text{there exists some } \xi \in M \text{ such that } T^{\xi}(\alpha) \text{ is not wellfounded} \\ &\Leftrightarrow \text{there exists some } \xi \in M \text{ such that} \\ &\qquad M \models (\exists \mathbf{S})(\exists \mathbf{T})[\varphi(\xi, \mathbf{T}) \mathbin{\&} \psi(\alpha, \mathbf{S}, \mathbf{T}) \mathbin{\&} \neg\chi(\mathbf{S})] \\ &\Leftrightarrow M \models (\exists \boldsymbol{\xi})(\exists \mathbf{S})(\exists \mathbf{T})[\varphi(\boldsymbol{\xi}, \mathbf{T}) \mathbin{\&} \psi(\alpha, \mathbf{S}, \mathbf{T}) \mathbin{\&} \neg\chi(\mathbf{S})]. \end{aligned}$$

The result for arbitrary Σ^1_2 pointsets $P \subseteq \mathfrak{X}$ follows easily by considering $A = f[P]$, where $f : \mathfrak{X} \to \mathcal{N}$ is a Δ^1_1 injection with Δ^1_1 inverse and applying 8E.11.

To prove the second assertion, take $A \subseteq \omega$ for simplicity of notation and suppose

$$n \in A \Leftrightarrow P(n, \alpha_0)$$

where P is Σ^1_2 and $\alpha_0 \in L$ and let $\psi(\mathbf{n}, \boldsymbol{\alpha})$ define P absolutely as in the first part, so that in particular

$$P(n, \alpha) \Leftrightarrow L \models \psi(n, \alpha).$$

The sentence

$$(\forall \boldsymbol{\alpha})(\exists \mathbf{x})[\mathbf{x} \subseteq \omega \mathbin{\&} (\forall \mathbf{n})[\mathbf{n} \in \mathbf{x} \leftrightarrow \psi(\mathbf{n}, \boldsymbol{\alpha})]]$$

is a theorem of ZF and hence it holds in L. Taking $\boldsymbol{\alpha} = \alpha_0$, this implies that there is some $x \in L$ such that $x \subseteq \omega$ and for all n,

$$L \models n \in x \Leftrightarrow \psi(n, \alpha_0),$$

so that

$$\begin{aligned} n \in x &\Leftrightarrow L \models \psi(n, \alpha_0) \\ &\Leftrightarrow P(n, \alpha_0) \\ &\Leftrightarrow n \in A; \end{aligned}$$

thus $x = A$ and $A \in L$.

If $\beta \in \Delta^1_2(\alpha_0)$, with $\alpha_0 \in L$, apply this to

$$A(n, m) \Leftrightarrow \beta(n) = m$$

to infer that $A \in L$ and hence $\beta \in L$. ⊣

To appreciate the significance of Shoenfield's Theorem, recall from the exercises of 8B that a formula $\theta(\boldsymbol{\alpha}_1, \ldots, \boldsymbol{\alpha}_m)$ of the language of second order arithmetic A^2 is Σ^1_n if

$$\theta(\boldsymbol{\alpha}_1, \ldots, \boldsymbol{\alpha}_m) \Leftrightarrow (\exists \boldsymbol{\beta}_1)(\forall \boldsymbol{\beta}_2)(\exists \boldsymbol{\beta}_3) \cdots (-\boldsymbol{\beta}_n)\varphi(\boldsymbol{\alpha}_1, \ldots, \boldsymbol{\alpha}_m, \boldsymbol{\beta}_1, \ldots, \boldsymbol{\beta}_n),$$

where $\varphi(\boldsymbol{\alpha}_1, \ldots, \boldsymbol{\alpha}_m, \boldsymbol{\beta}_1, \ldots, \boldsymbol{\beta}_n)$ has no quantifiers over $\mathcal{N}$. It is clear that we can interpret these formulas over standard models of *ZF* simply by putting (for $\alpha_1, \ldots, \alpha_m \in M$),

$$M \models \theta(\alpha_1, \ldots, \alpha_m) \Leftrightarrow (\omega, \mathcal{N} \cap M, +, \cdot, ap, 0, 1) \models \theta(\alpha_1, \ldots, \alpha_m),$$

i.e. by interpreting the quantifiers $\exists \boldsymbol{\beta}_i$, $\forall \boldsymbol{\beta}_j$ as ranging over the irrationals in M and using the standard interpretations for the operations $+, \cdot, ap$ (which are *ZF*-absolute by 8E.8) and the quantifiers $\exists n$, $\forall n$ (since ω is also *ZF*-absolute and hence a member of M).

8F.10. Shoenfield's Theorem (II) (Shoenfield [1961]). (i) *If $\theta(\boldsymbol{\alpha}_1, \ldots, \boldsymbol{\alpha}_m)$ is a Σ^1_2 or Π^1_2 formula of second order arithmetic, then for every standard model M of ZF such that $\aleph_1 \subseteq M$ and $\alpha_1, \ldots, \alpha_m \in M$,*

$$V \models \theta(\alpha_1, \ldots, \alpha_m) \Leftrightarrow M \models \theta(\alpha_1, \ldots, \alpha_m);$$

in particular, if $\alpha_1, \ldots, \alpha_m \in L$, then

$$V \models \theta(\alpha_1, \ldots, \alpha_m) \Leftrightarrow L \models \theta(\alpha_1, \ldots, \alpha_m).$$

(ii) *If we can prove a Σ^1_2 or Π^1_2 sentence θ by assuming in addition to the axioms of ZF the hypothesis $V = L$ (and its consequences AC and GCH, then θ is in fact true (i.e. $V \models \theta$).*

Proof. Take a Σ^1_2 sentence for simplicity of notation

$$\theta \Leftrightarrow (\exists \boldsymbol{\alpha})(\forall \boldsymbol{\beta})\varphi(\boldsymbol{\alpha}, \boldsymbol{\beta}),$$

and let

$$P(\alpha, \beta) \Leftrightarrow \mathsf{A}^2 \models \varphi(\alpha, \beta)$$

be the arithmetical pointset defined by the matrix of θ so that

$$V \models \theta \Leftrightarrow (\exists\alpha)(\forall\beta)P(\alpha, \beta),$$
$$M \models \theta \Leftrightarrow (\exists\alpha \in M)(\forall\beta \in M)P(\alpha, \beta).$$

Using the Basis Theorem for Σ^1_2, 4E.5,

$$\begin{aligned} V \models \theta &\Rightarrow (\exists\alpha)(\forall\beta)P(\alpha, \beta), \\ &\Rightarrow (\exists\alpha \in \Delta^1_2)(\forall\beta)P(\alpha, \beta) && \text{(by 4E.5)} \\ &\Rightarrow (\exists\alpha \in M)(\forall\beta)P(\alpha, \beta) && \text{(by 8F.9)} \\ &\Rightarrow (\exists\alpha \in M)(\forall\beta \in M)P(\alpha, \beta) && \text{(obviously)} \\ &\Rightarrow M \models \theta. \end{aligned}$$

Conversely, assuming that $M \models \theta$, choose some $\alpha_0 \in M$ such that

$$(\forall\beta \in M)P(\alpha_0, \beta)$$

and assume towards a contradiction that

$$(\exists\beta)\neg P(\alpha_0, \beta);$$

by the Basis Theorem 4E.5 again, we then have

$$(\exists\beta \in \Delta^1_2(\alpha_0))\neg P(\alpha_0, \beta)$$

so that by 8F.9,

$$(\exists\beta \in M)\neg P(\alpha_0, \beta)$$

contradicting our assumption and establishing $(\forall\beta)P(\alpha_0, \beta)$, i.e. $V \models \theta$.

The second assertion follows immediately, because if we can prove θ using the additional hypothesis $V = L$, then we know that $L \models \theta$ by 8F.2 and hence $V \models \theta$ by the first assertion. ⊣

This theorem is quite startling because so many of the propositions that we consider in ordinary mathematics are expressible by Σ^1_2 sentences—including all propositions of elementary or analytic number theory and most of the propositions of hard analysis. The techniques in the proof of 8F.1 allow us to prove that many set theoretic propositions are also equivalent to Σ^1_2 sentences. Theorem 8F.8 assures us then that the truth or falsity of these "basic" propositions does not depend on the answers to difficult and delicate questions about the nature of sets like the continuum hypothesis; we might as well assume $V = L$ in attempting to prove or disprove them.

Of course, in descriptive set theory we worry about propositions much more complicated than Σ_2^1 which may well have different truth values in L and in V.

EXERCISES

We should put down for the record one of the best known results that comes out of the theory of constructibility.

8F.11. Prove that the proposition $V = L$, the axiom of choice AC and the generalized continuum hypothesis GCH are consistent with ZF—i.e. we cannot prove the negation of any of these propositions in ZF. (Gödel [1938], [1940].)

Hint. All these propositions hold in L, so none of them can be refuted from the axioms of ZF which also hold in L. ⊣

A few remarks are in order on this proof.

We developed the theory of constructibility here with no appeal to the axiom of choice, so there is no use of choice in this argument. On the other hand, we have presented this proof in the same general framework in which we have proved all the other results in this book—i.e. we have assumed that we have the structure $(V, \in)$ which satisfies all the axioms of ZF (even though it may fail to satisfy AC) and that we can reason about this structure in the ordinary way. Granting these assumptions and these methods of proof, the hint outlines a valid argument for the consistency of ZF with $V = L$, AC and GCH.

It is customary in mathematical logic to give consistency proofs on the basis of minimal hypotheses and in fact the consistency of ZF with $V = L$, AC and GCH can be established assuming only that ZF is *formally consistent* and using only very basic, combinatorial arguments. To give that argument, we would have to give a precise definition of *formal logical consequence* (or *formal proof*) which would take us far afield from descriptive set theory. Suffice it to say that anyone who knows the rudiments of the theory of formal proofs will have no difficulty turning the hint above into a constructive, combinatorial demonstration of the consistency of ZF with $V = L$, AC and GCH, granting only the formal consistency of ZF.

We should also point out that at least as far as AC is concerned, the observation that we did not use the axiom of choice in 8F.11 is essential, if we are to have a nontrivial theorem. Because if we assume that $(V, \in)$ also satisfies AC, then it is obvious that $\neg AC$ cannot be a consequence of ZF—or else $(V, \in)$ would satisfy both AC and $\neg AC$.

Next we put down two results about the axiom of choice in the models $L(A)$.

8F.12. Prove that if A is a set of constructible sets $(A \subseteq L)$, then $L(A)$ satisfies the axiom of choice. (Notice that the hypothesis holds if A is any set of ordinals.)

Hint. Rework the proof of 8D.12 to show that there is a ZF-absolute relation $P(A, W, x, y)$ such that wehenever $W \subseteq A \times A$ is a wellordering of $TC(A)$, then $\{(x, y): P(A, W, x, y)\}$ is a wellordering of $L(A)$. Now argue that if $A \subseteq L$, we can find in $L(A)$ a wellordering W of $TC(A)$ and complete the proof as in 8F.3. ⊣

8F.13. Prove that the model $L(\mathcal{N})$ satisfies the axiom of dependent choices, DC. (You will need to use the fact that $V \models DC$.)

Hint. Show first that for every ordinal ξ, there is some ordinal λ_ξ and a surjection

$$\pi_\xi : \lambda_\xi \times \mathcal{N} \twoheadrightarrow L_\xi(\mathcal{N}),$$

such that $\pi_\xi \in L(\mathcal{N})$. (Use induction on ξ. In the successor case, it is enough to construct from any surjection $\pi : \lambda \times \mathcal{N} \twoheadrightarrow A$, another surjection $\rho : \lambda' \times \mathcal{N} \twoheadrightarrow \bigcup \{{}^n A : n \in \omega\}$. In the limit case, you may assume given surjections $\pi_\xi : \lambda' \times \mathcal{N} \twoheadrightarrow L_\xi(\mathcal{N})$, for each $\xi < \lambda$; put first

$$\pi(\xi, \eta, \alpha) = \pi_\xi(\eta, \alpha)$$

and define $\rho : \lambda^* \times \mathcal{N} \twoheadrightarrow L_\lambda(A)$ by $\rho(\zeta, \alpha) = \pi(\rho_1(\zeta), \rho_2(\zeta), \alpha)$, where $\zeta \mapsto (\rho_1(\zeta), \rho_2(\zeta))$ is a surjection of λ^* onto $\lambda' \times \lambda'$.)

Suppose now that $P \in L_\lambda(\mathcal{N})$, $A \in L_\lambda(\mathcal{N})$, $A \neq \emptyset$, $P \subseteq A \times A$ and $(\forall x \in A)(\exists y \in A)P(x, y)$; we must show that there is a function $f : \omega \to A$ in $L(\mathcal{N})$, such that $(\forall n)P(f(n), f(n+1))$.

By DC in V, there is some infinite sequence $x_0, x_1, x_2, \ldots$ in V such that $(\forall n)P(x_n, x_{n+1})$, with each $x_i \in A$. If $\pi : \lambda \times \mathcal{N} \twoheadrightarrow L_\lambda(\mathcal{N})$ is a surjection which lies in $L(\mathcal{N})$, we can further choose ordinals ζ_n, α_n (in V) such that $x_n = \pi(\zeta_n, \alpha_n)$. Now forget about the ordinals ζ_n, but check that the function $n \mapsto \alpha_n$ lies in $L(\mathcal{N})$, because it can be coded by a single irrational.

On $\omega\times\lambda$ define the binary relation

$$(k,\xi)<(m,\eta)\Leftrightarrow k=m+1\ \&\ P(\pi(\eta,\alpha_m),\pi(\xi,\alpha_k));$$

this relation lies in $L(\mathcal{N})$ and it is not wellfounded in V, because we have

$$(0,\zeta_0)>(1,\zeta_1)>(2,\zeta_2)>\dots.$$

By Mostowski's Theorem 8E.4 then, this relation is not wellfounded in $L(\mathcal{N})$, so that we have a function $n\mapsto\xi_n$ in $L(\mathcal{N})$ such that $(\forall n)P(\pi(\xi_n,\alpha_n),\pi(\xi_{n+1},\alpha_{n+1}))$ and we can finally take $f(n)=\pi(\xi_n,\alpha_n)$. ⊣

8F.14. Prove (without using the axiom of choice) that for every finite set T_0 of axioms of ZF, there is a countable transitive set M such that $M\models T_0$.

Hint. Imitate the proof of the Countable Reflection Theorem 8C.10 using the hierarch $\{L_\xi:\xi\in ON\}$ instead of $\{V_\xi:\xi\in ON\}$. ⊣

8F.15. Prove that there is a smallest standard model of ZF (Shepherdson [1951], Cohen [1963]).

Hint. If no set $M\models ZF$, then every standard model of ZF is an inner model and L is the least standard model by 8F.6.

Assume now that some transitive set $M\models ZF$. Since we proved that $L\models ZF$ using only the fact that $V\models ZF$, it follows that for every transitive set M and for every sentence $\varphi\in ZF$,

$$M\models ZF\Rightarrow M\models\text{"the collection of constructible sets satisfies }\varphi\text{"},$$

where for each φ, the expression in quotes can be easily transformed into a sentence of $\mathfrak{L}^{\in}$. Now argue using 8F.1 that for $M\models ZF$ and $x\in M$,

$$M\models\text{"}x\text{ is constructible"}\Leftrightarrow x\in M\cap L$$

and if λ is the least ordinal not in M and $M\models ZF$,

$$M\cap L=L_\lambda.$$

Thus the least model of ZF is L_λ, where λ is the least ordinal such that $L_\lambda\models ZF$. ⊣

8F.16. Prove that if $\mathcal{N}-L\neq\emptyset$, then the pointset $\mathcal{N}\cap L$ is not Π^1_2 (Shoenfield [1961]).

Hint. If it were, then $\mathcal{N}-L$ would be Σ^1_2 and thus have a Δ^1_2 member by 4E.5, contradicting Shoenfield's Theorem 8F.9. ⊣

8F.17. Prove that if $\mathcal{N}-L\neq\emptyset$, then Π^1_3 sentences of the language of second order arithmetic are not absolute for L—i.e. for some such sentence θ, $L \models \theta$ but $V \models \neg\theta$ (Shoenfield [1961]). ⊣

8F.18. Prove that not all Σ^1_2 pointsets are ZF-absolute as relations.

Hint. If every Σ^1_2 pointset were ZF-absolute, then (easily) there would be a finite $T_0 \subseteq ZF$ such that every standard model of T_0 would contain all Σ^1_2 sets of integers. Using the technique in the proof of 8F.7 argue that if

$$P(\beta) \Leftrightarrow \mathfrak{A}(8, \beta) \text{ is a wellfounded model of } T_0,$$

then every Σ^1_2 set of integers must be recursive in any β such that $P(\beta)$ and hence every Δ^1_2 irrational must be recursive in every β such that $P(\beta)$; but P is Σ^1_2 and by 8F.14 and the Basis Theorem 4E.5 there must exist some $\beta \in \Delta^1_2$ such that $P(\beta)$, which is absurd. ⊣

We now turn to the study of so-called *relative constructibility*.

Consider structures of the form

$$(M, \in, P),$$

where $P \subseteq M$. The language $\mathcal{L}^{\in,1}$ of these structures is obtained by adding a unary relation symbol $\mathbf{P}$ to $\mathcal{L}^{\in}$ and interpreting $\mathbf{P}(\mathbf{x})$ by $x \in P$.

8F.19. Prove that there is a ZF-absolute operation $\boldsymbol{Def}(A, P)$ such that for any two sets A, P,

$$\begin{aligned} x \in \boldsymbol{Def}(A, P) \Leftrightarrow x \subseteq A \;\&\; & \text{there is a formula} \\ & \varphi(\mathbf{v}_0, \dots, \mathbf{v}_{n-1}, \mathbf{v}_n) \text{ in the language} \\ & \mathcal{L}^{\in,1} \text{ and members } x_0, \dots, x_{n-1} \text{ of } A, \text{ such that} \\ & s \in x \Leftrightarrow (A, \in, P \cap A) \models \varphi(x_0, \dots, x_{n-1}, s). \end{aligned}$$

Hint. Look up the proof of 8D.1 and alter the definition of the satisfaction relation #24 to take account of the additional unary relation $P \cap A$. ⊣

For each set A, the hierarchy of *sets constructible relative to* A is defined by the transfinite recursion

$$\begin{aligned} L_0[A] &= \emptyset, \\ L_{\xi+1}[A] &= \boldsymbol{Def}(L_\xi[A], A) \\ L_\lambda[A] &= \textstyle\bigcup_{\xi<\lambda} L_\xi[A] \quad \text{if} \quad \lambda \text{ is a limit ordinal} \end{aligned}$$

and of course we put

$$L[A] = \bigcup_\xi L_\xi[A].$$

We collect in one theorem all the basic facts about the models $L[A]$.

8F.20. (i) Prove that the operation

$$(\xi, A) \mapsto L_\xi[A]$$

and the relation

$$P(x, \xi, A) \Leftrightarrow x \in L_\xi[A]$$

are ZF-absolute.

(ii) Prove that the relation

$$Q(x, A) \Leftrightarrow x \in L[A]$$

is absolute for inner models of ZF.

(iii) Prove that each $L_\xi[A]$ is a transitive set and

$$\xi \leq \eta \Rightarrow L_\xi[A] \subseteq L_\eta[A].$$

(iv) Prove that each $L[A]$ is an inner model of ZF.

(v) Prove that if $A \cap L[A] = B \cap L[A]$, then $L[A] = L[B]$; thus

$$L[A] = L[A \cap L[A]].$$

(vi) Prove that $A \cap L[A] \in L[A]$ and

$$L[A] = L(A \cap L[A]);$$

in particular,

$$A \subseteq L \Rightarrow L[A] = L(A).$$

Hint. (i)–(iv) are established exactly like the corresponding results for the L_ξ's.

To show (v), check by induction on ξ that if $A \cap L[A] = B \cap L[A]$, then $L_\xi[A] = L_\xi[B]$; then take $B = A \cap L[A]$.

For (vi) argue first that $A \cap L[A] \subseteq L_\xi[A]$ for some ξ, and then easily $A \cap L[A] \in L_{\xi+1}[A]$. The ZF-absoluteness of $(\xi, A) \mapsto L_\xi[A]$ implies immediately that $L[A] \subseteq L(A)$, so applying this to $A \cap L[A]$ we get

$$L[A \cap L[A]] \subseteq L(A \cap L[A]);$$

hence $L[A] = L[A \cap L[A]] \subseteq L(A \cap L[A])$. On the other hand, given

$A\cap L[A]\in L[A]$, the ZF-absoluteness of $(\xi, A)\mapsto L_\xi(A)$ implies that $L(A\cap L[A])\subseteq L[A]$. Finally, if $A\subseteq L\subseteq L[A]$, then $A\cap L[A]=A$. ⊣

Caution: Since relative constructibility is often studied in the case where A is a set of ordinals, when $L[A]=L(A)$, there is some confusion in the literature between the notations $L[A]$ and $L(A)$ and the clear distinction we have established here is not always observed.

Notice that by 8F.20, if $\alpha\in\mathcal{N}$, then

$$L[\alpha]=L(\alpha);$$

on the other hand,

$$L[\mathcal{N}]=L,$$

since for each transitive set A the membership relation in the set $\mathcal{N}\cap A$ is certainly definable in $(A,\in)$.

In general, $L[A]$ is a much better model than $L(A)$.

8F.21. Show that for each set A, $L[A]$ satisfies the axiom of choice.

Hint. Look up the proofs of 8D.12 and 8F.3, checking first that

$$L[A]\models\text{"every set is in } L[A\cap L[A]]\text{"}. \qquad ⊣$$

In the universe constructible from an irrational, we also have the full effect of the key Theorem 8F.4.

8F.22. Show that there is a fixed formula $\varphi_L(\boldsymbol{\alpha})$ with one free variable $\boldsymbol{\alpha}$ such that for every $\alpha\in\mathcal{N}$ the following hold.

(i) $L[\alpha]\models\varphi_L(\alpha)$.

(ii) If A is a transitive set, $\alpha\in A$ and $A\models\varphi_L(\alpha)$, then $A=L_\lambda[\alpha]$ for some limit ordinal λ.

(iii) For every infinite ordinal ξ and every set $x\in L[\alpha]$ such that $x\subseteq L_\xi[\alpha]$, there is some ordinal λ such that

$$\xi\le\lambda<\xi^+,$$

$L_\lambda[\alpha]\models\varphi_L(\alpha)$ and $x\in L_\lambda[\alpha]$.

Infer that $L[\alpha]$ satisfies GCH.

Hint. Copy over the proof of 8F.4. ⊣

8F.23. Prove that the pointset $\{(\alpha,\beta): \alpha\in L[\beta]\}$ is Σ^1_2. ⊣

8F.24. Prove that for each β, if $\mathcal{N}\subseteq L[\beta]$, then $\mathcal{N}$ admits a $\Sigma^1_2(\beta)$-good wellordering of rank $\aleph_1$. ⊣

8G. Regularity results and inner models

We already know from 6G.10–6G.12 that if there exists a measurable cardinal, then $\mathbf{\Sigma}^1_2$ pointsets are absolutely measurable, they all have the property of Baire and every uncountable $\mathbf{\Sigma}^1_2$ set has a perfect subset. In this section we will derive interesting metamathematical refinements of these regularity properties and extensions of them to the higher Lusin pointclasses.

In proving these results, we will introduce and study briefly some very interesting inner models of ZF which are much like L but which also satisfy various forms of definable determinacy.

We start with a beautiful and very useful result of Mansfield which we will establish by rethinking the proof of the Perfect Set Theorem 2C.2 in the light of the theory of absoluteness.

8G.1. MANSFIELD'S LEMMA (Mansfield [1970]). *Suppose M is a standard model of ZF, T is a tree on $\omega \times \varkappa$, $T \in M$ and*

$$A = \mathrm{p}[T] = \{\alpha \in \mathcal{N}: T(\alpha) \text{ is not wellfounded}\};$$

then either $A \subseteq M$ or A has a perfect subset.

PROOF (Solovay). Following the proof of 2C.2, assign to each tree S on $\omega \times \varkappa$ and each ordinal ξ the tree S^ξ by the recursion

$$S^0 = S,$$

$$S^{\xi+1} = \{u \in S^\xi : \mathrm{p}[S^\xi_u] \text{ has more than one (irrational) member}\},$$

$$S^\lambda = \textstyle\bigcap_{\xi<\lambda} S^\xi, \quad \text{if} \quad \lambda \text{ is a limit ordinal.}$$

Each S^ξ is a tree on $\omega \times \varkappa$ and

$$\eta < \xi \Longrightarrow S^\eta \supseteq S^\xi$$

so that easily, for some λ,

$$S^{\lambda+1} = S^\lambda;$$

we take

$$S^* = S^\lambda$$

for the least such λ.

The relation

$$R_1(S, \varkappa) \Longleftrightarrow S \text{ is a tree on } \omega \times \varkappa \;\&\; \mathrm{p}[S] \neq \emptyset$$

is ZF-absolute by the general properties of absoluteness and Mostowski's Theorem 8E.4, since

$$\mathfrak{p}[S] \neq \emptyset \leftrightarrow \{\langle u, v\rangle : u, v \text{ are finite sequences on } \omega \times \varkappa \text{ and } u, v \in S \text{ and } v \text{ is a proper initial segment of } u\} \text{ is not wellfounded.}$$

But also

$$\begin{aligned} &\mathfrak{p}[S] \text{ has more than one element} \\ &\leftrightarrow (\exists u)(\exists n)(\exists \xi)(\exists m)(\exists \eta)[u^\frown(n, \xi) \in S \ \&\ u^\frown(m, \eta) \in S \\ &\quad \&\ n \neq m \\ &\quad \&\ \mathfrak{p}[S_{u^\frown(n,\xi)}] \neq \emptyset \\ &\quad \&\ \mathfrak{p}[S_{u^\frown(m,\eta)}] \neq \emptyset, \end{aligned}$$

so that the relation

$$R_2(S, \varkappa) \Leftrightarrow S \text{ is a tree on } \omega \times \varkappa \text{ and } \mathfrak{p}[S] \text{ has more than one element}$$

is also ZF-absolute. Thus by the ZF-absoluteness of definition by transfinite recursion 8E.5, the operation

$$(\xi, S) \mapsto S^\xi$$

is ZF-absolute.

The assertion

$$(\forall S)(\exists \lambda)[S^{\lambda+1} = S^\lambda]$$

is a theorem of ZF which is easily expressible in $\mathcal{L}^\in$, so it must hold on all standard models of sufficiently may axioms in ZF. From this we infer easily that the operation

$$S \mapsto S^* = S^\lambda \text{ for the least } \lambda \text{ such that } S^{\lambda+1} = S^\lambda$$

is also ZF-absolute.

Assume now the hypotheses of the lemma for a specific $M \models ZF$ and a tree T on $\omega \times \varkappa$, $T \in M$. From the discussion above we know that

$$\xi \in M \rightarrow T^\xi \in M$$

and that for some $\lambda \in M$,

$$T^{\lambda+1} = T^\lambda = T^* \in M.$$

If $T^* \neq \emptyset$, then $\mathfrak{p}[T]$ has a perfect subset as in the proof of 2C.2. It remains to show that if $T^* = \emptyset$, then $\mathfrak{p}[T] \subseteq M$.

Assume $T^* = \emptyset$ and suppose that $\alpha \in \mathfrak{p}[T]$, so that for some $f: \omega \to \varkappa$, $(\alpha, f) \in [T]$, i.e. for all n

$$(\alpha(0), f(0), \alpha(1), f(1), \ldots, \alpha(n-1), f(n-1)) \in T;$$

as in the proof of 2C.2 again, there must be a ξ such that $(\alpha, f) \in [T^\xi] - [T^{\xi+1}]$ and consequently for some n,

$$u = (\alpha(0), f(0), \ldots, \alpha(n-1), f(n-1)) \in T^\xi - T^{\xi+1}.$$

For this ξ, we have by the definition that

$$\mathfrak{p}[T^\xi_u] \text{ has exactly one element,}$$

namely α.

Suppose now $\varphi(\mathbf{S}, \boldsymbol{\varkappa})$ and $\psi(\mathbf{S}, \boldsymbol{\beta})$ are formulas of $\mathcal{L}^\in$ which define absolutely for standard models of ZF the relations $R_1(S, \varkappa)$ and $\beta \in \mathfrak{p}[S]$. We have

$$V \models \varphi(T^\xi_u, \varkappa),$$

hence also

$$M \models \varphi(T^\xi_u, \varkappa).$$

But the sentence

$$(\forall \mathbf{S})(\forall \boldsymbol{\varkappa})[\varphi(\mathbf{S}, \boldsymbol{\varkappa}) \to (\exists \boldsymbol{\beta})\psi(\mathbf{S}, \boldsymbol{\beta})]$$

is obviously a theorem of ZF (at least if $\varphi(\mathbf{S}, \boldsymbol{\varkappa})$ and $\psi(\mathbf{S}, \boldsymbol{\beta})$ are chosen in the natural way), so that it must hold in M. In particular,

$$M \models \varphi(T^\xi_u, \varkappa) \to (\exists \boldsymbol{\beta})\psi(T^\xi_u, \boldsymbol{\beta})$$

and thus there is some $\beta \in M$ so that

$$M \models \psi(T^\xi_u, \beta)$$

which implies $\beta \in T^\xi_u$, i.e. $\beta = \alpha$ and $\alpha \in M$. ⊣

In order to put down an elegant version of this result, let us call a pointset $P \subseteq \mathfrak{X}$ *$\varkappa$-Suslin over* (a transitive class) M, if $\varkappa \in M$ and there is a tree T on $\omega \times \varkappa$ in M and a Δ^1_1 function

$$f: \mathfrak{X} \to \mathcal{N}$$

with Δ^1_1 inverse such that

$$P(x) \Leftrightarrow f(x) \in \mathfrak{p}[T].$$

8G.2. MANSFIELD'S PERFECT SET THEOREM. *If P is $\varkappa$-Suslin over a standard model M of ZF, then either $P \subseteq M$ or P has a non-empty perfect subset.*

PROOF. If $x_0 \in P - M$, then by the ZF-absoluteness of f^{-1}, $\alpha_0 = f(x_0) \in \mathfrak{p}[T] - M$ and hence $\mathfrak{p}[T]$ has a non-empty perfect subset F. Now $f^{-1}[F]$ is an uncountable Borel subset of P, so P has a non-empty perfect subset. ⊣

8G.3. THEOREM (Shoenfield [1961]). *Every $\Sigma^1_2(\beta_0)$ pointset in $\aleph_1$-Suslin over $L[\beta_0]$.*

PROOF. It is enough by the definition to prove the result for $A \subseteq \mathcal{N}$, so suppose

$$\alpha \in A \Leftrightarrow P(\alpha, \beta_0),$$

where P is Σ^1_2. Put

$$\gamma \in B \Leftrightarrow P(((\gamma(0))_0, (\gamma(1))_0, \dots), ((\gamma(0))_1, (\gamma(1))_1, \dots))$$

so that B is in Σ^1_2 and there is some $T \in L$, T a tree on $\omega \times \aleph_1$ such that $B = \mathfrak{p}[T]$. Compute:

$$\begin{aligned}
\alpha \in A &\Leftrightarrow P(\alpha, \beta_0) \\
&\Leftrightarrow (\langle \alpha(0), \beta_0(0)\rangle, \langle \alpha(1), \beta_0(1)\rangle, \dots) \in B \\
&\Leftrightarrow \text{for some } f: \omega \to \aleph_1 \text{ and all } n, \\
&\qquad ((\langle \alpha(0), \beta_0(0)\rangle, f(0)), (\langle \alpha(1), \beta_0(1)\rangle, f(1)), \dots, \\
&\qquad\qquad (\langle \alpha(n-1), \beta_0(n-1)\rangle, f(n-1))) \in T \\
&\Leftrightarrow \text{for some } f: \omega \to \aleph_1 \text{ and all } n, \\
&\qquad\qquad ((\alpha(0), f(0)), \dots, (\alpha(n-1), f(n-1))) \in S
\end{aligned}$$

where

$$\begin{aligned}
&((a_0, \xi_0), \dots, (a_{n-1}, \xi_{n-1})) \in S \\
&\qquad \Leftrightarrow ((\langle a_0, \beta_0(0)\rangle, \xi_0), \dots, (\langle a_{n-1}, \beta_0(n-1)\rangle, \xi_{n-1})) \in T
\end{aligned}$$

and clearly $S \in L[\beta_0]$. ⊣

8G.4. COROLLARY (Solovay [1969]). *If $P \subseteq \mathfrak{X}$ is $\Sigma^1_2(\beta_0)$ and P has a member $x_0 \notin L[\beta_0]$, then P has a non-empty perfect subset.*

In particular, if $card(\mathcal{N} \cap L[\beta_0]) = \aleph_0$, *then every uncountable* $\Sigma^1_2(\beta_0)$ *set has a non-empty perfect subset and*

$$\begin{aligned} \mathcal{N} \cap L[\beta_0] &= C_2(\beta_0) \\ &= \text{the largest countable } \Sigma^1_2(\beta_0) \text{ set.} \end{aligned} \qquad \dashv$$

We can consider this theorem a "lightface" improvement of 6G.10, since as we will show in 8H.15, 8H.7,

$$(\exists \varkappa)[\varkappa \to (\aleph_1)] \Rightarrow (\forall \beta)[card(\mathcal{N} \cap L[\beta]) = \aleph_0].$$

But the present result is much more appealing than 6G.10, because it seems to give an explanation of why some $\Sigma^1_2(\beta)$ sets are countable and others are not: at least granting that $card(\mathcal{N} \cap L[\beta]) = \aleph_0$, then for any $\Sigma^1_2(\beta)$ set P,

$$P \text{ is countable} \Leftrightarrow P \subseteq L[\beta].$$

We now turn to similar "lightface" versions of the other known regularity properties of $\mathbf{\Sigma}^1_2$ sets, on the basis of the hypothesis $(\forall \beta) card(\mathcal{N} \cap L[\beta]) = \aleph_0]$. The key result is a metamathematical theorem on approximating sets modulo a σ-ideal, a theorem quite similar to 2H.1.

First some preliminary definitions and absoluteness computations.

Recall the coding of Borel sets by irrationals which we introduced in 7B and suppose M is a transitive class, possibly an inner model of *ZF*. A Borel set $P \subseteq \mathcal{X}$ is *rational over* M if there is some $\alpha \in M$ which is a code for P.

For each ordinal $\lambda > \omega$, we define the sets C^ξ_λ by the recursion on ξ

$$\begin{aligned} C^0_\lambda &= \{\langle 1, s\rangle : s \in \omega\}, \\ C^{\xi+1}_\lambda &= C^\xi_\lambda \\ &\quad \cup \{\langle 2, a\rangle : a \in C^\xi_\lambda\} \\ &\quad \cup \{\langle 3, f\rangle : f : \eta \to C^\xi_\lambda \text{ is a function with domain} \\ &\qquad\qquad \text{some } \eta < \lambda \text{ and image in } C^\xi_\lambda\}, \\ C^\mu_\lambda &= \textstyle\bigcup_{\xi<\mu} C^\xi_\lambda \text{ if } \mu \text{ is a limit ordinal.} \end{aligned}$$

We let

$$C_\lambda = \textstyle\bigcup_\xi C^\xi_\lambda$$

and we call the members of the class C_λ *codes for the* λ*-Borel sets*. Notice that

$$\lambda \leq \lambda' \Rightarrow C_\lambda \subseteq C_{\lambda'}.$$

For each space $\mathcal{X}$ and each $a \in C_\lambda$ we define a pointset

$$B_a = B(a, \lambda, \mathcal{X}) \subseteq \mathcal{X}$$

by the recursion

$$B(\langle 1, s\rangle, \lambda, \mathcal{X}) = N_s(\mathcal{X}) = \text{the } s\text{'th basic nbhd of } \mathcal{X},$$
$$B(\langle 2, a\rangle, \lambda, \mathcal{X}) = \mathcal{X} - B(a, \lambda, \mathcal{X}),$$
$$B(\langle 3, f\rangle, \lambda, \mathcal{X}) = \bigcup_{\xi<\eta} B(f(\xi), \lambda, \mathcal{X}),$$

where in the last clause $\eta < \lambda$ and $f: \eta \to C_\lambda$. It is clear how to express this definition in the language of set theory by using recursion on ξ to define $B: C_\lambda \to V$ on each C_λ^ξ first. It is also immediate that

$$\lambda \leq \lambda' \ \&\ a \in C_\lambda \Longrightarrow B(a, \lambda, \mathcal{X}) = B(a, \lambda', \mathcal{X}),$$

so that the notation "B_a" is unambiguous once we fix $\mathcal{X}$ and we know $a \in C_\lambda$ for some λ.

It is easy to prove (using the axiom of choice) that a set $P \subseteq \mathcal{X}$ is B_a for some $a \in C_\lambda$ if and only if P is λ-Borel in the sense of Section 2E. We say that *P is λ-Borel over* (an inner model of *ZF*) M if

$$P = B_a = B(a, \lambda, \mathcal{X})$$

for some $a \in C_\lambda \cap M$.

8G.5. LEMMA. (i) *The relations*

$$C(\lambda, a) \Leftrightarrow a \in C_\lambda,$$
$$P(\lambda, a, x) \Leftrightarrow a \in C_\lambda \ \&\ x \in B(a, \lambda, \mathcal{X})$$

are ZF-absolute.

(ii) *There is a ZF-absolute operation*

$$F: C_{\omega+1} \to \mathcal{N}$$

such that for each $a \in C_{\omega+1}$, $F(a)$ is a Borel code of $B(a, \omega\ \mathcal{X})$; similarly, there is a ZF-absolute operation

$$G: \mathcal{N} \to V$$

such that if α is a Borel code of some $P \subseteq \mathcal{X}$, then $G(\alpha) \in C_{\omega+1}$ and $B(G(\alpha), \omega, \mathcal{X}) = P$.

(iii) *Let M be a standard model of ZF and let*

$$\lambda^* = \aleph_1^M$$
$$= \text{supremum}\{\xi \in M: \text{there is some bijection } f: \omega \rightarrowtail\!\!\!\!\twoheadrightarrow \xi, f \in M\}.$$

A set $P \subseteq \mathscr{X}$ is λ^*-Borel over M if and only if P is Borel rational over M.

(iv) *If M is an inner model of ZF and $P \subseteq \mathscr{X}$ is $\varkappa$-Suslin over M, then P is $(\varkappa^+ + 1)$-Borel over M, where*

$$\varkappa^+ = \textit{least cardinal greater than } \varkappa.$$

(v) *Every $\Sigma^1_2(\beta)$ pointset is $(\aleph_1 + 1)$-Borel over $L[\beta]$.*

PROOF. (i) and (ii) are proved by standard absoluteness arguments which we will omit.

(iii) The assertion

$$(\forall a)[a \in C_{\aleph_1} \Rightarrow \text{for some } \xi < \aleph_1,\ a \in C_{\xi+1}]$$

is easily expressible in $\mathcal{L}^\in$ and it is a theorem of ZF. Thus if M is a standard model of ZF and $\lambda^* = \aleph_1^M$, then (using (i))

$$a \in C_{\lambda^*} \cap M \Rightarrow \text{for some } \xi < \lambda^*,\ a \in C_{\xi+1} \cap M.$$

At the same time, it is easy to define a ZF-absolute operation

$$(f, \xi, a) \mapsto a^* = H(f, \xi, a),$$

such that whenever $f : \omega \twoheadrightarrow \xi$ is a bijection of ξ with ω and $a \in C_{\xi+1}$, then $a^* = H(f, \xi, a) \in C_{\omega+1}$ and $B_a = B_{a^*}$.

Now if $a \in C_{\lambda^*} \cap M$ with $\lambda^* = \aleph_1^M$, choose $\xi < \lambda^*$ such that $a \in C_{\xi+1} \cap M$, choose $f \in M$, $f : \omega \rightarrowtail\!\!\!\!\twoheadrightarrow \xi$ and take $a^* = F(f, \xi, a) \in M$. Using (ii), this shows that if $P \subseteq \mathscr{X}$ is λ^*-Borel over M, then P is Borel rational over M and the converse is immediate by (ii).

(iv) Suppose first that $T \in M$, T a tree on $\omega \times \varkappa$. By the Sierpinski formulas in 2F.1,

$$\mathcal{N} - \mathrm{p}[T] = \bigcup_{\lambda < \varkappa^+} B^\lambda_{\emptyset}$$

where the sets B^λ_u are defined by a simple recursion on λ with u, $\varkappa$, T as parameters, so that in fact

$$\alpha \in B^\lambda_u \Leftrightarrow R(\alpha, \lambda, u, \varkappa, T)$$

with some ZF-absolute R. It is easy to define a ZF-absolute operation $G(\lambda, u, \varkappa, \mu, T)$ so that for each $\mu > \varkappa$ and each $\lambda < \mu$, $G(\lambda, u, \varkappa, \mu, T)$ is a code in C_μ of B^λ_u and then

$$H(\mu, \varkappa, T) = \langle 3, \{\langle \lambda, \langle 2, G(\lambda, \emptyset, \varkappa, \mu, T)\rangle\rangle : \lambda < \mu\}\rangle$$

is ZF-absolute and gives a code of $\mathrm{p}[T]$ in $C_{\varkappa^+ + 1}$ when we substitute $\mu = \varkappa^+$. Since $M \models ZF$ and $\varkappa$, $\varkappa^+$, $T \in M$, we have $H(\varkappa^+, \varkappa, T) \in M$ and $\mathrm{p}[T]$ is $(\varkappa^+ + 1)$-Borel over M.

In order to prove the result for $\varkappa$-Suslin sets $P \subseteq \mathcal{X}$, fix come Δ^1_1 function

$$f : \mathcal{X} \to \mathcal{N}$$

with Δ^1_1 inverse so that the relation

$$Q(x, s) \Leftrightarrow f(x) \in N_s$$

is Δ^1_1. Now (ii) above and the Suslin–Kleene theorem imply easily that there is a ZF-absolute $G(s, \mu)$ such that for each $s \in \omega$ and $\mu > \omega$, $G(s, \mu)$ is a code of $f^{-1}[N_s]$ in C_μ and from this we can get a further ZF-absolute operation $G'(a, \mu)$ such that for each $a \in C_\mu$, $G'(a, \mu)$ is a code in C_μ of $f^{-1}[B(a, \mu, \mathcal{N})]$. If $P = f^{-1}[\mathfrak{p}[T]]$ with $T \in M$ and $\mathfrak{p}[T] = B(a, \varkappa^+, \mathcal{N})$, then $G'(a, \varkappa^+)$ is a code of P and hence P is $(\varkappa^+ + 1)$-Borel over M. ⊣

Recall from 2H that a collection J of subsets of a space $\mathcal{X}$ is a σ-ideal if it closed under subsets and countable unions.

A σ-ideal J is *definable over* an inner model M of ZF if there exists some formula $\varphi(\boldsymbol{\alpha}, \mathbf{c})$ of $\mathcal{L}^\in$ and some $c \in M$, such that for all $\alpha \in \mathcal{N} \cap M$,

$$\alpha \text{ is the Borel code of some } A \in J \Leftrightarrow M \models \varphi(\alpha, c).$$

8G.6. LEMMA. (i) *For each space $\mathcal{X}$ and each inner model M of ZF, the ideal of meager subsets of $\mathcal{X}$ is definable over M.*

(ii) *If μ is a σ-finite Borel measure on $\mathcal{X}$, then there is some $\alpha_\mu \in \mathcal{N}$ such that whenever $\alpha_\mu \in M \models ZF$, then the σ-ideal Z_μ of subsets of $\mathcal{X}$ of μ-measure 0 is definable over M. If μ is the Lebesgue measure on the reals, then Z_μ is definable over every inner model M of ZF.*

(iii) *If J is definable over M, then there is a formula $\psi(\mathbf{a}, \mathbf{c})$ of $\mathcal{L}^\in$ and some $c \in M$ such that if $\lambda^* = \aleph_1^M$ and $a \in C_{\lambda^*} \cap M$, then*

$$B_a \in J \Leftrightarrow M \models \psi(a, c).$$

PROOF. By 4F.19, the relation

$$P(\alpha) \Leftrightarrow \alpha \text{ codes a meager Borel subset of } \mathcal{X}$$

is easily Π^1_2 and hence ZF-absolute. This establishes (i) and (ii) follows by a similar computation.

To prove (iii), let $\varphi(\boldsymbol{\alpha}, \mathbf{c})$ and $c \in M$ be such that

$$\alpha \text{ codes a Borel set in } J \Leftrightarrow M \models \varphi(\alpha, c)$$

and take $\psi(\mathbf{a}, \mathbf{c})$ to be the formula expressing the condition

$$\begin{aligned}(\exists\xi)(\exists f)(\exists\alpha)\{f: \omega \twoheadrightarrow \xi &\text{ is a bijection}\\ &\& \ a \in C_{\xi+1}\\ &\& \ \alpha \text{ is the code of a Borel set in } J\\ &\& \ F(f, \xi, a) = \alpha\},\end{aligned}$$

where F is the ZF-absolute operation defined as in the proof of 8G.5 such that with f, ξ and a as above and $\alpha = F(f, \xi, a)$,

$$B_\alpha = B_a$$

and where we define "α is the code of a Borel set in J" using the formula $\varphi(\boldsymbol{\alpha}, \mathbf{c})$ supplied by the definition. ⊣

The Approximation Theorem 2H.1 had the additional hypothesis that J is regular from above. Here we will work with σ-ideals J which satisfy the *countable chain condition*, i.e. such that every uncountable collection of pairwise disjoint Borel sets must contain sets in J. We must be a bit careful in formulating this condition in arbitrary inner models of ZF.

A σ-ideal J *satisfies the CCC in an inner model M of ZF*, if there is a formula $\psi(\mathbf{a}, \mathbf{c})$ of $\mathcal{L}^\in$ and some $c \in M$ such that:

(i) for $\lambda^* = \aleph_1^M$ and $a \in C_{\lambda^*} \cap M$,

$$B_a \in J \Leftrightarrow M \models \psi(a, c),$$

(ii) $M \models$ "for every function $f: \mu \to C_{\aleph_1}$,

$$\begin{aligned}&\text{if } \xi, \eta < \mu \ \& \ \xi \neq \eta \Rightarrow B_{f(\xi)} \cap B_{f(\eta)} = \emptyset,\\ &\text{then } \{\xi: B_{f(\xi)} \notin J\} \text{ is countable"};\end{aligned}$$

in turning the expression in quotes into a formal assertion in $\mathcal{L}^\in$ here, we use the given $\psi(\mathbf{a}, \mathbf{c})$ to express $B_a \in J$ and we use a formula $\varphi(\mathbf{x}, \mathbf{a}, \lambda, \mathfrak{X})$ supplied by (i) of 8G.5 to express "$x \in B_a$," so that e.g.

$$B_{f(\xi)} \cap B_{f(\eta)} = \emptyset$$

is expressed by

$$\neg(\exists x)[\varphi(x, f(\xi), \aleph_1, \mathfrak{X}) \ \& \ \varphi(x, f(\eta), \aleph_1, \mathfrak{X})].$$

Also, by "$\aleph_1$" we mean the supremum of countable ordinals within M, so that a more detailed version of (ii) would read

$$M \models \text{"if } \lambda = \text{the least uncountable ordinal, then}$$
$$\text{for every } f : \mu \to C_\lambda,$$
$$\text{if } \xi,\ \eta < \mu \cdots \text{etc."}$$

Notice that if J satisfies the CCC in M, then J is definable over M.

8G.7. LEMMA. *For each $\mathfrak{X}$ and each inner model M of ZF, the collection of meager subsets of $\mathfrak{X}$ satisfies the CCC in M.*

(ii) *If μ is a σ-finite Borel measure on $\mathfrak{X}$, then there is some $\alpha_\mu \in \mathcal{N}$ such that whenever $\alpha_\mu \in M \models ZF$, then the σ-ideal Z_μ satisfies the CCC in M; if μ is the Lebesgue measure on the reals, then Z_μ satisfies the CCC in every inner model M of ZF.*

PROOF is immediate because it is a theorem of ZF that these ideals satisfy the CCC, so this assertion must hold in every model of ZF. (More precisely: write up a proof of the CCC for these ideals, check what properties of the relation

$$P(\alpha) \Leftrightarrow \alpha \text{ codes a Borel set in } J$$

you use in the proof, verify that each M must satisfy the formal sentences expressing these properties because of the absoluteness in the definition of $P(\alpha)$.) ⊣

Suppose now that M is an inner model of ZF and J is a σ-ideal on $\mathfrak{X}$ which satisfies the CCC in M. The set of points of $\mathfrak{X}$ which are *J-algebraic over M* is defined by

$$Alg(M, J) = Alg^{\mathfrak{X}}(M, J) = \{x \in \mathfrak{X} : \text{for some } a \in M \cap C_{\omega+1},\ B_a \in J \text{ and } x \in B_a\}.$$

The points in $Alg(M, J)$ can be *approximated* in M, in the sense that they belong to a "small" Borel set (i.e. a Borel set in J) with code in M.

Notice that if $card(\mathcal{N} \cap M) = \aleph_0$, then $Alg(M, J)$ is a countable union of sets in J, so it is in J. Of course, if $\mathcal{N} \subseteq M$, then $Alg(M, J)$ is the union of all the Borel sets in J and is most likely all of $\mathfrak{X}$.

Let

$$Trans(M, J) = Trans^{\mathfrak{X}}(M, J) = \mathfrak{X} - Alg(M, J)$$

be the set of points of $\mathfrak{X}$ which are *J-transcendental over M.*

8G.8. THE APPROXIMATION THEOREM OVER MODELS OF *ZF*. *Let M be an inner model of ZF, let J be a σ-ideal of subsets of $\mathfrak{X}$ which satisfies the CCC in M and suppose $P \subseteq \mathfrak{X}$ is λ-Borel over M, for any $\lambda > \omega$. Then there exists a Borel set P^* which is rational over M such that*

$$P \triangle P^* = (P - P^*) \cup (P^* - P) \subseteq Alg(M, J).^{(2)}$$

PROOF. Fix J and M which satisfy the hypotheses and let

$$\lambda^* = \aleph_1^M$$

as in 8G.5. The idea is to define an operation

$$a \mapsto a^*$$

such that if $a \in C_\lambda \cap M$, then $a^* \in C_{\lambda^*} \cap M$ and

$$B_a \triangle B_{a^*} \subseteq Alg(M, J).$$

To define the operation $a \mapsto a^*$ *we will work in M*, i.e. we will perform a set theoretic construction within the model M. All that this means is that in the definition below, by "set" we mean "set in M" and all notions of set theory are interpreted within M.

The definition of a^* is by recursion on the least ξ such that $a \in C_\lambda^\xi$, but of course it comes down to treating three cases.

(i) If $a = \langle 1, s \rangle$ for some $s \in \omega$, put $a^* = a$.

(ii) If $a = \langle 2, b \rangle$ for some b, put $a^* = \langle 2, b^* \rangle$.

(iii) Suppose now $a = \langle 3, f \rangle$, where f is a function, $f: \eta \to C_\lambda$ where $\eta < \lambda$ and we may assume that for each $\xi < \eta$, $(f(\xi))^*$ has already been defined.

Let us define first a function

$$g: \aleph_1 \to \eta$$

by the following subsidiary recursion. (*Caution.* Since we are working within M, actually

$$g: \lambda^* = \aleph_1^M \to \eta.)$$

Put

$$g(0) = 0,$$

$$g(\zeta + 1) = \begin{cases} \text{the least } \xi > g(\zeta) \text{ such that} \\ \qquad B_{(f(\xi))^*} - \bigcup_{\nu \le \zeta} B_{f(g(\nu))^*} \notin J, \\ \qquad \text{if one such } \xi \text{ exists,} \\ 0 \quad \text{otherwise.} \end{cases}$$

$$g(\mu) = 0 \quad \text{if } \mu \text{ is a limit.}$$

Caution: To make precise this definition, we first construct some *ZF*-absolute operation $H(a, h)$ such that whenever $a \in C_\lambda$ and $h : \zeta \to C_{\lambda^*}$, we have

$$B_{H(a,h)} = B_a - \bigcup_{\nu \leq \zeta} B_{h(\nu)}$$

and then we ask in the first case whether

$$(*) \qquad B_{H(f(\xi)^*, \{\langle \nu, f(g(\nu))^* \rangle : \nu \leq \zeta\})} \in J$$

by using the formula $\psi(\mathbf{d}, \mathbf{c})$ and the object $c \in M$ supplied by the hypothesis that J satisfies the *CCC* in M. As a result, the set in $(*)$ will be in J exactly when M thinks that it is in J.

Notice now that since J satisfies the *CCC* in M, there must be some ordinal $\zeta_0 < \lambda^*$, such that

$$\zeta \geq \zeta_0 \to g(\zeta) = 0,$$

otherwise we would get $\aleph_1$ disjoint Borel sets with none of them in J (as M sees things). Pick the least such ζ_0 and put

$$a^* = \langle 3, \{\langle \zeta, (f(g(\zeta)))^* \rangle : \zeta < \zeta_0\} \rangle.$$

This completes the definition of the operation $a \mapsto a^*$ in M and we obviously have for $a \in M$,

$$a \in C_\lambda \Rightarrow a^* \in C_{\lambda^*}.$$

We now move outside M and show by induction on ξ that

$$a \in C_\lambda^\xi \cap M \Rightarrow B_a \,\triangle\, B_{a^*} \subseteq Alg(M, J).$$

Again the proof comes down to three cases and the first two are completely trivial.

(iii) Suppose $a = \langle 3, f \rangle$ with $f : \eta \to C_\lambda$ with $\eta < \lambda$ and $a^* = \langle 3, f^* \rangle$ is defined as above.

If $x \in Trans(M, J)$ and $x \in B_{a^*}$, then for some $\zeta < \zeta_0$, $x \in B_{f^*(\zeta)} = B_{f(g(\zeta))^*}$ and $B_{f(g(\zeta))^*} \triangle B_{f(g(\zeta))} \subseteq Alg(M, J)$ by the induction hypothesis, so that $x \in B_{f(g(\zeta))} \subseteq B_a$.

Conversely, if $x \in B_a$, then let ξ be the least ordinal such that $x \in B_{f(\xi)}$ and assume towards a contradiction that $x \in Trans(M, J)$ and $x \notin B_{a^*}$. Now we obviously cannot have $\xi = g(\zeta)$ for some $\zeta < \zeta_0$, since in that case we would have by the induction hypothesis $x \in B_{f(g(\zeta))^*} = B_{f^*(\zeta)} \subseteq B_{a^*}$; hence by the definition of f^*, if

$$\zeta_1 = supremum\{\zeta : g(\zeta) < \xi\},$$

we must have $g(\zeta_1+1)\neq\xi$, i.e.

$$(**)\qquad M\models B_{f(\xi)^*}-\bigcup_{\nu\leq\zeta_1}B_{f(g(\nu))^*}\in J.$$

By the construction then, there is some $d\in C_{\lambda^*}\cap M$ such that

$$B_d=B_{f(\xi)^*}=\bigcup_{\nu\leq\zeta_1}B_{f(g(\nu))^*}$$

and

$$B_d\in J,$$

since the interpretation of $(**)$ was that $M\models\psi(d,c)$. But obviously $x\in B_d$, so that x is a member of some J-set with code in M contradicting our assumption that $x\in \mathit{Trans}(M,J)$. ⊣

The corollaries make the effort worthwhile.

8G.9. Theorem. (i) *Let J be the σ-ideal of all meager subsets of $\mathfrak{X}$ and suppose $P\subseteq\mathfrak{X}$ is $\Sigma^1_2(\beta_0)$; then there exists a Borel set $\tilde{P}$ which is rational over $L[\beta_0]$ and such that*

$$P\triangle\tilde{P}\subseteq Alg(L[\beta_0],J).$$

In particular, if $card(\mathcal{N}\cap L[\beta_0])=\aleph_0$, then P has the property of Baire.

(ii) *Suppose μ is a σ-finite Borel measure on $\mathfrak{X}$ which satisfies the CCC in $L[\beta_0]$ (e.g. if μ is the Lebesgue measure on $\mathbb{R}$) and suppose $P\subseteq\mathfrak{X}$ is $\Sigma^1_2(\beta_0)$; then there is a Borel set $\tilde{P}$ which is rational over $L[\beta_0]$ such that*

$$P\triangle\tilde{P}\subseteq Alg(L[\beta_0],Z_\mu).$$

In particular, if $card(\mathcal{N}\cap L[\beta_0])=\aleph_0$, then P is μ-measurable.

(iii) (Solovay) *If $(\forall\beta)[card(\mathcal{N}\cap L[\beta])=\aleph_0]$, then every $\mathbf{\Sigma}^1_2$ set has the property of Baire and is absolutely measurable.*[(2)]

Proof is immediate from 8G.8 and the preceding lemmas. ⊣

The applications of these results to the higher Lusin pointclasses depend on the connection between semiscales and trees which we established in 2B.1 and which we now reformulate in a suitable notation.

Suppose Γ is a Spector pointclass closed under $\forall^{\mathcal{N}}$ with ordinal

$$\boldsymbol{\delta}=supremum\{|\leq|:\ \leq\text{ is a prewellordering of }\mathcal{N}\text{ in }\mathbf{\Delta}\},$$

let $G\subseteq\omega\times\mathcal{N}$ be universal in Γ and let $\{\bar{\varphi}\}=\{\varphi_n\}_{n\in\omega}$ be a Γ-scale on G. By 4C.14 we know that the length $|\varphi_n|$ of each (regular) norm φ_n is $\boldsymbol{\delta}$, i.e.

$$\varphi_n:G\twoheadrightarrow\boldsymbol{\delta}.$$

Put

$$T_\Gamma = T_{\Gamma,G,\bar\varphi}$$
$$= \{((e, \varphi_0(e, \alpha)), (\alpha(0), \varphi_1(e, \alpha)), \dots, (\alpha(n-2), \varphi_{n-1}(e, \alpha))) : G(e, \alpha)\}.$$

It is important to notice that although we will use the notation "T_Γ" for this *tree associated with the pointclass* Γ, actually T_Γ depends on the particular choice of G and $\bar\varphi$.

In case

$$\Gamma = \Pi^1_{2n+1}$$

and under the hypothesis $Det(\mathbf{\Delta}^1_{2n})$, we use the simpler notation

$$T_{2n+1} = T_{\Pi^1_{2n+1}}.$$

When we think of trees as *sets* (members of V) as in the next theorem, we should (strictly speaking) use the notation

$$\langle \xi, \eta \rangle$$

for the set-theoretic pair, rather than the intuitive

$$(\xi, \eta)$$

which we have been using all along. It is not worth here to complicate notation by insisting on this pedantic difference.

8G.10. THEOREM. *Let Γ be a Spector pointclass which is closed under $\forall^{\mathcal{N}}$ and has the scale property and let T_Γ be the tree on $\omega \times \boldsymbol{\delta}$ associated with Γ and some fixed universal set $G \subseteq \omega \times \mathcal{N}$ and Γ-scale $\{\bar\varphi\}$ on G; suppose $P \subseteq \mathfrak{X}$ is in $\exists^{\mathcal{N}}\Gamma(\beta_0)$, i.e.*

$$P(x) \Leftrightarrow (\exists\alpha) R(x, \beta_0, \alpha)$$

where $R \in \Gamma$ and suppose M is an inner model of ZF such that T_Γ, $\beta_0 \in M$, e.g.

$$M = L[T_\Gamma, \beta_0] (= L(\langle T_\Gamma, \beta_0 \rangle)).$$

(i) *P is $\boldsymbol{\delta}$-Suslin over M.*

(ii) *Either $P \subseteq M$ or P has a non-empty perfect subset.*

(iii) *If J is a σ-ideal which satisfies the CCC in M, then there is a Borel set $\tilde P$ which is rational over M and such that*

$$P \triangle \tilde P \subseteq Alg(M, J);$$

if in addition card $(\mathcal{N} \cap M) = \aleph_0$, then P is J-measurable.

In particular with Γ and T_Γ as above, if

$$(\forall\beta)[card(\mathcal{N}\cap L[T_\Gamma, \beta]) = \aleph_0],$$

then every pointset in $\exists^{\mathcal{N}}\Gamma$ has the property of Baire, it is absolutely measurable and if uncountable, it has a non-empty perfect subset.

PROOF. It is enough to prove (i) from which the rest follow by what we have already established in this section.

Suppose first $A \subseteq \mathcal{N}$ and

$$\alpha \in A \Leftrightarrow (\exists\gamma)R(\alpha, \beta_0, \gamma)$$

with R in Γ, so that for a fixed $e^* \in \omega$, easily

$$\alpha \in A \Leftrightarrow (\exists\gamma)G(e^*, \alpha \circ \beta_0 \circ \gamma)$$

where

$$\alpha \circ \beta \circ \gamma = (\alpha(0), \beta_0(0), \gamma(0), \alpha(1), \beta_0(1), \gamma(1),\ldots, \alpha(n-1), \beta_0(n-1), \gamma(n-1),\ldots).$$

As in the proof of Shoenfield's Lemma 8F.9, let

$$S = \text{all initial segments of sequences of the form}$$

$$((a_0, \xi_0), (a_1, c_0), (a_2, \xi_1), (a_3, c_1), (a_4, \xi_2), \ldots (a_{2n+1}, c_n), (a_{2n+2}, \xi_{n+1}))$$

such that

$$((e^*, \xi_0), (a_0, \xi_1), (\beta_0(0), \xi_2), (c_0, \xi_3), \ldots (a_{n-3}, \xi_{n-2}), (\beta_0(n-3), \xi_{n-1}), (c_{n-3}, \xi_n)) \in T_\Gamma$$

so that immediately,

$$\beta_0, T_\Gamma \in M \Longrightarrow S \in M.$$

If $\alpha \in A$, then choose γ so that $G(e^*, \alpha \circ \beta_0 \circ \gamma)$ and take $c_n = \gamma(n)$, $\xi_n = \varphi_n(e^*, \alpha \circ \beta_0 \circ \gamma)$; clearly the infinite sequence

$$(*) \qquad ((\alpha(0), \xi_0), (\alpha(1), \beta_0(0), (\alpha(2), \xi_1),\ldots) \in [S]$$

and $\alpha \in \mathfrak{p}[S]$. Conversely, if for some c_0, ξ_0, c_1, ξ_1,... we have

$$((\alpha(0), \xi_0), (\alpha(1), c_0), (\alpha(2), \xi_1), (\alpha(3), c_1),\ldots) \in [S],$$

then by the definition we know that for each n, there are irrationals α_n, γ_n such that $G(e^*, \alpha_n \circ \beta_0 \circ \gamma_n)$,

$$\varphi_k(e^*, \alpha_n \circ \beta_0 \circ \gamma_n) = \xi_k \quad (k \leq n)$$

and

$$limit_{n\to\infty} \alpha_n = \alpha,$$
$$limit_{n\to\infty} \gamma_n = \gamma = (c_0, c_1, c_2, \ldots).$$

Since $\bar{\varphi}$ is a scale on G, it follows that $G(e^*, \alpha \circ \beta_0 \circ \gamma)$ holds so that $\alpha \in A$.

The result for arbitrary $P \subseteq \mathfrak{X}$ in $\exists^{\mathcal{N}}\mathbf{\Gamma}$ follows by the definition of $\boldsymbol{\delta}$-Suslin over M. ⊣

The important special cases here are

$$\Gamma = \Pi^1_{2n+1}$$

and

$$\Gamma = IND = \text{all absolutely inductive pointsets}$$

with the appropriate determinacy hypotheses. We will look at some of the basic facts about the corresponding models $L[T_\Gamma]$ in the exercises.

EXERCISES

Let us first use Mansfield's perfect set theorem to prove a simple converse to the proposition

$$\mathcal{N} \subseteq L \Rightarrow \mathcal{N} \text{ admits a } \Sigma^1_2\text{-good wellordering of rank } \aleph_1.$$

8G.11. Prove that if $\mathcal{N}$ admits a Σ^1_2-good wellordering of rank $\aleph_1$, then $\mathcal{N} \subseteq L$.

Hint. The results in Chapter 5 were all proved under the hypothesis "$\mathcal{N}$ admits a Σ^1_2-good wellordering of rank $\aleph_1$" which was indicated there by the symbolic notation "$\mathcal{N} \subseteq L$". By 5A.11 then, if $\mathcal{N}$ admits a Σ^1_2-good wellordering, then every α is recursive in some $\beta \in C_1$, where C_1 is the largest Π^1_1 set with no non-empty perfect subset; but by Solovay's Theorem 8G.4, $C_1 \subseteq L$, so that every irrational α is recursive in some $\beta \in L$ and hence $\alpha \in L$. ⊣

Mansfield [1975] has established the much nicer result that if $\mathcal{N}$ admits a Σ^1_2 wellordering (or any rank and not necessarily Σ^1_2-good), then $\mathcal{N} \subseteq L$; see also Kechris [198?] for a simple proof of this and some related results.

8G.12. Prove that if there is a function $f: \mathcal{N} \to \mathcal{N}$ whose graph is a thin, Π^1_1 subset of $\mathcal{R} \times \mathcal{R}$, then $\mathcal{N} \subseteq L$.

Hint. $Graph(f) \subseteq L$ by 8G.2, so that for each α, the pair $\langle \alpha, f(\alpha) \rangle \in L$ and $\alpha \in L$. ⊣

Let us call $\varepsilon \in \mathcal{N}$ a *code* of a closed set $F \subseteq \mathcal{N}$, if

$$F = \{\alpha : (\forall n)[\bar{\varepsilon}(\alpha(n)) = 1]\}.$$

It is worth putting down an alternative version of 8G.1 and 8G.2 in terms of these codes.

8G.13. Prove that if M is a standard model of *ZF*, T a tree on $\omega \times \kappa$, $T \in M$ and $\mathfrak{p}[T]$ has a member not in M, then $\mathfrak{p}[T]$ has a non-empty perfect subset with code in M.

Infer that every Σ^1_2 set $A \subseteq \mathcal{N}$ which has a nonconstructible member has a non-empty perfect subset with code in L.

Hint. We showed in the proof of 8G.1 that if $\mathfrak{p}[T]$ has a member not in M, then the tree T^* is not empty and $T^* \in M$. Now interpret the proof of 2C.2 after that *within* M, to get a perfect subset of $\mathfrak{p}[T]$; the code of this perfect subset is in M and it also codes a perfect subset of $\mathfrak{p}[T]$ in the world.

The second assertion follows easily. ⊣

It is worth pointing out here that the condition

$$(\forall \beta)[card(\mathcal{N} \cap L[\beta]) = \aleph_0]$$

follows from our familiar hypothesis $Det(\mathbf{\Delta}^1_2)$. Something stronger and more interesting is actually true.

8G.14. Assume $Det(\mathbf{\Delta}^1_2)$ and show that for each β,

$$\alpha \in L[\beta] \Rightarrow \alpha \in \Delta^1_3(\beta);$$

in particular,

$$Det(\mathbf{\Delta}^1_2) \Rightarrow (\forall \beta)[card(\mathcal{N} \cap L[\beta]) = \aleph_0].$$

Hint. Let C_1 be the largest thin Π^1_1 subset of 4F.4 and consider the

$*$-game for C_1 as we defined this in the exercises of 6A, in which I plays finite (non-empty) sequences

$$s_0 = a_0, \ldots, a_{k_0-1},$$
$$s_1 = a_{k_0+1}, \ldots, a_{k_1-1},$$

II plays integers

$$a_{k_0}, a_{k_1}, \ldots$$

and I wins if the play

$$\alpha = a_0, \ldots, a_{k_0-1}, a_{k_0}, a_{k_0+1}, \ldots .$$

is in C_1. The game is determined since C_1 is Π^1_1 and by 6A.10, I cannot win it, so II does; but then C_1 is countable and by 6E.1, II has a Δ^1_3 winning strategy τ. Now the proof of 6A.11 makes it clear that every irrational in C_1 is recursive in τ and hence lies in Δ^1_3. Finally, from 5A.11 it follows that each $\alpha \in L$ is recursive in some $\beta \in C_1$, so that each $\alpha \in L$ is Δ^1_3.

The argument for the relativized case is similar. ⊣

In 8H.15, 8H.16 we will show that this strong result also follows from the large cardinal hypothesis $(\exists\varkappa)[\varkappa \to (\aleph_1)]$.

We now turn to a brief study of the inner models associated with various nice pointclasses. First a few basic absoluteness facts about the models $L[T_{2n+1}]$; in the statements of these results, we always assume tacitly that T_{2n+1} is the tree on $\omega \times \boldsymbol{\delta}^1_{2n+1}$ associated with some universal Π^1_{2n+1} set $G \subseteq \omega \times \mathcal{N}$ and some Π^1_{2n+1}-scale $\bar{\varphi}$ on G-granting at least $Det(\boldsymbol{\Delta}^1_{2n})$.

8G.15. Assume $Det(\boldsymbol{\Delta}^1_{2n})$ and suppose M is an inner model of ZF such that $T_{2n+1} \in M$. Show that for each Σ^1_{2n+2} formula $\theta(\boldsymbol{\alpha}_1, \ldots, \boldsymbol{\alpha}_m)$ of second order arithmetic and $\alpha_1, \ldots, \alpha_m \in M$,

$$V \models \theta(\alpha_1, \ldots, \alpha_m) \Leftrightarrow M \models \theta(\alpha_1, \ldots, \alpha_m).$$

Hint. Assume the hypotheses on T_{2n+1} and M and prove by induction on $i \leq 2n+2$ that every Σ^1_i formula is absolute for M; use the Basis Theorem 6C.6 and 8G.10 as in the proof of Shoenfield's Theorem (II), 8F.10. ⊣

8G.16. Assume $Det(\boldsymbol{\Delta}^1_{2n})$ and suppose M is an inner model of ZF such that $T_{2n+1} \in M$. Show that

$$M \models \text{“}Det(\mathbf{\Delta}^1_{2n})\text{.”}$$

Show under the further hypothesis $Det(\mathbf{\Delta}^1_{2n+1})$ that

$$M \models \text{“}Det(\mathbf{\Delta}^1_{2n+1})\text{.”}$$

Hint. Take the second assertion which is a bit harder and for simplicity of notation take $n = 1$.

Clearly M satisfies the formal sentence of $\mathcal{L}^\epsilon$ expressing $Det(\mathbf{\Delta}^1_3)$ if and only if for every two Σ^1_3 formulas $\theta_1(\boldsymbol{\alpha}, \boldsymbol{\beta})$ and $\theta_2(\boldsymbol{\alpha}, \boldsymbol{\beta})$, M satisfies

$$(\forall \boldsymbol{\alpha})\{(\forall \boldsymbol{\beta})[\theta_1(\boldsymbol{\alpha}, \boldsymbol{\beta}) \leftrightarrow \neg\theta_2(\boldsymbol{\alpha}, \boldsymbol{\beta})] \\ \rightarrow [(\exists \boldsymbol{\sigma})(\forall \boldsymbol{\tau})\theta_1(\boldsymbol{\alpha}, \boldsymbol{\sigma} * \boldsymbol{\tau}) \vee (\exists \boldsymbol{\tau})(\forall \boldsymbol{\sigma})\theta_2(\boldsymbol{\alpha}, \boldsymbol{\sigma} * \boldsymbol{\tau})]\},$$

so assume $\alpha \in M$ and

(1) $$M \models (\forall \boldsymbol{\beta})[\theta_1(\alpha, \boldsymbol{\beta}) \leftrightarrow \neg\theta_2(\alpha, \boldsymbol{\beta})];$$

we must show that

(2) $$M \models (\exists \boldsymbol{\sigma})(\forall \boldsymbol{\tau})\theta_1(\alpha, \sigma * \tau) \vee (\exists \boldsymbol{\tau})(\forall \boldsymbol{\sigma})\theta_2(\alpha, \boldsymbol{\sigma} * \boldsymbol{\tau}).$$

From (1) and 8G.15 we know that

$$V \models (\forall \boldsymbol{\beta})[(\theta_1(\alpha, \boldsymbol{\beta}) \leftrightarrow \neg\theta_2(\alpha, \boldsymbol{\beta})]$$

so that the set

$$A = \{\beta : V \models \theta_1(\alpha, \beta)\}$$

is $\Delta^1_3(\alpha)$ and hence determined. Assume without loss of generality that I wins the game defined by A, so that

$$V \models (\exists \boldsymbol{\sigma}(\forall \boldsymbol{\tau})\theta_1(\alpha, \boldsymbol{\sigma} * \boldsymbol{\tau});$$

by 8G.15 again

$$M \models (\exists \boldsymbol{\sigma})(\forall \boldsymbol{\tau})\theta_1(\alpha, \boldsymbol{\sigma} * \boldsymbol{\tau}),$$

so that (2) holds. ⊣

This a typical application of the absoluteness result in 8G.15: the models $L[T_{2n+1}]$ reflect enough of the properties of V to satisfy just a bit more determinacy than we need to construct them.

The most important problem about these models is their *invariance*, i.e. the question whether $L[T_{2n+1}]$ depends on the particular choice of G and $\bar{\varphi}$ used in constructing T_{2n+1}. This is still open, but it is known that the set

$\mathcal{N} \cap L[T_{2n+1}]$ of irrationals in $L[T_{2n+1}]$ is independent of the particular choice of T_{2n+1}, in fact

$$\begin{aligned}\mathcal{N} \cap L[T_{2n+1}] &= C_{2n+2} \\ &= \text{the largest countable } \Sigma^1_{2n+2} \text{ subset of } \mathcal{N}.\end{aligned}$$

We will outline a proof of this result of Kechris and Harrington (Kechris and Martin for $n = 1$) in a sequence of lemmas which are quite important in themselves. It will pay to formulate these lemmas in a reasonably general context, since they have applications to the study of many pointclasses other than Π^1_{2n+1}.

Let us say for the remainder of this chapter that a pointclass Γ *resembles* Π^1_1 if the following two conditions hold.

(i) Γ is a Spector pointclass with the scale property and closed under $\forall^{\mathcal{N}}$.

(ii) For each $\alpha \in \mathcal{N}$, if $P \subseteq \mathfrak{X} \times \mathcal{Y}$ is in $\Delta(\alpha)$ and

$$Q(x) \Leftrightarrow P_x = \{y: P(x, y)\} \text{ is not meager},$$

then Q is also in $\Delta(\alpha)$.

It is clear that Π^1_1 resembles Π^1_1 by 4F.19 and that if Γ resembles Π^1_1 so does each relativization $\Gamma(z)$. The next result gives us a large stock of pointclasses which resemble Π^1_1.

8G.17. Suppose Γ is a Spector pointclass with the scale property and closed under $\forall^{\mathcal{N}}$ and assume that there is some adequate pointclass $\Gamma_1 \subseteq \Gamma$ such that

$$\Gamma = \forall^{\mathcal{N}} \Gamma_1,$$

Γ_1 is closed under $\exists^{\mathcal{N}}$ and $\mathbf{\Sigma}^1_1 \subseteq \mathbf{\Gamma}_1$. Assume also that $Det(\mathbf{\Gamma})$ holds and show that Γ resembles Π^1_1.

In particular, if $Det(\mathbf{\Pi}^1_{2n+1})$ holds then Π^1_{2n+1} resembles Π^1_1 and if $Det(\boldsymbol{IND})$ holds, then IND resembles Π^1_1. (Kechris [1973].)

Hint. Show first by a prewellordering argument that if $P \in \Delta$ and

$$P(x) \Leftrightarrow (\forall \alpha) R(x, \alpha)$$

for some $R \in \Gamma$, then actually $R \in \mathbf{\Delta}$. (Let

$$R(x, \alpha) \leftrightarrow G(\varepsilon_1, x, \alpha)$$

where G is universal in Γ, let φ be a Γ-norm on G and assuming towards a contradiction that R is not in $\mathbf{\Delta}$, check that

$$G(\varepsilon, y, \beta) \Leftrightarrow (\exists x)[P(x) \,\&\, (\exists \alpha)[\neg(\varepsilon_1, x, \alpha) <^*_\varphi (\varepsilon, y, \beta)]],$$

so that in that case $G \in \neg\Gamma$.) Similarly for $P \in \Delta(\alpha)$.

Suppose now $P \subseteq \mathfrak{X} \times \mathcal{Y}$ is in Δ and choose $R \subseteq \mathfrak{X} \times \mathcal{Y} \times \mathcal{N}$ in $\neg\Gamma_1$ such that

$$P(x, y) \Leftrightarrow (\exists\alpha)R(x, y, \alpha),$$

where Γ_1 is the pointclass given by the hypothesis. By the remark above we also know that $R \in \mathbf{\Delta}$.

Claim:

$$\begin{aligned} P_x \text{ is not meager} \\ &\Leftrightarrow (\exists A \subseteq \mathcal{Y} \times \mathcal{N})[A \text{ is } \mathbf{\Sigma}^1_1 \;\&\; (\forall y)(\forall\alpha)[A(y, \alpha) \Rightarrow R(x, y, \alpha)] \\ &\qquad \& \; \mathrm{p}[A] = \{y : (\exists\alpha)A(y, \alpha)\} \text{ is not meager}]. \end{aligned}$$

For the non-trivial direction ($\Rightarrow$) of the claim, notice that if P_x is not meager, then it has a non-meager Borel subset B. The relation

$$R'_x(y, \alpha) \Leftrightarrow y \in B \;\&\; R(x, y, \alpha)$$

is in $\mathbf{\Delta}$, so it can be uniformized by some $R^*_x \subseteq R'_x$ in $\mathbf{\Gamma}$ since Γ is scaled and then the function $f: B \to \mathcal{Y} \times \mathcal{N}$ with graph R^*_x is Baire-measurable by 6A.16. Using 2H.10, find some non-meager Borel set $C \subseteq B$ such that $f \upharpoonright C$ is continuous and take $A = R^*_x \cap C \times \mathcal{Y} \times \mathcal{N}$; this is the required $\mathbf{\Sigma}^1_1$ subset of R'_x, since easily $C = \mathrm{p}[A]$.

Now the claim is proved and together with the key hypothesis $\exists^{\mathcal{N}}\Gamma_1 \subseteq \Gamma_1$, it implies easily that the relation Q is in $\neg\Gamma$. Since also

$$\begin{aligned} Q(x) &\Leftrightarrow P_x \text{ is not meager} \\ &\Leftrightarrow (\exists s)[N_s - P_x \text{ is meager}], \end{aligned}$$

we easily get that Q is also in Γ.

The argument is similar for P in $\Delta(\alpha)$.

For the specific examples, take $\Gamma_1 = \Sigma^1_{2n}$ for Π^1_{2n+1} and $\Gamma_1 = \mathit{IND}$ for IND. ⊣

The next result is the key lemma that we need.

8G.18. Suppose Γ resembles Π^1_1, $A \subseteq \mathfrak{X}$ is in Δ and not meager and

$$\varphi : A \to \mathit{Ordinals}$$

is a Δ-norm on A and assume $\mathit{Det}(\mathbf{\Gamma})$. Prove that there is some λ^* and a non-empty perfect set F with code in Δ such that

$$x \in F \Rightarrow \varphi(x) = \lambda^*.$$

In particular, if $A \subseteq \mathfrak{X}$ is in Δ and not meager, then A has a non-empty perfect subset with code in $\boldsymbol{\Delta}$.

(The Kechris Perfect Set Lemma, Kechris [1973].)

Hint. The second assertion follows immediately from the first, taking $\varphi(x) = 0$.

For the first assertion, argue first using 5A.10 and the fact that all sets in $\boldsymbol{\Delta}$ have the property of Baire that if

$$\psi : B \to \mathit{Ordinals}$$

is any $\boldsymbol{\Delta}$-norm on a set $B \in \boldsymbol{\Delta}$ which is not meager, then for some λ the set

$$\{x \in B : \psi(x) = \lambda\}$$

must be non-meager.

It will be enough to prove the result for $A \subseteq \mathcal{N}$, easily, so suppose φ is a Δ-norm on A, put

$$\lambda^* = \text{least } \lambda \text{ such that } \{\alpha \in A : \varphi(\alpha) = \lambda\} \text{ is not meager,}$$
$$A^* = \{\alpha \in A : \varphi(\alpha) = \lambda^*\}$$

and notice that A^* is in Δ because Γ resembles Π^1_1.

Consider the $*$-game for A^* as we defined it in the exercises of 6A, in which I plays finite (non-empty) sequences from ω

$$s_0 = a_0, \dots, a_{k_0 - 1},$$
$$s_1 = a_{k_0+1}, \dots, a_{k_1 - 1},$$
$$\dots$$

II plays single integers $a_{k_0}, a_{k_1}, \dots$ and I wins if the play

$$\alpha = a_0, \dots, a_{k_0-1}, a_{k_0}, a_{k_0+1}, \dots, a_{k_1-1}, a_{k_1}, \dots$$

is in A^*. It is obviously enough to show that I has a winning strategy in Δ, since from any such strategy we can easily get a perfect subset of A^* with code in Δ.

Fix a Δ-scale $\bar{\psi} = \{\psi_m\}$ on A^* and put

$$\lambda_0 = \text{least } \lambda \text{ such that } \{\alpha \in A^* : \psi_0(\alpha) = \lambda\}$$
$$\text{is not meager,}$$
$$\bar{s}_0 = \text{least } s \text{ such that}$$
$$N(s) - \{\alpha \in A^* : \psi_0(\alpha) = \lambda_0\} \text{ is meager,}$$
$$A_0 = \{\alpha \in A^* : \alpha \in N(\bar{s}_0) \;\&\; \psi_0(\alpha) = \lambda_0\}$$

and have I start the game by playing the sequence s_0 with code $\bar{s}_0$. (Recal here that if

$$s = \langle a_0, \ldots, a_{n-1} \rangle,$$

then

$$N(s) = \{\alpha \in \mathcal{N}: \alpha(0) = a_0, \ldots, \alpha(n-1) = a_{n-1}\};$$

we are also using the fact that for each non-meager B with the property of Baire, there must be some s such that $N(s) - B$ is meager.) In general, having defined λ_i, $\bar{s}_i$, A_i, for $i < m$, I answers the move

$$a = a_{k_{m-1}}$$

of II by setting

$$\lambda_m = \text{least } \lambda \text{ such that } \{\alpha \in A_{m-1}: \alpha(k_{m-1}) = a \ \&\ \psi_m(\alpha) = \lambda\}$$
$$\text{is not meager,}$$
$$\bar{s}_m = \text{least } s \text{ such that}$$
$$N(s) - \{\alpha \in A_{m-1}: \alpha(k_{m-1}) = a \ \&\ \psi_m(\alpha) = \lambda_m\} \text{ is meager,}$$
$$A_m = \{\alpha \in A_{m-1}: \alpha(k_{m-1}) = a \ \&\ \alpha \in N(\bar{s}_m) \ \&\ \psi_m(\alpha) = \lambda_m\}$$

and playing the sequence s_m with code $\bar{s}_m$. It is clear (by induction) that each A_i is not meager and that the unique point $\alpha \in \bigcap_m A_m$ which is the play must be in A^*, using the fact that $\bar{\psi}$ is a scale. Finally, it is not hard to check that this strategy for I is in Δ, using the hypothesis that Γ resembles Π^1_1. ⊣

In the computation below we will actually use a corollary of this which is worth separate billing.

8G.19. Suppose Γ resembles Π^1_1 and

$$\varphi : \mathcal{N} \to \mathit{Ordinals}$$

is a Δ-norm on $\mathcal{N}$ and assume $\mathit{Det}(\mathbf{\Gamma})$. Prove that there is some λ^* with the following property: for every $\alpha \in \mathcal{N}$ there exists some $\gamma \in \mathcal{N}$ such that $\varphi(\gamma) = \lambda^*$ and $\alpha \in \Delta(\gamma)$.

Hint. By the Kechris Perfect Set Lemma we can find a perfect set $F \subseteq \mathcal{N}$ with code ε in Δ and a λ^* such that

$$\gamma \in F \Rightarrow \varphi(\gamma) = \lambda^*.$$

Using the method of 1A.2, 1A.3, we can easily get an injection

$$f : \mathcal{C} \rightarrowtail F$$

of the Cantor set into F which is recursive in ε, so that for each $\alpha' \in \mathcal{C}$ there is some $\gamma = f(\alpha') \in F$ such $\alpha' \in \Delta_1^1(\varepsilon, \gamma)$ – because

$$\alpha'(n) = m \Leftrightarrow (\exists \alpha)[f(\alpha) = \gamma \;\&\; \alpha(n) = m].$$

The result follows because every $\alpha \in \mathcal{N}$ is recursive in some $\alpha' \in \mathcal{C}$ and $\Delta_1^1(\varepsilon, \gamma) \subseteq \Delta(\gamma)$ when $\varepsilon \in \Delta$. ⊣

Suppose $P \subseteq \mathfrak{X} \times \mathcal{Y}$, $S \subseteq \mathfrak{X}$ and

$$\varphi : S \twoheadrightarrow \lambda$$

is any regular norm. We say that P is *uniform in x* (*relative to the norm* φ) if for all $x, x' \in S$ and all $y \in \mathcal{Y}$,

$$\varphi(x) = \varphi(x') \Rightarrow [P(x, y) \Leftrightarrow P(x', y)].$$

If P is uniform in x, it clearly defines a relation on $\lambda \times \mathcal{Y}$ which we will denote by the same symbol,

$$\begin{aligned} P(\xi, y) &\Leftrightarrow (\exists x \in S)[\varphi(x) = \xi \;\&\; P(x, y)] \\ &\Leftrightarrow (\forall x \in S)[\varphi(x) = \xi \Rightarrow P(x, y)]. \end{aligned}$$

8G.20. Suppose Γ resembles Π_1^1,

$$\varphi : S \twoheadrightarrow \lambda \leq \boldsymbol{\delta}$$

is a regular Γ-norm on some $S \subseteq \mathfrak{X}$, $S \in \Gamma$ and $P \subseteq \mathfrak{X} \times \mathcal{Y}$ is in $\exists^{\mathcal{N}}\Gamma$ and uniform in x. Assume $Det(\boldsymbol{\Gamma})$ and prove that the relations

$$\begin{aligned} Q_1(x, y) &\Leftrightarrow x \in S \;\&\; (\exists \xi < \varphi(x))P(\xi, y), \\ Q_2(x, y) &\Leftrightarrow x \in S \;\&\; (\forall \xi < \varphi(x))P(\xi, y), \end{aligned}$$

are both in $\exists^{\mathcal{N}}\Gamma$. (The Harrington–Kechris theorem. Harrington–Kechris [198?]).

Hint. That Q_1 is in $\exists^{\mathcal{N}}\Gamma$ is trivial (and uses very little of the hypotheses).

Rename $Q_2 = Q$, assume the hypotheses and suppose

$$(*) \qquad P(x, y) \Leftrightarrow (\exists \alpha) R(x, y, \alpha)$$

with R in Γ.

Let $\pi : \mathcal{N} \twoheadrightarrow \mathfrak{X}$ be a recursive surjection and put

$$\begin{aligned} R'(x, y, \alpha) \Leftrightarrow x \in S \;\&\; \pi((\alpha)_0) \in S \;\&\; \varphi(x) = \varphi(\pi((\alpha)_0)) \\ \&\; R(\pi((\alpha)_0), y, (\alpha)_1); \end{aligned}$$

now R' is clearly uniform in x and in Γ and

$$P(x, y) \Leftrightarrow (\exists\alpha)R'(x, y, \alpha),$$

so that we may assume (*) holds with R uniform in x.

For each x, y, consider the game $G(x, y)$ where I plays β, II plays γ and

$$\text{II wins} \Leftrightarrow T(\beta, \gamma, x, y),$$

where

$$\begin{aligned} T(\beta, \gamma, x, y) \Leftrightarrow {} & [\pi(\beta) \notin S \vee \varphi(x) \leq \varphi(\pi(\beta))] \\ & \vee [\pi(\beta) \in S \;\&\; \varphi(\pi(\beta)) < \varphi(x) \\ & \quad \& \; (\exists\alpha \in \Delta(\gamma))R(\pi(\beta), y, \alpha)], \end{aligned}$$

with $\pi : \mathcal{N} \twoheadrightarrow \mathfrak{X}$ as above. Clearly T is in Γ (by the theorem on restricted quantification 4D.3) so the game is determined.

We claim that for $x \in S$,

$$(\forall\xi < \varphi(x))P(\xi, y) \Leftrightarrow \text{II wins } G(x, y);$$

from this the result will follow immediately, since

$$\text{II wins } G(x, y) \Leftrightarrow (\exists\tau)(\forall\beta)T(\beta, [\beta]*\tau, x, y)$$

and the expression on the right defines a relation in $\exists^{\mathcal{N}}\Gamma$.

If II wins $G(x, y)$, then for any $\xi < \varphi(x)$ have I play some β with $\varphi(\pi(\beta)) = \xi$; if II responds with γ, then $(\exists\alpha \in \Delta(\gamma))R(\pi(\beta), y, \alpha)$ holds, so that we have $P(\xi, y)$ as required.

Assume towards a contradiction that I wins $G(x, y)$ with a strategy σ but $(\forall\xi < \varphi(x))P(\xi, y)$. Imagine II playing irrationals of the form $\langle\sigma, \delta\rangle$ for arbitrary δ to which I responds by $\sigma*[\langle\sigma, \delta\rangle] \in S$ and consider the norm on $\mathcal{N}$

$$\psi(\delta) = \varphi(\sigma*[\langle\sigma, \delta\rangle]).$$

This is obviously a $\Delta(\sigma)$-norm and the pointclass $\Gamma(\sigma)$ resembles Π^1_1, so by 8G.19 there must be some ξ with the following property: for each α, there is some δ with $\gamma(\delta) = \xi$ and $\alpha \in \Delta(\sigma, \delta)$. Now choose any x such that $\varphi(x) = \xi$ and since $P(x, y) \Leftrightarrow (\exists\alpha)R(x, y, \alpha)$ holds, choose α such that $R(x, y, \alpha)$, choose δ such that $\psi(\sigma) = \xi$ and $\alpha \in \Delta(\sigma, \delta)$ and have II play $\langle\sigma, \delta\rangle$. If $\beta = \sigma*[\langle\sigma, \delta\rangle]$ is I's response then, we know that $\pi(\beta) \in S$ and $\varphi(\pi(\beta)) = \xi$ and since $\alpha \in \Delta(\langle\sigma, \delta\rangle)$ and I wins, we have $\neg R(\pi(\beta), y, \alpha)$; this contradicts $R(x, y, \alpha)$, since R is uniform in x. ⊣

Let us immediately check out one important application of this very basic theorem. (We will include full determinacy among the hypotheses here, but in the applications we will be working with the model $L(\mathfrak{R})$, so only definable determinacy hypotheses will be needed.)

8G.21. Suppose Γ resembles Π^1_1, $S\subseteq\mathfrak{X}$, $T\subseteq\mathcal{Y}$ are in Γ and

$$\varphi : S \twoheadrightarrow \lambda \leq \boldsymbol{\delta}, \qquad \psi : T \twoheadrightarrow \mu \leq \boldsymbol{\delta}$$

are regular Γ-norms. Assume AD and show that the relations

$$P(x, y) \Leftrightarrow x \in S \;\&\; y \in T \;\&\; \varphi(x) = \psi(y)$$
$$Q(x, y) \Leftrightarrow x \in S \;\&\; y \in T \;\&\; \varphi(x) < \psi(y),$$

are both in $\exists^{\mathcal{N}}\Gamma$ (Harrington–Kechris [198?].)

Hint. Let $G\subseteq\mathcal{N}\times\mathfrak{X}\times\mathcal{Y}$ be universal in Γ, let $\chi : G \twoheadrightarrow \boldsymbol{\delta}$ be a Γ-norm on G and with each

$$z = (\varepsilon, \alpha, x, y) \in \mathcal{N}\times\mathcal{N}\times\mathfrak{X}\times\mathcal{Y}$$

such that $G(\alpha, x, y)$ associate the set $A^z\subseteq\mathfrak{X}\times\mathcal{Y}$ by

$$A^z(x', y') \Leftrightarrow G(\varepsilon, x', y') \;\&\; \chi(\varepsilon, x', y') \leq \chi(\alpha, x, y).$$

Each A^z is obviously in $\boldsymbol{\Delta}$, uniformly in z.

The relation

$$(1) \qquad \begin{aligned} R_1(z) &\Leftrightarrow R_1(\varepsilon, \alpha, x, y)\\ &\Leftrightarrow G(\alpha, x, y)\end{aligned}$$

is clearly in Γ and so is the relation

$$(2) \qquad \begin{aligned} R_2(z) \Leftrightarrow\; & R_1(z) \;\&\; A^z \text{ is uniform in } x'\\ & \text{(relative to the norm } \varphi) \text{ and in } y'\\ & \text{(relative to the norm } \psi)\\ \Leftrightarrow\; & R_1(z)\\ & \;\&\; (\forall x')(\forall y')[A^z(x', y') \rightarrow (x' \in S \;\&\; y' \in T)]\\ & \;\&\; (\forall x')(\forall x'')(\forall y')(\forall y'')\{[A^z(x', y')\\ & \qquad \;\&\; \varphi(x') = \varphi(x'') \;\&\; \psi(y') = \psi(y'')] \rightarrow A^z(x'', y'')]\}.\end{aligned}$$

With each z such that $R_2(z)$ holds, we associate the set of pairs or ordinals

$$B^z \subseteq \lambda\times\mu$$

by

$$B^z(\eta, \xi) \Leftrightarrow (\exists x)(\exists y)[x \in S \,\&\, \varphi(x) = \eta \,\&\, y \in T \,\&\, \psi(y) = \xi \,\&\, A^z(x, y)]$$
$$\Leftrightarrow (\forall x)(\forall y)\{[x \in S \,\&\, \varphi(x) = \eta \,\&\, y \in T \,\&\, \psi(y) = \xi] \Rightarrow A^z(x, y)\}.$$

Conversely, suppose $B \subseteq \lambda \times \mu$ is *bounded below* $\boldsymbol{\delta}$, i.e.

$$B \subseteq \lambda_1 \times \mu_1$$

where $\lambda_1 \leq \lambda$, $\lambda_1 < \boldsymbol{\delta}$ and $\mu_1 \leq \mu$, $\mu_1 < \boldsymbol{\delta}$; using *AD* now, the Coding Lemma (II) 7D.6 (applied to $\neg\Gamma$) implies directly that the set

$$B' = \{(x, y) : x \in S \,\&\, y \in T \,\&\, B(\varphi(x), \psi(y))\}$$

is in $\boldsymbol{\Delta}$ and hence by the Covering Lemma 4C.11, easily,

$$B' = A^z$$

with some z such that $R_2(z)$ holds and hence

$$B = B^z.$$

Proceeding with the computation, notice that the relation

$$R_3(x, y, z) \Leftrightarrow x \in S \,\&\, y \in T \,\&\, R_2(z) \,\&\, B^z \subseteq \varphi(x) \times \psi(y)$$
$$\Leftrightarrow x \in S \,\&\, y \in T \,\&\, R_2(z) \,\&\, (\forall x')(\forall y')[A^z(x', y') \Rightarrow [\varphi(x') < \varphi(x) \,\&\, \psi(y') < \psi(y)]]$$

is clearly in Γ and uniform in x (relative to φ) and in y (relative to ψ).

Now

$$R_4(x', y', z) \Leftrightarrow R_2(z) \,\&\, x' \in S \,\&\, y' \in T \,\&\, A^z(x', y')$$

is in Γ and uniform in x' (relative to φ) and in y' (relative to ψ), so by the Harrington–Kechris theorem,

$$R_5(x, y, z) \Leftrightarrow x \in S \,\&\, y \in T \,\&\, (\forall \xi < \varphi(x))(\exists \eta < \psi(y)) R_4(\xi, \eta, z)$$

is in $\exists^{\mathcal{N}}\Gamma$; clearly

$$R_5(x, y, z) \Leftrightarrow x \in S \,\&\, y \in T \,\&\, R_2(z) \,\&\, B^z \subseteq \varphi(x) \times \psi(y) \,\&\, (\forall \xi < \varphi(x))(\exists \eta < \psi(y)) B^z(\xi, \eta).$$

Proceeding in the same way, by successive applications of the Harrington–Kechris theorem we show that the following relation is in $\exists^{\mathcal{N}}\Gamma$:

$$\begin{aligned} R_6(x, y, z) \Leftrightarrow x \in S \ \&\ y \in T \ \&\ R_2(z) \\ \&\ B^z \subseteq \varphi(x) \times \psi(y) \\ \&\ B^z \text{ is the graph of an order-preserving one-to-one} \\ \text{function from } \varphi(x) \text{ onto } \psi(y). \end{aligned}$$

Using the remark above then, that by AD every subset of $\varphi(x) \times \psi(y)$ is B^z for some z such that $R_3(x, y, z)$, we finally get that

$$\begin{aligned} P(x, y) &\Leftrightarrow x \in S \ \&\ y \in T \ \&\text{ there exists a one-to-one order-preserving function from } \varphi(x) \text{ onto } \psi(y) \\ &\Leftrightarrow (\exists z)\{R_3(x, y, z) \ \&\ R_6(x, y, z)\} \end{aligned}$$

is in $\exists^{\mathcal{N}}\Gamma$.

The argument for $\varphi(x) < \psi(y)$ is similar. ⊣

The usefulness of this result will be apparent very soon, but the method of proof also has wide applicability, which is why we explained it in such considerable detail.

Fix now a pointclass Γ which resembles Π_1^1, let

$$\varphi : S \twoheadrightarrow \boldsymbol{\delta}$$

be a regular Γ-norm on some $S \subseteq \mathcal{X}$ in Γ which is onto the ordinal $\boldsymbol{\delta}$ associated with Γ and let $G \subseteq \omega \times \mathcal{X}$ be a good universal set in $\exists^{\mathcal{N}}\Gamma$ in the sense of 3H.4. For this φ, G, define

$$P_{\varphi,G} \subseteq \omega \times \boldsymbol{\delta}$$

by

$$P_{\varphi,G}(n, \xi) \Leftrightarrow (\exists x)[x \in S \ \&\ \varphi(x) = \xi \ \&\ G(n, x)];$$

intuitively, $P_{\varphi,G}(n, \xi)$ asks if the ordinal $\xi < \boldsymbol{\delta}$ has the $\exists^{\mathcal{N}}\Gamma$-property n, *in the coding of ordinals and properties determined by* φ *and* G.

8G.22. Suppose Γ resembles Π_1^1 and let $P_{\varphi,G}$, $P_{\psi,H}$ be defined as above for two different choices φ, G and ψ, H of Γ-norms and good universal sets in $\exists^{\mathcal{N}}\Gamma$. Assume AD and prove that there is a recursive function $f : \omega \to \omega$ such that

$$P_{\varphi,G}(n, \xi) \Leftrightarrow P_{\psi,H}(f(n), \xi).$$

In particular, if AD holds, then

$$L[P_{\varphi,G}] = L[P_{\psi,H}].$$

(Moschovakis.)

Hint. Suppose $\psi: T \twoheadrightarrow \boldsymbol{\delta}$ with $T \subseteq \mathcal{Y}$ and compute:

$$\begin{aligned} P_{\varphi,G}(n, \xi) &\Leftrightarrow (\exists x)[x \in S \;\&\; \varphi(x) = \xi \;\&\; G(n, x)] \\ &\Leftrightarrow (\exists y)[y \in T \;\&\; \psi(y) = \xi \;\&\; R(n, y)], \end{aligned}$$

where

$$R(n, y) \Leftrightarrow (\exists x)[x \in S \;\&\; \varphi(x) = \psi(y) \;\&\; G(n, x)].$$

Now R is in $\exists^{\mathcal{N}}\Gamma$ by 8G.21 and hence (using the fact that H is a good universal set, as in 3H.2. easily) there is a recursive $f: \omega \to \omega$ such that

$$R(n, y) \Leftrightarrow H(f(n), y),$$

so that

$$\begin{aligned} P_{\varphi,G}(n, \xi) &\Leftrightarrow (\exists y)[y \in T \;\&\; \psi(y) = \xi \;\&\; H(f(n), y)) \\ &\Leftrightarrow P_{\psi,H}(f(n), \xi). \end{aligned}$$ ⊣

If Γ resembles Π^1_1 and AD holds, we let

$$H_\Gamma = L[P_{\varphi,G}]$$

where φ, G are chosen as above. Notice that (by 4C.14) we can always find such φ and G and by 8G.22, which particular φ, G we choose is irrelevant. Notice also that if $\Gamma \subseteq L(\mathfrak{R})$ as in the case with Π^1_{2n+1} and *IND*, we need only assume that every set of irrationals in $L(\mathfrak{R})$ is determined, as the whole construction of H_Γ takes place within the model $L(\mathfrak{R})$ by very simple absoluteness considerations.

For $\Gamma = \Pi^1_{2n+1}$ we use the simpler notation

$$H_{2n+1} = H_{\Pi^1_{2n+1}}.$$

One extension of 8G.22 is worth putting down.

8G.23. Suppose Γ resembles Π^1_1 and

$$\psi: S_1 \twoheadrightarrow \boldsymbol{\delta}, \ldots, \psi_n: S_n \twoheadrightarrow \boldsymbol{\delta}$$

are regular Γ-norms on sets $S_1, \ldots, S_n$ in Γ, put

$$\begin{aligned} Q(\xi_1, \ldots, \xi_n) \Leftrightarrow\ & \xi_1, \ldots, \xi_n < \boldsymbol{\delta} \\ & \& \; (\exists x_1 \in S_1) \cdots (\exists x_n \in S_n)[\psi_1(x_1) = \xi_1 \;\&\; \cdots \\ & \& \; \psi_n(x_n) = \xi_n \;\&\; R(x_1, \ldots, x_n)], \end{aligned}$$

where R is an arbitrary pointset in $\exists^{\mathcal{N}}\Gamma$. Assume AD and prove that $Q \in H_\Gamma$. (Moschovakis).

Hint. The *Gödel wellordering* of pairs of ordinals is defined by

$$\begin{aligned}(\xi, \eta) \leq (\xi', \eta') \Leftrightarrow\ & max(\xi, \eta) < max(\xi', \eta') \\ & \vee [max(\xi, \eta) = max(\xi', \eta') \ \&\ \xi < \xi'] \\ & \vee [max(\xi, \eta) = max(\xi', \eta') \ \&\ \xi = \xi' \ \&\ \eta \leq \eta'].\end{aligned}$$

The initial segments of this wellordering of $ON \times ON$ are sets and hence by 8C.7, easily, there is a ZF-absolute operation

$$\pi_2 : ON \times ON \rightarrowtail\!\!\!\!\!\twoheadrightarrow ON$$

such that

$$(\xi, \eta) \leq (\xi', \eta') \Leftrightarrow \pi_2(\xi, \eta) \leq \pi_2(\xi', \eta').$$

Choose a regular Γ-norm $\varphi : S \twoheadrightarrow \boldsymbol{\delta}$ and on $S \times S$ define the norm

$$\psi(x, y) = \pi_2(\varphi(x), \varphi(y)).$$

It is easy to check that ψ is a regular Γ-norm, so if $G \subseteq \omega \times \mathfrak{X} \times \mathfrak{X}$ is universal in $\exists^{\mathcal{N}}\Gamma$ and

$$P(n, \xi) \Leftrightarrow (\exists x)(\exists y)[S(x) \ \&\ S(y) \ \&\ \psi(x, y) = \xi \ \&\ G(n, x, y)],$$

we know that

$$H_\Gamma = L[P]$$

by 8G.22. But obviously

$$\begin{aligned}Q(\xi_1, \xi_2) \Leftrightarrow\ & (\exists x)(\exists y)\{S(x) \ \&\ S(y) \ \&\ \psi(x, y) = \pi_2(\xi_1, \xi_2) \\ & \qquad \&\ (\exists x_1 \in S_1)(\exists x_2 \in S_2)[\varphi(x) = \psi_1(x_1) \\ & \qquad\qquad \&\ \varphi(y) = \psi_2(x_2) \\ & \qquad\qquad \&\ R(x_1, x_2)]\} \\ \Leftrightarrow\ & P(n^*, \pi_2(\xi_1, \xi_2))\end{aligned}$$

with a fixed n^* and $Q \in L[P]$.

The result for $n \geq 2$ follows by the same argument using an appropriate $\pi_n : ON^n \twoheadrightarrow ON$. ⊣

Let us now turn to the relation between H_Γ and the models $L[T_\Gamma]$ which we introduced in 8G.10.

8G.24. Suppose Γ resembles Π^1_1 and let

$$T_\Gamma = T_{\Gamma,G,\bar{\varphi}}$$

be the tree associated with an universal set $G \subseteq \omega \times \mathcal{N}$ in Γ and a Γ-scale $\bar{\varphi}$ on G as in 8G.10. Assume AD and prove that

$$T_\Gamma \in H_\Gamma$$

so that $L[T_\Gamma] \subseteq H_\Gamma$. (Moschovakis.)

Hint. This time take the Gödel wellordering of all tuples of ordinals defined by

$$\begin{aligned}(\xi_1,\dots,\xi_n) \leq (\eta_1,\dots,\eta_m) \Leftrightarrow\ & max(\xi_1,\dots,\xi_n) < max(\eta_1,\dots,\eta_m)\\ & \vee[max(\xi_1,\dots,\xi_n) = max(\eta_1,\dots,\eta_m)\ \&\ n < m]\\ & \vee[max(\xi_1,\dots,\xi_n) = max(\eta_1,\dots,\eta_m)\ \&\ n = m\\ & \quad \&\ \text{the tuple } (\xi_1,\dots,\xi_n) \text{ precedes}\\ & \quad (\eta_1,\dots,\eta_n) \text{ lexicographically}]\end{aligned}$$

and argue as before that for some ZF-absolute

$$\pi : V \twoheadrightarrow ON,$$

$$(\xi_1,\dots,\xi_n) \leq (\eta_1,\dots,\eta_m) \Leftrightarrow \pi(\langle\xi_1,\dots,\xi_n\rangle) \leq \pi(\langle\eta_1,\dots,\eta_m\rangle).$$

Choose a regular Γ-norm $\varphi : S \twoheadrightarrow \boldsymbol{\delta}$ where $S \subseteq \mathcal{N}$ for simplicity and on

$$S^* = \{(m, \alpha) : (\forall i < m)[(\alpha)_i \in S]\}$$

define the norm

$$\psi(m, \alpha) = \pi(\langle\varphi((\alpha)_0),\dots,\varphi((\alpha)_{m-1})\rangle).$$

Again it is a simple computation to check that ψ is a Γ-norm and it is regular, because $\boldsymbol{\delta}$ is a cardinal and π easily establishes a one-to-one correspondence of all tuples below $\boldsymbol{\delta}$ with $\boldsymbol{\delta}$.

Suppose now $\varphi_0, \varphi_1,\dots$ is a sequence of regular Γ-norms on some set A which are uniformly in Γ in the sense that both relations

$$Q_1(n, x, y) \Leftrightarrow x \leq^*_{\varphi_n} y,$$

$$Q_2(n, x, y) \Leftrightarrow x <^*_{\varphi_n} y,$$

are in Γ and suppose

$$\begin{aligned}Q(u, a) \Leftrightarrow\ & u \in \omega\ \&\ a = \langle\xi_0,\dots,\xi_{m-1}\rangle \text{ for some ordinals } \xi_0,\dots,\xi_{m-1}\\ & \&\ (\exists x)\{x \in A\ \&\ \varphi_0(x) = \xi_0\ \&\ \cdots\ \&\ \varphi_{m-1}(x) = \xi_{m-1}\\ & \qquad \&\ R(u, x)\}\end{aligned}$$

where R is in $\exists^{\mathcal{N}}\Gamma$. As in the previous argument,

$$(*)\quad Q(u,a) \leftrightarrow (\exists m)(\exists\alpha)\{\psi(m,\alpha)=\pi(a) \\ \& (\exists x)\{\varphi_0(x)=\varphi((\alpha)_0) \& \cdots \& \varphi_{m-1}(x)=\varphi((\alpha)_{m-1}) \\ \& R(u,x)\}\} \\ \leftrightarrow (\exists m)\{a \text{ is an } m\text{-tuple of ordinals } \& P(f(m,u),\pi(a))\}$$

with a recursive f, where

$$P(n,\xi) \leftrightarrow (\exists m)(\exists\alpha)\{\psi(m,\alpha)=\xi \& G(n,m,\alpha)\}$$

and hence $Q\in L[P]=H_\Gamma$.

Finally, if T_Γ is the tree that comes from a Γ-scale on some $G\subseteq\omega\times\mathcal{N}$ in Γ, put

$$Q(u,a) \leftrightarrow u\in\omega \& a=\langle\xi_0,\ldots,\xi_{m-1}\rangle \text{ for some } \xi_0,\ldots,\xi_{m-1} \\ \& (((u)_0,\xi_0),\ldots,((u)_{m-1},\xi_{m-1}))\in T$$

and check immediately that Q satisfies $(*)$ with a suitable recursive R, so that $Q\in H_\Gamma$. But of course T can be defined directly from Q, so also $T\in H_\Gamma$. ⊣

For the case $\Gamma=\Pi^1_1$, it is quite easy to check that

$$L[T_1]=H_1=L,$$

see Kechris–Moschovakis [1972]. In general, for an arbitrary Γ which resembles Π^1_1 (and assuming AD), it is not known whether

$$L[T_\Gamma]=H_\Gamma,$$

although there are some positive results of Martin for the special case $\Gamma=\Pi^1_3$. The question is important, because a positive answer implies the invariance of $L[T_\Gamma]$ from the arbitrary choices of G and $\bar\varphi$ in its definition—and this would be a very strong invariance property of scales.

Finally we are ready to show that for Γ as above,

$$(*)\qquad \mathcal{N}\cap L[T_\Gamma]=\mathcal{N}\cap H_\Gamma \\ = \text{the largest countable } \exists^{\mathcal{N}}\Gamma\text{-subset of } \mathcal{N}.$$

First the easy half.

8G.25. Suppose Γ resembles Π^1_1 and assume AD. Show that there is a largest countable set $C_\Gamma \subseteq \mathcal{N}$ in $\exists^{\mathcal{N}}\Gamma$ and for any choice of T_Γ

$$C_\Gamma \subseteq L[T_\Gamma] \subseteq H_\Gamma.$$

Hint. Show first that for each $\mathfrak{X}$, there is a largest countable subset of $\mathfrak{X}$ in Γ as in 6E.9 and then proceed as in 6E.10 to get a largest countable subset $C_\Gamma \subseteq \mathcal{N}$ in $\exists^{\mathcal{N}}\Gamma$. By 8G.10 then, $C_\Gamma \subseteq L[T_\Gamma]$. ⊣

To show (*) above it is enough to prove that the pointset $\mathcal{N} \cap H_\Gamma$ is in $\exists^{\mathcal{N}}\Gamma$. The computation is very similar to that in the proof of 8G.21, only a bit more elaborate.

Fix a pointclass Γ which resembles Π^1_1 and choose a regular Γ-norm

$$\varphi : S \twoheadrightarrow \boldsymbol{\delta}$$

where $S \subseteq \mathfrak{X}$. Fix also a good universal $G \subseteq \omega \times \mathfrak{X}$ in $\exists^{\mathcal{N}}\Gamma$, so that

$$H_\Gamma = L[P],$$

where

$$P(n, \xi) \Leftrightarrow (\exists x)[x \in S \ \& \ \varphi(x) = \xi \ \& \ G(n, x)].$$

We will obviously code ordinals below $\boldsymbol{\delta}$ using φ.

We also need to code *bounded n-ary subsets* of $\boldsymbol{\delta}$, where $B \subseteq {}^n\boldsymbol{\delta}$ is bounded if there is some $\lambda < \boldsymbol{\delta}$ such that

$$B(\xi_1, \ldots, \xi_n) \Rightarrow \xi_1, \ldots, \xi_n < \lambda.$$

To do this, choose for each n a universal set

$$G^n \subseteq \mathcal{N} \times \mathfrak{X}^n$$

in Γ, let

$$\chi^n : G^n \twoheadrightarrow \boldsymbol{\delta}$$

be a Γ-norm, let

$$S^n = \{(x_0, \varepsilon, \alpha, x_1, \ldots, x_n) : x_0 \in S \ \& \ G^n(\alpha, x_1, \ldots, x_n)\}$$

and for each

$$z = (x_0, \varepsilon, \alpha, x_1, \ldots, x_n)$$

put

$$A^z = \{(x'_1, \ldots, x'_n) : G(\varepsilon, x'_1, \ldots, x'_n) \ \& \ \chi^n(\varepsilon, x'_1, \ldots, x'_n) \leq \chi^n(\alpha, x_1, \ldots, x_n)$$
$$\& \ \varphi(x'_1) < \varphi(x_0) \ \& \cdots \& \ \varphi(x'_n) < \varphi(x_0)\}.$$

Clearly each A^z is in $\mathbf{\Delta}$ and the relation

$$\begin{aligned} U^n(z) \Leftrightarrow\ & A^z \text{ is uniform in } x_1',\ldots,x_n' \text{ relative to } \varphi \\ \Leftrightarrow\ & z \in S^n \\ & \& (\forall x_1')(\forall x_1'') \cdots (\forall x_n')(\forall x_n'') \\ & \{[A^z(x_1',\ldots,x_n') \ \&\ \varphi(x_1') = \varphi(x_1'') \ \&\ \cdots \ \&\ \varphi(x_n') = \varphi(x_n'')] \\ & \qquad \Rightarrow A^z(x_n'',\ldots,x_n'')\} \end{aligned}$$

is in Γ. With each $z \in U^n$ we associate the set

$$\begin{aligned} B^z = \{(\xi_1,\ldots,\xi_n) \colon (\exists x_1') \cdots (\exists x_n')[\varphi(x_1') = \xi_1 \ \&\ \cdots \ \&\ \varphi(x_n') = \xi_n \\ \&\ A^z(x_1',\ldots,x_n')]\} \end{aligned}$$

which is obviously bounded, since if $z = (x_0, \varepsilon, \alpha, x_1,\ldots,x_n)$, then

$$B^z(\xi_1,\ldots,\xi_n) \Rightarrow \xi_1,\ldots,\xi_n < \varphi(x_0).$$

8G.26. Fix Γ, φ, etc. as above, assume AD and prove that every bounded n-ary subset of $\boldsymbol{\delta}$ is B^z for some $z \in U^n$.

Hint. Use the Covering lemma 4C.11 and the Coding Lemma (II) 7D.6. ⊣

A *restricted second order relation on* $\boldsymbol{\delta}$ is a relation of the form

$$R(\xi_1,\ldots,\xi_n, B_1,\ldots,B_m),$$

where each ξ_i ranges over $\boldsymbol{\delta}$ and each B_j ranges over the bounded subsets of $\boldsymbol{\delta}^{k_j}$, for some integer k_j. With each such R we associate the pointset

$$\begin{aligned} R^{\#}(x_1,\ldots,x_n, z_1,\ldots,z_m) \Leftrightarrow\ & x_1,\ldots,x_n \in S \\ & \&\ z_1 \in U^{k_1} \ \&\ \cdots \ \&\ z_m \in U^{k_m} \\ & \&\ R(\varphi(x_1),\ldots,\varphi(x_n), B^{z_1},\ldots,B^{z_m}) \end{aligned}$$

relative to a coding of ordinals and bounded subsets as above. We say that R *is in* $\exists^{\mathcal{N}}\Gamma$ *in the codes* if $R^{\#}$ is in $\exists^{\mathcal{N}}\Gamma$.

8G.27. Suppose Γ resembles Π_1^1, fix φ, G, G^n, χ^n as above and assume AD. Prove that the collection of restricted second order relations on $\boldsymbol{\delta}$ which are $\exists^{\mathcal{N}}\Gamma$ in the codes includes the basic relations

$$\eta < \xi, \quad \eta = \xi, \quad (\xi_1,\ldots,\xi_n) \in B, \quad B \subseteq \lambda^n$$

and their negations and it is closed under &, ∨, additions, permutations and identifications of variables and the quantifiers

$$(\exists\xi<\lambda),\quad (\forall\xi<\lambda),$$
$$(\exists\xi_1,\dots,\xi_n)[B(\xi_1,\dots,\xi_n)\ \&\ \cdots],$$
$$(\forall\xi_1,\dots,\xi_n)[B(\xi_1,\dots,\xi_n)\Rightarrow\cdots],$$
$$(\exists B)[B \text{ is bounded } \&\ \cdots].$$

Hint. The computations are all trivial, using of course, the Harrington–Kechris Theorem 8G.20. ⊣

This is the general version of the Harrington–Kechris theorem which we need. In the results below we assume tacitly that the codings are relative to a fixed choice of $\varphi: S \twoheadrightarrow \boldsymbol{\delta}$, $G\subseteq\omega\times\mathfrak{X}$, G^n, etc. as above.

8G.28. Suppose Γ resembles Π^1_1, assume AD and prove that the relation

$$R(A,B)\Leftrightarrow A\in L(B)$$

on bounded n-ary subsets of $\boldsymbol{\delta}$ is in $\exists^{\mathcal{N}}\Gamma$ in the codes.

Hint. This is a set-theoretic computation, very much like that of $\mathcal{N}\cap L$ in 8F.7. Using unary A, B for simplicity, check first by a Skolem–Löwenheim argument that

$$A\in L(B)\Leftrightarrow(\exists\lambda<\boldsymbol{\delta})(\exists\mu<\boldsymbol{\delta})[A\subseteq\lambda\ \&\ B\subseteq\lambda\ \&\ \lambda<\mu\ \&\ A\in L_\mu(B)].$$

By 8F.1, next, choose a formula $\varphi_L(\mathbf{x},\boldsymbol{\mu},\mathbf{y})$ and a finite $T_0\subseteq ZF$ such that for all standard models M of T_0, if A, B, $\mu\in M$ then

$$A\in L_\mu(B)\Leftrightarrow M\models\varphi_L(A,\mu,B).$$

Suppose the axiom of extensionality is also in T_0. By Mostowski collapsing then, we easily have

$$\begin{aligned}
&A\subseteq\lambda\ \&\ B\subseteq\lambda\ \&\ A\in L_\mu(B)\\
&\quad\Leftrightarrow(\exists\nu<\boldsymbol{\delta})(\exists E\subseteq\nu\times\nu)\{\lambda<\mu<\nu\ \&\ (\forall\eta,\zeta\leq\mu)[\eta<\zeta\Leftrightarrow E(\eta,\zeta)]\\
&\qquad\&\ (\exists\xi_A)(\exists\xi_B)[(\forall\xi<\mu)[E(\xi,\xi_A)\Leftrightarrow\xi\in A]\\
&\qquad\&\ (\forall\xi<\mu)[E(\xi,\xi_B)\Leftrightarrow\xi\in B]\ \&\ (\nu,E)\models\varphi_L(\xi_A,\mu,\xi_B)]\}
\end{aligned}$$

which then implies that the relation $A\in L(B)$ is in $\exists^{\mathcal{N}}\Gamma$ in the codes by 8G.27. ⊣

8G.29. Suppose Γ resembles Π^1_1 and assume AD.

(i) Show that the relation

$$R(A) \Leftrightarrow A \in H_\Gamma$$

on bounded subsets of $\boldsymbol{\delta}$ is in $\exists^{\mathcal{N}}\Gamma$ in the codes.

(ii) Prove that $\mathcal{N} \cap H_\Gamma$ is a pointset in $\exists^{\mathcal{N}}\Gamma$.

(iii) Infer that

$$\begin{aligned}\mathcal{N} \cap H_\Gamma &= \mathcal{N} \cap L[T_\Gamma] \\ &= \text{the largest countable } \exists^{\mathcal{N}}\Gamma\text{-subset of } \mathcal{N}.\end{aligned}$$

(For $\mathcal{N} \cap L[T_\Gamma]$ and $\Gamma = \Pi^1_3$, Kechris–Martin [1978]; for $\mathcal{N} \cap L[T_\Gamma]$ in general Harrington–Kechris [198?]; for $\mathcal{N} \cap H_\Gamma$, Moschovakis.)

Hint. Fix a regular Γ-norm

$$\varphi : S \twoheadrightarrow \boldsymbol{\delta}$$

and a universal set $G \subseteq \omega \times \mathfrak{X}$ and put

$$P(n, \xi) \Leftrightarrow (\exists x)[x \in S \ \& \ \varphi(x) = \xi \ \& \ G(n, x)]$$

so that $H_\Gamma = L[P]$.

Use first a Skolem–Löwenheim argument and the regularity of $\boldsymbol{\delta}$ to argue that for bounded $A \subseteq \boldsymbol{\delta}^m$,

$$A \in L[P] \Leftrightarrow (\exists \lambda < \boldsymbol{\delta})[A \in L(P \cap \omega \times \lambda)]. \tag{1}$$

Suppose now

$$G(n, x) \Leftrightarrow (\exists \beta) G_1(n, x, \beta)$$

where G_1 is in Γ and fix a Γ-norm

$$\psi_1 : G_1 \twoheadrightarrow \boldsymbol{\delta}.$$

Put

$$T(n, x) \Leftrightarrow x \in S \ \& \ (\exists \beta) G_1(n, x, \beta)$$

and on T define the norm

$$\psi(n, x) = \mathit{infimum}\{\psi_1(n, x', \beta) : x' \in S \ \& \ \varphi(x') = \varphi(x) \ \& \ G_1(n, x, \beta)\}.$$

It is clear that ψ is an $\exists^{\mathcal{N}}\Gamma$-norm on T and it is *uniform* on $n \in \omega$, $x \in S$ relative to φ, so we can define

$$\psi(n, \xi) = \psi(n, x), \quad \text{for any} \quad x \in S \quad \text{such that} \quad \varphi(x) = \xi.$$

Finally put

$$P^{\lambda,\mu}(n,\xi) \Leftrightarrow \xi < \lambda < \mu \;\&\; \psi(n,\xi) < \mu$$

and let

$$Q(\lambda,\mu,n,\xi) \Leftrightarrow P^{\lambda,\mu}(n,\xi).$$

The key to the proof is that *both* Q *and* $\neg Q$ *are in* $\exists^{\mathcal{N}}\Gamma$ *in the codes.*

Proof that Q *is in* $\exists^{\mathcal{N}}\Gamma$ *in the codes* is quite easy. We must show that

$$\begin{aligned} Q^{\#}(u,v,a,x) \Leftrightarrow\; & u,v,a,x \in S \\ & \& \; \varphi(x) < \varphi(u) < \varphi(v) \\ & \& \; (\exists n)[\varphi(a) = n \;\&\; \psi(n,x) < \varphi(v)] \end{aligned}$$

is in $\exists^{\mathcal{N}}\Gamma$ and only the last clause here needs any computation,

$$\begin{aligned} \psi(n,x) < \varphi(v) \Leftrightarrow (\exists x')(\exists\beta)[\varphi(x') = \varphi(x) \;\&\; G_1(n,x',\beta) \\ \& \; \psi_1(n,x',\beta) < \varphi(v)]. \end{aligned}$$

(We will omit the computation of "$\varphi(a) = n$" which involves saying that exactly n points precede a in φ.)

Proof that $\neg Q$ *is in* $\exists^{\mathcal{N}}\Gamma$. Compute:

$$\begin{aligned} (\neg Q)^{\#}(u,v,a,x) \Leftrightarrow\; & u,v,a,x \in S \\ & \& \{\neg[\varphi(x) < \varphi(u) < \varphi(v) \;\&\; (\exists n)[\varphi(a) = n]] \\ & \vee (\exists n)[\varphi(a) = n \;\&\; \& \; \neg(\psi(n,x) < \varphi(v))]\}. \end{aligned}$$

For the non-trivial last clause, notice that for $v \in S$,

$$\begin{aligned} & \neg(\psi(n,x) < \varphi(v)) \\ & \Leftrightarrow (\exists m)(\exists z)(\exists\gamma)\{G_1(m,z,\gamma) \;\&\; \psi_1(m,z,\gamma) = \varphi(v) \\ & \qquad \& \; (\forall x')(\forall\beta)[\varphi(x) = \varphi(x') \\ & \qquad \Rightarrow (m,z,\gamma) \leq^*_{\psi_1} (n,x',\beta)]\} \end{aligned}$$

which establishes by 8G.21 that this clause and hence $(\neg Q)^{\#}$ is in $\exists^{\mathcal{N}}\Gamma$.

Claim: if A is bounded, then

(2) $$A \in L[P] \Leftrightarrow (\exists\lambda < \boldsymbol{\delta})(\exists\mu < \boldsymbol{\delta})[A \in L(P^{\lambda,\mu})].$$

To show this in the direction ($\Rightarrow$), choose λ by (1) so that $A \in L(P \cap \omega \times \lambda)$ and let

$$\mu = \mathit{supremum}\{\psi(n,\xi) : T(n,\xi) \;\&\; \xi < \lambda\};$$

since $\boldsymbol{\delta}$ is regular, we have $\mu < \boldsymbol{\delta}$ and then clearly

$$P \cap (\omega \times \lambda) = P^{\lambda,\mu}.$$

For the other direction notice that since Q is in $\exists^{\mathcal{N}}\Gamma$ in the codes, then 8G.23 implies that $Q \in H_\Gamma$ and hence each $P^{\lambda,\mu} \in H_\Gamma$ and $L(P^{\lambda,\mu}) \subseteq H_\Gamma$.

Finally, since both Q and $\neg Q$ are in $\exists^{\mathcal{N}}\Gamma$ in the codes, so is the relation on a bounded set $B \subseteq \omega \times \delta$

$$B = P^{\lambda,\mu} \Leftrightarrow (\forall n, \xi \in B) Q(\lambda, \mu, n, \xi)$$
$$\& \ (\forall n, \xi < \lambda)[Q(\lambda, \mu, n, \xi) \Rightarrow B(n, \xi)]$$

and hence so is the relation

$$A \in H_\Gamma \Leftrightarrow (\exists B)(\exists \lambda)(\exists \mu)[B = P^{\lambda,\mu} \ \& \ A \in L(B)].$$

This completes the proof of (i). To prove (ii) now, simply compute

$$\alpha \in H_\Gamma \Leftrightarrow (\exists A)[A \in L[P] \ \& \ (\forall n)(\forall m)[\alpha(n) = m \Leftrightarrow A(n, m)]]$$

and check that the definition of "in $\exists^{\mathcal{N}}\Gamma$ in the codes" easily implies now that this is in $\exists^{\mathcal{N}}\Gamma$.

Finally, $\mathcal{N} \cap \mathrm{H}_\Gamma$ is wellorderable by 8F.12, hence countable by 7D.4, hence contained in $L[T_\Gamma]$ by 8G.10. ⊣

The same method yields easily the next basic result about these models.

8G.30. Suppose Γ resembles Π^1_1 and assume *AD*.

(i) Show that the set

$$C_\Gamma = \mathcal{N} \cap H_\Gamma$$

admits an $\exists^{\mathcal{N}}\Gamma$-good wellordering.

(ii) Show that

$$H_\Gamma \models \text{“}2^{\aleph_0} = \aleph_1\text{”}.$$

(iii) In the special case $\Gamma = \Pi^1_{2n+1}$ and granting *DC*, show that C_{2n+2} admits a Σ^1_{2n+2}-good wellordering and

$$H_{2n+1} \models \text{“}\mathcal{N} \text{ admits a } \Sigma^1_{2n+2}\text{-good wellordering of rank } \aleph_1\text{”}.$$

(Kechris [1975]).

Hint. For (i), use 8F.12 to get a canonical wellordering $\leq^{\lambda,\mu}$ for each $L(P^{\lambda,\mu})$ in the notation of 8G.29 and on C_Γ define the wellordering

$$\begin{aligned}\alpha \leq_\Gamma \beta \Leftrightarrow\ & [\alpha \in L_\xi(P^{\lambda,\mu}) \text{ for some triple } \langle\lambda, \mu, \xi\rangle \\ & \text{such that for every smaller triple } \langle\lambda', \mu', \xi'\rangle \\ & \text{(in the lexicographic ordering) } \beta \notin L_{\xi'}(P^{\lambda',\mu'})] \\ & \vee [\text{for the least triple } \langle\lambda, \mu, \xi\rangle \text{ such that } L_\xi(P^{\lambda,\mu}) \\ & \text{contains both } \alpha \text{ and } \beta,\ L_\xi(P^{\lambda,\mu}) \text{ thinks that} \\ & \alpha \leq^{\lambda,\mu} \beta].\end{aligned}$$

Here we only look at triples $\langle\lambda, \mu, \xi\rangle$ such that $L_\xi(P^{\lambda,\mu})$ satisfies enough axioms of ZF to decide correctly the formula defining $\leq^{\lambda,\mu}$. The computation that this $\leq_\Gamma$ is in $\exists^{\mathcal{N}}\Gamma$ and $\exists^{\mathcal{N}}\Gamma$-good is similar to that in the proof of 8G.29. They key of course, is that

$$\alpha \leq_\Gamma \beta \ \& \ \beta \in L_\xi(P^{\lambda,\mu}) \Rightarrow \alpha \in L_\xi(P^{\lambda,\mu});$$

compare the proof of 8F.7.

To prove (ii), put

$$P(\beta, \gamma) \Leftrightarrow \beta \in C_\Gamma \ \& \ (\forall n)[(\gamma)_n <_\Gamma \beta] \ \& \ (\forall \delta <_\Gamma \beta)(\exists n)[\delta = (\gamma)_n]$$

where $\leq_\Gamma$ is the $\exists^{\mathcal{N}}\Gamma$-good wellordering of C_Γ. Clearly P is in $\exists^{\mathcal{N}}\Gamma$ and since (in V) C_Γ is countable, we have

$$(\forall \beta \in C_\Gamma)(\exists \gamma)P(\beta, \gamma).$$

Let $P^* \subseteq P$ uniformize P in $\exists^{\mathcal{N}}\Gamma$ and notice that $P^* \subseteq \mathcal{N} \times \mathcal{N}$ is countable so that $P^* \subseteq H_\Gamma$; thus for each $\beta \in C_\Gamma$ there is some $\gamma \in C_\Gamma$ which enumerates $\{\delta: \delta <_\Gamma \beta\}$, so that *within H_Γ every initial segment of the wellordering $<_\Gamma$ has countable length.* This means precisely that

$$H_\Gamma \models \text{"the length of } <_\Gamma \text{ is } \aleph_1\text{"},$$

so that in H_Γ we have the continuum hypothesis.

(iii) is immediate by a small absoluteness argument. ⊣

Kechris [1975] established this result about C_{2n+2} on the basis of the weaker hypothesis $Det(\mathbf{\Delta}^1_{2n})$ and by a different method, long before the proof of the Harrington–Kechris theorem. In this connection, the models

$$L_{2n+2} = L(C_{2n+2}) \subseteq L[T_{2n+1}]$$

have been studied quite adequately, see Kechris–Moschovakis [1978b], Kechris [1975] and Becker [1978].

One of the reasons for looking at these various "partially playful universes" is the desire to obtain consistency and independence results about determinacy hypotheses. Here is one example.

8G.31. Assume $Det(Power(\mathcal{N}) \cap L(\mathfrak{R}))$ and show that for each n, $Det(\mathbf{\Sigma}^1_{2n+1})$ is not a theorem of $ZFC + Det(\mathbf{\Delta}^1_{2n+1})$; in particular, for no n does $Det(\mathbf{\Delta}^1_n)$ imply PD. (Moschovakis, see Kechris–Moschovakis [1978b].)

Hint. By 8G.16, $H_{2n+1} \models ZFC + Det(\mathbf{\Delta}^1_{2n+1})$. On the other hand, in H_{2n+1} we have a Σ^1_{2n+2}-good wellordering of $\mathcal{N}$ of rank $\aleph_1$ and this

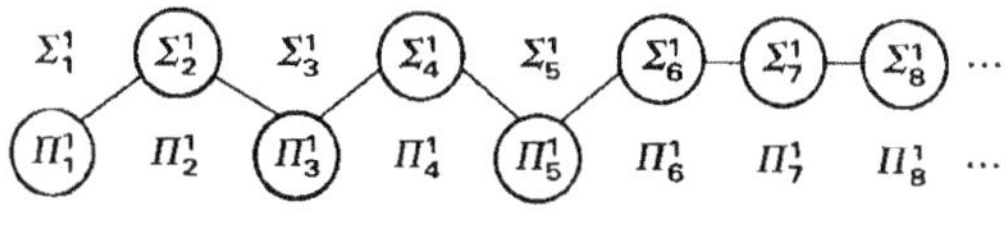

Diagram 8G.1. The normed Kleene pointclass in H_5.

implies easily as in 5A.6 that there is an uncountable, thin Π^1_{2n+1} set in this model, which violates $Det(\mathbf{\Sigma}^1_{2n+1})$ by 6A.12. ⊣

In this connection it is worth pointing out that Martin has shown (granting DC)

$$Det(\mathbf{\Delta}^1_{2n}) \Rightarrow Det(\mathbf{\Sigma}^1_{2n}),$$

see his forthcoming Martin [19?c].

Consider also the prewellordering property in the model H_Γ. It is clear from the proof of 5A.3, that the existence of a Σ^1_{2n+2}-good wellordering of rank $\aleph_1$ implies that for each $m \geq 2n+2$, Σ^1_m is normed, so that using $H_{2n+1} \models Det(\mathbf{\Delta}^1_{2n})$, we infer that in H_5 (for example) the circled pointclasses in Diagram 8G.1 are all normed.

The corresponding diagram for H_{IND} has infinitely many "teeth" since

$$H_{IND} \models PD;$$

but in H_{IND} we have, of course, a canonical wellordering of the universe and (it is easy to see) that the restriction of this wellordering to $\mathcal{N}$ is in fact definable (inductive) and of rank $\aleph_1$. This model, incidentally satisfies a good deal of absoluteness beyond the absoluteness of analytical statements, see Moschovakis [1978].

Using forcing and granting AD, one can show that for any pointclass Γ which resembles Π^1_1, if λ is any ordinal which is countable (in V), then

$$H_\Gamma \models 2^\lambda = \lambda^+.$$

A much easier consequence of constructibility theory is that

$$\lambda \geq \mathbf{\delta} \Rightarrow H_\Gamma \models 2^\lambda = \lambda^+.$$

It is not known whether these models satisfy the generalized continuum hypothesis.

It is also not hard to see that if

$$\lambda = \aleph_1 = \text{the least ordinal uncountable (in } V),$$

then for any such Γ,

$$H_\Gamma \models \text{“}\lambda \text{ is strongly inaccessible”}.$$

Actually,

$$H_\Gamma \models \text{“}\lambda \text{ is measurable”}$$

and the same is true for $\lambda = \aleph_2$ (Moschovakis).

The study of these fascinating inner models is one of the most intriguing research areas in the theory of determinacy and it is still in its infancy. We add here just one more result which is quite easy to show by the methods we have established.

8G.32. Suppose Γ resembles Π^1_1, assume AD and let $A \subseteq \boldsymbol{\delta}^n$ be bounded. Prove that $A \in H_\Gamma$ if and only if there is a regular Γ-norm $\varphi : S \twoheadrightarrow \boldsymbol{\delta}$, a pointset $R \subseteq \mathfrak{X}^{n+1}$ and an ordinal $\lambda < \boldsymbol{\delta}$, such that

$$[x \in S, x_1, \ldots, x_n \in S \;\&\; \varphi(x) = \lambda, \varphi(x_1) = \xi_1, \ldots, \varphi(x_n) = \xi_n]$$
$$\Rightarrow [A(\xi_1, \ldots, \xi_n) \Leftrightarrow R(x, x_1, \ldots, x_n)]. \quad \dashv$$

8H. On the theory of indiscernibles[(4)]

In this section we will develop just enough of the celebrated theory of indiscernibles in L to allow us to prove the following important theorem of Martin and Solovay (improved in part by Mansfield).

If there exists a measurable cardinal, then every Π^1_2 set admits a Δ^1_3-scale into an ordinal

$$u_\omega \leq \aleph_\omega;$$

if in addition the axiom of choice holds, then

$$u_\omega < \aleph_3,$$

so that every $\boldsymbol{\Sigma}^1_3$ set is $\aleph_2$-Suslin.

Together with results we have already proved and assuming the axiom of choice and that every set in $L(\mathcal{R})$ is determined, this implies easily

$$\boldsymbol{\delta}^1_3 = (\aleph_{\omega+1})^{L(\mathcal{R})} \leq \aleph_3,$$

i.e. $\boldsymbol{\delta}^1_3$ is the ordinal which appears within $L(\mathcal{R})$ to be the $\omega+1$'st cardinal—but in V, actually $\boldsymbol{\delta}^1_3 \leq \aleph_3$.

Another consequence (granting a measurable cardinal and the axiom of choice) is that *every* $\mathbf{\Sigma}^1_3$ *set is the union of* $\aleph_2$ *Borel sets* and (assuming further $Det(\mathbf{\Delta}^1_2)$) *every* $\mathbf{\Sigma}^1_4$ *set is the union of* $\aleph_3$ *Borel sets.*

Extension of these elegant results to the higher Lusin pointclasses is one of the most challenging and fascinating open problems of descriptive set theory.

The basic notions of the theory of indiscernibles are model-theoretic and it is useful to explain them in a general setting.

Consider first the theory

$$ZFL = ZF + \text{“}V = L\text{”}$$

which extends Zermelo–Fraenkel set theory by the (formal sentence of $\mathcal{L}^\epsilon$ expressing) the axiom of constructibility and let

$$\mathbf{x} \leq \mathbf{y}$$

be a formula of $\mathcal{L}^\epsilon$ which defines the canonical wellordering of L,

$$\mathbf{x} < \mathbf{y} \Leftrightarrow \mathbf{x} \leq \mathbf{y} \ \& \ \neg(\mathbf{x} = \mathbf{y}).$$

It is clear that if $\mathfrak{A} = (A, E)$ is any model of ZFL, not necessarily a standard model, then

(1) $$\mathfrak{A} \models \text{“}\{(\mathbf{x}, \mathbf{y}) : \mathbf{x} \leq \mathbf{y}\} \text{ is an ordering”}$$

and for any formula $\psi(\mathbf{z}_1, \dots, \mathbf{z}_n, \mathbf{y})$,

(2) $$\begin{aligned} \mathfrak{A} \models (\forall \mathbf{z}_1) \cdots (\forall \mathbf{z}_n)\{(\exists \mathbf{y})\psi(\mathbf{z}_1, \dots, \mathbf{z}_n, \mathbf{y}) \\ \rightarrow (\exists \mathbf{y})[\psi(\mathbf{z}_1, \dots, \mathbf{z}_n, \mathbf{y}) \\ \& \ (\forall \mathbf{x})[\mathbf{x} < \mathbf{y} \rightarrow \neg\psi(\mathbf{z}_1, \dots, \mathbf{z}_n, \mathbf{x})]]\}; \end{aligned}$$

this is simply because these formal sentences are all theorems of ZFL and hence they are true in every model of ZFL.

Let us call a theory (a set of sentences) T in a language $\mathcal{L} = \mathcal{L}^u$ *good* (relative to a formula $\mathbf{x} < \mathbf{y}$) if T has a model and if every model $\mathfrak{A}$ of T satisfies (1) and (2) above. We will assume for simplicity that there are no function symbols or constants in $\mathcal{L}$.

Another good theory which we will use is

$$\begin{aligned} ZFL[\dot{\alpha}] = ZF + \text{“}\dot{\alpha} \text{ is the graph of a function on } \omega \text{ to } \omega\text{”} \\ + V = L[\dot{\alpha}]; \end{aligned}$$

this is in the language $\mathcal{L}^\epsilon[\dot{\alpha}]$ obtained by adding to $\mathcal{L}^\epsilon$ a *binary relation symbol* $\dot{\alpha}$ meant to represent the graph of some irrational α and its axioms express precisely the assertion that every set is constructible from

α. It is clear that $ZFL[\dot{\alpha}]$ is good relative to the formula defining the canonical wellordering of $L[\alpha]$. The models of $ZFL[\dot{\alpha}]$ are of the form

$$\mathfrak{A} = (A, E, \bar{\alpha})$$

where $E \subseteq A \times A$ and $\bar{\alpha} \subseteq A \times A$ interpret $\in$ and $\dot{\alpha}$ respectively.

If $\mathfrak{A} = (A, \text{-})$ is any model of a good theory T, let

$$x \leq^{\mathfrak{A}} y \Leftrightarrow \mathfrak{A} \models x \leq y$$

be *the canonical ordering* on A and call $\mathfrak{A}$ wellfounded if $\leq^{\mathfrak{A}}$ is a wellordering. It is clear from 8F.4 and 8F.22 (using Mostowski collapsing) that every wellfounded model of ZFL is isomorphic to some

$$(L_\lambda, \in)$$

and every wellfounded model of $ZFL[\dot{\alpha}]$ is isomorphic to some

$$(L_\lambda[\alpha], \in, \alpha),$$

where $\alpha \in \mathcal{N}$ and λ is limit.

Let us now fix a good theory T. With each formula $\varphi(\mathbf{v}_0, \mathbf{v}_1, \ldots, \mathbf{v}_n)$ whose free variables are among the first $n+1$ variables in our standard list $\mathbf{v}_0, \mathbf{v}_1, \ldots$ and in which $\mathbf{v}_n$ actually occurs free, we associate the *formal term*

$$\tau(\mathbf{v}_1, \ldots, \mathbf{v}_n) = \mu \mathbf{v}_0 \varphi(\mathbf{v}_0, \mathbf{v}_1, \ldots, \mathbf{v}_n)$$

which intuitively defines *the least* $\mathbf{v}_0$ *such that* $\varphi(\mathbf{v}_0, \mathbf{v}_1, \ldots, \mathbf{v}_n)$. Of course, these terms are not in the language $\mathfrak{L}$ of T, but we can easily interpret them by partial functions on each model $\mathfrak{A} = (A, \text{-})$ of T:

$$[\mu \mathbf{v}_0 \varphi(\mathbf{v}_0, x_1, \ldots, x_n)]^{\mathfrak{A}}\downarrow \Leftrightarrow \mathfrak{A} \models (\exists \mathbf{v}_0)\varphi(\mathbf{v}_0, x_1, \ldots, x_n),$$

$$[\mu \mathbf{v}_0 \varphi(\mathbf{v}_0, x_1, \ldots, x_n)]^{\mathfrak{A}} = \text{the unique } y \in A \text{ such that}$$
$$\mathfrak{A} \models \varphi(y, x_1, \ldots, x_n) \,\&\, (\forall \mathbf{z} < y) \neg \varphi(\mathbf{z}, x_1, \ldots, x_n).$$

We will say that the partial function

$$\tau^{\mathfrak{A}}(x_1, \ldots, x_n) = [\mu \mathbf{v}_0 \varphi(\mathbf{v}_0, x_1, \ldots, x_n)]^{\mathfrak{A}}$$

is *definable by a term in* $\mathfrak{A}$.

Moreover, we can easily interpret substitutions of these terms into formulas of $\mathfrak{L}$ as abbreviations of formulas,

$$\psi(\mu \mathbf{v}_0 \varphi(\mathbf{v}_0, \mathbf{v}_1, \ldots, \mathbf{v}_n), \mathbf{x}_1 \ldots, \mathbf{x}_m)$$
$$\Leftrightarrow (\exists \mathbf{y})\{\varphi(\mathbf{y}, \mathbf{v}_1, \ldots, \mathbf{v}_n) \,\&\, (\forall \mathbf{z} < \mathbf{y}) \neg \varphi(\mathbf{z}, \mathbf{v}_1, \ldots, \mathbf{v}_n)$$
$$\&\, \psi(\mathbf{y}, \mathbf{x}_1, \ldots, \mathbf{x}_m)\},$$

where the variable $\mathbf{y}$ is chosen different from all the variables occurring in $\psi(\mathbf{w}, \mathbf{x}_1,\ldots, \mathbf{x}_m)$ to avoid conflicts in interpretation. As another example,

$$\mu\mathbf{v}_0\varphi(\mathbf{v}_0, \mathbf{v}_1,\ldots, \mathbf{v}_n) = \mu\mathbf{v}_0\psi(\mathbf{v}_0, \mathbf{v}_1,\ldots, \mathbf{v}_m)$$

is an abbreviation of

$$(\exists\mathbf{v}_0)\{\varphi(\mathbf{v}_0, \mathbf{v}_1,\ldots, \mathbf{v}_n) \mathbin{\&} (\forall\mathbf{z}<\mathbf{v}_0)\neg\varphi(\mathbf{z}, \mathbf{v}_1,\ldots, \mathbf{v}_n) \\ \mathbin{\&} \psi(\mathbf{v}_0, \mathbf{v}_1,\ldots, \mathbf{v}_m) \mathbin{\&} (\forall\mathbf{z}<\mathbf{v}_0)\neg\psi(\mathbf{z}, \mathbf{v}_1,\ldots, \mathbf{v}_m)\}$$

where $\mathbf{z}$ is any variable not occurring in $\varphi(\mathbf{v}_0, \mathbf{v}_1,\ldots, \mathbf{v}_n)$, $\psi(\mathbf{v}_0, \mathbf{v}_1,\ldots, \mathbf{v}_m)$.

If $\mathfrak{A} = (A, \text{-})$ is a model of T and $B \subseteq A$, put

$$B^* = \{\tau^{\mathfrak{A}}(x_1,\ldots, x_n)\colon x_1 <^{\mathfrak{A}} x_2 <^{\mathfrak{A}} \cdots <^{\mathfrak{A}} x_n,\ x_1,\ldots, x_n \in B \\ \text{and } \tau(\mathbf{v}_1,\ldots, \mathbf{v}_n) = \mu\mathbf{v}_0\varphi(\mathbf{v}_0, \mathbf{v}_1,\ldots, \mathbf{v}_n) \text{ is a term} \\ \text{such that } \tau^{\mathfrak{A}}(x_1,\ldots, x_n)\downarrow\}$$

and let $\mathfrak{B}^* = (B^*, \text{-})$ be the substructure of $\mathfrak{A}$ obtained by restricting all the relations of $\mathfrak{A}$ to B^*.

We say that B *generates* $\mathfrak{A}$ if

$$\mathfrak{A} = \mathfrak{B}^*.$$

8H.1. LEMMA. *If $\mathfrak{A} = (A, \text{-})$ is a model of a good theory T and $\emptyset \neq B \subseteq A$, then $\mathfrak{B}^*$ is an elementary substructure of $\mathfrak{A}$.*

PROOF. As in the proof of the Skolem–Löwenheim Theorem 8A.4, we show by induction on the length of a formula $\varphi(\mathbf{x}_1,\ldots, \mathbf{x}_n)$, that if $x_1,\ldots, x_n \in B^*$, then

$$\mathfrak{B}^* \models \varphi(x_1,\ldots, x_n) \Leftrightarrow \mathfrak{A} \models \varphi(x_1,\ldots, x_n).$$

The only non-trivial case is when we consider formulas that start with a quantifier, say the formula

$$(\exists\mathbf{w})\chi(\mathbf{s}, \mathbf{t}, \mathbf{w}),$$

where we have taken only two free variables to simplify notation. We must show that if $s, t \in B^*$ and

$$\mathfrak{A} \models (\exists\mathbf{w})\chi(s, t, \mathbf{w}),$$

then there is some $w \in B^*$ such that $\mathfrak{A} \models \chi(s, t, w)$, so that by the induction hypothesis,

$$\mathfrak{B}^* \models (\exists\mathbf{w})\chi(s, t, \mathbf{w}).$$

We have

$$s=\tau^{\mathfrak{A}}(x_1,\ldots,x_n)=[\mu\mathbf{v}_0\varphi(\mathbf{v}_0,x_1,\ldots,x_n)]^{\mathfrak{A}},$$
$$t=\sigma^{\mathfrak{A}}(y_1,\ldots,y_m)=[\mu\mathbf{v}_0\psi(\mathbf{v}_0,y_1,\ldots,y_m)]^{\mathfrak{A}}$$

where $x_1<^{\mathfrak{A}}\cdots<^{\mathfrak{A}}x_n$, $y_1<^{\mathfrak{A}}\cdots<^{\mathfrak{A}}y_m$ are increasing sequences in B. Let $z_1<^{\mathfrak{A}}z_2<^{\mathfrak{A}}\cdots<^{\mathfrak{A}}z_k$ be an increasing enumeration of the finite set $\{x_1,\ldots,x_n,y_1,\ldots,y_m\}$, so that

$$x_i=z_{a_i},\qquad y_j=z_{b_j}$$

for suitable integers a_i $(1\le i\le n)$, b_j $(1\le j\le m)$ and put

$$\rho(\mathbf{v}_1,\ldots,\mathbf{v}_k)=\mu\mathbf{v}_0\chi(\mu\mathbf{v}_0\varphi(\mathbf{v}_0,\mathbf{v}_{a_1},\ldots,\mathbf{v}_{a_n}),\ \mu\mathbf{v}_0\psi(\mathbf{v}_0,\mathbf{v}_{b_1},\ldots,\mathbf{b}_{b_m}),\mathbf{v}_0),$$

where of course, the formula defining $\rho(\mathbf{v}_1,\ldots,\mathbf{v}_k)$ must be "unabbreviated" as above, using variables that will not conflict with $\mathbf{v}_0,\mathbf{v}_1,\ldots,\mathbf{v}_k$. Clearly now

$$\begin{aligned}w&=\rho^{\mathfrak{A}}(z_1,\ldots,z_k)\\&=[\mu\mathbf{v}_0\chi([\mu\mathbf{v}_0\varphi(\mathbf{v}_0,x_1,\ldots,x_n)]^{\mathfrak{A}},[\mu\mathbf{v}_0\psi(\mathbf{v}_0,y_1,\ldots,y_m)]^{\mathfrak{A}},\mathbf{v}_0)]^{\mathfrak{A}}\\&=[\mu\mathbf{v}_0\chi(s,t,\mathbf{v}_0)]^{\mathfrak{A}}\end{aligned}$$

is in B^* and $\mathfrak{A}\vDash\chi(s,t,w)$. ⊣

In the future we will often omit the details of such arguments where one has to fuss with renumbering variables, as they are quite routine.

Suppose again that $\mathfrak{A}$ is a model of T and $I\subseteq A$ is a subset of the domain A. We say that I is *homogeneous in* $\mathfrak{A}$ (a *set of indiscernibles* in $\mathfrak{A}$) if for each formula $\varphi(\mathbf{x}_1,\ldots,\mathbf{x}_n)$ and every two n-tuples of increasing members of I,

$$x_1,\ldots,x_n,\qquad y_1,\ldots,y_n\in I,$$
$$x_1<^{\mathfrak{A}}x_2<^{\mathfrak{A}}\cdots<^{\mathfrak{A}}x_m,\qquad y_1<^{\mathfrak{A}}y_2<^{\mathfrak{A}}\cdots<^{\mathfrak{A}}y_n,$$

we have

$$\mathfrak{A}\vDash\varphi(x_1,\ldots,x_n)\Leftrightarrow\mathfrak{A}\vDash\varphi(y_1,\ldots,y_n).$$

This is the key notion of the theory of indiscernibles and it is obviously related to the notion of a *homogeneous set of ordinals* (relative to a partition) with which we worked in section 6G.

We will be dealing with pairs

$$(\mathfrak{A},I),$$

where $\mathfrak{A} = (A, \text{-})$ is a model of T and $I \subseteq A$ is homogeneous in $\mathfrak{A}$. For each such pair, let

$$\|I\|^{\mathfrak{A}} = \text{the order type of the ordering } \{(x, y): x, y \in I, x \leq^{\mathfrak{A}} y\}$$

and define *the character* of $(\mathfrak{A}, I)$ by

$$\Phi = Char(\mathfrak{A}, I) = \{\varphi(\mathbf{v}_1,\ldots,\mathbf{v}_n): \varphi(\mathbf{v}_1,\ldots,\mathbf{v}_n) \text{ is a formula whose free variables are among } \mathbf{v}_1,\ldots,\mathbf{v}_n \text{ and for some } x_1 <^{\mathfrak{A}} x_2 <^{\mathfrak{A}} \cdots <^{\mathfrak{A}} x_n,\ x_1,\ldots,x_n \in I,\ \mathfrak{A} \models \varphi(x_1,\ldots,x_n)\}.$$

We are allowing here $n = 0$, so that the collection of formulas Φ contains all the sentences true in $\mathfrak{A}$.

Mostly, we will work with pairs $(\mathfrak{A}, I)$ where I is wellordered by $\leq^{\mathfrak{A}}$, so that $\|I\|^{\mathfrak{A}}$ is some ordinal λ. One should be careful however, because it may happen that $\|I\|^{\mathfrak{A}}$ is an ordinal but $\mathfrak{A}$ is not a wellfounded structure, i.e. the entire set A is not wellordered by $\leq^{\mathfrak{A}}$.

We are now ready to state and outline a proof of the basic result in the so-called Ehrenfeucht–Mostowski theory.

8H.2. THEOREM. *Suppose Φ is a collection of formulas in the language of a good theory T and assume that there exists a pair $(\mathfrak{A}, I)$ such that $\mathfrak{A}$ is a model of T, I is an infinite homogeneous set in $\mathfrak{A}$ and*

$$\Phi = Char(\mathfrak{A}, I).$$

Then for each infinite, limit ordinal λ, *we can define a pair $(\mathfrak{A}_\lambda, I_\lambda)$ (which depends only on Φ) so that the following hold:*

(i) *$\mathfrak{A}_\lambda$ is a model of T, I_λ is homogeneous in $\mathfrak{A}_\lambda$, I_λ generates $\mathfrak{A}_\lambda$ and*

$$\|I_\lambda\|^{\mathfrak{A}_\lambda} = \lambda.$$

(ii) *$Char(\mathfrak{A}_\lambda, I_\lambda) = \Phi$.*

(iii) *If $(\mathfrak{B}, J)$ is any pair where $\mathfrak{B} = (B, \text{-})$ is a model of T, $J \subseteq B$ is homogeneous in $\mathfrak{B}$ and generates $\mathfrak{B}$ and $Char(\mathfrak{B}, J) = \Phi$ and if*

$$f : J \rightarrowtail I_\lambda$$

is any order-preserving injection, then there is a unique extension

$$f^* : B \to A_\lambda$$

which is an elementary imbedding of $\mathfrak{B}$ into $\mathfrak{A}_\lambda$, i.e. for each formula

$\varphi(\mathbf{x}_1,\dots,\mathbf{x}_n)$ and $x_1,\dots,x_n \in B$,

$$\mathfrak{B} \models \varphi(x_1,\dots,x_n) \Leftrightarrow \mathfrak{A}_\lambda \models \varphi(f^*(x_1),\dots,f^*(x_n)).$$

(v) *With the same hypotheses as in* (iii), *if $f: J \twoheadrightarrow I_\lambda$ is in fact an order-preserving bijection of J with I_λ, then f^* is an isomorphism of $\mathfrak{B}$ with $\mathfrak{A}_\lambda$.*

(v) *The following three conditions are equivalent:*

(a) *Each $\mathfrak{A}_\lambda$ is a wellfounded model of T (i.e. $\leq^{\mathfrak{A}_\lambda}$ is a wellordering).*

(b) *For each countable λ, $\mathfrak{A}_\lambda$ is wellfounded.*

(c) *For some uncountable λ, $\mathfrak{A}_\lambda$ is wellfounded.*

PROOF. The idea is to think of the ordinals below λ as the indiscernibles in the model which we will construct, so every member of this model will be of the form

$$x = \tau^{\mathfrak{A}_\lambda}(\xi_1,\dots,\xi_n) = [\mu\mathbf{v}_0\varphi(\mathbf{v}_0, \xi_1,\dots,\xi_n)]^{\mathfrak{A}_\lambda}$$

for some formula $\varphi(\mathbf{v}_0, \mathbf{v}_1,\dots,\mathbf{v}_n)$ and some $\xi_1 < \xi_2 < \cdots < \xi_n < \lambda$. We can then think of the tuple

$$(\xi_1,\dots,\xi_n, \mu\mathbf{v}_0\varphi(\mathbf{v}_0, \mathbf{v}_1,\dots,\mathbf{v}_n))$$

as a *name* of $[\mu\mathbf{v}_0\varphi(\mathbf{v}_0, \xi_1,\dots,\xi_n)]^{\mathfrak{A}_\lambda}$ (still to be constructed) and define first the collection of names

$$B_\lambda = \{(\xi_1,\dots,\xi_n, \mu\mathbf{v}_0\varphi(\mathbf{v}_0, \mathbf{v}_1,\dots,\mathbf{v}_n)) : \xi_1 < \xi_2 < \cdots < \xi_n < \lambda$$
$$\text{and the formula } (\exists\mathbf{v}_0)\varphi(\mathbf{v}_0, \mathbf{v}_1,\dots,\mathbf{v}_n) \in \Phi\}.$$

There is an obvious equivalence relation $\sim$ on B_λ (names are equivalent when they name the same object) which is easiest to define in the special case when the ordinals of the names are *far apart*. For example, if

$$\xi_1 < \xi_2 < \eta_1 < \eta_2 < \eta_3,$$

put

$$\begin{aligned}
&(\xi_1, \xi_2, \mu\mathbf{v}_0\varphi(\mathbf{v}_0, \mathbf{v}_1, \mathbf{v}_2)) \sim (\eta_1, \eta_2, \eta_3, \mu\mathbf{v}_0\psi(\mathbf{v}_0, \mathbf{v}_1, \mathbf{v}_2, \mathbf{v}_3))\\
&\Leftrightarrow \mu\mathbf{v}_0\varphi(\mathbf{v}_0, \mathbf{v}_1, \mathbf{v}_2) = \mu\mathbf{v}_0\psi(\mathbf{v}_0, \mathbf{v}_3, \mathbf{v}_4, \mathbf{v}_5) \in \Phi\\
&\Leftrightarrow \text{whenever } \mathfrak{A} = (A, \text{-}) \models T,\ I \subseteq A \text{ is homogeneous}\\
&\quad \text{in } \mathfrak{A},\ \mathit{Char}(\mathfrak{A}, I) = \Phi \text{ and } x_1 <^{\mathfrak{A}} x_2 <^{\mathfrak{A}} y_1 <^{\mathfrak{A}} y_2 <^{\mathfrak{A}} y_3\\
&\quad \text{are all in } I, \text{ then}\\
&\quad \mathfrak{A} \models \mu\mathbf{v}_0\varphi(\mathbf{v}_0, x_1, x_2) = \mu\mathbf{v}_0\psi(\mathbf{v}_0, y_1, y_2, y_3).
\end{aligned}$$

In the general case, let $\lambda_1 < \lambda_2 < \cdots < \lambda_k$ be an increasing enumeration of the set $\{\xi_1, \ldots, \xi_n, \eta_1, \ldots, \eta_m\}$ so that

$$\xi_i = \lambda_{s_i}, \qquad \eta_j = \lambda_{t_j},$$

and put

$$\begin{aligned}
&(\xi_1, \ldots, \xi_n, \mu \mathbf{v}_0 \varphi(\mathbf{v}_0, \mathbf{v}_1, \ldots, \mathbf{v}_n)) \sim (\eta_1, \ldots, \eta_m, \mu \mathbf{v}_0 \psi(\mathbf{v}_0, \mathbf{v}_1, \ldots, \mathbf{v}_m)) \\
&\quad \Leftrightarrow \mu \mathbf{v}_0 \varphi(\mathbf{v}_0, \mathbf{v}_{s_1}, \ldots, \mathbf{v}_{s_n}) = \mu \mathbf{v}_0 \psi(\mathbf{v}_0, \mathbf{v}_{t_1}, \ldots, \mathbf{v}_{t_m}) \in \Phi \\
&\quad \Leftrightarrow \text{whenever } \mathfrak{A} = (A, \text{-}) \models T,\ I \subseteq A \text{ is homogeneous in } \mathfrak{A}, \\
&\qquad Char(\mathfrak{A}, I) = \Phi \text{ and } z_1 <^{\mathfrak{A}} z_2 <^{\mathfrak{A}} \cdots <^{\mathfrak{A}} z_k \text{ are all in } I, \text{ then} \\
&\qquad \mathfrak{A} \models \mu \mathbf{v}_0 \varphi(\mathbf{v}_0, z_{s_1}, \ldots, z_{s_n}) = \mu \mathbf{v}_0 \psi(\mathbf{v}_0, z_{t_1}, \ldots, z_{t_m}).
\end{aligned}$$

It is now easy to prove that $\sim$ is an equivalence relation on B_λ, using the hypothesis that there exists some pair $(\mathfrak{A}, I)$ with $\mathfrak{A} \models T$, I infinite and homogeneous in $\mathfrak{A}$ and $Char(\mathfrak{A}, I) = \Phi$.

The domain of $\mathfrak{A}_\lambda$ will be then

$$\begin{aligned}
A_\lambda &= B_\lambda / \sim \\
&= \text{all equivalence classes of members of } B_\lambda \text{ under } \sim.
\end{aligned}$$

Similarly,

$$I_\lambda = \{[(\xi, \mu \mathbf{v}_0(\mathbf{v}_0 = \mathbf{v}_1))]: \xi < \lambda\},$$

where [-] is the equivalence class of -, so that I_λ is naturally ordered with order type λ and we can identify it with λ for the proof.

For each unary relation symbol $\mathbf{R}$ in the language, define on A_λ

$$\begin{aligned}
&R^\lambda([(\xi_1, \ldots, \xi_n, \mu \mathbf{v}_0 \varphi(\mathbf{v}_0, \mathbf{v}_1, \ldots, \mathbf{v}_n))]) \Leftrightarrow \mathbf{R}(\mu \mathbf{v}_0 \varphi(\mathbf{v}_1, \ldots, \mathbf{v}_n)) \in \Phi \\
&\qquad \Leftrightarrow \text{whenever } \mathfrak{A} = (A, I) \models T,\ I \subseteq A \text{ is homogeneous in } \mathfrak{A}, \\
&\qquad\quad Char(\mathfrak{A}, I) = \Phi \text{ and } x_1 <^{\mathfrak{A}} x_2 <^{\mathfrak{A}} \cdots <^{\mathfrak{A}} x_n \text{ are all in } I, \\
&\qquad\quad \text{then } \mathfrak{A} \models \mathbf{R}_0(\mu \mathbf{v}_0 \varphi(\mathbf{v}_0, x_1, \ldots, x_n)).
\end{aligned}$$

The same idea will work for relation symbols of more arguments, except that we will have to indulge in some renumbering of variables in the definition, as we did (in effect) with $=$ above. These relations are easily well-defined on A_λ, although we define them to begin with via representatives of equivalence classes—that argument too uses the fact that $\Phi = Char(\mathfrak{A}, I)$ for some suitable pair.

At this point we have a structure $\mathfrak{A}_\lambda = (A_\lambda, \text{-})$ and a set $I_\lambda \subseteq A_\lambda$ and it is obvious that the truth of formal sentences in $\mathfrak{A}_\lambda$ can be computed by determining whether various formulas are in Φ. The precise fact that we need is as follows.

Lemma. Suppose $\varphi(\mathbf{x}_1,\ldots,\mathbf{x}_n)$ is a formula and

$$x_1 = (\xi_1^1,\ldots,\xi_{m_1}^1, \mu\mathbf{v}_0\varphi_1(\mathbf{v}_0,\mathbf{v}_1,\ldots,\mathbf{v}_{m_1})),\ldots,$$
$$x_n = (\xi_1^n,\ldots,\xi_{m_n}^n, \mu\mathbf{v}_0\varphi_n(\mathbf{v}_0,\mathbf{v}_1,\ldots,\mathbf{v}_{m_n}))$$

is a sequence of names in B_λ with corresponding equivalence classes $\bar{x}_1,\ldots,\bar{x}_n$. Let $\lambda_1 < \lambda_2 < \cdots < \lambda_k$ be an increasing enumeration of the finite set $\{\xi_j^i : 1 \le i \le n, 1 \le j \le m_i\}$ so that with a suitable choice of integers $a(i,j)$,

$$\xi_j^i = \lambda_{a(i,j)} \quad (1 \le i \le n, 1 \le j \le m_i).$$

Then

$$\mathfrak{A}_\lambda \models \varphi(\bar{x}_1,\ldots,\bar{x}_n)$$
$$\Leftrightarrow \varphi(\mu\mathbf{v}_0\varphi_1(\mathbf{v}_0,\mathbf{v}_{a(1,1)},\ldots,\mathbf{v}_{a(1,m_1)}),\ldots,\mu\mathbf{v}_0\varphi_n(\mathbf{v}_0,\mathbf{v}_{a(n,1)},\ldots,\mathbf{v}_{a(n,m_n)})) \in \Phi.$$

This is easy to check by induction on φ and from it (i) and (ii) of the theorem follow routinely.

To prove (iii), given $f: J \rightarrowtail I_\lambda$ which is order preserving and given $x \in B$, find first an increasing sequence $x_1 <^{\mathfrak{B}} x_2 <^{\mathfrak{B}} \cdots <^{\mathfrak{B}} x_m$ of members of J and a formula $\varphi(\mathbf{v}_0,\mathbf{v}_1,\ldots,\mathbf{v}_n)$ such that

$$x = [\mu\mathbf{v}_0\varphi(\mathbf{v}_0, x_1,\ldots,x_n)]^{\mathfrak{B}}$$

and set

$$f^*(x) = \text{equivalence class of } (f(x_1),\ldots,f(x_n), \mu\mathbf{v}_0\varphi(\mathbf{v}_0,\mathbf{v}_1,\ldots,\mathbf{v}_n)).$$

It is easy to check that the definition of $f^*(x)$ is independent of the choice of $x_1,\ldots,x_n$, $\varphi(\mathbf{v}_0,\mathbf{v}_1,\ldots,\mathbf{v}_n)$ using the homogeneity of J and the fact that f^* is an elementary imbedding follows directly from the lemma. Also (iv) follows immediately, because if f is onto I_λ, then f^* is clearly onto A_λ.

Finally, to prove (v), suppose first that $\varkappa$ is uncountable and $\mathfrak{A}_\varkappa$ is wellfounded and let λ be any countable ordinal, let $J \subseteq I_\varkappa$ be any subset of $I_\varkappa$ of order-tuple λ and consider the structure $\mathfrak{B} = (J^*, \text{-})$ which is an elementary substructure of $\mathfrak{A}_\varkappa$ by 8H.1. Now J is homogeneous in $\mathfrak{B}$ and generates $\mathfrak{B}$, and

$$Char(\mathfrak{B}, J) = Char(\mathfrak{A}_\varkappa, I_\varkappa) = \Phi;$$

thus if we let $f: J \twoheadrightarrow I_\lambda$ be the unique order-preserving bijection of J with I_λ, the extended map $f^*: J^* \twoheadrightarrow A_\lambda$ is an isomorphism, in particular it takes the ordering $\leq^{\mathfrak{B}}$ in an order-preserving way onto $\leq^{\mathfrak{A}_\lambda}$. But $\leq^{\mathfrak{B}}$ is a wellordering since it is the restriction of $\leq^*$ to J^*, and so $\leq^{\mathfrak{A}_\lambda}$ is also a wellordering.

This shows that in (v), (c) $\Rightarrow$ (b). To show that (b) $\Rightarrow$ (a) and complete the proof, suppose that for some $\varkappa$, $\mathfrak{A}_\varkappa$ is not wellfounded and let

$$x_1 >^{\mathfrak{A}_\varkappa} x_2 >^{\mathfrak{A}_\varkappa} \cdots$$

be an infinite descending chain—we can find one even without using the axiom of choice, because the domain $A_\varkappa$ is obviously a wellorderable set. Now each $x_i = [\mu \mathbf{v}_0 \varphi_i(\mathbf{v}_0, y_1^i, \ldots, y_{n_i}^k)]^{\mathfrak{A}_\varkappa}$ for a suitable $\varphi_i(\mathbf{v}_0, \mathbf{v}_1, \ldots, \mathbf{v}_{n_i})$ and members $y_1^i, \ldots, y_{n_i}^i$ of $I_\varkappa$, so there is a countable subset $J \subseteq I_\varkappa$ such that for each i, $x_i \in J^*$; this simply means that the elementary substructure $\mathfrak{B} = (J^*, \text{-})$ is not wellfounded. Finally, if λ is the countable order-type of the set J, then as above $\mathfrak{B}$ is isomorphic with $\mathfrak{A}_\lambda$ and $\mathfrak{A}_\lambda$ is not wellfounded. ⊣

It is very important in this model-theoretic construction that the pairs $(\mathfrak{A}_\lambda, I_\lambda)$ were constructed directly from the set of formulas Φ, assuming only that $\Phi = Char(\mathfrak{A}, I)$ *for some* pair $(\mathfrak{A}, I)$, but not using any particular $\mathfrak{A}$, I in the construction. Thu ve have an operation

$$\Gamma(\lambda, \Phi) = (\mathfrak{A}_\lambda, I_\lambda)$$

which assigns to each ordinal λ and each character Φ a pair $(\mathfrak{A}_\lambda, I_\lambda)$ with the appropriate properties.

We now turn to apply this basic result from model theory to *ZFL* and *ZFL*$[\dot{\alpha}]$ following Silver [1971].

It will be convenient to abuse notation slightly by using the symbol "$L_\varkappa[\alpha]$" to refer both to the set $L_\varkappa[\alpha]$ and to the structure $(L_\varkappa[\alpha], \in, \alpha)$, for any given $\alpha \in \mathcal{N}$. One should keep in mind that in the language of these structures we have a relation symbol which defines α, although this will be suppressed in our notation.

We should also point out that *in any model of* *ZFL*$[\dot{\alpha}]$, *the ordinals are cofinal in the canonical ordering*. This is because by (the relativized version of) 8F.1 and 8D.10

$$y \in L_\xi[\alpha] \;\&\; x \leq y \Rightarrow x \in L_\xi[\alpha],$$

$$\xi \notin L_\xi[\alpha],$$

so that

$$y \in L_\xi[\alpha] \Rightarrow y < \xi;$$

now this is a theorem of *ZF*, so its formal version holds in every model of

$ZFL[\dot{\alpha}]$ which also satisfies "$(\forall y)(\exists\xi)[y \in L_\xi[\alpha]]$" and which hence also satisfies "$(\forall y)(\exists\xi)[y < \xi]$".

8H.3. THEOREM (Silver [1971]). *Suppose $\alpha \in \mathcal{N}$, $\varkappa$ is a measurable cardinal and $\mathfrak{U}$ is a normal ultrafilter on $\varkappa$.*

(i) $L_\varkappa[\alpha] \models ZFL[\dot{\alpha}]$.

(ii) *There exists some set $I \subseteq \varkappa$ such that $I \in \mathfrak{U}$ and I is homogeneous in $L_\varkappa[\alpha]$.*

(iii) *If $I \subseteq \varkappa$, $I \in \mathfrak{U}$ and I is homogeneous in $L_\varkappa[\alpha]$, then the following hold.*

(a) *For each formula $\varphi(\mathbf{v}_0, \mathbf{v}_1, \ldots, \mathbf{v}_n)$ and for each increasing sequence $\xi_1 < \xi_2 < \cdots < \xi_n < \lambda$ of ordinals in I,*

$$L_\varkappa[\alpha] \models (\exists \mathbf{v}_0)\varphi(\mathbf{v}_0, \xi_1, \ldots, \xi_n) \rightarrow \mu\mathbf{v}_0\varphi(\mathbf{v}_0, \xi_1, \ldots, \xi_n) < \lambda.$$

(b) *For each formula $\varphi(\mathbf{v}_0, \mathbf{v}_1, \ldots, \mathbf{v}_n, \mathbf{v}_{n+1}, \ldots, \mathbf{v}_{n+m})$, if $\xi_1 < \xi_2 < \cdots < \xi_n$, $\eta_1 < \eta_2 < \cdots < \eta_m$, $\eta'_1 < \eta'_2 < \cdots < \eta'_m$ are increasing sequences in I and if $\xi_n < \eta_1$, $\xi_n < \eta'_1$, then*

$$L_\varkappa[\alpha] \models \mu\mathbf{v}_0\varphi(\mathbf{v}_0, \xi_1, \ldots, \xi_n, \eta_1, \ldots, \eta_m) < \eta_1$$
$$\rightarrow \mu\mathbf{v}_0\varphi(\mathbf{v}_0, \xi_1, \ldots, \xi_n, \eta_1, \ldots, \eta_m) = \mu\mathbf{v}_0\varphi(\mathbf{v}_0, \xi_1, \ldots, \xi_n, \eta'_1, \ldots, \eta'_m).$$

PROOF. To check (i), go back to the proof of 6G.9 and check that the following was actually shown, without appeal to the axiom of choice: if $\varkappa$ is measurable, $\lambda < \varkappa$ and $\{X_\xi\}_{\xi<\mu}$ is a wellordered sequence of distinct subsets of λ, then $\mu < \varkappa$. This implies easily that

$$L[\alpha] \models \text{“}\varkappa \text{ is a strongly inaccessible cardinal”}$$

and hence by 8C.11, since $L_\varkappa[\alpha]$ satisfies the axiom of choice,

$$L[\alpha] \models \text{“}L_\varkappa[\alpha] \text{ is a model of } ZFC\text{”}.$$

Now simple absoluteness considerations imply that in fact

$$L_\varkappa[\alpha] \models ZFC$$

and again, easily, using 8F.1, $L_\varkappa[\alpha] \models$ "$V = L(\alpha)$", so that $L_\varkappa[\alpha] \models ZFL[\dot{\alpha}]$.

(ii) For each formula $\varphi(\mathbf{x}_1, \ldots, \mathbf{x}_n)$ with n free variables define

$$F_\varphi : \varkappa^{[n]} \rightarrow \{0, 1\}$$

by setting first

$$F_1(\xi_1, \ldots, \xi_n) = \begin{cases} 1 & \text{if } L_\varkappa[\alpha] \models \varphi(\xi_1, \ldots, \xi_n) \\ 0 & \text{if } L_\varkappa[\alpha] \models \neg\varphi(\xi_1, \ldots, \xi_n) \end{cases}$$

and for each n-element subset X of $\varkappa$, putting

$$F_\varphi(X) = F_1(\xi_1,\ldots,\xi_n)$$

where $\xi_1 < \xi_2 < \cdots < \xi_n$ is an increasing enumeration of X. Since there are only countably many formulas, we can use 6G.4 to get a set $I \in \mathfrak{U}$ which is homogeneous for each partition F_φ and hence homogeneous for $L_\varkappa[\alpha]$.

(iii) (a) If $L_\varkappa[\alpha] \models (\exists \mathbf{v}_0)\varphi(\mathbf{v}_0, \xi_1,\ldots,\xi_n)$ with $\xi_1 < \cdots < \xi_n$ in I, then for some x we have $L_\varkappa[\alpha] \models \varphi(x, \xi_1,\ldots,\xi_n)$ and since I is cofinal in $\varkappa$ and hence in the canonical ordering of $L_\varkappa[\alpha]$, we have some $\mu \in I$, $\xi_1 < \cdots < \xi_n < \mu$ and $x < \mu$. Thus

$$L_\varkappa[\alpha] \models \mu \mathbf{v}_0 \varphi(\mathbf{v}_0, \xi_1,\ldots,\xi_n) < \mu$$

and by homogeneity, for any $\lambda > \xi_n$, $\lambda \in I$,

$$L_\varkappa[\alpha] \models \mu \mathbf{v}_0 \varphi(\mathbf{v}_0, \xi_1,\ldots,\xi_n) < \lambda.$$

(iii) (b) Assume

$$L_\varkappa[\alpha] \models \mu \mathbf{v}_0 \varphi(\mathbf{v}_0, \xi_1,\ldots,\xi_n, \eta_1,\ldots,\eta_m) < \eta_1$$

and let

$$x = [\mu \mathbf{v}_0 \varphi(\mathbf{v}_0, \xi_1,\ldots,\xi_n, \eta_1,\ldots,\eta_m)]^{L_\varkappa[\alpha]}.$$

Since $x < \eta_1$, we have $x \in L_{\eta_1}[\alpha]$ by the basic properties of the canonical wellordering on $L_\varkappa[\alpha]$ and hence $x \in L_\xi[\alpha]$ for some $\xi < \eta_1$. (The set of limit ordinals below $\varkappa$ is easily of normal measure 1, so that η_1 is limit.) Thus

$$(*) \qquad L_\varkappa[\alpha] \models \text{“}(\exists \xi)[\mu \mathbf{v}_0 \varphi(\mathbf{v}_0, \xi_1,\ldots,\xi_n, \eta_1,\ldots,\eta_m) \in L_\xi[\alpha] \;\&\; \xi < \eta_1]\text{”}$$

and hence the same formula must be satisfied by $L_\varkappa[\alpha]$ if we replace $\eta_1,\ldots,\eta_m$ by any other increasing tuple $\zeta_1 < \cdots < \zeta_m$ from I with $\xi_n < \zeta_1$.

We now define $F(\zeta)$ for $\zeta \in I$, $\zeta > \xi_n$ as follows: let $\zeta_1 = \zeta$, let $\zeta_2,\ldots,\zeta_m$ be the first $m-1$ members of I above ζ and put

$$F(\zeta) = \text{least } \xi \text{ such that } L_\varkappa[\alpha] \models \text{“}\mu \mathbf{v}_0 \varphi(\mathbf{v}_0, \xi_1,\ldots,\xi_n, \zeta_1,\ldots,\zeta_m) \in L_\xi[\alpha]\text{”}.$$

We have $F(\zeta) < \zeta$ for all $\zeta \in I$, $\zeta > \xi_n$, so by normality there is a fixed ζ^* such that $F(\zeta) = \zeta^*$ for all ζ in some set $J \in \mathfrak{U}$.

Again define $G(\zeta)$ for $\zeta \in J$, $\zeta > \xi_n$ by choosing $\zeta_1,\ldots,\zeta_m$ as above and putting

$$G(\zeta) = [\mu \mathbf{v}_0 \varphi(\mathbf{v}_0, \xi_1,\ldots,\xi_n, \zeta_1,\ldots,\zeta_m]^{L_\varkappa[\alpha]}.$$

Now $G(\zeta)\in L_{\zeta^*}[\alpha]$ for all $\zeta\in J$, $\zeta>\xi_n$ and since $card(L_{\zeta^*}[\alpha])=card(\zeta^*)<\varkappa$ we have a fixed element $x^*\in L_{\zeta^*}[\alpha]$ such that for all ζ in some $H\in\mathfrak{U}$, $G(\zeta)=x^*$.

If ζ and ζ'' are far apart in H so that

$$\zeta_1<\zeta_2<\cdots<\zeta_m<\zeta_1''<\zeta_2''<\cdots<\zeta_m'',$$

the equation $G(\zeta)=G(\zeta'')$ simply means that

$$(**)\qquad L_\varkappa[\alpha]\models \mu\mathbf{v}_0\varphi(\xi_1,\ldots,\xi_n,\zeta_1,\ldots,\zeta_m)=\mu\mathbf{v}_0\varphi(\xi_1,\ldots,\xi_n,\zeta_1'',\ldots,\zeta_m'')$$

so by homogeneity, (**) actually holds for all

$$\xi_1<\cdots<\xi_n<\zeta_1<\cdots<\zeta_m<\zeta_1''<\cdots<\zeta_m''$$

in I. Finally given $\xi_1<\cdots<\xi_n<\eta_1<\cdots<\eta_m$ and $\xi_1<\cdots<\xi_n<\eta_1'<\cdots<\eta_m'$ in I, choose ordinals $\zeta_1''<\cdots<\zeta_m''$ in I above η_m and η_m' and apply (**) with $\zeta_1'',\ldots,\zeta_m''$ and taking $\zeta_i=\eta_i$ first and $\zeta_i=\eta_i'$ afterwards to obtain the desired result

$$L_\varkappa[\alpha]\models \mu\mathbf{v}_0\varphi(\mathbf{v}_0,\xi_1,\ldots,\xi_n,\eta_1,\ldots,\eta_m)=\mu\mathbf{v}_0,\xi_1,\ldots,\xi_n,\eta_1',\ldots,\eta_m'). \qquad \dashv$$

Fix now a measurable cardinal $\varkappa$, an irrational α and a set $I\subseteq\mathfrak{U}$ of normal measure 1 which is homogeneous in $L_\varkappa[\alpha]$ and let

$$\begin{aligned}\Phi &= Char(L_\varkappa[\alpha], I)\\ &=\{\varphi(\mathbf{v}_1,\ldots,\mathbf{v}_n): \varphi(\mathbf{v}_1,\ldots,\mathbf{v}_n) \text{ is in the language of}\\ &\qquad ZFL[\dot\alpha] \text{ and for } \xi_1<\cdots<\xi_n \text{ in } I,\ L_\varkappa[\alpha]\models\varphi(\xi_1,\ldots,\xi_n)\}.\end{aligned}$$

It is clear from the last two theorems that Φ satisfies the following conditions, where for each n, m, $\varphi_{n,m}$ is the sentence of $ZFL[\dot\alpha]$ which expresses formally the assertion

$$\langle n,m\rangle\in\dot\alpha.$$

(R1) There exists a pair $(\mathfrak{A}, I)$, where $\mathfrak{A}\models ZFL[\dot\alpha]$, I is an infinite homogeneous set in $\mathfrak{A}$ and

$$\Phi=Char(\mathfrak{A}, I).$$

(R2) The formula $Ordinal(\mathbf{v}_1)$ is in Φ, and for each n, m,

$$\alpha(n)=m \Leftrightarrow \varphi_{n,m}\in\Phi.$$

(R3) For each formula $\varphi(\mathbf{v}_0,\mathbf{v}_1,\ldots,\mathbf{v}_n)$ in the language of $ZFL[\dot\alpha]$, the formula

$$(\exists \mathbf{v}_0)\varphi(\mathbf{v}_0, \mathbf{v}_1,\ldots, \mathbf{v}_n) \to \mu \mathbf{v}_0\varphi(\mathbf{v}_0, \mathbf{v}_1,\ldots, \mathbf{v}_n) < \mathbf{v}_{n+1}$$

is in Φ.

(R4) For each formula $\varphi(\mathbf{v}_0, \mathbf{v}_1,\ldots, \mathbf{v}_n, \mathbf{v}_{n+1},\ldots, \mathbf{v}_{n+m})$, the formula

$$\begin{aligned}\mu \mathbf{v}_0\varphi(\mathbf{v}_0, \mathbf{v}_1,\ldots, \mathbf{v}_n, \mathbf{v}_{n+1},\ldots, \mathbf{v}_{n+m}) < \mathbf{v}_{n+1} \\ \to \mu \mathbf{v}_0\varphi(\mathbf{v}_0, \mathbf{v}_1,\ldots, \mathbf{v}_n, \mathbf{v}_{n+1},\ldots, \mathbf{v}_{n+m}) \\ = \mu \mathbf{v}_0\varphi(\mathbf{v}_0, \mathbf{v}_1,\ldots, \mathbf{v}_n, \mathbf{v}_{n+m+1},\ldots, \mathbf{v}_{n+m+m})\end{aligned}$$

is in Φ.

(R5) For each countable limit ordinal λ, if $\Gamma(\lambda, \Phi) = (\mathfrak{A}_\lambda, I_\lambda)$ is the canonical pair associated with Φ by 8H.2, then $\mathfrak{A}_\lambda$ is wellfounded.

The remarkable thing about these conditions is that *they do not refer at all to $\varkappa$ or I*—whether they hold or not depends only on the set of formulas Φ and the irrational α—and yet as we will see in the next theorem they determine $Char(L_\varkappa[\alpha], I)$ completely, no matter which measurable cardinal $\varkappa$ and which set I of normal measure 1 in $\varkappa$ we choose.

If Φ is any collection of formulas satisfying (1)–(5), we will call it a *remarkable character for α*. For the record:

8H.4. COROLLARY (Silver [1971]). *If there exists a measurable cardinal, then for each $\alpha \in \mathcal{N}$ there exists a remarkable character for α.* ⊣

We next come to the main theorem in the theory of indiscernibles for L and $L[\alpha]$.

8H.5. THEOREM (Silver [1971]). *If $\alpha \in \mathcal{N}$ and Φ is a remarkable character for α, then there exists exactly one class of ordinals I with the following two properties.*

(i) *I is closed and unbounded.*

(ii) *For each uncountable cardinal λ, $I \cap \lambda$ is a homogeneous set in $L_\lambda[\alpha]$ which generates $L_\lambda[\alpha]$ and satisfies*

$$\Phi = Char(L_\lambda[\alpha], I \cap \lambda).$$

Moreover, I contains all uncountable cardinals and for any uncountable cardinal $\varkappa \geq \aleph_\omega$

$$\Phi = \{\varphi(\mathbf{v}_1,\ldots, \mathbf{v}_n) : L_\varkappa[\alpha] \models \varphi(\aleph_1,\ldots, \aleph_n)\}.$$

In particular, there is at most one remarkable character for α.

PROOF. Fix Φ and let

$$\Gamma(\lambda, \Phi) = (\mathfrak{A}_\lambda, I'_\lambda)$$

be the canonical pair associated with Φ and each ordinal λ by 8H.2. By (R5) and 8H.2 each $\mathfrak{A}_\lambda$ is isomorphic with some $L_{\lambda^*}[\alpha']$ and we must have $\alpha' = \alpha$ because of (R2). Also, if I_λ is the image of the homogeneous set I'_λ under the isomorphism of $\mathfrak{A}_\lambda$ with $L_{\lambda^*}[\alpha]$, then I_λ is homogeneous in $L_{\lambda^*}[\alpha]$, it generates $L_{\lambda^*}[\alpha]$ and by (R2) again, it is a set of ordinals of order type λ, so in particular $\lambda \leq \lambda^*$.

We will eventually set

$$I = \bigcup_\lambda I_\lambda$$

after some lemmas. To simplify notation, let us put

$$\tau^\lambda_\varphi(\xi_1, \ldots, \xi_n) = [\mu \mathbf{v}_0 \varphi(\mathbf{v}_0, \xi_1, \ldots, \xi_n)]^{L_{\lambda^*}[\alpha]},$$

for each formula $\varphi(\mathbf{v}_0, \mathbf{v}_1, \ldots, \mathbf{v}_n)$ such that $(\exists \mathbf{v}_0)\varphi(\mathbf{v}_0, \mathbf{v}_1, \ldots, \mathbf{v}_n) \in \Phi$ and for each $\xi_1 < \cdots < \xi_n$ in I_λ.

Lemma 1. I_λ is cofinal and closed in $L_{\lambda^*}[\alpha]$.

Proof. Each $x \in L_{\lambda^*}[\alpha]$ is $\tau^\lambda_\varphi(\xi_1, \ldots, \xi_n)$ for some $\xi_1 < \cdots < \xi_n$ in I_λ, and by (R3), if $\xi_n < \eta$ and $\eta \in I_\lambda$, we have $\tau^\lambda_\varphi(\xi_1, \ldots, \xi_n) < \eta$; thus I_λ is cofinal in $L_{\lambda^*}[\alpha]$.

To prove that I_λ is closed, suppose towards a contradiction that $\mu < \lambda^*$, $I_\lambda \cap \mu$ is cofinal in μ but $\mu \notin I_\lambda$, so that

$$\mu = \tau^\lambda_\varphi(\xi_1, \ldots, \xi_n, \eta_1, \ldots, \eta_m)$$

for some τ^λ_φ and ordinals in I_λ with

$$\xi_1 < \cdots < \xi_n < \mu < \eta_1 < \cdots < \eta_m.$$

By (R4) then, taking $\eta'_1 < \cdots < \eta'_m < \mu$ in I_λ with $\xi_n < \eta'_1$, we have

$$\mu = \tau^\lambda_\varphi(\xi_1, \ldots, \xi_n, \eta_1, \ldots, \eta_m) = \tau^\lambda_\varphi(\xi_1, \ldots, \xi_n, \eta'_1, \ldots, \eta'_m)$$

and hence by (R3)

$$\mu = \tau^\lambda_\varphi(\xi_1, \ldots, \xi_n, \eta'_1, \ldots, \eta'_m) < \mu$$

which is absurd.

Lemma 2. If $\lambda < \varkappa$ are limit ordinals, then $\lambda^* \in I_\varkappa$,

$$I_\lambda = I_\varkappa \cap \lambda^*$$

and $L_{\lambda^*}[\alpha]$ is an elementary submodel of $L_{\varkappa^*}[\alpha]$.

Proof. Let

$$f: I_\lambda \rightarrowtail I_\varkappa$$

be the unique order-preserving map of I_λ onto an initial segment J of $I_\varkappa$, say

$$J = I_\varkappa \cap \mu$$

where $\mu \in I_\varkappa$ since $I_\varkappa$ is closed. Let J^* be the elementary substructure of $L_{\varkappa^*}[\alpha]$ generated by J, so that J is homogeneous in J^* and $Char(J^*, J) = \Phi$. Finally by 8H.2, let

$$f^*: L_{\lambda^*}[\alpha] \rightarrowtail\!\!\!\!\rightarrow J^*$$

be the canonical isomorphism induced by f. We will prove that

$$(*) \qquad J^* = L_\mu[\alpha];$$

from this it follows immediately that f^* takes the ordinals below λ^* in an order-preserving fashion onto the ordinals below μ, so that f^* is the identity map on λ^*, $\mu = \lambda^*$ and

$$I_\lambda = J = I_\varkappa \cap \mu = I_\varkappa \cap \lambda^*.$$

Moreover, since the identity

$$f^*: L_{\lambda^*}[\alpha] \rightarrowtail L_{\varkappa^*}[\alpha]$$

is an elementary embedding by 8H.2, it follows that $L_{\lambda^*}[\alpha]$ is an elementary submodel of $L_{\varkappa^*}[\alpha]$.

To prove (*), choose $\psi(\mathbf{x}, \boldsymbol{\xi})$ in the language of $ZFL[\dot\alpha]$ so that for each $L_\zeta[\alpha] \models ZFL[\dot\alpha]$ and $x, \xi \in L_\zeta[\dot\alpha]$

$$x \in L_\xi[\alpha] \Leftrightarrow L_\zeta[\alpha] \models \psi(x, \xi)$$

and if $x \in J^*$, choose $\xi_1 < \cdots < \xi_n \in J$ and $\varphi(\mathbf{v}_0, \mathbf{v}_1, \ldots, \mathbf{v}_n)$ so that

$$x = \tau_\varphi^\varkappa(\xi_1, \ldots, \xi_n).$$

Now for some $\eta \in I_\varkappa$,

$$L_{\varkappa^*}[\alpha] \models \psi(\mu \mathbf{v}_0 \varphi(\mathbf{v}_0, \xi_1, \ldots, \xi_n), \eta)$$

and hence by homogeneity and the fact that $\xi_n < \mu \in I_\varkappa$,

$$L_{\varkappa^*}[\alpha] \models \psi(\mu \mathbf{v}_0 \varphi(\mathbf{v}_0, \xi_1, \ldots, \xi_n), \mu)$$

and $x \in L_\mu[\alpha]$.

On the other hand if $x \in L_\mu[\alpha]$ and

$$x = \tau_\varphi^\varkappa(\xi_1,\ldots,\xi_n,\eta_1,\ldots,\eta_m)$$

with $\xi_1 < \cdots < \xi_n$ in J and $\eta_1 < \cdots < \eta_m$ in $I_\varkappa$ above J, then

$$L_{\varkappa^*}[\alpha] \models \mu\mathbf{v}_0\varphi(\mathbf{v}_0,\xi_1,\ldots,\xi_n,\eta_1,\ldots,\eta_m) < \mu < \eta_1$$

so that by (R4), if $\eta_1' < \cdots < \eta_m' \in J$ with $\xi_n < \eta_1'$ we have

$$L_{\varkappa^*}[\alpha] \models \mu\mathbf{v}_0\varphi(\mathbf{v}_0,\xi_1,\ldots,\xi_n,\eta_1,\ldots,\eta_m) = \mu\mathbf{v}_0\varphi(\mathbf{v}_0,\xi_1,\ldots,\xi_n,\eta_1',\ldots,\eta_m'),$$

i.e.

$$x = \tau_\varphi^\varkappa(\xi_1,\ldots,\xi_n,\eta_1',\ldots,\eta_m') \in J^*.$$

Lemma 3. If λ is an uncountable cardinal, then

$$\lambda^* = \lambda.$$

Proof. Let $\varkappa$ be any limit ordinal above λ and assume towards a contradiction that

$$card(I_\varkappa \cap \lambda) \leq \zeta < \lambda$$

with ζ infinite. Now every ordinal $\xi < \lambda$ is of the form

$$\xi = \tau_\varphi^\varkappa(\xi_1,\ldots,\xi_n,\eta_1,\ldots,\eta_m)$$

with some $\varphi(\mathbf{v}_0,\mathbf{v}_1,\ldots,\mathbf{v}_n,\mathbf{v}_{n+1},\ldots,\mathbf{v}_{m+1})$ and $\xi_1 < \cdots < \xi_n$ in $I_\varkappa \cap \lambda$, $\eta_1 < \cdots < \eta_m$ in $I_\varkappa$ above λ and by (R4), the value of $\tau_\varphi^\varkappa(\xi_1,\ldots,\xi_n,\eta_1,\ldots,\eta_m)$ depends only on the tuple

$$\langle \varphi(\mathbf{v}_0,\mathbf{v}_1,\ldots,\mathbf{v}_n,\mathbf{v}_{n+m}),\xi_1,\ldots,\xi_n \rangle,$$

since $\xi < \eta_1$. But there are only $\aleph_0 \cdot \zeta^n = \zeta$ such tuples for each n and hence only ζ such tuples altogether, which is absurd since there are $\lambda > \zeta$ ordinals below λ.

We now have $card(I_\varkappa \cap \lambda) = \lambda$ and hence the order type of $I_\varkappa \cap \lambda$ is exactly λ. Since $I_\varkappa \cap \lambda$ is an initial segment of $I_\varkappa \cap \lambda^*$ which also has order type λ, we must have $I_\varkappa \cap \lambda = I_\varkappa \cap \lambda^* = I_\lambda$ and since I_λ is cofinal in λ^*, we have $\lambda = \lambda^*$.

Now (i) and (ii) follow immediately with

$$I = \bigcup_\lambda I_\lambda.$$

If J is another class of ordinals which also satisfies (i) and (ii), then for each uncountable λ let

$$f : J \cap \lambda \rightarrowtail\!\!\!\!\twoheadrightarrow I \cap \lambda$$

be the unique order-preserving bijection and let

$$f^*: L_\lambda[\alpha] \rightarrowtail\!\!\!\!\twoheadrightarrow L_\lambda[\alpha]$$

be the canonical extension of f to an isomorphism. Clearly f^* is the identity on λ and hence f is the identity and

$$J \cap \lambda = I \cap \lambda.$$

Since λ was arbitrary uncountable, $J = I$.

Finally, if λ is an uncountable cardinal, then

$$\lambda = \lambda^* \in I$$

by Lemmas 2 and 3 and the characterization of Φ follows immediately. ⊣

If Φ is the unique remarkable character for α (assuming it exists), put

$$\alpha^{\#}(n) = \begin{cases} 1 & \text{if } n \text{ is the code of some formula } \varphi(\mathbf{v}_1, \ldots, \mathbf{v}_{k_n}) \in \Phi, \\ 0 & \text{otherwise,} \end{cases}$$

where of course, we are using the recursive coding of formulas of Section 8A. It is convenient (and traditional) to treat the unrelativized case of L as a special case of this, noticing that

$$L = L[t \mapsto 0]$$

and putting

$$0^{\#} = (t \mapsto 0)^{\#},$$

assuming of course, that there exists a remarkable character for the constant function $t \mapsto 0$. It is also traditional to abbreviate the ponderous-sounding hypothesis

"there exists a remarkable character for α"

by the simpler if somewhat sloppier

"$\alpha^{\#}$ exists".

By 8H.4 then, if there exists a measurable cardinal, then $(\forall\alpha)[\alpha^{\#}$ exists].

8H.6. THEOREM (Solovay [1967]). *The relation*

$$P(\alpha, \beta) \Leftrightarrow \alpha^{\#} \text{ exists and } \beta = \alpha^{\#}$$

is Π^1_2.

Thus, if $\alpha^{\#}$ exists, then

$$\gamma \in L[\alpha] \Rightarrow \gamma \textit{ is recursive in } \alpha^{\#} \Rightarrow \gamma \in \Delta_3^1(\alpha)$$

and $\alpha^{\#}$ is an irrational in $\Delta_3^1(\alpha)$ which is not in $L[\alpha]$.

PROOF. Compute,

$$\begin{aligned} P(\alpha, \beta) \Leftrightarrow (\forall n)[\beta(n) \le 1 \;\&\; [\beta(n) = 1 \Rightarrow (&n \text{ is the code of some} \\ &\text{formula } \varphi(\mathbf{v}_1,\dots,\mathbf{v}_{k_n}) \text{ in the} \\ &\text{language of } ZFL[\dot{\alpha}])]] \\ \&\text{ if } \Phi_\beta = \{\varphi(\mathbf{v}_1,\dots,\mathbf{v}_k)&: \text{if } n \text{ is the code of} \\ \varphi(\mathbf{v}_1,\dots,\mathbf{v}_k)&, \text{ then } \beta(n) = 1\}, \text{ then} \\ \Phi_\beta \text{ satisfies (R1)–(R5) in the}& \text{ definition of} \\ \text{“remarkable character”.}& \end{aligned}$$

It is now obvious that

$$P_1(\alpha, \beta) \Leftrightarrow \Phi_\beta \text{ satisfies (R1)}$$

is Σ_1^1. (Actually P_1 is arithmetical, but the proof of this requires Gödel's *Completeness Theorem* from logic which we have not proved.)

Similarly,

$$P_2(\alpha, \beta) \Leftrightarrow \Phi_\beta \text{ satisfies (R2), (R3) and (R4)}$$

is obviously arithmetical. Thus to complete the proof, it will be enough to find a Π_2^1 relation $P_3(\alpha, \beta)$ such that

$$\begin{aligned} &P_1(\alpha, \beta) \;\&\; P_2(\alpha, \beta) \\ &\qquad \Rightarrow [P_3(\alpha, \beta) \Leftrightarrow \Phi_\beta \text{ satisfies (R5)}] \end{aligned}$$

and for this we use Theorem 8H.2.

From that result we know that if $P_1(\alpha, \beta)$ hold, then for each countable ordinal λ there exists a pair $(\mathfrak{A}_\lambda, I_\lambda)$ with certain properties, that any two such pairs are isomorphic and that (R5) holds exactly when in every such pair, $\mathfrak{A}_\lambda$ is a wellfounded model. Using the irrational codes for structures with domain in ω that we introduced in the exercises of 8A, put

$$\begin{aligned} P_4(\alpha, \beta, \gamma, \delta) \Leftrightarrow P_1(\alpha, \beta) \ \&\ & P_2(\alpha, \beta) \\ &\ \&\ Str(\langle 2, 2\rangle, \gamma) \\ &\ \&\ [\text{the set } I_\delta = \{n: \delta(n) = 1\} \text{ is homogeneous in} \\ &\qquad \mathfrak{A}(\langle 2, 2\rangle, \gamma) \text{ and generates } \mathfrak{A}(\langle 2, 2\rangle, \gamma)] \\ &\ \&\ Char(\mathfrak{A}(\langle 2, 2\rangle, \gamma), I_\delta) = \Phi_\beta \end{aligned}$$

and compute easily that P_4 is Σ^1_1, using the fact that the satisfaction relation is Δ^1_1, 8A.6. But then the relation

$$\begin{aligned} P_5(\alpha, \beta, \gamma, \delta, \varepsilon) \Leftrightarrow P_4(\alpha, \beta, \gamma, \delta) &\\ &\ \&\ \varepsilon \in WO \\ &\ \&\ [\text{the order-type of } I_\delta \text{ under the canonical} \\ &\qquad \text{ordering of } \mathfrak{A}(\langle 2, 2\rangle, \gamma) \text{ is } |\varepsilon|] \end{aligned}$$

is easily Δ^1_2 and hence

$$\begin{aligned} P_3(\alpha, \beta) \Leftrightarrow P_1(\alpha, \beta) \ \&\ P_2(\alpha, \beta) &\\ &\ \&\ (\forall \varepsilon)(\forall \gamma)(\forall \delta)[P_5(\alpha, \beta, \gamma, \delta, \varepsilon) \\ &\qquad \rightarrow [\text{the canonical ordering of} \\ &\qquad \mathfrak{A}(\langle 2, 2\rangle, \gamma) \text{ is a wellordering}] \end{aligned}$$

is in Π^1_2 and the proof is complete.

If $\alpha^{\#}$ exists, then obviously

$$\begin{aligned} \alpha^{\#}(n) = i &\Leftrightarrow (\exists \beta)[P(\alpha, \beta) \ \&\ \beta(n) = i] \\ &\Leftrightarrow (\forall \beta)[P(\alpha, \beta) \rightarrow \beta(n) = i] \end{aligned}$$

so that $\alpha^{\#} \in \Delta^1_3(\alpha)$.

Also, if $\gamma \in L[\alpha]$, then

$$\gamma = [\mu \mathbf{v}_0 \varphi(\mathbf{v}_0, \xi_1, \ldots, \xi_n)]^{L_\varkappa[\alpha]}$$

with any $\varkappa \geq \aleph_1$, $\varkappa$ a cardinal and $\xi_1 < \cdots < \xi_n$ in the canonical homogeneous set $I \cap \varkappa$ for $L_\varkappa[\alpha]$, so that if Φ is the remarkable character for α,

$$\begin{aligned} \gamma(s) = t \Leftrightarrow &\text{ the sentence} \\ &(\exists \mathbf{v}_0)[\varphi(\mathbf{v}_0, \mathbf{v}_1, \ldots, \mathbf{v}_n) \ \&\ (\forall \mathbf{x} < \mathbf{v}_0) \neg \varphi(\mathbf{x}, \mathbf{v}_1, \ldots, \mathbf{v}_n) \\ &\qquad \&\ \text{“}\langle s, t\rangle \in \mathbf{v}_0\text{”}] \\ &\text{is in } \Phi \end{aligned}$$

where of course,

$$\text{“}\langle s, t\rangle \in \mathbf{v}_0\text{”}$$

is some formula with code recursively obtained from s, t and which expresses that $\langle s, t\rangle \in \mathbf{v}_0$. Thus γ is clearly recursive in $\alpha^{\#}$. ⊣

8H.7. COROLLARY. *If $\alpha^{\#}$ exists, then card$(\mathcal{N} \cap L[\alpha]) = \aleph_0$; in particular, if there exists a measurable cardinal, then card$(\mathcal{N} \cap L[\alpha]) = \aleph_0$. (Silver [1971]; Rowbottom [1971] for the inference from the existence of a measurable cardinal.)* ⊣

Assuming that $(\forall \alpha)[\alpha^{\#}$ exists], let I_α be the canonical, closed and unbounded class of *Silver indiscernibles for* $L[\alpha]$ supplied by 8H.5 and let

$$I^* = \bigcap \{I_\alpha : \alpha \in \mathcal{N}\}$$

be the class of *uniform indiscernibles.*

8H.8. LEMMA. *Assume $(\forall \alpha)[\alpha^{\#}$ exists].*

(i) *There is a fixed formula $\varphi(\boldsymbol{\alpha}, \boldsymbol{\xi})$ of $\mathcal{L}^{\epsilon}$ such that for every inner model M of ZF such that $\alpha^{\#} \in M$,*

$$\xi \in I_\alpha \leftrightarrow M \models \varphi(\alpha, \xi).$$

(ii) *If $\alpha \in L[\beta]$, then $I_\beta \subseteq I_\alpha$.*

(iii) *The class I^* of uniform indiscernibles is closed and unbounded, it contains all uncountable cardinals and for any $\alpha \in \mathcal{N}$ and any $u \in I^*$, the order-type of $I_\alpha \cap u$ is u.*

PROOF. (i) Using 8H.6, we easily get a definition of $\alpha^{\#}$ in terms of α and which is absolute for all inner models of *ZF* which contain $\alpha^{\#}$; again, using the construction in 8H.5 we can get a definition of I_α from $\alpha^{\#}$ which is absolute for all inner models of *ZF*.

(ii) If $\alpha \in L[\beta]$, then $\alpha = [\mu \mathbf{v}_0 \varphi(\mathbf{v}_0, \xi_1, \ldots, \xi_n)]^{L[\beta]}$ for some $\varphi(\mathbf{v}_0, \mathbf{v}_1, \ldots, \mathbf{v}_n)$ and some $\xi_1 < \cdots < \xi_n$ in I_β and then using (R4), easily,

$$\alpha = [\mu \mathbf{v}_0 \varphi(\mathbf{v}_0, \xi_1, \ldots, \xi_n)]^{L[\beta]}$$

for some $\varphi(\mathbf{v}_0, \mathbf{v}_1, \ldots, \mathbf{v}_n)$ and *any* increasing sequence $\xi_1 < \cdots < \xi_n$ in I_β. Using this and an induction on formulas, we easily assign to each $\psi(\mathbf{x}_1, \ldots, \mathbf{x}_k)$ in the language of $ZFL[\dot{\alpha}]$ another formula $\psi^*(\mathbf{x}_1, \ldots, \mathbf{x}_k, \mathbf{y}_1, \ldots, \mathbf{y}_n)$ so that for $x_1, \ldots, x_k \in L[\alpha]$ and any $\xi_1 < \cdots < \xi_n$ in

I_β,

$$L[\alpha] \models \psi(x_1, \ldots, x_k) \leftrightarrow L[\beta] \models \psi^*(x_1, \ldots, x_k, \xi_1, \ldots, \xi_n),$$

Thus for any ordinals $\eta_1, \ldots, \eta_k$,

$$(*) \qquad [\mu \mathbf{v}_0 \psi(\mathbf{v}_0, \eta_1, \ldots, \eta_k)]^{L[\alpha]} = [\mu \mathbf{v}_0 \psi^*(\mathbf{v}_0, \eta_1, \ldots, \eta_k, \xi_1, \ldots, \xi_n)]^{L[\beta]}.$$

where $\xi_1 < \cdots < \xi_n$ are any cardinals above η_k.

Suppose now towards a contradiction that λ is the least ordinal in $I_\beta - I_\alpha$ so that

$$\lambda = [\mu \mathbf{v}_0 \chi(\mathbf{v}_0, \eta_1, \ldots, \eta_l, \eta_1', \ldots, \eta_m')]^{L[\alpha]}$$

with $\eta_1 < \cdots < \eta_l < \lambda$ in I_α (and hence in I_β) and $\eta_1' < \cdots < \eta_m'$ in I_α above λ. Using (R4) we may replace the η_i' by cardinals above λ, so they too are in I_β and then using $(*)$ with cardinals $\xi_1 < \cdots < \xi_n$ above η_m' we have

$$(**) \qquad \lambda = [\mu \mathbf{v}_0 \chi^*(\mathbf{v}_0, \eta_1, \ldots, \eta_l, \eta_1', \ldots, \eta_m', \xi_1, \ldots, \xi_n)]^{L[\beta]} < \eta_1'.$$

Since $\lambda \in I_\beta$, by (R4) we then have

$$\lambda' = [\mu \mathbf{v}_0 \chi^*(\mathbf{v}_0, \eta_1, \ldots, \eta_l, \lambda, \ldots, \eta_m', \xi_1, \ldots, \xi_n)]^{L[\beta]} < \lambda$$

which is absurd.

(iii) Let $\{c_\xi^\alpha\}_{\xi \in ON}$ be an increasing inumeration of I_α and let $\lambda < u \in I^*$. By (i), the operation

$$\xi \mapsto c_\xi^\alpha$$

is definable from α in $L[\alpha^\#]$, so that there is some formula $\psi(\mathbf{v}_0, \boldsymbol{\alpha}, \mathbf{v}_1)$ and

$$c_\lambda = [\mu \mathbf{v}_0 \psi(\mathbf{v}_0, \alpha, \lambda)]^{L[\alpha^\#]} = \mathit{supremum}\{c_\xi^\alpha : \xi < \lambda\}$$

and then, since $u \in I_{\alpha^\#}$, easily $a_\lambda < u$. Thus there are λ distinct members of I_α below u for each $\lambda < u$ and hence the order-type of $I_\alpha \cap u$ is u.

The other parts of (iii) are trivial. ⊣

We now come to the Martin–Solovay theorem for Π_2^1 which we declared our aim in the beginning of this section. We let

$$u_1, u_2, u_3, \ldots, u_\omega$$

be the first $\omega + 1$ uniform indiscernibles so that (easily)

$$u_1 = \aleph_1$$

and for each n, $u_n \leq \aleph_n$.

8H.9. THEOREM. *If* $(\forall\alpha)[\alpha^{\#}$ *exists*], *then every* $\Pi^1_2(\alpha)$ *pointset admits a* $\Delta^1_3(\alpha)$*-scale into* u_ω. (*Martin–Solovay* [1969] *as improved by Mansfield* [1971] *and Martin* [19?a].)

PROOF. We will prove the result for a Π^1_2 set $P\subseteq\mathcal{N}$, since the relativized case is proved similarly and then the result follows for subsets of an arbitrary product space by 4E.6.

By 8F.8, there is a *ZF*-absolute operation

$$\lambda \mapsto T(\lambda)$$

such that for each infinite ordinal λ, $T(\lambda)$ is a tree on $\omega\times\lambda$ and for $\lambda\geq\aleph_1$,

$$P(\alpha)\Leftrightarrow T(\lambda,\alpha) \text{ is wellfounded,}$$

where $T(\lambda,\alpha)$ is the tree on λ determined by α,

$$T(\lambda,\alpha)=\{(\eta_0,\ldots,\eta_{k-1}):\eta_0,\ldots,\eta_{k-1}<\lambda \\ \& \ (\alpha(0),\eta_0,\ldots,\alpha(k-1),\eta_{k-1})\in T(\lambda)\}.$$

For each $(\eta_0,\ldots,\eta_{k-1})$ below λ, we define *the rank of* $T(\lambda,\alpha)$ *at* $(\eta_0,\ldots,\eta_{k-1})$ as in 2D,

$$\rho(T(\lambda,\alpha),(\eta_0,\ldots,\eta_{k-1})) \\ =\mathit{supremum}\{\rho(T(\lambda,\alpha),(\eta_0,\ldots,\eta_{k-1},\eta))+1:\eta<\lambda\}.$$

Let $\varphi_0(\mathbf{v}_0,\mathbf{v}_1,\ldots,\mathbf{v}_{k(0)})$, $\varphi_1(\mathbf{v}_0,\mathbf{v}_1,\ldots,\mathbf{v}_{k(1)}),\ldots$ be an enumeration of all formulas in the language of set theory such that

$$L\models\text{“}\mu\mathbf{v}_0\varphi_n(\mathbf{v}_0,\aleph_1,\ldots,\aleph_{k(n)}) \text{ is a sequence of ordinals”}.$$

It is clear that we can enumerate these formulas recursively in $0^{\#}$ and that if I_0 is the class of Silver indiscernibles for $t\mapsto 0$ and λ is any limit point of I_0, then the following two things hold.

(1) If $\xi_1<\cdots<\xi_{k(n)}<\lambda$ are all in I_0, then

$$\tau^L_n(\xi_1,\ldots,\xi_{k(n)})=[\mu\mathbf{v}_0\varphi_n(\mathbf{v}_0,\xi_1,\ldots,\xi_{k(n)})]^L$$

is a finite sequence of ordinals below λ.

(2) If $(\eta_0,\ldots,\eta_{m-1})$ is any sequence of ordinals below λ, then

$$(\eta_0,\ldots,\eta_{m-1})=\tau^L_n(\xi_1,\ldots,\xi_{k(n)})$$

for some n and some $\xi_1<\cdots<\xi_{k(n)}$ below λ in I_0.

We now define a sequence of norms on P by

$$\psi_n(\alpha)=\rho(T(u_{k(n)+1},\alpha),\tau^L_n(u_1,\ldots,u_{k(n)}))$$

and we claim that $\bar{\psi} = \{\psi_n\}$ is the required Δ_3^1-scale on P. The motivation for this definition will become clear from the argument to follow.

To check first that $\bar{\psi}$ is a scale on P, suppose $\alpha_0, \alpha_1, \ldots$ are all in P, $limit_{i\to\infty} \alpha_i = \alpha$ and for each n and all large enough i,

$$\psi_n(\alpha_i) = \lambda_n;$$

we must show that $\alpha \in P$ and for each n, $\psi_n(\alpha) \le \lambda_n$.

Let β be any irrational such that each α_i is recursive in β and by the lemma

$$I_\beta \subseteq I_{\alpha_i}$$

and fix an increasing enumeration $\{c(\xi) : \xi \in ON\}$ of I_β. By the lemma,

$$\xi < u_\omega \Rightarrow c(\xi) < u_\omega.$$

Since the relation

$$\rho(T(\lambda, \alpha_i), \tau_n^L(x_1, \ldots, x_{k(n)})) = \rho(T(\lambda, \alpha_j), \tau_n^L(x_1, \ldots, x_{k(n)}))$$

is easily definable in $L[\beta]$ and since it is true for all large enough i and j when

$$x_1 = u_1, \ldots, x_{k(n)} = u_{k(n)}, \qquad \lambda = u_{k(n)+1},$$

it is also true for all large enough i and j when

$$x_1 = c(\xi_1), \ldots, x_{k(n)} = c(\xi_n), \qquad \lambda = u_\omega$$

for any increasing sequence $\xi_1 < \cdots < \xi_{k(n)}$ of ordinals below u_ω.

For each sequence $v = (\eta_0, \ldots, \eta_{m-1})$ in $T(u_\omega, \alpha)$, choose n and $\xi_1 < \cdots < \xi_{k(n)}$ in I_0 such that

$$v = \tau_n^L(\xi_1, \ldots, \xi_{k(n)})$$

and let

$$f(v) = limit_{i\to\infty} \rho(T(u_\omega, \alpha_i), \tau_n^L(c(\xi_1), \ldots, c(\xi_{k(n)}))).$$

The definition of $f(v)$ is independent of the choice of n and $\xi_1 < \cdots < \xi_{k(n)}$ in I_0, since

$$(*) \qquad \tau_n^L(\xi_1, \ldots, \xi_{k(n)}) = \tau_m^L(\zeta_1, \ldots, \zeta_{k(m)})$$

is a formal assertion in L about the indiscernibles $\xi_1, \ldots, \xi_{k(n)}, \zeta_1, \ldots, \zeta_{k(m)}$ and the ordinals $c(\xi_1), \ldots, c(\xi_{k(n)}), c(\zeta_1), \ldots, c(\zeta_{k(m)})$ are also indiscernibles and with the same ordering, so (*) implies

$$\tau_n^L(c(\xi_1), \ldots, c(\xi_{k(n)})) = \tau_m^L(c(\zeta_1), \ldots, c(\zeta_{k(m)})).$$

Also

$$(\eta_0,\ldots,\eta_{m-1})\in T(u_\omega,\alpha)\Leftrightarrow(\alpha(0),\eta_0,\ldots,\alpha(m-1),\eta_{m-1})\in T(u_\omega)$$

so that the assertion

$$(**)\qquad \tau_n^L(\xi_1,\ldots,\xi_{k(n)})\in T(u_\omega,\alpha)$$

depends only on the first m values of α, it is true for α_i when i is large enough and it is a formal assertion in L about the indiscernibles $\xi_1,\ldots,\xi_{k(n)}$, u_ω, so that (**) implies

$$\tau_n^L(c(\xi_1),\ldots,c(\xi_{k(n)}))\in T(u_\omega,\alpha_i)$$

for all large i. Thus—and this is the key point—*if $v=\tau_n^L(\xi_1,\ldots,\xi_{k(n)})$ and $v'=\tau_m^L(\xi_1',\ldots,\xi_{k(m)}')$ are both sequences in $T(u_\omega,\alpha)$ and v is an initial segment of v', then for all large i, $\tau_n^L(c(\xi_1),\ldots,c(\xi_{k(n)}))$, $\tau_m^L(c(\xi_1'),\ldots,c(\xi_{k(m)}'))$ are both sequences in $T(u_\omega,\alpha_i)$ and the first of these is an initial segment of the second*; it follows that for all large i

$$\rho(T(u_\omega,\alpha_i),\tau_n^L(c(\xi_1),\ldots,c(\xi_{k(n)})))>\rho(T(u_\omega,\alpha_i),\tau_m^L(c(\xi_1'),\ldots,c(\xi_{k(m)}'))).$$

At this point we have produced a map

$$(1)\qquad f:T(u_\omega,\alpha)\to \textit{Ordinals}$$

such that

$$(2)\qquad \text{if } v \text{ is an initial segment of } v', \text{ then } f(v)>f(v').$$

It is immediate then that $T(u_\omega,\alpha)$ is wellfounded and hence $\alpha\in P$. Moreover (easily, using just (1) and (2))

$$\rho(T(u_\omega,\alpha),v)\le f(v)$$

for each $v\in T(u_\omega,\alpha)$, so writing out what this means: for arbitrary n, $\xi_1<\cdots<\xi_{k(n)}$ in I_0 and all large i,

$$\rho(T(u_\omega,\alpha),\tau_n^L(\xi_1,\ldots,\xi_{k(n)}))\le\rho(T(u_\omega,\alpha_i),\tau_n^L(c(\xi_1),\ldots,c(\xi_{k(n)}))).$$

By the lemma now,

$$c(u_n)=u_n.$$

so for all n and all large i

$$\rho(T(u_\omega,\alpha),\tau_n^L(u_1,\ldots,u_{k(n)}))\le\rho(T(u_\omega,\alpha_i),\tau_n^L(u_1,\ldots,u_{k(n)}))$$

and replacing u_ω by $u_{k(n)+1}$, we have finally for all large i,

$$\begin{aligned}\psi_n(\alpha) &= \rho(T(u_{k(n)+1}, \alpha), \tau_n^L(u_1, \ldots, u_{k(n)}))\\ &\leq \rho(T(u_{k(n)+1}, \alpha_i), \tau_n^L(u_1, \ldots, u_{k(n)})) = \lambda_n\end{aligned}$$

as required.

To verify that $\bar{\psi}$ is a Δ_3^1-scale on P, we argue as in the proof of (ii) of 8H.8 (and using the fact that $\lambda \mapsto T(\lambda)$ is ZF-absolute) that

$$\begin{aligned}\rho(T(\lambda, \alpha), \tau_n^L(\xi_1, \ldots, \xi_{k(n)})) &\leq \rho(T(\lambda, \beta), \tau_n^L(\xi_1, \ldots, \xi_{k(n)}))\\ &\Leftrightarrow L[\langle\alpha, \beta\rangle] \models \varphi(\xi_1, \ldots, \xi_{k(n)}, \lambda)\end{aligned}$$

where $\varphi(\mathbf{v}_1, \ldots, \mathbf{v}_{k(n)}, \mathbf{v}_{k(n)+1})$ is some fixed formula in the language of $ZFL[\dot{\alpha}]$. Thus

$$\begin{aligned}\alpha \leq^*_{\psi_n} \beta \Leftrightarrow\ & P(\alpha)\ \&\ \neg P(\beta)\\ & \vee [P(\alpha)\ \&\ P(\beta)\ \&\ \text{if } m \text{ is the code of}\\ & \quad \varphi(\mathbf{v}_1, \ldots, \mathbf{v}_{k(n)}, \mathbf{v}_{k(n)+1}), \text{ then}\\ & \quad \langle\alpha, \beta\rangle^{\#}(m) = 1]\end{aligned}$$

and the relation on the right is obviously in Δ_3^1. ⊣

Let us collect the corollaries.

8H.10. COROLLARY. *If $(\forall\alpha)[\alpha^{\#}$ exists], then every Π_2^1 set can be uniformized by a Π_3^1 set and every nonempty Σ_3^1 set has a Δ_4^1 member. (Martin–Solovay* [1969] *as improved by Mansfield* [1971].)

PROOF is immediate by 8H.9 and 4E.3, easily. ⊣

8H.11. COROLLARY. *Assume $AD+DC$; then*

$$\begin{aligned}&u_\omega = \aleph_\omega,\\ &\boldsymbol{\delta}_3^1 = \aleph_{\omega+1}, \boldsymbol{\delta}_4^1 = \aleph_{\omega+2},\end{aligned}$$

every $\boldsymbol{\Sigma}_3^1$ set is $\aleph_\omega$-Suslin and then every $\boldsymbol{\Sigma}_4^1$-set is $\aleph_{\omega+1}$-Suslin. (Martin [19?a]; *for the parts about $\boldsymbol{\delta}_4^1$ and $\boldsymbol{\Sigma}_4^1$ also Kunen.)*

PROOF. Clearly $u_\omega \leq \aleph_\omega$ since $\aleph_1, \aleph_2, \ldots, \aleph_\omega$ are all uniform indiscernibles and every $\boldsymbol{\Sigma}_3^1$ set is u_ω-Suslin. For the converse, assume towards a contradiction that $u_\omega < \aleph_\omega$, let n be the least integer such that every $\boldsymbol{\Sigma}_3^1$ set is $\aleph_n$-Suslin and notice that by 8H.9, $\aleph_n$ is certainly the order-type of a $\mathbf{\Delta}^1_3$ prewellordering, since $\aleph_n$ is $\leq$ the length of the scale on u_ω we constructed. Now the choiceless version of 2F.4 explained in 7F and 7D.9 implies that every $\mathbf{\Sigma}^1_3$ set is $\mathbf{\Delta}^1_3$ which is absurd.

From $u_\omega = \aleph_\omega$, the Kunen–Martin Theorem 2G.2 and 7D.8 imply immediately that $\boldsymbol{\delta}^1_3 = \aleph_{\omega+1}$, $\boldsymbol{\delta}^1_4 = \aleph_{\omega+2}$ and that every $\mathbf{\Sigma}^1_4$ set is $\aleph_{\omega+1}$-Suslin follows from 6C.2 and 6C.4. ⊣

Solovay and Kunen have shown that actually

$$u_n = \aleph_n \quad n = 1, 2, 3, \ldots$$

granting *AD*, but we will not prove this here; see Kechris [1978] or Kleinberg [1977]. We need another of Solovay's lemmas however, to derive the corollaries of 8H.9 which depend on the axiom of choice.

8H.12. LEMMA (Solovay). *If* $(\forall \alpha)[\alpha^\# \text{ exists}]$, *then for each* n,

$$\textit{cofinality}(u_{n+1}) = \textit{cofinality}(u_2) \leq \aleph_2.$$

In particular, if the axiom of choice holds, then $u_3, u_4, \ldots$ *are all singular and* $u_\omega < \aleph_3$.

PROOF. It is easy to show by induction on ξ, that for each $\xi < u_\omega$ there is some α and some $\varphi(\mathbf{v}_0, \mathbf{v}_1, \ldots, \mathbf{v}_n)$ in the language of $ZFL[\dot{\alpha}]$ such that for any cardinal $\varkappa \geq u_\omega$,

$$\xi = [\mu \mathbf{v}_0 \varphi(\mathbf{v}_0, u_1, \ldots, u_n)]^{L_\varkappa[\alpha]};$$

assuming the result for all $\xi' < \xi$ and assuming $\xi \neq u_n$, then

$$\xi = [\mu \mathbf{v}_0 \psi(\mathbf{v}_0, \xi_1, \ldots, \xi_n, \eta_1, \ldots, \eta_m)]^{L_\varkappa[\alpha]}$$

with $\xi_1 < \cdots < \xi_n < \xi < \eta_1 < \cdots < \eta_m$ and the ξ_i, η_j in I_α, each ξ_i is definable in some $L_\varkappa[\alpha_i]$ in terms of the uniform indiscernibles, we can substitute larger uniform indiscernibles for the η_j and ξ is easily definable from uniform indiscernibles in any $L_\varkappa[\beta]$ where $\alpha, \alpha_1, \ldots, \alpha_n$ are all recursive in β.

Now put for each ordinal λ

$$\textit{next}(\lambda, \alpha) = \text{the least member of } I_\alpha \text{ greater than } \lambda$$

and notice that for each n, α by (iii) of 8H.8

$$u_n < \textit{next}(u_n, \alpha) < u_{n+1}.$$

Claim:

$$u_{n+1} = supremum\{next(u_n, \alpha) : \alpha \in \mathcal{N}\}.$$

To prove this suppose

$$\lambda = supremum\{next(u_n, \alpha) : \alpha \in \mathcal{N}\} < u_{n+1}$$

so that λ is not a uniform indiscernible and for some α,

$$\lambda = [\mu \mathbf{v}_0 \varphi(\mathbf{v}_0, u_1, \ldots, u_n, u_{n+1}, \ldots, u_k)]^{L[\alpha]} < u_{n+1}$$

by what we proved above. But we can then substitute $next(u_n, \alpha)$ for u_{n+1} in this formula and we get $\lambda < next(u_n, \alpha)$ which is absurd.

Finally, notice that

$$next(u_1, \alpha) = next(u_1, \beta) \Rightarrow next(u_n, \alpha) = next(u_n, \beta)$$

since the assertion on the left can be expressed formally as

$$L[\langle \alpha, \beta \rangle^{\#}] \models \varphi(u_1)$$

with a formula $\varphi(\mathbf{v}_1)$ in the language of $ZFL[\dot{\alpha}]$ and so we should also have

$$L[\langle \alpha, \beta \rangle^{\#}] \models \varphi(u_n).$$

Thus the set of ordinals

$$A = \{\xi < u_2 : \text{for some } \alpha,\ \xi = next(u_1, \alpha)\}$$

is cofinal in u_2 and the map that sends

$$\xi = next(u_1, \alpha)$$

to

$$f(\xi) = next(u_n, \alpha)$$

is well-defined and established that $cofinality(u_{n+1}) = cofinality(u_2)$.

If the axiom of choice holds, then $\aleph_3$ is a regular cardinal so we must have $u_n < \aleph_3$ for each n and finally $u_\omega < \aleph_3$, since u_ω has cofinality ω. ⊣

8H.13. COROLLARY (Martin [19?a]). *If* $(\forall \alpha)[\alpha^{\#}$ *exists*] *and the axiom of choice holds, then*

$$\boldsymbol{\delta}_3^1 \leq \aleph_3$$

and every $\boldsymbol{\Sigma}_3^1$ *set is the union of* $\aleph_2$ *Borel sets. If in addition* $Det(\boldsymbol{\Delta}_2^1)$ *holds, then*

$$\boldsymbol{\delta}^1_4 \leq \aleph_4$$

and every $\boldsymbol{\Sigma}^1_4$ *set is the union of* $\aleph_3$ *Borel sets.*

PROOF. Since $u_\omega < \aleph_3$, $card(u_\omega) = card(u_2) \leq \aleph_2$ and hence by 8H.9, every $\boldsymbol{\Sigma}^1_3$ set is $\aleph_2$-Suslin; now the Kunen–Martin Theorem 2G.2 and 2F.4 imply the results about $\boldsymbol{\delta}^1_3$ and $\boldsymbol{\Sigma}^1_3$. If $Det(\boldsymbol{\Delta}^1_2)$ holds we also know that every $\boldsymbol{\Sigma}^1_4$ set is $\boldsymbol{\delta}^1_3$-Suslin, hence $\aleph_3$-Suslin, and again 2G.2 and 2F.4 apply. ⊣

A good deal of effort has gone into attempts to generalize these beautiful results about Π^1_2, Σ^1_3 and Σ^1_4 to the higher levels of the projective hierarchy, but without success so far, although there has been some progress; see Kechris [1978] and Solovay [1978b] in particular where earlier work of Kunen is also described. The obvious obstruction is that the proofs in this section depend on very special properties of L which we do not know how to extend to the higher analogs of L described in 8G—and there are even doubts whether these models are the correct vehicles for generalizing the present arguments.

EXERCISES

8H.14. Prove that if $\alpha^\#$ exists and I_α is the class of Silver indiscernibles for α, then for any uncountable cardinal $\varkappa$ and $\xi_1 < \cdots < \xi_n < \varkappa$ in I_α and for any formulas $\varphi(\mathbf{v}_1, \ldots, \mathbf{v}_n)$ in the language of $ZFL[\dot{\alpha}]$,

$$L_\varkappa[\alpha] \models \varphi(\xi_1, \ldots, \xi_n) \Leftrightarrow L[\alpha] \models \varphi(\xi_1, \ldots, \xi_n).$$

In particular, for each uncountable $\varkappa$, $L_\varkappa[\alpha]$ is an elementary substructure of $L[\alpha]$.

Hint. Use 8H.4 and the Reflection Theorem 8C.4. ⊣

8H.15. Prove that if there exists a Ramsey cardinal, then $(\forall\alpha)[\alpha^\#$ exists$]$. (Silver [1971].)

Hint. By $\varkappa \to (\aleph_1)$, easily there exists a set $I \subseteq \varkappa$ of order type $\aleph_1$ which is homogeneous in $L_\varkappa[\alpha]$. For each such I let

$$f(I) = \text{the } \omega\text{'th ordinal in an increasing enumeration of } I,$$

and let

$$\lambda^* = \mathit{infimum}\{f(I): I \text{ is homogeneous in } L_\varkappa[\alpha] \text{ with order-type } \aleph_1\}.$$

Finally check that if I is homogeneous with order type $\aleph_1$ and such that $f(I)=\lambda^*$, then $Char(L_\varkappa[\alpha], I)$ is remarkable. (For example in checking (R4), suppose that for some increasing sequence

$$\xi_1 < \cdots < \xi_n < \eta_1 < \cdots < \eta_m < \eta'_1 < \cdots < n'_m$$

of ordinals in I and some term $\tau_\varphi(\mathbf{v}_1, \ldots, \mathbf{v}_n, \mathbf{v}_{n+1}, \ldots, \mathbf{v}_{n+m})$ we have

$$\tau^\varkappa_\varphi(\xi_1, \ldots, \xi_n, \eta_1, \ldots, \eta_m) < \eta_1,$$

$$\tau^\varkappa_\varphi(\xi_1, \ldots, \xi_n, \eta_1, \ldots, \eta_m) < \tau^\varkappa_\varphi(\xi_1, \ldots, \xi_n, \eta'_1, \ldots, \eta'_m),$$

let $\{c_\xi: \xi < \aleph_1\}$ be an increasing enumeration of I and choose for each $\xi < \aleph_1$ an m-tuple $\bar{\eta}_\xi = \eta^\xi_1, \ldots, \eta^\xi_m$ from I such that

$$\xi < \zeta \Rightarrow \eta^\xi_m < \eta^\zeta_1, \qquad \xi < \omega \Rightarrow \eta^\xi_i = c_j$$

for some $j < n$ and $\bar{\eta}_\omega = c_\omega, c_{\omega+1}, \ldots, c_{\omega+m-1}$. Now let

$$J = \{\tau^\varkappa_\varphi(c_1, \ldots, c_n, \bar{\eta}_\xi): \xi < \aleph_1\}$$

and show that J is also homogeneous in $L_\varkappa[\alpha]$ of order-type $\aleph_1$ and ω'th member less than λ^*.) ⊣

8H.16. Show that

$$(\forall\alpha)[\alpha^\# \text{ exists}] \Rightarrow Det(\Sigma^1_1)$$

(Martin [1970].)

Hint. Go back to the proof of 6G.7 and suppose $A \subseteq \mathcal{N}$ is Σ^1_1, the relativized case having a similar proof. Define the auxilliary game A^* on any uncountable cardinal $\varkappa$, say $\varkappa = \aleph_1$ and check that the set of finite sequences which defines the open set A^* lies in L, since it has an absolute definition (in terms of $\varkappa$). Thus either I or II *wins the game* A^* *in* L and we must show that the same player wins A in V.

In the non-trivial case when I wins, suppose $\sigma^* \in L$ is a winning strategy for I within L, let $I_\varkappa = I \cap \varkappa$ be the Silver indiscernibles for L below $\varkappa$ and let

$$\sigma(a_0, a_1, \ldots, a_{2t-1}) = \sigma^*(a_0, (a_1, \xi_{n(1)}), a_2, (a_3, \xi_{n(2)}), \ldots, a_{2t-2}, (a_{2t-1}, \xi_{n(t)}))$$

where $\xi_{n(1)}, \ldots, \xi_{n(t)}$ are distinct indiscernibles in $I_\varkappa$ and ordered so that II has not lost in the position

$$\begin{array}{lllll} \text{I} & a_0 & & \cdots & a_{2t-2} \\ \text{II} & & a_1, \xi_{n(1)} & \cdots & & a_{2t-1}, \xi_{n(t)}. \end{array}$$

It is clear that the definition of σ is independent of the particular choice of $\xi_{n(1)}, \ldots, \xi_{n(t)}$ in I_κ, so we have a strategy for I in A. To show that it is winning argue by contradiction as in the proof of 6G.7: if α is the play in some run of A in which I follows σ but $\alpha \notin A$, then $\bigcup_t D(\bar{\alpha}(2t))$ is a wellordering of countable rank, so there is an order-preserving map

$$x_t \mapsto \xi_t \in J_\kappa$$

into the indiscernibles and in every position of the run of A^*

$$\begin{array}{lllll} \text{I} & a_0 & & a_2 & & \cdots \\ \text{II} & & a_1, \xi_1 & & a_3, \xi_2 & & \cdots \end{array}$$

I has not lost while I is following a winning strategy, which is absurd. ⊣

Improving on earlier results of Martin, Harrington [1978] has proved the converse of this, so that

$$(\forall \alpha)[\alpha^\# \text{ exists}] \Leftrightarrow Det(\mathbf{\Sigma}_1^1)$$

and the metamathematical hypothesis $(\forall \alpha)[\alpha^\# \text{ exists}]$ is equivalent to the very natural, game-theoretic hypothesis $Det(\mathbf{\Sigma}_1^1)$. Martin [19?c] extends this beautiful result and shows that for many natural pointclasses $\Lambda \subseteq \mathbf{\Delta}_2^1$, $Det(\Lambda)$ is equivalent to "the analytical content" of various large cardinal assumptions.

8I. Some remarks about strong hypotheses

The need to consider strong set theoretic hypotheses has been explained in the introduction to this monograph and again in Chapter 5 and in the introduction to Chapter 6: simply put, *ZFC* is just not strong enough to decide the most natural and basic questions about definable sets of real numbers. Here we will discuss very briefly some of the serious foundational questions which arise in the study of strong hypotheses.

There are essentially three propositions whose consequences we have considered in some detail, Gödel's axiom of constructibility ($V = L$), the assumption that measurable cardinals exist (*MC*) and various determinacy hypotheses among which the strongest (and most natural) is

$$Det\,L(\mathbb{R}) \Leftrightarrow \text{every game in } L(\mathbb{R}) \text{ is determined}$$

$$\Leftrightarrow L(\mathbb{R}) \models AD.$$

We take it for granted that neither $V=L$ nor $Det\,L(\mathbb{R})$ is "obviously true" on the basis of our current understanding of the notion of set.

MC is in a different category as it follows from strong "large cardinal axioms" whose truth can be supported by some a priori arguments, see Solovay–Reinhardt–Kanamori [1978]. If we accept these large cardinal axioms as (at least) highly plausible on the basis of their meaning, then the chief foundational problem of descriptive set theory becomes simply *to prove* the fruitful hypothesis $Det\,L(\mathbb{R})$; Martin [19?b] has recently broken new ground in this important program by showing that $Det(\mathbf{\Sigma}^1_2)$ follows from "the existence of a non-trivial, iterable elementary imbedding of some V_κ into itself."

We cannot do justice here to this foundational position that takes (some) large cardinal axioms as evident without going into the technical results of that part of set theory, so we will defer to the forthcoming Martin [19?c]. Suffice it to say that the surprising connection between large cardinals and determinacy hypotheses is quite fascinating independently of any philosophical considerations.

It is also fair to remark that not all set theorists accept the intrinsic plausibility of large cardinal axioms.

Going one step further, many set theorists do not adhere to the realistic approach towards mathematics which we have adopted throughout this book and are uncomfortable with references to the "truth" or "falsity" of complicated set theoretic propositions like $V=L$, *MC* and $Det\,L(\mathbb{R})$. Without embarking in one of those shallow and fruitless discussions of "formalism" versus "realism," we will make just a few remarks to help clarify our position.

The main point in favor of the realistic approach to mathematics is the instinctive certainty of most everybody who has ever tried to solve a problem that he is thinking about "real objects," whether they are sets, numbers, or whatever; and that these objects have intrinsic properties above and beyond the specific axioms about them on which he is basing his thinking for the moment. Nevertheless, most attempts to turn these strong feelings into a coherent foundation for mathematics invariably lead to vague discussions of "existence of abstract notions" which are quite repugnant to a mathematician.

Contrast with this the relative ease with which formalism can be explained in a precise, elegant and self-consistent manner and you will have the main reason why most mathematicians claim to be formalists (when pressed) while they spend their working hours behaving as if they were completely unabashed realists.

It is not unreasonable to accept naively that the universe of sets exists (and conforms substantially to the description we gave in the introduction to this chapter) and that we can reason about sets much as physicists reason about elementary particles or astronomers reason about stars—while conceding immediately that we are in no position now to make precise (or even talk eloquently about) the kind of "existence" we have in mind. And it is certainly natural (and useful) for a mathematician to behave *as if* the universe of sets existed and conformed to his common-sense ideas about sets.

One of the main features of this consciously naive realistic approach is that it forces us to abandon any claims of absolute certainty for our assertions about sets. Instead of "axioms" (taken as unassailable, by definition, in the formalist approach), we must speak of "hypotheses" to be tested against each other and against our basic intuitions about sets, perhaps to be adopted temporarily and discarded later in the light of new evidence. To be sure, we have a great deal of confidence in the truth of some propositions (like the axioms of *ZFC*) which appear to be evident on the basis of the (incomplete and vague) description of the universe of sets we gave in the introduction to this chapter; for others, we must weigh carefully the evidence before we accept them (tentatively) as true or we reject them.

The most serious foundational problem in this naive realistic approach to set theory is to determine what kind of evidence we may hope to find for favoring a hypothesis over its negation and how we should weigh such evidence. Before we comment briefly on this difficult question as it affects the hypotheses we have been studying, we should discuss the equally important problem of *consistency* for extensions of *ZF*.

In 8F.11 we proved the consistency of the theory $ZFC + V = L$ (within our standard framework of realistic set theory or even constructively from the formal consistency of the weaker theory *ZF* according to the remarks following 8F.11); can we do the same for $ZFC + MC$ or $ZFC + Det\,L(\mathfrak{R})$?

To formulate this question precisely in a general setting, suppose T is any *axiomatic set theory*, i.e. any set of sentences of $\mathcal{L}^{\in}$ such that the corresponding set of (number) codes

$$T^{\#} = \{[\chi]^{8} : \chi \in T\}$$

(as we defined these in 8A) is recursive. Since $T^{\#}$ is a definable set of numbers, we can talk about T within the language $\mathcal{L}^{\in}$. The precise definition of *formal proof from the axioms in* T to which we have often alluded is naturally effective, since we should be able to recognize when a sequence of assertions constitutes a correct proof. Proofs can be coded by numbers, so that for a given axiomatic set theory T, the relation

$$\begin{aligned} Proof_T(n, m) \Leftrightarrow\ & n \text{ is the code of some sentence } \chi \\ & \text{and } m \text{ is the code of a proof of} \\ & \chi \text{ from the axioms in } T \end{aligned}$$

is recursive and hence definable in $\mathcal{L}^{\in}$, in fact ZF-absolute. Suppose then that $\varphi_T(\mathbf{x}, \mathbf{y})$ is a formula of $\mathcal{L}^{\in}$ such that for all standard models M of some fragment of ZF,

$$Proof_T(n, m) \Leftrightarrow M \models \varphi_T(n, m),$$

let n_0 be the code of some self-contradictory assertion (like $(\exists \mathbf{z})[\mathbf{z} \neq \mathbf{z}]$) and put

$$Consis(T) \Leftrightarrow (\forall \mathbf{y}) \neg \varphi_T(n_0, \mathbf{y});$$

now this formal sentence $Consis(T)$ clearly expresses within $\mathcal{L}^{\in}$ (and ZF-absolutely) the consistency of the axiomatic theory T.

To simplify notation, put

$$\begin{aligned} T \vdash \psi \Leftrightarrow\ & \psi \text{ is a theorem of } T \\ \Leftrightarrow\ & \psi \text{ can be proved from the axioms of } T. \end{aligned}$$

By the remarks after the proof of 8F.11,

$$ZF \vdash Consis(ZF) \rightarrow Consis(ZFC + V = L).$$

Unfortunately, the corresponding *relative consistency* assertions for $ZFC + MC$ or $ZFC + Det\, L(\mathbb{R})$ cannot be proved, because of the following fundamental theorem of Gödel.

8I.1. THE SECOND INCOMPLETENESS THEOREM OF GÖDEL. *If T is a consistent axiomatic set theory at least as strong as ZF (i.e. $ZF \subseteq T$), then $Consis(T)$ is not a theorem of T.* ⊣

This is a very weak version of Gödel's celebrated theorem whose natural, general version does not refer explicitly to set theory—it expresses an inherent weakness of sufficiently strong axiomatic systems in any language. For our purposes here, however, this version is strong enough, in fact the following corollary suffices.

8I.2. COROLLARY. *Suppose T is a consistent axiomatic set theory at least as strong as ZF and such that*

$$T \vdash Consis(ZF);$$

then the assertion of relative consistency

$$Consis(ZF) \rightarrow Consis(T)$$

is not a theorem of T—a fortiori it is not a theorem of ZF.

PROOF. If both $Consis(ZF) \rightarrow Consis(T)$ and $Consis(ZF)$ were theorems of T, so would $Consis(T)$ be a theorem of T, contradicting the second incompleteness theorem. ⊣

8I.3. COROLLARY. *If $ZF + MC$ is consistent, then the relative consistency assertion*

$$Consis(ZF) \rightarrow Consis(ZF + MC)$$

cannot be proved in ZF; similarly, if $ZF + Det\, L(\mathcal{R})$ is consistent, then

$$Consis(ZF) \rightarrow Consis(ZF + Det\, L(\mathcal{R}))$$

cannot be proved in ZF.

OUTLINE OF PROOF. Consider first the easier case of the theory $ZFC + MC$ and recall from 6G.9 that (using choice) every measurable cardinal is strongly inaccessible, so that in $ZFC + MC$ we can prove the existence of a strongly inaccessible κ; but for such κ, V_κ is a model of ZF (by 8C.11) and the existence of a model for a theory implies by elementary means that the theory is consistent, so that

$$ZFC + MC \vdash Consis(ZF)$$

and 8I.2 applies.

For the subtler case without choice, recall from 8H.3 that if κ is measurable, then L_κ is a model of ZF; thus again

$$ZF + MC \vdash Consis(ZF)$$

and 8I.2 applies.

Finally, by 7D.18

$$ZF + AD \vdash MC,$$

so that

$$ZF + AD \vdash Consis(ZF)$$

and

$$Consis(ZF) \rightarrow Consis(ZF + AD)$$

cannot be established in ZF. But

$$Consis(ZF + Det\, L(\mathcal{R})) \rightarrow Consis(ZF + AD)$$

can be established easily in ZF once the notion of proof is made precise—by using the inner model $L(\mathcal{R})$ just as we use L to show the implication

$$Consis(ZF) \rightarrow Consis(ZF + V = L). \quad \dashv$$

Much stronger results are known in this direction, for example

$$Consis(ZF) \rightarrow Consis(ZF + Det(\Sigma^1_1))$$

cannot be established in ZFC and

$$Consis(ZFC + MC) \rightarrow Consis(ZF + Det(\Delta^1_2))$$

cannot be established in $ZFC + MC$, assuming that the theories involved are in fact consistent (the second result is due to Solovay).

Thus large cardinal and determinacy hypotheses are quite different from $V = L$ in this respect; if they are consistent, then their consistency (relative to ZF) cannot be established, unless we are willing to accept methods of proof which go beyond ZFC and which may be suspect themselves. What are we to make of this?

A formalist would conclude that the hypothesis of constructibility $V = L$ is by far the best way to strengthen ZF, as it is safe (from the point of view of consistency) and it answers completely (and almost trivially) all the interesting questions of descriptive set theory.

In the naive, realistic approach consistency is very weak evidence for truth—after all the theory $ZF + \neg Consis(ZF)$ is consistent (by the second incompleteness theorem) and it is obviously false. We must look for criteria other than consistency to evaluate the plausibility of a new hypothesis.

The key argument against accepting $V = L$ (or even $\mathcal{N} \subseteq L$) is that the axiom of constructibility appears to restrict unduly the notion of *arbitrary* set of integers; there is no a priori reason why every subset of ω should be definable from ordinal parameters, much less by an elementary definition over some countable L_ξ. Some would go further and claim disbelief that the real line can be *definably* wellordered on any rank—it is quite plausible that the only sets of reals which admit definable wellorderings are countable.

We are arguing here that there are some (perhaps weak) direct intuitions which make $V = L$ look implausible, just as there are some direct intuitions which lend credibility to large cardinal hypotheses like *MC*.

No one claims direct intuitions of this type either for or against determinacy hypotheses—those who have come to favor these hypotheses as plausible, argue from their consequences as we developed them in the last three chapters. In addition to the richness and internal harmony of these consequences, two aspects of the theory we have developed deserve explicit mention.

One is the *naturalness* of the proofs from determinacy—in each instance where we prove a property of Π^1_3 (say from $Det(\mathbf{\Delta}^1_2)$), the same argument gives a new proof of the same (known) property for Π^1_1, using only the determinacy of clopen sets (which is a theorem of *ZF*). Thus the new results appear to be natural generalizations of known results and their proofs shed new light on classical descriptive set theory. (This is not the case with the proofs from $V = L$ which all depend on the Σ^1_2-good wellordering of $\mathcal{N}$ and shed no light on Π^1_1.)

The second point is the surprising connection between determinacy and large cardinal hypotheses on which we have commented many times and which lends credence to both. To take one example, the fact that Π^1_2 sets can be uniformized by Π^1_3 sets follows both from *MC* and from $Det(\mathbf{\Delta}^1_2)$, by proofs which (at least on the surface) are totally unrelated; one tends to believe the result then and consequently to take both proofs seriously and to feel a little more sympathetic towards their respective hypotheses.

At the present state of knowledge only few set theorists accept $Det\, L(\mathcal{R})$ as highly plausible and no one is quite ready to believe it beyond a reasonable doubt; and it is certainly possible that someone will simply refute *Det* $L(\mathcal{R})$ (or even $Det(\Sigma_1^1)$) in *ZFC*. On the other hand, it is also possible that the web of implications involving determinacy hypotheses and relating them to large cardinals will grow steadily until it presents such a natural and compelling picture that more will succumb to its beauty.

We should end by quoting directly a paragraph from Gödel [1947] which was written primarily about large cardinals but which is perhaps even more relevant to determinacy hypotheses.

> "There might exist axioms so abundant in their verifiable consequences, shedding so much light upon a whole discipline, and furnishing such powerful methods for solving given problems (and even solving them, as far as that is possible, in a constructivistic way) that quite irrespective of their intrinsic necessity they would have to be assumed at least in the same sense as any well established physical theory."

8J. Historical remarks

[1] The brief introduction to logic in 8A–8C and our development of Gödel's theory of constructibility in 8D–8F follow quite standard lines and we will not attempt to provide specific references here.

[2] The present, general version of the approximation Theorem 8G.8 is due to Moschovakis and first appeared in Kechris–Moschovakis [1978b], but the result is implicit in Solovay [1969], [1970]. The original proofs of Solovay used forcing techniques.

[3] For references to early work on inner models of set theory which reflect some determinacy, see Becker [1978]. Our development in the exercises of 8G (starting with 8G.17) is based on the recent and still unpublished work of Harrington and Kechris.

[4] Finally, our brief development of the theory of indiscernibles in 8H follows closely Silver's fundamental work in his Ph.D. Thesis, Silver [1971]. See also the expository Silver [1973] which has references to earlier work, particularly Rowbottom [1971], Gaifman, and Scott [1961] where it was first shown that the existence of measurable cardinals contradicts $V = L$.

This page intentionally left blank

REFERENCES AND INDEX TO REFERENCES

The numbers after a name or mathematical work indicate the pages where there are references to this mathematician or work; e.g. something is said about John Addison on pages 9, 118, 189, etc. and his 1954 Doctoral Dissertation is mentioned on pages 7, 189 and 270.

J. W. ADDISON: *9, 118, 189, 204, 205, 270, 272, 380.*

[1954] *On some points of the theory of recursive functions*, Doctoral Dissertation, University of Wisconsin (1954): *7, 189, 270.*

[1959a] Separation principles in the hierarchies of classical and effective descriptive set theory, *Fund. Math.* 46 (1959) 123–135: *7, 64, 189, 270, 271.*

[1959b] Some consequences of the axiom of constructibility, *Fund. Math.* 46 (1959), 337–357: *4, 278, 281, 285, 519.*

J. W. ADDISON and S. C. KLEENE

[1957] A note on function quantification, *Proc. Amer. Math. Soc.* 8 (1957) 1002–1006: *114.*

J. W. ADDISON and Y. N. MOSCHOVAKIS

[1968] Some consequences of the axiom of definable determinateness, *Proc. Nat. Acad. Sci.*, USA 59 (1968) 708–712: *9, 380.*

P. ALEXANDROFF

[1916] Sur la puissance des ensembles mesurables B, *Comptes Rendus Acad. Sci.* 162 (1916) 323–325: *113.*

W. J. ARSENIN

[1940] Sur les projections de certain ensembles measurable B, *C. R. (Doklady) Acad. Sci.* USSR (N.S.) 27 (1940) 107–109: *259.*

R. BAIRE: *1.*

[1899] Sur les fonctions de variables reelles, *Ann. Matematica Pura ed Applicata, Serie III^a*, 3 (1899) 1–122: *1, 62.*

S. BANACH: *299, 379.*

[1931] Über analytisch darstellebare Operationen in abstrakten Räume, *Fund. Math.* 17 (1931) 283–295: *62.*

J. BAUMGARTNER: *379.*

H. BECKER

[1978] Partially playful universes, in *Cabal Seminar* 76–77 ed. by A. S. Kechris and Y. N. Moschovakis, Lecture Notes in Mathematics #689, Springer (1978) 55–90: *572, 611.*

D. BLACKWELL

[1967] Infinite games and analytic sets, *Proc. Nat. Acad. Sci.* USA 58 (1967) 1836–1837: *8, 380, 442.*

E. BOREL: *1.*

L. E. J. BROUWER: *270.*

[1924] Beweiss, dass jede volle Funktion gleichmässig stetig ist, *Konin. Neder. Akad. van Weten. te Amst. Proc.* 27 (1924) 189–193; reprinted in L. E. J. Brouwer, Collected Works vol. 1, North Holland, Amsterdam, 1975: *270.*

L. CARLESON

[1967] Selected problems on exceptional sets, Van Nostrand Math. Studies, Princeton, 1967: *115.*

G. CHOQUET

[1955] Theory of capacities, *Annales de l'Institute de Fourier,* 5 (1955) 131–292: *115.*

J. P. R. CHRISTENSEN

[1974] Topology and Borel structure, *North-Holland Mathematical Studies* #10, (North-Holland, Amsterdam, 1974): *62.*

A. CHURCH: *5, 188, 269.*

[1936] An unsolvable problem in elementary number theory, *Amer. J. Math.* 58 (1936) 345–363: *6.*

P. J. COHEN: *65, 275, 283.*

[1963a] A minimal model for set theory, *Bull. Amer. Math. Soc.* 69 (1963) 537–540: *530.*

[1963b] The independence of the continuum hypothesis, I, II, *Proc. Nat. Acad. Sci.* USA, 50 (1963) 1143–1148, 51 (1964), 105–110: *4.*

MARTIN DAVIS

[1950] *On the theory of recursive unsolvability,* Doctoral Dissertation, Princeton University (1950): *6, 188.*

MORTON DAVIS

[1964] Infinite games of perfect information, *Annals of Math. Studies,* 52 (1964) 85–101: *296, 297, 379, 424.*

C. DELLACHERIE: *147.*

F. R. DRAKE

[1974] Set theory, an introduction to large cardinals (North-Holland 1974): *370, 380.*

P. ERDÖS and A. HAJNAL

[1958] On the structure of set mappings, *Acta Math. Acad. Sci. Hungar,* 9 (1958) 111–131: *370.*

S. FEFERMAN

[1965] Some applications of forcing and generic sets, *Fund. Math.* 56 (1965) 325–345: *199.*

C. FREILING: *vii.*

H. FRIEDMAN: *442, 444.*

[1971] Axiomatic recursive function theory, in *Logic Colloquium* 69, Gandy and Yates eds., (North-Holland 1971): *188.*

D. GALE and F. M. STEWART

[1953] Infinite games with perfect information, *Ann. Math. Studies,* 28 (1953) 245–266: *289, 293, 379.*

R. O. GANDY

[1960] Proof of Mostowski's conjecture, *Bull. Acad. Polon. Sci. Math. Astron. Phys.*, 8 (1960) 571–575: *245.*

K. GÖDEL: *4, 5, 65, 274, 281, 285, 607, 608, 611.*

[1931] Über formal unentscheidbare Sätze der Principia Mathematica und verwandter Systeme I, *Monatshefte für Mathematik und Physik*, 38 (1931) 173–198. (English translation in van Heijehoort, J. (editor), From Frege to Gödel, A source book in mathematical logic 1879–1931, Cambridge, Mass. (Harvard Univ. Press), 1967): *5.*

[1938] The consistency of the axiom of choice and the generalized continuum hypothesis, *Proc. Nat. Acad. Sci.* 24 (1938) 556–557 and Consistency proof for the generalized continuum hypothesis, *Proc. Nat. Acad. Sci.* 25 (1939) 220–224: *4, 285, 516, 518, 528.*

[1940] The consistency of the axiom of choice and of the generalized continuum hypothesis with the axioms of set theory, *Annals of Math. Studies*, Study 3, Princeton University Press, Princeton, New Jersey 1940: *285, 516, 518, 519, 528.*

[1947] What is Cantor's continuum problem? *Amer. Math. Monthly* 54 (1947) 515–525: *611.*

D. GUASPARI

[19??] The largest constructible Π^1_1 set, to appear: *247.*

P. HALMOS

[1960] *Naive Set Theory*, Van Nostrand, Princeton 1960: *xii.*

G. H. HARDY and E. M. WRIGHT

[1960] An introduction to the theory of numbers, Fourth edition, (Oxford, 1960): *12.*

L. HARRINGTON: *vii, 300, 553, 611.*

[1977] Long projective wellorderings, *Ann. of Math. Logic* (1977) 1–24: *284.*

[1978] Analytic determinacy and $0^{\#}$, *J. Symbolic Logic* 43 (1978) 685–693: *284, 604.*

L. HARRINGTON and A. S. KECHRIS

[19??] In preparation: *557, 559, 569.*

J. HARRISON

[1967] Doctoral Dissertation, Stanford University (1967): *243.*

F. HAUSDORFF

[1916] Die Mächtigkeit der Borelschen Mengen, *Math. Annalen* 77 (1916) 430–437: *113.*

[1919] Über halbstetige Funktionen und deren Verallgemeinerung, *Math. Zeit.* 5 (1919) 292–309: *62.*

[1957] Set Theory, Chelsea Publishing Co., New York, NY 1957 (Translated from the (1937) third edition of Hausdorff's *Mengenlehre*): *61.*

P. G. HINMAN

[1969] Some applications of forcing to hierarchy problems in arithmetic, *Z. Math. Logik Grundlagen Math.* 15 (1969) 341–352: *264.*

J. HOFFMANN-JORGENSEN

[1970] The theory of analytic spaces, *Various Publication Series* nr. 10, Matematisk Institut Århus, Danmark: *62.*

T. JECH
[1968] ω_1 can be measurable, *Israel J. Math.* 6 (1968) 363–367: *374, 442.*
[1971] Lectures in set theory with particular emphasis on the method of forcing, Lecture Notes in Mathematics 217 (1971), Springer-Verlag: *285.*

T. JOHN: *414.*

L. KANTOROVITCH and E. LIVENSON
[1932] Memoir on the analytical operations and projective sets (I), *Fund. Math.* 18 (1932) 214–279: *75, 114.*

A. S. KECHRIS: *vii, 227, 270, 271, 321, 331, 342, 377, 432, 553, 611.*
[1973] Measure and category in effective descriptive set theory, *Annals of Math. Logic* 5 (1973) 337–384: *9, 249, 262, 264, 266, 342, 357, 376, 553, 555.*
[1974] On projective ordinals, *J. Symbolic Logic* 39 (1974) 269–282: *9, 444.*
[1975] The theory of countable analytical sets, *Trans. Amer. Math. Soc.* 202 (1975) 259–297: *9, 247, 249, 250, 318, 344, 346, 347, 571, 572.*
[1978] AD and projective ordinals, in *Cabal Seminar* 76–77, Lecture Notes in Mathematics #689, Springer-Verlag (1978) 91–132: *437, 442, 600, 602.*
[198?] The perfect set theorem and definable wellorderings of the continuum, to appear: *550.*

A. S. KECHRIS and D. A. MARTIN
[1978] On the theory of Π^1_3 sets of reals, *Bull. Amer. Math. Soc.* 84 (1978) 149–151: *569.*

A. S. KECHRIS and Y. N. MOSCHOVAKIS
[1972] Two theorems about projective sets, *Israel J. Math.* 12 (1972) 391–399: *346, 565.*
[1978a] editors, *Cabal Seminar* 76–77, Lecture Notes in Mathematics #689, Springer-Verlag (1978): *422.*
[1978b] Notes on the theory of scales, manuscript circulated in 1971 and published in *Cabal Seminar* 76–77, Lecture notes in Mathematics #689, Springer-Verlag (1978) 1–54: *572, 611.*

S. C. KLEENE: *5, 118, 164, 188, 189, 193, 205, 269, 270, 272, 381, 382, 383, 386, 388, 397, 400, 409, 444.*
[1938] On notation for ordinal numbers, *J. Symbolic Logic* 3 (1938) 150–155: *443.*
[1943] Recursive predicates and quantifiers, *Trans. Amer. Math. Soc.* 53 (1943) 41–73: *6, 7, 63, 171, 188.*
[1944] On the form of predicates in the theory of constructive ordinals, *Amer. J. Math.* 66 (1944) 41–58: *443.*
[1950] A symmetric form of Gödel's theorem, Kon. Nederlandsche Akad. van Weterschappen, *Proc. of the Section of Sciences*, vol. 53 (1950) 800–802: *271.*
[1952a] Introduction to Metamathematics, (Amsterdam, North-Holland); Groningen (Noordhoff) and New York and Toronto (Van Nostrand), 1952: *6, 188.*
[1952b] Recursive functions and intuitionistic mathematics, *Proc. of the International Congress of Mathematicians*, (Cambridge, Mass. 1950), 1952: *6, 188.*
[1955a] On the form of predicates in the theory of constructive ordinals (second paper), *Amer. J. Math.* 77 (1955) 405–428: *6, 269, 270, 444.*
[1955b] Arithmetical predicates and function quantifiers, *Trans. Amer. Math. Soc.* 79 (1955) 312–340: *6, 7, 188, 238, 269, 271, 444.*
[1955c] Hierarchies of number theoretic predicates, *Bull. Amer. Math. Soc.* 61 (1955) 193–213: *6, 7, 188, 224, 269, 401, 444.*

[1959a] Countable functionals, *Constructivity in Mathematics*, A. Heyting, Ed., (North-Holland, Amsterdam, 1959) 81–100: *188*.
[1959b] Quantification of number-theoretic functions, *Compositio Mathematica* 14 (1959) 23–40: *220, 271*.

E. M. KLEINBERG
[1977] Infinitary combinatorics and the axiom of determinateness, Lecture notes in mathematics, 612 (1977), Springer-Verlag: *422, 600*.

M. KONDO: *272*.
[1938] Sur l'uniformization des complementaires analytiques et les ensembles projectifs de la second classe, *Jap. J. Math.* 15 (1938) 197–230: *4, 235, 272*.

G. KREISEL
[1959] Interpretation of analysis by means of constructive functionals of finite types, *Constructivity in Mathematics*, A. Heyting, Ed. (North-Holland, Amsterdam), (1959) 101–128: *188*.
[1962] The axiom of choice and the class of hyperarithmetic functions, *Indag. Math.* 24 (1962) 307–319: *202, 203*.

K. KUNUGUI: *255*.
[1940] Sur un problème de M. E. Szpilrajn, *Proc. Imp. Acad. Tokyo XVI* No. 3 (1940) 73–78: *259*.

C. KURATOWSKI: *4, 204, 285*.
[1931] Evaluation de la classe borélienne ou projective d'un ensemble de points à l'aide des symboles logiques, *Fund. Math.* 17 (1931) 249–272: *63*.
[1936] Sur les théorèmes de separation dans la théorie des ensembles, *Fund. Math.* 26 (1936) 183–191: *64, 271*.
[1966] Topology, Vol. 1 (Academic Press, New York and London, 1966) (translated from the French *Topologie*, Vol. 1, published by PWN, Warsaw, 1958): *61, 63, 64, 115, 269*.

C. KURATOWSKI and A. TARSKI
[1931] Les Opérations logiques et les ensembles projectifs, *Fund. Math.* 17 (1931) 240–248: *63*.

D. LACOMBE
[1955] Extension de la notion de fonction récursive aux fonctions d'une ou plusieurs variables reélles, *Comptes Rendus Acad. Sci.* 240 (1955) 2478–2480 and 241 (1955) 13–14, 151–153: *188*.
[1959] Quelques procedes de definition en topologie recursive, *Constructivity in Mathematics*, A. Heyting, Ed. (North-Holland, Amsterdam, 1955) 129–158: *188*.

D. G. LARMAN
[1972] Projecting and uniformizing Borel sets with K_σ sections I, II, *Mathematika* 19 (1972) 231–244 and 20 (1973) 233–246: *259*.

M. LAVRENTIEFF
[1925] Sur les sous-classes de la classification de M. Baire, *Comptes Rendus Acad. Sci.* 180 (1925) 111–114: *64*.

H. LEBESGUE: *1*.
[1905] Sur les fonctions représentables analytiquement, *Journal de Math.* 6e serie 1 (1905) 139–216: *1, 62, 63, 114, 443*.

A. LEVY: *283.*
[1965] Definability in axiomatic set theory I, in *Logic, Methodology and Philosophy of Science*, Y. Bar-Hillel, ed. (North-Holland, Amsterdam, 1965) 127–151: *284.*

A. LOUVEAU: *260, 273.*

N. LUSIN: *2, 3, 4, 86, 91, 92, 193, 196, 205, 269, 270, 272.*
[1917] Sur la classification de M. Baire, *Comptes Rendus Acad. Sci. Paris* 164 (1917) 91–94: *2, 63, 113, 114, 115.*
[1925a] Sur un probleme de M. Emile Borel et les ensembles projectifs de M. Henri Lebesgue: les ensembles analytiques, *Comptes Rendus Acad. Sci.* 180 (1925) 1318–1320: *3, 63.*
[1925b] Sur les ensembles projectifs de M. Henri Lebesgue, *Comptes Rendus Acad. Sci.* 180 (1925) 1572–1574: *3, 63.*
[1925c] Les proprietes des ensembles projectifs, *Comptes Rendus Acad. Sci.* 180 (1925) 1817–1819: *3, 63.*
[1925d] Sur les ensembles non measurables B et l'emploi de la diagonale de Cantor, *Comptes Rendus Acad. Sci.* 181 (1925) 95–96: *63.*
[1927] Sur les ensembles analytiques, *Fund. Math.* 10 (1927), 1–95: *114, 271.*
[1930a] Sur le problème de M. J. Hadamard d'uniformisation des ensembles, *Comptes Rendus Acad. Sci.* Paris 190 (1930) 349–351: *3, 64, 250, 259, 267, 271.*
[1930b] Leçons sur les ensembles analytiques et leurs applications, Collection de monographies sur la theorie des fonctions, Paris, Gauthier-Villars 1930: *61, 114.*

N. LUSIN and P. NOVIKOV
[1935] Choix effectiv d'un point dans un complémentaire analytique arbitraire, donné par un crible, *Fund. Math.* 25 (1935) 559–560: *272.*

N. LUSIN and W. SIERPINSKI
[1918] Sur quelques proprietés des ensembles (A), *Bull. Int. Acad. Sci. Cracovie,* Série A: Sciences Mathematiques (1918) 35–48: *114, 115, 269, 270.*
[1923] Sur un ensemble non mesurable B, Journal de Mathématiques, 9e serie, 2 (1923) 53–72: *114, 115, 268, 269, 270.*

R. MANSFIELD: *65, 79, 534, 537, 574.*
[1970] Perfect subsets of definable sets of real numbers, *Pacific J. Math.* 35 (1970), 451–457: *113, 114, 243, 534.*
[1971] A Souslin operation for Π^1_2, *Israel J. Math.* 9 (1971) 367–379: *596, 599.*
[1975] The nonexistence of Σ^1_2 wellorderings of the Baire space, *Fund. Math.* 86 (1975) 279–282: *549.*

A. A. MARKOV: *188.*

D. A. MARTIN: *vii, 8, 9, 99, 101, 114, 267, 304, 318, 319, 320, 341, 342, 370, 373, 375, 379, 380, 433, 437, 439, 440, 442, 444, 553, 565, 573, 574, 595, 604.*
[1968] The axiom of determinateness and reduction principles in the analytical hierarchy, *Bull. Amer. Math. Soc.* 74 (1968) 687–689: *9, 380, 441, 442.*
[1970] Measurable cardinals and analytic games, *Fund. Math.* 66 (1970) 287–291: *364, 371, 379, 603.*
[1975] Borel determinacy, *Annals of Math.* 102 (1975) 363–371: *357, 358, 379, 441.*
[19?a] Projective sets and cardinal numbers, to appear in J. Symbolic Logic: *8, 9, 75, 97, 103, 113, 115, 433, 596, 599, 601.*

[19?b] Infinite games, to appear in the Proceedings of the International Congress of Mathematicians held in Helsinki, 1978: *605.*

[19?c] Borel and projective games, monograph in preparation: *358, 374, 442, 573, 604, 605.*

D. A. MARTIN and R. M. SOLOVAY

[1969] A basis theorem for Σ^1_3 sets of reals, *Annals of Math.* 89 (1969) 138–160: *8, 596, 599.*

I. MAXIMOFF

[1940] Sur le système de Souslin d'ensembles dans l'espace transfini, *Bull. Amer. Math. Soc.* 46 (1940), 543–550: *113.*

MAZUR: *379.*

A. E. MOSCHOVAKIS: *vii.*

N. R. MOSCHOVAKIS: *vii.*

Y. N. MOSCHOVAKIS: *9, 202, 209, 270, 272, 304, 321, 325, 331, 347, 351, 380, 411, 433, 562, 563, 564, 569, 572, 574, 611.*

[1967] Hyperanalytic predicates, *Trans. Amer. Math. Soc.* 129 (1967) 249–282: *271.*

[1969] Abstract first order computability I and II, *Trans. Amer. Math. Soc.* 138 (1969) 427–504: *271, 444.*

[1970] Determinacy and prewellorderings of the continuum, *Mathematical Logic and Foundations of Set Theory*, Y. Bar-Hillel, Ed. (North-Holland, Amsterdam, 1970), 24–62: *217, 271, 426, 429, 432, 434, 442, 443, 444.*

[1971a] Uniformization in a playful universe, *Bull. Amer. Math. Soc.* 77 (1971) 731–736: *9, 113, 310, 317, 321.*

[1971b] The game quantifier, *Proc. Amer. Math. Soc.* 31 (1971) 245–250: *411, 444.*

[1973] Analytical definability in a playful universe, *Logic, Methodology and Philosophy of Science IV*, P. Suppes et al., eds. (North-Holland, Amsterdam-London 1973) 77–83: *9, 335, 339, 342, 349.*

[1974a] Elementary induction on abstract structures, (North-Holland, Amsterdam, 1974): *271, 419, 444.*

[1974b] Structural characterizations of classes of relations, in *Generalized recursion theory*, J. E. Fenstad and P. G. Hinman, eds. (North-Holland, Amsterdam 1974) 53–81: *271, 414.*

[1974c] New methods and results in descriptive set theory, Proceedings of the International Congress of Mathematicians held in Vancouver in 1974, Vol. 1 251–257: *9.*

[1978] Inductive scales on inductive sets, in *Cabal Seminar* 76–77, Lecture Notes in Mathematics #689, Springer-Verlag, (1978) 185–192: *411, 573.*

A. MOSTOWSKI: *4, 189, 270, 285.*

[1946] On definable sets of positive integers, *Fund. Math.* 34 (1946) 81–112: *6, 63, 188.*

[1951] A classification of logical systems, *Studia Philosphica* 4 (1951) 237–274: *6, 188.*

J. MYCIELSKI

[1964] On the axiom of determinateness, *Fund. Math.* 53 (1964) 205–224: *380, 424.*

[1966] On the axiom of determinateness II, *Fund. Math.* 59 (1966) 203–212: *380, 424.*

J. MYCIELSKI and H. STEINHAUS

[1962] A mathematical axiom contradicting the axiom of choice, *Bull Pol. Acad.* 10 (1962) 1–3: *379, 380, 422.*

J. MYCIELSKI and S. SWIERCZKOWSKI

[1964] On the Lebesgue measurability and the axiom of determinateness, *Fund. Math.* 54 (1964) 67–71: *302, 380, 424.*

O. NIKODYM

[1925] Sur une propriété de l'operation $\mathscr{A}$, Fund. Math. 7 (1925) 149–154: *115.*

P. NOVIKOFF: *4, 202, 205, 272, 285.*

[1931] Sur les fonctions implicites measurables B, *Fund. Math.* 17 (1931) 8–25: *250, 259, 272.*

[1935] Sur las séparabilité des ensembles projectifs de seconde classe, *Fund. Math.* 25 (1935) 459–466: *270.*

J. C. OXTOBY

[1957] The Banach-Mazur game and Banach category theorem, *Annals of Math. Studies,* 39 (1957) 159–163: *299, 379.*

[1971] Measure and category, Springer-Verlag, New York-Heidelberg 1971: *281.*

J. B. PARIS

[1972] $ZF\vdash\Sigma_4^0$ determinateness, *J. Symbolic Logic* 37 (1972) 661–667: *379.*

E. POST: *5, 188.*

H. J. ROGERS

[1967] Theory of recursive functions and effective computability, McGraw-Hill Book Company, New York 1967: *270.*

F. ROWBOTTOM: *8.*

[1971] Some strong axioms of infinity incompatible with the axiom of constructibility, *Annals Math. Logic* 3 (1971), 1–44: *368, 594, 611.*

G. E. SACKS

[1969] Measure-theoretic uniformity in recursion theory and set theory, *Trans. Amer. Math. Soc.* 142 (1909), 381–420: *266.*

[1976] Countable admissible ordinals and hyperdegrees, *Advances in Math.* 19 (1976) 213–262: *247.*

J. SAINT RAYMOND: *260.*

D. SCOTT: *8.*

[1961] Measurable cardinals and constructible sets, *Bull. Acad. Polon. Sci.*, Ser. des Sc. Ast. et Phys. 9 (1961), 521–524: *611.*

J. C. SHEPHERDSON

[1951] Inner models of set theory Parts I and II, *J. Symbolic Logic* 16 (1951) 161–190 and 17 (1952) 225–237: *530.*

J. R. SHOENFIELD: *84, 522.*

[1961] The problem of predicativity, Essays on the foundations of Mathematics, Magnes Press, Hebrew Univ. Jerusalem (1961) 132–139: *64, 114, 524, 526, 530, 531, 537.*

W. SIERPINSKI: *3, 4, 66, 94, 96, 193, 196, 269, 270, 272.*

[1924] Sur une propriété des ensembles ambigus, *Fund. Math.* 6 (1924) 1–5: *64.*

[1925] Sur une classe d'ensembles, *Fund. Math.* 7 (1925) 237–243: *3, 63.*

[1926] Sur une propriété des ensembles (A), *Fund. Math.* 8 (1926) 362–369: *114.*

[1927] Sur la puissance des ensembles d'une certaine classe, *Fund. Math.* 9 (1927) 45–49: *113.*

[1928] Sur les produits des images continue des ensembles C(A), *Fund. Math.* 11 (1928) 123–126: *63.*

[1930] Sur l'uniformisation des ensembles mesurables (B), *Fund. Math.* 16 (1930) 136–139: *272.*

[1950] Les ensembles projectifs et analytiques, Mémorial des Sciences Mathematiques, CXII Paris, Gauthier-Villars, (1950): *2, 61, 64.*

[1956] General Topology, University of Toronto Press, Toronto 1956: *61.*

J. SILVER: *8.*

[1971] Some applications of model theory in set theory, *Annals of Math. Logic* 3 (1971) 45–110: *583, 584, 587, 594, 602, 611.*

[1973] The bearing of large cardinals on constructibility, *Studies in Model Theory*, M. Morley ed., M.A.A. Studies in Mathematics, 1973: *611.*

R. M. SOLOVAY: *vii, 8, 65, 114, 284, 318, 319, 320, 334, 376, 377, 378, 380, 404, 414, 432, 442, 443, 444, 534, 546, 574, 596, 600, 609, 611.*

[1967] A non-constructible Δ^1_3 set of integers, *Trans. Amer. Math. Soc.* 127 (1967) 58–75: *591.*

[1969] On the cardinality of $\mathbf{\Sigma}^1_2$ sets of reals, *Foundations of Mathematics*, Bullof et al eds., Springer-Verlag, Berlin-Heidelberg-New York (1969) 58–73: *8, 286, 375, 380, 537, 611.*

[1970] A model of set theory in which every set is Lebesgue measurable, *Annals of Math.* 92 (1970) 1–56: *4, 284, 611.*

[1978a] The independence of DC from AD, in *Cabal Seminar* 76–77, Lecture Notes in Mathematics #689, Springer-Verlag (1978) 171–184: *423.*

[1978b] A Δ^1_3 coding of the subsets of ${}^{\omega}\omega$, in *Cabal Seminar* 76–77, Lecture Notes in Mathematics #689, Springer-Verlag (1978) 133–150: *437, 602.*

R. M. SOLOVAY, W. N. REINHARDT and A. KANAMORI

[1978] Strong axioms of infinity and elementary embeddings, *Ann. of Math. Logic* 13 (1978) 73–116: *605.*

C. SPECTOR: *196, 269, 270.*

[1955] Recursive wellorderings, *J. Symbolic Logic* 20 (1955) 151–163. *197, 269, 271.*

[1960] Hyperarithmetical quantifiers, *Fund. Math.* 48 (1960) 313–320: *245.*

[1961] Inductively defined sets of natural numbers, *Infinitistic methods*, Pergamon, New York, (1961) 97–102: *409, 444.*

J. STEEL: *267.*

[1977] Ph.D. Thesis, University of California, Berkeley: *440.*

A. H. STONE

[1962] Non-separable Borel sets, *Rozprawy Matematyczne*, 23 (1962), 40 pages: *113.*

M. SUSLIN: *2, 3, 79, 81, 90, 91, 114, 444.*

[1917] Sur une definition des ensembles mesurables B sans nombres transfinis, *Comptes Rendus Acad. Science*, Paris 164 (1917) 88–91: *2, 63, 113, 114.*

E. SZPILRAJN-MARCZEWSKI: *115.*

H. TANAKA

[1968] A basis theorem for Π^1_1 sets of positive measure, *Comment. Math. Univ. of St. Paul*, 16 (1968) 115–127: *266.*

S. K. THOMASON

[1967] The forcing method and the upper semilattice of hyperdegrees, *Trans. Amer. Math. Soc.* 129 (1967) 38–57: *264*.

A. TURING: *5, 156, 188.*

R. VAN WESEP

[1977] Ph.D. Thesis, University of California, Berkeley: *440.*

[1978] Wadge degrees and descriptive set theory, in *Cabal Seminar* 76–77, Lecture Notes in Mathematics #689, Springer-Verlag (1978) 151–170: *440.*

[1978] Separation principles and the axiom of determinateness, *J. Symbolic Logic*, 43 (1978) 77–81: *440.*

J. VON NEUMANN

[1949] On rings of operators, reduction theory, *Annals of Math.* 50 (1949) 448–451: *240, 272.*

W. WADGE

[19??] Ph.D. Thesis, University of California, Berkeley: *424, 440.*

P. WOLFE

[1955] The strict determinateness of certain infinite games, *Pacific J. Math.* 5 (1955) 841–847: *290, 379.*

SUBJECT INDEX

In alphabetizing entries we have adopted the following conventions.

(1) Greek letters are read as if they were spelled out in English.

(2) Hyphenated expressions are read separated, as if the hyphen were a blank space.

(3) Subscripts and superscripts do not count.

For example, "Δ^1_k-hull" is alphabetized as if it read "delta hull" and precedes "Δ^1_1 sets" which is placed as if it read "delta sets".

INDEX OF SYMBOLS

This page intentionally left blank